Biology and Breeding of Food Legumes

Biology and Breeding of Food Legumes

Edited by

Aditya Pratap and Jitendra Kumar

Crop Improvement Division, Indian Institute of Pulses Research, Kanpur, INDIA

www.cabi.org

CABI is a trading name of CAB International

CABI
Nosworthy Way
Wallingford
Oxfordshire OX10 8DE
UK

Tel: +44 (0)1491 832111
Fax: +44 (0)1491 833508
E-mail: cabi@cabi.org
Website: www.cabi.org

CABI
875 Massachusetts Avenue
7th Floor
Cambridge, MA 02139
USA

Tel: +1 617 395 4056
Fax: +1 617 354 6875
E-mail: cabi-nao@cabi.org

A catalogue record for this book is available from the British Library, London, UK.

Library of Congress Cataloging-in-Publication Data

Biology and breeding of food legumes / edited by Aditya Pratap and Jitendra Kumar.
p. cm.
Includes bibliographical references and index.
ISBN 978-1-84593-766-9 (alk. paper)
1. Legumes--Breeding. 2. Food crops--Breeding. 3. Legumes as food. I. Pratap, Aditya, 1976- II. Kumar, Jitendra, 1973- III. Title.

SB177.L45B56 2011
583'.74--dc22 2011008615

ISBN-13: 978 1 84593 766 9

Commissioning editor: Meredith Carroll
Editorial assistant: Gwenan Spearing
Production editor: Fiona Chippendale

Typeset by SPi, Pondicherry, India.
Printed and bound by CPI Group (UK) Ltd, Croydon, CR0 4YY.

Contents

Contributors

Angenon, G. *Laboratory of Plant Genetics, Institute for Molecular Biology and Biotechnology, Vrije Universiteit Brussel (VUB), Pleinlaan 2, B-1050 Brussels, Belgium; E-mail: Geert.Angenon@vub.ac.be*

Asibuo, J.Y. *CSIR-Crops Research Institute, P.O. Box 3785, Kumasi, Ghana; E-mail: jyasibuo@gmail.com*

Basandrai, Ashwani K. *CSK Himachal Pradesh Krishi Vishvavidyalaya, Hill Agricultural Research and Extension Centre, Dhaulakuan, District Sirmour (HP)-173001, India; E-mail: ashwanispp@gmail.com*

Basandrai, Daisy *CSK Himachal Pradesh Krishi Vishvavidyalaya, Hill Agricultural Research and Extension Centre, Dhaulakuan, District Sirmour (HP)-173001, India; E-mail: bunchy@rediffmail.com*

Bebee, S. *International Center for Tropical Agriculture, A.A. 6713, Cali, Colombia; E-mail: s.beebe@cgiar.org*

Blair, M. *International Center for Tropical Agriculture (CIAT), Bean Project, A.A. 6713, Cali, Colombia, South America; E-mail: m.blair@cgiar.org*

Burstin, J. *UMR-102 Legume Ecophysiology and Genetics, INRA, 17 rue de Sully, 21065 Dijon Cedex, France; E-mail: burstin@dijon.inra.fr*

Cannon, S.B. *United States Department of Agriculture – Agricultural Research Service, Corn Insects and Crop Genetics Research Unit, Ames, IA 50011, USA; E-mail: steven.cannon@ars.usda.gov*

Chamarthi, S.K. *International Crops Research Institute for the Semi-Arid Tropics (ICRISAT), Patancheru-502 324, Andhra Pradesh, India*

Chaturvedi, S.K. *Division of Crop Improvement, Indian Institute of Pulses Research, Kanpur-208024, India; E-mail: sushilk.chaturvedi@gmail.com*

Chaubey, B.K. *Division of Crop Improvement, Indian Institute of Pulses Research, Kanpur-208024, India; E-mail: brijchaturvedi.chaubey@gmail.com*

Chaudhary, H.K. *Molecular Cytogenetics and Tissue Culture Laboratory, CSK Himachal Pradesh Agricultural University, Palampur, H.P. India–176062; E-mail: ctlhkc@rediffmail.com*

Chimwamurombe, P.M. *Department of Biological Sciences, University of Namibia, Namibia; E-mail: pchimwa@gmail.com*

Croser, J.S. *Centre for Legumes in Mediterranean Agriculture (CLIMA), University of Western Australia, 35 Stirling Hwy, Crawley, WA 6009, Australia, E-mail: jcroser@clima.uwa.edu.au*

Czyczyło-Mysza, I. *Polish Academy of Sciences, Franciszek Górski Institute of Plant Physiology, Niezapominajek 21, 30-239 Kraków, Poland; E-mail: czyczylo-mysza@ifr-pan.krakow.pl*

Dalvi, V.A. *Guangxi Crop Genetic Improvement and Biotechnology Laboratory, Nanning, People's Republic of China; E-mail: vijay_dalvi79@rediffmail.com*

De La Fuente, M. *Misión Biológica de Galicia-CSIC, P.O. Box 28, 36080, Pontevedra, Spain; E-mail: mfuente@mbg.cesga.es*

De Ron, A.M. *Misión Biológica de Galicia-CSIC, P.O. Box 28, 36080, Pontevedra, Spain*

Duc, G. *UMR-102 Legume Ecophysiology and Genetics, INRA, 17 rue de Sully, 21065 Dijon cedex, France; E-mail: Gerard.Duc@dijon.inra.fr*

Duraimurugan, P. *Crop Protection Division, Indian Institute of Pulses Research, Kanpur–208024, Uttar Pradesh, India; E-mail: duraimuruganp@rediffmail.com*

Gai, J. *National Centre for Soybean Improvement, Nanjing Agricultural University, Nanjing, Jingsu Province, 210095, People's Republic of China; E-mail: sri@njau.edu.in*

Gallardo, K. *UMR-102 Legume Ecophysiology and Genetics, INRA, 17 rue de Sully, 21065 Dijon cedex, France; E-mail: Karine.Gallardo@dijon.inra.fr*

Gaur, P.M. *International Crops Research Institute for the Semi-Arid Tropics (ICRISAT), Patancheru-502 324, Andhra Pradesh, India; E-mail: p.gaur@cgiar.org*

Gnanasambandam, Annathurai *Grains Innovation Park, Department of Primary Industries, Private Bag 260, Horsham, Victoria 3401, Australia; E-mail: annathurai.gnanasambandam@dpi.vic.gov.au*

Gupta, D.S. *Crop Improvement Division, Indian Institute of Pulses Research, Kanpur-208024, India; E-mail: debgpb@gmail.com*

Gupta, M. *School of Food Science and Environmental Health, Dublin Institute of Technology, Dublin 1, Ireland; E-mail: mahesh.Gupta@DIT.ie*

Gupta, Sanjeev *Division of Crop Improvement, Indian Institute of Pulses Research, Kanpur-208024, India; E-mail: saniipr@rediffmail.com*

Haq, Nazmul *Centre for Underutilised Crops, Environment Division, School of Civil Engineering and the Environment, Southampton University, Southampton SO17 1BJ, UK; E-mail: N.N.Haq@soton.ac.uk*

Hobson, Kristy *Grains Innovation Park, Department of Primary Industries, Private Bag 260, Horsham, Victoria 3401, Australia; E-mail: Kristi.hobson@dpi.vic.gov.au*

Imtiaz, Muhammad *Biodiversity and Integrated Gene Management, International Centre for Agricultural Research in the Dry Areas, P.O. Box 5466, Aleppo, Syria; E-mail: M.Imtiaz@cgiar.org*

Ishitani, M. *International Centre for Tropical Agriculture, A.A. 6713, Cali, Colombia; E-mail: m.ishitani@cgiar.org*

Jain, Rashmi *Division of Crop Improvement, Indian Institute of Pulses Research, Kanpur-208024, India; E-mail: rashmijain_25@yahoo.com*

Kaila, V. *Molecular Cytogenetics and Tissue Culture Lab., CSK Himachal Pradesh Agricultural University, Palampur, H.P. India–176062; E-mail: vineetakaila@yahoo.com*

Khulbe, Rajesh *Department of Genetics and Plant Breeding, GB Pant University of Agriculture & Technology, Pantnagar, India; E-mail: rkkhulbe@gmail.com*

Kumar, A. *International Crops Research Institute for the Semi-Arid Tropics (ICRISAT), Patancheru-502324, Andhra Pradesh, India*

Kumar, J. *Division of Crop Improvement, Indian Institute of Pulses Research, Kanpur-208024, India; E-mail: jitendra73@gmail.com*

Kumar, Shiv *Biodiversity and Integrated Gene Management, International Centre for Agricultural Research in the Dry Areas, P.O. Box 5466, Aleppo, Syria; E-mail: sk.agrawal@cgiar.org*

Leonforte, Antonio *Grains Innovation Park, Department of Primary Industries, Private Bag 260, Horsham, Victoria 3401, Australia; E-mail: tony.leonforte@dpi.vic.gov.au*

Lulsdorf, M.M. *Crop Development Centre (CDC), University of Saskatchewan, 51 Campus Drive, Saskatoon SK S7N 5A8, Canada; E-mail: monika.lulsdorf@usask.ca*

Materne, Michael *Grains Innovation Park, Department of Primary Industries, Private Bag 260, Horsham, Victoria 3401, Australia; E-mail: Michael.Materne@dpi.vic.gov.au*

May, G.D. *National Center for Genome Resources, 2935 Rodeo Park Drive East, Santa Fe, NM 87505, USA*

Mir, R.R. *International Crops Research Institute for the Semi-Arid Tropics (ICRISAT), Patancheru-502324, Andhra Pradesh, India; E-mail: imrouf2006@gmail.com*

Mutlu, N. *Faculty of Agriculture, Akdeniz University, TR-07070 Antalya, Turkey; E-mail: severmutlu@hotmail.com*

Nguyen, H.T. *National Center for Soybean Biotechnology (NCSB), University of Missouri, 40 Agriculture Building, Columbia, MO 65211-7140, USA*

Norton, T. *Department of Food Engineering, Harper Adams University College, TF10 8NB, UK; E-mail: tnorton@harper-adams.ac.uk*

Ochatt, S. *Laboratoire de Physiologie Cellulaire, Morphogenèse et Validation (PCMV), Unité Mixte de Recherches en Génétique et Ecophysiologie des Légumineuses à Graines (UMRLEG), Centre de Recherches, INRA de Dijon, B.P. 86510, 21065 Dijon Cedex, France; E-mail: ochatt@epoisses.inra.fr*

Ohmido, Nobuko *Graduate School of Human Development and Environment, Kobe University, Kobe 657-8501, Japan; E-mail: ohmido@kobe-u.ac.jp*

Oldach, K.H. *South Australia Research Development Institute, Plant Genomics Centre, Waite Research Precinct, Hartley Grove, Urrbrae SA, 5064, Australia; E-mail: Klaus.Oldach@sa.gov.au*

Palmer, R.G. *USDA-ARS, Agronomy Department, Iowa State University, Ames, IA 50011, USA; E-mail: Reid.Palmer@ars.usda.gov*

Paull, Jeffrey *School of Agriculture, Food and Wine, University of Adelaide, Waite Campus, Glen Osmond, South Australia 506, Australia; E-mail: jeffrey.paull@adelaide.edu.au*

Pereira, E.A. *Faculdade de Agronomia e Medicina Veterinária, Universidade de Brasília, Campus Universitário Darcy Ribeiro, Asa Norte, Instituto Central de Ciências Ala Sul, Caixa Postal 4.508 - CEP: 70.910-970 Brasília, DF, Brazil; E-mail: everaldo@unb.br*

Pratap, Aditya *Division of Crop Improvement, Indian Institute of Pulses Research, Kanpur-208024, India; E-mail: adityapratapgarg@gmail.com*

Rajan, Neha *Division of Crop Improvement, Indian Institute of Pulses Research, Kanpur-208024, India; E-mail: neha_rajan96@yahoo.com*

Rane, J. *International Centre for Tropical Agriculture, A.A. 6713, Cali, Colombia; E-mail: j.rane@cgiar.org*

Rao, I.M. *International Centre for Tropical Agriculture, A.A. 6713, Cali, Colombia; E-mail: i.rao@cgiar.org*

Rodiño, A.P. *Misión Biológica de Galicia-CSIC, P.O. Box 28, 36080, Pontevedra, Spain; E-mail: aprodino@mbg.cesga.es*

Safari, S. *Department of Bioinformatics and Genomics, University of North Carolina at Charlotte, 9201 University City Blvd., Charlotte, NC 28223, USA; E-mail: ssafari@uncc.edu*

Sankaran, M. *Central Agricultural Research Institute, Port Blair, A & N Islands, India; E-mail: kmsankaran@gmail.com*

Santalla, M. *Misión Biológica de Galicia-CSIC. P.O. Box 28, 36080, Pontevedra, Spain; E-mail: msantalla@mbg.cesga.es*

Sardana, S. *National Bureau of Plant Genetic Resources, New Delhi, 110 012, India*

Sato, Shusei *Kazusa DNA Research Institute, 2-6-7 Kazusa-kamatari, Kisarazu, Chiba 292-0818, Japan; E-mail: ssato@kazusa.or.jp*

Schlueter, J.A. *Department of Bioinformatics and Genomics, University of North Carolina at Charlotte, 9201 University City Blvd., Charlotte, NC 28223, USA; E-mail: jschluet@uncc.edu*

Sharma, S.K. *National Bureau of Plant Genetic Resources, New Delhi, 110 012, India; E-mail: skspbg@yahoo.co.in*

Singh, B.B. *Additional Director General (Oilseeds and Pulses), Indian Council of Agricultural Research, Krishi Bhawan, New Delhi-110001, India; E-mail: singhbb_55@yahoo.com*

Singh, Mohar *National Bureau of Plant Genetic Resources, New Delhi, 110 012, India; E-mail: mmohar26@yahoo.co.in*

Skrzypek, E. *Polish Academy of Sciences, Franciszek Górski Institute of Plant Physiology, Niezapominajek 21, 30-239 Kraków, Poland; E-mail: skrzypek@ifr-pan.krakow.pl*

Solanki, R.K. *Division of Crop Improvement, Indian Institute of Pulses Research, Kanpur-208024, India; E-mail: rks.iipr@gmail.com*

Sood, A. *Molecular Cytogenetics and Tissue Culture Lab., CSK Himachal Pradesh Agricultural University, Palampur, H.P., India-176062; E-mail: archit_sh@rediffmail.com*

Sood, V.K. *Molecular Cytogenetics and Tissue Culture Lab., CSK Himachal Pradesh Agricultural University, Palampur, H.P., India-176062; E-mail: vkspbg23@rediffmail.com*

Souza, L.A.C. *Ministério do Desenvolvimento Agrário, Ed. Palácio do Desenvolvimento, 10° andar, Brasília, CEP: 71.000-000 Brasília DF, Brazil; E-mail: gutocopati@yahoo.com*

Spehar, C.R. *Faculdade de Agronomia e Medicina Veterinária, Universidade de Brasília, Campus Universitário Darcy Ribeiro, Asa Norte, Instituto Central de Ciências Ala Sul, Caixa Postal 4.508 - CEP: 70.910-970 Brasília, DF, Brazil; E-mail: spehar@brturbo.com.br*

Srinivasan, T. *Coconut Research Station, Tamil Nadu Agricultural University, Aliyar Nagar-642 101, Tamil Nadu, India; E-mail: srini_vasant@yahoo.com*

Suso, María José *Instituto de Agricultura Sostenible (CSIC), Apdo. 4084, 14080 Córdoba, Spain; E-mail: ge1susom@uco.es*

Tabata, Satoshi *Kazusa DNA Research Institute, 2-6-7 Kazusa-kamatari, Kisarazu, Chiba 292-0818, Japan; E-mail: tabata@kazusa.or.jp*

Tayeng, T. *Molecular Cytogenetics and Tissue Culture Lab., CSK Himachal Pradesh Agricultural University, Palampur, H.P., India-176062; E-mail: tisutayeng@rediffmail.com*

Thu, T.T. *Laboratory of Plant Genetics, Institute for Molecular Biology and Biotechnology, Vrije Universiteit Brussel (VUB), Pleinlaan 2, B-1050 Brussels, Belgium; E-mail: tthu@vub.ac.be*

Tiwari, B.K. *Manchester Food Research Centre, Manchester Metropolitan University, M14 6HR, UK; E-mail: brijesh.tiwari@ucd.ie*

Toker, C. *Faculty of Agriculture, Akdeniz University, TR-07070 Antalya, Turkey; E-mail: toker@akdeniz.edu.tr*

Varshney, R.K. *International Crops Research Institute for the Semi-Arid Tropics (ICRISAT), Patancheru-502324, Andhra Pradesh, India; E-mail: r.k.varshney@cgiar.org*

Verma, P. *Agricultural Research Station, MP University of Agriculture and Technology, Kota 324001, India; E-mail: preetiarskota2005@hotmail.com*

Vuong, T.D. *National Center for Soybean Biotechnology (NCSB), University of Missouri, 40 Agriculture Building, Columbia, MO 65211-7140, USA*

Wędzony, M. *Pedagogical University of Kraków, Podchorążych 2, 30-084 Kraków, Poland; E-mail: mwedzony@ap.krakow.pl*

Young, N.D. *Department of Plant Pathology, 495 Borlaug Hall, University of Minnesota, St. Paul, MN55108, USA; E-mail: neviny@umn.edu*

Foreword

Food legumes, comprising dry bean, dry pea, soybean, groundnut, chickpea, pigeon pea, lentil, mung bean, urd bean, lathyrus and cowpea, have considerable global area under cultivation, and these crops are important constituents of cereal-based vegetarian diets. With their high protein content and ability to fix nitrogen, which reduces fertilizer use in agriculture, grain legumes have become important targets for agricultural, environmental and biotechnological research. However, over the last five decades, global food legume production involving major grain legume crops except soybean and groundnut has witnessed only a marginal annual increase of 0.77%, with fluctuation only from 40.78 to 55.85 million t. This slow growth in production, along with a rising population, diversified uses for end products and improved purchasing capacity, has put tremendous pressure on the per capita availability of pulses. Several constraints such as drought, pest and disease problems and unavailability of quality seeds of improved varieties have made the situation more complex. The influence of abiotic stresses on cultivation of pulses on marginal lands increases these difficulties under the present scenario of climate change.

However, the present global production of legumes could easily be increased by 30–40% if: (i) losses caused by several biotic and abiotic stresses were prevented; and (ii) genotypes less influenced by environment were developed. The scientific community has responded positively to these challenges by directing a greater amount of research towards increasing production and improving the quality of pulses for both edible and industrial purposes. To sustain this progress and accelerate the development of better and superior varieties, crop breeding and biotechnology play a vital role in transferring economically important traits from distant/wild species to the cultivated backgrounds. A synergy of conventional and modern crop improvement tools has opened up new avenues of target-oriented research for legume scientists.

This book, *Biology and Breeding of Food Legumes*, represents to date the most modern and comprehensive volume compiled by two young scientists from this institute, who deserve appreciation for their efforts. This volume offers an extensive reference on the recent developments made in major food legumes. It offers exhaustive information on various aspects related to history, origin and evolution, botany, breeding objectives and methods, hybrid technology, doubled-haploid breeding and *in vitro* techniques; and on recent developments made through biotechnology, genetic engineering and molecular approaches. Contributions to all the chapters in this book have been made by renowned scientists whose research contributions are acknowledged globally. I am hopeful that the information contained in this book will further motivate the research efforts of breeders to promote the productivity and yield stability of food legumes, and that the book will be a useful knowledge resource for those involved in the teaching, extension and production of these important crops.

N. Nadarajan
Director, IIPR, Kanpur, India
April, 2011

Preface

In terms of agricultural importance, after cereals food legumes represent the most valued food source because of their importance for humans and animals, soil ameliorative values and ability to thrive under harsh and fragile environments. Bearing in mind their key role in the diversification and intensification of contemporary agriculture, systematic national and international efforts towards their genetic improvement began in the1960s using classical breeding tools. With the advent of modern techniques and the creation of new selection opportunities in the form of alien variations, global scientific research has been directed towards precise and target-oriented goals and remarkable results have been obtained in developing high-yielding, input-responsive, early-maturing and high-nutrition varieties in pulses.

However, despite the tremendous advances made in the breeding of food legumes, the need and opportunities to further improve their production, productivity and protein and nutritional quality, are as great today as they have ever been. There is an urgent need to search for new gene pools with special reference to wild species and to update the knowledge gained through recent technological advancements. Over the years, a greater portion of food legume breeders' efforts has been directed towards developing improved plant types and technologies while working in concert with the conventional techniques of crop improvement. Consequently, voluminous literature has been generated on different aspects of legume improvement but is scattered over numerous journals and books. However, to date no single publication has provided a comprehensive insight into this literature with a focus on the breeding aspects of food legumes. This book has been edited with the objective of addressing this issue.

Biology and Breeding of Food Legumes comprises 26 chapters contributed by eminent legume scientists around the world. The first two chapters present the historical and evolutionary aspects, while the third chapter deals with the biology of food legumes. The subsequent five chapters (4 to 8) deal with breeding methods, with special reference to distant hybridization and breeding for warm and cool season food legumes and resistance to stresses. This is followed by a section on specific technologies, i.e. polyploidy, cytology and molecular cytogenetics, *in vitro* techniques, haploidy breeding, transgenesis, male sterility and mutagenesis (Chapters 9 to 16). Chapter 17 deals with cultivation of food legumes in the problem soils of the savannahs, and is followed by two chapters on more recent techniques involving molecular markers. The next chapter covers protein content and nutritional quality. The subsequent three chapters (19 to 21) deal with underutilized food legumes, legumes as models and plant genetic resources, these being followed by a chapter on seed dormancy and viability. Postharvest technology, value addition and international trade are dealt with in the last two chapters.

A review of the entire gamut of published work was not possible in this single volume, nor was this the aim. However, the contributors of individual chapters have tried to provide important references on significant work published to date on different aspects of legume improvement. Bearing in mind the scope of the book, slight overlapping in subject matter is possible albeit all chapters having been dealt with in depth by various experts. We are extremely grateful to all our experienced authors who, despite great demands on their time while writing these chapters, completed the task with the utmost responsibility and great care.

We are highly indebted to Dr S. Ayyappan, Director-General and Secretary, Indian Council of Agricultural Research (ICAR), Department of Agricultural Research and Education, Government of India for providing necessary support and guidance in the preparation of this publication. Professor Swapan Datta, Deputy Director-General (Crop Science), ICAR and Dr V.D. Patil, retired Additional Director-General (Oilseed and Pulses), ICAR deserve our heart-felt thanks for providing us with state-of-the-art facilities at IIPR to carry out pulses research. In addition, Dr N. Nadarajan, Director, Dr Masood Ali, Ex-Director and Dr S.K. Chaturvedi, Head, Crop Improvement Division, all globally recognized pioneer pulses researchers at IIPR, deserve special mention for their encouragement to us in undertaking this endeavour.

Many others have also rendered invaluable help in bringing this publication to life, and they deserve our heartfelt appreciation and gratitude: Dr B.B. Singh, Project Coordinator, Mungbean, Urdbean, Lentil, Lathyrus, Rajmash and Pea Crops, IIPR (now ADG (O & P), ICAR) for providing the cover image and technical comments; Dr Shiv Kumar, Lentil Breeder, ICARDA, Syria and scientists from the Crop Improvement Division, IIPR for their valuable technical input during the course of editing the various chapters; Mr Debjyoti Sen Gupta for editorial corrections; Mr Rakesh Agrawal, Senior Technical Assistant, Mr Brijesh Kumar and Miss Neha Rajan, Senior Research Fellows for typographical help; and CAB International for shepherding the book through the editorial process with a thoroughly professional approach. The first editor owes so very much to the late Sh Surinder Kumar Mittal, who always inspired us to strive for better, but unfortunately left for his heavenly abode before he could see this book through to print. Thanks are also due to our lovely kids Puranjay, Neha and Gunika, whose time we have compromised in order to complete this task. And lastly, Dr Rakhi Gupta and Mrs Renu Rani, our better halves, deserve special thanks for their unstinting help, patience and emotional support during the preparation of this manuscript.

Aditya Pratap
Jitendra Kumar
IIPR, Kanpur, India
April, 2011

1 History, Origin and Evolution

Aditya Pratap and Jitendra Kumar

1.1 Introduction

The Latin word *legumen*, which is believed to have come from the verb *legere* (to gather) is supposed to be the origin of the term legume. However, the English language borrowed this term from the French word *légume*, that refers to any kind of vegetable. Written documentary records have been found only for the legume, soybean, in the books of Shen Nung, dating back to 2800 BC in China. Its high protein content was used to flavour and enrich their basic food grains. Methods to extract oil from soybean had been devised by 400 BC.

Theophrastus (a Greek botanist) wrote in 300 BC that leguminous plants 'reinvigorate' the soil and could be used as manure. The Romans also emphasized the use of leguminous plants for this purpose, and historically they became important for enriching the soil fertility of the nutrient-poor Mediterranean soil (Ladock, 2010).

The ancient Egyptians had a high regard for lentils, and the Romans appreciated them, during the reign of Caligula they transported 840 t of lentils to Rome. However, during this period the use of beans as a foodstuff was negligible in Egypt. Although peas had been a staple food in Rome for some time, their use became popular in the green form in the 17th century, when it was a fashionable dish of the rich. Madame de Maintenon (from the court of Louis XIV) mentioned pea as 'a fashion and a madness' (FDM, 1996). The grasspea was described as an aphrodisiac in the 17th century in a Moroccan medical compendium, *Tuhfat al-ahbāb*, because when eaten in quantity without other foods it led to a disease known as lathyrism, which resulted in a permanent paralysis of the lower limbs (Wright, 2011).

In the 16th century, the bean was brought from North America (where it had been grown since ancient times) to Europe and it became a specialized luxury dish there due to its accessibility to only the rich. When visiting the West Indies, Columbus was impressed to see the cultivation of peanuts along with other crops. The Mediterranean peoples ate beans, which have the highest protein content among all plant foods and have been shown nutritionally important for those too poor to afford or choose to eat meat. The amino acids found in beans are perfectly complemented by those in cereals, and these two foods were the first to be found preserved at archaeological sites. In general, Mediterranean dishes are a combination of wheat and beans, or rice and lentils, or maize and peas, which basically fulfil the protein needs in the human diet. In the words of the botanist Charles B. Heiser Jr, the use of legumes with cereals was a 'happy accident' for primitive people with regard to balancing dietary protein, because they did

©CAB International 2011. *Biology and Breeding of Food Legumes* (eds A. Pratap and J. Kumar)

not know about the importance of amino acids or proteins (Wright, 2011).

1.2 Origin and Distribution of Legumes

The history of legumes starts with human civilization and their evolution throughout many different regions of the world. According to Harlan (1971), agriculture originated independently in three 'real' centres (South-west Asia, China and Meso-America) along with other 'non-centre' regions such as Africa, South-east Asia and South America, in which domestic activities were dispersed over a span of 5000–10,000 km from the 'real' centres. Subsequently, Harlan *et al.* (1976) referred to the Near East as the 'centre of agricultural innovation' where pea, lentil, vetch and faba bean had become domesticated food legumes. This entire system of agriculture was moved out along the shores of the Mediterranean to the Danube and Rhine Rivers, eastward to the Indus and northern India and southward across Arabia, the Yemen and into the Ethiopian plateau, although it did not advance further into tropical Africa. It reached China in the second half of the 2nd millennium BC (Harlan *et al.*, 1976). Figure 1.1 shows the distribution of different legume crops, while a discussion on the origin of agriculture in the three major regions listed above based on historical data – including archaeological and recent molecular evidence – now follows.

South-west Asia region (Mesopotamia)

Legumes, accompanied by cereals, were the first plants cultivated by man in Mesopotamia (south-west Asian region), where ancient agriculture evolved. In this region, lupins and lentils have been identified as the oldest cultivated legumes on the basis of archaeobotanical remains found from the Epipalaeolithic (17,000 BC; Hopf and Bar-Yosef, 1987). Lentil from the later phase of Interstadial to the end of the Younger Dryas has been discovered at the Ohalo II site in the Levant area. It is assumed that this period is associated with a wetter and warmer climate toward the Holocene (starting around 10,000 BC), a time

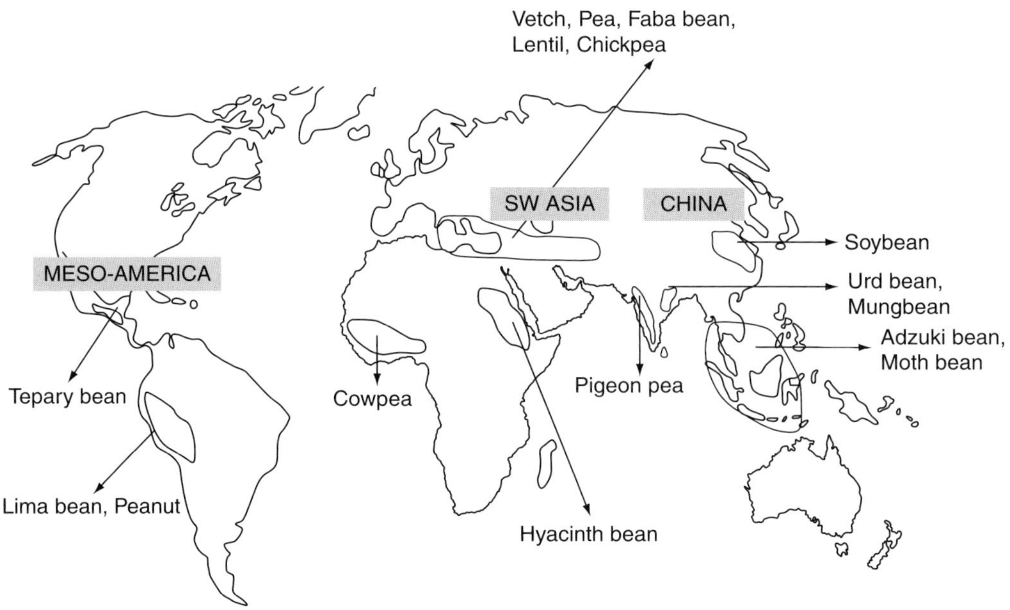

Fig. 1.1. The main centres of agricultural origin (Harlan, 1971) and distribution of major food legumes.

of forest expansion. Recovery of domesticated cereals from most sites during this period suggests the beginning of widespread cultivation of associated legumes. Subsequently from the Pre-Pottery Neolithic A (8500–7500 BC), food legumes such as bitter vetch, pea and faba bean have been identified at different sites including Jericho and Iraq-el-Dubb in the Levant and Tell Aswad in Syria (Colledge, 1994). Grasspea has been identified from sites in both Turkey and Syria, and, during this period, chickpea also appeared for the first time. The small-seeded legumes in particular were found toward the end of the Pre-Pottery Neolithic (Late Pre-Pottery Neolithic B (6600–5500 BC)). However, among food legumes, lentil was still predominant during this period, while other food legumes were probably of less importance (Butler, 2007).

Meso-America and South America region

The process of agriculture in Meso-America (the New World) probably started around 8000 BC, which roughly coincides with the beginning of domestication in the Old World (Piperno *et al.*, 2009). Archaeological evidence from grinding stones indicates the existence of beans, along with starch grains, by 7000 BC. The extreme North-west Balsas–Jalisco region of Meso-America has been identified as a possible area of domestication of beans with maize, where their wild ancestors have been found in abundance (Zizumbo-Villarreal and Colunga-Garcia Marin, 2010). Other authors have also suggested that maize, beans and squash were domesticated in different regions and periods (Harlan, 1995; Kwak *et al.*, 2009). Later, beans spread to the rest of Meso-America via existing biological–cultural corridors (Perry *et al.*, 2007). In these regions, different types of bean were domesticated: the scarlet runner bean and the tepary bean, originally from Mexico, and the small- and large-seeded lima bean (also known as the butter or sieva bean), originally from Peru (Kaplan, 1965). However, beans were unknown in the Old World until 1493, after the return of Columbus from his second journey to the New World. Here, people knew vigna beans as phaseolus beans. The spread of beans from Central America to Spain and Portugal occurred in 1506 after the discovery of the New World, and beans had reached Europe from the Andes by 1532, while a description in a German herbal tome indicates their arrival there by at least 1543.

South America has been seen as a separate region for the domestication of legumes, where the groundnut was known to indigenous people over 4000 years ago. Many pre-Columbian cultures, such as the Moche, depicted groundnuts (peanuts) in their art (Katherine and Museum, 1997). The oldest specimens of groundnuts found in Peru have been estimated to be around 7600 years old (Dillehay, 2007). This legume might have first been domesticated in Paraguay or Bolivia, because wild strains of peanut are found in abundance in these regions, where some are still being cultivated.

North and South China regions

China represents the third region where agriculture has evolved independently. Evidence indicates that two Neolithic cultures – the Yang-shao (centred around the middle course of the Huangho River in Honan and Shansi) and Lungshan (in East and North China) – were predominant in ancient times. Soybean, one of oldest cultivated food legumes, has been known to man here for over 5000 years, and this region represents a candidate location of its domestication (Hymowitz, 1970). Molecular diversity studies conducted on soybean populations collected from both North and South China suggest that this food legume crop was also domesticated from ancient times in South China (Ding *et al.*, 2008). See Section 1.6 for more detail on the origins of this food crop.

1.3 Timeline Origin of Leguminosae

The origin of leguminous plants is largely speculative, and fossil records do not provide much help in judging the exact time of origin of the Leguminosae. However, evidence obtained from fossils and phylogenetic records (Schrire *et al.*, 2005a, b) suggests that members of the legume family originally evolved in arid and/or semi-arid regions along the Tethys seaway during the early Tertiary (Herendeen, 1992). The West Gondwanan hypothesis for the origin of the family also supports a 'moist equatorial megathermal' origin for legumes during the mid- to late Cretaceous (Raven and Axelrod, 1974; Polhill and Raven, 1981). Tertiary legume diversification immediately followed the origin of the family. Legumes are now highly diverse in tropical to subtropical Africa and South America, and therefore, these regions may indicate possible candidates for the origin of this family (Pan *et al.*, 2010). A minimum point of 84 million years ago (MYA) has been suggested as being the split between Fagales and Cucurbitales as an internal calibration point, and an age of 74–79 MYA has been estimated for Fabaceae (Soltis *et al.*, 2000). The fossil record of the Fabaceae is abundant and diverse, particularly in the Tertiary.

Figure 1.2 shows the timeline evolution of the legume family and its subsequent divergences in subfamilies and clades on the basis of archaeological and molecular data. Lavin *et al.* (2005) used the tertiary macrofossils of the Leguminosae as time constraints and molecular data and estimated the ages of the earliest branching clades of subfamilies. They proposed that the first definitive legumes appeared during the Late Paleocene (~56 MYA) (Herendeen and Wing, 2001; Wing *et al.*, 2004). The oldest caesalpinioid, mimosoid and papilionoid clades evolved during approximately the same time range of 39–59 MYA. These traditionally recognized subfamilies of legumes and other taxonomically large clades within these subfamilies (genistoids) are recorded from fossil records soon afterward, beginning around 50–55 MYA (Herendeen, 1992). A prediction derived from the legume fossil record is that there should be little difference between the estimated age of the origin of legumes and their subsequent diversification. The fossil record of legumes predicts this genetic finding of a rapid diversification of extant lineages.

1.4 Taxonomic History

Details on the history depicting the taxonomic classification of legumes have been reviewed by Cronk (1990). Linnaeus (1753) has given an account based on the sexual system, and he grouped genera belonging to the family Leguminosae into three orders, Diadelphia Decandria (the precursor of the Papilionoideae), Polyandria Monogynia (the precursor of the Mimosoideae) and Decandria Monogynia (the precursor of the Caesalpinioideae). Using the same system, other genera were also included in order to update Linnaeus' inventory (Persoon, 1805, 1807). Linnaeus continued to use the sexual system, although a natural system with 94 genera in the Leguminosae was published by De Jussieu (1789). The beginning

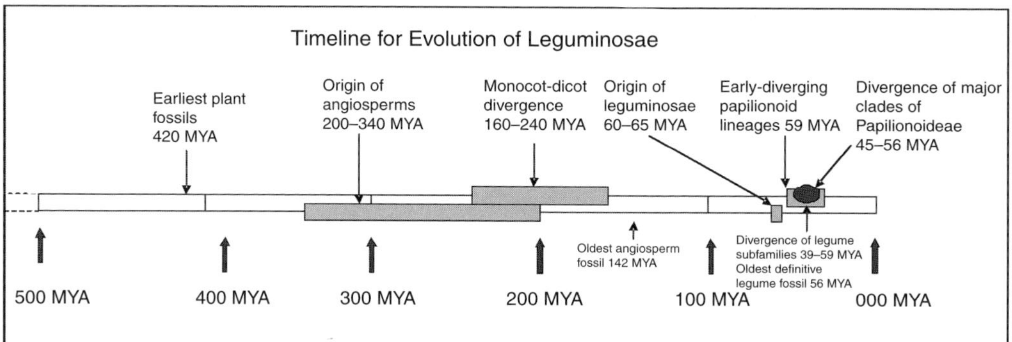

Fig. 1.2. Timeline evolution of the legume family based on archaeological and molecular data (sources: Doyle and Luckow, 2003; Lavin *et al.*, 2005; cover page of *Annual Wheat News Letter*, 2010; Pan *et al.*, 2010). MYA, million years ago.

of generic reform was shown by de Candolle (1825), which was subsequently followed by Bentham. Bentham (1865) gave an estimate of the number of species in each genus, based partly on the available literature and partly on the large collection of specimens in the herbarium of the Royal Botanic Gardens at Kew, London. The three main classes of legume in the sexual system have now become suborders of the order Leguminosae in this more natural system. Bentham achieved a remarkable system by the intuitive use of the natural method. At each step in the construction of a classification, he reassessed the relative value of characters in order to weed out artificial ones (Bentham, 1861). Although Taubert (1894) surveyed the plant kingdom in the light of Engler's new system, it did not affect the generic classification of the Leguminosae. Later, Hutchinson (1964) included a very large number of new genera although this was largely regarded as a revision of Bentham's work. Polhill and Raven (1981) and Gunn (1983) provided another overview of the family, which was a supplement to Hutchinson (1964).

Fabaceae are placed in the order Fabales according to most taxonomic systems, including the APG III, a modern system of plant taxonomy for flowering plant classification. The Fabaceae traditionally have three subfamilies, Caesalpinioideae, Mimosoideae and Faboideae. Polhill and Raven (1981) and Polhill (1994) have described Papilionoideae for Faboideae. Some taxonomists have also recognized these three subfamilies as separate families (Hutchinson, 1964; Cronquist, 1981). The last of these families, which includes most of the food legumes, has loosely been divided into four groups of tribes (Kirkbride *et al.*, 2003): (i) the basal Swartzieae and Sophoreae tribes; (ii) temperate tribes; (iii) tropical tribes; and (iv) temperate herbaceous tribes. The tribe Swartzieae has been placed in the subfamily Caesalpinioideae or even considered to be a fourth subfamily, but the general consensus of opinion among legume taxonomists is that it should be in the Faboideae (Cowan, 1981). Cladistic studies (Herendeen, 1995) and rubidium chloride (RbCl) data (Doyle *et al.*, 1997) indicate that Swartzieae and Sophoreae should be merged into a single tribe in the Faboideae. A general consensus on the tribal and generic classification of the legumes was taken at the First International Legume Conference at the Royal Botanic Gardens, Kew, in 1978. This conference was attended by Charles R. (Bob) Gunn, who recognized that this would enable sweeping family-wide studies of many aspects of the legumes (Polhill and Raven, 1981). He prepared a nomenclature of legume genera on the basis of seed and fruit characteristics (Gunn, 1983). These legume fruits and seeds collected from institutions and individuals throughout the world were incorporated into the US National Seed Herbarium (BARC), Beltsville, MD. The systematic of genus was further classified on the basis of crosses among the species and subspecies by establishing the biological barriers to the access to a common gene pool. The recent developments made in taxonomic classification of legumes based on further molecular data are discussed in detail in Chapter 22.

1.5 Concept of Centre of Origin

The question regarding the origin of crops was first considered in 1807, by Alexander von Humboldt in his work *Essai sur la Géographie des Plantes*. However, Alphonse de Candolle was the one who first recognized the significance of the relationship between plant domestication and the development of man. He incorporated taxonomic, archaeological, historical and philological data into a geographical framework to postulate regions such as China, South-west Asia (including Egypt) and Tropical Asia as being regions of plant domestication, documented in his classical book *Origin of Cultivated Plants*, published in 1882. Much later, his ideas were again taken up (Harlan, 1992). Thus the overlap between wild ancestors and cultigens, and archaeological remains connecting both, are *sine qua non* conditions in establishing such a centre of origin. Subsequently, the concept of centres of origin of crop plants was first discussed by Vavilov at the Fifth International Genetics Congress held in Berlin in 1926, where he recognized China, India, Indo-Malaya, Central Asia, the Near East, the Mediterranean, Ethiopia, southern Mexico and Central America, South America and Chile as centres of origin of crop plants. These independent regions were nominated on the basis of the greatest diversity of types occurring for a particular crop, and hence for the first cultivation of various plants including legumes. Vavilov (1926) divided all cultivated plants into two groups: (i) those known in cultivation or the wild state; and (ii) those derived from weeds. Those plants belonging to the latter category were placed in the primary group, with those in the former being put in the secondary group. He worked on this theory until his death in 1943. These centres of origin were further modified to three 'real' centres and three 'non-centres' (Harlan, 1971). These centre–non-centre combinations have been recorded as: (i) Near East–Africa; (ii) North China, South-east Asia and the South Pacific; and (iii) Meso-America–South America.

1.6 Cultivated Species: History, Origin and Evolution

The origin of a cultivated food legume species has been considered in those regions where both archaeological remains and wild species coexist. Natural forces including mutation, migration, hybridization and genetic drift have led to alteration in wild species resulting in the evolution of the wild ancestors of cultivated species. These were recruited by man, both knowingly and unknowingly, as 'landraces' compatible with the farming methods of ancient times. Subsequently, cultivars (i.e. cultivated varieties) of known species evolved from these landrace progenitors. The alterations that occurred with regard to specific traits during the domestication process are discussed in Chapter 2.

These observations suggest that the primary temperate legumes, including the garden pea (*Pisum sativum*), field pea (*Pisum arvense*), winged pea (*Tetragonolobus purpureus*), green bean (*Phaseolus vulgaris*), runner bean (*Phaseolus coccineus*), butter bean (*Phaseolus lunatus*), lima bean (*Phaseolus limensis*), soybean (*Glycine max*), lentil (*Lens culinaris*) and broad bean (*Vicia faba*) originated in humid, sub-humid, cool, subtropical, semiarid and temperate areas in diverse regions of the world, ranging from South-west Asia and East Asia to the Mediterranean, Peru, Mexico and Guatemala (Muehlbauer, 1993). The common tropical legumes, including the winged bean (*Psophocarpus tetragonolobus*), jicama (*Pachyrhizus erosus* and *Pachyrhizus tuberosus*), chickpea (*Cicer arietinum*), black-eyed pea (*Vigna unguiculata*) and peanut (*Arachis hypogaea*) originated in the humid, semi-arid, cool, subtropical and tropical climates of South America, South-west Asia, Ethiopia, India, Japan, China and West Africa (Hymowitz, 1990). Details on the history, origin and evolution of the major food legumes are now summarized.

Pigeon pea (*Cajanus cajan*)

The name pigeon pea was first reported in Barbados, where the seeds of these plants were used as pigeon feed (Plukenet, 1962). In India, many Sanskrit names have their modern equivalents, and the common name *Arhar* used in the north of the country is considered to be derived from the word *Adhaki* or *Adhuki*. In southern India, the name *Tur* is believed to be derived from the Dravidian *Tovarai* or *Tuvari*, used in Sanskrit since AD 300–400 (De, 1974; van der Maesen, 1990). The Portuguese *Guandu* and Spanish *Gandul* would appear to be derived from the Telugu word *Kandulu* (van der Maesen, 1986), while some consider it to be a corruption of the word *Cajan* (Royes, 1976).

The presence of several wild relatives, the large diversity of the crop gene pool, linguistic evidence and a few archaeological remains, as well as its wide usage in daily cuisine, make India a fitting candidate for the place of origin of the pigeon pea (Vavilov, 1951; van der Maesen, 1990). However, according to another group of scientists (Krauss, 1932; Purseglove, 1968), pigeon pea originated in Africa and from there it was introduced to the West Indies, Brazil and also India (Tothill, 1948), and then from India to Australia, Sri Lanka, Jamaica and Zambia (FAO, 1959). However, it was observed by van der Maesen (1979) that only a single close wild relative of the pigeon pea, *Cajanus kerstingii* (Harms), was widespread in Africa while another, *Cajanus scarabaeoides* (L.) Thourars, was limited to the coastal areas and therefore appeared to have arrived only recently. Others also agreed with Vavilov's view that it must have spread eastwards to Malaysia from India around 200 BC and was perhaps carried subsequently to China, later reaching Australia via Indonesia (De, 1974; Royes, 1976; van der Maesen, 1980). The occurrence of the greatest diversity of *Cajanus cajan* and its wild relatives in Western Ghats and the Malabar Coast of India supported the view that India is the centre of origin of pigeon pea (De, 1974). Some *Atylosia* (*Cajanus*) species bear a very striking resemblance to *Cajanus*, while during exploration trips to Western Ghats during the years 2009/2010, a vast amount of diversity was observed for *Cajanus lineatus* and *Rhyncosia* (Pratap and John, 2010, unpublished data).

Chickpea (*Cicer arietinum*)

Chickpea is a member of the West Asian Neolithic crop assemblage, associated with the origin of agriculture in the Fertile Crescent some 10,000 years ago (Zohary and Hopf, 2000; Abbo *et al.*, 2003). It most probably originated in an area of present-day south-eastern Turkey and adjoining Syria. The wild progenitor of chickpea, *Cicer reticulatum* Lad. (Ladizinsky and Adler, 1976) is currently reported from only 18 narrowly distributed locations in south-eastern Turkey, while the other two wild annual species of *Cicer* closely related to chickpea, *Cicer bijugum* and *Cicer echinospermum*, are also found distributed in Turkey and Syria. The earliest record of chickpea from the Middle East dates back to 6250 BC.

The use of chickpea may date back to the early Neolithic period (8000–7000 BC) and is evidenced in the archaeological remains of carbonized chickpea reported from Cajoni in Turkey (van Zeist, 1972) and Tell Abu Hureyra in Syria (Hillman, 1975). Another authentic record for chickpea comes from the Hacilar site near Burdur in Turkey, radiocarbon-dated to 5450 BC (Helbaek, 1970). A bowl of chickpea seed dated to 1400 BC as a grave gift was found in Dier-el-Medineh in Ancient Egypt (Darby *et al.*, 1977). The biological remains of chickpea have been unearthed at various archaeological sites in Israel and Jordan, and dated to 3000–1000 BC (Hopf, 1969, 1978; Ellison *et al.*, 1978; McGreery, 1979; van der Maesen, 1987). Archaeological evidence indicates the spread of chickpea in Greece, at the earliest, from 800 BC (Kroll, 1981), in southern France from about 1000 BC (Cowtin and Erroux, 1974) and in Ethiopia via the Mediterranean by AD 1000 (Ramanujam, 1976a). In India, the earliest occurrence of chickpea, dating back to 2000 BC, has been reported from Atranjikhera in Uttar Pradesh (Chowdhury *et al.*, 1971; Vishnu-Mittre, 1974). In the 16th century, Spanish and Portuguese travellers took it to New World, most notably Mexico (Ramanujam, 1976b). Vavilov (1926, 1949) designated two primary centres of origin, South-west Asia and the Mediterranean, with Ethiopia being designated as a secondary centre. De Candolle (1883) traced the origin of chickpea to an area south of the Caucasus and northern Persia (now Iran).

Vigna spp.

Mung bean (*Vigna radiata* var. *radiata*) and urd bean (*Vigna mungo*) originated in the Indian subcontinent (de Candolle, 1884; Vavilov, 1926; Zukovskij, 1962). India contains a wide range of diversity of cultivated as well as weedy wild types of mung bean and is considered as the region of first domestication (Baudoin and Maréchal, 1988). Himachal Pradesh and Western Ghats in India are noted as centres of diversity of wild mung bean (Chandel, 1981), and maximum diversity among related species is limited to the upper Western Ghats and the Deccan hills (Pratap and John, 2010, unpublished data). A secondary centre of diversity exists in Bihar State in India. The progenitors of mung bean (*V. radiata* var. *sublobata*) and urd bean (*V. mungo* var. *sylvestris*) are seen in abundance as weeds in cultivated and wasteland areas of India (Singh *et al.*, 1974; Chandel *et al.*, 1984, Lawn and Cottell, 1988), as well as in wetlands in subtropical regions of northern and eastern Australia (Lawn and Cottell, 1988). Mention of mung bean in Vedic texts, such as *Charak Samhita*, indicates an origin far beyond the Christian era (Jain and Mehra, 1980) and the occurrence of archaeological records is unknown from anywhere outside India (Kajale, 1974). Charred grains of mung bean have been reported from Chalcolithic Navdatoli (1500–1000 BC), Neolithic Chairand, Bihar (1800 BC to AD 200), while carbonized grains of wild types have been reported by Kajale (1977) from Daimabad in Ahmednagar District in Maharashtra. The closeness between mung bean and urd bean is so prominent that they appear to be variants of a single species (Verdcourt, 1970). However, Watt and Maréchal (1977) discriminated them on the basis of free dipeptides in their seeds and confirmed them to be distinct. There is some archaeological evidence indicating the use of urd bean as early as around 1660–1440 BC (Vishnu-Mittre, 1974) and 2200–1000 BC (Kajale, 1977).

Adzuki bean (*Vigna angularis*) originated in the Chinese centre (Vavilov, 1926). The presumed wild ancestor of cultivated adzuki bean is *V. angularis* var. *nipponensis* (Yamaguchi, 1992; Kaga *et al.*, 2008). This wild species is distributed across a wide area from Japan, the Korean Peninsula and China to Nepal and Bhutan (Tomooka *et al.*, 2002). It exists as a crop complex in Japan where cultivated, wild and weedy adzuki bean are found (Vaughan *et al.*, 2004). Archaeological evidence in Japan, in the form of carbonized remains, trace it back to around 4000 years ago (Maeda, 1987; Yano *et al.*, 2004), predating archaeological remains in China and Korea (Crawford, 2006).

Moth bean (*Vigna aconitifolia*) is one of the most primitive *Vigna* species with respect to its evolution (Smartt, 1985). According to de Candolle (1886), moth bean grows wild in India. Vavilov (1926) also mentioned India as being the centre of origin due to the abundance of both wild and cultivated forms. However, Maréchal *et al.* (1978) reported Sri Lanka and Pakistan as being the centres of diversity of this crop. Rachie and Roberts (1974) concluded that moth bean is native to India, Pakistan and Burma. Its earliest mention is in the ancient Hindu text *Taitriya Brahmana*, a commentary in *Yajurveda* (*c.*7000 BC). *Kautilya* (321–296 BC) mentions it as a rainy season crop, while Watt (1889) described it as a drought-resistant crop widespread throughout the entire Indian subcontinent.

Regarding cowpea (*Vigna unguiculata*), it is now generally accepted that this crop originated in Africa. The origin and domestication of cowpea is believed to be West or Central Africa, very likely in Nigeria, where an abundance of both wild and weedy types flourishes in both savannah and forested zones (Harlan, 1971; Rawal, 1975). Archaeological evidence from West Africa dates the existence of cowpea to 3000 BC. Vavilov (1928, 1949), however, recognized India as the main centre of origin of this crop, while Africa and China were considered as secondary centres of origin. Cowpea reached India more than 2000 years ago (Ng and Maréchal, 1985) and two cultigroups, *biflora* and *sesquipedalis*, evolved from *V. unguiculata* in India and South-east Asia, respectively, as a result of intensive human selection. Excavations at Harappa (Indus–Saraswati civilization, 3200–2000 BC), have revealed that cowpea was one of the major grain legumes of those times (Mehra, 2002). The evolution of cowpea has been associated with changes in pod structure and seed coat, as well as with an increase in the rate of inbreeding (Maréchal *et al.*, 1978). In India, ssp. *cylindrica* evolved from ssp. *unguiculata* while ssp. *sesquipedalis* evolved in South-east Asia from vegetable types selected from ssp. *unguiculata*. This species reached Europe from Asia, and perhaps from North Africa, before 300 BC, and the Spanish took the crop to the West Indies in the 17th century. Later, more cultivars reached the New World from West Africa with the slave trade in the 16th century (Singh, 1991), reaching the southern part of the modern USA in the early eighteenth century (Steele and Mehra, 1980). Its wild forms are distributed all over tropical Africa and Madagascar, but are not seen in Asia. The wild forms of *V. unguiculata* are polymorphic and tentatively subdivided in subspecies – *dekindtiana* (Harms) Verdc., *tenuis* (E. Mey) M.M.& S. and *stenophylla* (Harvey) M.M.& S.

Rice bean (*Vigna umbellata*) is found in both wild and cultivated form in tropical areas of the Indian subcontinent, from the Himalayas to Sri Lanka, and it is very similar to *Phaseolus* (Hooker, 1879). Vavilov (1926) designated India as the centre of origin of both cultivated and wild forms of rice bean, inclusive of Assam and Burma but exclusive of north-west India. According to Chandel and Pant (1982), the cultivated forms seem to have originated from the wild populations growing in the Indian subcontinent. The species grows wild in the Himalayas (Chandel, 1981) and central China, extending its lower latitudinal limits to Malaysia and thus showing a diverse distributional and adaptive range from humid subtropical to warm and temperate climates (Chandel and Pant, 1982). The wild form, var. *gracilis*, is likely to be an ancestor of the rice bean. *Vigna mimima* (Roxb.) Ohwi & Ohashi and *Vigna delzelliana* display similarities to var. *gracilis* (Maréchal *et al.*, 1978). *V. minima* was considered a wild relative of the rice bean located in Western Ghats and Kerala (Gopinathan and Babu, 1986).

Common bean (*Phaseolus*)

The genus *Phaseolus* is of American origin and comprises over 30 species (Westphal, 1974; Debouck, 1999). However, only five of these species – *Phaseolus acutifolius* A. Gray (tepary bean), *Phaseolus coccineus* L. (scarlet runner bean), *Phaseolus lunatus* L. (lima bean), *Phaseolus polyanthus* Greenman (year-long bean) and *Phaseolus vulgaris* L. (common bean) have been used in cultivation (Gepts and Debouck, 1991; Debouck, 1999, 2000), the common bean being the most widely grown among these. Other species – *Phaseolus formosus* H.B.K. and *P. polystachyus* (L.) B.S.P. – are now also under cultivation or have been gathered from their habitat in the tropical areas of the American continent (Evans, 1980).

Wild populations of *Phaseolus* are distributed from northern Mexico to north-western Argentina (Gepts *et al.*, 1986; Koenig *et al.*, 1990). Common bean has multiple domestication sites through the distribution range in Middle and Andean South America (Harlan and de Wet, 1971; Gepts *et al.*, 1986). Archaeological evidence from South America indicates the domestication of *P. vulgaris* as far back as 6500–5000 BC (Kaplan *et al.*, 1973; Evans, 1976). The large-seeded lima bean (*P. lunatus*) is believed to have been domesticated in Peru around 4000 BC. *Phaseolus acutifolius* was domesticated around 1000 BC, while *P. coccineus* has a comparatively recent origin of the last millennium or so in the Tahau Can valley in Mexico (Kaplan, 1965). *Phaseolus vulgaris* and *P. lunatus* are likely to have travelled from America via the Philippines to Asia and from Brazil to Africa (Wanjari, 2005). Evans (1980) reported that they were widespread in Italy, Turkey, Iran and Greece during the 17th century. In the eastern USA, they were introduced only in the 19th century.

Pea (*Pisum*)

Archaeological evidence from the Near East dates the existence of peas back to 7000 BC (Baldev, 1988). In Europe, it has been grown since the Bronze Age, and in America it was introduced in the 16th century (Wanjari, 2005). In India, the earliest references to pea are found in the dictionary of Amarsimha (Amarcosa, c. 200 BC), where it was named *khandika* or *harenu* in Sanskrit. It also found mention by both Varahamihira (6th century AD) and Bhavaprakash (16th century AD).

Neither the wild progenitor nor the early history of the pea is known. However, Vavilov (1949) recognized Ethiopia, the Mediterranean and Central Asia regions as the probable centre of origin of this crop, while the north-eastern centre is a secondary centre of diversity where other related types such as *Pisum elatius*, *Pisum humile* and *Pisum fulvum* are abundant. de Candolle (1886) believed that the progenitor of *Pisum* existed in northern India. Purseglove (1968) opined that wild forms found in Georgia and Russia are very similar to the cultivated variety.

Therefore, *P. elatius* may be an ancestral form in the evolution of field peas created through the introgression of genes. However, an independently derived cultivated type known as *Pisum sativum* ssp. *abyssinicum*, which is restricted to highland regions of Ethiopia and southern Yemen and shows a greater affinity to *P. fulvum* (Vershinin *et al.*, 2003). *Pisum fulvum* is found abundantly in Syria, Lebanon, Israel, Palestine and Jordan (Maxted and Ambrose, 2001). It was also indicated that *P. sativum* probably originated in medieval times through a mutation to white flowers and large seeds in the cultivated form of *Pisum arvense* (Smartt, 1976). However, more recent studies based on molecular data suggest that *P. sativum* is nested within the diversity of *P. elatius* and may even be paraphyletic

(Vershinin *et al.*, 2003; Baranger *et al.*, 2004; Taran *et al.*, 2005), suggesting that cultivated *P. sativum* was derived mainly from *P. elatius* (Jing *et al.*, 2010). There is also support for the view that cultivated species are not cross-compatible with *P. fulvum*. Other claimed wild species, such as *Pisum humile* and *Pisum jomardii*, have received little support from molecular studies (Jing *et al.*, 2010). However, on the other hand, extensive sharing of molecular markers among *Pisum* species has suggested significant outcrossing and introgression between species, although it is a predominantly inbreeding genus by nature (greater than 99%; Jing *et al.*, 2010). These studies, therefore, favour *Pisum* as a species complex with multiple subspecies, with interbreeding having occurred in varying degrees (Vershinin *et al.*, 2003).

Lentil (*Lens*)

The word 'lentil' comes from the Latin *lens*, and Medikus, a German botanist and physician, gave it its scientific name *Lens culinaris* in 1787 (see Cubero, 1981). Most probably, it is one of the oldest cultivated crops adapted to the most inhospitable agricultural environments in the cooler temperate zones of the world, or to the winter season in Mediterranean climates (Harlan, 1992). Archaeological evidence indicates the lentil as being native to south-western Asia (Turkey–Syria–Iraq region).

Lentils were known in India before the 1st century AD, where it is known as *masura*, which means pillow in Sanskrit. The word *masura* has also been mentioned by later authors in the *Brahadaranyaka* (c. 5500 BC), in the *Yajurveda* (c. 7000 BC), in a commentary in the *Rigveda* (c. 8000 BC) and in the *Charaka* (c. 700 BC), *Susruta* (c. 400 BC) and *Kautilya* (c. 321–296 BC). Interestingly, the Turkic word *mercimek* and an Old Persian word *marjunak* for lentil are both phonetically close to *masura*. The 15th-century *Hortus Sanitatis* lists some medicinal properties of lentil by collecting information from the *Dioscordies* and other ancient references. In Spain, Gabriel-Alonso de Herrera writes in 1502 that the best ones are the biggest, as they are large, white and do not produce a black tint in water (Cubero *et al.*, 2009).

Based on a rural 'cyclopaedia' of the mid-19th century, lentil had been introduced into England from France during the 15th century, and by the middle of the 19th century the UK had four varieties that were succinctly described as big, small, red and yellow. Thus it seems that the European variety structure remained largely unchanged from the 16th to the 19th century, and was probably the same as that in the Middle Ages. It was introduced in the early 1900s to the USA. The archaeological data, the distribution of wild species and overlapping of both wild and cultivated lentils in the same regions suggest that the Near East and central Asia, i.e. the Turkey–Cyprus region (south-west Asia or Near East or Mediterranean area), is the obvious candidate for the origin of the cultivated species *Lens culinaris* (Cubero, 1981). This region is the likely site of lentil domestication, where some populations of *Lens orientalis* were unconsciously subjected to automatic selection, leading to a new species, *L. culinaris* (see Cubero *et al.*, 2009 for details). Previously, the eastern border of south-west Asia (i.e. the region between Afghanistan, India and Turkistan) has been considered as the possible centre of origin due to the presence of the highest proportion of endemic varieties (Barulina, 1930). However, more recently this region has become better explained as a secondary centre of diversity.

The most detailed and complete study of the cultivated lentil was made by Barulina (1930), who described *Lens microsperma* and *Lens macrosperma* as two subspecies of cultivated species on the basis of seed size. She also considered the geographical distribution and defined six different regional groups or *greges* (i.e. *pilosae*, *subspontanea*, *aethiopicae*, *europeae*, *asiaticae* and *intermediate*) within former subspecies and no geographical group within later subspecies. Distributions of Barulina's *greges* and wild lentils have better explained the evolution of cultivated species and its varietal *facies* in lentil. Three *greges* having only a distinct character are restricted to very concrete regions: *pilosae* to the Indian subcontinent (a strong pubescence), *aethiopicae* to Ethopia and Yemen (pods with a characteristically elongated apex) and *subspontanea* to the Afghan regions closest to the Indian subcontinent (very dehiscent pods, purple coloured before maturity). All those characters distinguishing *greges* from others are seen together in the closely related species *orientalis*.

However, the unique characteristics of each *grege* mentioned above are shown together with a cluster of primitive characteristics of closely related to *orientalis*. The distribution of *subspontanea* also overlaps with that of the wild species *orientalis*, and both *orientalis* and *culnaris* forms are found together in the south Turkey– north Syria region. Thus *orientalis* has played a leading role in the evolution of eastern small-seeded lentils, while the wild species *Lens ervoides* has spread southwards and overlaps with the short-calyx, *Lens aethiopicae*, suggesting its contribution to the evolution of this small-seeded *grege*. The *microsperma* and *macrosperma* varieties overlap to a greater or lesser extent with known wild lentils and are clearly intermixed. However, the easy cross-compatibility of *Lens odemensis* with *Lens culnaris* may have generated the genetic raw material for the western lentils with their larger seeds, high number of large leaflets and calyx teeth longer than the corolla. The westward spread of *Lens nigricans* and *L. ervoides* implies their role in the evolution of western lentils, because of the probability of survival of some crosses in natural environments despite their cross-incompatibility with cultigens due to hybrid embryo abortion. Thus *L. orientalis* and *L. odemensis* forms are most likely candidates as companion weeds of the cultigen, and *L. microsperma* and *L. macrosperma* have evolved simply through disruptive selection (Cubero *et al.*, 2009).

Faba bean (*Vicia faba*)

Contrasting views have been reported on the origin and domestication of faba bean (Maxted *et al.*, 1991). Earlier studies postulated the Near East as the centre of origin (Cubero, 1973, 1974), with several different routes possibly having led to its spread to Europe: along the north African coast to Spain, along the Nile to Ethiopia and from Mesopotamia to India. However, later studies suggest that central Asia (Ladizinsky, 1975) or south-eastern Europe and south-western Asia (Muratova, 1931; Maxted, 1995) were the centres of origin for the genus *Vicia*.

The small- and large-seeded forms of faba bean are predominant in nature. The former type is very ancient compared with the latter as it has been traced back to the Neolithic culture and its remains have been found in an archaeological excavation in Israel, dating it at 6800–6500 BC (Kislev, 1985; Garfinkel, 1987). The small-seeded group is found over a large area (from Spain to the Himalayas), and also has the greatest number of endemics and diversity with many specific traits that are lacking in other groups (Muratova, 1931). Therefore south-western Asia, where the small-seeded faba bean is predominant, is considered the principal centre origin of *V. faba* and the Mediterranean region, with its concentration of large-seeded forms, is considered a secondary centre (Muratova, 1931). Another secondary centre of diversity for genetic resources of faba bean is probably China, where the faba bean gene pool, especially the winter gene pool, has been reproductively isolated from the European and West Asian gene pools (Zong *et al.*, 2009). However, the timing of the introduction of faba bean to China is uncertain and various views have been reported (Zheng *et al.*, 1997; Ye *et al.*, 2003). A detailed account on the origin of various types of faba bean can be found in a recent review by Duc *et al.* (2010).

There are different views regarding the probable progenitors of cultivated species. Earlier, *Vicia pliniana* (Trabut) Murat from Algeria, used for cooking (Trabut, 1911), was considered the closest wild relative (Muratova, 1931). However, differences from *V. faba* in morphological characters including a broad arillus, the anatomical structure of the seed coat and weak swelling properties have allowed it as an independent species, *Vicia pliniana*. *Vicia faba paucijuga* is presumed to be another ancestor, which has a short stem, small number of leaflets per leaf and very small seeds (Cubero and Suso, 1981). On the basis of many morphological similarities and coincidence in their distribution, Hopf (1973) proposed *Vicia narbonensis* L. as a probable wild ancestor, although he later argued against this species being a progenitor (Ladizinsky, 1975). Although *V. narbonensis*, *Vicia johannis* and

Vicia bithynica all cross well with each other, many attempts to cross *V. faba* with any of its relatives have failed (Cubero, 1982; Hanelt and Mettin, 1989). Neither has the crossing compatibility of cultivated species been observed with other wild species such as *V. narbonensis*, *Vicia melanops*, *Vicia lutea* and *V. johannis*, and the phenomenon of embryo abortion has been observed in hybrids (Ramsay and Pickersgill, 1986; Raupakias, 1986).

Soybean (*Glycine*)

The history of soybean is well documented (Hymowitz, 1970; Guo, 1993; Guo *et al.*, 2010). Evidence suggests that soybean emerged as a domesticate during the Zhou dynasty in the eastern half of northern China. The oldest records appear in bronze inscriptions and in early writings that date not much earlier than 1100 BC. Because domestication is a process of trial and error and is not a time-datable event, this process probably took place during the Shang dynasty.

The current evidence for the antiquity of the soybean lies in the pictographical analysis of the archaic Chinese word for soybean (*Shu*) that first appeared in *The Book of Odes* (*Shih-ching*) during the Zhou dynasty and on bronze inscriptions. The word *Shu* pictographically depicts the horizontal line in the middle as earth; the upper and lower parts represent the stem and root, while around the root the three teardrop-like lines illustrate the nodules. With the expansion of the Zhou dynasty, trading in soybean moved to South China, Korea, Japan and South-east Asia. By the 1st century AD, soybean had probably been distributed throughout China by trade missions and, with time, to other Asian countries. The earliest Japanese reference to soybean is found in the *Kojiki* (Records of Ancient Matters), completed in AD 712. In the 16th and 17th centuries, there are several references to native soy foods in the diaries of European visitors to China and Japan. The first soybeans were brought to the USA in 1765 by Samuel Bowen, a seaman employed by the East India Company, and planted by Henry Yonge on his plantation 'Greenwich' located at Thunderbolt, a few miles east of Savannah, Georgia. Mr. Bowen used the soybean to produce soy sauce and a soybean noodle for export to England (Soybean Meal Information Centre, 2011).

Molecular and morphological data on genetic diversity among the wild and cultivated types collected from both South and North China favour South China as the place of origin of cultivated species. The late-type soybean from South China was found closer to the wild type and it is expected that the wild soybean is the common ancestor for the cultivated type of South China, from which early cultivated types were originated during the process of dissemination to North China (Gai *et al.*, 2000). The higher genetic diversity among the South China population compared with that of North China also supports the origin of soybean as being South China (Ding *et al.*, 2008). These contrasting pieces of evidence therefore support the earlier study that domestication of soybean occurred simultaneously in several regions of China (Lu, 1978).

Studies show that the genus *Glycine* is of ancient polyploid origin (Qiu and Chang, 2010), and its genome has passed through two major rounds of duplication events during speciation (Schlueter *et al.*, 2004; Van *et al.* 2008; see Chapter 7, this volume). *Glycine* (perennial) and *Soja* (annual) are two subgenera. Twenty-four species of *Glycine*, including both annual and perennial, are known, but taxonomically annual species (both wild and cultivated) are grouped under the subgenus *Soja*. The wild and cultivated annual species are known as *Glycine soja* and *Glycine max*, respectively. Most molecular studies show that the cultivated species *G. max* has close a phylogenetic relationship to the wild species *G. soja*, which is known as the progenitor of this species (see Qiu and Chang, 2010 for more details).

References

Abbo, S., Shtienberg, D., Lichtenzveig, J., Lev Yadun, S. and Gopher, A. (2003) The chickpea, summer cropping, and a new model for pulse domestication in the ancient Near East. *The Quarterly Review of Biology* 78, 435–448.

Annual Wheat News Letter (2010) Timeline for wheat evolution, 56. Available at http://wheat.pw.usda.gov/ggpages/awn/56/ (accessed 14 March 2011).
Baldev, B. (1988) Origin, distribution, taxonomy and morphology. In: Baldev, B., Ramanujam, S. and Jain, H.K. (eds) *Pulse Crops (Grain Legumes)*. Oxford & IBH Publishers, New Delhi, India, pp. 3–51.
Baranger, A., Aubert, G., Arnau, G., Lainé, A.L., Deniot, G., Potier, J. *et al.* (2004) Genetic diversity within *Pisum sativum* using protein- and PCR-based markers. *Theoretical Applied Genetics* 108, 1309–1321.
Barulina, H. (1930) Lentils of the USSR and other countries. *Bulletin of Applied Botany, Genetics and Plant Breeding* 40, 265–304 [in Russian with English summary].
Baudoin, J.P. and Maréchal, R. (1988) Taxonomy and evolution of the genus *Vigna*. In: *Mungbean. Proceedings of Second International Symposium*, November 16–20, 1987, Bangkok. AVRDC, Shanhua, Taiwan.
Bentham, G. (1861) On the species and genera of plants considered with reference to the practical application to systematic botany. *Natural History Review* 1861, 133–151.
Bentham, G. (1865) Ordo LVII Leguminosae. In: Bentham, G. and Hooker, J.D. (eds) *Genera Plantarum, vol. (2)*. Reeve, London, pp. 434–600.
Butler, A. (2007) Grain legumes: Evidence of these important ancient food resources from early pre-agrarian and agrarian sites in southwest Asia. In: Damania, A.B., Valkoun, J., Willcox, G. and Qualset, C.O. (eds) *The Origins of Agriculture and Crop Domestication*. ICARDA, Aleppo, Syria, pp. 345.
Chandel, K.P.S. (1981) Wild *Vigna* species in Himalayas. *Plant Genetic Resources Newsletter* 45, 17–19.
Chandel, K.P.S. and Pant, K.C. (1982) Genetic resources of *Vigna* species in India: their distribution, diversity and utilization in crop improvement. *Annals of Agriculture Research* 3, 19–34.
Chandel, K.P.S., Lester, R.N. and Starling, R.J. (1984) The wild ancestors of urd and mung beans (*Vigna mungo* (L.) Hepper and *V. radiata* (L.) Wilczek). *Botanical Journal of the Linnean Society* 89, 85–96.
Chowdhury, K.A., Saraswat, K.S., Hasan, S.N. and Gaur, R.G. (1971) 4.000–3.5000 year old barley, rice and pulses from Atranjikhera. *Science and Culture* 37, 531–533.
Colledge, S.M. (1994) Plant exploitation on Epipalaeolithic and early Neolithic sites in the Levant. PhD Thesis, University of Sheffield, UK.
Cowan, R.S. (1981) Swartzieae. In: Polhill, R.M. and Raven, P.H. (eds) *Advances in Legume Systematics, Part 1*. Royal Botanic Gardens, Kew, UK, pp. 209–212.
Cowtin, J. and Erroux, J. (1974) Apercusur. I. Agriculture prehistaridue dansle sudest dela Frances. *Bulletin of the Society of Prehistory in France* 71, 321–334.
Crawford, G.W. (2006) East Asian plant domestication, In: Stark, M.T. (ed.) *Archaeology of Asia*. Blackwell Publishing, Oxford, UK, pp. 77–95.
Cronk, Q.C.B. (1990) The name of the pea: a quantitative history of legume classification. *New Phytology* 116, 163–175.
Cronquist, A. (1981) *An Integrated System of Classification of Flowering Plants*. Columbia University Press, New York.
Cubero, J.I. (1973) Evolutionary trends in *Vicia faba*. *Theoretical and Applied Genetics* 43, 59–65.
Cubero, J.I. (1974) On the evolution of *Vicia faba*. *Theoretical and Applied Genetics* 45, 47–51.
Cubero, J.I. (1981) Origin, taxonomy and domestication. In: Webb, C. and Hawtin, G.C. (eds) *Lentils*. Commonwealth Agricultural Bureaux, Farnham Royal, UK, pp. 15–21.
Cubero, J.I. (1982) Interspecific hybridization in *Vicia*. In: Hawtin, G. and Webb, C. (eds) *Faba Bean Improvement*. Martinus Nijhoff Publishers, The Hague, The Netherlands, pp. 91–108.
Cubero, J.I. and Suso, M. (1981) Primitive and modern forms of *Vicia faba*. *Kulturflanze* XXIX, 137–145.
Cubero, J.I., Perez de la Vega, M. and Fratini, R. (2009) Origin, phylogeny, domestication and spread. In: Erskine, W., Muehlbauer, F.J., Sarker, A. and Sharma, B. (eds) *The Lentil: Botany, Production and Uses*. CAB International, Wallingford, UK, pp. 13–33.
Darby, W., Ghalioungui, P. and Grivetti, L. (1977) *Food: The Gift of Osiris*, 2 vols, Academic Press, New York and San Francisco, California.
De, D.N. (1974) Pigeon pea. In: Hutchinson, J. (ed.) *Evolutionary Studies of World Crops: Diversity and Change in the Indian Subcontinent*. Cambridge University Press, London, pp. 79–87.
Debouck, D.G. (1999) Diversity in *Phaseolus* species in relation to the common bean. In: Singh, S.P. (ed.) *Common Bean Improvement in the Twenty-First Century*. Kluwer, Dordrecht, The Netherlands, pp. 25–52.
Debouck, D.G. (2000) Biodiversity, ecology and genetic resources of *Phaseolus* beans – seven answered and unanswered questions. In: *Proceedings of the 7th MAFE International Workshop on Genetic Resources Part 1. Wild Legumes*. AFFRC and NIAR, Japan, pp. 95–123.
de Candolle, A. (1883) *Origine des Plantes Cultivées*. Nabu Press, Paris, pp. 250–260.
de Candolle, A. (1884) *Origine des Plantes Cultivées*. Nabu Press, USA, pp. 424.

de Candolle, A. (1886) *Origin of Cultivated Plants*. 2nd edn. Hafner Publishing Company, New York.
de Candolle, A. and De, P. (1825) *Prodromus Systematis Naturalis*. Treuttel and Wurtz, Paris.
de Jussieu, A.L. (1789) *Genera Plantarum, Secundum Ordines Naturales Disposita Juxta Methodum in Horto Regio Parisiensi Eexaratam* (available online at http://gallica.bnf.fr/).
Dillehay, T.D. (2007) *Earliest Known Evidence of Peanut, Cotton and Squash Farming Found*. Available at http://www.eurekalert.org/pub_releases/2007-06/vu-eeo062507.php (accessed 14 March 2011).
Ding, Y.L., Zhao, T.J. and Gai, J.Y. (2008) Genetic diversity and ecological differentiation of Chinese annual wild soybean (*Glycine soja*). *Biodiversity Science* 16, 133–142.
Doyle, J.J. and Lucknow, M.A. (2003) The rest of the iceberg. Legume diversity and evolution in a phylogenetic context. *Plant Physiology* 131, 900–910.
Doyle, J.J., Doyle, J.L., Ballenger, J.A., Dickson, E.E., Kajita, T. and Ohashi, H. (1997) A phylogeny of the chloroplast gene *rbcL* in the Leguminosae: taxonomic correlations and insights into the evolution of nodulation. *American Journal of Botany* 84, 541–554.
Duc, G., Bao, S., Baum, M., Redden, B., Sadiki, M., Suso, M.J. *et al*. (2010) Diversity maintenance and use of *Vicia faba* L. genetic resources. *Field Crops Research* 115, 270–278.
Ellison, R., Renfrew, J., Brothwell, D. and Seeley, N. (1978) Some food offerings from *Ur*, excavated by Sir Leonard Woolley, and previously unpublished. *Journal of Archeological Science* 5, 167–177.
Evans, A.M. (1976) Beans – *Phaseolus* spp. In: Simmonds, N.W. (ed.) *Evolution of Crop Plants*. Longman, London, pp. 168–172.
Evans, A.M. (1980) Structure, variation, evolution and classification in *Phaseolus*. In: Summerfield, R.J. and Bunting, A.H. (eds) *Advances in Legume Science*. Royal Botanic Gardens, Kew, UK, pp. 337–347.
FAO (1959) *Tabulated Information on Tropical and Subtropical Grain Legumes*. Food and Agricultural Organization of the United Nations, Rome, pp. 45–62.
FDM (1996) *Mediterranean Diet Foundation* (available at: http://fdmed.org/en/mediterranean-diet/food-pyramid/legumes).
Gai, J.Y., Xu, D.H., Gao, Z., Abe, Y.S.J., Fukushi, H. and Kitajima, S. (2000) Studies on the evolutionary relationship among eco-types of *G. max* and *G. soja* in China. *Acta Agronomica Sinica* 26, 513–520.
Garfinkel, Y. (1987) Yiftahel: a neolithic village from the seventh millennium BC in lower Galilee. *Israel Journal of Field Archaeology* 14, 199–212.
Gepts, P. and Debouck, D. (1991) Origin, domestication and evolution of the common bean (*Phaseolus vulgaris* L.). In: van Schoonhoven, A. and Voysest, O. (eds) *Common Beans: Research for Crop Improvement*. CAB International, Wallingford, UK and CIAT, Cali, Colombia, pp. 7–53.
Gepts, P., Osborn, T.C., Rashka, K. and Bliss, F.A. (1986) Phaseolin protein variability in wild forms and landraces of the common bean (*Phaseolus vulgaris*): evidence for multiple centres of domestication. *Economic Botany* 40, 451–468.
Gopinathan, M.C. and Babu, C.R. (1986) A unique growth pattern associated with cliestogamy in a tropical legume, *Vigna minima* (Roxb.) Ohwi and Ohashi (Leguminosae). *Botanical Journal of the Linnean Society* 92, 263–268.
Gunn, A. (1983) A Nomenclator of legume (Fabaceae) genera. US Department of Agriculture Technical Bulletin No. 168.
Guo, J., Wang, Y., Song, C., Zhou, J., Qiu, L., Huang, H. *et al*. (2010) A single origin and moderate bottleneck during domestication of soybean (*Glycine max*): implications from microsatellites and nucleotide sequences. *Annals of Botany*, doi: 10.1093/aob/mcq125.
Guo, W.T. (1993) *The History of Soybean Cultivation*. Hehai University Press, Nanjing, China.
Hanelt, P. and Mettin, D. (1989) Biosystematics of the genus *Vicia* L. (Leguminosae). *Annual Review of Ecology and Systematics* 20, 199–223.
Harlan, J.R. (1971) Agricultural origins: centres and noncentres. *Science* 174, 468–474.
Harlan, J.R. (1992) *Crop and Man*. American Society of Agronomy, Madison, Wisconsin.
Harlan, J.R. (1995) *The Living Fields*. Cambridge University Press, Cambridge, UK.
Harlan, J.R. and de Wet, J.M.J. (1971) Toward a rational classification of cultivated plants. *Taxon* 20, 509–517.
Harlan, J.R., de Wet, J.M.J. and Stemler, A.B.L. (1976) Plant domestication and indigenous African agriculture. In: Harlan, J.R., de Wet, J.M.J. and Stemler, A.B.L. (eds) *Origins of African Plant Domestication*. Mouton, The Hague, The Netherlands, pp. 3–19.
Helbaek, H. (1970) The plant husbandry at Hacilar. In: Mellaart, J. (ed.) *Excavation at Hacilar*. Edinburgh University Press, Edinburgh, UK, pp. 189–244.
Herendeen, P.S. (1992) The fossil history of leguminosae from the eocene of southeastern North America. In: Herendeen, P.S. and Dilcher, D.L. (eds) *Advances in Legume Systematics; Part 4, The Fossil Record*. Royal Botanic Gardens, Kew, UK. pp. 85–160.

Herendeen, P.S. (1995) Phylogenetic relationships of the tribe Swartzieae. In: Crisp, M.D. and Doyle, J.J. (eds) *Advances in Legume Systematics, Part 7: Phylogeny*. Royal Botanic Gardens, Kew, UK, pp. 123–132.
Herendeen, P.S. and Wing, S. (2001) Papilionoid legume fruits and leaves from the Paleocene of northwestern Wyoming. *Botany 2001 Abstracts*, Botanical Society of America, St Louis, Missouri (http://www.botany2001.org/).
Hillman, G. (1975) The plant remains from Tell Abv. Hureyra. A preliminary report. *Proceedings of the Prehistory Society* 41, 70–73.
Hooker, J.D. (1879) *The Flora of British India. Vol. II. Sabiaceae to Cornacra*. L. Reeve and Company Ltd, Ashford, UK.
Hopf, M. (1969) Plant remains and early farming in Jericho. In: Ucko, U. and Dimbley, G.W. (eds) *The Domestication and Exploitation of Plants and Animals*. Duckworth, London, pp. 355–359.
Hopf, M. (1973) Fruhe kulturpflanzen aus Bulgarien. *Jahrbuch des Romisch-Germanischer Zentralmuseums Mainz* 20, 1–47.
Hopf, M. (1978) Plant remains: Early arad I. In: Amerian, R. (ed.) *The Chalcolithic Settlement and Early Bronze Age City*. Israel Exploration Society, Jerusalem, Israel, pp. 64–82.
Hopf, M. and Bar-Yosef, O. (1987) Plant remains from Hayonim Cave, Western Galilee. *Paléorient* 13, 115–120.
Hutchinson, J. (1964) *The Genera of Flowering Plants, Vol. 1, Order 7, Leguminales*. Clarendon Press, Oxford, UK, pp. 221–289.
Hymowitz, T. (1970) On the domestication of the soybeans. *Economic Botany* 23, 408–421.
Hymowitz, T. (1990) Grain legumes. In: Janick, J. and Simon. J.E. (eds) *Advances in New Crops*, Timber Press, Portland, Oregon.
Jain, H.K. and Mehra, K.L. (1980) Evolution, adaptation, relationships and uses of the species of *Vigna* cultivated in India. In: Summerfield, R.J. and Bunting, A.H. (eds) *Advances in Legume Science*. Royal Botanic Gardens, Kew, UK, pp. 459–468.
Jing, R., Vershinin, A., Grzebyta1, J., Shaw, P., Smýkal, P., Marshall, D. *et al.* (2010) The genetic diversity and evolution of field pea (*Pisum*) studied by high throughput retrotransposon based insertion polymorphism (RBIP) marker analysis. *BMC Evolutionary Biology* 10, 44–53.
Kaga, A., Isemura, T., Tomooka, N. and Vaughan, D.A. (2008) The genetics of domestication of the azuki bean. *Genetics* 178, 1013–1036.
Kajale, M.D. (1974) Plant economy at Bhokardan. Appendix A. In: Dev, S.B. and Gupta, R.S. (eds), *Excavations at Bhokardan* (*Bhogavardana*) 1973. Nagpur University and Maharashtra Marathwada University, Maharashtra, India, pp. 7–224.
Kajale, M.D. (1977) On the botanical findings from excavations at Diamabad, a Chalecolithic site in Western Maharashtra, India. *Current Science* 46, 818–819.
Kaplan, L. (1965) Archaeology and domestication in America. *Phaseolus. Economic Botany* 19, 358–368.
Kaplan, L., Lynch, T.F. and Smith, C.E. Jr. (1973) Early cultivated beans (*Phaseolus vulgaris*) from an intermontane Peruvian valley. *Science* 179, 76–77.
Katherine, B. and Museum, L. (1997) *The Spirit of Ancient Peru: Treasures from the Museo Arqueológico Rafael Larco Herrera*. Thames and Hudson, New York, USA.
Kirkbride Jr., J.H., Gunn, C.R. and Weitzman, A.L. (2003) Fruits and seeds of genera in the subfamily Faboideae (Fabaceae). U.S. Department of Agriculture, Technical Bulletin No. 1890, 1212 pp.
Kislev, M.E. (1985) Early neolithic horsebean from Yiftahel, Israel. *Science* 228, 319–320.
Koenig, R.L., Singh, S.P. and Gepts, P. (1990) Novel phaseolin types in wild and cultivated common bean (*Phaseolus vulgaris*: Fabaceae). *Economic Botany* 44, 50–60.
Krauss, F.G. (1932) The pigeon pea (*Cajanus indicus*): its improvement, culture, and utilization in Hawaii. *Hawaii Agricultural Experiment Station Bulletin* 64, 1–46.
Kroll, H. (1981) Thessalische Kulturpflanzen. *Z. Arachaol* 15, 97–103.
Kwak, M., Kami, J. and Gepts, P. (2009) The putative Mesoamerican centre of domestication of *Phaseolus vulgaris* L. is located in the Rio Lerma-Santiago basin of Mexico. *Crop Science* 49, 554–563.
Ladizinsky, G. (1975) On the origin of the broad bean *Vicia faba* L. *Israel Journal of Botany* 24, 80–88.
Ladock, J. (2010) *History of Legumes: Man's Use of Legumes* (available at: http://www.healthguidance.org/entry/9933/1/History-of-Legumes-Mans-Use-of-Legumes.html).
Ladizinsky, G. and Adler, A. (1976) The origin of chickpea *Cicer arietinum* L. *Euphytica* 25, 211–217.
Lavin, M., Herendeen, P.S. and Wojciechowski, M.F. (2005) Evolutionary rates analysis of Leguminosae implicates a rapid diversification of lineages during the Tertiary. *Systematic Biology* 54, 530–549.
Lawn, R.J. and Cottell, A. (1988) Wild mung bean and its relative in Australia. *Biologist* 35, 267–273.

Linnaeus, C. (1753) *Species Plantarum* (edn 1). L. Salvus, Stockholm. Facsimile in 2 vols, 1957, Ray Society, London.

Lu, S. (1978) Discussion on the original region of cultivated soybean in China. *Scienta Agricultura Sinica* 4, 90–94.

Maeda, K. (1987) *Legumes and Humans: A 10,000 Year History*. Kokonshin, Tokyo, Japan [in Japanese].

Maréchal, R., Masherpa, J.M. and Stainer, F. (1978) Etude taxonomique d'un groupe complexe d'especes des genres *Phaseolus* et *Vigna* (*Papilionaceae*) sur la base de donnees morphologiques et polliniques, traitees par l'analyse informatique. *Boissiera* 28, 1–273.

Maxted, N. (1995) An ecogeographical study of *Vicia* subgenus *Vicia*. In: *Systematic and Ecogeographical Studies on Crop Genepools, vol. 8*. IPGRI, Rome.

Maxted, N., Callimassia, M.A. and Bennet, M.D. (1991) Cytotaxonomic studies of eastern Mediterranean *Vicia* species (Leguminosae). *Plant Systematic and Evolution* 77, 221–234.

Maxted, N. and Ambrose, M. (2001) Peas (*Pisum* L.). In: Maxted, N. and Bennett, S.J. (eds) *Plant Genetic Resources of Legumes in the Mediterranean*. Kluwer Academic Publishers. Dordrecht, The Netherlands, pp. 81–190.

McGreery, D.W. (1979) Flotation of the Badedh-Dhra and Numeria plant remains. Annual American schools. *Oriental Research* 46, 165–169.

Mehra, K.L. (2002) Agricultural foundation of Indus-Saraswati civilization. In: Nene, Y.L. and Choudhary, S.L. (eds) *Proceedings of the National Conference* on *Agriculture Heritage of India*, 10–13 February 2002. Rajsthan College of Agriculture, MPUAT, Udaipur, Rajasthan, India, pp. 1–21.

Muehlbauer, F.J. (1993) Food and grain legumes. In: Janick, J. and Simon, J.E. (eds) *New Crops*. John Wiley & Sons, New York.

Muratova, V.S. (1931) Common beans (*Vicia faba* L.). *Bulletin of Applied Botany, Genetics and Plant Breeding Suppl*. 50, 1–298.

Ng, N.Q. and Maréchal, R. (1985) Cowpea taxonomy, origin and germplasm. In: Singh, S.R. and Rachie, K.O. (eds) *Cowpea Research, Production and Utilization*. John Wiley & Sons, New York, pp. 11–21.

Pan, A.D., Jacobs, B.F. and Herendeen, P.S. (2010) Detarieae *sensu lato* (Fabaceae) from the Late Oligocene (27.23 Ma) Guang River flora of north-western Ethiopia. *Botanical Journal of the Linnean Society* 163, 44–54.

Perry, L., Dickau, R., Zarrillo, S., Holst, I., Persall, D.M., Piperno, D. *et al*. (2007) Starch fossils and the domestication and dispersal of chili peppers (*Capsicum* spp. L.) in the Americas. *Science* 315, 986–988.

Persoon, C. (1805) *Synopsis Plantarum* (*Part 1*). C.F. Cramer, Paris and J.G. Cottam, Tubingen, Germany, 546 pp.

Persoon, C. (1807) *Synopsis Plantarum, Part 2*. Treuttel & Wurtz, Paris and J.G. Cottam, Tubingen, Germany, 657 pp.

Piperno, D.R., Ranere, A.J., Holst, I., Dickau, R. and Iriarte, J. (2009) Starch grain and phytolith evidence for early ninth millennium B.P. maize in the central Balsas river valley, Mexico. *Proceedings of the National Academy of Sciences U.S.A.* 106, 5019–5024.

Plukenet, L. (1962) *Phytographia* 3, Table 213, Figure 3.

Polhill, R.M. (1994) Classification of the Leguminosae. In: Bisby, F.A., Buckingham, J. and Harborne, J.B. (eds), *Phytochemical Dictionary of the Leguminosae*. Chapman and Hall, New York, pp. xxxv–xlviii.

Polhill, R.M. and Raven, P.H. (1981) *Advances in Legume Systematics, Part 1*. Royal Botanic Gardens, Kew, UK.

Purseglove, J.W. (1968) *Tropical Crops – Dicotyledons 1*. Longman, London, pp. 236–237.

Qiu, L.J. and Chang, R.Z. (2010) The origin and history of soybean. In: Singh, G. (ed.) *The Soybean: Botany, Production and Uses*. CAB International, Wallingford, UK, pp. 1–23.

Rachie, K.O. and Roberts, L.M. (1974) Grain legumes of the lowland tropics. *Advances in Agronomy* 26, 1–132.

Ramanujam, S. (1976a) Chickpea. In: Simmonds, N.W. (ed.) *Evolution of Crop Plants*. Longman, London, pp. 167–176.

Ramanujam, S. (1976b) Chickpea. In: Simmonds, N.W. (ed.) *Evolution of Crop Plants*. Longman, London, pp. 157–159.

Ramsay, G. and Pickersgill, B. (1986) Interspecific hybridisation between *Vicia faba* and other species of *Vicia*: approaches to delaying embryo abortion. *Biology Zentralbla* 105, 171–179.

Raupakias, D.G. (1986) Interspecific hybridization between *Vicia faba* (L.) and *Vicia narbonensis* (L.). Early pod growth and embryo-sac development. *Euphytica* 35, 175–183.

Raven, P.H. and Axelrod, D.I. (1974) Angiosperm biogeography and past continental movements. *Annals of the Missouri Botanic Garden* 61, 539–657.

Rawal, K.M. (1975) Natural hybridization among wild, weedy and cultivated *Vigna unguiculata* (L.) Walp. *Euphytica* 24, 699–707.
Royes, W.W. (1976) Pigeon pea. In: Simmonds, N.W. (ed.) *Evolution of Crop Plants*. Longman, London, pp. 154–156.
Schlueter, J.A.. Dixon, P., Granger, C., Grant, D., Clark, L. *et al.* (2004) Mining EST databases to resolve evolutionary events in major crop species. *Genome* 47, 868–876.
Schrire, B.D., Lewis, G.P. and Lavin, M. (2005a) Biogeography of the leguminosae. In: Lewis, G., Schrire, G., Mackinder, B. *et al.* (eds), *Legumes of the World*. Royal Botanic Gardens, Kew, UK, pp. 21–54.
Schrire, B.D., Lavin, M. and Lewis, G.P. (2005b) Global distribution patterns of the Leguminosae: insights from recent phylogenies. In: Friis, I. and Balslev, H. (eds), *Plant Diversity and Complexity Patterns: Local, Regional and Global Dimensions*. Biologiske Skrifter 55, Special-Trykkeriet Viborg A/S, Viborg, Denmark, pp. 375–422.
Singh, D.P. (1991) *Genetics and Breeding of Pulse Crops*. Oxford & IBH Publications, New Delhi, India.
Singh, H.B., Joshi, B.S., Chandel, K.P.S., Pant, K.C. and Saxena, R.K. (1974) Genetic diversity in some Asiatic *Phaseolus* species and its conservation. *Indian Journal of Genetics and Plant Breeding* 34A, 52–57.
Smartt, J. (1976) Comparative evolution of pulse crops. *Euphytica* 25, 139–143.
Smartt, J. (1985) Evolution of grain legumes. III. Pulses in the genus *Vigna*. *Experimental Agriculture* 21, 87–100.
Soltis, D.E., Soltis, P.S., Chase, M.W., Mort, M.E., Albach, D.C. *et al.* (2000) Angiosperm phylogeny inferred from 18S rDNA, *rbcL*, and *atpB* sequences. *Botanical Journal of the Linnean Society* 133, 381–461.
Soybean Meal Information Centre (2011) Soybeans – History and Future (available at http://www.soymeal.org/pdf/HistorySoybeanUse.pdf, accessed 2010).
Steele, W.M. and Mehra, K.L. (1980) Structure, evolution and adaptation to farming systems and environments in *Vigna*. In: Summerfield, R.J. and Bunting, A.H. (eds), *Advances in Legume Science*. Royal Botanic Gardens, Kew, UK, pp. 393–404.
Taran, B., Zhang, C., Wankertin, T., Tullu, A. and Vandenberg, A. (2005) Genetic diversity among varieties and wild species accessions of pea (*Pisum sativum* L.) based on molecular markers, and morphological and physiological characters. *Genome* 4, 257–272.
Taubert, P. (1894) Leguminosae. In: Engler, A. and Prantl, K. (eds) *Die Naturlichen Pflanzenfamilien*, 1st edn, vol. 3(3). Leipzig, Germany, pp. 70–388.
Tomooka, N., Vaughan, D.A., Moss, H. and Maxted, N. (2002) *The Asian Vigna: Genus Vigna Subgenus Ceratotropis Genetic Resources*. Kluwer Academic Publishers, Dordrecht, The Netherlands.
Tothill, J.D. (1948) *Agriculture in Sudan*. Oxford University Press, London, pp. 974.
Trabut, L. (1911) L'indegenat de la Flore en Algérie. *Bulletin of the National African Society* 7/15, 1–7.
Van, K., Kim, D.H., Cai, C.M., Kim, M.Y., Shin, J.H., Graham, M.A. *et al.* (2008) Sequence level analysis of recently duplicated regions in soybean [*Glycine max* (L.) Merr.] genome. *DNA Research* 15, 93–102.
van der Maesen L.J.G. (1987) Origin, history and taxonomy of chickpea. In: Saxena, M.C. and Singh, K.B. (eds), *The Chickpea*. CAB International, Wallingford, UK, pp. 11–34.
van der Maesen, L.J.G. (1979) Wild pigeon peas in Africa. *Plant Genetic Resources Newsletter* 40, 8–10.
van der Maesen, L.J.G. (1980) India is the native home of the pigeon pea. In: Arends, J.C., Boelema, G., de Groot, C.T. and Leeuwenberg, A.J.M. (eds) *Libergratulatorius in honorem H.C.D. de Wit*. Landbouwhogeschool Miscellaneous Paper no. 19. H. Veenman and B.V. Zonen, Wageningen, The Netherlands, pp. 257–262.
van der Maesen, L.J.G. (1986) Cajanus DC. and Atylosia W. & A. (Leguminosae). Agricultural University of Wageningen Papers 85-4 (1985). Agricultural University, Wageningen, The Netherlands, pp. 225.
van der Maesen, L.J.G. (1990) *The Pigeonpea*. CAB International, Wallingford, UK, pp. 15–46.
van Zeist, W. (1972) Palaeobotanical results of the 1970 season at Cayonii, Turkey. *Helinium* 12, 3–19.
Vaughan, D.A., Tomooka, N. and Kaga, A. (2004) Azuki bean in genetic resources, chromosome engineering and crop improvement. In: *Grain Legumes*, CRC Press, Boca Raton, Florida, pp. 341–353.
Vavilov, N.I. (1926) *Studies on the Origin of Cultivated Plants*. Leningrad, 1951.
Vavilov, N.I. (1928) Geographical centres of our cultivated plants. In: *Proceedings of the V International Genetic Congress*, New York, pp. 342–369.
Vavilov, N.I. (1949) The origin, variation, immunity and breeding of cultivated plants. *Chronica Botanica* 13(1/6), 26–151.
Vavilov, N.I. (1951) The origin, variation, immunity and breeding of cultivated plants. *Chronica Botanica* 13(1/6), 1–366.
Verdcourt, B. (1970) Studies in the leguminosae-papilionoideae for the flora of tropical east Africa IV. *Kew Bulletin* 24, 507–569.

Vershinin, A.V., Alnutt, T.R., Knox, M.R., Ambrose, M.R. and Ellis, T.H.N. (2003) Transposable elements reveal the impact of introgression, rather than transposition, in *Pisum* diversity, evolution and domestication. *Molecular Biology and Evolution* 20, 2067–2075.

Vishnu-Mittre (1974) The beginning of agriculture: paleaobotanical evidence in India. In: Hutchinson, J.B. (ed.), *Evolutionary Studies in World Crops: Diversity and Change in the Indian Sub-continent*. Cambridge University Press, Cambridge, UK.

Wanjari, K.B. (2005) Origin and history of pulses In: Singh, G., Sekhon, H.S. and Kour, J.S. (eds), *Pulses*. Geeta Somani Agrotech Publishing Company, New Delhi, India, pp. 59–78.

Watt, E.E. and Maréchal, R. (1977) The differences between mung and urid beans. *Tropical Grain Legume Bulletin* 7, 31–33.

Watt, G. (1889) *A Dictionary of Economic Products of India*. Cosmo Publications, Delhi, India [reprinted 1972].

Westphal, E. (1974) Pulses in Ethiopia, their taxonomy and agricultural significance. Centre for Agricultural Publishing and Documentation, PUDOC, Agricultural Research Reports No. 815, Wageningen, The Netherlands.

Wing, S.L., Herrera, F. and Jaramillo, C. (2004) A Paleocene flora from the Cerrajón Formation, Guajíra Peninsula, northeastern Colombia. In: *VII International Organization of Paleobotany Conference Abstracts*, 21–26 March, Museo Egidio Feruglio, Trelew, Argentina, pp. 146–147.

Wright, C. (2011) *The World of Legumes* (available at www.cliffordawright.com/caw/food/entries/display.php/topic_id/6/id/103/, accessed 2010).

Yamaguchi, H. (1992) Wild and weed azuki beans in Japan. *Economic Botany* 46, 384–394.

Yano, A., Yasuda, K. and Yamaguchi, H. (2004) A test for molecular identification of Japanese archaeological beans and phylogenetic relationship of wild and cultivated species of subgenus *Ceralolropis* (genus *Vigna*, Papilionaceae) using sequence variation in two non-coding regions of the *lrn*L and *lrn*F genes. *Economic Botany* 58 (Suppl.), S135–S146.

Ye, Y., Lang, L., Xia, M. and Tu, J. (2003) *Faba Beans in China*. China Agriculture Press, Beijing, China [in Chinese].

Zheng, Z., Wang, S. and Zong, X. (1997) *Food Legume Crops in China*. China Agriculture Press, Beijing, China, pp. 53–92 [in Chinese].

Zizumbo-Villarreal, D. and Colunga-GarcıaMarın, P. (2010) Origin of agriculture and plant domestication in West Mesoamerica. *Genetic Resources and Crop Evolution* 57, 813–825.

Zohary, D. and Hopf, M. (2000) *Domestication of Plants in the Old World. The Origin and Spread of Cultivated Plants in West Asia, Europe and the Nile Valley. Edn 3*. Oxford University Press Inc., New York.

Zong, X., Liu, X., Guan, J., Wang S., Liu, Q., Paull, J.G. *et al.* (2009) Molecular variation among Chinese and global winter faba bean germplasm. *Theoretical and Applied Genetics* 118, 971–978.

Zukovskij, P.M. (1962) *Cultivated Plants and Their Wild Relatives*. Commonwealth Agriculture Bureau, London.

2 Domestication

P.M. Chimwamurombe and R.K. Khulbe

2.1 Introduction

Food legumes, cultivated for their highly nutritious seeds, accompanied cereals in most regions of grain agriculture and formed an important dietary component of many early civilizations. Each major civilization developed not only its staple cereals, but also its characteristic companion legumes. Wheat and barley agriculture in West Asia and Europe had pea, lentil, faba bean and chickpea. Maize in Meso-America was accompanied by *Phaseolus* beans; and in South America by groundnut. Pearl millet and sorghum cultivation in the African savanna belt was associated with cowpea and bambara groundnut. Soybean was added to cereal cultivation in China, and hyacinth bean, black gram and green gram in India (Zohary and Hopf, 2000).

All food legumes belong to the family Fabaceae, which possesses the greatest number of domesticated crops of any family with 41 domesticated species (Harlan, 1992; Weeden, 2007). The available archaeological evidence indicates that pea, lentil, chickpea, bitter vetch and grass pea were taken into cultivation more or less together with the principal cereals. The establishment of this set of 'first wave' pulses was followed by several other legumes, prominent among which were the faba bean and fenugreek. They were followed, apparently later, by the lupin. The cowpea was domesticated in Africa, south of the Sahara, and reached the Mediterranean basin only in classical times (Zohary and Hopf, 2000). The recovery of large quantities of pulses from storage structures – for example, 7.4 kg of lentils and 2850 seeds of faba beans from Yiftah'el in Israel, dated to the middle Pre-Pottery Neolithic B (PPNB) (Kislev, 1985; Garfinkel *et al.*, 1988), 500 chickpeas at the pottery Neolithic site of Hoyucek in Turkey (Nesbitt, 2002) – prompts suggestions of the domestication of pulses predating that of cereals (Kislev and Bar-Yosef, 1988).

The domestication syndrome refers to all modifications occurring in a crop plant when it becomes cultivated from the wild form and is therefore dependent on man (Hammer 1984, 2003). For pulses, this applies equally as well, with increases in seed size, reduced pod-shattering and, importantly, loss of germination inhibition (Plitman and Kislev, 1989; Smartt, 1990; Zohary and Hopf, 2000). Additionally, the wild-type chemical defences have been selected against. Many wild legumes contain potent toxins and anti-metabolites in their seeds to protect them against animal predation. Cultivars frequently lack or contain only reduced amounts of these toxic compounds. In others, fermentation (soybean) or cooking (common bean) is necessary to render the seed safe for human consumption (Zohary and Hopf, 2000). Self-pollination seems to have been a major asset in the domestication of food legumes, on account of the advantages conferred by it in the establishment of a barrier between wild and cultivated populations, and the automatic fixation of the desired genotypes (Zohary and Hopf, 2000).

The 'domestication syndrome' traits for legume crops are generally common to all (Table 2.1). The rate and the order in which the domestication of these traits occurred, however, differ somewhat (Fuller, 2007). The processes and the accompanying genetic changes leading to the evolution of the wild progenitors into the present-day domesticates are discussed briefly in this chapter for the major food legumes.

2.2 Pea (*Pisum sativum* L.)

The carbonized remains of *Pisum sativum* found at many historical sites in the Fertile Crescent of the Middle East dating back to the sixth or seventh millennium BC suggest the occurrence of domestication during that era (Zohary and Hopf, 1973). Cultivation of *Pisum sativum* then spread from the Fertile Crescent to Russia and westward into Europe and eastward into China and India, and into the Western Hemisphere upon the discovery of the New World.

Pisum sativum ssp. *elatius*, the presumed wild ancestor of the cultivated pea, is sufficiently close to the wild ancestor to provide a reasonable starting point for the domestication process (Weeden, 2007). Several apparently intermediate stages for the domestication of pea are available in germplasm collections. The germplasm identified by the taxonomic label *P. sativum* ssp. *abyssinicum* appears to be a primitive landrace that displays several traits (indehiscent pods, smooth pods, thin testa) that are usually associated with initial steps in the domestication process. This landrace that is presumed to have been isolated some 3000 years ago may provide insight into the progress made in domestication of peas after some 5000 years of cultivation. Another divergent and distinct landrace described as the 'Afghanistan' type (Weeden and Wolko, 1988) is found in the foothills and higher slopes of Afghanistan, Nepal, Iran and Pakistan.

By about 5000 years ago, four traits in pea had been at least partly changed (Weeden, 2007). The indehiscent pod trait appears to have been fixed in domesticated germplasm by that time, and the longer-term seed dormancy appears to have been eliminated. Gigantism in the form of seed weight had increased about twofold. Earliness may also have been selected before the split of *Pisum sativum* ssp. *abyssinicum* from the main track of pea domestication, because it flowers relatively early. However, the allele responsible for earliness in this subspecies appears not to be present in the remaining domesticated germplasm, and an alternative explanation would be that the allele was selected after the divergence.

In *Pisum sativum*, at least 11 loci involved in the domestication syndrome have been identified. The *Dpo* locus was identified relatively early as a primary factor controlling pod dehiscence (Blixt, 1972). Flowering time is controlled by at least six loci (Murfet and Reid, 1985), although not all of these are important in domestication. Numerous genes or QTL have been identified that influence plant habit and seed quality (Blixt, 1972) and seed size (Timmerman-Vaughan *et al.*, 1996). More recent studies have added to this list of genes controlling the traits of the domestication syndrome (Weeden *et al.*, 2002; Timmerman-Vaughan *et al.*, 2005), and molecular studies have now identified the coding sequences of many of these genes.

Approximately 20 genes or QTL are responsible for the modifications of plant form and function that accompanied the domestication of pea. Thus, we know that the substitution of 'a' for 'A' in peas improved seed quality and reduced seed dormancy. Loss of *Np* increased seed size (at least under certain conditions) but also reduced tolerance to bruchid attack. The recessive 'r' allele improves seed quality (sweetness) but appears to reduce seed size. Homozygosity for the dwarfing gene may increase root mass, and two of the photoperiod response genes, *Sn* and *Hr*, are either closely linked to genes that influence root/shoot ratio or are directly involved themselves.

Table 2.1. The domestication syndrome traits of important legume crops.

Crop	Domestication syndrome traits	Wild	Cultivated	Reference(s)
Pea (*Pisum sativum* L.)	Pod dehiscence	Present	Absent	Blixt (1972); Vaughan *et al.* (1996); Weeden and Muehlbauer (2004); Timmerman-Vaughan *et al.* (1996)
	Dormancy	Present	Absent	
	Plant height	Tall	Dwarf	
	Branches	Many basal	Few basal	
	Seed size	Small	Large	
	Seed quality	Poor	Good	
	Flowering	Long-day	Day-neutral	
Common bean (*Phaseolus vulgaris* L.)	Seed dispersal	Present	Absent	Koinange *et al.* (1996)
	Dormancy	Present	Absent	
	Growth habit			
	determinacy	Indeterminate	Determinate	
	twining	Twining	Non-twining	
	Gigantism			
	Pod length (cm)	5.7	9.8	
	100-seed wt (g)	3.5	19.5	
	Earliness			
	days to flowering	69	46	
	days to maturity	107	80	
	Photoperiod sensitivity	>60	0	
	Harvest index	0.42	0.62	
	Seed pigmentation	Present	Absent	
Chickpea (*Cicer arietinum* L.)	Dormancy	Present	Absent	Cobos *et al.* (2009)
	Pod dehiscence	Present	Absent	
	Flower colour	Purple	White	
	Growth habit	Prostrate	Erect to semi-erect	
	Seed size	Small	Medium–large	
	Seed colour	Blackish brown	Brown/creamish	
	Seed coat texture	Reticulated	Rough–smooth	
Cowpea (*Vigna unguiculata* L.)	Pods per peduncle	5 or more	2–3	Lush and Evans (1981)
	Pod disposition at maturity	Erect	Pendent	
	Pod dehiscence	Present	Absent	
	Flowering	Later than cultivated	Earlier than wild	
	Mature seed	Hard, i.e. impermeable to water	Rough- and smooth-coated imbibe water readily	
	Germination	Germinate less rapidly than domesticates outside 20–30°C	Germinate more rapidly than wild accession outside at 20–30°C or at high temperature	

Continued

Table 2.1. Continued.

Crop	Domestication syndrome traits	Wild	Cultivated	Reference(s)
Lentil (*Lens culinaris* L.)	Pod dehiscence	Present	Absent	Ladizinsky (1979); Sonnante *et al.* (2009)
	Flower colour	Purple	White	
	Growth habit	Prostrate	Erect	
	Epicotyl colour	Bluish	Green or purple	
	Seed coat spotting	Dark + brown spots	Yellowish-grey	
Adzuki bean (*Vigna angularis* L.)	Seed dormancy (% germination in field)	0.0	73.3	Kaga *et al.* (2008)
	Pod dehiscence (no. of twists)	2.8	0.6	
	Increase in organ size			
	pod length (cm)	5.8	11.1	
	seeds/pod	8.5	6.0	
	100-seed wt. (g)	2.5	24.0	
	Twining (%)	100.0	0.0	
	Days to 100% pod maturity	112.3	79.3	
	Epicotyl colour	Purple	Green	
	Seed colour	No red	Red	
	Black mottle	Present	Absent	
Bambara groundnut (*Vigna subterranea*)	Germination	30 days or longer; erratic/ staggered	15 days; uniform	Hepper (1963); Pasquet and Fotso (1997); Swanevelder (1998); Pasquet *et al.* (1999); Basu *et al.* (2007a)
	Root	No clear tap root	Compact, well-developed tap root	
	Plant type	Spreading	Compact/bunch type	
	Stem	Limited number of lateral stems	Many short lateral stems	
	Internodes (cm)	Long (6.5–10.0)	Short (1.3–3.4 cm)	
	Leaves	Small (4.5–6.5 × 1.9–2.8)	Large (7.5–9.4 × 2.8–3.6)	
	Pods	Borne along the length of the elongated stems	Clustered at the base	
		Thin and smooth		
	Pod testa	Small (9–11 mm) and vary in size	Thick and wrinkled	
	Seed size		Larger (11–15 mm) and quite uniform in size	

2.3 Common Bean (*Phaseolus vulgaris* L.)

The common bean originated in the Americas, and a variety of studies over many years indicate that there are two major gene pools that diverged prior to its domestication (Gepts, 1998), generally referred to as the Meso-American and Andean gene pools. Domestication also appears to have occurred independently in the two gene pools, with the Andean domestication 4000 years ago pre-dating the Meso-American by 2000 years (Kaplan and Lynch, 1999; Piperno and Dillehay, 2008). Gepts (1998), using phaseolin-S, identified a well-circumscribed area in west-central Mexico as the putative domestication centre for the common bean. This area is located relatively close to the area proposed for the domestication of maize, although it does not match it.

The common bean is a non-centric crop that has had multiple domestications throughout the range of wild populations (Harlan, 1975; Gepts *et al.*, 1986). Among the array of domestication traits in common bean, the two most important attributes of the domestication syndrome in this crop are the loss of seed dispersal ability and seed dormancy, because these are crucial for adaptation to a cultivated environment. The former is conditioned by the presence of fibres in the pods, both in their sutures and their walls. Loss of these fibres leads to indehiscence of the pods and lack of seed dispersal at maturity. Cultivated beans display a more compact growth habit compared with their wild progenitor. In its most evolved form under domestication, this growth habit is characterized by a combination of traits comprising determinacy, non-twining branches, few vegetative nodes and long internodes. Selection by humans has also led to pods and seeds that are larger and show different or no anthocyanin pigmentation. The dissemination of cultivated beans from their domestication centres in the tropics to new areas at higher altitudes has led to a selection of genotypes that are insensitive to day length compared with the wild progenitor, which will only flower under short days. In concert with the changes in growth habit and photoperiod sensitivity, common bean cultivars flower and mature generally earlier than their wild ancestors.

The two most important distinguishing characteristics between wild and cultivated beans are seed dispersal, conditioned by the presence of fibres in the pods, and dormancy, conditioned by impermeability of the seed coat. The lack of pod suture fibres is conditioned by a single gene (*St*) on linkage group D2, and tightly linked or identical to the gene *St* controlling the presence of pod suture fibres. Four unlinked QTL were identified for seed dormancy (DO). A majority of the major genes controlling the domestication syndrome in common bean were concentrated on few (3) of the 11 linkage groups of the genome. Of the 16 qualitative or quantitative traits of the domestic syndrome analysed in the study, 11 were controlled partially by factors on linkage group D1 (principally growth habit and phenology), 4 on linkage group D2 (principally seed dispersal and dormancy) and 2 on linkage group D7 (size of the harvested organs). The results of chi-square tests suggest that the factors involved in the domestication syndrome are not distributed proportionately to the genetic length of the linkage groups.

2.4 Chickpea (*Cicer arietinum* L.)

Chickpea was one of the first grain legumes to be domesticated in the Old World (van der Maesen, 1987). Domestication of chickpea occurred in a small core area within the Fertile Crescent, in present-day south-eastern Turkey – northern Syria, near the springs of the Tigris and Euphrates Rivers. This area is supposed to be the real 'cradle of agriculture' (Lev-Yadun *et al.*, 2000). The species *Cicer reticulatum* Ladiz. is the wild progenitor of the domesticated chickpea (Ladizinsky and Adler, 1976a, b; Redden and Berger, 2007; van der Maesen *et al.*, 2007), and crosses readily with cultivated chickpea.

Conventionally, cultivated chickpea is divided into two types, *Kabuli* and *Desi*. *Kabuli*, which has large, ram-shaped cream- or beige-coloured seeds, is predominantly distributed in Mediterranean countries and the Near East. *Desi*, which has small, angular dark-coloured seeds, prevails in the eastern and southern parts of the distribution area of the crop (Van der Maesen, 1972; Zohary and Hopf, 1993). It is commonly accepted that the large-seeded domestic *Kabuli* chickpeas originated from the small-seeded *Desi* chickpeas, but the induced mutant (white flower and cream seed coat colour) of *C. reticulatum* may suggest an additional path for the evolution of *Kabuli* chickpea. Based on historical records records and the induced mutants obtained from the study, the domestic *kabuli* chickpea could have emerged directly from *C. reticulatum* in south-eastern Turkey and adjoining Syria (Toker, 2009).

In chickpea, changes accompanying domestication initially included the loss of dormancy, followed by reduced pod dehiscence, larger seed size, larger plant size and variants with more erect habit and reduced anthocyanin pigmentation (Smartt, 1984; Ladizinsky, 1987). However, the key to chickpea domestication was the change from a winter habit with an autumn sowing to a spring habit, which avoided or reduced the threat of lethal infestation of the endemic *Ascochyta* pathogen complex (Abbo *et al.*, 2003). However, both annual and perennial wild relatives in the region are adapted to winter cropping whereas domestic chickpea is spring-sown for summer cropping, and the main differentiation between the wild and domestic species is the loss of responsiveness to vernalization, a polygenic trait. Domestic chickpea is also characterized by larger plant and seed size than the *C. reticulatum* wild progenitor. In contrast to other grain legumes, loss of seed dormancy and reduction in pod shattering do not appear to be key traits for domestication in chickpea (Ladizinsky, 1987).

2.5 Lentil (*Lens culinaris* L.)

Domestication of lentil is now generally accepted to have occurred in the same core area as chickpea (see Section 2.4). Recent molecular and biochemical evidence confirms that the ssp. *orientalis* is the taxon from which the crop was domesticated. The analysis of variation of intronic regions of a cytosolic glutamine synthetase gene and two paralogous genes coding for Bowman–Birk protease inhibitors by Sonnante *et al.* (2005) supports the idea that lentils derive from a specific stock of the ssp. *orientalis*, the one where the mutants that triggered domestication first appeared.

In lentil, seed dispersal is considered to be the first character of selection (Zohary, 1999), with the selection for increased seed size being later (Sonnante *et al.*, 2009). Recent evidence tends to suggest that cultivation was carried out by man far before domestication traits were fixed (Pringle, 1998; Balter, 2007). According to the weedy/dump-heap hypothesis (Abbo *et al.*, 2005), humans brought wild seeds to their villages and unconsciously dispersed them to the proximities or to dump areas: in these areas, due to the better soil fertility, stronger plants were observed by the inhabitants, triggering the idea of cultivation.

For lentils, two main traits were involved in the domestication process: pod dehiscence and seed dormancy, both of which were reported to be under the control of single recessive genes. A third major trait, seed size, appears to be under a more complex control (Sonnante *et al.*, 2009). According to Ladizinsky (1979), the domestication of lentils was accomplished in a single-step event, due to a single mutation.

The genetics of traits involved in the domestication of lentil has not received the same attention as in other legumes (Sonnante *et al.*, 2009). Ladizinsky (1979) analysed the inheritance of seed colour (*Scp*), epicotyl colour (*Gs*), growth habit (*Gh*), flower colour and pod dehiscence in a lentil × ssp. *orientalis* cross. Of these traits, the white flowers, erect growth and pod indehiscence are typical of the cultivated lentil. While seeds of cultivated lentil can germinate shortly after maturation, wild lentil seeds undergo seed dormancy due to a hard seed coat (Ladizinsky, 1985). The hard seed coat in ssp. *orientalis* is controlled by a single recessive gene in the homozygous condition. Together with pod dehiscence, the breakdown of seed dormancy is one of the first traits implied in lentil domestication. As this trait is governed by one recessive gene in ssp. *orientalis*, a mutant with a soft coat must have appeared during domestication in a relatively short space of time (Ladizinsky, 1985).

Tahir and Muehlbauer (1993) found that three morphological traits involved in the domestication syndrome of lentil (epicotyl colour, pod indehiscence and growth habit) were associated with genes or factors that gave a selective advantage to cultivated lentil alleles during the development of recombinant inbred lines. A map obtained from a cross of lentil × ssp. *orientalis* showed that each of the five morphological loci (seed colour pattern, cotyledon colour, stem pigment, pod dehiscence–indehiscence, seed ground colour), except for pod dehiscence, was found to be linked to one or more molecular markers. In one of the first linkage maps based on a population derived from lentil × ssp. *orientalis* (Harvey and Muehlbauer, 1989), the authors found linkage between some isozymes and morphological characters, and the linkage *Pi-Gall-Pdp* was particularly interesting because pod dehiscence (*Pi*) and pigmentation (*Pdp*) are also linked in pea.

2.6 Cowpea (*Vigna unguiculata* L.)

For cowpea, two domestication areas have been proposed in western and north-eastern Africa, respectively (Baudoin and Maréchal, 1985; Ng and Maréchal, 1985; Vaillancourt and Weeden, 1992; Ng, 1995; Pasquet, 2000). Cowpea was probably domesticated by farmers in West Africa, which is also a major centre of diversity of cultivated cowpea (Ng and Padulosi, 1988). However, studies based on amplified fragment length polymorphism (AFLP) markers by Coulibaly *et al.* (2002) furnish evidence of occurrence of domestication in north-eastern Africa. Domestication of cowpea could have occurred simultaneously with domestication of sorghum (*Sorghum bicolor*) and pearl millet (*Pennisetum typhoides*) in the third millennium BC (Steele, 1976). The wild cowpea, *Vigna unguiculata* ssp. *unguiculata* var. *spontanea* is the likely progenitor of cultivated cowpea (Pasquet, 1999). The loss of a BamHI restriction site in chloroplast DNA differentiates all domesticated accessions and a few wild (*V. u.* ssp. *U.* var. *spontanea*) accessions (Feleke *et al.*, 2006). The morphology and growth habit of the wild cowpea are very similar to that of cowpea landraces, but it also possesses wild-like attributes such as shattering pods with small seeds. Despite the wide distribution of var. *spontanea* throughout sub-Saharan Africa, molecular studies point to a unique domestication event (Panella and Gepts, 1992; Pasquet, 1999; Ba *et al.*, 2004).

The domesticates of cowpea rarely developed more than two or three pods per peduncle, which are pendent at maturity, but five or more can mature in succession on wild plants and often these remain erect (Lush and Evans, 1980a, b). However, domesticates tend to flower earlier then wild accessions (Lush *et al.*, 1980). The pods of wild cowpea are dehiscent whereas the pods of domesticates are indehiscent. In cowpea, pod dehiscence is said to be controlled by a single dominant gene (Rawal, 1975). Most of the mature seed of wild cowpea is hard, i.e. impermeable to water. The seeds of rough-coated domesticates all imbibe water readily, as do most smooth-coated domesticates. Cowpea germinates rapidly between 20 and 30°C, but outside this range domesticates tend to germinate more rapidly than wild accessions, particularly at high temperatures (Lush *et al.*, 1980).

All changes that characterize the evolution of most seed crops have not occurred in cowpea, as the wild subsp. *dekindtiana* (also referred to as ssp. *spontanea*) already possessed the appropriate attributes, for example, annuality. Other changes have not occurred in cowpea domesticates, perhaps because they were of no value in the traditional agricultural conditions under which most cowpeas are grown. Plant attributes in the last category are photoperiodic controls on reproduction, which appear to have adaptive value not only in natural conditions but also under traditional conditions (Lush *et al.*, 1980), and an indeterminate plant habit, which may be better suited to intercropping and weed control and associated with the ability to recover from drought stress.

2.7 Adzuki Bean (*Vigna angularis* L.)

It is not known where adzuki bean was domesticated. However, adzuki bean exists as a crop complex in Japan where its cultivated, wild and weedy forms can be found (Vaughan *et al.*, 2004). In addition, carbonized adzuki bean seeds have been found from archaeological sites in Japan dated to 4000 years ago (Maeda, 1987; Yano *et al.*, 2004), pre-dating archaeobotanical remains of adzuki bean in China and Korea (Crawford, 2006). Thus Japan is one possible place where this crop was domesticated. The presumed wild ancestor of cultivated adzuki bean is *Vigna angularis* var. *nipponensis* (Yamaguchi, 1992). In various parts of Japan where wild and cultivated adzuki beans are sympatric, plants with variable phenotype are commonly found (Kaga *et al.*, 2004). The occurrence of plants in wild populations having genes from cultivated adzuki bean (Wang *et al.*, 2004) suggests that natural crossing among components of this crop complex is a regular occurrence (Yamamoto *et al.*, 2006), which may have resulted in landraces accumulating alleles as a result of natural introgression and farmer selection. Domestication of adzuki has been also involved a trade-off between yield and seed size, with fewer but longer pods and fewer but larger seeds on plants with shorter stature in cultivated adzuki bean being at the expense of overall seed yield (Kaga *et al.*, 2008).

Adzuki bean shows numerous differences in morphological and physiological traits associated with domestication compared with its closely related wild relatives. Domestication of adzuki bean has resulted in a conspicuous increase in seed and pod size, non-twining growth habit and loss of seed dormancy and seed dispersal ability. In addition, seed colour variation that is not found in its wild relatives is present in adzuki bean cultivars. Among the domesticated Asian *Vigna* and their presumed wild ancestors, seed size, seed colour and life history traits differ markedly (Isemura *et al.*, 2007). For example, cultivated mung bean generally has green seeds that are about five times the size of the wild mung bean, while cultivated adzuki bean usually has red seeds more than eight times the size of the wild adzuki bean (Tomooka *et al.*, 2000). Kaga *et al.* (2008) identified a reciprocal translocation between cultivated and wild adzuki bean parents on the basis of the linkage map having a pseudo-linkage group and clustering of seed productivity-related QTL with large effect near the presumed break points.

In adzuki bean a few domestication-related traits are controlled by a single major gene, and most of these are controlled by a small number of QTL. Pod dehiscence in adzuki bean is controlled by a single gene, and three QTL for twining habit were detected. For the majority of traits measured, between two and nine QTL on two or more linkage groups were detected. The genes controlling domestication-related traits are not randomly distributed across crop genomes (Doebley and Stec, 1991, 1993; Poncet *et al.*, 2000). Particularly important linkage groups with major QTL for domestication-related traits were groups 1, 2, 4, 7 and 9.

A broad array of domestication-related traits in adzuki bean have been analysed and their QTL mapped on a molecular linkage map. Most traits are controlled by two to nine QTL that occur on different linkage groups. QTL for domestication-related traits are not evenly distributed across the adzuki bean linkage map, and 5 of the 11 linkage groups in adzuki bean (groups 1, 2, 4, 7 and 9) possess 80% of the QTL detected. In addition, within a linkage group QTL are clustered (Isemura *et al.*, 2007).

2.8 Pigeon Pea (*Cajanus cajan* L.) Millsp.

Pigeon pea (*Cajanus cajan*) is an important legume crop with cultivation taking place primarily in the semi-arid tropics of the world. Despite its importance, the understanding of the domestication history of this species, or the relationships between domesticated and wild species, is limited (Mulualem *et al.*, 2010). Pigeon pea is likely to have evolved by interspecific hybridization of *Cajanus cajanifolia* and *Cajanus scarabaeoides* (Nadimpalli *et al.*, 1992) somewhere on the Indian subcontinent (van der Maesen, 1980; for details, see Chapter 1). Restriction fragment length polymorphism (RFLP) analysis (Nadimpalli *et al.*, 1992) and single-nucleotide polymorphism (SNP) genotyping (Muluelam *et al.*, 2010) support *C. cajanifolia* as the progenitor of cultivated pigeon pea. It is likely that India was also the centre of domestication sometime before 2000 BC, as evidenced by the presence of several wild species of pigeon pea including the progenitor species, high morphological diversity among varieties, ample linguistic evidence and variety of use in daily cuisine (van der Maesen, 1990). East Africa is considered a secondary centre of diversity of pigeon pea (Smartt, 1990; van der Maesen, 1990). The genetic analysis of Indian and African accessions using simple sequence repeat (SSR) markers supports this hypothesis (Songok *et al.*, 2010). A further centre of diversity occurs in Australia (Nene and Sheila, 1990). After domestication, pigeon pea is believed to have travelled from India to Malaysia and then to East Africa (van der Maesen, 1990).

No wild form of pigeon pea is known, and the few reports of such forms apparently refer to types that have escaped from cultivation. On the other hand, various lines of evidence indicate that the genus *Atylosia* is closely related to *Cajanus* (Ladizinsky and Hamel, 1980). Pigeon pea has been successfully crossed with both Indian and Australian wild species, and also with two native Australian species, *Atylosia acutifolia* and *Atylosia pluriflora*. These hybrids showed high levels of sterility (Dundas *et al.*, 1987), however. On the basis of the appearance of specific *Atylosia* bands in some of the electrophoretic variants of *Cajanus*, Ladizinsky and Hamel (1980) suggested that the gene flow is still effective between pigeon pea and various *Atylosia* species. Mulualem *et al.* (2010), in a study on 31 wild and 79 cultivated genotypes of pigeon pea by using high-throughput SNP genotyping, observed genetic admixture between wild and cultivated genomes, which suggested the involvement of successive rounds of gene flow during domestication.

In pigeon pea, besides shortening of maturity duration and increase in pod and seed size, change in the content and composition of protein and anti-metabolites appears to have occurred during domestication. The poor solubility of the *Atylosia* seed protein in comparison with *Cajanus* indicates that domestication of *Cajanus* was coupled with increased solubility and perhaps a better nutritional value (Ladizinsky and Hamel, 1980). Aruna *et al.* (2007) observed variations in the trypsin inhibitors and lectin content in the developing pods of *C. scarabaeoides* and pigeon pea accessions. The protein and trypsin inhibitor contents were higher in the wild accessions than the cultivated genotypes. The occurrence of very high broad-sense heritability estimates indicated involvement of few genes in the inheritance of these biochemical components. Loss of proteinase inhibitor (PI) activity has also occurred during domestication. The PIs that constitute pigeon pea's defence machinery exhibited monomorphism in pigeon pea cultivars in terms of TI (trypsin inhibitor) and CI (chymotrypsin inhibitor) isoforms, contrary to the diverse inhibitory profiles of the pigeon pea wild relatives.

2.9 Bambara Groundnut (*Vigna subterranea* L.)

Bambara groundnut is closely related to cowpea (*V. unguiculata*), with which it shares much of its area of cultivation and origins of genetic diversity (Basu *et al.*, 2007b). The centre of origin of bambara groundnut is believed to be in north-eastern Nigeria and northern Cameroon (Hepper, 1970). Bambara groundnut consists of two botanical forms; *Vigna subterranea* var. *subterranea* and var. *spontanea*. The cultivated form var. *subterranea* exists as landraces and is grown extensively in sub-Saharan Africa. The wild forms comprise var. *spontanea* and are restricted to an area from Nigeria to Sudan, with a centre of diversity around Cameroon. The chromosome number in both wild and cultivated plants is 2n=22 (Frahm-Leliveld, 1953). High genetic identity between wild and domesticated forms suggests that wild bambara groundnut (*V. subterranea* var. *spontanea*) is the true progenitor of domesticated Bombara groundnut (Doku and Karikari, 1971; Pasquet *et al.*, 1999; Massawe *et al.*, 2002; Ntundu *et al.*, 2004).

Domestication of bambara groundnut primarily involved a change from a spreading/trailing growth habit to a compact/bushy plant type, which was mainly brought about by shortening of internodes, increase in the number of lateral branches and shortening of the lateral branches. An increase in leaf size and a slight increase in flower size accompanied these changes. The change in plant type led to clustering of the pods at the base, which in the wild forms are borne along the length of the stems. The pod testa is thickened in the domesticated forms compared with the thin testa of the wild forms. As a result, the pods of the wild plant do not wrinkle upon drying, while the thick, fleshy pods of the freshly dug domesticated fruit wrinkle on drying. As in other legumes, domestication resulted in increase in size and uniformity of bambara groundnut seed. Another change accompanying bambara groundnut domestication is the uniformity in germination as compared with the staggered seed germination in the wild forms (Hepper, 1963; Basu *et al.*, 2007b).

Initial investigation into bambara groundnut domestication by Basu *et al.* (2007b) suggests that the major morphological difference between *spontanea* and *subterranea* types (spreading or compact plant habit) is under the control of a relatively limited numbers of genes. The major components of compact plant habit, i.e. internode length and stems per plant, both showed monogenic inheritance in a cross between DipC (var. *subterranea*) and VSSP11 (var. *spontanea*). A single co-dominant gene for stems per plant and a single dominant gene for long internodes were postulated to explain the majority of the variation present. Early emergence is postulated to be largely controlled by a single dominant gene, whereas leaf area and 100-seed weight were clearly multigenic. A linkage map based on the wide cross consists of 81 AFLP markers and 2 microsatellites (Basu *et al.*, 2007c) distributed across 20 linkage groups. Development of a genetic linkage map for bambara groundnut will allow the dissection of traits through linkage and QTL analysis, besides establishing linkages between bambara groundnut and other more characterized legume genomes such as soybean and *Medicago* (Mayes *et al.*, 2009).

2.10 Genome Conservation and Synteny among Legumes

A comparison of linkage maps of the common and adzuki bean shows that QTL for seed length and pod length on LG 7 of the common bean are present in almost the same region on LG of adzuki bean. QTL for pod and growth habit detected on LG7 in adzuki bean were, however, not detected on LG B5 of common bean. Using populations derived from crosses between cowpea and wild cowpea and mung bean and wild mung bean, two and four QTL for seed weight, respectively, were reported (Fatokun *et al.*, 1992), and a significant correspondence was observed between linkage groups in the two crops. In this study, QTL for seed weight was detected on linkage group 1 at a location corresponding to that of a QTL for this trait on linkage group II in cowpea and mung bean. Thus seed weight QTL appears to be conserved among these three species. QTL for seed weight were also detected at similar locations on adzuki bean linkage group 9 and mung bean linkage group I. Although the QTL with the largest effect for seed weight was detected on the LG2 in adzuki bean, no QTL was detected on the linkage groups corresponding to this linkage group in cowpea and mung bean suggesting that QTL on LG VI of cowpea, III and VI of mung bean and 8 of adzuki bean appear to be specific to these crops. These results suggest that the main genome regions related to increased seed weight under domestication do not correspond among these related species, despite high homology between the linkage groups. In adzuki bean, seed weight in cultivated taxa is about eight times that of the wild parent. In contrast, seed weight in cultivated and wild parents of crosses analysed for both cowpea and mung bean exhibited only a fivefold difference (Fatokun *et al.*, 1992). Adzuki bean has the largest seed for the cultivated Asian *Vigna* (Tomooka *et al.*, 2000). It seems that increase in seed size compared with cowpea and mung bean involves different loci.

In soybean (tribe Phaseolae) a QTL detected for seed weight by Maughan *et al.* (1996) corresponds to LG1 in adzuki bean. However, this RFLP marker was well separated from the molecular makers associated with seed weight variation in adzuki bean, mung bean and cowpea.

In *Pisum sativum* L. (tribe Vicieae), a QTL for seed weight was also detected in the region that corresponds to the region with seed weight QTL on LG1 of adzuki bean and II of cowpea and mung bean based on RFLP comparison (Timmerman-Vaughan *et al.*, 1996). Therefore, it seems that this region has been conserved across the Leguminosae and plays an important role in increasing seed size.

Weeden *et al.* (1992), in an intercross of *L. ervoides* × lentil, found that in eight regions linkage among marker loci appeared to be conserved between lentil and pea. The observed synteny between lentils and pea could foster genetic studies in lentils. Microsyntenic relationships between lentils and the model legume *Medicago truncula*ta were established by Phan *et al.* (2006). The integration of present knowledge on lentil genetic maps in a consensus map, also including information from other legumes such as pea (Weeden *et al.*, 1992), could serve as a groundwork for future studies in lentil genetics and genomics (Ford *et al.*, 2007). This knowledge would surely provide a powerful tool for filling the gap in lentil breeding and at the same time provide more information on the genetics of lentil domestication, and thus insight into origins of this crop that the present fragmented knowledge is unable to do. It was revealed that, despite many parallels in the modifications during domestication between pea and common bean, no genes that were involved in the domestication of both crops were identified. Problems with seed dispersal, growth habit, earliness, seed quality and seed pigmentation all appear to involve different suites of genes in pea compared with bean. The case for seed dormancy, gigantism and particularly the loss of photoperiod sensitivity is less clear, and may involve homologous or orthologous sequences. Resolution of these issues will probably require the identification of the coding sequence of the gene affected in one crop followed by mapping of that sequence in the other. However, it is encouraging from a breeder's perspective to find that there are at least several ways to modify unwanted characters such a pod dehiscence and plant habit, and possibly avoid some of the detrimental effects accompanying the substitution of certain alleles for others.

2.11 Conclusion

In this chapter it has been made clear that the domestication of food legumes has been a long journey for some of the legumes such as soybean, pea, adzuki bean, common bean and cowpea, which applies to most crops in general. This has been the case primarily because of the lack of tools that could quicken the process. In future, the domestication and evolution of pulses is envisaged as being shorter, due to the availability of research tools and the immense pressure being exerted by climate change effects and the ever-increasing demand for more food resources. Furthermore, there is always a need to do research on the little-known legume plants of the world, as these may hold the key to solving some of the problems of inhabitants of harsh environments. However, the availability of funding for such programmes remains a real challenge. One of the broader impacts of the domestication of legumes will be the availability of a new crop alternative for resource-poor farmers in southern Africa and other arid regions of the world.

References

Abbo, S., Shtienberg, D., Lichtenzveig, J., Lev-Yadun, S. and Gopher, A. (2003) Chickpea, summer cropping, and a new model for pulse domestication in the ancient Near-East. *The Quarterly Review of Biology* 78, 435–448.

Abbo, S., Gopher, A., Rubin, B. and Lev-Yadun, S. (2005) On the origin of Near Eastern founder crops and the "dump-heap hypothesis". *Genetic Resources and Crop Evolution* 352, 491–495.

Aruna, R., Rao, D.M., Reddy, L.J., Ramakrishnan, S.S. and Upadhyaya, H.D. (2007) Influence of pod maturity and level of domestication on biochemical components in wild and cultivated pigeon pea (*Cajanus cajan*). *Annals of Applied Biology* 151, 25–32.

Ba, F.S., Pasquet, R.S. and Gepts, P. (2004) Genetic diversity in cowpea [*Vigna unguiculata* (L.) Walp.] as revealed by RAPD markers. *Genetic Resources and Crop Evolution* 51, 539–550.

Balter, M. (2007) Seeking agriculture's ancient roots. *Science* 316, 1830–1835.

Basu, S., Roberts, J.A., Azim-Ali, S.N. and Mayes, S. (2007a) Bambara groundnut. In: Kole, C.M. (ed.) *Genome Mapping and Molecular Breeding in Plants: Pulses, Sugar and Tuber Crops. 3.* Springer, New York, pp. 159–173.

Basu, S., Mayes, S., Davey, M., Roberts, J.A., Azam-Ali, S.N., Mithen, R. *et al.* (2007b) Inheritance of 'domestication' traits in bambara groundnut (*Vigna subterranea* (L.) Verdc.). *Euphytica* 157, 59–68.

Basu, S., Roberts, J.A., Azim-Ali, S.N. and Mayes, S. (2007c) Development of microsatellite marker for Bambara groundnut (*Vigna subterranea* L. Verdc.). *Molecular Ecology Notes* 7, 1326–1328.

Baudoin, J.P. and Maréchal, R. (1985) Genetic diversity in *Vigna*. In: Singh, S.R. and Rachie, K.O. (eds) *Cowpea Research, Production and Utilization*. John Wiley and Sons, Chichester, UK, pp. 3–9.

Blixt, S. (1972) Mutation genetics in *Pisum. Agriculture Horticulture Genetica* 30, 1–293.

Cobos, M.J., Winter, P., Kharrat, M., Cubero, J.I., Gil, J., Millan, T. *et al.* (2009) Genetic analysis of agronomic traits in a wide cross of chickpea. *Field Crops Research* 111, 130–136.

Coulibaly, S., Pasquet, R.S., Papa, R. and Gepts, P. (2002) AFLP analysis of the phenetic organization and genetic diversity of *Vigna unguiculata* L. Walp. reveals extensive gene flow between wild and domesticated types. *Theoretical and Applied Genetics* 104, 358–366.

Crawford, G.W. (2006) East Asian plant domestication. In: Stark, M.T. (ed.) *Archaeology of Asia*. Blackwell Publishing, Oxford, UK, pp. 77–95.

Doebley, J. and Stec, A. (1991) Genetic analysis of the morphological differences between maize and teosinte. *Genetics* 129, 285–295.

Doebley, J. and Stec, A. (1993) Inheritance of morphological differences between maize and teosinte: comparison of results for two F2 populations. *Genetics* 134, 559–570.

Doku, D.V. and Karikari, S.K. (1971) Operational selection in wild bambara groundnut. *Ghana Journal of Science* 11, 45–56.

Dundas, I.S., Britten, E.J., Byth, D.E. and Gordon, G.H. (1987) Meiotic behavior of hybrids of pigeon pea and two Australian native Atylosia species. *Journal of Heredity* 78, 261–265.

Fatokun, C.A., Menancio-Hautea, D., Danesh, D. and Young, N.D. (1992) Evidence for orthologous seed weight genes in cowpea and mung bean based on RFLP mapping. *Genetics* 132, 841–846.

Feleke, Y., Pasquet, R.S. and Gepts, P. (2006) Development of PCR-based chloroplast DNA markers that characterize domesticated cowpea (*Vigna unguiculata* ssp. *unguiculata* var. *unguiculata*) and highlight its crop-weed complex. *Plant Systematic and Evolution* 262, 75–87.

Ford, R., Redden, R.J., Materne, M.M. and Taylor, P.W.J. (2007) Lentil. In: Kole, C.M. (ed.) *Genome Mapping and Molecular Breeding in Plants: Pulses, Sugar and Tuber Crops. 3.* Springer, New York, pp. 91–108.

Frahm-Leliveld, J.A. (1953) Some chromosome numbers in tropical leguminous plants. *Euphytica* 2, 46–48.

Fuller, D.Q. (2007) Contrasting patterns in crop domestication rates: recent archaeobotanical insights from the Old World. *Annals of Botany* 100, 903–924.

Garfinkel, Y., Kislev, M.E. and Zohary, D. (1988) Lentil in the pre-pottery neolithic B Yiftah'el: additional evidence of its early domestication. *Israel Journal of Botany* 37, 49–51.

Gepts, P. (1998) Origin and evolution of common bean: past events and recent trends. *Horticultural Science* 33, 1124–1130.

Gepts, P., Osborn, T.C., Rashka, K. and Bliss, F.A. (1986) Phaseolin-protein variability in wild forms and landraces of the common bean (*Phaseolus vulgaris*): evidence for multiple centers of domestication. *Economic Botany* 40, 451–468.

Hammer, K. (1984) The domestication syndrome. *Kulturpflanze* 32, 11–34.
Hammer, K. (2003) Evolution of cultivated plants and biodiversity. *Nova Acta Leopoldina NF* 87, 133–146.
Harlan, J.R. (1975) Geographic patterns of variation in some cultivated plants. *Journal of Heredity* 66, 84–191.
Harlan, J.R. (1992) *Crops and Man*. American Society of Agronomy, Madison, Wisconsin.
Harvey, M.J. and Muehlbauer, F.J. (1989) Variability for restriction fragment lengths and phylogenies in lentil. *Theoretical and Applied Genetics* 77, 839–843.
Hepper, F.N. (1963) Plants of the 1957–58 West Africa expedition II: The bambara groundnut (*Voandzeia subterranea*) and Kersting's groundnut (*Kerstingiella geocarpa*) wild in West Africa. *Kew Bulletin* 16, 395–407.
Hepper, F.N. (1970) Bambara groundnut (*Voandzeia subterranea*). *Field Crop Abstracts* 23, 1–6.
Isemura, T., Kaga, A., Knoishi, S., Ando, T., Tomooka, N., Han, O.K. *et al.* (2007) Genome dissection of traits related to domestication in azuki bean (*Vigna angularis*) and comparison with other warm-season legumes. *Annals of Botany* 100, 1053–1071.
Kaga, A., Han, O.K., Hirashima, S., Sarvankumar, P. and Kumari, H.M.P.S. (2004) Collecting and monitoring of the azuki bean (*Vigna angularis*) complex populations in Tottori prefecture, Japan. Annual report on exploration and introduction of plant genetic resources. *National Institute of Agrobiological Sciences* 20, 61–74.
Kaga, A., Isemura, T., Tomooka, N and Vaughan, D.A. (2008) The genetics of domestication of the azuki bean (*Vigna angularis*). *Genetics* 178, 1013–1036.
Kaplan, L. and Lynch, T.F. (1999) *Phaseolus* (Fabaceae) in archaeology: AMS radiocarbon dates and their significance for pre-Columbian agriculture. *Economic Botany* 53, 261–272.
Kislev, M.E. (1985) Early neolithic horsebean from Yiftah'el, Israel. *Science* 279, 302–303.
Kislev, M.E. and Bar-Yosef, O. (1988) The legumes: the earliest domesticated plants in the Near East? *Current Anthropology* 29, 175–179.
Koinange, E.M.K., Singh, S.P. and Gepts, P. (1996) Genetic control of the domestication syndrome in common bean. *Crop Science* 36, 1037–1045.
Ladizinsky, G. (1979) The genetics of several morphological traits in lentil. *Journal of Heredity* 70, 135–137.
Ladizinsky, G. (1985) The genetics of hard seed coat in the genus *Lens*. *Euphytica* 34, 539–543.
Ladizinsky, G. (1987) Pulse domestication before cultivation. *Economic Botany* 41, 60–65.
Ladizinsky, G. and Adler, A. (1976a) The origin of chickpea *Cicer arietinum* L. *Euphytica* 25, 211–217.
Ladizinsky, G. and Adler, A. (1976b) Genetic relationships among the annual species of *Cicer* L. *Theoretical and Applied Genetics* 48, 197–203.
Ladizinsky, G. and Hamel, A. (1980) Seed protein profiles of pigeon pea (*Cajanus cajan*) and some *Atylosia* species. *Euphytica* 29, 313–317.
Lev-Yadun, S., Gopher, A. and Abbo, S. (2000) The cradle of agriculture. *Science* 288, 1602–1603.
Lush, W.M. and Evans, L.T. (1980a) The seed coats of cowpeas and other grain legumes: structure in relation to function. *Field Crops Research* 3, 267–286.
Lush, W.M. and Evans, L.T. (1980b) The domestication and improvement of cowpeas (*Vigna unguiculata* (L.) WALP.). *Euphytica* 30, 579–587.
Lush, W.M. and Evans, L.T. (1981) Domestication and improvement of cowpea. *Euphytica* 30, 579–587.
Lush, W.M., Evans, L.T. and Wien, H.C. (1980) Environmental adaptation of wild and domesticated cowpeas (*Vigna unguiculata* (L.) Walp.). *Field Crops Research* 3, 173–187.
Maeda, K. (1987) *Legumes and Humans: a 10,000-Year History*. Kokonshin, Tokyo.
Massawe, F.J., Dickinson, M., Roberts, J.A. and Azam-Ali, S.N. (2002) Genetic diversity in bambara groundnut (*Vigna subterranea* (L) Verdc.) landraces revealed by AFLP markers. *Genome* 45, 1175–1180.
Maughan, P.J., Saghai Maroof, M.A. and Buss, G.R. (1996) Molecular-marker analysis of seed-weight: genomic locations, gene action, and evidence for orthologous evolution among three legume species. *Theoretical and Applied Genetics* 93, 574–579.
Mayes, S., Stadler, F., Basu, S., Murchie, E., Massawe, F., Kilian, A. *et al.* (2009) BAMLINK – a cross-disciplinary programme to enhance the role of bambara groundnut (*Vigna subterranea* L. Verdc.) for food security in Africa and India. *Acta Horticulturae* 806, 137–149.
Mulualem, T., Kassa, R., Penmetsa, V., Farmer, A.D., Carrasquilla-Garcia, N., Datta, S. *et al.* (2010) Single nucleotide polymorphism (SNP) genotyping in diverse genotypes of cultivated pigeon pea and wild

relatives with the Illumina Goldengate Assay. In: *Plant & Animal Genomes XVIII Conference*, 9–13 January 2010, Town & Country Convention Center, San Diego, California, pp. 182.

Murfet, I.C. and Reid, J.B. (1985) The control of flowering and internode length in *Pisum*. In: Hebblethwaite, P.D., Heath, M.C. and Dawkins, T.C.K. (eds) *The Pea Crop*. Butterworths, London, pp. 67–80.

Nadimpalli, B.G., Jarret, R.L., Pathak, S.C. and Kochert, G. (1992) Phylogenetic relationships of pigeon pea (*Cajanus cajan*) based on nuclear restriction fragment length polymorphisms. *Genome* 36, 216–223.

Nene, Y.L. and Sheila, V.K. (1990) Pigeon pea: geography and importance. In: Nene, Y.L., Hall, S.D. and Sheila, V.K. (eds) *The Pigeonpea*. CAB International, Wallingford, UK, pp. 1–14.

Nesbitt, M. (2002) When and where did domesticated cereals first occur in southwest Asia? In: Cappers, R.T.J. and Bottema, S. (eds) *The Dawn of Farming in the Near East*. Ex Oriente, Berlin, pp. 113–132.

Ng, Q. (1995) Cowpea, *Vigna unguiculata* (Leguminosae: Papilionoideae). In: Smartt, J. and Simmonds, N.W. (eds) *Evolution of Crop Plants*, 2nd edn. Longman, New York, pp. 326–332.

Ng, Q. and Maréchal, R. (1985) Cowpea taxonomy, origin and germplasm. In: Singh, S.E. and Bachie, K.O. (eds) *Cowpea Genetic Resources*. International Institute of Tropical Agriculture, Ibadan, Nigeria, pp. 11–21.

Ng, Q. and Padulosi, S. (1988) Cowpea gene pool distribution and crop improvement. In: Ng, Q., Perrino, P., Attere, F. and Zedan, H. (eds) *Crop Genetic Resources of Africa*, vol. II. IBPGR, Rome, pp. 161–174.

Ntundu, W.H., Bach, I.C., Christiansen, J.C. and Andresen, S.B. (2004) Analysis of genetic diversity in Bambara groundnut (*Vigna subterrenea* (L) Verdc.) landraces using amplified fragment length polymorphism (AFLP) markers. *African Journal of Biotechnology* 3, 220–225.

Panella, L. and Gepts, P. (1992) Genetic relationships within *Vigna unguiculata* (L.) Walp. based on isozyme analyses. *Genetic Resources and Crop Evolution* 39, 71–88.

Pasquet, R.S. (1999) Genetic relationships among subspecies of *Vigna unguiculata* (L.) Walp. based on allozyme variation. *Theoretical and Applied Genetics* 98, 1104–1119.

Pasquet, R.S. (2000) Allozyme diversity of cultivated cowpea *Vigna unguiculata* (L.) Walp. *Theoretical and Applied Genetics* 101, 211–219.

Pasquet, R.S. and Fotso, M. (1997) The ORSTOM bambara groundnut collection. In: Heller, J., Begemann, F. and Mushonga, J. (eds) *Bambara Groundnut Vigna subterranea* (L.) IPGRI, Rome, pp. 119–123.

Pasquet, R.S., Schwedes, S. and Gepts, P. (1999) Isozyme diversity in bambara groundnut. *Crop Science* 39, 1228–1236.

Phan, H.T.T., Ellwood, S.R., Ford, R., Thomas, S. and Oliver, R. (2006) Differences in syntenic complexity between *Medicago Truncatula* with *Lens culinaris* and *Lupinus albus*. *Functional Plant Biology* 33, 775–782.

Piperno, D.R. and Dillehay, T.D. (2008) Starch grains on human teeth reveal early broad crop diet in northern Peru. *Proceedings of the National Academy of Sciences USA* 105, 19622–19627.

Plitman, U. and Kislev, M.E. (1989) Reproductive changes induced by domestication. In: Stirton, C.H. and Zarucchi, J.L. (eds) *Advances in Legume Biology*. Botanical Garden, St. Louis, Missouri, pp. 487–503.

Poncet, V., Lamy, F., Devos, K.M., Gale, M.D., Sarr, A. and Robert, T. (2000) Genetic control of domestication traits in pearl millet (*Pennisetum glaucum* L., Poaceae). *Theoretical and Applied Genetics* 100, 147–159.

Pringle, H. (1998) Neolithic agriculture: the slow birth of agriculture. *Science* 282, 1446.

Rawal, K.M. (1975) Natural hybridization among wild, weedy and cultivated *Vigna unguiculata* (L.) Walp. *Euphytica* 24, 699–707.

Redden, R.J. and Berger, J.D. (2007) History and origin of chickpea. In: Yadav, S.S., Redden, R.J., Chen, W. and Sharma, B. (eds) *Chickpea Breeding and Management*. CAB International, Wallingford, UK.

Smartt, J. (1984) Evolution of pulse legumes. 1. Mediterranean pulses. *Experimental Agriculture* 20, 275–296.

Smartt, J. (1990) *Grain Legumes: Evolution and Genetic Resources*. Cambridge University Press, Cambridge, UK, pp. 278–293.

Songok, S., Ferguson, M., Muigai, A.W. and Silim, S. (2010) Genetic diversity in pigeon pea [*Cajanus cajan* (L.) Millsp.] landraces as revealed by simple sequence repeat markers. *African Journal of Biotechnology* 9, 3231–3241.

Sonnante, G., De Paolis, A. and Pignone, D. (2005) Bowman-Birk inhibitors in *Lens*: identification and characterization of two paralogous gene classes in cultivated lentil and wild relatives. *Theoretical and Applied Genetics* 110, 596–604.

Sonnante, G., Hammer, K. and Pignone, D. (2009) From the cradle of agriculture a handful of lentils: History of domestication. *Rendiconti Lincei* 20, 21–37.
Steele, W.M. (1976) Cowpeas, *Vigna unguiculata* (Leguminosae-Papillionatae). In: Simmonds N.W. (ed.) *Evolution of Crop Plants*. Longman, London, pp. 183–185.
Swanevelder, C.J. (1998) *Bambara Groundnut*. Department of Agriculture, Pretoria, Republic of South Africa.
Tahir, M. and Muehlbauer, F. (1993) Gene mapping in lentil with recombinant inbred lines. *Journal of Heredity* 85, 306–310.
Timmerman-Vaughan, G.M., McCallum, J.A., Frew, T.J., Weeden, N.F. and Russell, A.C. (1996) Linkage mapping of quantitative trait loci controlling seed weight in pea (*Pisum sativum* L.). *Theoretical and Applied Genetics* 93, 431–439.
Timmerman-Vaughan, G.M., Mills, A., Whitfield, C., Frew, T., Butler, R. and Murray, S. (2005) Linkage mapping of QTL for seed yield, yield components and developmental traits in pea. *Crop Science* 45, 1336–1344.
Toker, C. (2009) A note on the evolution of kabuli chickpeas as shown by induced mutations in *Cicer reticulatum* Ladizinsky. *Genetic Resources and Crop Evolution* 56, 7–12.
Tomooka, N., Kashiwaba, K., Vaughan, D.A., Ishimoto, M. and Egawa, Y. (2000) The effectiveness of evaluating wild species: searching for sources of resistance to bruchid beetle in the genus *Vigna subgenus* Ceratotropis. *Euphytica* 115, 27–41.
Vaillancourt, R.E. and Weeden, N.F. (1992) Chloroplast DNA polymorphism suggests Nigerian center of domestication for the cowpea, *Vigna unguiculata* (Leguminosae). *American Journal of Botany* 79, 1194–1199.
van der Maesen, L.J.G. (1972) *Cicer L., a Monograph of the Genus, with Special Reference to the Chickpea (Cicer arietunum L.), its Ecology and Distribution*. Mendelingen Landbouhogeschool, Wageningen, The Netherlands, pp. 1–341.
van der Maesen, L.J.G. (1980) India is the native home of the pigeon pea. In: Arends, J.C., Boelema, G., de Groot, C.T. and Leeuwenberg, A.J.M. (eds) *Libergratulatorius in Honorem H.C.D. de Wit*. Wageningen, The Netherlands, pp. 257–262.
van der Maesen, L.J.G. (1987) Origin, history and taxonomy of chickpea. In: Saxena, M.C. and Singh, K.B. (eds) *The Chickpea*. CAB International, Wallingford, UK, pp. 11–34.
van der Maesen, L.J.G. (1990) Pigeon pea: origin, history, evolution, and taxonomy. In: Nene, Y.L., Hall, S.D. and Sheila, V.K. (eds) *The Pigeonpea*. CAB International, Wallingford, UK, pp. 15–46.
van der Maesen, L.J.G., Maxted, N., Javadi, F., Coles, S. and Davies, A.M.R. (2007) Taxonomy of the genus *Cicer* revisited. In: Yadav, S.S., Redden, B., Chen, W. and Sharma, B. (eds) *Chickpea Breeding and Management*. CAB International, Wallingford, UK, pp. 14–46.
Vaughan, D.A., Tomooka, N. and Kaga, A. (2004) Azuki bean. In: Singh, R.J. and Jauhar, P.P. (eds) *Genetic Resources, Chromosome Engineering, and Crop Improvement: Grain Legumes*. CRC Press, Boca Raton, Florida, pp. 341–353.
Wang, X.W., Kaga, A., Tomooka, N. and Vaughan, D.A. (2004) The development of SSR markers by a new method in plants and their application to gene flow studies in azuki bean [*Vigna angularis* (Willd.) Ohwi & Ohashi]. *Theoretical and Applied Genetics* 109, 352–360.
Weeden, N.F. (2007) Genetic changes accompanying the domestication of *Pisum sativum*: is there a common genetic basis to the "domestication syndrome" for legumes? *Annals of Botany* 100, 1017–1025.
Weeden, N.F. and Muehlbauer, F.J. (2004) Genomics and genetic improvement in the cool season pulse crops pea, lentils and chickpea. In: Wilson, R.F., Stalker, H.T. and Brummer, E.C. (eds) *Legume Crop Genomics*. AOCS Press, Champaign, Illinois, pp. 83–96.
Weeden, N.F. and Wolko, B. (1988) *Measurement of Genetic Diversity in Pea Accessions Collected Near the Center of Origin of Domesticated Pea*. IPBGR, Rome.
Weeden, N.F., Muehlbauer, F.J. and Ladizinsky, G. (1992) Extensive conservation of linkage relationships exists between pea and lentil genetic maps. *Journal of Heredity* 83, 123–129.
Weeden, N.F., Brauner, S. and Przyborowski, J.A. (2002) Genetic analysis of pod dehiscence in pea (*Pisum sativum* L.). *Cellular and Molecular Biology Letters* 7, 657–663.
Yamaguchi, H. (1992) Wild and weedy azuki beans in Japan. *Economic Botany* 46, 384–394.
Yamamoto, Y., Sano, C.M., Tatsumi, Y. and Sano, A. (2006) Field analysis of horizontal gene flow among *Vigna angularis* complex plants. *Plant Breeding* 125, 156–160.
Yano, A., Yashuda, K. and Yamaguchi, H. (2004) A test for molecular identification of Japanese archaeological beans and phylogenetic relationship of wild and cultivated species of subgenus

Ceratotropis (genus *Vigna*, Papilionaceae) using sequence variation in two non-coding regions of the *trnL* and *trnF* genes. *Economic Botany* 58, S135–S146.
Zohary, D. (1999) Monophyletic vs. polyphyletic origin of the crops on which agriculture was founded in the Near East. *Genetic Resources and Crop Evolution* 46, 133–142.
Zohary, D. and Hopf, M. (1973) Domestication of pulses in the old world. *Science* 182, 887–894.
Zohary, D. and Hopf, M. (1993) *Domestication of Plants in the Old World – the Origin and Spread of Cultivated Plants in West Asia, Europe, and the Nile Valley*. Clarendon Press, Oxford, UK.
Zohary, D. and Hopf, M. (2000) *Domestication of Plants in the Old World*, 3rd edn. Oxford University Press, Oxford, UK.

3 Biology of Food Legumes

S.K. Chaturvedi, Debjyoti Sen Gupta and Rashmi Jain

3.1 Introduction

Food legumes, because of their most prominent biological features and ability to fix atmospheric nitrogen due to the presence of bacteria in their root nodules, provide ample justification for their significant involvement in major crop improvement programmes throughout the world. This group of crops is important for sustainable agricultural production in areas where double cropping has become a must to provide nutritional and food security to an increasing human population. With some 20,000 species, the legumes are the third largest family of higher plants. Fabaceae/Leguminosae is a large family (about 700 genera and 18,000 species), and is nearly ubiquitous over temperate and tropical parts of the world (Polhill and Raven, 1981). Many agronomically important plants are members of this family and are second only to cereal crops in agricultural importance with regard to area coverage and total production. In 2004, more than 300 million t of grain legumes were produced on 190 million ha (or about 13% of total land under cultivation, including arable land and land under permanent crops (FAOSTAT, 2011).

In contrast with other botanical families, wind-pollinated species are extremely rare in the Fabaceae, which are largely self-fertilized or insect-pollinated. Although not unique to the legumes, insect pollination is accompanied by adaptations in the plant host such as the development of specific morphological traits and the production of volatile attractants. Morphological traits include specific inflorescence types, such as racemes and pseudoracemes and a zygomorphic (bilateral) flower (Tucker, 2003).

Grain legumes are important in human nutrition in several parts of the world, and they contribute substantially to the total protein intake, mainly of vegetarian diets. The subfamily Papilionoideae is the most important of all, containing most of the cultivated food grain legumes with 30 tribes, 455 genera and about 12,000 species. It is a specialized monophyletic group derived from within the Caesalpinioideae subfamily, based on morphological (Chappill, 1995) and molecular evidence (Doyle, 1995; Doyle *et al.*, 2000). Its monophyly is supported by imparipinnate leaves, petal claws, a lateral seed hilum, the presence of a hilar fissure and unidirectional sepal initiation (Doyle *et al.*, 2000). This subfamily includes herbs, shrubs and trees that generally have alternate, compound, pinnate or trifoliate leaves with stipules and often with stiples (Cobley and Steele, 1976). The inflorescence is generally a raceme and flowers are typically known as papilionaceous, from which the subfamily name is derived. Stamens are usually ten and mostly diadelphous. The superior ovary is enclosed by a staminal tube that matures into a dry, dehiscent fruit, known as a pod. The seeds vary greatly in shape, size and colour. Since large variation in reproductive biology is present in members of the Papilionoideae, understanding the biology and floral morphology will help in formulating appropriate research strategies for development of suitable plant types, as well for applying breeding methods for improvement. This chapter discusses the biology and floral morphology of legumes in general and major food legume crops in particular.

3.2 Reproductive Biology

The success of a hybridization-based crop improvement programme relies heavily upon the reproductive behaviour of the species. Breeding methods differ in cross- and self-pollinated species, which greatly depend upon the floral morphology and pollination behaviour. Species of the flowering plants are most reliably identified by their flowers, the sexually reproductive organs (Tucker, 2003). The family Fabaceae/ Leguminoseae comprises mainly three subfamilies: Caesalpiniaceae, Mimosoideae and Papilionioideae, all differing greatly in floral symmetry. The subfamilies Mimosoideae and Papilionoideae are monophyletic and have been derived from the third subfamilily, Caesalpiniodeae, which is basal and paraphyletic (Doyle, 1995; Doyle *et al.*, 2000; Bruneau *et al.*, 2001). Although many food grain legumes have a typical papilionaceous type of flower and people consider them to be the representatives of legumes, many legume taxa differ markedly from this type of flower (Fig. 3.1; Tucker, 1987, 2003). Flowers of most of the grain legumes belonging to Papilionoideae have a pentamerous ground plan. The inflorescences of Papilionoideae are generally racemes or panicles, although two other kinds of inflorescence, pseudoracemes and cymes, are also rarely found in the subfamily (Tucker, 1987). The flowers are specialized zygomorphic. The calyx consists of five sepals and the corolla comprises a standard, two wings and two lower petals that lie inside the wings and are united at the lower margins to form a keel (Fig. 3.1). There are ten stamens surrounding the pistil, which is superior in position and differentiates into a gynoecium with stigma, style and ovary. Therefore, papilionaceous flowers comprise 21 organs in all and the members of each whorl alternate with those of the preceding whorl (Tucker, 2003). The anthers open lengthwise and shed their pollen directly on to the stigma. After anther dehiscence and pollination is completed, the ovary elongates.

During the process of floral development the petals are similar, although they differentiate late in ontogeny (Fig. 3.1, F). All floral organs are initiated in a successive, unidirectional order in each whorl starting on the abaxial side (Tucker, 2003). However, timing of development of one whorl may overlap with that of the next in some papilionoids. For example, stamens start to initiate before the last petals have been initiated (Fig. 3.2, C; Tucker, 1989, 2003; Ferrándiz *et al.*, 1999). Unisexual flowers are rare in this subfamily, although male sterility has been reported in some species of *Vigna*, *Lathyrus* and *Lupinus* (Karlin Arroyo, 1981). The flowers are generally cleistogamous, although they are also well adapted for pollination by insects. There is minimal cross-pollination in pea (Gritton, 1980), lentil (Wilson and Law, 1972) and chickpea (Niknejad and Khosh-Khui, 1972); however, cross-pollination can be more than 50% in faba bean (Hanna and Lawes, 1967) and pigeon pea. Cross-pollination is believed to be high in grass pea, but actual data have not been reported to date.

The subfamily Caesalpinioideae is highly diverse and consists of 170 genera and about 3000 species (Doyle *et al.*, 2000; Tucker, 2003). The Caesalpinioid inflorescences are racemes and the floral symmetry is highly variable within the members of this subfamily, reflecting the fact that the family is polyphyletic, as revealed by molecular phylogenies (Doyle *et al.*, 2000). The whorls of sepals, petals and the two whorls of stamens are each pentamerous and alternating. Although most legumes have 21 floral organs, many caesalpinioid taxa have undergone complete loss of some organs such as sepals, petals or stamens, or

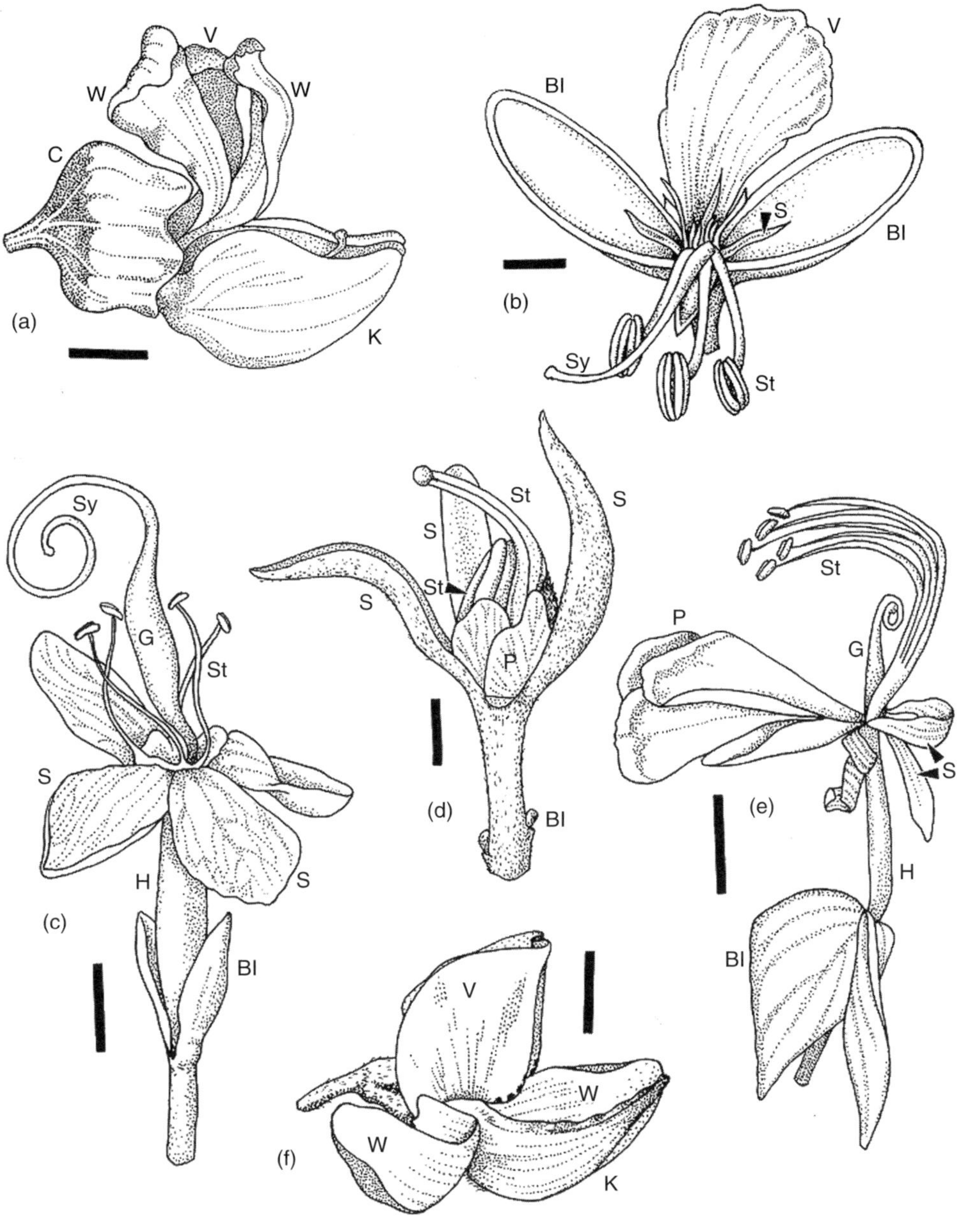

Fig. 3.1. A–F, legume flowers. A, papilionaceous flower of redbud (*Cercis canadensis*) with three forms of petal: standard or vexillum, wing and keel. B, *Paramacrolobium caeruleum*, zygomorphic flower with large bracteoles, five tiny sepals, one large petal, one carpel and three stamens. C, *Saraca declinata*, radially symmetrical flower with sepals, no petals, one carpel and only four stamens. D, *Labichea lanceolata*, asymmetric flower with sepals, four reduced petals, one carpel (not shown) and only two stamens. E, strongly zygomorphic flower of *Amherstia nobilis*, with petalloid bracteoles, four sepals, three large petals, ten stamens and an elongate hypanthium. F, papilionoid flower of *Lupinus succulentus*, with standard or vexillum, wings and keel. Bl, bracteole; C, calyx; G, gynoecium; H, hypanthium; K, keel petal; P, petal; V, standard or vexillum petal; S, sepal; St, stamen; Sy, style; W, wing petal. Scale bars: 4 mm for A–C, E, F; 2 mm for D. (Photograph adapted from Tucker, 2003; copyrighted by the American Society of Plant Biologists and reprinted with permission.)

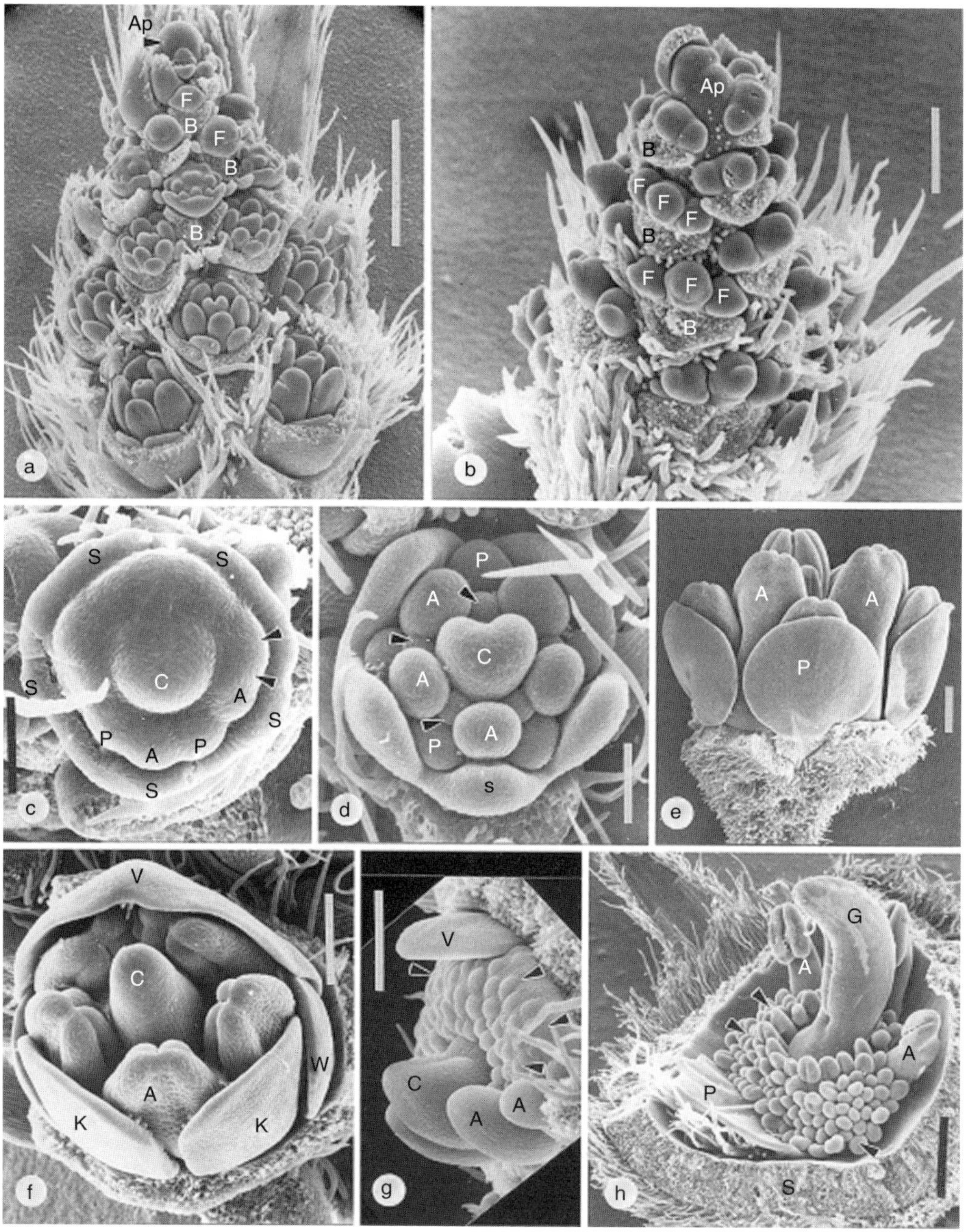

Fig. 3.2. A–H, Floral initiation and specialization in Papilionoideae (SEM micrographs). The abaxial side is at the base in C, D and F. A and B, inflorescences with most bracts removed. A, raceme of *Lupinus affinis* with flower buds developing successively and acropetally; each flower is subtended by a bract. B, pseudoraceme inflorescence of *Psoralea macrostachys* with three flowers in each bract axil. C, floral bud of garden pea (*Pisum sativum*) showing overlap in time of initiation among whorls of sepals, petals, stamens and carpel. All organ types have initiated on the abaxial side but only sepals on the adaxial side; a common primordium (arrowheads) has initiated one stamen primordium and would have initiated two more primordia. D, polar view of floral bud of *Genista tinctoria* at mid-stage with all organs initiated; three of the

even entire whorls may be missing (Tucker, 1998, 2000, 2003).

Mimosoideae contains about 65 genera and about 3000 species (Doyle *et al.*, 2000). The group is believed to be derived from among the Caesalpinioids based upon morphological (Chappill, 1995) and molecular data (Doyle *et al.*, 2000; Lucknow *et al.*, 2000; Tucker, 2003). This subfamily includes four tribes – Mimosae, Acacieae, Ingeae and Parkieae. Its flowers are usually borne in a raceme, or a panicle inflorescence. The flowers have a successive acropetal initiation, although formation of floral organ takes place simultaneously in all the buds in an inflorescence. This leads to a synchronous development in mimosoid inflorescence. The flowers are radially symmetrical in all taxa of the mimosoid subfamily (Tucker, 2003). These flowers usually have four or five organs in a whorl, and all the members of a whorl are similar. Flowers of Mimoseae and Parkieae tribes have eight or ten stamens in two whorls, while the Acacieae may have multistaminate species. Most of the members of Mimosoideae have a single carpel per flower, although some taxa may also have multicarpellary flowers.

3.3 Biology of Major Food Legume Crops

Chickpea

The biology of chickpea (*Cicer arietinum* L.) has been described by a number of researchers (Muller, 1973; Pate and Kuo, 1981; Cubero, 1987). *Cicer* includes both annuals as well as perennials. The plants are 0.2–1.0 m tall, olive to dark bluish green in colour and shrubby, but never reach the size of an authentic bush (Cubero, 1987). Plants rarely attain > 1 m in height and are pubescent, with both glandular and aglandular hairs. Stems are branched, straight or flaxous, erect to prostrate, sometimes shrubby and much branched and strong with more or less pronounced ribs (Cubero, 1987). Some varieties are semi-erect with main stem and only a few branches, while others are semi-spreading types with profuse branching. The branches are usually quadrangular, ribbed, green and densely coated with glandular hair. The main stem is round and sometimes divaricates from the base. Its stipules are generally toothed, and concrescent with the stem but not with the leaves.

The leaves of *Cicer* are imparipinnate, glandular-pubescent with 3–8 pairs of leaflets and a top leaflet (rachis ending in a leaflet). Leaflets are ovate to elliptic, 0.6–2.0 cm long, 0.3–1.4 cm wide with serrate margin and acuminate to aristate apex, cuneate base; stipules 2–5-toothed. They are serrated, somewhat sticky, pinnate reticulate and without stipules, strongly veined. The stipules are also toothed and furrowed on the upper surface. Under good conditions, plants grow to 30–65 cm bearing a taproot of 15–30 cm along with about four rows of lateral roots. The primary root is long and strong and it branches very quickly. Being generally tolerant to drought, the plant is known to thrive in winter, cold and dew. Deep and prolific root systems in chickpea have been associated with enhanced avoidance of terminal moisture stress.

Flowers are typically zygomorphic, solitary, sometimes 2 per inflorescence, axillary,

inner stamen primordia are indicated by arrowheads. The median sagittal sepal is on the abaxial (lower) side, while the median petal is on the adaxial (upper) side. E, lateral view of flower bud of *Cadia purpurea* showing all petals of same size, none overlapping at this stage. F, near-polar view of large bud of *Genista tinctoria*, sepals removed, to show descending cochleate aestivation of petals. G, lateral view of flower bud of *Swartzia sericea*, showing single petal and ring meristem (arrowheads), on which numerous stamen primordia have initiated. H, older flower bud of *Swartzia aureosericea*, sepals removed. The flower has a single petal, three large stamens, about 100 small stamens (some at arrowheads) and a gynoecium. A, outer-whorl stamen; a, inner-whorl stamen; Ap, inflorescence apical meristem; B, bract; C, carpel; F, flower bud/floral apex; G, gynoecium; K, keel petal; P, petal; S, sepal/calyx tube; V, standard or vexillum petal; W, wing petal. Scale bars: 100 μm in C, D, G; 200 μm in B, F; 500 μm in A, E, H. (Photograph adapted from Tucker *et al.*, 2003; copyrighted by the American Society of Plant Biologists and reprinted with permission.)

polypetalous with a vexillary aestivation. Peduncles are 0.6–3.0 cm long, pedicels 0.5–1.3 cm long, bracts triangular or tripartite; calyx 7–10 mm long; corolla white, pink, purplish (fading to blue) or blue, 0.8–1.2 cm long. The flowers are borne on short, jointed peduncles arising from the leaf axil and are situated opposite the leaves. Chickpea is characterized by a semi-prostrate bushy plant habit and by single flowers per peduncle (rarely double or triple), and a low number of seeds per pod. The calyx tube is oblique, gamosepalous, lanceolate and densely covered with glandular hair persistent with anterior, two lateral, two posterior subconnate, sublanceolate lobes. The corolla varies in colour from white to purple/pink or blue, the standard petal being ovate with a number of coloured, forking veins running from the centre to the edge of the petal. The wings are almost half as broad as the standard petal, clawed and spurred. The keels are nearly half as broad as the wing and clawed and free. The staminal column is diadelphous (9 + 1). The anthers are bicelled, orange in colour and basifixed. The ovary is superior, sessile, pubescent (Duke, 1981; Cubero, 1987; van der Maesen, 1987) and oval, with a terminal, slightly bent style and a blunt stigma. Pods are rhomboid-ellipsoid, having 1–2 seeds, 3 at a maximum, inflated, glandular-pubescent. A great variability in shape, size and colour of seeds is observed, which may be cream, yellow, brown, black or green, rounded to angular, with smooth or wrinkled or tuberculate seed coat, laterally compressed with a median groove around two-thirds of the seed, anterior beaked; germination cryptocotylar (Duke, 1981; Cubero, 1987; van der Maesen, 1987). *Cicer* is hopogeal and there are no hypocotyls.

Pigeon pea

Pigeon pea (*Cajanus cajan* (L.) Millsp.) is a vigorous, drought-tolerant legume widely grown in subtropical and tropical regions as an edible and forage legume. It is an erect annual or short-lived perennial usually reaching a height of 1–3 m. The plants grow into woody shrubs, 1.0–2.5 m tall when harvested annually, but may attain a height of 3–4 m when grown as a perennial plant in fence rows or agroforestry plots. Its seeds do not possess dormancy and germination is hypogeal.

This plant possesses a deep, strong and woody taproot system with well-developed lateral branches. It is well adapted to dry conditions, because it can penetrate plough layers and sparingly take up soluble sources of phosphate. Normally, root depth ranges from 30 to 90 cm, although under certain conditions the roots can grow more than 2 m. However, the most extensive development takes place in the upper 60 cm portion (Natarajan and Willey, 1980; Reddy, 1990). Compact varieties produce more deeply penetrating roots, while the spreading types produce shallower, spreading and denser root systems (Pathak, 1970).

The pigeon pea plant is erect and branching. The stems are ribbed and up to 12–15 cm in diameter. Young stems are angular and pubescent. The stem is woody; leaves are trifoliate, compound. The first two leaves are simple, opposite and caducous and are narrowly ovate with a cordate-truncate base, and an acute-acuminate apex (Reddy, 1990). Subsequent leaves are compound, pinnately trifoliate and spirally arranged. The leaflets are entire and deeply silky on the lower surface. Petioles are short, slender, grooved and subtended by small stipules. Petiole length ranges from 2.5 to 6.4 cm and it is prominently grooved on the adaxial side. Terminal leaflets have a longer stalk and are mostly symmetrical and longer than laterals; leaflets are 5–10 cm long. Terminal leaflets are usually bigger than lateral leaflets. The leaves are pubescent, more so on the lower than the upper surface (Bisen and Sheldrake, 1981). Simple and glandular hairs are also seen on all aerial parts of the plant, with the exception of floral organs such as petals and stamens.

Inflorescence is small recemes, mostly axillary, sometimes terminal, 4–10 cm long. The flowers are clustered at the top of the peduncles, which are 1–8 cm long. Flowers are mostly yellow, sometimes tinged red or purple. The bracts are small with a thick middle nerve. They are ovate-lanceolate with hairy margins and curved inwards to form a boat-like structure to enclose 1–3 young lateral buds. The pedicel is thin, 5–15 mm long and covered with hair.

Flowers are self-compatible and are frequently self-pollinated. Many cleistogamous lines are available in germplasm. The flowers are visited by insects and, depending on the frequency of visits, outcrossing can be observed in 5–40% of cases. The calyx is gamosepalous with five lobes. The calyx tube is campanulate (bell-shaped) with nerved teeth. The upper two teeth are subconnate. The lower three are free and spreading. The upper lobes are paired, free or partly free, with the lower one the longest. The corolla is zygomorphic and bright yellow. The petals are imbricate and of three prominent types: standard, wings and keel. The standard is broad, large, auricled and erect. The wings are obliquely obovate with an incurved claw. The keel petals are obtuse (round), inwardly curved and boat-shaped. The keel covers the androecium (stamens) and gynoecium (female organs) of the flower. Normally the standard and wings are bright yellow; the keel is greenish yellow. Aestivation is a descending imbricate or one whorl outside is free and the one inside has both margins overlapped. The other whorls overlap by only one margin. Stamens are 10, diadelphous. The free stamen filament (4–7 mm) is attached at the base. The other filaments are fused together for the greater part and enclose the gynoecium. The upper free portion of the filaments terminates in anthers. The anthers are uniform, about 1 mm long. The two halves of the anthers are joined by a relatively large, sterile connective tube that is basifixed. The anthers are light or dark yellow, dorsifixed. Of the ten stamens, four have short filaments and six, including a posterior one, have long filaments.

The short anthers have blunt lobes and the long ones pointed lobes. The pollen produced by short stamens is generally used for self-fertilization (Bahadur *et al.*, 1981). The ovary is superior, subsessile, flattened dorsoventrally with a long style. It has a very short stalk, densely pubescent and glandular punctate (dotted or pitted) with two to nine ovules, marginal placentation, monocarpellary and unilocular. The style is long, filiform, upturned beyond the middle region and glabrous. It is attached to a thickened, incurved and capitate (swollen) stigma.

In general, pods are green and pointed with a little reddish mottling, but purplish pods are also found. Several pods are produced in clusters on an upright stem. The pod is 7 cm long and 1.3–1.4 cm broad. The seeds are smooth and green. The pods are compressed with a diagonal depression between the seeds up to 8 in number, up to 8 cm long, and 1.0–1.5 cm broad and non-shattering. Seed orbicular and oval with one flattened edge, testa colour is white, grey, red, brown, purple, etc.

Lentil

The botanical features of lentil (*Lens culinaris* Medik.) can be described as annual bushy herb, slender, almost erect or sub-erect, much branched, softly hairy with slender and angular stems, and 15–75 cm height (Duke, 1981; Muehlbauer *et al.*, 1985; Saxena, 2009). The lentil plant has a slender taproot system with a mass of fibrous lateral roots (Saxena, 2009). The taproot and the lateral roots in the upper soil layer carry numerous small, round, elongated nodules when a plant grows on a medium that contains appropriate strains of *Rhizobium*. The nodules may start appearing 15 days after emergence, but the peak growth in number and mass occurs when the plant reaches peak vegetative growth and it starts to decline with the onset of flowering.

Ten to sixteen leaflets are subtended on the rachis (40–50 mm); upper leaves have simple tendrils while lower leaves are mucronate (Muehlbauer *et al.*, 1985). The leaves are alternate, compound, pinnate, usually ending in a tendril; leaflets 4–7 pairs, alternate or opposite; oval, sessile, 1–2 cm long; stipules small or absent. Flowers are small, pale blue, purple, white or pink, in axillary 1–4-flowered racemes; 1–4 flowers are borne on a single peduncle and a single plant can produce up to 10–150 peduncles, each being 2.5–5.0 cm long (Muehlbauer *et al.*, 1985). Flowering proceeds acropetally. The flowers are hermaphrodite and cleistogamous.

Pods are oblong, flattened or compressed, smooth, up to 1.3 cm long, 1–2-seeded with biconvex, rounded and small seeds that are lens-shaped, green, greenish-brown or light red speckled with black. Cotyledons are red, orange, yellow or green, bleaching to yellow, often showing through the testa, influencing its apparent colour (Kay, 1979; Duke, 1981; Muehlbauer *et al.*, 1985). The size of seed is greater in the types grown in eastern regions to those in western areas. Accordingly, there are two types, namely, macrosperma, found mainly in the Mediterranean region and the New World (seed size ranging from 6 to 9 mm in diameter and yellow cotyledons with little or no pigmentation), and microsperma (2–6 mm with red-orange or yellow cotyledons) found on the Indian subcontinent, and Near East and East Africa, respectively (Muehlbauer *et al.*, 1985). The first type includes the Chilean or yellow cotyledon ones while the latter includes the small-seeded Persian or red cotyledon lentils (Kay, 1979).

Mung bean

The mung bean (*Vigna radiata* (L.) Wilczek) is an erect to sub-erect, deep-rooted, much-branched and somewhat hairy annual herb with a height ranging from 30 to 130 cm. Plants are generally branched and habit can vary from erect to sub-erect in the cultivated types to prostrate in wild progenitors. It may have a tendency of twining. The root system is an extensive taproot, while the stem is hollow, furrowed, squarish and hairy with green and sometimes purple pigmentation. Roots bear nodules that fix atmospheric nitrogen via a symbiotic association with the bacterium *Rhizobium*.

Leaves are alternate, compound, mostly trifoliate, even quadra- and pentafoliate, and covered with hairs. Stipules are broad and ovate. Petiole and rachis are grooved, pubescent, two lower leaflets are opposite and asymmetrical, terminal symmetrical, leaflets are large, ovate and entire. These are palmately three-veined, cuneate at the base and acuminate at the distal end. Flowers are in an axillary or terminal raceme, peduncle up to 13 cm in length with clusters of 10–20 flowers. The corolla is yellow, sometimes curved, 5–10 cm long. Small flowers are borne in capitate clusters on the end of long, hairy peduncles. The flowers are produced in short axillary recemes in clusters of 9–15. The flower is typically papilionaceous having one standard, two wings, two keels, a diadelphous androecieum and a gynoecium. The gynoecium is monocarpellary with a superior unilocular ovary. The style is twisted below the stigmatic surface. The stigma is hairy and placentation is marginal. The calyx comprises 5 sepals, 3 large and free, 2 small and fused. The keel encloses the reproductive organs, 10 stamens and 1 gynoecium. The number of seeds per pod ranges from 10 to 15. The seeds are oblong, green or olive green in colour, sometimes yellow, brown or blackish.

Urd bean

Black gram (*Vigna mungo* (L.) Hepper) is an erect, herbaceous, well-branched and hairy annual that can attain a height of 30–90 cm. The stems are slightly ridged with brownish hairs. Leaves are large, trifoliate, compound and hairy, generally green in colour with a purplish tinge. Leaflets are 5–10 cm long, broad, hairy, ovate and entire with small stipules. The plants have a well developed taproot system with good number of nodules for fixing atmospheric nitrogen.

The inflorescence is axillary raceme which may be branched with capitate clusters of 5–6 flowers on a short hairy peduncle which elongates later. There are five sepals and five petals. Stamens are 9 and 1, style hairy and spirally twisted. The flowers are axillary, recemose, complete, self pollinated and bright or pale yellow in colour. Calyx segments are ovate, corolla is papilionaceous, yellow, stamens 10, diadelphous, with vexillary stamen free. The pods are 4–6cm long, slender round, covered with small hairs, with short hooked beak black or greenish in colour and they contain 6–14 seeds in them. Seeds are globular, generally black, olive green or grey, germination is epigeal.

Field pea

Field pea (*Pisum sativum* L.) is an annual herbaceous legume adapted to cool and humid climates. The plant is semi-erect but has a tendency to climb on support if available. Pea roots can grow to a depth of three to four feet, however, over 75% of the root biomass is within two feet of the soil surface. A relatively shallow root system and high water use efficiency make field pea an excellent rotational crop with small grains, especially in arid areas where soil moisture conservation is critical. The stems grow to a length of 2 to 4 ft and these are slender, hollow and succulent. Leaves are pinnately compound, consist of one to three pairs of leaflets with a terminal, branched tendril. These are pale green with a whitish bloom on the surface. At maturity, the plant is a prostrate vine.

Flowers are borne in the axil of leaf always in pairs. Each consists of five petals i.e. one standard, two wings and two keels that are fused except at their base. They cover the pistils and the stamens. The standard has a notch in the center. It is composed of five sepals in gamosepalous condition. Two sepals are behind the standard, 2 subtending the wings and fifth anterior one subtending the keel. Androecium consists 10 stamens in 9+1 arrangement. The filaments of 9 stamens are joined much of their length to form a staminal tube around the ovary. In white seeded types, usually number of seeds per pod vary from 4–12 but in vegetable types, seeds per pod vary from 5–18. The stamen is free. When young, the filaments are shorter than the style but elongate by the time of pollen shedding. Ovary is superior, green and flattened containing 5–12 ovules. The style is slightly flattened, cylindrical and bends at right angle to ovary. It recurs towards the ovary near its tips. The tip has a brush of stylar hairs. Stigma is elliptical, viscous and sticky.

Cowpea

Cowpea (*Vigna unguiculata* (L.) Walp.) is a very diverse, usually glabrous, annual herb which is twining to sub-erect and rarely erect. It has a deep taproot system with many lateral branches in the surface soil and many globular nodules. The root nodules are smooth and spherical, about 5 mm in diameter, numerous on the main taproot and its branches but sparse on the smaller roots. The stem is ridged, almost glabrous but hairy at the nodes. Leaves are compound, glabrous, alternate, stipulate, long petioled, trifoliate with the lower leaflets opposite and asymmetrical, top leaflet symmetrical with a short petiole. The terminal leaflet of the trifoliate leaves is commonly around 12 cm long and larger than the lateral leaflets. The stipules are large and spurred at the base while the stiples are inconspicuous. The flowers occur in alternate pairs on a long axillary peduncle, and these are large, showy, white or yellow or pink, bracteates with short pedicels and 2 bracteoles. Flowers are pentamerous and cyclic. Calyx tube has 5 lobes, subequal, campanulate, fleshy at the base, corolla is papilionaceous with 5 petals, polypetalous, stamens 10, diadelphous, filaments alternately winged, long and short, anthers uniform, yellow and style upturned, laterally compressed, stigma beaked, globular. Many flowers may be produced in each inflorescence, but only 2–4 produce the fruit. The fruit is a pod which is long cylindrical and slightly compressed and their colour varies from pale straw to brown, red or dark purple, depending upon the subspecies. In subsp. *unguiculata*, the pods are 10–30 cm long, pendent while the seeds are 5–12 mm long. In subsp. *cylindrica*, the pods are 7.5–13 cm long and usually erect. The seeds are 5–6 mm long. In subsp. *sesquipedalis*, the pods are longer than 30 cm, flabby and are shrinking between seeds before drying. The seeds are usually 8–12 mm long and elongated kidney shaped. Seed germination is epigeal, very quick and very high.

Rice bean

Rice bean (*Vigna umbellata* (Thung.) Ohwi and Ohashi) is highly branched, flaxous annual growing 1–4 m in height. The plants are erect during early growth stage, which tend to become viny and tendrillous with the progress of growth. The younger vegetative parts are covered with fine deciduous deflexed hairs. The taproot system bearing small nodules is very extensive with a number of fine deep rooting branches. The plant produces a large number of spreading and intertwining branches, glaucous though the younger branches have short hairs. Leaves are pinnately trifoliate, leaflets broad ovate, sub-glabrous, entire or with lobes, tip acute to acuminate and the terminal leaflet cuneate. The inflorescence is axillary raceme with linear bracteoles. Flowers are bright yellow in colour and occur in clusters, papilionaceous, calyx deltoid and shortly toothed, ovary with upturned style and stigma. Pods are slender, somewhat curved, and pubescent with a prominent blunt beak. Seeds are 6–10 in a pod, oblong, 6–8 mm long, different coloured ranging from yellow to brown to black and mottled, and germination is epigeal.

Grass pea

Grass pea (*Lathyrus sativus* L.) in an annual plant with a spreading to prostrate habit and main axis about 15 to 30 cm. The stems are slender, quadrangular, hairy and with small internodes. The leaves are alternate and trifoliate with deeply lobed leaflets. The leaflets are 2–4, sessile, linear, lanceolate, with acuminate tip and cuneate base. The leaf is supported by a ridged petiole and subtended by lobed stipules. The inflorescence is axillary, long peduncled capitates racemes and flowers are solitary, white to reddish purple, calyx 5-lobed, corolla typical of papilionaceous flowers. The basal ovary is minutely hirsute having a twisted style, bearded on the lower side and a flat papillate stigma. The fruits are a pod which is oblong, flat, about 2.5 to 5 cm long, 5 mm wide. They have a short curved beak and there are short stiff bristles. Seeds are 3–5 in a pod, angled, yellow to brownish grey in colour with yellow to reddish yellow cotyledons. Germination is hypogeal.

Soybean

Soybean (*Glycine max* (L.) Merrill.) is a hairy annual with an extensive taproot system, most of it in the top 15 cm of the soil. The taproot may grow as deep as 2 m and adventitious roots grow from the hypocotyls. Aloni *et al.* (2006) found that the average length of soybean main roots that had grown for six days was 104 mm. Few or no lateral roots are indicative of a strong apical dominance. The modern cultivars of soybean are erect, bushy, 20–180 cm tall, usually with a few primary branches and no secondary branches. Exceptionally prostrate and freely branching forms are also found.

Soybean leaves are trifoliate and alternate with long petioles and small stipules and stipules; the leaflets are ovate to lanceolate with mucronate tip. The flowers are white or pale purple, very typical of Papilionadeae with a tubular calyx of five unequal sepal lobes and a five-member corolla that consists of a posterior standard petal, two lateral wing petals and two anterior keel petals (Guard, 1931). The androecium is diadelphous (9+ 1) arrangement. The single pistil is unicarpellate and has one to four campylotropous ovules (Palmer *et al.*, 2001). The style curves back toward the posterior stamen and surrounded by a knob-like stigma (Carlson and Lersten, 1987). Each flower is subtended by two bracteoles and has a hairy calyx of five pointed sepals united for about half of their length. The flowers are normally self pollinated but around 1% of cross pollination aided by insects does occur. The pods are short stalked and occur in groups of 3–15, 3–7 cm long, hairy. Light brown at maturity and slightly constricted between the seeds. The seeds vary greatly in shape, size and colour though these are most often round and yellowish, brown or black with epigeal germination.

Common bean

Three main kinds of the common bean (*Phaseolus vulgaris* L.) are recognized. The 'bushy' type cultivars are day-neutral, early maturing dwarf plants with a height of 20–60 cm with lateral and terminal inflorescences and determinate growth. The 'semi-pole' are runner types having 4–8 internodes in their main axis and are longer than the bushy types. The 'pole' types are climbing and indeterminate, may grow 2–3 m tall if provided with a support to grow by twining. The internode is longer than the bush types. The optimal plant growth habit and architecture of common bean is dependent on environmental conditions. 'Bush' type beans produce a crop in as little as 65 days and the climbing beans, on the other hand, have a longer growing season 100–120 days; some even up to 240 days and have higher yield potential (Checa *et al.*, 2006). Shoot growth habit plays a complex and important role in adaptation to P-deficiency where indeterminate types were found to be more tolerant. Common beans generally have compound leaves, with three smooth edged oval leaflets that taper to a point.

Common bean has a taproot system with many branches in the upper soil. The stem is slender, twisted, angled and ribbed, more or less square and often streaked with purple colour. The leaves are alternate, trifoliate and large. The terminal leaflet is subtended by a pair of tiny stipules while the lateral symmetrical leaflets by a single stipule. The inflorescence is axillary raceme, which may bear up to 12 flowers that may be white, pink, purple or variegated. Flowers are smaller, short-stalked, papilionaceous with 10 diadelphous stamens, long ovary, coiled style and hairy stigma. Pods are slender, cylindrical or flattened, 10–20 cm long, straight or curved and terminated by a prominent beak containing 4–10 seeds. Depending on the variety or genotype, the pods can be green, yellow, black or purple. Seeds are borne alternately, non-endospermic and vary greatly in size and colour. Multiple commercial seed types exist based on seed colour with white, yellow, cream, brown, pink, red, purple, black and mottled, pinto or striped seed types popular in different regions of the world and with different cultures (Voysest and Dessert, 1991; Voysest *et al.*, 1994). Ibarra Perez *et al.* (1997) reported the incidence of multiple paternity in common bean, where they found that most multiplied pods ($\approx 70\%$) were filled by non-hybrid seeds plus a single hybrid seed. On average, hybrid seeds occurred more frequently in ovules in positions 7 (most basal) and 1 (most stylar) than in ovules in the middle positions of the pod. Seed germination is epigeal in common bean.

Groundnut

Groundnut (peanut) (*Arachis hypogea* L.) is an annual herbaceous plant growing 30–50 cm tall. The leaves are alternate, pinnate with four leaflets (two opposite pairs; no terminal leaflet), each leaflet 1–7 cm long and 1–3 cm broad.

Inflorescences are borne in the axils of leaves on both primary and secondary branches. They are simple or compound and each has up to five flowers, only one flower per inflorescence usually opening on any given day. Flowers are papilionaceous and sessile, but appear to be stalked because of an elongated tubular hypanthium or calyx tube. Styles are contained within the calyx tube, and both the style and calyx tubes rapidly elongate 12–24 h prior to anthesis. The ovary is superior, to which the hypanthium is attached at the base. The flower ranges in colour from deep orange to light yellow, and in rare cases it may be white. A central crescent area exists on the face of the standard that is usually darker in colour, or in some cases a different colour than the remainder of the standard (Moss and Rao, 1995). Flowers generally have 10 androecia, with 5 anthers being elongated and the remaining 5 being more globular and small. The few anthers are usually sterile and difficult to observe. Sterility is more common in Spanish and Valencia types than in Virginia types (Maeda, 1972). Both the stigma and anthers are enclosed by the keel, which induces self-fertilization.

After pollination, the fruit develops into a pod 3–7 cm long containing 2–3 (rarely 1 or 4) seeds, the stalks at the bases of the ovaries, called pegs, elongate rapidly and turn downward to bury the fruits several centimetres underground to complete their development. The pro-embryo divides three to four times (resulting in an 8–16-nucleate egg) and then becomes quiescent at the time when a meristem located adjacent to the basal ovule becomes active. A carpophore (or gynophore, but commonly called a 'peg') begins to elongate by positive geotropism into the soil (Zamski and Ziv, 1976). After the peg enters the soil, it becomes diageotropic (i.e. begins to grow horizontally), ceases to elongate and at the same time it swells, and the embryos resume cell division. Pods generally develop in the absence of light (Ziv, 1981), but aerial pods can occur. In *A. hypogaea*, pod development generally begins 16–17 days after pollination, but in other species the process may be delayed until 23–25 days (Halward and Stalker, 1987). Pegs of the domesticated species are relatively short and do not break easily, but pegs of wild *Arachis* species may be 1 m or more in length and are fragile. The seed has two cotyledons, a hypocotyl, epicotyl and radicle. The cotyledons comprise nearly 96% of the seed weight and are the major storage tissue for the developing seedlings (Moss and Rao, 1995). The mature seeds resemble other legume seeds such as beans, but they have paper-thin seed coats as opposed to the usual, hard legume seed coats.

References

Aloni, R., Aloni, E., Langhans, M. and Ullrich, C.I. (2006) Role of cytokinin and auxin in shaping root architecture, regulating vascular differentiation lateral root initiation root apical dominance and root gravitropism. *Annals of Botany* 97, 883–893.

Bahadur, B., Madhusudana Rao, M. and Lokendar Rao, K. (1981) Studies on dimorphic stamens and pollen (SEM) and its possible role in pollination biology of *Cajanus cajan* (L.) Millsp. *Indian Journal of Botany* 4, 122–129.

Bisen, S.S. and Sheldrake, A.R. (1981) *The Anatomy of the Pigeonpea*. Research Bulletin No. 5. ICRISAT, Patancheru, AP, India, pp. 24.

Bruneau, A., Forest, F., Herendeen, P.S., Klitgaard, B.B. and Lewis, G.P. (2001) Phylogenetic relationship in the Caesdalpinioideae (Leguminosae) as inferred from chloroplast *trn*L intron sequences. *Systematic Botany* 26, 487–514.

Carlson, J.B. and Lersten, N.R. (1987) Reproductive morphology. In: Wilcox, J.R. (ed.) *Soybean, Improvement, Production and Uses*. Agronomy Monographs 2nd edn, No. 16, American Society of Agronomy (ASA), Madison, Wisconsin, pp. 303–416.

Chappill, J.A. (1995) Cladistic analysis of the leguminosae, the development of an explicit hypothesis. In: Crisp, M. and Doyle, J.J. (eds) *Advances in Legume Systematic, Part 7. Phylogeny*. Royal Botanical Garden, Kew, UK, pp. 1–10.

Checa, O., Ceballos, H. and Blair, M.W. (2006) Generation means analysis of climbing ability in common bean (*Phaseolus vulgaris* L.). *Journal of Heredity* 97, 456–465.

Cobley, L.S. and Steele, W.M. (1976) *The Botany of Tropical Crops*. Longman, London.

Cubero, J.I. (1987) Morphology of chickpea. In: Saxena, M.C. and Singh, K.B. (eds) *The Chickpea*. CAB International, Wallingford, UK, pp. 35–66.

Doyle, J.J. (1995) DNA data and legume phylogeny: a progress report. In: Crisp, M. and Doyle, J.J. (eds) *Advances in Legume Systematic, Part 7. Phylogeny*. Royal Botanical Gardens, Kew, UK, pp. 11–30.

Doyle, J.J., Chappill, J.A., Bailey, D.C. and Kajita, T. (2000) Towards a comprehensive phylogeny of legumes evidence from *rbc*L sequences and non-molecular data. In: Herendeen, P.S. and Bruneau, A. (eds) *Advances in Legume Systematic*. Royal Botanical Gardens, Kew, UK, pp. 1–20.

Duke, J.A. (1981) *Handbook of Legumes of World Economic Importance*. Plenum Press, New York, pp. 52–57.

FAOSTAT (2011) Available at http://faostat.fao.org/faostat/collections?subset5 (accessed 30 October 2010).

Ferrandiz, C., Navarro, C., Gomez, M.D., Canas, L.A. and Betran, J.P. (1999) Flower development in *Pisum sativum*, from the war of the whorls to the battle of the common primordia. *Developmental Genetics* 25, 280–290.

Gritton, E.T. (1980) Field pea. In: Fehr, W.R. and Hadley, H.H. (eds) *Hybridization of Crop Plants*. American Society of Agronomy, and Crop Science Society of America, Madison, Wisconsin, pp. 347–356.

Guard, A.T. (1931) Development of floral organs of the soybean. *Botany Gazette* 91, 97–102.

Halward, T.M. and Stalker. H.T. (1987) Comparison of embryo development in wild and cultivated *Arachis* species. *Annals of Botany* 59, 9–14.

Hanna, A.S. and Lawes, D.A. (1967) Studies on pollination and fertilization in the field bean (*Vicia faba* L.). *Annals of Applied Biology* 59, 289–295.

Ibarra Perez, F.J., Ehdaie, B. and Waines, J.G. (1997) Estimation of outcrossing rate in common bean. *Crop Science*, 37, 60–65.

Karlin Arroyo, M.T. (1981) Breeding systems and pollination biology in Leguminosae. In: Polhill, R.M. and Raven, P.H. (eds) *Advances in Legume Systematic*, Part 2. Royal Botanical Gardens, Kew, UK, pp. 723–769.

Kay, D. (1979) *Food Legumes*. Tropical Products Institute Crop and Products Digest No. 3. Tropical Products Institute, London, pp.48–71.

Lucknow, M., White, P.L. and Bruneau, A. (2000) Relationships among the basal genera of mimosoid legumes. In: Herendeen, P.S. and Bruneau, A. (eds) *Advances in Legume Systematics*, Part 9. Royal Botanical Garden, Kew, UK, pp.165–180.

Maeda, K. (1972) Growth analysis on the plant type in peanut varieties, *Arachis hypogoea* L. IV. Relationship between the varietal difference of the progress of leaf emergence on the mainstem during preflowering period and the degree of morphological differentiation of leaf primordia in the embryo. *Proceedings of the Crop Science Society of Japan* 41, 179–186.

Moss, J.P. and Ramanatha Rao, V. (1995) The peanut-reproductive development to plant maturity. In: Pattee, H.E. and Stalker, H.T. (eds) *Advances in Peanut Science*. American Peanut Research and Education Society, Stillwater, Oklahoma, pp. 1–13.

Muehlbauer, F.J., Cubero, J.I. and Summerfield, R.J. (1985) Lentil (*Lens culinaris* Medik.). In: Summerfield, R.J. and Roberts, E.I.I. (eds) *Grain Legume Crops*. Collins, London, pp. 266–311.

Muller, C. (1973) La tige feuilée et les cotyledons des Vicieas a germination hypogée. *La Cellule* 46, 195–354.

Natarajan, M. and Willey, R.W. (1980) Sorghum–pigeon pea intercropping and the effects of plant population density. 1. Growth and yield. *Journal of Agricultural Sciences* 95, 51–58.

Niknejad, M. and Khosh-Khui, M. (1972) National cross pollination in gram (*Cicer arietinum* L.) *Indian Journal of Agricultural Science* 42, 273–274.

Palmer, R.G., Gai, J., Sun, H. and Burton, J.W. (2001) Production and evaluation of hybrid soybean. *Plant Breeding Reviews* 21, 263–307.

Pate, J.S. and Kuo, J. (1981) Anotomical studies of legume pods. A possible tool in taxonomic research. In: Polhill, R.M. and Raven, P.H. (eds) *Advances in Legume Systematics*, Part 1. Royal Botanical Gardens, Kew, UK, pp. 903–925.

Pathak, G.N. (1970) Red gram. In: *Pulse Crops of India*. Indian Council of Agricultural Research, New Delhi, India, pp. 14–53.

Polhill, R.M. and Raven P.H. (1981) *Advances in Legume Systematics*, Parts 1 and 2. Royal Botanical Gardens, Kew, UK.

Reddy, L.J. (1990) Pigeon pea morphology. In: Nene, Y.L., Hall, S.D. and Shiela, V.K. (eds) *The Pigeonpea*. CAB International, Wallingford, UK, pp. 44–87.

Saxena, M.C. (2009) Plant morphology, anatomy and growth habit. In: Erskine, W., Muehlbauer, F.J., Sarker, A. and Sharma, B. (eds) *The Lentil: Botany, Production and Uses*. CABI, Wallingford, UK, pp. 34–46.

Tucker, S.C. (1987) Pseudoracemes in papilionoid legumes, their nature, development, and variation. *Botanical Journal of the Linnean Society* 95, 181–206.

Tucker, S.C. (1989) Overlapping organ initiation in common primordia in flowers of *Pisum sativum* (Leguminosae, Papilionoideae). *American Journal of Botany* 76, 714–729.

Tucker, S.C. (1998) Floral ontogeny in legume genera Petalostylis, Labichea and Dialium (Caesalpinioidae, Cassieae), a series in floral reduction. *American Journal of Botany* 85, 184–208.

Tucker, S.C. (2000) Evolutionary loss of sepals and/or petals in detarivoid taxa Aphanocalyx, Brachistegia and Monopetalanthus (Leguminosae, Caesalpinioidae). *American Journal of Botany* 87, 608–624.

Tucker, S.C. (2003) Floral development in legumes. *Plant Physiology* 131, 911–926.

van der Maesen, L.J.G. (1987) *Cicer* L.: origin, history and taxonomy of chickpea In: Saxena, M.C. and Singh, K.B. (eds) *The Chickpea*. CAB International, Wallingford, UK, pp. 11–34.

Voysest, O. and Dessert M. (1991). Bean cultivars: classes and commercial seed types. In: van Schoohoven, A. and Vosest, O. (eds) *Common Beans: Research for Crop Improvement*. CAB International and CIAT, Wallingford, UK. pp. 119–162.

Voysest, O., Valencia, M.C. and Amezquita, M.C. (1994) Genetic diversity among Latin American, Andean, and Mesoamerican common bean cultivars. *Crop Science* 34, 1100–1110.

Wilson, V.E. and Law, A.G. (1972) Natural crossing in *Lens esculenta* Moench. *Journal of the American Society of Horticultural Science* 87, 142–143.

Zamski, E. and Ziv, M. (1976) Pod formation and its geotropic orientation in the peanut, *Arachis hypogoea* L., in relation to light and mechanical stimulus. *Annals of Botany* 40, 631–636.

Ziv, M. (1981) Photomorphogenesis of the gynophore, pod and embryo in peanut, *Arachis hypogoea* L. *Annals of Botany* 48, 353–359.

4 Breeding for Improvement of Cool Season Food Legumes

Michael Materne, Antonio Leonforte, Kristy Hobson, Jeffrey Paull and Annathurai Gnanasambandam

4.1 Introduction

The main cool season food legumes cultivated around the world are lentil (*Lens culinaris* Medik.), chickpea (*Cicer arietinum* L.), field pea (*Pisum sativum* L.) and faba bean (*Vicia faba* L.). These are among the world's oldest cultivated plants. Breeding of these pulses is relatively recent and limited compared with cereals, even though the father of genetics, Gregor Mendel, used peas in his classical genetics studies in the mid-1800s. Focused efforts in breeding pulses began only in the 1970s with the establishment of the International Centre for Agricultural Research in Dry Areas (ICARDA) in Syria and the International Crops Research Institute for Semi-Arid Tropics (ICRISAT) in India, supported by the Consultative Group in International Agricultural Research (CGIAR), as well as through strengthening of the agricultural research systems of different conditions. Both ICARDA and ICRISAT have: (i) established, characterized and distributed landraces (traditional farmers' varieties); (ii) initiated breeding programmes that involve more diverse hybridizations; and (iii) distributed segregating populations and inbred lines to partner countries for selection and release to farmers. While ICARDA stimulated breeding of lentil, *Kabuli* chickpea and faba bean, ICRISAT stimulated *desi* chickpea breeding internationally. The development of modern, semi-leafless dwarf field pea in Europe provided a major breakthrough in field pea breeding globally. Achievements in pulse breeding are demonstrated through the successful delivery of cultivars that have established or secured production in many countries of the world.

4.2 Development and Utilization of Genetic Resources for Breeding

Genetic resources for use in cool season food legume breeding are maintained at ICARDA, ICRISAT and also by other national programmes, particularly in the USA, Canada, Australia, India and a number of other important repositories. These are discussed in detail in Chapter 23. These genetic resources contain mostly landraces, breeding materials and a limited number of wild species. Although the number of germplasm accessions of cool season food legumes available in genebanks throughout the world ranges from 23,000 in lentil to 49,000 in field pea, this is still small in comparison with world cereal collections, which include more than 410,000 wheat accessions and 210,000 rice accessions (Tanksley and McCouch, 1997). Additional collection from regions underrepresented in germplasm collections are required to capture available allelic variations for various traits.

The landraces and wild species have several useful traits that are being exploited in breeding programmes (Redden *et al.*, 2005; Singh *et al.*, 2008; Furman *et al.*, 2009; Duc *et al.*, 2010). For example, wild species have been used to develop resistance to anthracnose in lentil (Fiala *et al.*, 2009) and phytophthora root rot in chickpea (Knights *et al.*, 2008).

Utilization of wild species in breeding has been hampered by crossability barriers. Only the wild species in the primary gene pool (see Chapter 6) are readily crossable with the cultivated species. Improvements in tissue culture technologies are needed to access valuable genes in wild relatives, such as those that exist for ascochyta blight in wild chickpea species. Fertile hybrids between lentil cultigen and *Lens ervoides* were successfully obtained with the aid of embryo rescue to develop recombinant inbred populations and to transfer resistance to anthracnose to the cultivated background (Fiala *et al.*, 2009).

The wild ancestor of faba bean has not been discovered yet, or has become extinct. Hence, collection and preservation of faba bean germplasm is more crucial for present and future breeding programmes (Duc *et al.*, 2010). Also, due to the technical difficulties of achieving interspecific crosses and the political sensitivities of producing transgenic lines of faba bean, exploitation of natural variability within the cultivated species and induced mutagenesis are the only options currently available to breeders.

4.3 Breeding Methodologies

Cultivated lentil, chickpea, field pea and faba bean are all diploids with varying chromosome numbers (Singh, 2005). While faba bean generally exhibits a high percentage of outcrossing, lentil, chickpea, field pea are predominantly self-pollinated. Hence, the breeding methods adopted for lentil, chickpea and field pea have been similar to other self-pollinated crops, and generally have involved hybridization among cultivars or between cultivars, landraces and primitive forms, followed by combinations of pedigree, bulk, backcross or single-seed descent methods of selection (Ahmad *et al.*, 2005; Muehlbauer and McPhee, 2005; Redden *et al.*, 2005; Materne and McNeil, 2007). The presence of partial allogamy should be considered for faba bean breeding, which normally uses bulk selection and recurrent selection and the separation of lines during seed production to prevent outcrossing (Cubero and Nadal, 2005).

Breeding strategies

Although the basis for selection in breeding programmes has obviously varied according to the trait and species targeted, in a broad sense it has focused on combining desirable variation for major yield-limiting traits across and within environments. The selection for major genes for adaptation has been essential in establishing new breeding programmes. Genes such as those that control flowering time provide basic adaptation to an environment and can create a bottleneck to the introduction of diversity into a breeding programme. The introgression of single or a few genes into an adapted background can be achieved through backcrossing, pedigree systems of selection (e.g. single-seed descent) and effective phenotyping. In Canada and Australia, complex crosses have been used effectively to explore diversity within the lentil gene pool.

For more complex traits, maintaining segregating populations as bulk lines has been an important strategy to increase frequency combinations of minor genes (e.g. improved lodging resistance) that additively contribute to the desired variation, followed by cycles of recurrent selection. Mass selection has been a useful strategy for eliminating highly deleterious genes in relation to poor adaptation caused by high disease susceptibility (e.g. ascochyta blight) or high sensitivity to specific stress factors (e.g. cold, herbicide damage, soil boron toxicity) and for improving grain quality.

Progenies are normally tested in rows or mini-plots grown from the individual plant selections for observational purposes.

Targeted progeny testing is sometimes used to expose germplasm to high disease or abiotic stress pressure. In addition, out of season seed increase in the field or glasshouse has been an effective breeding strategy to accelerate generations, but more so for short season climates. Selections are usually grown over several years to permit observations of performance (e.g. grain yield) under different environmental conditions to enable the selection of lines that are more broadly adapted over years and environments. Selected inbred lines in most programmes are comprehensively compared to existing commercial varieties in their yielding performance, quality and other aspects of agronomic importance in advanced regional testing. In this respect, statistical analysis of genotype by environment interactions has been a useful tool for identifying sources of variation for improving both regional and general adaptation.

Mutation breeding

A number of spontaneous mutations have been very important for the development of erect-growing field pea varieties across a number of countries (Redden *et al.*, 2005). However, as spontaneous mutations occur at a low frequency in natural populations, they have therefore been induced by physical or chemical agents or by insertion of DNA to disrupt the gene (Tadege *et al.*, 2009). Induced mutations are highly useful to create variability when: (i) a desired trait may not be available in existing germplasm; and (ii) suitable screening methods are available that can be adapted to evaluate large mutagenized populations.

By the end of 2000 at least 32 mutant varieties had been reported in pea, 13 in faba bean, 11 in chickpea and 2 in lentil (Maluszynski *et al.*, 2000). Some of these varieties have produced a significant impact financially, and also on food legume production. For example, two mutant chickpea cultivars (CM-88 and CM-98) with disease resistance were grown in 350,000 ha in Pakistan, resulting in an additional estimated income of US$9.6 million per year to farmers (Ahloowalia *et al.*, 2004). Induced mutation was used in Canada to identify a lentil line with tolerance to imidazolinone herbicides. The trait was patented (US Patent 7232942) and licensed for use in Clearfield® lentil varieties, which are now widely grown in Canada and the USA. The trait has been transferred to cultivars of all market classes, resulting in the release of a series of herbicide-tolerant cultivars (Muehlbauer *et al.*, 2009). Mutant lentil lines with resistance to imidazolinone have also been developed in Australia.

4.4 Breeding Priorities

Abiotic stresses

Pulse crops are an important component of rotations in farming systems, but are considered more sensitive than cereals to a wide range of abiotic stresses, including drought, heat, frost, chilling, waterlogging, salinity and mineral toxicities (Dita *et al.*, 2006). As the majority of the world's pulse production occurs under rainfed conditions, the most common abiotic limitations to grain production occur within the reproductive development phase, as pods and developing seeds are highly sensitive to abortion and physical damage. While direct selection for abiotic stress tolerance during reproductive development has proved difficult in the field, as multiple stresses typically occur in combination and to varying degrees, long-term targeted selection for grain yield over a number of years has effectively led to the pyramiding of genes for higher general adaptation. The selection for yield under rainfed conditions has been the major strategy for selecting lentil cultivars with adaptation to variable climatic and soil factors, leading to increased water use efficiency, principally through an increased response to moisture availability (Materne and McNeil, 2007; Muehlbauer *et al.*, 2009).

Matching a crop's phenology to an environment, including the avoidance of drought and heat, is a key part of improving adaptation and increasing crop yields, and has been a major global focus in breeding for local and broad adaptation of all the cool season food legumes (Materne and Siddique, 2009; Khan *et al.*, 2010). One of the major achievements of ICARDA's collaborative lentil research is broadening the narrow genetic base of lentil in South Asia through introgression of genes from ICARDA germplasm. Extra-early and extra-bold lentil lines have been developed in India for different cropping systems, and the cultivar Shekher (ILL 4404) is being grown in the mid-hills region of Nepal, a new area for lentils (see references in Materne and McNeil, 2007).

In field pea, specific phenology traits such as time to flowering, flowering duration, flower number per inflorescence, seeds per pod and inherent rate of ovule and seed abortion have been researched. Relative timing and duration of flowering (Alcalde *et al.*, 2000) have been the main phenology traits manipulated by field pea breeders. In typically short (both winter and spring sown) growing season climates, selection for earlier-flowering genotypes has been an important trait for avoidance of late season abiotic stress (e.g. terminal drought and high temperature). Early flowering and maturing faba bean varieties have enabled expansion of production in the subtropical region of Australia (Rose and van Leur, 2006). In contrast, a longer growing season or variable rainfall climates require a longer duration of flowering to ensure optimal response to rainfall and available soil moisture. For chickpea, a large global breeding effort has targeted early maturity to avoid drought. Whilst the *Kabuli* type is generally considered more drought sensitive than *Desi* types (Leport *et al.*, 2006), ICRISAT developed an extra-short-duration *Kabuli* variety (ICCV 2), which improved yields and expanded production. Since the release of this cultivar, even earlier-maturing germplasm has been developed and combined with a double-podding trait (Ahmad *et al.*, 2005).

Cold tolerance has been an important trait for improvement in crop adaptation in many countries. In the USA and Turkey, large yield increases have been achieved by sowing lentil in winter rather than spring, using genotypes tolerant to cold temperatures during winter (Materne and McNeil, 2007). Similarly, very high tolerance of seedlings to cold temperatures has been identified in faba bean (Link *et al.*, 2010) and field pea. This has led to the development of winter types of both crops, including peas that have a longer photoperiod requirement for flowering (Lejeune-Hénaut *et al.*, 2008) in Europe and North America. To overcome frost damage during the reproductive cycle, indeterminate pod and seed development may be an effective strategy to reduce damage, particularly on developing ovules (Leonforte, unpublished data). For chickpea, chilling temperatures at the reproductive phase often result in pod abortion, and Clarke *et al.* (2004) successfully used pollen selection methods to develop and release two cultivars that produce pods under lower temperatures than other cultivars.

Soil constraints, such as salinity, are attracting greater attention from researchers and breeding programmes internationally. Breeding for improved tolerance to soil factors (e.g. high soil boron, salinity and sodicity), which limit water availability late in the growing season, are likely to contribute to higher drought tolerance per se (Leonforte *et al.*, 2010). Lentil cultivars with improved tolerance to NaCl have been released already in Australia (Materne and Siddique, 2009). The recent review by Flowers *et al.* (2010) gives a comprehensive overview of studies conducted to explore genetic variation to salt sensitivity in chickpea. Greater efforts have also been focused on quantifying thresholds, and it was recently reported that subsoil chloride (Cl) concentration was the most effective indicator of reduced grain yields rather than salinity, and that growing chickpea on soils with Cl > 600 mg should be avoided due to high yield losses (Dang *et al.*, 2010). Similarly, faba bean has been reported to be more sensitive to Cl^- than Na^+, and genetic variation for tolerance to the individual ions was observed (Tavakkoli *et al.*, 2010). Screening methodologies range from pot-based to field methods. More recently, attention has been focused on improving genetic knowledge that could provide molecular markers for salt tolerance in the near future (Varshney *et al.*, 2009).

In the subsoils of Australia's southern grain belt, boron (B) toxicity often occurs in tandem with soil salinity. In Australia, lentil breeding lines with improved tolerance to B have been developed that could improve yields by up to 91% in the target region, based on controlled environment experiments (Hobson *et al.*, 2006). Whilst genetic variation has been identified in chickpea (Hobson *et al.*, 2009), only limited research in this crop has been undertaken. Genetic variation has been identified in both field pea (Redden *et al.*, 2005) and faba bean (Paull, unpublished), and the overall level of tolerance of both crops is greater than in lentil and chickpea. Screening for B tolerance involves growing plants in soil that is high in B and rating symptom expression. In contrast, B deficiency has been identified as a limitation to lentil production in Nepal, and cultivars must be efficient in the uptake of B (Srivastava *et al.*, 2000). Similarly, cultivars that are efficient in the uptake of iron (Fe) are required on the alkaline soils of Syria and Australia (Materne and Siddique, 2009).

Biotic stresses

Lentil

Ascochyta blight caused by *Ascochyta lentis* is a major disease of lentil in Canada, India, Australia and Pakistan, where it devastates production and product quality. Many sources of resistance to ascochyta blight have been identified, particularly ILL5588, Indianhead and ILL7537, and cultivars have been released (Tivoli *et al.*, 2006). In Australia, the cultivar 'Nipper' has been released, having good resistance to ascochyta blight and botrytis grey mould, caused by *Botrytis fabae*. Anthracnose (*Colletotrichum truncatum*) is another significant disease in Canada, and cultivars such as Robin have been released that have resistance to ascochyta blight and moderate resistance to anthracnose derived from Indianhead (Vandenberg *et al.*, 2002). Improved resistance to anthracnose is now being transferred from *Lens ervoides* (Fiala *et al.*, 2009). Bari-Masur varieties with stemphyllium blight (*Stemphyllium botryosum*) resistance (developed through collaborative efforts between ICARDA and the Bangladesh government) are making a major impact in Bangladesh (Materne and McNeil, 2007). The rust (*Uromyces viciae-fabae*)-resistant varieties Bakria (ILL4605), Bichette (ILL5562) and Hamira (ILL6238) were released in Morocco (Sarker and Erskine, 2002). Similarly, in Ethiopia, varieties like Adaa and Alemaya have been released that have a high level of resistance to rust and the wilt root rot complex (Sarker and Erskine, 2002). Rust is also a breeding objective in subtropical areas of the Indian subcontinent and South America. Fusarium wilt (*Fusarium oxysporum* f. sp. *lentis*) is the major soil-borne disease of lentil internationally and the major disease of lentil in the Middle East. Long-term breeding at ICARDA has successfully delivered resistant cultivars to farmers, such as Talia 2, based on resistance from ILL5588 (Materne and McNeil, 2007).

Chickpea

The major biotic constraints to chickpea production globally include diseases such as fusarium wilt (*Fusarium oxysporum* f. sp. *Ciceri*), ascochyta blight (*Ascochyta rabiei*), botrytis grey mould (*Botrytis cinerea*) and phytophthora root rot (*Phytophthora medicaginis*) (Ahmad *et al.*, 2005; Knights *et al.*, 2008; Singh *et al.*, 2008). Several varieties with durable and stable resistance to fusarium wilt have been released in India and a number of other countries, and recent advances in the understanding of the genetic control of resistance are likely to result in successful pyramiding of resistance genes (Singh *et al.*, 2008). Varieties with improved ascochyta blight resistance have been released and widely adopted by growers in India, Pakistan, Syria, the USA, Canada and Australia (Ahmad *et al.*, 2005).

Viral diseases have become an important constraint in countries such as Australia, and these are mainly caused by the luteovirids. Plant-parasitic nematodes (root-knot *Meloidogyne* spp., root-lesion *Pratylenchus* spp., cyst-forming *Heterodera* spp. and reniform nematode *Rotylenchulus reniformis*) are reported in the major chickpea-growing areas and estimated to cause an annual yield loss of 14% (Castillo *et al.*, 2008). The major pests include helicoverpa pod borer (*Helicoverpa armigera* and *Helicoverpa punctigera*) and leaf miner (*Liriomyza cicerina*) (Ahmad *et al.*, 2005). Whilst genetic variation has been exploited for most traits, scarcity of sources of resistance is a problem, especially for ascochyta blight. Genetic variation in the wild relatives has been utilized for traits such as botrytis grey mould (Pande *et al.*, 2006), rust, phytophthora root rot, nematodes and helicoverpa, but is still considered underutilized (Singh *et al.*, 2008).

Field pea

Peas are adversely affected by a large number of fungal and viral diseases, bacterial blight and pests. Of the foliar fungal diseases, extensive efforts in breeding have focused on combining minor genes for resistance to ascochyta blight (caused by *Mycosphorella pinnodes*, *Phoma medicaginis* var. *pinodella*, *Ascochyta pisi* and *Phoma koolunga*), as single genes with major effect have not been identified. However, progress in early season whole-plant resistance (McMurray *et al.*, 2010) has been achieved in Australia and an erect architecture appears to be important (Le May *et al.*, 2005). Detached leaf assay methodology (Onfroy *et al.*, 2007) identified significant pathogen-specific resistance within adapted Australian bred germplasm (Richardson *et al.*, 2009). Downy mildew caused by *Peronospora viciae* is also widely distributed, but is more prevalent in cool and wet growing regions. This fungus causes systemic infection of seedlings, local infections on leaves and pod infections. Major genes for resistance have been identified and effective screening established (Davidson *et al.*, 2004). However, rapid pathogen specialization has been a widespread problem. Powdery mildew caused by *Erysiphe pisi* can be a serious disease of field pea, particularly in warm and humid growing climates. Two major genes for resistance, *er1* and *er2*, confer high and stable resistance to this disease (Katoch *et al.*, 2010) and have been extensively used to develop resistant varieties globally. A third major gene (*er3*) conferring resistance has also been identified from *Pisum fulvum* (Fondevilla *et al.*, 2008). Other regionally important foliar fungal diseases for which high phenotypic resistance has been identified include pea rust (*Uromyces pisi*) (Barilli *et al.*, 2009) and septoria blotch (*Septoria pisi*).

Bacterial blight (*Pseudomonas syringe* pv. *pisi* and *Psuedomonas syringae* pv. *syringae*) is a localized but very devastating disease in cool temperate regions. Breeding has mostly focused on pyramiding available race-specific resistance to pv. *pisi* (i.e. from seven races) (Hollaway and Bretag, 1995; Elvira-Recuenco *et al.*, 2003). Recently in Australia pv. *syringae* has proved damaging, and field-based screening has identified major variation for resistance and led to rapid release of resistant varieties. A large number of aphid-transmitted viruses can produce a range of disease symptoms individually or in combination. These include cucumber mosaic virus (CMV), pea early browning virus (PEBV), pea enation mosaic virus (PEMV), luteo viruses: pea leaf roll virus (PLRV) and bean leaf roll virus (BLRV), poty viruses: bean yellow mosaic virus (BYMV) and pea seedborne mosaic virus (PSbMV), alfalfa mosaic virus (AMV), pea streak virus (PeSV) and red clover vein mosaic virus (RCVMV). Root rot diseases are widespread and may be caused by one or a combination of several common soil fungal pathogens: aphanomyces root rot (*Aphanomyces euteiches*), pythium tip blight (*Pythium ultimum*), fusarium root rot (*Fusarium solani* f. sp. *pisi*), rhizoctonia root rot (*Rhizoctonia solani*) and fusarium wilt (*Fusarium oxysporum*). Whilst high resistance is found only to fusarium wilt, effort is focusing on developing resistance to aphanomyces root rot. Resistance to *Aphanomyces* is partial and controlled by several quantitative trait loci (QTL) (Pilet-Nayel *et al.*, 2002, 2005), but major gene resistance in the model legume species *Medicago truncatula* was recently identified (Pilet-Nayel *et al.*, 2009). Useful resistance to pests has been identified only to pea weevil (*Bruchus pisorum* L.) in the secondary gene pool (*Pisum fulvum*), which is a widespread problem (Clement *et al.*, 2002), and transfer of resistance from *P. fulvum* appears feasible (Clement *et al.*, 2009).

Faba bean

Faba bean is infected by many pathogens and pests worldwide (see review by Sillero *et al.*, 2010). While genetic variation has been identified in response to many of these pathogens and pests, relatively few are major objectives in breeding programmes. The major fungal pathogens that are targeted in breeding programmes include ascochyta blight (*Ascochyta fabae*), chocolate spot (*Botrytis fabae and B. cinerea*) and rust (*Uromyces viciae-fabae*), with more localized selection for cercospora leaf spot (*Cercospora zonata*) and downy mildew (*Peronospora viciae*). Screening at ICARDA in the 1980s identified resistance to ascochyta blight and chocolate spot (Hanounik and Robertson, 1988) and further screening, under both field and controlled conditions, has identified more sources of disease resistance. Resistance, or partial resistance, to ascochyta blight has been identified in germplasm from diverse locations. It would appear that there are a number of genes that control resistance to ascochyta blight, as the reported genetic control of resistance differs depending on the source of resistance studied and the combination of parents (Sillero *et al.*, 2010). In contrast, resistance to chocolate spot is partial at best and genetic control is poorly understood. Resistant germplasm appears to be concentrated in the Andean region (Hanounik and Robertson, 1988; Sillero *et al.*, 2010), although other resistant germplasm has been identified (Bouhassan *et al.*, 2004; Villegas-Fernández *et al.*, 2009).

Viruses, including bean leaf roll virus (BLRV), bean yellow mosaic virus (BYMV), faba bean necrotic yellows virus, broad bean stain virus and pea seed-borne mosaic virus, affect a range of pulse crops, including faba bean. Resistance to BLRV and BYMV has been reported at ICARDA (Makkouk and Kumari, 1995; Makkouk *et al.*, 2002). Field screening with inoculation of faba bean plants with viruliferous aphids, combined with tissue blot immunosorbent assay (TBIA), has successfully introduced BLRV resistance from germplasm originating from Yunnan, China (van Leur *et al.*, 2000) to advanced breeding lines in Australia. The parasitic weed broomrape (*Orobanche* spp.) is a major pest of faba bean in the Mediterranean region; partial resistance has been identified and improved varieties released (see Nadal *et al.*, 2004a; Sillero *et al.*, 2010).

Quality

Lentil

Traditionally, lentil consumers have sourced local product and this has dictated preferences in terms of seed size, shape and colour. Breeding for quality has focused on seed characteristics, as these are most relevant in terms of how lentil is primarily traded. Inheritance and selection is also relatively simple, enabling breeders to concentrate on agronomic traits that limit profitability. Larger size in green lentil is preferred in many markets, except in areas such as in North Africa, where a medium-round green lentil is desired. Good colour and blemish-free seed is also important. Depending on region, the preferred size of red lentil ranges from very small (< 3 g per 100 seeds, e.g. Bangladesh) to medium-large (> 5 g per 100 seeds, e.g. Sri Lanka), with a general preference for round seed that can be de-hulled and split or retained whole (footballs; Vandenberg, 2009). Increasingly, breeders are selecting for characteristics that improve milling and cooking qualities; however, place of cultivation and farm management, have a large impact on quality.

Chickpea

Seed size, shape and colour are important traits for both *desi* and *Kabuli* types of chickpea. For *desi*, milling characteristics such as de-hulling efficiency are considered very important. The Australian *desi* variety, Jimbour, has a good reputation in the subcontinent for the whole seed and split markets due to its size, seed colour and the ease with which the seed coat is removed. For *Kabuli* types, large, white-coloured seeds are preferred for premium markets but there is also a large global demand for 8 mm *Kabulis*, particularly for the canning market. Laboratory-scale quality testing is common in breeding programmes in developed countries where there is a heavy reliance on cultivars meeting the requirements of export markets. Common tests include seed size, colour, hydration capacity, de-hulling and splitting efficiency and cooking time (Wood *et al.*, 2008).

Field pea

Dry pea grain is used extensively for both human consumption markets and stockfeed. Pea grain types for human consumption can be classified into those for yellow split, green split, whole green, snack food and sprouting markets. The main trade of grain for human consumption is of yellow split peas used directly in cooking or for producing flour. The focus of breeding has been to deliver grain that is highly spherical and has high splitting efficiency. Most countries have focused variety development on yellow peas, which have a clear seed coat and are tannin free. However, Australia has specialized in the development of split yellow peas that have a coloured and non-patterned seed coat (e.g. Dun types and Kaspa types) aimed at higher-value niche markets in Asia. For green split and whole green pea grain markets, colour and hydration (e.g. for canning) are the main trait targets in breeding. Whole pea grain is used in a variety of roasted snack foods, mostly within Asia. In this market there is differential preference for taste (e.g. Kaspa type), coat colour (e.g. green seed coat) and grain size. For the sprouting market, non-tendril seedlings and production of anthocyanin (e.g. dun types) appears to be preferred. All grain types are suitable for stockfeed; however, it is the lowest-value dry pea grain commodity traded, with the exception of niche speciality types for stockfeed (i.e. maple types for pigeon feed).

Faba bean

Faba beans range in size from 200 g to more than 2000 g per 1000 seeds, and are classified as either *Vigna faba minor* (small), *V. faba equine* (medium) or *V. faba major* (large or broad beans). There is regional variation in preferred seed type, with the small seeds dominant in northern Europe, medium in the Middle East, North Africa and Australia, and broad beans in Southern Europe and areas of China. Faba beans are used for food, particularly in the Middle East and North Africa, and for feed in developed countries. Major breeding objectives for the food markets include seed colour, de-hulling efficiency, hydration and cooking time. Faba bean contains a number of anti-nutritional factors, the two most important for breeding being tannins, which reduce protein utilization, and the glycosides vicine and convicine, which can cause favism in humans lacking the enzyme glucose-6-phosphate dehydrogenase and also reduce feeding efficiency in pigs and poultry. Both these anti-nutritional factors are controlled by major genes, with the 'zero' types being homozygous recessive.

Important agronomic traits

Lentil

The large-scale production of lentil in developed countries has only been achieved with mechanized harvesting systems, whereas in many traditional lentil-producing countries it continues to be harvested by hand. However, hand-harvesting is increasingly being considered a major constraint to lentil production, and the development of taller, lodging-resistant cultivars that retain their pods and seed at maturity is a prime breeding goal of lentil programmes in many parts of the world. Plant height is correlated with higher pods, maturity and tendency to lodge, but lines have been identified that are tall and early maturing. Idlib 1 and Idlib 2 were released in Syria, Rachyya in Lebanon, IPA 98 in Iraq and Sayran 96 in Turkey for use in combination with mechanized harvesting (Sarker and Erskine, 2002). Cultivars combining tall height, lodging resistance, yield and optimum maturity are being released in Canada and Australia and will potentially expand production into drier areas in Australia. Natural selection within bulk-segregating populations by delaying harvest decreased pod dehiscence, and delayed harvest was suggested as a suitable method for breeding with selection for height and lodging resistance.

Lentils compete poorly with weeds due to their slow growth during winter and short stature. Hence, weed control is a major limitation to growing lentil worldwide. Improved weed control has been achieved through the development of lentil cultivars with resistance to imidazolinone herbicides in Canada and Australia and early maturity for crop topping in Australia (Materne and McNeil, 2007).

Chickpea

As more chickpea-producing countries move towards mechanized harvesting, harvestability has become a trait of greater importance (Whish *et al.*, 2007). A tall lodging-resistant growth habit has been targeted to improve the efficiency of harvesting and reduce harvest losses. The achievement of this plant architecture has resulted in chickpea becoming a favourable legume option for wide-row and no-till farming systems (e.g. Canada and Australia). There are very few reports of pod drop and shattering in the literature, but both can occur if harvest is delayed due to unfavourable conditions at crop maturity.

Weed management of chickpea crops is extremely important, as chickpea also competes very poorly with weeds. Chickpea is slow to emerge and obtain canopy closure, which allows weeds to grow rapidly without suppression by the crop. Grass weeds are usually successfully controlled using selective herbicides, but broadleaf weeds generally pose the greatest challenges and the least weed control options. Whilst there are herbicides registered for use in chickpea, many have a narrow safety margin and crop damage can be substantial under certain environmental conditions (Datta *et al.*, 2009). More recently, research has been aimed to develop herbicide-tolerant cultivars (Tar'an *et al.*, 2009).

Field pea

The main breakthrough in field pea variety development globally has been the release of erect semi-dwarf types with the afila leaf trait (Redden *et al.*, 2005). The level of dwarfism is closely linked with adaptation, particularly to differential climates such that taller dwarfs (e.g. Kaspa type) are better suited to winter-sown Mediterranean-type climates such as Australia and shorter dwarfs (e.g. spring types such as cultivar 'Baccarra') are better suited for spring–summer sowing in the long-day, short-season climates of Europe and North America. Breeding for lodging resistance at harvest has required targeted selection, particularly in longer-growing season climates (Leonforte *et al.*, 2006). High resistance to lodging appears reasonably heritable and consistent across growing season climates (Tar'an *et al.*, 2003). Height of pod set has also been an important characteristic in reducing late season ascochyta disease infection (Le May *et al.*, 2005) and in improving harvesting efficiency and reducing contamination. The use of genes conferring reduced pod parchment layers in the pod wall has been successfully used in Australia to develop highly pod-shattering-resistant cultivars (e.g. cultivar 'Kaspa') for low-humidity climates.

Faba bean

Faba bean production in Europe, Australia and North America is highly mechanized and specific plant traits are selected for these management systems. Harvesting ability is very significant, and traits that contribute include height of the lowest pod, standing ability, time of maturity relative to optimum weather for harvesting and non-shattering pods. There is inherent variation for height of lowest pods, but this trait is also affected by time of flowering and time of sowing. A stiff straw mutant has been identified (Frauen and Sass, 1989), while reduced internode length and semi-determinate growth habit also contribute to standing ability and time of ripening. A mutant with a terminal inflorescence and determinate growth has been identified (Sjödin, 1971), and although yield potential of determinate varieties for broad acre crops has been less than for semi-determinate varieties, the trait has been incorporated in a variety for mechanical harvesting of green pods (Nadal *et al.*, 2004b). In Mediterranean-type environments, such as Australia, there is a very significant relationship between early sowing and yield potential (Adisarwanto and Knight, 1997), and varieties grown in this system require a high level of disease resistance to withstand the higher disease pressure associated with early sowing.

4.5 Utilization of New Tools and Technologies in Cultivar Development

Continued technological improvements and innovations across a range of fields are essential to improve efficiency and accuracy in breeding cool season food legumes. Some examples of current utilization of technologies for cultivar development in developed countries such as Australia and Canada include the use of: (i) satellite guidance and automatic steering to improve accuracy of sowing and spraying, and also to reduce labour costs; (ii) modified harvesters with floating fronts to ensure consistent cutting heights; and (iii) specific plant-breeding relational databases (e.g. Agrobase II) for data management and experimental design.

Induced polyploidy could be useful to increase both grain and plant size and to create new genetic variability. Molecular tools that will accelerate crop improvement, such as trait-linked DNA markers, doubled haploids, genomics and genetic transformation, are in the developmental phase or being increasingly used in cool season food legumes (Popelka *et al.*, 2004; Dita *et al.*, 2006). It is expected that these molecular tools are likely to become more applicable to crop improvement over the next decade. The use of molecular markers and identification of QTL (Table 4.1) could accelerate the selection process by alleviating time-consuming approaches of direct screening under greenhouse and field conditions, particularly in the quest to combine genes for many traits. Functional and comparative genomics and post-genomic tools would greatly help the identification of genes and pathways and functional analysis of these genes. The transfer of important genes could be achieved through genetic modification (GM). Currently available GM traits such as herbicide tolerance and insect and virus resistance could have immediate impact in pulses. ICRISAT is investigating the potential to produce chickpea resistant to *Helicoverpa* using the *Bt* gene used widely in other crops, including cotton. However, potential limitations to the use of GM technology include large costs and difficulties in taking genetically modified crops to market, hesitant adoption by consumers and lack of financial returns and therefore limited investment by private companies.

4.6 Conclusions

The introduction and release of germplasms around the world and increased breeding efforts are overcoming biotic and abiotic constraints to production. Success has been commendable considering the short period of breeding and

Table 4.1. List of some QTL identified in cool season food legumes.

Crop	Biotic/abiotic stress/traits of interest	Gene(s)/QTL identified[1]
Lentil	*Ascochyta lentis*	Ral2, AbR1
	Fusarium oxysporum f. sp. *cicer*	FW
	(Cold)	Frt
Chickpea	*Ascochyta rabiei*	Ar19
Pea	*Erysiphe pisi*	er
	Orobanche crenata	Ocp1, Ocp2
	Pea seed-borne mosaic virus	sbm-1, sbm-2
Faba bean	*Orobanche crenata*	Oc1, Oc2, Oc3, Oc4, Oc5
	Ascochyta fabae	Af1, Af2, Af3, Af4
	Uromyces viciae-fabae	Uvf-1
	Frost tolerance	U_AUSPC-1, U_AUSPC-2, U_AUSPC-3
	Zero tannins	Zt-1, Zt-2

[1]Sources: Dita *et al.*, 2006; Torres *et al.*, 2010.

low level of investment compared with larger crops such as wheat, maize and rice. However, systematic evaluation and characterization of germplasm accessions for various agronomic and morphological characteristics, biotic/abiotic stresses, grain yield and quality is still required to effectively utilize these genetic resources for future crop improvement.

In many cases pulse-breeding programmes must combine genes for many traits to develop cultivars that provide reliable and profitable production compared with cereals. This is being achieved with focused phenotyping efforts, but the development and uptake of reliable cost-effective markers is essential to fast-track this process. Fortunately, advancements in the technology and international collaborative efforts will provide genetic tools to breeders over the next 5 years. Genetic modification is achievable and offers great potential for pulses, but sensitivities associated with consumer demand must be addressed before cultivars are developed and released. Similarly, efforts towards improving tissue culture techniques may expand access to genes in wild relatives and the use of double haploids in research and breeding.

International collaboration has been the foundation of pulse breeding and remains a priority into the future if pulses are to compete with cereals for production area and maintain food markets. The effective use of resources and intellectual property (IP) globally is essential to provide the technologies and germplasm required to develop cultivars that increase productivity and reduce cost. This would increase profitability and expand production to meet the expanding demands for high-quality protein.

Quality will become increasingly important as markets and consumers have more choice and become more sophisticated in their specifications. A greater focus will be given to quality traits as cultivars are released that address disease and agronomic limitations to production. Pulses are traded on the physical characteristics of the grain, and this will remain the focus of breeding until users and processors recognize and pay for improvements in processing, cooking or taste characteristics. Supply of pulses is unlikely to exceed demand due to increasing populations, greater consumption (as standards of living rise, especially in target regions for pulses), need for protein feed for animals and, potentially, a decrease in cropping area as a result of degradation and competition from alternate industries, agriculture, environment and urbanization.

Breeding of cool season food legumes has been undertaken by private companies in Europe, but the number of companies has declined due to a lack of returns based primarily on seed sales. In Australia, end-point royalties are established but breeding programmes are still publicly funded by farmers' levies, federal and state governments and universities, as they are not yet viable as private entities. In Canada, there is a very good relationship between the grower-funded bodies such as the Saskatchewan Pulse Growers and research providers, and varieties are released without end-point royalties. In most developing countries pulse breeding and research is government funded with the international centres having a major impact; adoption of varieties and availability of technology is still a major limitation in many of these countries. Collaborative research, utilizing the resources of developed countries particularly in technology development, in combination with targeted research at international centres and local research and breeding efforts, will provide much-needed advances in these countries. Fortunately, goodwill within the small pulse-breeding community will foster such relationships to benefit both developing and developed countries.

References

Adisarwanto, T. and Knight, R. (1997) Effect of sowing date and plant density on yield and yield components in the faba bean. *Australian Journal of Agricultural Research* 48, 1161–1168.

Ahloowalia, B.S., Maluszynski, M. and Nichterlein, K. (2004) Global impact of mutation-derived varieties. *Euphytica* 135, 187–204.

Ahmad, F., Gaur, P.M. and Croser, J.S. (2005) Chickpea (*Cicer arietinum* L.). In: Singh, R.J. and Jauhar, P.P. (eds) *Genetic Resources, Chromosome Engineering, and Crop Improvement: Grain Legumes*. Vol. 1, CRC Press, Boca Raton, Florida, pp. 187–217.

Alcalde, J.A., Wheeler, T.R. and Summerfield, R.J. (2000) Genetic characterization of flowering of diverse cultivars of pea. *Agronomy Journal* 92, 772–779.

Barilli, E., Sillero, J.C., Moral, A. and Rubiales, D. (2009) Characterization of resistance response of pea (*Pisum* spp.) against rust (*Uromyces pisi*). *Plant Breeding* 128(6), 665–670.

Bouhassan, A., Sadiki, M. and Tivoli, B. (2004) Evaluation of a collection of faba bean (*Vicia faba* L.) genotypes originating from the Maghreb for resistance to chocolate spot (*Botrytis fabae*) by assessment in the field and laboratory. *Euphytica* 135, 55–62.

Castillo, P., Navas-Cortes, J.A., Landa, B.B., Jimenez-Diaz, R.M. and Vovlas, N. (2008) Plant-parasitic nematodes attacking chickpea and their *in planta* interactions with rhizobia and phytopathogenic fungi. *Plant Disease* 92, 840–853.

Clarke, H.J., Khan, T.N. and Siddique, K.H.M. (2004) Pollen selection for chilling tolerance at hybridisation leads to improved chickpea cultivars. *Euphytica* 139, 65–74.

Clement, S.L., Hardie, D.C. and Elberson, L.R. (2002) Variation among accessions of *Pisum fulvum* for resistance to pea weevil. *Crop Science* 42, 2167–2173.

Clement, S.L., McPhee, K.E., Elberson, L.R. and Evans, M.A. (2009) Pea weevil, *Bruchus pisorum* L. (Coleoptera: *Bruchidae*), resistance in *Pisum sativum* × *Pisum fulvum* interspecific crosses. *Plant Breeding* 128, 478–485.

Cubero, J.I. and Nadal, S. (2005) Faba bean (*Vicia faba* L.) In: Singh, R.J. and Jauhar, P.P. (eds) *Genetic Resources, Chromosome Engineering, and Crop Improvement: Grain Legumes*. Vol. 1, CRC Press, Boca Raton, Florida, pp. 163–186.

Dang, Y.P., Dalal, R.C., Buck, S.R., Harms, B., Kelly, R., Hochman, Z. *et al.* (2010) Diagnosis, extent, impacts, and management of subsoil constraints in the northern grains cropping region of Australia. *Australian Journal of Soil Research* 48, 105–119.

Datta, A., Sindel, B.M., Kristiansen, P., Jessop, R.S. and Felton, W.L. (2009) Effect of isoxaflutole on the growth, nodulation and nitrogen fixation of chickpea (*Cicer arietinum* L.). *Crop Protection* 28, 923–927.

Davidson, J.A., Krysinska-Kaczmarek, M., Kimber, R.B.E. and Ramsey, M.D. (2004) Screening field pea germplasm for resistance to downy mildew (*Peronospora viciae*) and powdery mildew (*Erysiphe pisi*). *Australasian Plant Pathology* 33, 413–417.

Dita, M.A., Rispail, N., Prats, E., Rubiales, D. and Singh, K.B. (2006) Biotechnology approaches to overcome biotic and abiotic stress constraints in legumes. *Euphytica* 147, 1–24.

Duc, G., Bao S., Baum, M., Redden, B., Sadiki, M., Suso, M.J. *et al.* (2010) Diversity maintenance and use of *Vicia faba* L. genetic resources. *Field Crops Research* 115, 270–278.

Elvira-Recuenco, M., Bevan, J.R. and Taylor, J.D. (2003) Differential responses to pea bacterial blight in stems, leaves and pods under glasshouse and field conditions. *European Journal of Plant Pathology* 109, 555–564.

Fiala, J.V., Tullu, A., Banniza, S., Seguin-Swartz, G. and Vandenberg, A. (2009) Interspecies transfer of resistance to anthracnose in lentil (*Lens culinaris* Medic.). *Crop Science* 49, 825–830.

Flowers, T.J., Gaur, P.M., Gowda, C.L.L., Krishnamurthy, L., Samineni, S., Siddique, K.H.M. *et al.* (2010) Salt sensitivity in chickpea. *Plant, Cell and Environment* 33, 490–509.

Fondevilla, S., Rubiales, D., Moreno, M.T. and Torres, A.M. (2008) Identification and validation of RAPD and SCAR markers linked to the gene *Er3* conferring resistance to *Erysiphe pisi* DC in pea. *Molecular Breeding*, 22, 193–200.

Frauen, M. and Sass, O. (1989) Inheritance and performance of the stiff-strawed mutant in *Vicia faba* L. *XII Eucarpia Congress*, 13–8, p. 15.

Furman, B.J., Coyne, C., Redden, B., Sharma, S.K. and Vishnyakova, M. (2009) Genetic resources: collection, characterization, conservation and documentation. In: Yadav, S.S., McNeil, D.L. and Stevenson, P.C. (eds) *Lentil: An Ancient Crop for Modern Times*. Springer, Dordrecht, The Netherlands, pp. 64–75.

Hanounik, S.B. and Robertson, L.D. (1988) New sources of resistance in *Vicia faba* to chocolate spot caused by *Botrytis fabae*. *Plant Disease* 72, 596–698.

Hobson, K.B., Armstrong, R.D., Nicolas, M., Connor, D.J. and Materne, M. (2006) Response of lentil (*Lens culinaris*) germplasm to high concentrations of soil boron. *Euphytica* 151, 371–382.

Hobson, K., Materne, M. and Noy, D. (2009) Evaluation of chickpea (*Cicer arietinum*) germplasm for tolerance to high soil boron. In: *Proceedings of the 14th Australasian Plant Breeding Conference and 11th SABRAO Congress*, 10–14 August 2009, Cairns, Australia.

Hollaway, G.J. and Bretag, T.W. (1995) Occurrence and distribution of races of *Pseudomonas syringae* pv. *pisi* in Australia and their specificity towards various field pea (*Pisum sativum*) cultivars. *Australian Journal of Experimental Agriculture* 35, 629–632.

Katoch, V., Sharma, S., Pathania, S., Banayal, D.K., Sharma, S.K. and Rathour, R. (2010) Molecular mapping of pea powdery mildew resistance gene *er2* to pea linkage group III. *Molecular Breeding* 25, 229–237.

Khan, H.R., Paull, J.G., Siddique, K.H.M and Stoddard, F.L. (2010) Faba bean breeding for drought-affected environments: A physiological and agronomic perspective. *Field Crops Research* 115, 279–286.

Knights, E.J., Southwell, R.J., Schwinghamer, M.W. and Harden, S. (2008) Resistance to *Phytophthora medicaginis* Hansen and Maxwell in wild *Cicer* species and its use in breeding root rot resistance chickpea (*Cicer arietinum L.*). *Australian Journal of Agricultural Research* 59, 383–387.

Lejeune-Hénaut, I., Hanocq, E., Béthencourt, L., Fontaine, V., Delbreil, B., Morin, J. *et al.* (2008) The flowering locus *Hr* colocalizes with a major QTL affecting winter frost tolerance in *Pisum sativum* L. *Theoretical and Applied Genetics* 116, 1105–1116.

Le May, C., Schoeny, A., Tiroli, B. and Ney, B. (2005) Improvement and validation of a pea crop growth model to simulate the growth of cultivars infected with Ascochyta blight. *European Journal of Plant Pathology* 112, 1–12.

Leonforte, A., Armstrong, E., McMurray, L., Regan, K. and Moore, S. (2006) Breeding reliable and lodging resistant semi-dwarf field peas for Australia. *Proceedings of the 13th Australasian Plant Breeding Conference*, May 2006, Christchurch, New Zealand.

Leonforte, A., Noy, D., Forster, J. and Salisbury, P. (2010) Evaluation for higher tolerance to NaCl in *Pisum sativum* L. *Proceedings of the 5th International Research Conference*, April 2010, Antalya, Turkey.

Leport, L., Turner, N.C., Davies, S.L. and Siddique, K.H.M. (2006) Variation in pod production and abortion among chickpea cultivars under terminal drought. *European Journal of Agronomy* 24, 236–246.

Link, W., Balko, C. and Stoddard, F.L. (2010) Winter hardiness in faba bean: Physiology and breeding. *Field Crops Research* 115, 287–296.

Makkouk, K.M. and Kumari, S.G. (1995) Screening and selection of faba bean (*Vicia faba* L.) germplasm for resistance to bean yellow mosaic potyvirus. *Journal of Plant Diseases and Protection* 102, 461–466.

Makkouk, K.M., Kumari, S.G. and van Leur, J.A.G. (2002) Screening and selection of faba bean (*Vicia faba* L.) germplasm resistant to Bean leafroll virus. *Australian Journal of Agricultural Research* 53, 1077–1082.

Maluszynski, M., Nichterlein, K., van Zanten, L. and Ahloowalia, B.S. (2000) Officially released mutant varieties – the FAO/IAEA Database. *Mutation Breeding Review* 12, 1–84.

Materne, M. and McNeil, D.L. (2007) Breeding methods and achievements. In: Yadav, S.S., McNeil, D.L. and Stevenson, P.C. (eds) *Lentil: An Ancient Crop for Modern Times*. Springer, Dordrecht, The Netherlands, pp. 241–253.

Materne, M. and Siddique, K.H.M. (2009) Agroecology and crop adaptation. In: Erskine, W., Muehlbauer, F., Sarker, A. and Sharma, B. (eds) *The Lentil: Botany, Production and Uses*. CAB International, Wallingford, UK, pp. 47–63.

McMurray, L.S., Davidson, J.A., Lines, M.D. and Leonforte, A. (2010) Combining pathological, agronomic and breeding advances to maximise *Pisum sativum* yields under changing climatic conditions in South-Eastern Australia. *Proceedings of the 5th International Research Conference*, April 2010, Antalya, Turkey.

Muehlbauer, F.J. and McPhee (2005) Lentil (*Lens culinaris* Medik.). In: Singh, R.J. and Jauhar, P.P. (eds) *Genetic Resources, Chromosome Engineering, and Crop Improvement: Grain Legumes*. Vol. 1, CRC Press, Boca Raton, Florida, pp. 219–230.

Muehlbauer, F.J., Mihov, M., Vandenberg, A., Tullu, A. and Materne, M. (2009) Improvements in developed countries. In: Erskine, W., Muehlbauer, F., Sarker, A. and Sharma, B. (eds). *The Lentil: Botany, Production and Uses*. CAB International, Wallingford, UK, pp. 137–154.

Nadal, S., Moreno, M.T. and Cubero, J.I. (2004a) Registration of 'Baraca' faba bean. *Crop Science* 44, 1864.

Nadal, S., Moreno, M.T. and Cubero, J.I. (2004b) Registration of 'Retaca' faba bean. *Crop Science* 44, 1865.

Onfroy, C., Baranger, A. and Tivoli, B. (2007) Biotic factors affecting the expression of partial resistance in pea to ascochyta blight in a detached stipule assay. *European Journal of Plant Pathology* 119, 13–27.

Pande, S., Galloway, G., Gaur, P.M., Siddique, K.H.M., Tripathi, H.S., Taylor, P. *et al.* (2006). Botrytis grey mould of chickpea: A review of biology, epidemiology and disease management. *Australian Journal of Agricultural Research* 57, 1137–1150.

Pilet-Nayel, M.L., Muehlbauer, F.J., McGee, R.J., Kraft, J.M., Baranger, A. and Coyne, C.J. (2002) Quantitative trait loci for partial resistance to *Aphanomyces* root rot in pea. *Theoretical and Applied Genetics* 106, 28–39.

Pilet-Nayel, M.L., Muehlbauer, F.J., McGee, R.J., Kraft, J.M., Baranger, A. and Coyne, C.J. (2005) Consistent quantitative trait loci in pea for partial resistance to *Aphanomyces euteiches* isolates from the United States and France. *Phytopathology* 95, 1287–1293.

Pilet-Nayel, M.L., Prospéri, J.M., Hamon, C., Lesné, A., Lecointe, R., Le Goff, I. *et al.* (2009) *AER1*, a major gene conferring resistance to *Aphanomyces euteiches* in *Medicago truncatula*. *Phytopathology* 99, 203–208.

Popelka, J.C., Terryn, N. and Higgins, T.J.V. (2004) Gene technology for grain legumes: can it contribute to the food challenge in developing countries? *Plant Science* 167, 195–206.

Redden, B., Leonforte, A., Ford, R., Croser, J. and Slattery, J. (2005) Pea (*Pisum sativum* L.). In: Singh, R.J. and Jauhar, P.P. (eds) *Genetic Resources, Chromosome Engineering, and Crop Improvement: Grain Legumes*. Vol. 1, CRC Press, Boca Raton, Florida, pp. 49–83.

Richardson, H.J., Leonforte, A. and Smith, A.J. (2009) Response of field pea varieties to the fungal components of the Ascochyta complex. In: *Ascochyta 2009: The Second International Ascochyta Workshop*, Pullman, Washington, p. 60.

Rose, I.A. and van Leur, J.A.G. (2006) Breeding faba beans (*Vicia faba*) for adaptation to short season environments in Australia. In: Mercer, C.F. (ed.) *Proceedings of the 13th Australasian Plant Breeding Conference*, Christchurch, New Zealand, pp. 34–38.

Sarker, A. and Erskine, W. (2002) Lentil production in the traditional lentil world. In: Brouwer J.B. (ed.) *Proceedings of Lentil Focus 2002*, Horsham, Victoria, Australia, pp. 35–40.

Sillero, J.C., Villegas-Fernandez, A.M., Thomas, J., Rojas-Molina, M.M., Emeran, A.A., Fernandez-Aparicio, M. *et al.* (2010) Faba bean breeding for disease resistance. *Field Crops Research* 115, 297–307.

Singh, R.J. (2005) Landmark research in grain legumes. In: Singh, R.J. and Jauhar, P.P. (eds) *Genetic Resources, Chromosome Engineering, and Crop Improvement: Grain Legumes*. Vol. 1, CRC Press, Boca Raton, Florida, pp. 1–9.

Singh, R., Sharma, P., Varshney, R.K., Sharma, S.K. and Singh, N.K. (2008) Chickpea improvement: Role of wild species and genetic markers. *Biotechnology and Genetic Engineering Reviews* 25, 267–314.

Sjödin, J. (1971) Induced morphological variation in *Vicia faba* L. *Hereditas* 67, 155–180.

Srivastava, S.P., Bhandari, T.M.S., Yadav, C.R., Joshi, M. and Erskine, W. (2000) Boron deficiency in lentil: yield loss and geographic distribution in a germplasm collection. *Plant and Soil* 219, 147–151.

Tadege, M., Wang, T.L., Wen, J., Ratet, P. and Mysore, K.S. (2009) Mutagenesis and beyond! Tools for understanding legume biology. *Plant Physiology* 151, 978–984.

Tanksley, S.D. and McCouch, S.R. (1997) Seed banks and molecular maps: unlocking genetic potential from the wild. *Science* 277, 1063–1066.

Tar'an, B., Warkentin, T., Somers, D.J., Miranda, D., Vandenberg, A., Blade, S. *et al.* (2003) Quantitative trait loci for lodging resistance, plant height and partial resistance to mycosphaerella blight in field pea (*Pisum sativum* L.). *Theoretical and Applied Genetics* 8, 1482–1491.

Taran, B., Warkentin, T.D., Vandenberg, A. and Holm, F.F. (2009) Variation in chickpea germplasm for tolerance to imazethapyr and imazamox herbicides. *Canadian Journal of Plant Science* 90, 139–142.

Tavakkoli, E., Rengasamy, P. and McDonald, G.K. (2010) High concentrations of Na^+ and Cl^- ions in soil solution have simultaneous detrimental effects on growth of faba bean under salinity stress. *Journal of Experimental Botany* (DOI: 10.1093/jxb/erq422).

Tivoli, B., Baranger, A., Avila, C.M., Banniza, S., Barbetti, M., Chen, W. *et al.* (2006) Screening techniques and sources of resistance to foliar diseases caused by major necrotrophic fungi in grain legumes. *Euphytica* 147, 223–253.

Torres, A.M., Avila, C.M., Gutierrez, N., Palomino, C., Moreno, M.T. and Cubero, J.I. (2010) Marker-assisted selection in faba bean (*Vicia faba* L.). *Field Crops Research* 115, 243–252.

Vandenberg, A. (2009) Postharvest processing and value addition. In: Erskine, W., Muehlbauer, F., Sarker, A. and Sharma, B. (eds) *The Lentil: Botany, Production and Uses*. CAB International, Wallingford, UK, pp. 391–407.

Vandenberg, A., Kiehn, F.A., Vera, C., Gaudiel, R., Buchwaldt, L., Dueck, S. *et al.* (2002) CDC Robin lentil. *Canadian Journal of Plant Science* 82, 111–112.

van Leur, J.A.G., Marcellos, H., Makkouk, K.M., Paull, J. and Rose, I.A. (2000) Identification of resistance to bean leaf roll luteo virus in faba bean. *Biological and Cultural Tests for Control of Plant Diseases* 15, 27.

Varshney, R.K., Hiremath, P., Lekha, P., Kashiwagi, J., Balaji, J., Deokar, A.A. *et al.* (2009) A comprehensive resource of drought- and salinity- responsive ESTs for gene discovery and marker development in chickpea (*Cicer arietinum* L.). *BMC Genomics* 10, 523.

Villegas-Fernandez, A.M., Sillero, J.C., Emeran, A.A., Winkler, J., Raffiot, B., Tay, J. *et al.* (2009) Identification and multi-environment validation of resistance to *Botrytis fabae* in *Vicia faba*. *Field Crops Research* 114, 84–90.

Whish, J.P.M., Castor, P., Carberry, P.S. and Peake, A. (2007) On-farm assessment of constraints to chickpea (*Cicer arietinum* L.) production in marginal areas of northern Australia. *Experimental Agriculture* 43, 505–520.

Wood, J.A., Knights, E.J., and Harden, S. (2008) Milling performance in desi-type chickpea (*Cicer arietinum* L.): effects of genotype, environment and seed size. *Journal of the Science of Food and Agriculture* 88, 105–115.

5 Breeding for Improvement of Warm Season Food Legumes

B.B. Singh, R.K. Solanki, B.K. Chaubey and Preeti Verma

5.1 Introduction

The warm season food legumes including soybean, pigeon pea, mung bean, urd bean and cowpea are mainly grown in hot and humid climatic conditions. These crops hold prime importance as they cover a maximum area under rainfed cultivation, alhough most of them can also be grown in spring and summer seasons. Warm season food legumes are popular in different parts of world; for example, soybean (*Glycine max*) is an important crop in the USA, pigeon pea is mainly grown in India and African countries, while mung bean and urd bean are important crops in South-east Asian countries, particularly in the Indian subcontinent. In addition to this, cowpea is an important crop in the USA and African countries.

All these crops have immense importance in vegetarian diets as a source of protein, and therefore tremendous breeding efforts have been made worldwide to improve yield and quality using both conventional and modern approaches (Singh *et al*., 2005; Gupta and Kumar, 2006; Pathan and Sleper, 2008; Dupare *et al*., 2009). Focused efforts on the breeding of warm season food legumes have been made in different international centres supported by the Consultative Group in International Agricultural Research (CGIAR). Among these centres, the International Crops Research Institute for the Semi-Arid Tropics (ICRISAT), located in India, has focused research on pigeon pea and the International Institute of Tropical Agriculture (IITA) has a global mandate for cowpea improvement. The Asian Vegetable Research and Development Centre (AVRDC) was established for the improvement of mung bean worldwide. Besides, the US Department of Agriculture (USDA) has focused research activities on soybean. The Indian Institute of Pulses Research, Kanpur, a leading centre of the Indian Council of Agriculture Research and other Agriculture Universities in India are also involved in genetic improvements in warm season legume crops, including pigeon pea, mung bean and urd bean. These national and international centres are involved in collection, evaluation and sharing of germplasm, and also undertake breeding programmes for genetic improvement. The international centres also distribute the segregating populations and inbred lines to partner countries for selection and their release as varieties, resulting in stimulated breeding internationally. Hall *et al*. (1997) and Singh *et al*. (1997, 2002) have described cowpea breeding programmes in different regions of the world. The bean/cowpea CRSP (Cowpea Collaborative Research Program) is also catalysing and supporting research on cowpea improvement in the USA, Cameroon and Senegal. Significant research on various aspects of cowpea improvement is also being carried out in Brazil, Nigeria, Burkina Faso, Senegal, Mali and India and, to a lesser extent, in a number of other countries. These efforts have led to the development of different types of cowpea cultivar, including *Vigna unguiculata*, *Vigna biflora* (or catjang) and *Vigna sesquipedalis* (Hall *et al.*, 1997). This chapter focuses on significant breeding efforts and research achievements in major warm season food legumes that have been accomplished over recent years.

5.2 Important Constraints

The productivity of any crop depends upon its genetic make-up and the environment in which it grows; favourable environment helps the plant to express its genetic potential maximally. Besides the prevailing climatic conditions, favourable environment refers to biotic, abiotic and edaphic factors existing at different stages of growth. The major biotic and abiotic constraints in warm season food legumes are elaborated in Chapters 15 and 16, and are briefly mentioned below.

Biotic stresses

Bacterial pustules, frog eye leaf spot, purple seed stain, soybean mosaic, bud blight, collar rot, *Rhizoctonia* aerial blight, rust and powdery mildew are important biotic stressors in soybean. It has been shown recently in the central southern part of India that rust and powdery mildew cause a yield loss of 10–100% (Rao *et al.*, 1995). Though fusarium wilt, sterility mosaic and phytophthora blight (*Phytophthora drechsleri*) are the most economically important diseases in pigeon pea, fusarium wilt causes heavy losses (Kannaiyan *et al.*, 1984). Sterility mosaic has also caused severe yield loss in India, which was around 100,000 t in the period 1975–1980 (Kannaiyan *et al.*, 1984). Phytopthora blight, first reported by Williams *et al.* (1968), is more common in short-duration (120–150 days' maturity) pigeon pea varieties as compared with medium- and long-duration varieties. In *Vigna* species, yellow mosaic virus, powdery mildew, cerscospora leaf spot and root disease caused by *Pythium* and *Fusarium* spp. cause significant losses. The problem of stored grain pests, i.e. bruchids, is a major factor causing damage after harvesting in almost all legume crops. In cowpea, the diseases cowpea yellow mosaic, blackeye cowpea mosaic, cowpea aphid-borne mosaic, cercospora leaf spot, ascochyta blight and bacterial blight are all of economic importance. The aphids, thrips and bruchids that commonly affect food legume crops are also important pests of cowpea (Hall *et al.*, 1997).

Abiotic stresses

Abiotic factors affecting yield are very common among all warm season legumes, as these crops are grown mainly during the rainy season; therefore waterlogging conditions, particularly in the early stage of growth and especially in areas receiving high rainfall, greatly affect yield potential. Besides this, moisture stress is also responsible for yield loss in areas of low rainfall. Salinity and other abiotic factors affecting a favourable soil environment are also important. Poor seed dormancy is one of the major concerns in mung bean, as it leads to preharvest sprouting causing high yield loss if rains or conditions of high humidity arise during the harvesting/maturity period.

5.3 Genetic Resources

The success of any crop improvement scheme depends on the availability of genetic resources, because this provides the opportunity for genetic manipulation involving diverse parents in hybridization programmes. Thus collection, evaluation, documentation and utilization of germplasm are important activities for enhancing crop improvement programmes (see Chapter 23 for details). Genetic resources under investigation in warm season legumes are maintained at various repositories in the USA, China, India, Taiwan and other countries linked to the international network of USDA, ICRISAT and AVRDC, or in national programmes within different countries. The largest collection of germplasm is of soybean, representing around 170,000 accessions maintained in over the 70 countries; China holds 26,000, followed by the USA with 19,000 (Carter *et al.*, 2004). Pigeon pea research and cultivation is concentrated mainly in South-east Asia and some parts of Africa. The global collection of nearly 24,938 accessions of pigeon pea is maintained at ICRISAT and the National Bureau of Plant Genetic Resources (NBPGR), both in India. Other national institutes of developing countries also maintain germplasm of pigeon pea, although stocks of mung bean and urd bean are limited; a base collection of nearly 5600 mung bean and 480 urd bean accessions is maintained at AVRDC, while NBPGR holds a stock of 3497 mung bean and 1200 urd bean accessions. A collection of over 15,000 cowpea accessions of cultivated varieties from over 100 countries and 560 accessions of wild cowpea is maintained at IITA, Nigeria. These have been characterized and evaluated for desirable traits, and are preserved and used in breeding programmes (Ng and Singh, 1997; Singh, 2005). Extreme cowpea genotypes have been observed with respect to many traits, and genetic studies have identified several desirable genes (Fery and Singh, 1997; Singh, 2002).

5.4 Breeding Methods and Strategies

Improving crops according to human need requires knowledge of floral biology, genome diversity, cross-compatibility between cultivated and other species and the genetics underlying target traits. Using this information, plant-breeding programmes are designed in such a way that high seed yield with minimum quality standards essential for human dietary needs can be harvested. Ranalli and Cubero (1997) discussed the basis of genetic improvement in legumes and the application of breeding methods, including introduction, hybridization, early generation selection and mutation, along with molecular markers that offer opportunity to enhance precision.

Introduction

This is a primary approach in crop improvement, in which a variety or a genotype is introduced directly for commercial cultivation into a new environment. This method has been used successfully in India and the USA for improving warm season legumes, resulting in the introduction and development of a large number of soybean lines (Bragg, Clark-33, Davis, Hardee, Improved Pelican, KM-1), mung bean (Pusa 105, Pusa 9531, Pant Moong 5, Pusa Vishal, SML 668) and pigeon pea (Hy 3C; a selection from PI 2812-2). One particular line, Brazil 1-1, an early pigeon pea line, was introduced from Brazil and has been involved in a breeding programme aimed at transferring the earliness trait, resulting in the development of early-maturing varieties like Mukta, Sharad and Pusa Ageti (Singh *et al.*, 2005). In the USA, the introduction of various lines has contributed significantly to genetic improvement of the yield potential of soybean (Pathan and Sleper, 2008).

Pureline breeding

Pureline selection is the step preceding introduction of a line, in which the selection of better plant types is made from an already existing genetically heterogeneous population or landrace. These superior plant types are identified as the result of natural selection pressure, which helps to evolve new plant types with strong genetic potential. These variants are fixed by breeders through a continuous cycle of selfing and selection. The use of this method has often been more successful in cross-pollinated, warm season legumes, because heterogeneity in the gene pool helps to release new genetic variability in nature. Using this method in pigeon pea, a number of varieties, e.g. T 7, UPAS 120, Bahar, BDN 2 and Narendran Arhar 1 have been developed in India, and some of these are still popular among farmers; more than 60 improved cultivars of mung bean and urd bean have been developed using this breeding method (Gupta and Kumar, 2006; Tickoo *et al.*, 2006).

Recombination breeding

Hybridization, which is also known as recombination breeding, is one of the important techniques involved in breeding programmes, and it refers to the development of better recombinants using intra- or interspecific variability. Exotic collections, primitive forms and landraces are important sources of rare alleles. Knowledge of parental performance, their combining ability and selection for yield per se is essential for the breeding of high-yielding genotypes. Before performing hybridization between desirable parents, the breeder should be aware of the component traits association and its affect on economic yield, as this helps in directing phenotypic selection in advancing the segregation of generations. In general, soybean, seed yield in pigeon pea, mung bean and urd bean is positively associated with pods per plant, seeds per pod and seed weight; therefore, selection for these component traits may be beneficial. Hybridization is generally followed by pedigree, bulk, recurrent, backcross or single-seed methods of selection (Ranalli and Cubero, 1997).

The pedigree method is the most commonly used for improving yield and other major component traits, leading to the development of many legume varieties. Besides, recurrent selection and population improvement methods have been suggested as ways to accumulate desirable traits and to break undesirable linkages. A modified version of the early-generation testing method has been found to be efficient and successful in soybean (Cooper, 1990). In this crop, various recurrent selection methods have been used or proposed, including mass selection for oil (Burton and Brim, 1981) and seed weight (Tinius *et al.*, 1991); half-sib family selection for seed yield (Burton and Carver, 1993) and oil quality (Carver *et al.*, 1986); and S 1 family selection for yield (Kenworthy and Brim, 1979; Rose *et al.*, 1992) and protein (Brim and Burton, 1979). Successful application of recurrent selection in soybean could be due to the availability of sterile lines, and this has been employed for yield improvement (Tinius *et al.*, 1991), oil and protein content (Burton and Brim, 1981) and fatty acid content (Carver *et al.*, 1986). Early-generation testing, which was developed in Canada as a modification of the bulk method, has also been shown as being very feasible for improving those characters showing additive and additive × additive genetic components of variance. It holds an advantage over late-generation testing due to the reduction in population load, as inferior lines are discarded in early generations. However, F_2, F_3 and even F_4 families are subjected to early-generation selection depending upon the target trait and environmental condition (Burton, 1997). Soybean breeding in the USA has been viewed as a process of cyclic recurrent selection. Breeding populations are often developed by two-way, three-way or four-way crosses of cultivars and/or breeding lines. If unadapted germplasm is used, at least one backcross to the adapted parent is often used (Burton, 1997). In cowpea, recombination breeding has focused on the development of improved cultivars having high yielding lines in the intercropping system. For this purpose, the standard pedigree method has been followed to select desirable plants/progenies (Singh *et al.*, 1996). It has been observed that breeding lines selected under intercropping are significantly better than those selected under sole-crop selection, which might be due to greater stress and selective pressure under intercropping. For improving the yield and yield components in cowpea, the single-seed descent method has been found more effective than that of progenies developed via single plant selection (Mehta and Zaveri, 1997). In addition, populations developed through the single-seed descent selection method have been shown to have high broad-sense heritability (Hall *et al.*, 1997).

Although successful interspecific crosses between *Vigna unguiculata* and *Vigna vexillata* have been reported, it has not been confirmed through backcross breeding whether the F_1 so developed are true F_1 hybrids (Gomathinayagam *et al.*, 1998). Tyagi and Chawla (1999) also reported successful crosses between *Vigna radiata* and *V. unguiculata* using *in vitro* culture techniques. Gibberellic acid treatment sustained the pods for 9–10 days, which were then used for embryo culture; around 10% of total embryos resulted in plantlet formation. However, the authors did not report further growth and culture of these plantlets and, therefore, it is not certain whether the crosses were true hybrids. There is a need to continue efforts to cross *V. vexillata* and other *Vigna* species with cowpea to broaden the genetic base using new, emerging techniques. Successful interspecific *Vigna radiata* × *Vigna mungo* crosses have resulted in the development of four mung bean (Pant M 4, HUM 1, Meha, PM 6) and one urd bean (Mash 1008) variety with improved plant types. A large number of novel traits in both mung bean and urd bean have been developed. The variability generated through these crosses for different agronomic traits is unique, as such extreme types are not available in the existing collections of either mung bean or urd bean (Singh and Singh, 1998; Singh and Dixit, 2002).

Hybrid breeding

The success of the hybrid breeding approach is better established in those crops where hybrid seed production is easy, i.e. those showing a sufficient level of cross-pollination, including pigeon pea, which is frequently cross-pollinated. Studies show that pigeon pea genotypes have a high degree of hybrid vigor in their genetic background that can be exploited commercially. In this crop, different male sterility systems have been identified and used in the development of hybrids – and for other warm season crops (see Chapter 13). In India, extensive research has been undertaken in hybrid technology on pigeon pea, and the world's first hybrid (ICPH 8) was released by ICRISAT in 1991. This hybrid has shown a yield advantage of 30.5% over the nearest line, UPAS 120 (Saxena, 2008). Many more hybrids have subsequently been developed but, due to their high seed production cost, farmers did not adopt these, and so efficient cytoplasmic nuclear male sterility systems have been identified. Presently, interspecific hybridization with available resources is being followed rigorously for the development of line CGMS in pigeon pea (Saxena, 2008, 2009), which has resulted in the identification of two cms lines, GT288A and 67A, with 100% sterility that have been extensively used to exploit their hybrid vigour (Singh *et al.*, 2005). The possibilities for the development of hybrid varieties in soybean have also been explored, and efforts have been made toward the identification of male sterility. Studies show that male sterility in soybean is controlled by a single recessive gene (Palmer and Lewers, 1998), but local conditions need to be addressed to maximize opportunities for pollination and pollination vectors for hybrid seed production (Perez *et al.*, 2008).

Mutation breeding

If desirable variability is not available for a target trait within a gene pool, mutation is the ultimate means of creating new genetic variation. Mutations may occur spontaneously or can be induced artificially. Several morphological and other mutants have been isolated in different legume crops, including warm season food legumes (Micke, 1984; Gopalakrishna and Reddy, 2009; Table 5.1). Studies show that the effect and efficiency of mutagens depends largely on genotype, and this varies with the dosage and nature of mutagen used (see Chapter 14 for details). In pigeon pea, gamma rays have been found to be more effective in generating a high frequency of chlorophyll mutants (Venkateswarlu *et al.*, 1981). Streptomycin sulfate and sodium azide (SA) induced male sterile plants at concentrations of 0.5 M and 0.025%, respectively (Pandey *et al.*, 1996). Sodium azide has been found to be more effective than ethyl methyl sulfonate (EMS) (Potdukhe and Narkhede, 2002). However, in the case of mung bean, EMS showed the highest mutagenic efficiency compared with other mutagens such as methyl methanesulfonate (MMS) and SA (Khan and Wani, 2006). A high EMS concentration increased fertile branches, pods per plant and plant height in mutants (Wani and Khan, 2004). Moreover, in urd bean, the effectiveness of EMS was shown to be high compared with mung bean (Rakshit *et al.*, 2001). In urd bean, mutation with gamma rays and EMS induced early mutants with increased pod numbers, number of seeds per pod, 100-seed weight and protein content (Sharma

Table 5.1. Some selected mutants reported with regard to different traits in warm season food legumes.

Crop	Key traits	Reference(s)
Cowpea	Fasciated mutant	Adu-Dapaah *et al.* (1999)
	Partial or complete male sterile mutants	Odeigah *et al.* (1996); Singh and Adu-Dapaah (1998)
Mung bean	High protein content and yield	Chakraborty *et al.* (1998)
	High for pods per plant, seeds per pod, 100-seed weight and seed yield	Singh *et al.* (2001)
	Leaf mutants, pod mutants and semi-dwarf plants	Srinives *et al.* (2000); Tah (2006)
	Branchless and multifoliate	Singh and Kole (2006)
	Resistance to YMV and synchronous maturity	
Soybean	Yellow seeded	Bhatnagar *et al.* (1990)
	Shatter resistant	Misra *et al.* (1981)
	Low linolenic acid	Brossman and Wilcox (1984)
	Leaf and floral modifications	Dwivedi and Pandey (1981)
	Prolonged stability of soya oil	Rahman *et al.* (1997)
Urd bean	100-seed weight, YMV resistant, drought tolerant, early maturing (70 days) and high yielding	Dixit *et al.* (2000)

et al., 2007). However, the use of a lower dose of mutagen has been observed to be more effective and efficient in urd bean (Sharma *et al.*, 2005).

John (1999) reported a 50 Kr dose of gamma rays to be the most effective for inducing mutations in cowpea. Using gamma rays, EMS and SA, several male sterile mutants have been obtained in this crop (Odeigah *et al.*, 1996). Although the use of gamma rays and ethidium bromide generated a reasonable level of variation for different agronomic traits, the former has been observed as being more effective in inducing mutation than the latter (Gunasekaran *et al.*, 1998). In mutation breeding, the M_1 plant-to-row method has been suggested as being efficient but, when dealing with bigger populations, the M_1 seed bulk method should be adopted (Balyan and Khan, 1995). These workers also suggested that the M_1 single-seed bulk method needs higher skill levels in identifying mutants. In mung bean, M_2 generation selection can give high potential gains for plant height, days to flowering and maturity (Khan and Wani, 2006).

Mutation breeding has been used to develop improved cultivars in warm season crops developed either through mutation breeding directly or by involving mutants as a parent in crossing programmes (Ahloowalia *et al.*, 2004; Gopalakrishna and Reddy, 2009). For example, in soybean, the mutant MACS 111 derived from Kalitur has been used to develop the elite cultivar MACS 450 (Raut *et al.*, 2000). In India, the pigeon pea varieties Trombay Vishakha 1, CO-3 (bold-seeded, high-yielding), CO-5 (early photo-insensitive), TAT-10 (extra-early) and CO-6 (intermediate type) have been developed through irradiation, and most of these are still popular among farmers. In mung bean, CO-4, Pant Moong-2, TAP-7, MUM 2, BM 4, LGG 407, LGG 450, CO-4, TT 9E and Pant Mung-1 are among the important mutant varieties released in India (Ahloowalia *et al.*, 2004). Most mutant cultivars are early maturing, high yielding and tolerant/resistant to YMV. Another variety, 'SML 668', has been developed through selection in a mutant line NM 94 for resistance to yellow mosaic virus (YMV) and synchronous maturity (Brar *et al.*, 2006). This variety is very popular in the Punjab, Haryana, Himachal Pradesh, Rajasthan and Bihar states of India. NIAB Mung 92 and NIAB Mung 98 mutant varieties, popular in Pakistan, are high yielding and resistant to YMV and cercospora leaf spot. In urd bean, mutant cultivars such as Vamban 2, TU 94-2, CO 4, Sarla, TAU 1, TAU 2, TPU 4, TAU 5 and TU 94-2 have been released as early-maturing cultivars. Among these varieties, TAU 1, TAU 2 and TPU 4 have been developed through crosses with the large-seeded neutron-induced mutants UM 196 and UM 201, which showed 5.6–6.9 g/100 seeds. Similarly, the mutant cultivars 'Vamban 2' and 'Sarla B-14-4' have been developed from the susceptible cultivar 'T 9' as being YMV resistant (Dixit *et al.*, 2000).

Molecular marker technology

Recently, the use of molecular markers has become important in conventional breeding programmes for several purposes, including the assessment of genetic diversity, confirmation of hybridity of F_1, mapping of important traits and marker-assisted selection for indirect selection of desirable alleles in segregating generations. Therefore, genomic resources have been developed for warm season food legumes (Choi *et al.*, 2004; Cannon *et al.*, 2009; Muchero *et al.*, 2009; Sato *et al.*, 2010). For example, in pigeon pea polymerase chain reaction (PCR)-based SSR and SNP markers have been developed for genetic mapping and marker-assisted improvement (Burns *et al.*, 2001; Odeny *et al.*, 2007; Datta *et al.*, 2010; Saxena *et al.*, 2010). Furthermore, a pigeon pea genomics initiative (PGI) programme has resulted in the development of 25 different mapping populations and genomic resources, including a BAC library of 69,120 clones, 16 cDNA libraries for wild and sterility mosaic diseases, 6590 primer pairs for SSRs identified from BAC end sequences, SSR (> 3000), DArT (> 15,000 features) and 66,345 SNP (from 1206 high-quality sequences) markers (Varshney *et al.*, 2010a). A consensus molecular map based on SNP markers has also been developed for cowpea (Muchero *et al.*, 2009). The development of molecular markers and the establishment of a marker–trait association for agronomically important traits in these crops have recently been reviewed (Varshney *et al.*, 2010b). The progress made in the use of marker-assisted selection (MAS) has been highlighted in recent reviews and in Chapter 19, emphasizing trait mapping and molecular breeding in legumes, including warm season food legumes (Varshney *et al.*, 2010b).

5.5 Important Target Traits for Improvement

Abiotic stresses

Warm season food legume crops encounter unpredictable environmental conditions such as waterlogging, terminal drought, high temperature, heavy rains, etc. These factors taken together affect yield. Therefore, the development of plant types that can survive under different environmental conditions will be required to boost the crop production and productivity (Pennisi, 2008). Some important target traits in breeding programmes for improving the genotypes of these crops against abiotic stress are discussed below.

Short duration and photo-thermal insensitivity

These are important traits in soybean, mung bean and urd bean, because the development of short-duration and photo-thermally insensitivite genotypes creates plants suitable for different cropping systems, and also avoids terminal drought (Singh, 2010, unpublished report). In cowpea, photosensitive cultivars not only flower early but also become extremely dwarf in habit when day length is under 12.5 h (Ishiyaku and Singh, 2001), and a complete association of photosensitivity has been observed with dwarfing, which is controlled by a monogenic recessive gene (Ishiyaku and Singh, 2001). In urd bean, earliness and photo-thermosensitivity are recessive traits and are controlled by major genes (Sinha, 1988). Thus selection of genotypes with early vigour holds tremendous importance in breeding programmes. As a result, some of the very popular early varieties, such as Narendra Urd 1, KU 300, Sarla, Vamban, and Urd 3, have been developed in India for commercial cultivation. Since urd bean is also cultivated in the spring/summer season, Pant U 19, T 9, KM 1 and TMV 1 have been developed as photo-thermoinsensitive varieties (Gupta and Kumar, 2006).

Leaf pubescence density

Suitability for soybean cultivation is improved by this trait in drought-prone areas, as it reduces leaf temperature and water loss by transpiration and enhances photosynthesis and vegetative vigour (Du *et al.*, 2009). Two additive genes control this trait in soybean (Pfeiffer and Pilcher, 2006). This is also an important trait of mung bean and urd bean; some lines of mung bean developed at AVRDC, e.g. V 2013, V 1281, V 3372, VC 1163D, VC 2750A, VC 2754A and VC 2768A, can withstand moisture stress (Tickoo *et al.*, 2006), including long spells of rainfall causing flooding.

Seed dormancy

Reduced seed dormancy is found in mung bean, resulting in preharvest sprouting during the maturity phase in the monsoon (*kharif*) season, and therefore the identification of lines with tolerance to preharvest sprouting is highly desirable in this crop (Tickoo *et al.*, 2006) and in urd bean.

Deep root system

Pigeon pea is cultivated mostly in rainfed zones, the deep and dense root system providing inherent potential to counteract drought or water stress during the critical growth phases.

Biotic stress

Warm season crops are also affected by a number of important diseases, insect pests and nematodes, now discussed below. Therefore, the development of cultivars resistant to these biotic stresses remains a target of breeding programmes in these crops.

Diseases

RUST. A devastating disease of soybean caused by *Phakpspora pachyrhizi*, yield losses of up to 95% have been reported in Brazil (Hartman *et al.*, 1997), 75% in Argentina (Yorinori *et al.*, 2005) and 50% in the USA (Hartman, 2005). Inheritance studies suggest that four single dominant genes control this trait (Hartman, 2005). Although genotype PI 459025, having a single dominant gene for resistance to all three rust isolates, has been identified, its use has been shown to be problematic due to the rapid breakdown of resistance. Therefore, the development of genotypes having multiple genes of resistance is an important target of soybean breeding programmes.

BACTERIAL PUSTULES. In soybean this condition is caused by *Xanthomonas axonopodis* pv. *glycines*, which is very much favoured by hot and humid conditions. Studies have shown that a single recessive gene controls resistance to this disease (Hartwig and Lehman, 1951). Molecular breeding has also been conducted, and SSR markers tightly linked to BLP resistance have been identified for using in breeding programmes (Kim *et al.*, 2010).

FUSARIUM WILT (FW). In pigeon pea, FW is an important biotic stress causing significant yield losses of up to 20–25% in India (Dhar and Reddy, 1999) and Africa (ICRISAT, 1983). Resistance to FW is a complex phenomenon, studies suggesting variously that it is governed by multiple genes (Pal, 1934), two complementary genes (Shaw, 1936; Pathak, 1970) and a single dominant gene (Pawar and Mayee, 1986; Singh, I.P., *et al.*, 1998). Many wilt-resistant varieties have been developed in India through pedigree and bulk-pedigree methods, e.g. Pusa 33, C 11, BDN 1, BDN 2, ICPL 8863, Jawahar Arhar 4, Birsa Arhar 1, ICPL 87119, KM 7 and MAL 13.

PYTOPHTHORA BLIGHT (PB). Caused by the fungus *Phytophthora drechsleri* f. sp. *cajani*, no resistant variety is available for pigeon pea (Singh *et al.*, 2005). Studies have variously claimed that resistance is governed by a single dominant gene (Sharma *et al.*, 1982) and two homozygous recessive genes (Singh *et al.*, 2003a). Some tolerant lines, e.g. KPBR 80-2-1, KPBR 80-2-2, GAUT 82-55 and ICP 8103 have been developed. Some level of resistance has been found among accessions of *Cajanus platycarpus* against PB.

POWDERY MILDEW (PM). Of importance in mung bean and urd bean, in the former PM is caused by *Erysiphe polygoni* DC, and can cause yield losses of up to 20–40% in India (Grewal, 1978). The status of PM resistance in this crop has been reviewed (Reddy *et al.*, 2008), an inheritance for resistance to PM has variously been reported as monogenic and polygenic (Yong *et al.*, 1993; Sorajjapinun *et al.*, 2005; Reddy, 2009). The TARM 1 and TARM 18 lines are well-known varieties showing a high level of resistance to PM. It is also a serious disease in urd bean, causing 20–25% yield losses. Resistance is controlled by a single recessive gene (Kaushal and Singh, 1989). Limited resistance sources are available for PM in mung bean and urd bean, e.g. Pant U 30, P 115, Line 6203 and LBG 642. Cultivar LBG 17, resistant to PM, is very popular in rice-fallow areas of India (Gupta and Kumar, 2006).

CERCOSPORA LEAF SPOT (CLS). In mung bean, CLS caused by *Cercospora canescens* Ell. and Mart. and *Cercospora cruenta* Sacc. is an important disease. Warm and humid weather conditions are very favourable for its appearance. It has been variously reported that resistance to CLS is governed by one or two genes (Singh and Patel, 1977; Mishra *et al.*, 1998) and a single recessive gene (Yadav *et al.*, 1981). ML 613 is a cultivated variety bearing resistance to CLS.

VIRAL MOSAICS. Viruses cause a number of diseases in warm season legume crops, including sterility mosaic virus (SMV) in pigeon pea, soybean mosaic virus (SMV) in soybean and mung bean yellow mosaic virus (MYMV) in mung bean. Inheritance studies have been conducted on these diseases; in soybean, SMV resistance is controlled by three independent genes (Moon *et al.*, 2009). Bud blight disease of soybean is caused by a strain of groundnut bud necrosis virus (GBNV), and is an important viral disease in major soybean-growing areas of India. Some lines such as MACS 754, NRC 55, VLS 55 and JS-SH-96-04 have been identified as resistant to bud blight (Lal *et al.*, 2002).

Sterility mosaic virus is an important viral disease of pigeon pea carried by an arthropod vector (Kumar *et al.*, 2000). Inheritance to this disease in pigeon pea has been reported to be monogenic to oligogenic (Singh, B.V. *et al.*, 1983; Srinivas *et al.*, 1997; Singh, I.P. *et al.*, 2003b). Some of the popular varieties in India such as Hy 3C, Bahar, Pusa 9, Narender Arhar 1, MA 3, MAL 13 and Asha have resistance to SMV.

Predominantly found across India, especially in the rainy season, MYMV is spread by the vector white fly (*Bemisia tabaci* Genn.). Resistance to MYMV is reported variously to be governed by a single recessive gene (Singh and Patel, 1977) and two recessive genes (Verma, 1985; Reddy, 1986). In India, a large number of varieties, e.g. Pant Moong 2, Narendra Moong 1, Meha, Samrat, IPM2-3, HUM 1 and PM 6 have considerable resistance to MYMV. MYMV is also the most common threat to the urd bean. Under severe conditions, yield loss has been observed up to 100%. Resistance to this disease has variously been reported to be monogenic dominant (Dahiya *et al.*, 1977) and digenic recessive (Singh, A., *et al.*, 1998). Pant U 84, UPU 2, Pant U-19, UH 81-7, UG-700 and IPU 94-1 are among the most important genotypes resistant to MYMV.

Insect pests

POD BORER (*HELICOVERPA ARMIGERA*, *MARUCA TESTULALIS*, *MARUCA VITRATA*) AND PODFLY (*MELANAGROMYZA OBTUSE*). For pigeon pea, these are the most important insect pests. Pod borers cause damage in all mature groups, while podfly is prevalent in late-duration genotypes. High-density trichomes on the pod wall surface and their associated exudates play a major role in resistance to pod borers. The inheritance of trichomes is governed by single dominant gene (Verulkar *et al.*, 1997, Rupakula *et al.*, 2005; Banu *et al.*, 2007) in *Cajanus scarabaeoides*. *C. scarabaeoides* shows resistance to podfly due to trichomes, their expression governed by a single dominant gene (Verulkar *et al.*, 1997), whereas for podfly resistance in cultivated species, two genes behave in both dominant and recessive fashion based on allelic interactions (Singh and Lal, 2002). Annual losses due to *M. vitrata* have been estimated at US$30 million (ICRISAT, 1992). Very limited efforts have been made to identify a source for its resistance. Recently, ICPL 98003 and ICPL 98008 have been identified as donors for use in breeding programmes (Sunitha *et al.*, 2008).

The pod borers *M. testulalis* and *H. armigera*) also cause heavy losses in mung bean. At AVRDC (Taiwan), resistant lines V 2019, V 4270, V 2106 and V 2135 have been used in breeding programmes. Only low levels of resistance have been observed for *Maruca* pod borers in cowpea; in this crop the P120 and C11 lines have been reported to be the least damaged (Jagginavan *et al.*, 1995), and TV × 7 line has been shown to be the genotype most resistant to these insects (Veeranna and Hussain, 1997).

BRUCHIDS (*CALLOSOBRUCHUS CHINENSIS AND CALLOSOBRUCHUS MACULATES*). Bruchids are the most important pests of stored grain in *Vigna* spp. Multiple seed factors are responsible for resistance against bruchids, i.e. the presence of α-amylase inhibitors, trypsin inhibitors, polyphenol and and tannin content (Ishimoto and Kitamura, 1989). Inheritance of resistance is variously reported as due to monogenic dominant (Tickoo *et al.*, 2006) and digenic dominant duplicate (Souframanien and Gopalakrishna, 2007) gene actions. No effective resistant source has been reported for mung bean, whereas in urd bean lines Mash 59, VM 2011 and VM 2166, some resistance has been documented (Gupta and Kumar, 2006). Resistance to multiple insects has been found in cowpea, and several improved cowpea varieties with combined resistance to aphids, thrips and bruchids have been developed (Singh *et al.*, 1996). The varieties IT97K-207-15, IT95K-398-14 and 98K-506-1 have a high level of bruchid resistance (Singh 1999), and the 7s-storage protein 'vicillin' has been reported to be responsible for bruchid resistance in cowpea lines related to TVu 2027 (Yunes *et al.*, 1998).

THRIPS AND STEMFLY. These are major pests in urd bean, with yield losses under severe attack amounting to up to 40%. A large degree of genotypic variation has been observed for resistance. Important donors against thrips include PDU 5, KB 63, UG 567 and UH 804. The genotypes UG 218, PDU 1, PDU 5, LBG 707 and CO 305 are suitable donors for stem fly resistance (Gupta and Kumar, 2006).

APHID (*APHIS GLYCINES MATSUMURA*). For soybean, this is a major pest. Genotypes PI 200538 and PI 243540 have strong resistance to aphids, and a single dominant gene governs resistance to this insect (Kang *et al.*, 2008; Hill *et al.*, 2009).

Nematodes

The nematodes also are responsible for major problems in some warm season legume crops. In Nigeria, nematode attack in cowpea is very severe in the dry season when planting with irrigation. Several resistance sources have been identified for nematodes (Singh, 1998), of which IT89KD-288 was found to be resistant to four strains of *Meloidogyne incognita* in the USA (Ehlers *et al.*, 2000); this genotype was found to be very effective against nematodes, and showed high yielding potential in trials conducted in areas highly prone to nematode attack in Nigeria (Singh *et al.* 2002). IT89KD-288 was taken by one farmer in 1994 and, through farmer to farmer diffusion, it has become a popular variety because of its nematode resistance and high yield in the dry season. Roberts *et al.* (1996) identified the IT84S-2049 cowpea line from IITA as being completely resistant to diverse populations of the root-knot nematodes *M. incognita* and *Meloidogyne javanica*. Systematic genetic studies have indicated that resistance in IT84S-2049 was conferred by a single dominant gene, which was allelic to either the *Rk* gene or another gene very closely linked to *Rk*; therefore, the symbol *Rk*2 was proposed to designate this new resistance factor. Rodriguez *et al.* (1996) screened nine cowpea varieties for resistance to the root-knot nematode *M. incognita*; they observed that IITA-3, Habana 82, Incarita-1, IT86D-364, IT87D-1463-8, Vinales 144, P902 and IITA-7 were highly resistant, whereas the local variety Cancharro was highly susceptible.

Seed quality traits

Warm season food legumes are well known for their high seed protein and oil content. The most important limiting amino acids in food legumes, such as the sulfur-containing amino acids (methionine and cystine), are importance targets in protein quality improvement programmes. Efforts to increase cystine and methionine levels in soy proteins have been primarily aimed at increasing the concentration of protein subunits, which are known to have higher levels of these two amino acids. Increasing the protein and oil content is also an important target in warm season food legume crops for improving seed quality along with yield. However, a negative correlation between yield and protein content or between yield and oil content is well documented in these crops (Dahiya *et al.*, 1977; Wilcox and Shibles, 2001). Increasing both protein and oil concentration in seeds is an important breeding goal in soybean, but these are negatively correlated (Brim and Burton, 1979). It has been reported that soybean oil content is governed by additive gene effects, additive × additive epistatic interaction and complementary epistasis (Rahangdale and Raut, 2002), and therefore use of recurrent selection schemes could be the most effective means of increasing oil content (Burton and Brim, 1981).

Protein content is governed by considerable non-additive gene action in mung bean, thus making it a complex trait to transfer (Chandra and Tickoo, 1998). Rotundo *et al.* (2009) suggested that this negative association could be overcome by increasing the supply of assimilates per seed without sacrificing reproductive efficiency. In India, Naik *et al.* (2002) developed a local pureline, BSN 1 from *Nagpuri*, having a high yield and 27.8% seed protein. In urd bean, a positive association has been observed between protein content, seed yield, 100-seed weight and pods/plant (Kole *et al.*, 2002). Urd bean seeds contain 25% protein, but only limited efforts have so far been made to study the extent of genotypic variation for protein content in relation to other yield components (Kole *et al.*, 2002). Dark green colour, shiny and bold seeds are important quality factors for mung bean consumers in India; however, in Bangladesh and adjoining regions, yellow and small grains are commonly consumed.

High phytic acid (PA) levels in soybean seeds cause mineral malnutrition in humans, and to investigate this problem systematic studies have been conducted. Recently, it has been observed that total phosphorus (P) and phytate P (PhyP) are controlled by dominant recessive epistasis, which may be of assistance in developing low-phytate varieties (Sompong *et al.*, 2010). The quality of soybean oil is also determined on the basis of the ratios of polysaturated fatty acids, saturated fatty acids and mono-unsaturated fatty acids, and essential fatty acids such as linoleic/linolenic. High linolenic acid levels in soybean oil have poor oxidative stability (Patil *et al.*, 2004). Isoflavon in soybean oil is another important target for improvement in oil quality. For this trait, epistatic interactions have been observed, apart from malonyldiadzin (MDZ). To obtain the largest selection gains for this trait, priority should be given to exploiting either the additive genetic variances in superior lines or the cytoplasmic effect and the epistatic interactions between cytoplasmic and nuclear genes (Chiari *et al.*, 2006). Lutein is a major carotenoid in soybean seed, and is beneficial for maintenance of eye health; this component is positively correlated with oleic acid and negatively correlated with linoleic and linolenic acid (Lee *et al.*, 2009).

Agronomic traits

In mung bean, yield is correlated with leaf area index (LAI), number of branches per plant, pods per plant and seeds per pod (Makeen *et al.*, 2007). Multiple leaflet traits give a greater leaf area, thus intercepting more sunlight to help increase yield. This trait is controlled by single recessive gene (Sripisut and Srinives, 1986), whereas leaflet number is controlled by two loci (Soehendi *et al.*, 2007). A recent study suggests that leaflet size is more important than leaflet number in relation to seed yield (Sriphadet *et al.*, 2010). Determinate growth habit and compact plant type are also preferred traits for the development of varieties suitable for intercropping in mung bean (Tickoo *et al.*, 2006). Large differences of 68–140 days exist for maturity time in urd bean; however, due to the intensification of multiple cropping systems, early varieties are required to suit this system (Sinha, 1988; Chadha *et al.*, 2009).

Lodging resistance is the main target characteristic for soybean cultivars. An erect growth habit, which reduces mechanical harvesting loss and allows maximum light penetration through the plant canopy, is a target trait in the USA in soybean for improved plant type. Soybean breeders have used several other traits with mixed results, including narrow leaflets, brachytic stems (short internode), stem termination to alter height, and more fibrous rooting (Wells *et al.*, 1993). Change in the length of the reproductive period has been focused in soybean on adaptation to specific environments. However, in practice, lengthening the pod-filling period and/or changing the rate of dry matter accumulation in pods have allowed minor improvement in yield, with a positive correlation having been between these two traits (Smith and Nelson, 1986).

References

Adu-Dapaah, H.K., Singh B.B. and Fatokun C.A. (1999) A fasciated mutant in cowpea (*Vigna unguiculata* (L.). *Acta Agronomica Hungarica* 47, 371–376.

Ahloowalia, B.S., Maluszynski, M. and Nichterlein, K. (2004) Global impact of mutation-derived varieties. *Euphytica* 135, 187–204.

Balyan, H.S. and Khan, M.N. (1995) Comparison of three methods of handling the M_1 and M_2 generations in Urd bean. *Indian Journal of Pulses Research* 8, 109–112.

Banu M.R., Muthiah, A. and Ashok, S. (2007) Inheritance of podborer (*Helicoverpa armigera*) tolerance in pigeon pea. *International Journal of Botany* 3(1), 125–127.

Bhatnagar, P.S., Tiwari, S.P. and Singh, P. (1990) Application of mutagenesis for improvement of indigenous soybean variety of India. *Mutation Breeding Newsletter* 36, 8.

Brar, J.S., Bains, T.S., Shanmugasundaram, S. and Singh, S. (2006) Developing short duration mung bean genotypes suitable for rice-wheat cropping system. In: S. Shanmugasundaram (ed.) *Proceedings of the Final Workshop and Final Meeting of the DFID-Mung Bean Project*, 27–31 May 2004. Punjab Agricultural University, Ludhiana, India, pp. 61–81.

Brim, C.A. and Burton, J.W. (1979) Recurrent selection in soybeans II. Selection for increased percent protein in seeds. *Crop Science* 19, 494–498.

Brossman, G.D. and Wilcox, J.R. (1984) Induction of genetic variation for oil properties and agronomic characteristics of soybean. *Crop Science* 24, 783–787.

Burns, M.J., Edwards, K.J., Newbury, H.J., Ford-Lloyd, B.V. and Baggott, C.D. (2001) Development of simple sequence repeat (SSR) markers for the assessment of gene flow and genetic diversity in pigeon pea (*Cajanus cajan*). *Molecular Ecology Notes* 1, 283–285.

Burton, J.M. and Brim, C.A. (1981) Recurrent selection in soybeans III. Selection for increased oil in seeds. *Crop Science* 21, 31–34.

Burton, J.W. (1997) Soybean (*Glycine max* (L.) Merr.). *Field Crops Research* 53, 171–186.

Burton, J.W. and Carver, B.F. (1993) Selection among S_1 families vs selfed half-sib or full-sib families in autogamous crops. *Crop Science* 33, 21–28.

Cannon, S.B., May, G.D. and Jackson, S.A. (2009) Three sequenced legume genomes and many crop species: rich opportunities for translational genomics. *Plant Physiology* 151, 970–977.

Carter, T.E. Jr., Nelson, R.L., Sneller, C.H. and Cui, Z. (2004) Genetic diversity in soybean. In: Boerma, H.R. and Specht, J.E. (eds), *Soybeans: Improvement, Production, and Uses.* Agronomy Monographs 3rd ed. No. 16, ASA-CSSA-SSSA, Madison, WI, pp. 303–416.

Carver, B.F., Burton, J.W., Willson, R.F. and Carter, J.E. Jr. (1986) Cumulative response to various recurrent selections schemes in soybean: oil quality and correlated agronomic traits. *Crop Science* 26, 853–858.

Chadha, M.L., Bains, T.S., Sekhon, H.S. and Sain, S.K. (2009) Short duration mung bean for diversification of rice wheat systems. In. Ali, M. and Kumar, S. (eds) *Milestones in Food Legumes Research.* Indian Institute of Pulses Research, Kanpur, India, pp. 151–177.

Chakraborty, R.K., Bhowmik, A., Hossain, T. and Mian, M.A.K. (1998) Induction of mutation in mung bean (*Vigna radiata*) through gamma-irradiation. *Annals of Bangladesh Agriculture* 8, 129–136.

Chandra, N. and Tickoo, J.L. (1998) Genetic analysis of protein content in mung bean (*Vigna radiata* L. Wilczek). *Indian Journal of Genetics and Plant Breeding* 58, 475–478.

Chiari, L., Naoe, L.K., Piovesan, N.D., José, I.C., Cruz, C.D., Moreira, M.A. *et al.* (2006) Inheritance of isoflavone contents in soybean seeds. *Euphytica* 150, 141–147.

Choi, H.K., Kim, D., Uhm, T., Limpens, E., Lim, H., Mun, J.H. *et al.* (2004) A sequence-based genetic map of *Medicago truncatula* and comparison of marker colinearity with *Medicago sativa*. *Genetics* 166, 1463–1502.

Cooper, R.L. (1990) Modified early generation testing procedure for yield selection in soybean. *Crop Science* 30, 417–419.

Dahiya, B.S., Singh, K. and Brar, J.S. (1977) Incorporation of resistance to mung bean yellow mosaic virus in blackgram (*Vigna mungo* L). *Tropical Grain Legume Bulletin* 9, 28–32.

Datta, S., Kaashyap, M. and Kumar, S. (2010) Amplification of chickpea-specific SSR primers in *Cajanus* species and their validity in diversity analysis. *Plant Breeding* (doi:10.1111/j.1439-0523.2009.01678.x).

Dhar, V. and Reddy, M.V. (1999) Disease management strategies for increasing pulses production. In: *Proceedings of the Brain Storming Meeting on Pulses Production*, 26–27 March 1999. National Bureau of Plant Genetic Resources, New Delhi, India.

Dixit, G.P., Tripathi, D.P., Chandra, S., Tewari, T.N. and Tickoo, J.L. (2000) MULLaRP crops: varieties developed during the last fifty years. All India Coordinated Research Project on MULLaRP (ICAR), Indian Institute of Pulses Research, Kanpur, India.

Du, W.J., Fu, S.X. and Yu, D.Y. (2009) Genetic analysis of leaf pubescence density and water status traits in soybean (*Glycine max* (L) Merr). *Plant Breeding* 128, 259–265.

Dupare, B.U., Joshi, O.P., Billore, S.D. and Husain, S.M. (2009) Soybean improvement and development in India. In. Ali, M. and Kumar, S. (eds) *Milestones in Food Legumes Research*. Indian Institute of Pulses Research, Kanpur, India, pp. 115–133.

Dwivedi, A.K. and Pandey, M.P. (1981) Evaluation of induced quantitative mutants in soybean. *Indian Journal of Agricultural Sciences* 51, 715.

Ehlers, J.D., Matthews, W.C., Hall, A.E. and Roberts, P.A. (2000) Inheritance of a broad-based form of nematode resistance in cowpea. *Crop Science* 40, 611–618.

Fery, R.L. and Singh, B.B. (1997) Cowpea genetics: a review of recent literature. In: Singh, B.B., Mohan Raj D.R., Dashiell K.E. and Jackai, L.E.N. (eds) *Advances in Cowpea Research*. Co-publication of IITA and JIRCAS, IITA, Ibadan, Nigeria, pp. 13–29.

Gomathinayagam, P., Ram, S.G., Rathnaswamy, R. and Ramaswamy, N.M. (1998) Interspecific hybridization between *Vigna unguiculata* and *V. vexillata* through in vitro embryo culture. *Euphytica* 102, 203–209.

Gopalakrishna, T. and Reddy, K.S. (2009) Mutation breeding of food legumes. In: Ali, M. and Kumar, S. (eds) *Milestones in Food Legumes Research*. Indian Institute of Pulses Research, Kanpur, India, pp. 206–224.

Grewal, J.S. (1978) Disease of mung bean in India. In: *Proceedings of the 1st International Mung Bean Symposium*, University of Philippines, Philippines, pp. 165–168.

Gunasekaran, M., Selvaraj, U. and Raveemdram, T.S. (1998) Induced polygenic mutations in cowpea (*Vigna unguiculata* L. Walp). *South-Indian Horticulture* 46, 13–17.

Gupta, S. and Kumar, S. (2006) Urd bean breeding. In: Ali, M. and Kumar, S. (eds) *Advances in Mung Bean and Urd Bean*. Indian Institute of Pulses Research, Kanpur, India, pp. 149–168.

Hall, A.E., Singh, B.B. and Ehlers, J.D. (1997) Cowpea breeding. *Plant Breeding Reviews* 15, 215–274.

Hartman, G.L. (2005) Breeding for resistance to soybean rust. *Plant Disease* 89, 664–669.

Hartman, G.L., Wang, T.C. and Shanmugasundaram, S. (1997) Soybean rust research: progress and future prospects. In: *Proceedings of the World Soybean Research Conference V*, Kasetsart, Thailand, pp. 180–186.

Hartwig, E.E. and Lehman, S.G. (1951) Inheritance of resistance to the bacterial pustules disease in soybean. *Agronomy Journal* 43, 226–229.

Hill, C.B., Kim, S., Crull, L., Diers, B.W. and Hartman, G.L. (2009) Inheritance of resistance to the soybean aphid in soybean PI 200538. *Crop Science* 49, 1193–1200.

ICRISAT (1983) *Annual Report, 1982*. International Crops Research Institute for the Semi-Arid Tropics (ICRISAT), Patancheru, Andra Pradesh, India, pp. 131.

ICRISAT (1992) *The Medium Term Plan*. International Crops Research Institute for the Semi-Arid Tropics (ICRISAT), Patancheru, Andra Pradesh, India.

Ishimoto, M. and Kitamura, K. (1989) Growth inhibitory effects of an alpha-amylase inhibitor from the kidney bean, *Phaseolus vulgaris* (L.) on three species of bruchids (Coleoptera: Bruchidae). *Applied Entomology and Zoology* 24, 281–286.

Ishiyaku, M.F. and Singh, B.B. (2001) Inheritance of shortday-induced dwarfing in photosensitive cowpea. *African Crop Science Journal* 9(2), 1–8.

Jagginavan, S.B., Kulkarni, K.A. and Lingappa, S. (1995) Reaction of cowpea genotypes to the damage of pod borer complex. *Karnataka Journal of Agricultural Science* 89(1), 90–93.

John, S.A. (1999) Mutation frequency and chlorophyll mutations in parents and hybrid of cowpea following gamma irradiation. *Indian Journal of Genetics and Plant Breeding* 59, 357–361.

Kang, S., Rouf Mian, M.A. and Hammond, R.B. (2008) Soybean aphid resistance in PI 243540 is controlled by a single dominant gene. *Crop Science* 48(5), 1744–1748.

Kannaiyan, J., Nene, Y.L., Reddy, M.V., Ryan, J.G. and Raju, T.N. (1984) Prevalance of pigeon pea diseases and associated crop losses in Asia, Africa and the Americas. *Tropical Pest Management* 30, 62–71.

Kaushal, R.P. and Singh, B.M. (1989) Evaluation of black gram (*Phaseolus mungo*) germplasm for multiple disease resistance. *Indian Journal of Agricultural Sciences* 59, 726–727.

Kenworthy, W.J. and Brim, C.A. (1979) Recurrent selection in soybeans I. Seed yield. *Crop Science* 19, 315–318.

Khan, S. and Wani, M.R. (2006) MMS and SA induced genetic variability for quantitative traits in mung bean. *Indian Journal of Pulses Research* 19, 50–52.

Kim, D.H., Kim, K.H., Van, K., Kim, M.Y. and Lee, S.H. (2010) Fine mapping of a resistance gene to bacterial leaf pustules in soybean. *Plant Breeding* 120, 1443–1450.

Kole, C., Mohanty, S.K. and Pattanayak, S.K. (2002) Selection of protein rich genotypes in urd bean (*Vigna mungo* (L.) Hepper). *Indian Journal of Genetics and Plant Breeding* 62, 345–346.

Kumar, P.L., Jones, A.T., Sreenivasulu, P. and Reddy, D.V.R. (2000) Breakthrough in the identification of the causal virus of pigeon pea sterility mosaic disease. *Journal of Mycology and Plant Pathology* 30, 249.

Lal, S.K., Bhat, A.I., Rana, V.K.S., Sapra, R.L. and Kumar, A. (2002) Identification of resistant sources against bud-blight disease of soybean (*Glycine max* (L.) Merrill.). *Indian Journal of Genetics and Plant Breeding* 62, 287–290.

Lee, J.D., Shannon, J.G., So, Y.S., Sleper, D.A., Nelson, R.L., Lee, J.H. *et al.* (2009) Environment effects on lutein content and relationship of lutein and other compounds in soybean. *Plant Breeding* 128, 97–100.

Makeen, K., Abrahim, G., Jan, A. and Singh, A. (2007) Genetic variability and correlation studies on yield and its component in mung bean (*Vigna radiata*). *Journal of Agronomy* 6, 216–218.

Mehta, D.R. and Zaveri, P.P. (1997) Single seed versus single plant selection in cowpea. *Legume Research* 20, 130–132.

Micke, A. (1984) Mutation breeding of grain legumes. *Plant and Soil* 82, 337–357.

Mishra, S.P., Asthana, A.N. and Yadav, L. (1988) Inheritance of *Cercospora* leaf spot resistance in mung bean, *Vigna radiata* (L.) Wilczek. *Plant Breeding* 100, 228–229.

Misra, R.K., Singh C.B., Sharma S.M. and Mehta, S.K. (1981) Note on induced variation in shattering habit of soybean. *Indian Journal of Agricultural Sciences* 51, 678.

Moon, J.K., Jeong, S.C., Van, K., Maroof, S.M.A. and Lee, S.H. (2009) Marker assisted identification of resistance gene to soybean mosaic virus in soybean lines. *Euphytica* 169, 375–385.

Muchero, W., Diop, N.N., Bhat, P.R., Fenton, R.D., Wanamaker, S., Pottorff, M. *et al.* (2009) A consensus genetic map of cowpea [*Vigna unguiculata* (L) Walp.] and synteny based on EST-derived SNPs. *Proceeding of the National Academy of Sciences U.S.A.* 106, 18159–18164.

Naik, B.S., Singh, B. and Kole, C. (2002) A promising mung bean (*Vigna radiata* (L.) Wilczek) genotype with high protein content and seed yield. *Indian Journal of Genetics and Plant Breeding* 62, 342–344.

Ng, N.Q. and Singh, B.B. (1997) Cowpea. In: Fuccillo, D., Sears, L. and Stapleton, P. (eds) *Biodiversity and Trust*. Cambridge University Press, Cambridge, UK, pp. 89–99.

Odeigah, P.G.C., Osanyin, P.A.O. and Myers, G.O. (1996) Induced male sterility in cowpea. *Journal of Genetics and Plant Breeding* 50, 171–175.

Odeny, D.A., Jayashree, B., Ferguson, M., Hoisington, D., Crouch, J. and Gebhardt, C. (2007) Development characterization and utilization of microsatellite markers in pigeon pea [*Cajanus cajan* (L.) Millsp.]. *Plant Breeding* 126, 130–137.

Pal, B.P. (1934) Recent progress in plant breeding at Pusa. *Agricultural and Livestock in India* 4, 505–515.

Palmer, R.G. and Lewers, K.S. (1998) Registration of 68 soybean germplasm lines segregating for male sterility. *Crop Science* 58, 560–562.

Pandey, N., Ojha, C.B., Jha, V.B. and Singh, N.B. (1996) Effect of chemical mutagens on the rate of germination, seedling mortality, and induced sterility in pigeon pea. *International Chickpea and Pigeonpea Newsletter* 3, 65–67.

Pathak, G.N. (1970) Red gram. In: *Pulse Crops of India*. Indian Council of Agricultural Research, New Delhi, India, pp. 14–53.

Pathan, M.S. and Sleper, D.A. (2008) Advances in soybean breeding. In: Stacey, G. (ed.) *Genetics and Genomics of Soybean*. Springer-Verlag, New York, pp. 113–133.

Patil, A., Taware, S.P., and Raut, V.M. (2004) Quality of Indian soybean (*Glycine max* (L.) Merrill) varieties in relation to fatty acids composition. *Indian Journal of Genetic and Plant Breeding* 64, 245–246.

Pawar, N.B. and Mayee, C.D. (1986) Reaction of pigeon pea genotypes and their crosses to *Fusarium* wilt. *Indian Phytopathology* 39, 70–74.

Pennisi, E. (2008) Plant genetics: the blue revolution, drop by drop gene by gene. *Science* 320, 171–173.

Perez, E.O., Mian, R.M.A., Mendiola, C.T., Tew, J., Horner, H.T., Hanlin, S.J. *et al.* (2008) Seed set evaluation of male sterile, female fertile soybean lines using alfalfa leaf cutting bees and honey bees as pollinators. *Journal of Agricultural Sciences* 146, 461–469.

Pfeiffer, T.W. and Pilcher, D.L. (2006) Registration of KY98-2047 and KY98-2932 extra dense pubesence soybean germplasm. *Crop Science* 46, 480.

Potdukhe, N.R. and Narkhede, M.N. (2002) Induced mutations in pigeon pea (*Cajanus cajan* (L.) Millsp.). *Journal of Nuclear Agriculture and Biology* 31, 41–46.

Rahangdale, S.R. and Raut, V.M. (2002) Gene effects for oil content and other quantitative traits in soybean (*Glycine max* (L.) Merill). *Indian Journal of Genetics and Plant Breeding* 62(4), 322–327.

Rahman, S.M., Takagi, Y. and Kinoshita, T. (1997) Genetic control of high stearic acid content in seed oil of two soybean mutants. *Theoretical and Applied Genetics* 95, 772–776.

Rakshit, S., Singh, V.P. and Rakshit, S. (2001) Chemosensitivity studies in mung bean and urd bean. *Indian Journal of Pulses Research* 14(2), 112–115.

Ranalli, P. and Cubero, J.I. (1997) Bases for genetic improvement of grain legumes. *Field Crop Research* 53, 69–82.

Rao, V.G., Raut, V.M. and Patil, V.P. (1995) Out-break of soybean rust in Maharashtra. *Journal of Maharashtra Agriculture Universities* 20, 479–480.

Raut, V.M., Taware, S.P., Halvankar, G.B. and Philips, V. (2000) Development of high yielding variety MACS-450 by using Kalitur mutant MACS-111. In: *DAE-BRNS Symposium*. BARC-Bombay, India, pp. 103–110.

Reddy, K.R. (1986) Introgression of qualitative and quantitative genes from wild progenitors into greengram and blackgram. PhD thesis, GB Pant University of Agriculture and Technology, Pantnagar, India.

Reddy, K.S. (2009) Identification and inheritance of a new gene for powdery mildew resistance in mung bean (*Vigna radiata* L. Wilczek). *Plant Breeding* 128, 521–523.

Reddy, K.S., Dhanasekar, P. and Dhole, V.J. (2008) A review on powdery mildew disease resistance in mung bean. *Journal of Food Legumes* 21, 151–155.

Roberts, P.A., Matheswand, W.C. and Ehlers, J.D. (1996) New resistance to virulent root-knot nematodes linked to *Rk* locus in cowpea. *Crop Science* 36, 889–894.

Rodriguez, I., Rodriguez, M.G., Sanchez, L. and Iglesias, A. (1996) Expression of resistance to *Meloidogyne incognita* in cowpea cultivars. *Revista de Proteccion Vegetal* 11, 63–65.

Rose, J.L., Butler, D.G. and Ryley, M.J. (1992) Yield improvement in soybeans using recurrent selection. *Australian Journal of Agricultural Research* 43, 135–144.

Rotundo, J.L., Borras, L., Westgate, M.E. and Orf, J.H. (2009) Relationship between assimilate supply per seed during seed filling and soybean seed composition. *Field Crop Research* 112, 90–96.

Rupakula, A., Manohar, R., Reddy, L.J., Upadhyaya, H.D. and Sharma, H.C. (2005) Inheritance of trichomes and resistance to pod borer (*Helicoverpa armigera*) and their association in interspecific crosses between cultivated pigeon pea (*Cajanus cajan*) and its wild relative *C. scarabaeoides*. *Euphytica* 145, 247–257.

Sato, S., Isobe, S. and Tabata, S. (2010) Structural analyses of the genomes in legumes. *Current Opinion in Plant Biology* 13, 1–17.

Saxena, K.B. (2008) Genetic improvement of pigeon pea – a review. *Tropical Plant Biology* 1, 159–178.

Saxena, K.B. (2009) Evolution of hybrid breeding technology in pigeon pea. In: Ali, M. and Kumar, S. (eds) *Milestones in Food Legumes Research*. Indian Institute of Pulses Research, Kanpur, India, pp. 82–114.

Saxena, R.K., Prathima, C., Saxena, K.B., Hoisington, D.A., Singh, N.K. and Varshney, R.K. (2010) Novel SSR markers for polymorphism detection in pigeon pea (*Cajanus* spp.). *Plant Breeding* 129, 142–148.

Sharma, A.K., Singh, V.P. and Sharma, M.K. (2007) Induced seed and pod colour mutations in urd bean [*Vigna mungo* (L.) Hepper]. *Indian Journal of Genetics and Plant Breeding* 67(3), 270–271.

Sharma, D., Gupta, S.C., Rai, G.S. and Reddy, L.J. (1982) Inheritance of resistance to blight in pigeon pea. *Plant Disease* 66, 22–25.

Sharma, S.K., Sood, R. and Pandey, D.P. (2005) Studies on mutagen sensitivity, effectiveness and efficiency in urd bean [*Vigna mungo* (L.) Hepper]. *Indian Journal of Genetics and Plant Breeding* 65, 20–22.

Shaw, F.J.F. (1936) Studies in Indian pulses: The inheritance of morphological characters and wilt resistance in arhar (*Cajanus indicus* Spreng.). *Indian Journal of Agricultural Sciences* 6, 139–187.

Singh, A., Sirohi, A. and Panwar, K.S. (1998) Inheritance of mung bean yellow mosaic virus resistance in urd bean (*V. mungo* L. Hepper). *Indian Journal of Virology* 14, 89–90.

Singh, B.B. (1998) Sources of resistance to septoria, scab, bacterial blight and *Cercospora* leaf spot. *IITA Annual Report 1998*, Project 11. IITA, Ibadan, Nigeria, pp. 24–27.

Singh, B.B. (1999). Improved breeding lines with resistance to bruchid. *IITA Annual Report*, Project 11. IITA, Ibadan, Nigeria, pp. 29–30.

Singh, B.B. (2002) Recent genetic studies in cowpea. In: Fatokun, C.A., Tarawali, S.A., Singh, B.B., Kormawa, P.M. and Tamo, M. (eds) *Challenges and Opportunities for Enhancing Sustainable Cowpea Production*. IITA, Ibadan, Nigeria, pp. 3–13.

Singh, B.B. (2005) Cowpea [*Vigna unguiculata* (L.) Walp.] In: Singh, R.J. and Jauhar, P.P. (eds) *Genetic Resources, Chromosome Engineering and Crop Improvement*, vol 1. CRC Press, Boca Raton, Florida, pp. 117–162.

Singh, B.B. and Adu-Dapaah, H.K. (1998) A partial male sterile mutant in cowpea. *African Crop Science Journal* 6, 97–101.

Singh, B.B. and Dixit, H.K. (2002) Possibilities and limitations of interspecific hybridization involving greengram and blackgram. *Indian Journal of Agricultural Sciences* 72, 676–678.

Singh, B.B. and Singh, D.P. (1998) Variation for yield and yield components in early segregating generation of a wide cross between mung bean and urd bean. *Indian Journal of Genetic and Plant Breeding* 58, 113–115.

Singh, B.B., Asante, S.K., Jackai, L.E.N. and Hughes, J.D. (1996) Screening for resistance to parasitic plants, virus, aphid and bruchid. *IITA Annual Report*, Project 11. IITA, Ibadan, Nigeria, pp. 24.

Singh, B.B., Cambliss, O.I. and Sharma, B. (1997) Recent advances in cowpea breeding. In: Singh B.B., Mohan Raj, D.R., Dashiell, K. and Jackai, L.E.N. (eds) *Advances in Cowpea Research*. IITA, Ibadan, Nigeria, pp. 30–49.

Singh, B.B., Ehlers, J.D., Sharma, B. and Freire Filho, F.R. (2002) Recent progress in cowpea breeding In: Fatokun, C.A., Tarawali, S.A., Singh, B.B., Kormawa, P.M. and Tamo, M. (eds) *Challenges and Opportunities for Enhancing Sustainable Cowpea Production*. IITA, Ibadan, Nigeria, pp. 22–40.

Singh, B.V., Pandya, B.P., Gautam, P.L., Beniwal, S.P.S. and Pandey, M.P. (1983) Inheritance of resistance of sterility mosaic virus in pigeon pea. *Indian Journal of Genetics and Plant Breeding* 43, 487–493.

Singh, D. and Patel, P.N. (1977) Studies on resistance in crops to bacterial diseases in India, Part VIII. Investigations on inheritance of reactions to bacterial leaf spot and yellow mosaic diseases and linkage, if any, with other characters in mung bean. *Indian Phytopathology* 30, 202–206.

Singh, G., Sareen, P.K., Saharan, R.P., Singh, A., Singh, G. and Singh, A. (2001) Induced variability in mung bean [*Vigna radiata* (L.) Wilczek]. *Indian Journal of Genetics and Plant Breeding* 61, 281–282.

Singh, I.P. and Lal, S.S. (2002) Inheritance of resistance to podfly in pigeon pea (*Cajanus cajan* (L.) Millsp.). *Indian Journal of Genetics and Plant Breeding* 56, 85–88.

Singh, I.P., Dhar, V. and Chaudhary, R.G. (1998) Inheritance of resistance to *Fusarium* wilt in pigeon pea. *Indian Journal of Agricultural Sciences* 68(11), 729–731.

Singh, I.P., Chaudhary, R.G., Katiyar, P.K. and Dua, R.P. (2003a) Inheritance of resistance to the 'Kanpur' race of *Phytophthora drechsleri* in pigeon pea. *Plant Breeding* 122, 453–455.

Singh, I.P., Dhar, V. and Dua, R.P. (2003b) Inheritance of resistance to sterility mosaic in pigeon pea. In: *Proceedings of Satellite Symposium on Grain Legumes*, 9–11 Februrary. *Indian Agricultural Research Institute* (IARI), New Delhi, India. pp. 291–293.

Singh, N.B., Singh, I.P. and Singh, B.B. (2005) Pigeon pea breeding. In: Ali, M. and Kumar, S. (eds) *Advances in Pigeon Pea Research*. Indian Institute of Pulses Research, Kanpur, India. pp. 67–95.

Singh, R. and Kole, C.R. (2006) Delineation of EMS induced genetic variability in some agronomic traits in mung bean (*Vigna radiata* L. Wilczek). *Crop Research*. 32, 94–96.

Sinha, R.P. (1988) Early maturity, dwarf mutant of urd bean (*V. mungo* (L.) Hepper). *Journal of Nuclear Agriculture and Biology* 17, 61–62.

Smith, J.R. and Nelson, R.L. (1986) Relationship between seed filling period and yield among soybean breeding lines. *Crop Science* 26, 469–472.

Soehendi, R., Chanprame, S., Toojinda, T., Ngamponsai, S. and Srinives, P. (2007) Genetics, agronomics and molecular study of leaflet mutant in mung bean (*Vigna radiata*). *Journal of Crop Science and Biotechnology* 10, 193–200.

Sompong, U., Kaewprasit, C., Nakasathien, S. and Srinives, P. (2010) Inheritance of seed phytate in mung bean (*Vigna radiata*). *Euphytica* 171, 389–396.

Sorajjapinun, W., Rewthougchum, S., Koizumin, M. and Srinives, P. (2005) Quantitative inheritance of resistance to powdery mildew disease in mung bean (*Vigna radiata* (L.) Wilczek). *SABRAO Journal of Breeding and Genetics* 37, 91–96.

Souframanien, J. and Gopalakrishna, T. (2007) Source for bruchid resistance and its inheritance in Trombay wild urd bean (*Vigna mungo* var. *silvestris*). *Journal of Food Legumes* 20, 19–21.

Srinivas, T., Reddy, M.V., Jain, K.C. and Reddy, M.S. (1997) Inheritance of resistance to two isolates of sterility mosaic pigeon pea (*Cajanus cajan* (L.) Millsp.). *Euphytica* 97, 45–52.

Srinives, P., Hual, A. N., Saengchot, S., Ngampongsai, S. and Srinives, P. (2000) The use of wild relatives and gamma radiation in mung bean and blackgram breeding. In: *The Seventh International Workshop on Genetic Resources. Part 1: Wild Legumes*, 13–15 October 1999. Ministry of Agriculture, Forestry and Fisheries (MAFF), Ibaraki, Japan, pp. 205–218.

Sriphadet, S., Kasemsap, P. and Srinives, P. (2010) Effect of leaflet size and number on agronomic and physiological traits of mung bean. *Journal of Agricultural Science* 148, 353–361.

Sripisut, W. and Srinives, P. (1986) Inheritance of lobed leaflets and multiple leaflets in mung bean (*Vigna radiata*). *Thai Agricultural Research Journal* 4, 192–197.

Sunitha, V., Lakshmi, K. and Ranga Rao, G.V. (2008) Screening of pigeon pea genotypes against *Maruca vitrata* (Geyer). *Journal of Food Legumes* 21, 137–139.

Tah, P.R. (2006) Studies on gamma ray induced mutations in mung bean [*Vigna radiata* (L.) Wilczek]. *Asian Journal of Plant Sciences* 5(1), 61–70.

Tickoo, J.L., Lal, S.K., Chandra, N. and Dikshit, H.K. (2006) Mung bean breeding. In: Ali, M. and Kumar, S. (eds) *Advances in Mung Bean and Urd Bean*. Indian Institute of Pulses Research, Kanpur, India, pp. 110–148.

Tinius, C.N., Burton, J.W. and Carter, T.E. Jr. (1991) Recurrent selection for seed size in soybean I. Response to selection in replicate populations. *Crop Science* 31, 1137–1141.

Tyagi, D.K. and Chawla, H.S. (1999) Effect of season and hormones on crossability barriers and *in vitro* hybrid development between *Vigna radiata* and *V. unguiculata*. *Acta Agronomica Hungarica* 47, 147–154.

Varshney, R.K., Penmetsa, R.V., Dutta, S., Kulwal, P.L., Saxena, R.K., Datta, S. *et al.* (2010a) Pigeon pea genomics initiative (PGI): an international effort to improve crop productivity of pigeon pea (*Cajanus cajan* L.). *Molecular Breeding* 26, 393–408.

Varshney, R.K., Mahendar, T., May, G.D. and Jackson, S.A. (2010b) Legume genomics and breeding. *Plant Breeding Reviews* 33, 257–304.

Veeranna, R. and Hussain, M.A. (1997) Trichomes as physical barriers for cowpea pod borer *Maruca testulalis* (Geyer) (Lepidoptera: Pyralidae). *Insect Environment* 3, 15.

Venkateswarlu, S., Singh, R.M. and Reddy, L.J. (1981) Induced mutagenesis in pigeon pea with gamma rays, ethyl methane sulfonate (EMS) and hydroxylamine. *Proceedings of the International Workshop on Pigeonpeas* vol. 2, 15–19 December, 1980, ICRISAT Center, Patancheru, India.

Verma, R.P.S. (1985) Inheritance of resistance to mung bean yellow mosaic virus in the interspecific and intervarietal crosses of greengram and blackgram. PhD thesis. GB Pant University of Agriculture and Technology, Pantnagar, India.

Verulkar, S.B., Singh, D.P. and Bhattacharya, A.K. (1997) Inheritance of resistance to podfly and pod borer in the interspecific cross of pigeon pea. *Theoretical and Applied Genetics* 95, 506–508.

Wani, M.R. and Khan, S. (2004) Ethylmethane sulphonate induced quantitative variability in mung bean. *Bionotes* 6, 111.

Wells, R., Burton, J.W. and Kilen, T.C. (1993) Soybean growth and light interception: Response to differing leaf and stem morphology. *Crop Science* 33, 520–524.

Wilcox, J.R. and Shibles, R.M. (2001) Interrelationships among seed quality attributes in soybean. *Crop Science* 41, 11–14.

Williams, F.J., Grewal, J.S. and Amin, K.S. (1968) Serious and new diseases of pulse crops in India. *Plant Disease Reporter* 52, 300–304.

Yadav, I.C., Chand, J.N. and Saharan, G.S. (1981) Inheritance of resistance in mung bean to bacterial leaf spot. In: Lozano, J.C. (ed.) *Proceedings of the Fifth International Conference on Plant Pathogenic Bacteria.* Centro Internacional de Agricultura Tropical, Cali, Colombia, pp. 580–583.

Yong, N.D., Danesh, D., Menancio, H.D and Kumar, L. (1993) Mapping oligogenic resistance to powdery mildew in mung bean with RFLPs. *Theoretical and Applied Genetics* 87, 243–249.

Yorinori, J.T., Paiva, W.M., Frederick, R.D., Costamilan, L.M., Bertagnoli, P.F., Hartman, G.L. *et al.* (2005) Epidemics of soybean rust (*Phakopsora pachyrhizi*) in Brazil and Paraguay from 2001 to 2003. *Plant Disease* 89, 675–677.

Yunes, A.N., Andrade, T.M., Sales, P.M., Morais, R.A., Fernandez, V.S., Gomes, V.M. *et al.* (1998) Legume seed vicilins interfere with the development of the cowpea weevil (*Callosobruchus maculatus*). *Journal of Agricultural and Food Chemistry* 76, 111–116.

6 Distant Hybridization and Alien Gene Introgression

Shiv Kumar, Muhammad Imtiaz, Sanjeev Gupta and Aditya Pratap

6.1 Introduction

Chickpea (*Cicer arietinum* L.), lentil (*Lens culinaris* Medik.), pigeon pea (*Cajanus cajan* L. Millsp.), green gram (*Vigna radiata* L. Wilczek), black gram (*Vigna mungo* L. Hepper), common bean (*Phaseolus vulgaris* L.) and grass pea (*Lathyrus sativus* L.) are among the important food legume crops grown on 74 million ha area with 64 million tons of global output (FAO, 2010). These crops are an integral part of subsistence agriculture with significant contributions to dietary protein supply, atmospheric nitrogen fixation and agricultural sustainability (Ali and Kumar, 2009). The average productivity of these crops is 846 kg/ha, which is dismally low compared with their potential harvestable yield. This is attributed to their cultivation on poor soils under rainfed conditions by marginal farmers with minimum care and, consequently, these crops suffer severe yield losses not only due to edaphic, abiotic and socio-economic factors but also to confounding effects of various biotic stresses. Yield losses caused by various fungal, bacterial and viral diseases are enormous, besides parasitic weed menace at various growth stages (Dita *et al.*, 2006). Being rich in protein, several insect pests also cause yield losses to food legumes both under field conditions and in storage (Clement *et al.*, 1994, 1999). Among abiotic stresses, drought, temperature extremities and edaphic problems (salinity and mineral toxicities) have great bearing on their harvestable yield (Stoddard *et al.*, 2006). Since plant breeding in practice is an option for crop improvement, efforts have been made to search for genes imparting resistance to these stresses within the cultivated species and, to a limited extent, among their wild relatives, but success has been limited to a few diseases and insect pests, and is confined to major gene(s) from the primary gene pool in few food legume crops (Knott and Dvorak, 1976; Stalker, 1980; Prescott-Allen and Prescott-Allen, 1986, 1988; Ladizinsky *et al.*, 1988; Hajjar and Hodgkin, 2007). To diversify and broaden the genetic base of cultivated germplasm, introgression of alien genes from wild species needs to be pursued vigorously, not only to minimize the risk of stress epidemics but also to make discernible yield advances in these legume crops. Therefore, pre-breeding efforts are urgently required involving particularly those wild species that carry useful alien genes for improving yield, quality and stress resistance. In this chapter we review the information on the present status of wild gene pools, their evaluation, introgression through distance hybridization and future crossing potential, crossability barriers and means to overcome them, strategies for successful introgressions, and future prospects in the selected legume crops.

6.2 Wild Gene Pool: Present Status

Wild species are a rich reservoir of useful alien genes that are no longer available within the cultivated gene pool (Hawkes, 1977; Doyle, 1988; Tanksley and McCouch, 1997). Continuous efforts have been under way to collect and conserve wild relatives of various food legume crops in national and international gene banks (Plucknett *et al.*, 1987; FAO, 1996). Over the years, ICARDA has collected and conserved, in its global germplasm repository, 587 accessions representing 6 wild *Lens* species from 26 countries, 270 accessions of 12 wild *Cicer* species from ten countries and 1555 accessions of 45 wild *Lathyrus* species from 45 countries. Similarly, the ICRISAT gene bank is reported to have 308 accessions of 18 *Cicer* species from 19 countries, 555 accessions of 57 *Cajanus* species from 41 countries and 478 accessions of 47 *Arachis* species from 7 countries in its wild gene pool (Upadhyaya, personal communication). The US Department of Agriculture, Agricultural Research Service (USDA-ARS), Western Regional Plant Introduction Station (WRPIS), Pullman, Washington also has a collection of 4602 accessions of chickpea (Hannon *et al.*, 2001). In spite of being the largest collections, these have major germplasm gaps at species and genotype levels (Ferguson and Erskine, 2001), and a continuum in our efforts is very much required to fill these gaps in wild gene pools from the unrepresented areas of diversity in the gene banks.

The gene pool concept of Harlan and De Wet (1971) has been very helpful to plant breeders for initiating a pre-breeding programme for directed crop improvement. Various species of major food legume crops have been grouped into primary, secondary and tertiary gene pools on the basis of crossability, cytogenetic, phylogenetic and molecular data (Table 6.1). The useful genes identified in the primary gene pool are readily usable for crop improvement. However, occurrence of useful genes is much more frequent in the secondary and tertiary gene pools of various food legume crops (Kaiser *et al.*, 1994; Collard *et al.*, 2001; Mallikarjuna *et al.*, 2006; Tullu *et al.*, 2006). This requires the deployment of much more effort and novel techniques for integrating this invaluable resource of nature into crop improvement programmes.

6.3 Evaluation of Wild Gene Pool

Sporadic efforts have been made in the past to screen wild species of food legume crops under field and controlled conditions in order to identify useful alien genes for desired traits. These efforts have resulted in identification of valuable sources of resistance to key diseases and insect pests in addition to useful traits such as protein content, cytoplasmic male sterility, fertility restoration and yield attributes (Table 6.2).

Chickpea

Annual *Cicer* species have been evaluated for reaction to ascochyta blight, fusarium wilt, cyst nematode, leaf miner, seed beetle and cold tolerance at ICARDA (International Centre for Agricultural Research in the Dry Areas), and a high level of resistance to each stress has been identified (Table 6.2). Kumar and Dua (2006) presented a list of possible wild species as a source of useful alien genes for chickpea improvement. *Cicer judaicum* is reported to have resistance genes for ascochyta blight, fusarium wilt and botrytis grey mould (van der Maesen and Pundir, 1984). Greco and Di Vito (1993) reported valuable sources of resistance to cyst nematode in *Cicer bijugum*, *Cicer pinnatifidum* and *Cicer reticulatum*. Some wild accessions have shown resistance to more than one stress (Singh *et al.*, 1994; Ahmad *et al.*, 2005). For example, ILWC 7-1 of *C. bijugum* showed resistance to ascochyta blight, fusarium wilt, leaf miner, cyst nematode and cold, and ILWC 33/S-4 of *C. pinnatifidum* to ascochyta blight, fusarium wilt, seed beetle and cyst nematode. Kaur *et al.* (1999) reported significantly lower larval density of helicoverpa pod borer on some of the accessions of *Cicer echinospermum*, *C. judaicum*, *C. pinnatifidum* and *C. reticulatum*. Recently, 150 accessions of wild chickpea have been evaluated for resistance to helicoverpa pod borer under field and greenhouse conditions

Table 6.1. Different gene pools of selected legume crops.

Crop	Primary gene pool	Secondary gene pool	Tertiary gene pool	References
Chickpea	*Cicer arietinum, C. reticulatum, C. echinospermum*	*C. bijugum, C. pinnatifidum, C. judaicum,*	*C. cuneatum, C. chorassanicum, C. yamashitae*	Ladizinsky and Adler (1976a, 1976b); Ahmad *et al.* (1988, 2005); van der Maesen *et al.* (2007)
Lentil	*Lens culinaris ssp. culinaris, L. culinaris ssp. orientalis, L. odemensis*	*L. ervoides, L. nigricans*	*L. lamottei, L. tomentosus*	Ladizinsky *et al.* (1984); Ladizinsky (1999); Muehlbauer and McPhee (2005)
Pigeon pea	*Cajanus cajan, C. cajanifolius*	*C. acutifolius, C. albicans, C. confertiflorus, C. lanceolatus, C. latisepalous, C. lineatus, C. reticulatus, C. scarabaeoides, C. sericeus, C. trinervius*	*C. goensis, C. heynei, C. kerstingii, C. mollis, C. platycarpus, C. rugosus, C. volubilis and other species*	Smartt (1990); Singh *et al.* (2006)
Mung bean	*Vigna radiata* var. *radiata, V. radiata* var. *sublobata, V. radiata var setulosa*	*V. mungo* var. *mungo, V. mungo* var. *var silvestris, V. aconitifolia, V. trilobata*	*V. angularis, V dalzelliana, V. glabrescens, V. grandis, V. umbellata, V. vexillata*	Smartt (1981, 1985); Dana and Karmakar (1990); Chandel and Lester (1991); Kumar *et al.* (2004)
Urd bean	*V. mungo* var. *mungo, V. mungo var sylvestris*	*Vigna radiata* var. *radiata, V. radiata* var. *sublobata, V. radiata* var. *setulosa, V. aconitifolia, V. trilobata*	*V. angularis, V. dalzelliana, V. glabrescens, V. grandis, V. umbellata, V. vexillata*	Dana and Karmakar (1990); Chandel and Lester (1991); Kumar *et al.* (2004)
Common bean	*Phaseolus vulgaris*	*P. coccineus, P. costaricensis, P. polyanthus*	*P. acutifolius, P. lunatus, other Phaseolus* spp.	Debouck and Smartt (1995); Debouck (1999, 2000)
Grass pea	*Lathyrus sativus*	*L. chrysanthus, L. cicera, L. gorgoni, L. marmoratus, L. pseudocicera, L. amphicarpus, L. blepharicarpus, L. chloranthus, L. hierosolymitanus, L. hirsutus*	*Remaining Lathyrus species*	Jackson and Yunus, (1984); Yunus and Jackson (1991); Kearney (1993); Kearney and Smartt (1995)

(Sharma, 2004). Potential accessions of *C. reticulatum* that can provide genes for high yield have also been reported by various workers (Jaiswal and Singh, 1989; Singh and Ocampo, 1997; Singh *et al.*, 2005).

Lentil

The *Lens* gene pool consists of many wild relatives offering resistance to biotic (Ahmad *et al.*, 1997a, b) and abiotic stresses

Table 6.2. Useful wild germplasm for introgression of alien genes in food legume crops

Crop	Useful trait(s)	Wild species	Reference(s)
Chickpea	Ascochyta blight resistance	*C. judaicum*, *C. montbretii*, *C. pinnatifidum*	van der Maesen and Pundir (1984); Singh and Reddy (1993)
	Fusarium wilt resistance	*C. bijugum*, *C. judaicum*, *C. reticulatum*	van der Maesen and Pundir (1984); Kaiser *et al.* (1994); Infantino *et al.* (1996)
	Botrytis grey mould resistance	*C. pinnatifidum*, *C. judaicum*	Singh *et al.* (1982); van der Maesen and Pundir (1984)
	Cyst nematode resistance	*C. bijugum*, *C. pinnatifidum*, *C. reticulatum*	Greco and Di Vito (1993); Di Vito *et al.* (1996)
	Phytophthora root rot resistance	*C. echinospermum*, *C. bijugum*, *C. reticulatum*, *and C. pinnatifidum*	Knights *et al.* (2008)
	Cold tolerance	*C. bijugum*, *C. echinospermum and C. reticulatum*	Singh *et al.* (1990)
	Helicoverpa pod borer tolerance	*C. bijugum*, *C. echinospermum*, *C. judaicum*, *C. pinnatifidum*, *C. reticulatum*, *C. cuneatum*	Kaur *et al.* (1999); Sharma (2004)
	Drought tolerance	*C. anatolicum*, *C. microphyllum*, *C. montbretii*, *C. oxydon and C. songaricum*	Toker *et al.* (2007)
	Yield attributes	*C. reticulatum*	Jaiswal and Singh (1989); Singh and Ocampo (1997); Singh *et al.* (2005)
Lentil	Anthracnose resistance	*Lens ervoides*, *L. lamottei*, *L. nigricans*	Tullu *et al.* (2006)
	Ascochyta blight resistance	*L. ervoides*, *L. culinaris* ssp. *orientalis*, *L. odemensis*, *L. nigricans*, *L. montbretti*	Bayaa *et al.* (1994)
	Fusarium wilt resistance	*L. culinaris* ssp. *orientalis*, *L. ervoides*	Bayaa *et al.* (1995); Gupta and Sharma (2006)
	Powdery mildew resistance	*L. culinaris* ssp. *orientalis*, *L. nigricans*	Gupta and Sharma (2006)
	Rust resistance	*L. culinaris* ssp. *orientalis*, *L. ervoides*, *L. nigricans*, *L. odemensis*	Gupta and Sharma (2006)
	Drought tolerance	*L. odemensis*, *L. ervoides*, *L. nigricans*	Hamdi and Erskine (1996) Gupta and Sharma (2006)
	Cold tolerance	*L. culinaris* ssp. *orientalis*	Hamdi *et al.* (1996)
	Yield attributes	*L. culinaris* ssp. *orientalis*	Gupta and Sharma (2006)
	Resistance to orobanche	*Lens ervoides*, *L. odemensis*, *L. orientalis*	Fernández-Aparicio *et al.* (2009)
	Resistance to sitona weevils	*L. odemensis*, *L. ervoides*, *L. nigricans*, *L. culinaris* ssp. *orientalis*	El-Bouhssini *et al.* (2008)
Grass pea	Low ODAP content	*L. cicera*	Aletor *et al.* (1994); Siddique *et al.* (1996); Hanbury *et al.* (1999); Kumar *et al.* (2010)

Continued

Table 6.2. Continued.

Crop	Useful trait(s)	Wild species	Reference(s)
Pigeon pea	Cytoplasmic male sterility	*Cajanus cajanifolius, C. sericeus, C. scarabaeoides, C. acutifolius*	Rathnaswamy *et al.* (1999) Ariyanayagam *et al.* (1993, 1995); Tikka *et al.* (1997); Saxena and Kumar (2003); Kalaimagal *et al.* (2008)
	High protein content	*Cajanus cajanifolius, C. sericeus*	Akinola *et al.* (1975); Dalvi *et al.* (2008)
	Sterility mosaic disease resistance	*C. sericeus, C. albicans*	Akinola *et al.* (1975); Singh *et al.* (1993, 2005)
	Phytophthora blight resistance	*C. sericeus, C. acutifolius, C. platycarpus*	Akinola *et al.* (1975); Mallikarjuna and Saxena (2002)
	Helicoverpa pod borer resistance	*C. scarabaeoides*	Verulkar *et al.* (1997)
	Salinity tolerance	*C. albicans*	Subba Rao (1990)
	Earliness	*C. platycarpus*	Saxena (2008)
Vigna	MYMV resistance	*V. umbellata, V. trilobata, V. mungo*	Singh and Dikshit (2002); Pandiyan *et al.* (2008)
Common bean	Common blight resistance	*P. acutifolius*	Singh and Munoz (1999)
	BGYMV resistance	*P. coccineus*	Osorno *et al.* (2003)
	Resistance to root rot, anthracnose and angular leaf spot	*P. coccineus*	Silbernagel and Hannan (1992); Mahuku *et al.* (2003)
	Heat tolerance	*P. acutifolius*	Federici *et al.* (1990)
	Drought tolerance	*P. acutifolius*	Parsons and Howe (1984); Markhart (1985)
	Freezing tolerance	*P. angustissimus*	Balsubramanian *et al.* (2004)
	Salt tolerance	*P. filiformis*	Bayuelo-Jimenez *et al.* (2002)

ODAP, β-N-oxalyl-L-α,β-diaminopropionic acid; MYMV, mung bean yellow mosaic virus; BGYMV, bean golden yellow mosaic virus.

(Hamdi *et al.*, 1996). A few attempts have been made at ICARDA and advanced research institutions to evaluate wild *Lens* taxa for agro-morphological traits besides key biotic and abiotic stresses (Erskine and Saxena, 1993; Bayaa *et al.*, 1994, 1995; Hamdi and Erskine, 1996; Hamdi *et al.*, 1996; Ferguson and Robertson, 1999; Tullu *et al.*, 2006; see also Table 6.2). The wild gene pool of lentil showed drought tolerance in *Lens odemensis* and *Lens ervoides* (Hamdi and Erskine, 1996; Gupta and Sharma, 2006), and cold tolerance and earliness in *Lens culinaris* ssp. *orientalis* (Hamdi *et al.*, 1996). Some of the wild accessions of *Lens* showing combined resistance to ascochyta blight, fusarium wilt (ILWL 138) and anthracnose disease (IG 72653, IG 72646, IG 72651) have also been identified (Bayaa *et al.*, 1995; Tullu *et al.*, 2006). Gupta and Sharma (2006) evaluated 70 accessions representing four wild species/subspecies (*L. culinaris* ssp. *orientalis*, *L. odemensis*, *L. ervoides* and *Lens nigricans*) for yield attributes and biotic and abiotic stresses. This resulted in identification of donors for resistance to powdery mildew in *L. c.* ssp. *orientalis* (ILWL 200) and *L. nigricans* (ILWL 37); rust and wilt resistance in all four species; drought tolerance in *L. nigricans*; and seeds per plant in *L. c.* ssp. *orientalis* (ILWL 90). Some accessions of *L. nigricans* (ILWL 37) and *L. c.* ssp. *orientalis* (ILWL 77) have multiple disease resistance and can be very useful sources of alien resistance genes. El-Bouhssini *et al.* (2008) identified increased resistance to sitona weevil in *L. odemensis*, followed by *L. ervoides*, *L. c.* ssp. *orientalis* and *L. nigricans*.

Grass pea

The wild gene pool is a rich reservoir of rare alleles for grass pea improvement, which have been evaluated sporadically to identify zero/low ODAP (β-N-oxalyl-L-α,β-diaminopropionic acid) lines (Jackson and Yunus, 1984). A total of 1082 accessions belonging to 30 species were evaluated for 21 descriptors and agronomic traits at ICARDA (Robertson and Abd-El-Moneim, 1997). Assessment of ODAP content in wild species of *Lathyrus* indicated that in none of the species is it absent (Aletor *et al.*, 1994; Siddique *et al.*, 1996; Hanbury *et al.*, 1999). On average, the ODAP concentration in *Lathyrus cicera* was lowest, followed by *Lathyrus sativus* and *Lathyrus ochrus* (Aletor *et al.*, 1994; Hanbury *et al.*, 1999). Evaluation of 142 accessions of *L. cicera* at ICARDA showed a range of 0.073–0.513% for ODAP content, which is much lower than that in cultivated species (Kumar *et al.*, 2010). The accessions of *L. cicera* are also a good source of earliness, orobanche tolerance and cold tolerance (Robertson *et al.*, 1996).

Pigeon pea

Evaluation of wild species of pigeon pea has shown many desirable characteristics that can be introgressed into cultivated species to make them more adapted and productive. The species with useful traits are listed in Table 6.2. These species have been reported to carry genes for high protein content, salinity tolerance, pod borer tolerance, sterility mosaic resistance, wilt resistance, phytophthora blight resistance and cytoplasmic male sterility. *Cajanus sericeus* and *Cajanus albicans* are rich in protein content, *Cajanus reticulatus* var. *grandifolius* is hardy and fire tolerant (Akinola *et al.*, 1975) and *C. albicans* is tolerant to soil salinity (Subba Rao, 1988).

Vigna crops

A wild accession of *Vigna radiata* var. *sublobata*, PLN 15, has been found to be the potential donor for pods per plant and seeds per pod (Reddy and Singh, 1990). Resistance to mung bean yellow mosaic virus (MYMV) has been reported in *Vigna umbellata*, *Vigna trolibata* and *Vigna mungo* (Nagaraj *et al.*, 1981; Singh and Dikshit, 2002).

Common bean

Wild species of *Phaseolus* have been characterized for biotic stresses. Wilkinson (1983) reported *Phaseolus coccineus* as a potential source of high yield for common bean. Resistance to angular leaf spot (Busogoro *et al.*, 1999), anthracnose (Hubbeling, 1957), ascochyta blight (Schmit and Baudoin, 1992), bean golden mosaic virus (BGMV) (CIAT, 1986; Beebe and Pastor-Corrales, 1991; Singh *et al.*, 1997), bean yellow mosaic virus (BYMV) (Baggett, 1956), common bean blight (CBB) (Mohan, 1982; Schuster *et al.*, 1983; Singh and Munoz, 1999), root rot (Yerkes and Freytag, 1956; Azzam, 1957; Hassan *et al.*, 1971), white mould (Abawi *et al.*, 1978; Hunter *et al.*, 1982) and cold (Bannerot, 1979) are found in the secondary gene pool. Some sources of resistance have also been identified in the tertiary gene pool. Resistance to ashy stem blight (*Macrophoma phaseolina*) and fusarium wilt (*Fusarium oxysporum* f. sp. *phaseoli*) (Miklas *et al.*, 1998b), BGMV (Miklas and Santiago, 1996), bruchids (Shade *et al.*, 1987; Dobie *et al.*, 1990; CIAT, 1995, 1996), CBB (Coyne *et al.*, 1963; Schuster *et al.*, 1983; Singh and Munoz, 1999), drought (Thomas *et al.*, 1983; Parsons and Howe, 1984; Markhart, 1985; Federici *et al.*, 1990; Rosas *et al.*, 1991), leafhopper (CIAT, 1995,1996) and rust (Miklas and Stavely, 1998) are found in *Phaseolus acutifolius*.

6.4. Distant Hybridization

Crosses between species of the same or different genera have contributed immensely to crop improvement, gene and genome mapping, understanding of chromosome behaviour and evolution in crops like rice, wheat, maize, sugar cane, cotton, tomato, etc. (Sharma, 1995). The ultimate goal of distant hybridization is to transfer useful genes from alien species into cultivated species, and this has been very successful in a few crops but not very encouraging for legume crops. Stalker (1980) discussed the gaps between hybridization and utilization, along with approaches for the utilization of wild species in food legumes. However, it is well recognized that gene transfer through wide crosses is a long and tedious process, due to lack of homology between chromosomes of participating species in the cross and pre- and post-zygotic crossability barriers between wild and cultivated species. Utilizing the wild gene pool in breeding programmes may also be constrained by collection gaps in wild species, with no information on genome relationships, poor/limited screening of wild species, linkage drag and genetic complexity of the traits. Therefore, improvement through distant hybridization often takes longer in order to recover genotypes associated with acceptable agronomic background, and thus requires a long-tem approach.

Crossability potential

The crossability of cultivars with wild species is a prerequisite for alien gene introgression. A large proportion of wild species are not crossable with cultivated species, and consequently of no use for crop improvement through sexual manipulation. However, variability for crossability has been observed not only among genotypes of cultivated species but also among those of alien species in several crops (Sirkka *et al.*, 1993; Sharma, 1995). Environmental factors can also influence embryo development of interspecific hybrids, and thereby the crossability potential (Percy, 1986; Sirkka *et al.*, 1993; Tyagi and Chawla, 1999). Therefore, an understanding of the extent of crossability is essential for successful production of hybrids and their derivatives. The early work on interspecific hybridization in grain legumes has been reviewed by Smartt (1979). Singh (1990) reviewed a wide spectrum of hybridization work in the genus *Vigna*, and Ocampo *et al.* (2000) in cool season legume crops. During the past two decades much information relating to possible gene flow between legume crops and their wild relatives, crossability barriers and methods of overcoming them has been generated. This has greatly enhanced the interest of breeders in distant hybridization. This section summarizes the crossability potential of different food legume crops using various wild and cultivated species.

Chickpea

Of the eight annual wild species, only *Cicer reticulatum* and *Cicer echinospermum* have been successfully crossed with chickpea (Ladizinsky and Alder, 1976a; Ahmad *et al.*, 1988, 2005; Verma *et al.*, 1990; Singh and Ocampo, 1993), a technique regularly utilized in the ICARDA chickpea breeding programme (Imtiaz, personal communication). Conventional crossing has been successful in producing interspecific hybrids between *Cicer arietinum* and *C. reticulatum* and between *C. arietinum* and *C. echinospermum*. Due to the presence of post-zygotic barriers, abortion of the immature embryo occurs for other interspecific crosses involving species from the tertiary gene pool such as *C. bijugum* and *C. judaicum* (Ahmad *et al.*, 1988; Clarke *et al.*, 2006). The availability of novel tissue culture techniques and biotechnological tools for circumventing crossing barriers has brightened the prospects of transferring useful traits from the tertiary gene pool (Shiela *et al.*, 1992; Mallikarjuna, 1999; Clarke *et al.*, 2006) and, as a result, hybrids were obtained between *C. pinnatifidum* and *C. bijugum* (Mallikarjuna, 1999).

Lentil

Many successful attempts have been made to develop interspecific hybrids, but still many cross combinations are yet to be attempted successfully. As far as the crossability status of wild *Lens* taxa is concerned, *L. c.* ssp. *orientalis* and *L. odemensis* are crossable with cultivated lentil (Ladizinsky *et al.*, 1984; Abbo and Ladizinsky, 1991, 1994; Fratini *et al.*, 2004; Fratini and Ruiz, 2006; Muehlbauer *et al.*, 2006), although the fertility of hybrids depends on the chromosome arrangement of the wild parent (Ladizinsky, 1979; Ladizinsky *et al.*, 1984). Most accessions of *L. c.* ssp. *orientalis* cross readily with *L. culinaris*, and both are genetically isolated from other species. *Lens nigricans* and *L. ervoides* are not readily crossable with the cultivated lentil using conventional crossing methods, due to hybrid embryo breakdown (Abbo and Ladizinsky, 1991, 1994; Gupta and Sharma, 2005). Crosses are possible between *L. culinaris* and the remaining species, but they are characterized by a high frequency of hybrid embryo abortion, albino seedlings and chromosomal rearrangements that result in hybrid sterility, if these seedlings reach maturity (Abbo and Ladizinsky, 1991, 1994; Ladizinsky, 1993; Gupta and Sharma, 2005). Only four crosses have not resulted in hybrids to date: *L. c.* ssp. *orientalis* × *L. ervoides; L. c.* ssp. *orientalis* × *L. nigricans* (Ladizinsky *et al.*, 1984); *L. c.* ssp. *tomentosus* × *L. lamottei* (Van Oss *et al.*, 1997); and *L. c.* ssp. *odemensis* × *L. ervoides* (Ladizinsky *et al.*, 1984), although viable hybrids have been reported between cultivated species and *L. ervoides*, *L. odemensis* and *L. nigricans* with the use of GA3 (Ahmad *et al.*, 1995). Fratini *et al.* (2006) reported a high correlation between crossing success and phenotypic similarity based on pollen morphology and *in vitro* pollen length, together with pistil and style length, indicating a good predictor of hybridization success between different species.

Grass pea

Interspecific hybridization has been successful between *L. sativus* and two wild *Lathyrus* species (*L. cicera* and *L. amphicarpus*) with viable seeds (Davies, 1957, 1958; Khawaja, 1985; Yunus, 1990). Yunus (1990) crossed 11 wild species with *L. sativus* and found viable seeds with *L. cicera* and *L. amphicarpus* only. Other species formed pods but did not give fully developed viable seeds (Yamamoto *et al.*, 1989; Yunus, 1990; Kearney, 1993). Some other successful interspecific hybrids reported in the genus *Lathyrus* were *L. annuus* with *L. hierosolymilanus* (Yamamoto *et al.*, 1989; Hammett *et al.*, 1994, 1996); *L. articulatus* with *L. clymenus* and *L. ochrus* (Davies, 1958; Trankovskij, 1962); *L. cicera* with *L. blepharicarpus, L. gorgoni, L. marmoratus* and *L. pseudocicera* (Yamamoto *et al.*, 1989; Kearney, 1993); *L. gorgoni* with *L. pseudocicera* (Yamamoto *et al.*, 1989; Kearney, 1993); *L. hirsutus* with *L. odoratus* (Davies, 1958; Trankovskij, 1962; Khawaja, 1988; Yamamoto *et al.*, 1989); *L. marmoratus* with *L. blepharicarpus* (Yamamoto *et al.*, 1989; Kearney, 1993); *L. odoratus* with *L. belinenesis* (Hemmett *et al.*, 1994, 1996); *L. rotundifolius* with *L. tuberosus* (Marsden-Jones, 1919); and *L. sylvestris* with *L. latifolius* (Davies, 1957).

Pigeon pea

Hybridization studies have shown that *C. cajan* can be successfully crossed with *C. albicans*, *C. cajanifolius*, *C. sericeus*, *C. scarabaeoides*, and *C. lineatus* (Reddy, 1981; Reddy and De, 1983; Kumar *et al.*, 1985; Pundir and Singh, 1985). Reddy *et al.* (1981) reported that five species of *Cajanus* (*C. sericeus, C. scarabaeoides, C. albicans, C. trinervius* and *C. cajanifolius*) were crossable with pigeon pea cultivars. However, *C. crassus* var. *crassus* and *C. platycarpus* cannot be crossed. With the help of *in vitro* embryo rescue technique, a *C. cajan* × *C. platycarpus* cross has also been successfully engineered (Dhanuj and Gill, 1985; Kumar *et al.*, 1985; Mallikarjuna and Moss, 1995; Mallikarjuna *et al.* 2006; Saxena *et al.*, 1996). Shahi *et al.* (2006) attempted crosses between *C. cajan* and *C. platycarpus* to diversify the existing gene pool. Since the pollen of *C. platycarpus* failed to germinate on the stigma of *C. cajan*, the former was used as the female parent. However, hybrids of *C. platycarpus* with two cultivars of *C. cajan* var. Bahar and Pant A3 survived through embryo culture. Mallikarjuna *et al.* (2006) were also able successfully to cross *C. platycarpus* with cultivated pigeon pea by hormone-aided pollinations, rescuing the hybrid embryos *in vitro* and treating the hybrids with colchicines as these were 100% sterile. Nevertheless, *Cajanus scarabaeoides* has several undesirable characteristics (Upadhyaya, 2006), but is cross-compatible with cultivated pigeon pea and interspecific gene transfer is possible through conventional hybridization. *C. acutifolius* can also be successfully crossed with pigeon pea as a one-way cross (Mallikarjuna and Saxena, 2005).

Vigna species

A number of studies undertaken on crossability among different *Vigna* species have been reviewed by Dana and Karmakar (1990) and Singh (1990). Most reports indicate that *V. radiata* produced successful hybrids as seed parent with *V. mungo*, *V. umbellata* and *V. angularis*, although their reciprocal cross hybrids were not viable. However, by using sequential embryo rescue methods, the reciprocal hybrids between *V. mungo* and *V. radiata* could be successfully produced (Gosal and Bajaj, 1983a; Verma and Singh, 1986). *V. mungo* was also successfully crossed with *V. delzelliana* (Chavan *et al.*, 1966), *V. glabrescens* (Dana, 1968; Krishnan and De, 1968) and *V. trilobata* (Dana, 1966). In some cases, hybrid plants could be obtained only through embryo rescue technique, e.g. *V. mungo* × *V. umbellata* (Biswas and Dana, 1975; Chen *et al.*, 1983). Mung bean × rice bean crosses were generated to incorporate MYMV resistance and other desirable traits into mung bean (Verma and Brar, 1996). However, genotypic differences were observed in successful crosses. Furthermore, four amphidiploids of mung bean (ML 267 and K 851) × rice bean (RBL 33 and RBL 140) crosses were successfully produced and evaluated for different characters (Dar *et al.*, 1991). Singh *et al.* (2003) also produced successful hybrids between *V. radiata* and *V. umbellata*, and the hybrids possessed intermediate morphology with MYMV resistance. Similarly, Pal *et al.* (2005) were also successful in producing interspecific crosses between *V. mungo* and *V. umbellata*. Interspecific hybridizations between cultivated cowpea (*V. unguiculata* ssp. *unguiculata* and *V. u.* ssp. *biflora*) and wild forms of cowpea (*V. u.* var. *spontanea*, *V. u.* ssp. *alba*, *V. u.* ssp. *stenophylla*, *V. u.* ssp. *pawekiae* and *V. u.* ssp. *baoulensis*) were attempted by Kouadio *et al.* (2007), and the highest success rate was obtained in crosses between cultivated and annual inbred forms, although hybridization between cultivated and wild allogamous forms gave an intermediate rate of success. The success rate was lower when *V. u.* ssp. *baoulensis* was crossed with cultivated forms.

Crossability barriers

Crossability barriers developed during the process of speciation frustrate breeders' efforts in successful hybridization between species of different gene pools. Reproductive isolation, embryo or endosperm abortion, hybrid sterility and limited levels of genetic recombination are significant obstacles to the greater use of wild germplasm. These obstacles are in addition to those of undesirable linkages to non-agronomic traits once gene flow has been achieved. These barriers can prevent fertilization, reduce the number of hybrid seeds, retard the normal development of hybrid endosperm leading to embryo death or can cause hybrid sterility. In nature, there is selection bias towards strengthening these barriers to avoid extinction of the species by chaotic hybridization. In food legume crops several crossability barriers have been reported, the most common being cross incompatibility, embryo abortion at early growth stage, inviability of F_1 hybrids and sterility of F_1 hybrid and subsequent progenies (Kumar *et al.*, 2007). The pre-fertilization cross incompatibility between parent species arises when pollen grains do not germinate, the pollen tube does not reach the ovary or the male gametes do not fuse with the female (Chowdhury and Chowdhury, 1983; Shanmugam *et al.*, 1983).

Chickpea

Both pre-zygotic and post-zygotic barriers to interspecific hybridization in chickpea have been reported (Croser *et al.*, 2003). In the case of pre-zygotic barriers, Mercy and Kakar (1975) attempted to clarify incompatibility barrier(s) present among *Cicer* genus. They found the evidence of a low molecular weight inhibitory substance, possibly a protein present in the stylar and stigmatic tissues, inhibiting the germination and tube growth of the pollen. One of the reasons reported for the failure of interspecific crosses is the presence of localized sticky stigmatic secretion at the time pollen needs to be placed directly on the most receptive part of the stigma (Croser *et al.*, 2003). However, Ahmed *et al.* (1988) and Ahmed and Slinkard (2004) demonstrated a post-zygotic barrier(s) to crossing incompatibility rather than a pre-zygotic. They used seven of the eight wild annual *Cicer* species, belonging to the secondary and tertiary gene pools in reciprocal crosses with cultivated chickpea, and confirmed that the zygote was formed in all interspecific crosses. The embryos showed continued and retarded growth at different rates in various crosses but eventually aborted at an early pro-embryo stage in all crosses, except for *C. arietinum* × *C. echinospermum*. There is thus clear evidence confirming post-zygotic barriers in interspecific hybridization; however, further research is required to establish the exact causes of endosperm breakdown leading to embryo abortion, which might now be more feasible with the availability of new tools.

Lentil

Strong crossability barriers exist among *Lens* species that limit the utilization of the wild gene pool for lentil improvement. In some crosses, such as *L. culinaris* × *L. tomentosus, the* problem of chromosome pairing was observed between the participating genomes (Ladizinsky, 1979). In some *L. culinaris* × *L. culinaris* ssp. *orientalis* crosses, the hybrid embryo ceased growing but the endosperm shows no sign of disintegration (Ladizinsky, 1993). In contrast, Abbo and Ladizinsky (1991) observed that the endosperm was either abnormal or lacking in *L. culinaris* × *L. c.* ssp. *orientalis* crosses. Hybrids showed varying degrees of fertility, usually due to chromosome translocations and subsequent problems with chromosome pairing at meiosis, in *Lens culinaris* × *L. nigricans* (Goshen *et al.*, 1982; Ladizinsky *et al.*, 1984). Fertility is often very low, with little viable pollen produced in anthers, and varies depending on the accession in *L. culinaris* × *L. c.* ssp. *orientalis* crosses from 2% to 69% (Ladizinsky *et al.*, 1984). These problems can occur in the F_1 and also persist in later generations, causing partial or complete sterility. Albino seedlings can also occur in the F_1 generation and thus prevent hybridization success (Ladizinsky and Abbo, 1993). Another common problem is that hybrid embryos cease to grow about 7–14 days after pollination due to endosperm degeneration, and thus need rescuing in order to obtain viable hybrids (Ladizinsky *et al.*, 1985; Ahmad *et al.*, 1995). Hence, *L. culinaris* × *L. ervoides* or *L. culinaris* × *L. nigricans* crosses need embryo rescue techniques in order to develop mature hybrid plants (Cohen *et al.*, 1984; Abbo and Ladizinsky, 1991).

Vigna *crops*

In *Vigna* crops a slow rate of pollen growth, in addition to abnormalities in stigmatic and stylar regions, could be one of the major causes for low percentage of pod set in *V. radiata* × *V. umbellata* and *V. mungo* × *V. umbellata* crosses (Thiyagu *et al.*, 2008). However, the ploidy level and style length difference may not be major barriers in the case of *Vigna* species, as the long-styled female parent *V. radiata* could be successfully crossed with the short-styled male parent *V. trilobata*. Crosses between diploid × tetraploid (*V. radiata* × *V. glabrescens*) (Krishnan and De, 1968; Chen *et al.*, 1989) and tetraploid × diploid (*V. glabrescens* × *V. umbellata*) were also successful. In many studies crossability was genotype dependent (Rashid *et al.*, 1988). It was observed that strong pre-fertilization barriers were present in the cross between *V. radiata* and *V. umbellata*, and growth and lethality of interspecific hybrid seedlings were influenced by the genotypes of both parental species (Kumar *et al.*, 2007). Male sterility in F_1 plants and subsequent generations in interspecific crosses of *Vigna* could be attributed to meiotic irregularities: for example, unequal separation of tetrads and female sterility to degeneration of megaspores during megasporogenesis (Pandiyan *et al.*, 2008). One fertile pod with two hybrid seeds was obtained when *V. angularis* was used as a male parent; consequently, a partly fertile interspecific hybrid was obtained. Among the post-fertilization barriers, production of shrivelled hybrid seed with reduced or no germination (hybrid inviability), development of dwarf and non-vigorous plants and death of F_1 plants at critical stages of development (hybrid lethality) are the most common crossability barriers (Biswas and Dana, 1975). These barriers were of varying degrees in most of the interspecific crosses (Dana, 1964; Al-Yasiri and Coyne, 1966; Biswas and Dana, 1976; Chowdhury and Chowdhury, 1977; Machado *et al.*, 1982; Chen *et al.*, 1983; Gopinathan *et al.*, 1986). Sidhu (2003) produced interspecific hybrids of *V. radiata* with *V. mungo* and *V. trilobata*. Although the crosses between *V. radiata* and *V. trilobata* were successful, the seeds produced between *V. mungo* and *V. trilobata* had very poor germination and the germinated seedlings did not survive. Cytological analysis revealed irregular chromosome behaviour at diakinesis/metaphase I. In some of the interspecific crosses of *Vigna*, hybrid sterility has been observed to be of segregational type and was due mainly to interchange, inversion and possibly the duplication-deficiency type of structural heterozogosities in the F_1 individuals (Dana, 1964; Biswas and Dana, 1975; Karmakar and Dana, 1987).

Strategy to overcoming crossability barriers

With better understanding of the processes involved in pollen germination, pollen tube growth and fertilization, the opportunities to manipulate these processes toward the development of viable and fertile interspecific hybrids have improved considerably. Various measures to crossability barriers were reviewed by various workers (Sharma and Satija, 1996; Singh and Munoz, 1999), and are summarized in Table 6.3.

Embryo rescue protocols

The advent of *in vitro* techniques such as embryo and ovule culture, coupled with *in vivo* hormonal treatments, has greatly increased the scope of distant hybridization in food legume crops where post-fertilization barriers (zygotic abortion mechanisms) are common (Gupta and Sharma, 2005; Clarke *et al.*, 2006; Fratini and Ruiz, 2006; Mallikarjuna *et al.*, 2006). In wide crosses where few embryos are produced, the efficiency of recovering viable hybrid plants may also be enhanced by callus induction from the embryo and subsequent regeneration of plantlets. These procedures are also directed towards obtaining more efficient survival of embryos in situations where very immature embryos are to be cultured. Wide crosses that do not produce viable seeds could also be obtained through embryo callus production and subsequent regeneration and rooting of the callus. The possibility of increasing crossability also exists by predisposing crop embryos to alien endosperm and then using plants raised from those embryos to cross with the alien species. Hybridization of cultivated lentil with *L. ervoides* and *L. nigricans* results in pod development that is arrested within 10–16 days after pollination and finally yields shrivelled, non-viable seeds (Ladizinsky *et al.*, 1985), but can be rescued by a two-step *in vitro* method of embryo–ovule rescue to obtain successful distant hybrids (Cohen *et al.*, 1984). However, Ahmad *et al.* (1995) and Gupta and Sharma (2005) could not produce hybrids using the same technique. Fratini and Ruiz (2006) developed a protocol in which hybrid ovules were rescued 18 days after pollination. Fiala (2006) also obtained *L. culinaris* × *L. ervoides* hybrids using the Cohen *et al.* (1984) protocol. In addition, one viable *L. culinaris* ssp. *culinaris* × *L. lamottei* hybrid was also produced in this study. In chickpea, Clarke *et al.* (2006) suggested that the appropriate time to rescue *C. arietinum* × *C. bijugum* hybrids is the early globular stage of embryogenesis (2–7 days). In contrast, *C. arietinum* × *C. pinnatifidum* hybrids abort later (15–20 days) at the heart-shaped or torpedo stages, and are easier to rescue *in vitro*. Genotype also plays a significant role in the ability of immature selfed ovules to germinate *in vitro*. Thus the development of appropriate and efficient in vitro protocols for rescuing immature hybrid embryos is a necessity for these legume crops to secure alien gene resources available for their improvement.

Chromosome doubling

Colchicine-induced allopolyploids have been raised from most of the semi-fertile and completely seed-sterile F_1 hybrids in *Vigna* having high pollen fertility and seed set (Dana, 1966; Pande *et al.*, 1990), and some of these allopolyploids were used as a bridge species in wide crosses. In pigeon pea, Mallikarjuna and Moss (1995) attempted chromosome doubling of diploid F_1 hybrids of *Cajanus platycarpus* × *C. cajan* to obtain tetraploid F_1 hybrids. Selfing in successive generations had given rise to mature seeds with introgression of a resistance gene to phytophthora blight

Table 6.3. Methods of overcoming crossability barriers in food legumes.

Method	Cross combination	Reference(s)
Reciprocal crosses	*Vigna radiata* × *V. mungo*	Verma and Singh (1986); Ravi *et al.* (1987)
	Phaseolus vulgaris × *P. coccineus*	Rabakoarihanta *et al.* (1979)
	P. vulgaris × *P. lunatus*	Leonard *et al.* (1987)
Growth regulators	*V. radiata* × *V. umbellata*	Gupta *et al.* (2002)
	V. mungo × *V. umbellata*	Chen *et al.* (1978)
Embryo rescue	*V. radiata* × *V. unguiculata*	Tyagi and Chawla (1999)
	V. mungo × *V. radiata*	Gosal and Bajaj (1983a,b)
	V. radiata × *V. trilobata*	Sharma and Satija (1996)
	V. radiata × *V. radiata* var. *sublobata*	Sharma and Satija (1996)
	V. marina × *V. luteola*	Palmer *et al.* (2002)
	V. glabrescens × *V. radiata*	Chen *et al.* (1990)
	V. vexillata × *V. unguiculata*	Gomathinayagam *et al.* (1998)
	V. unguiculata × *V. mungo*	Shrivastava and Chawla (1993)
	Cajanus cajan × *C. cajanifolius*	Singh *et al.* (1993)
	C. cajan × *C. platycarpus*	Singh *et al.* (1993); Shahi *et al.* (2006)
	C. cajan × *Rhynchosia aurea*	Singh *et al.* (1993)
	C. platycarpus × *C. cajan*	Shahi *et al.* (2006); Mallikarjuna and Moss (1995); Mallikarjuna *et al.* (2006)
	C. cajan × *C. scarabaeoides*	–
	C. cajan × *C. acutifolius*	–
	P. vulgaris × *P. lunatus*	Kobuyama *et al.* (1991)
	P. vulgaris × *P. acutifolius*	Harlan and de Wet (1971)
	P. vulgaris × *P. acutifolius*	Cabral and Crocomo (1989); Andrade-Aguilar and Jackson (1988)
	Lens culinaris × *L. orientalis*	Ladizinsky *et al.* (1985); Ahmad *et al.* (1995)
	L. culinaris × *L. odemensis*	Goshen *et al.* (1982); Fratini and Ruiz (2006)
	L. culinaris × *L. tomentosus*	Ladizinsky and Abbo (1993)
	L. culinaris × *L. ervoides*	Cohen *et al.* (1984); Ahmad *et al.* (1995); Fiala (2006); Fratini and Ruiz (2006)
	L. culinaris × *L. lamottei*	Fiala (2006)
	L. culinaris × *L. nigricans*	Cohen *et al.* (1984); Fratini and Ruiz (2006)
	L. orientalis × *L. odemensis*	Ladizinsky *et al.* (1985); Goshen *et al.* (1982)
	L. orientalis × *L. tomentosus*	Ladizinsky and Abbo (1993); van Oss *et al.* (1997)
	Cicer arietinum × *C. reticulatum*	Ladizinsky and Adler (1976a, b)
	C. arietinum × *C. echinospermum*	Pundir and Mengesha (1995)
	C. arietinum × *C. pinnatifidum*	Mallikarjuna (1999)
	C. arietinum × *C.bijugum*	Clarke *et al.* (2006)
Chromosome doubling using colchicine	*V. radiata* × *V. mungo*	Pande *et al.* (1990)
	V. radiata × *V. trilobata*	Dana (1966)
Use of bridge species	*(V. mungo* × *V. radiata)* × *V. angularis*	Gupta *et al.* (2002)

disease from *C. platycarpus*. In cases where cultivated species cannot tolerate a large portion of alien chromosome, irradiation techniques have been successfully used. Among food legumes, irradiation techniques have been successful in recovering fertile plants in F_1 and subsequent generations in interspecific crosses in *Vigna*. Pandiyan *et al.* (2008) reported increased pod set in interspecific *V. radiata* × *V. umbellata* crosses developed from gamma ray-irradiated parental lines.

Reciprocal crossing

Reciprocal differences in wide crosses are also very common, and can be due to chromosomal imbalance in the endosperm, the role of the sperm nucleus in differential endosperm development or the alteration of endosperm development by pollen through the effects of antipodal cells, which are assumed to supply nutrients during early endosperm development (Beaudry, 1951). If disharmony between the genome of one species and cytoplasm of the other is a cause of a fertilization barrier, reciprocal crosses can be successful in recovery of hybrids. For example, while a *V. mungo* × *V. radiata* cross was unsuccessful, its reciprocal cross, *V. radiata* × *V. mungo*, produced successful hybrids (Verma and Singh, 1986; Ravi *et al.*, 1987). Interspecific hybridization between *V. nakashimae* and *V. angularis* was successful in both directions and viable seeds were produced, while *V. riukinensis* produced successful hybrids when used as male parent only with *V. angularis* and *V. umbellata* (Siriwardhane *et al.*, 1991). In general, using a female parent with higher chromosome number is more successful than the reciprocal method.

Use of bridge species

When useful genes are available in secondary and tertiary gene pools and direct hybridization between cultivated and wild species does not result in fertile hybrids, involvement of a third species as a bridge species has often been used for introgression of alien genes. For example, attempts at hybridizing *Lens culinaris* with *L. lamottei* and *L. nigricans* have not yielded fertile hybrids. This offers the possibility of transferring the genes for resistance to ascochyta blight and anthracnose to *L. culinaris* by using *L. ervoides* as a bridge species, with the embryo rescue technique as a means of broadening the resistance gene base in the cultivated species (Ye *et al.*, 2002; Tullu *et al.*, 2006). Transfer of bruchid resistance from wild *Vigna* species is difficult due to cross incompatibility. By using the bridge species *V. nakashimae*, the bruchid resistance of *V. umbellata* is transferred to adzuki bean (Tomooka *et al.*, 1992, 2000). However, bridge crosses will work only under the condition where species A hybridizes with species B but not with species C, and species B and C form a viable hybrid. Based on the close relationship reported in perennial *Cicer anatolicum*, *C. reticulatum* and *C. echinospermum*, the bridge-crossing approach deserves further attention.

Growth hormones

In wide crosses, if the hybrid seeds die when their embryos are too small to be cultured, post-pollination application of growth regulators such as gibberellic acid, naphthalene acetic acid, kinetin or 2, 4-D (dimethylamine), singly or as in combination, may be helpful in maintaining the developing seeds by facilitating division of the hybrid zygote and endosperm. Mallikarjuna (1999) observed that the only way to obtain interspecific hybrid in chickpea is by the application of growth regulators to pollinated pistils, to prevent initial pod abscission and to save the aborting hybrid embryos by embryo rescue techniques. Some interspecific crosses have been successful in *Phaseolus* (Stalker, 1980), *Cajanus* (Singh *et al.*, 1993) and *Cicer* (Shiela *et al.*, 1992) by application of growth regulators after pollination. This suggests that further breakthroughs in wide crossing may be possible through the exploitation of growth regulators followed by embryo rescue. *In vivo* hormonal treatments have also greatly helped in recovery of interspecific hybrids in *Vigna*. A true-breeding *Vigna mungo* × *V. radiata* derivative was reciprocally crossed with *V. angularis*, and the pollinated pistils were treated with GA3 after 24 and 78 h of pollination.

Backcrossing

In wide crosses, plants in initial generations are generally of inferior nature with poor expression of desired traits. This requires advancing the cross populations up to F_8/F_9 generations for recovery of desired types. In many cases the crosses are abandoned midway due various reasons, in spite of reports that useful recombinants could be recovered in later generations (F_{10}–F_{12}) of an interspecific cross (Singh and Dikshit, 2002). Therefore, delayed segregation often causes problems in identification and utilization of useful recombinants in interspecific crosses. This problem can be overcome through backcrossing of F_1 hybrids with cultivated species in early generations. Mallikarjuna *et al.* (2006) introgressed the *Cajanus platycarpus* genome into cultivated pigeon pea by backcrossing embryo-rescued F_1 hybrids with cultivated pigeon pea followed by *in vitro* culture of aborting embryos of BC_1 progeny. Similarly, one or more backcrosses to the recurrent parent are often required in common bean to restore fertility of hybrids when crossed with *Phaseolus acutifolius* and *P. parvifolius*. Using *P. acutifolius* as female parent of the initial F_1 cross, and/or first backcrossing *P. vulgaris* × *P. acutifolius* hybrid on to *P. acutifolius*, is often more difficult than using *P. vulgaris* as the female parent of the initial cross and backcrossing the interspecies hybrid on to *P. vulgaris* (Mejia-Jimenez *et al.*, 1994). The choice of parents (Parker and Michaels, 1986; Federici and Waines, 1988; Mejia-Jimenez *et al.*, 1994) and use of the congruity backcross (i.e. backcrossing alternately to each species) over recurrent backcrossing (Haghighi and Ascher, 1988; Mejia-Jimenez *et al.*, 1994) facilitate interspecific crosses of common and tepary beans, in addition to recovery of fertility and more hybrid progenies.

6.5 Successful Examples of Alien Gene Introgression in Food Legumes

Successful examples of alien gene introgressions in food legumes are limited to a few, for various reasons (Table 6.4). Genes for disease and insect resistance, male sterility and fertility restoration and yield attributes have been transferred into cultivated species of various legume crops. For example, successful introgression of drought tolerance from *Cicer reticulatum* (Hajjar and Hodgkin, 2007), yield genes from *C. reticulatum* (Singh *et al.*, 2005) and tolerance to ascochyta blight, cyst nematode and leaf miner have been documented. In lentil, some progress has been made in introgression of alien genes for resistance to ascochyta blight, anthracnose and cold in cultivated lentil (Hamdi *et al.*, 1996; Ye *et al.*, 2002; Fiala, 2006). Successful examples of using crossable wild species in pigeon pea breeding include development of a highly cleistogamous line (Saxena *et al.*, 1992); genetic dwarfs (Saxena and Sharma, 1995); phytophthora blight resistance (Reddy *et al.*, 1996; Mallikarjuna and Saxena,

Table 6.4. Successful examples of introgression in food legumes.

Crop	Wild relatives	Character	Reference(s)
Chickpea	*Cicer reticulatum*	Cyst nematode	Di Vito *et al.* (1996)
	C. reticulatum	Yield	Jaiswal and Singh (1989); Singh *et al.* (2005)
	C. reticulatum	Cold tolerance	Singh *et al.* (1995)
Lentil	*Lens orientalis*	Cold tolerance	Hamdi *et al.* (1996)
		Agronomic traits	Abbo *et al.* (1992); ICARDA (1995)
	Lens ervoides	Anthracnose resistance	Fiala (2006); Tullu *et al.* (2006)
Pigeon pea	*Cajanus sericeus*	Male sterility	Ariyanayagam *et al.* (1995)
	C. scarabaeoides	Male sterility	Tikka *et al.* (1997)
Mung bean	*Vigna mungo*	YMV resistance, plant type traits	Singh and Dikshit (2002)

2002); high-protein lines (Saxena *et al.*, 2002); cytoplasmic male sterile (CMS) lines (Saxena *et al.*, 2006); cyst nematode resistance (Saxena *et al.*, 1990); salinity resistance (Subba Rao *et al.*, 1990); and helicoverpa tolerance (Reed and Lateef, 1990). Some successful examples of alien gene introgression in food legume crops are described below.

Yield genes

The notion that wild relatives are a prospective source of genes for biotic stress tolerance only has been dismantled with convincing evidence of introgression of yield QTLs from the wild progenitors in some crops, including oats (Frey *et al.*, 1983), rice (Xiao *et al.*, 1996) and tomato (Tanksley *et al.*, 1996; Fulton *et al.*, 2000). The possibilities of introgression of desirable alien genes from wild to cultivated chickpea have been explored (Jaiswal and Singh, 1989; Verma *et al.*, 1990; Singh *et al.*, 2005). Studies have shown that, besides disease resistance and drought tolerance, wild *Cicer* species have genes for desirable yield components such as high number of fruiting branches and pods per plants (Singh *et al.*, 1994). In chickpea, alien genes for productivity have been transferred from *Cicer echinospermum*, *C. reticulatum* (Singh and Ocampo, 1997) and *C. reticulatum* (Singh *et al.*, 2005). Singh and Ocampo (1997) transferred some genes from *C. echinospermum* and *C. reticulatum* into cultivated chickpea and observed up to 39% increase in seed yield following the pedigree method. Singh *et al.* (2005) also reported introgression of yield genes and disease resistance genes from *C. reticulatum* to cultivated variety L550, with interspecific derivatives showing 6–17% yield advantage. A cross between Pusa 256 and *C. reticulatum* was made and their F_1 was again crossed with the wilt-resistant variety Pusa 362. Further selection concluded with the development of Pusa 1103, which is a high-yielding early variety with resistance to wilt, root rot and stunt virus and tolerance to drought and heat (Hajjar and Hodgkin, 2007; Kumar *et al.*, 2010). Singh and Dikshit (2002) introgressed yield genes in mung bean from urd bean with 15–60% yield advantage. The derivatives from mung bean × urd bean crosses exhibit many other desirable features such as lodging resistance, synchrony in podding and non-shattering (Reddy and Singh, 1990).

Disease resistance

In chickpea, introgression of resistance to cyst nematode from *Cicer reticulatum* has been reported, with promising lines under evaluation at ICARDA (Di Vito *et al.*, 1996; Ocampo *et al.*, 2000). Recently, resistance to anthracnose found in *Lens ervoides* germplasm has been exploited in Canada by introgressing resistance genes into cultivated backgrounds (Fiala, 2006; Tullu *et al.*, 2006). This successful use of *L. ervoides* holds promise as a source of genes for resistance to other diseases, and possibly for plant habit, biomass production and other important agronomic and marketing traits. Further exploitation of *L. ervoides* and the other wild *Lens* species is warranted. Derivatives from mung bean × urd bean crosses exhibit a higher level of MYMV resistance (Gill *et al.*, 1983). A few mung bean × ricebean and mung bean × *Vigna radiata* var. *sublobata* crosses having a high degree of resistance to MYMV were also recovered (Verma and Brar, 1996). Three mung bean cultivars, HUM 1, Pant Moong 4 and IPM99-125, and one urd bean cultivar, Mash 1008 (Sandhu *et al.*, 2005) have been developed from mung bean × urd bean crosses. These cultivars have improved plant types, in addition to higher MYMV resistance and synchronous maturity. In common bean, successful introgressions of alien genes imparting CBB (Freytag *et al.*, 1982; Park and Dhanvantari, 1987; Miklas *et al.*, 1994a, b), fusarium root rot (Wallace and Wilkinson, 1965) and white mould (Abawi *et al.*, 1978; Dickson *et al.*, 1982; Lyons *et al.*, 1987; Miklas *et al.*, 1998a) from *Phaseolus coccineus* have been reported. In contrast, resistance to halo blight from the common bean was incorporated into *P. coccineus* (Ockendon *et al.*, 1982). A high level of resistance to CBB was transferred from tepary to common bean (Coyne *et al.*, 1963; McElroy, 1985; Scott and Michaels, 1992; Singh and Munoz, 1999).

Insect pest resistance

The major production constraint of food legumes is susceptibility to bruchids (*Callosobruchus chinensis* L.) that eat seeds in storage. One accession of wild mung bean (*Vigna radiata* var. *sublobata*) exhibited complete resistance to adzuki bean weevils and cowpea weevils (Fujii *et al.*, 1989), which has successfully been used in breeding programmes (Tomooka *et al.*, 1992). *Vigna mungo* var. *silvestris*) is also reported to be immune to bruchids (Fujii *et al.*, 1989; Dongre *et al.*, 1996). Recently, rice bean (*V. umbellata*) has been identified as being of use because many accessions show complete resistance to bruchids and it is a cultivated species. Efforts are in progress at AVRDC to utilize *V. r.* var. *sublobata* for resistance to bruchids. Similarly, sources of resistance to leaf miner were used successfully in a chickpea breeding programme at ICARDA to develop promising breeding lines with leaf miner resistance for North Africa and West Asia (Singh and Weigand, 1996).

Male sterility and fertility restoration

Several wild relatives were used in hybridization with *Cajanus cajan*, and male sterile plants were isolated from the segregating populations. Ariyanayagam *et al.* (1995) crossed *C. sericeus* with *C. cajan* and isolated male sterile plants from the BC_3F_1 population. Tikka *et al.* (1997) developed a CMS line using *C. scarabaeoides* cytoplasm. Male sterile plants were also isolated from an interspecific cross of *C. cajanifolius* with *C. volubilis*. Saxena and Kumar (2003) developed a CMS sterile line, cms 88039A, using *C. scarabaeoides* (ICPW 89) and an early-maturing line of *C. cajan* (ICPL 88039). Similarly, two CMS lines, CORG 990052A and CORG 990047A, were developed by interspecific hybridization of *C. cajan* and *C. scarabaeoides* (Kalaimagal *et al.*, 2008). Experimental hybrids based on cytoplasmic male sterility derived from *C. scarabaeoides* and *C. sericeus* in pigeon pea are currently being evaluated in multi-environment trials. One recently released hybrid, GTH 1, has male sterile cytoplasm from *C. scarabaeoides*.

6.6 Future Strategy for Alien Gene Introgression

Advanced backcross-QTL strategy

Since the mid-1990s, convincing evidence at both morphological and molecular levels has accumulated for the utility of wild progenitors and related species as donors of productivity alleles. Productivity-enhancing genes/QTLs (quantitative-trait loci) have been introgressed in oats from *Avena sterilis* (Frey *et al.*, 1983), in tomato from *Lycopersicon pimpinellifolium* and *L. parviflorum* (Tanksley *et al.*, 1996; Fulton *et al.*, 2000), in rice from *Oryza rufipogon* (Xiao *et al.*, 1996) and in chickpea from *Cicer reticulatum* (Singh *et al.*, 2005). Novel breeding strategies such as AB-QTL (advanced backcross-QTL) have been deployed to exploit the worth of the progenitor and related species as this helps minimize the negative effect of linkage drag associated with alien gene introgression (Tanksley and Nelson, 1996). The related species of mung bean, such as *Vigna umbellata* and *V. angularis*, have comparatively higher productivity and their relationship with mung bean offers an opportunity for the introgression of some productivity alleles using AB-QTL strategy. Another related species, *V. mungo*, and the wild progenitor of mung bean, *V. radiata* var. *sublobata*, may also contribute some productivity alleles to the elite mung bean lines using the same approach.

Looking for genes based on molecular maps

The traditional approach in utilizing exotic germplasm is to screen the phenotype of entries from a gene bank for a clearly defined character and to use them in a crossing programme in order to introduce the genes into cultivated germplasm. Although effective for qualitative traits, only a small proportion of the genetic variation has been exploited for crop improvement as a result of this strategy (Tanksley and McCouch, 1997). Availability of genetic linkage maps based on molecular markers has opened up new opportunities in the utilization of hitherto unexploitable exotic germplasm. This requires a paradigm shift from selecting potential parents on the basis of phenotype to evaluating them directly for the presence of useful genes, through the integration of molecular tools. A gene-based approach to screening exotic germplasm has already been successfully used in rice and tomato for improving yield levels (Tanksley *et al.*, 1996; Xiao *et al.*, 1996). Recently, good progress has been made in generating genomic resources for food legume crops that will be very useful in genetic mapping and QTL analysis in these crops (Varshney *et al.*, 2009). With the use of DNA profiles, the genetic uniqueness of each accession in a gene bank can be determined and quantified. Molecular marker technology allows a targeted approach to the selection and introgression of valuable genes from a range of genetic resources while retaining the integrity of valuable genetic background through forward and background selection.

Recombination DNA technology

Transgenic approaches provide new options for broadening the genetic base in those cases where current options are lacking in their efficacy or existence. Plant genetic transformation techniques such as *Agrobacterium*-mediated transformation and direct gene delivery system (biolistics) allow the precise transfer of genes from any organism into either plant nuclear or chloroplast genomes. Many isolated plant genes are now being transferred between sexually incompatible plant species. In chickpea and pigeon pea, helicoverpa pod borer is a major insect pest for which no genetic solution exists. This requires development of transgenics having *Cry* genes from the soil bacterium *Bacillus thuringiensis* to combat the menace of helicoverpa pod borer. The recent report of a *Bt.* chickpea is an encouraging step towards improvement of food legumes for difficult traits such as pod borer resistance (Acharjee *et al.*, 2010). Similar is the case for botrytis grey mould in chickpea, where efforts are under way to construct a resistance against this disease. For gene introgression purposes, difficult species falling in tertiary and quaternary gene pools may turn out to be important sources of alien genes. For example, identification and cloning useful genes from *Phaseolus filiformis*, *P. angustissimus* and *P. lunana* and successful regeneration and transformation of common bean may facilitate gene introgression in the future.

Protoplast technology

Somatic hybridization using protoplast fusion has potential to overcome pre- and post-zygotic barriers to interspecific hybridization (Powers *et al.*, 1976; Davey *et al.*, 2005). It is possible to regenerate plants from a number of legume species, including *Pisum* (Ochatt *et al.*, 2000), *Trifolium* (Gresshoff, 1980), *Lotus* (Ahuja *et al.*, 1983) and *Melilotus* (Luo and Jia, 1998), and asymmetric protoplast fusion has been used for *Medicago* improvement (Tian and Rose, 1999; Yuko *et al.*, 2006). However, only a few reports of successful regeneration of plantlets are available in legumes (Li *et al.*, 1995). Initially, protoplast-derived tissues in rice bean were obtained although no shoot regeneration could be obtained. Shoot regeneration from protoplasts of *Vigna sublobata* has more recently been reported by Bhadra *et al.* (1994), with the maximum protoplast yield being obtained from 5-day-old seedlings. There are no reports at the time of writing of successful growth or regeneration of protoplasts from *Lens* species. Rozwadowski *et al.* (1990) cultured protoplasts from lentil epicotyl tissue, and around 6% of protoplasts developed into cell colonies.

Doubled haploids

Doubled haploid breeding is an important approach in many crop species, including wheat, barley, rice, maize and canola, to fix the hybrid immediately. Implementation of doubled haploids increases selection efficiency and allows new varieties to be bred up to 5 years faster than with conventional breeding methods alone. Haploids may be produced from either immature pollen cells, immature egg cells or following asymmetric chromosome elimination after interspecific hybridization. Several attempts have been made to develop anther and microspore culture systems for chickpea (Huda *et al.*, 2001; Vessal *et al.*, 2002; Croser *et al.*, 2006), common bean (Peters *et al.*, 1977; Munoz-Florez and Baudoin 1994a, b), field pea (Croser *et al.*, 2006) and pigeon pea (Pratap *et al.*, 2009). In chickpea, cultivars responsive to isolated microspore cultures have been identified and the induction of sporophytic development achieved in uninucleate microspores via the application of heat stress (32.5°C) pre-treatment to the buds (Croser *et al.*, 2006). Due to difficulty in derivation of green haploid regenerants these species have been defined as recalcitrant to androgenesis, although some progress has been made towards standardizing callus induction media and culture conditions in some of these crops. However, the production of a successful double haploid system in chickpea has been reported (Grewal *et al.*, 2009). A review of the literature on doubled haploid production in Fabaceae (Croser *et al.*, 2006) indicated that none of these approaches had been successful in producing haploid plants in food legumes, but the early stages of isolated microspore division have been observed.

6.7 Prospects

Productivity of food legume crops is affected by various biotic and abiotic stresses. There is thus an urgent need to widen the cultivated gene pool of these crops by incorporating genes for economically important traits from diverse sources. Wild species have proved to be an important reservoir of useful genes, and offer great potential for the incorporation of such genes into commercial cultivars. Many of the useful alien genes are expected to be different from those of the cultivated species, and are thus useful in broadening the base of resistance to various stresses. Recently, QTLs (oligogenic traits) that have been identified for yield traits in wild species of pulse crops may enhance agronomic and market values of cultivated varieties. The molecular marker technique can also be used for authentication of interspecific hybrids (Yamini *et al.*, 2001). There is a need to identify high-crossability genes in food legumes, as has been identified in wheat cultivars such as Chinese Spring (Luo *et al.*, 1993; Sharma, 1995). Identification of such genes in food legumes can bring non-crossable species within the ambit of alien gene transfer technology. There are major gaps in germplasm collections of wild species and their evaluation in food legumes that need to be filled, in order to progress further inroads in alien gene introgression. Continuing advances in wide-crossing techniques, such as embryo culture and development of novel crossing strategies, are creating greater accessibility in wild gene pools of many crops. The success rate of gene transfer in such wide crosses can be increased by knowledge of chromosome pairing mechanisms and their genetic control. The modern tools of molecular biology, such as monoclonal antibodies and *in situ* hybridization using various DNA probes, may soon make it possible to study the switching on and off of various genes in diverse tissues of the fertilized ovule, and control over the levels and movements of both exogenous and endogenous growth substances within the developing seed. It is likely that continuing advances in structural genomics and genetic engineering will result in new strategies for alien gene introgression.

References

Abawi, G.S., Provvidenti, R., Crosier, D.C. and Hunter. J.E. (1978) Inheritance of resistance to white mold disease in *Phaseolus coccineus*. *Journal of Heredity* 69, 200–202.

Abbo, S. and Ladizinsky, G. (1991) Anatomical aspects of hybrid embryo abortion in the genus *Lens* L. *Botany Gazette* 152(3), 316–320.

Abbo, S. and Ladizinsky, G. (1994) Genetical aspects of hybrid embryo abortion in the genus *Lens* L. *Heredity* 72, 193–200.

Abbo, S., Ladizinsky, G. and Weeden, N.F. (1992) Genetic analysis and linkage studies of seed weight in lentil. *Euphytica* 58, 259–266.

Acharjee, S., Sarmah, B.K., Ananda Kumar, P., Olsen, K., Mahon, R., Moar, W.J. *et al.* (2010) Transgenic chickpeas (*Cicer arietinum* L.) expressing a sequence-modified *cry2Aa* gene. *Plant Science* 178, 333–339.

Ahmad, F. and Slinkard, A.E. (2004) The extent of embryo and endosperm growth following interspecific hybridization between *Cicer arietinum* L. and related annual wild species. *Genetic Resources and Crop Evolution* 51, 765–772.

Ahmad, F., Slinkard, A.E. and Scoles, G.J. (1988) Investigation into the barrier(s) to interspecific hybridization between *Cicer arietinum* L. and eight other annual *Cicer* species. *Plant Breeding* 100, 193–198.

Ahmad, M., Fautrier, A.G., McNeil, D.L., Burritt, D.J. and Hill, G.D. (1995) Attempts to overcome post-fertilization barrier in interspecific crosses of the genus *Lens*. *Plant Breeding* 114, 558–560.

Ahmad, M., Fautrier, A.G., McNeil, D.L., Hill, G.D. and Burritt, D.J. (1997b) *In vitro* propagation of *Lens* species and their F_1 interspecific hybrids. *Plant Cell Tissue Organ Culture* 47, 169–176.

Ahmad, M., McNeil, D.L. and Sedcole, J.R. (1997a) Phylogenetic relationships in *Lens* species and their interspecific hybrids as measured by morphological characters. *Euphytica* 94, 101–111.

Ahmad, F., Gaur, P.M. and Croser, J.S. (2005) Chickpea (*Cicer arietinum* L.). In: Singh, R.J. and Jauhar P.P. (eds) *Genetic Resources, Chromosome Engineering and Crop Improvement, Volume 1, Grain Legumes*. CRC Press, Boca Raton, Florida, pp. 187–217.

Ahuja, P.S., Hadiuzzaman, S., Davey, M.R. and Cocking, E.C. (1983) Prolific plant regeneration from protoplast derived tissues of *Lotus corniculatus* L. (birdsfoot trefoil). *Plant Cell Reports* 2, 101–104.

Akinola, J.O., Whiteman, P.C. and Wallis, E.S. (1975) *The Agronomy of Pigeon pea (Cajanus cajan). Review Series no.1/1975*. CAB International, Wallingford, UK, pp. 57.

Aletor, V.A., Abd-El-Moneim, A.M. and Goodchild, A.V. (1994) Evaluation of the seeds of selected lines of three *Lathyrus* spp. for b-N-oxalyl amino-L-alanine (BOAA), tannins, trypsin inhibitor activity and certain *in vitro* characteristics. *Journal of the Science of Food and Agriculture* 65, 143–151.

Ali, M. and Kumar, S. (2009) Major technological advances in pulses, Indian scenario. In: Ali, M. and Kumar, S. (eds) *Milestones in Food Legumes Research*. Indian Institute of Pulses Research, Kanpur, India, pp. 1–21.

Al-Yasiri, S.A. and Coyne, D.P. (1966) Interspecific hybridization in the genus *Phaseolus*. *Crop Science* 6, 59–60.

Andrade-Aguilar, J.A. and M.T. Jackson (1988) Attempts at interspecific hybridization between *Phaseolus vulgaris* L. and *P. acutifolius* A. Gray using embryo rescue. *Plant Breeding* 101, 173–180.

Ariyanayagam, R.P., Rao, A.N. and Zaveri, P.P. (1993) Gene-cytoplasmic male sterility in pigeon pea. *International Pigeonpea Newsletter* 18, 7–11.

Ariyanayagam, R.P., Rao, A.N. and Zaveri, P.P. (1995) Cytoplasmic genic male sterility in interspecific matings of *Cajanus*. *Crop Science* 35, 981–985.

Azzam, H.A. (1957) Inheritance of resistance to *Fusarium* root rot in *Phaseolus vulgaris* L. and *Phaseolus coccineus* L. (Diss. Abstract 18, 32–33). Oregon State University, Corvallis, Oregon.

Baggett, J.R. (1956) The inheritance of resistance to strains of bean yellow mosaic virus in the interspecific cross *Phaseolus vulgaris* × *P. coccineus*. *Plant Disease Reporter* 40, 702–707.

Balasubramanian, P., Vanderberg, A., Hucl, P. and Gusta, L. (2004) Resistance of *Phaseolus* species to ice crystallization at subzero temperature. *Plant Physiology* 120, 451–457.

Bannerot, H. (1979) Cold tolerance in beans. *Annuual Report of Bean Improvement Cooperative* 22, 81–84.

Baum, M., Laguda, E.S. and Appels, R. (1992) Wide crosses in cereals. *Annual Review of Plant Physiology and Molecular Biology* 43, 117–143.

Bayaa, B., Erskine, W. and Hamdi, A. (1994) Response of wild lentil to *Ascochyta fabae* f. sp. *lentis* from Syria. *Genetic Resources and Crop Evolution* 41, 61–65.

Bayaa, B., Erskine, W. and Hamdi, A. (1995) Evaluation of a wild lentil collection for resistance to vascular wilt. *Genetic Resources and Crop Evolution* 42, 231–235.

Bayuelo-Jimenez, J.S., Debouck, D.G. and Lynch, J. (2002) Salinity tolerance in *Phaseolus* species during early vegetative growth. *Crop Science* 42, 2184–2192.

Beaudry, J.R. (1951) Seed development following the mating of *Elymus virginicus* L. and *Agropyron repens* L. Beauv. *Genetics* 36, 109–126.

Beebe, S. and Pastor-Corrales, M.A. (1991) Breeding for disease reisistance. In: van Schoonhoren, A. and Voyset, O. (eds) *Common Beans: Research for Crop Improvement*. CAB International and CIAT, Wallingford, UK, pp. 561–617.

Bhadra, S.K., Hammatt, N., Paner, J.B. and Davey, M.R. (1994) A reproducible procedure for plant regeneration from seedling hypocotyls protoplast of *Vigna sublobata* L. *Plant Cell Reports* 14, 175–179.

Biswas, M.R. and Dana, S. (1975) Black gram × rice bean cross. *Cytologia* 40, 787–795.

Biswas, M.R. and Dana, S. (1976) *Phaseolus aconitifolius* × *P. trilobatus* cross. *Indian Journal of Genetics and Plant Breeding* 36, 125–131.

Busogoro, J.P., Jijakli, M.H. and Lepoivre, P. (1999) Identification of a novel source of resistance to angular leaf spot disease of common bean within the secondary gene pool. *Plant Breeding* 118, 417–423.

Cabral, J.B. and Crocomo, O.J. (1989) Interspecific hybridization of *Phaseolus vulgaris*, *P. acutifolius* and *P. lunatus* using in vitro technique. *Turrialba* 39, 243–246.

Chandel, K.P.S. and Lester, R.N. (1991) Origin and evolution of Asiatic *Vigna* species. In: Sharma, B. and Mehra, R.B. (eds) *Golden Jubilee Celebration Symposium on Grain Legumes*, 9–11 February 1991, IARI, New Delhi, India, pp. 25–45.

Chavan, V.M., Patil, G.D. and Bhapkar, D.G. (1966) Improvement of cultivated *Phaseolus* species – need for interspecific hybridization. *Indian Journal of Genetics and Plant Breeding* 26, 152–154.

Chen, H.K., Mok, M.C., Shanmugasundaram, S. and Mok, D.W.S. (1989) Interspecific hybridization between *Vigna radiata* (L.) Wilczek and *V. glabrescens*. *Theoretical and Applied Genetics* 78, 641–647.

Chen, H.K., Mok, M.C. and Mok, D.W.S. (1990) Somatic embryogenesis and shoot organogenesis from interspecific hybrid embryos of *Vigna glabrescens* and *V. radiata*. *Plant Cell Reports* 9, 77–79.

Chen, N.C., Parrot, J.F., Jacobs, J., Baker, L.R. and Carlson, P.S. (1978) Interspecific hybridization of food legumes by unconventional methods of plant breeding. In: *International Mungbean Symposium*, 1977. Asian Vegetable Research and Development Centre, Shanhua, Taiwan, pp. 247–252.

Chen, N.C., Baker, R.L. and Honma, S. (1983) Interspecific crossability among four species of *Vigna* food legumes. *Euphytica* 32, 925–937.

Chowdhury, R.K. and Chowdhury, J.B. (1977) Intergeneric hybridization between *Vigna mungo* and *Phaseolus calcaratus*. *Indian Journal of Agricultural Sciences* 47, 117–121.

Chowdhury, R.K. and Chowdhury, J.B. (1983) Compatibility between *Vigna radiata* (L.) Wilczek and *Vigna umbellata* (Thumb) Ohwi and Ohashi. *Genetica Agraria* 37, 257–266.

CIAT (1986) *Bean Programme Annual Report*. CIAT, Cali. Colombia.

CIAT. (1995) *Bean Programme Annual Report*. CIAT, Cali. Colombia.

CIAT. (1996) *Bean Programme Annual Report*. CIAT, Cali. Colombia

Clarke, H.J., Wilson, J.M., Kuo, I., Lulsdorf, M., Mallikarjuna, N. and Siddique, K.H.M. (2006) Embryo rescue and plant regeneration in vitro of selfed chickpea (*Cicer arietinum* L.) and its wild annual relatives. *Plant Cell Tissue Organ Culture* 85, 197–204.

Clement, S.L., Sharaf El-Din N., Weigand, S. and Lateef, S.S. (1994) Research achievements in plant resistance to insect pests of cool season food legumes. *Euphytica* 73, 41–50.

Clement, S.L., Cristofaro, M., Cowgill, S.E. and Weigand, S. (1999) Germplasm resources, insect resistance, and grain legume improvement. In: Clement, S.L. and Quisenberry, S.S. (eds) *Global Plant Genetic Resources for Insect Resistant Crops*. CRC Press, Boca Raton, Florida, pp. 131–148.

Cohen, D., Ladizinsky, G., Ziv, M. and Muehlbauer, F.J. (1984) Rescue of interspecific *Lens* hybrids by means of embryo culture. *Plant Cell Tissue Organ Culture* 3, 343–347.

Collard, B.C.Y., Ades, P.K., Pang, E.C.K., Brouwer, J.B. and Taylor, P.W.J. (2001) Prospecting for sources of resistance to ascochyta blight in wild *Cicer* species. *Australian Journal of Plant Pathology* 30, 271–276.

Coyne, D.P., Schuster, M.L. and Al-Yasiri, S. (1963) Reaction studies of bean species and varieties to common blight and bacterial wilt. *Plant Disease Reporter* 47, 534–537.

Croser, J.S., Ahmad, F., Clarke, H.J. and Siddique, K.H.M. (2003) Utilization of wild *Cicer* in chickpea improvement – progress, constraints and prospects. *Australian Journal of Agricultural Research* 54, 429–444.

Croser, J.S., Lulsdorf, M., Davies, P.A., Clarke, H., Bayliss, K., Mallikarjuna, N. *et al.* (2006) Towards doubled haploid production on the fabaceae, progress and constraints. *Critical Reviews in Plant Science* 25, 139–157.

Dalvi, V.A., Saxena, K.B. and Madrap, I.A. (2008) Fertility restoration in cytoplasmic-nuclear male-sterile line derived from 3 wild relatives from pigeon pea. *Journal of Heredity* 99, 671–673.

Dana, S. (1964) Interspecific cross between tetraploid *Phaseolus* species and *P. ricciardianus*. *Nucleus* 7, 1–10.

Dana, S. (1966) Cross between *Phaseolus aureus* and *P. mungo*. *Genetica* 37, 259–274.

Dana, S. (1968) Hybrid between *Phaseolus mungo* and tetraploid *Phaseolus species*. *Japan Journal of Genetics* 43, 153–155.

Dana, S. and Karmakar, P.G. (1990) Species relation in *Vigna* subgenus *Ceratotropis* and its implications in breeding. *Plant Breeding Reviews* 8, 19–42.

Dar, G.M., Verma, M.M., Gosal, S.S. and Brar, J.S. (1991) Characterization of some interspecific hybrids and amphiploids in *Vigna*. In: Sharma, B. and Mehra, R.B. (eds) *Golden Jubilee Celebration Symposium on Grain Legumes*. Indian Society of Genetics and Plant Breeding, New Delhi, India, pp. 73–78.

Davey, M.R., Anthony, P., Power, J.B. and Lowe, K.C. (2005) Plant protoplasts, status and biotechnological perspectives. *Biotechnology Advances* 23, 131–171.

Davies, A.J.S. (1957) Successful crossing in the genus *Lathyrus* through stylar amputation. *Nature* 180, 612.

Davies, A.J.S. (1958) A cytogenetic study in the genus *Lathyrus*. PhD thesis, University of Manchester, UK.

Debouck, D.G. (1999) Diversity in *Phaseolus* species in relation to the common bean. In: Singh, S.P. (ed.) *Common Bean Improvement in the Twenty-first Century*. Kluwer Academic Publishers, Dordrecht, The Netherlands, pp. 25–52.

Debouck, D.G. (2000) Biodiversity, ecology and genetic resources of *Phaseolus* beans – seven answered and unanswered questions. In: Oono, K. (ed.) *Wild Legumes*. National Institute of Biological Resources, Tsukuba, Japan, pp. 95–123.

Debouck, D.G. and Smartt, J. (1995) Beans, *Phaseolus* spp. (Leguminosae–Papilionoideae). In: Smartt, J. and Simmonds, N.W. (eds) *Evolution of Crop Plants, 2nd edn*, Longman, London, pp. 287–294.

Dhanuj, M.S. and Gill, B.S. (1985) Intergeneric hybridization between *Cajanus cajan* and *Atylosia platycarpa*. *Annals of Biology* 1, 229–231.

Dickson, M.H, Hunter, J.E., Boettger, M.A and Cigna, J.A. (1982) Selection for resistance in *Phaseolus vulgaris* L. to white mold disease caused by *Sclerotinia sclerotiorum* (Lib.) de Bary. *Journal of the American Society of Horticultural Science* 107, 231–234.

Dita, M.A., Rispail, N., Prats, E., Rubiales, D. and Singh, K.B. (2006) Biotechnology approaches to overcome biotic and abiotic stress constraints in legumes. *Euphytica* 147, 1–24.

Di Vito, M., Singh, K.B., Greco, N. and Saxena, M.C. (1996) Sources of resistance to cyst nematode in cultivated and wild *Cicer* species. *Genetic Resources and Crop Evolution* 43, 103–107.

Dobie, P., Dendy, J., Sherman, A., Padgham, J., Wood, J.A. and Gatehouse, A.M.R. (1990) New sources of resistance to *Acanthoscelides obtectus* (Say) and *Zabrotes subfasciatus* Boheman (Coleopter: Bruchidae) in mature seeds of five species of *Phaseolus*. *Journal of Stored Products Research* 26, 177–186.

Dongre, T.K., Pawar, S.E., Thakare, R.G. and Harwalkar, M.R. (1996) Identification of resistant source to cowpea weevil (*Callosobruchus maculatus* (F.)) in *Vigna* sp. and inheritance of their resistance in blackgram (*Vigna mungo* var. *mungo*). *Journal of Stored Products Research* 32, 201–204.

Doyle, J.J. (1988) 5S ribosomal gene variation in the soybean and its progenitor. *Theoretical and Applied Genetics* 75, 621–624.

El-Bouhssini, M., Sarker, A., Erskine, W. and Joubi, A. (2008) First sources of resistance to Sitona weevil (*Sitona crinitus* Herbst) in wild Lens species. *Genetic Resources and Crop Evolution* 55, 1–4.

Erskine, W. and Saxena, M.C. (1993) Problems and prospects of stress resistance breeding in lentil. In: Singh, K.B. and Saxena, M.C. (eds) *Breeding for Stress Tolerance in Cool Season Food Legumes*. ICARDA/ Wiley, Chichester, UK, pp. 51–62.

FAO (1996) *State of the World's Plant Genetic Resources for Food and Agriculture*. Food and Agriculture Organization, Rome.

FAO (2010) *Production Statistics*. Food and Agriculture Organization, Rome.

Federici, C.T. and Waines, J.G. (1988) Interspecific hybrid compatibility of selected *Phaseolus vulgaris* L. lines with *P. acutifolius* A. Gray, P. *lunatus* L., and P. *filiformis* Bentham. *Annual Reporter Bean Improvement Cooperation* 31, 201–202.

Federici, C.T., Ehdaie, B. and Waines, J.C. (1990) Domesticated and wild tepary bean, Field performance with and without drought-stress. *Agronomy Journal* 82, 896–900.

Ferguson, M.E. and Erskine, W. (2001) Lentils (*Lens* L.). In: Maxted, N. and Bennett, S.J. (eds) *Plant Genetic Resources of Legumes in the Mediterranean*. Kluwer Academic Publishers, Dordrecht, The Netherlands, pp. 125–131.

Ferguson, M.E. and Robertson, L.D. (1999) Morphological and phenological variation in the wild relatives of lentil. *Genetic Resources and Crop Evolution* 46, 3–12.

Fernández-Aparicio, M., Sillero, J.C. and Rubiales, D. (2009) Resistance to broomrape in wild lentils (*Lens* spp.). *Plant Breeding* 128, 266–270.

Fiala, J.V. (2006) Transferring resistance to *Colletotrichum truncatum* from wild lentil species to cultivated lentil species (*Lens culinaris* subsp *culinaris*). MSc thesis, University of Saskatchewan, Saskatoon, Canada, pp. 131.

Fratini, R. and Ruiz, M.L. (2006) Interspecific hybridization in the genus *Lens* applying *in vitro* embryo rescue. *Euphytica* 150, 271–280.

Fratini, R., Ruiz, M.L. and Perez de la Vega, M. (2004) Intra-specific and inter-sub-specific crossing in lentil (*Lens culinaris* Medik.) *Canadian Journal of Plant Science* 84, 981–986.

Fratini, R., Garcia, P. and Ruiz, M.L. (2006) Pollen and pistil morphology, *in vitro* pollen grain germination and crossing success of *Lens* cultivars and species. *Plant Breeding* 125, 501–505.

Frey, K.J., Cox, T.S., Rodgers, D.M. and Bramel-Cox, P. (1983) Increasing cereal yields with genes from wild and weedy species. In: *Proceedings of the 15th International Genetics Congress*. Oxford and IBH Publishing, New Delhi, India, pp. 51–68.

Freytag, G.F., Bassett, M.J. and Zapata, M. (1982) Registration of XR-235-1-1 bean germplasm. *Crop Science* 22, 1268–1269.

Fujii, K., Ishimoto, M. and Kitamura, K. (1989) Pattern of resistance to bean weevil (Bruchidae) in Vigna radiate–Mungo-sublobata complex inform the breeding of new resistant varieties. *Applied Entomology and Zoology* 24, 126–132.

Fulton, T.M., Grandillo, S., Beck-Bunn, T., Fridman, E., Framton, A., Lopez, J. *et al.* (2000) Advanced back-cross QTL analysis of *Lycopersicon esculatum* × *L. parviflorum* cross. *Theoretical and Applied Genetics* 100, 1025–1042.

Gill, A.S., Verma, M.M., Dhaliwal, H.S. and Sandhu, T.S. (1983) Interspecific transfer of resistance to mung bean yellow mosaic virus from *Vigna mungo* to *Vigna radiata*. *Current Science* 52, 31–33.

Gomathinayagam, P., Ram, S.G., Rathnaswamy, R. and Ramaswamy, N.M. (1998) Interspecific hybridization between *Vigna unguiculata* and *V. vexillata* through *in vitro* embryo culture. *Euphytica* 102, 203–209.

Gopinathan, M.C., Babu, C.R. and Shivanna, K.R. (1986) Interspecific hybridization between rice bean (*Vigna umbellata*) and its wild relative (*V. minima*), fertility sterility relationships. *Euphytica* 35, 1017–1022.

Gosal, S.S. and Bajaj, Y.P.S. (1983a) *In vitro* hybridization in an incompatible cross – black gram × green gram. *Current Science* 52, 556–557.

Gosal, S.S. and Bajaj, Y.P.S. (1983b) Interspecific hybridization between *Vigna mungo* and *Vigna radiata* through embryo culture. *Euphytica* 32, 129–137.

Goshen, D., Ladizinsky, G. and Muehlbauer, F.J. (1982) Restoration of meiotic regularity and fertility among derivatives of *Lens culinaris* × *L. nigricans* hybrids. *Euphytica* 31, 795–799.

Greco, N. and Di Vito, M. (1993) Selection for nematode resistance in cool season food legumes. In: Singh, K.B. and Saxena, M.C. (eds) *Breeding for Stress Tolerance in Cool Season Food Legumes*. John Wiley & Sons/ICARDA, Chichester, UK, pp. 157–166.

Gresshoff, P.M. (1980) *In vitro* culture of white clover, callus, suspension, protoplast culture and plant regeneration. *Botany Gazette* 141, 157–164.

Grewal, R.K., Lulsdorf, M., Croser, J., Ochatt, S., Vandenberg, A. and Warkentin, T.D. (2009) Doubled-haploid production in chickpea (*Cicer arietinum* L.), role of stress treatments. *Plant Cell Reports* 28, 1289–1299.

Gupta, D. and Sharma, S.K. (2005) Embryo-ovule rescue technique for overcoming post-fertilization barriers in interspecific crosses of *Lens*. *Journal of Lentil Research* 2, 27–30.

Gupta, D. and Sharma, S.K. (2006) Evaluation of wild *Lens* taxa for agro-morphological traits, fungal diseases and moisture stress in northwestern Indian hills. *Genetic Resources and Crop Evolution* 53, 1233–1241.

Gupta, P.V., Plaha, P. and Rathore, P.K. (2002) Partially fertile interspecific hybrids between a black gram × green gram derivative and adzuki bean. *Plant Breeding* 121, 182–183.

Haghighi, K.R. and Ascher, P.D. (1988) Fertile intermediate hybrid between *Phaseolus vulgaris* L. and *P. acutifolius* from congruity backcrossing. *Sexual Plant Reproduction* 1, 51–58.

Hajjar, R. and Hodgkin, T. (2007) The use of wild relatives in crop improvement: A survey of developments over the last 20 years. *Euphytica* 156, 1–13.

Hamdi, A. and Erskine, W. (1996) Reaction of wild species of the genus *Lens* to drought. *Euphytica* 91, 173–179.

Hamdi, A., Küsmenoglu, I. and Erskine, W. (1996) Sources of winter hardiness in wild lentil. *Genetic Resources and Crop Evolution* 43, 63–67.

Hammett, K.R.W., Murray, B.G., Markham, K.R. and Hallett, I.C. (1994) Interspecific hybridization between *Lathyrus odoratus* and *L. belinensis*. *International Journal of Plant Science* 155, 763–771.

Hammett, K.R.W., Murray, B.G., Markham, K.R., Hallett, I.C. and Osterloh, I. (1996) New interspecific hybrids in *Lathyrus* (Leguminosae), *Lathyrus annuus* × *L. hierosolymitanus*. *Botanical Journal of the Linnean Society* 122, 89–101.

Hanbury, C.D., Siddique, K.H.M., Galwey, N.W. and Cocks, P.S. (1999) Genotype-environment interaction for seed yield and ODAP concentration of *Lathyrus sativus* L. and *L. cicera* L. in Mediterranean-type environments. *Euphytica* 110, 45–60.

Hannon, R., Açikgöz, N. and Robertson, L.D. (2001) Chickpeas (*Cicer* L.) In: Maxted, N. and Bennett, S.J. (eds) *Plant Genetic Resources of Legumes in the Mediterranean*. Kluwer Academic, Dordrecht, The Netherlands, pp. 115–124.

Harlan, J.R. and De Wet, M.J. (1971) Towards a rational classification of crop plants. *Taxonomy* 20, 509–517.

Hassan, A.A., Wilkinson, R.E. and Wallace, D.H. (1971) Genetics and heritability of resistance to *Fusarium solani* f. sp. *phaseoli* in beans. *Journal of the American Society for Horticultural Science* 96, 623–627.

Hawkes, J.G. (1977) The importance of wild germplasm in plant breeding. *Euphytica* 26, 615–621.

Hubbeling, N. (1957) New aspects of breeding for disease resistance in beans (*Phaseolus vulgaris* L.). *Euphytica* 6, 111–141.

Huda, S., Islam, R., Bari, M.A. and Asaduzzaman, M. (2001) Anther culture of chickpea. *International Chickpea and Pigeonpea Newsletter* 8, 24–26.

Hunter, J.E., Dickson, M.H., Boettger, M.A. and Cigna, J.A. (1982) Evaluation of plant introduction of *Phaseolus* spp. for resistance to white mold. *Plant Disease* 66, 320–322.

ICARDA (1995) *Legume Program Annual Report*. International Center for Agricultural Research in the Dry Areas, Aleppo, Syria.

Infantino, A., Porta-Puglia, A. and Singh, K.B. (1996) Screening of wild *Cicer* species for resistance to *Fusarium* wilt. *Plant Disease* 80, 42–44.

Jackson, M.T. and Yunus, A.G. (1984) Variation in the grass pea (*L. sativus* L.) and wild species. *Euphytica* 33, 549–559.

Jaiswal, H.K. and Singh, B.D. (1989) Analysis of gene effects for yield traits in crosses between *C. arietinum* and *C. reticulatum*. *Indian Journal of Genetics and Plant Breeding* 49, 9–17.

Kaiser, W.J., Alcala-Jimenez, A.R., Hervas-Vargas, A., Trapero-Casas, J.L. and Jimenez-Diaz, R.M. (1994) Screening of wild *Cicer* species for resistance to races 0 to 5 of *Fusarium oxysporum* f. sp. *ciceris*. *Plant Disease* 78, 962–967.

Kalaimagal, T., Muthaiah, A., Rajarathinam, S., Malini, S., Nadarajan, N. and Pechiammal, I. (2008) Development of new cytoplasmic genetic male-sterile line in pigeon pea from crosses between *Cajanus cajan* (L.) Millsp. and *C. scarabaeoides* (L.) Thouars. *Journal of Applied Genetics* 49(3), 221–227.

Karmakar, P.G. and Dana, S. (1987) Cytogenetic identification of a *Vigna sublobata* collection. *Nucleus* 30, 47–50.

Kaur, S., Chhabra, K.S. and Arora, B.S. (1999) Incidence of *Helicoverpa armigera* (Hubner) on wild and cultivated species of chickpea. *International Chickpea and Pigeon pea Newsletter* 6, 18–19.

Kearney, J.P. (1993) Wild *Lathyrus* species as genetic resources for improvement of grasspea (*L. Sativus*). PhD thesis, University of Southampton, UK.

Kearney, J.P. and Smartt, J. (1995) The grass pea *Lathyrus sativus* (Leguminosae – Papilionoideae). In: Smartt, J. and Simmonds, N.W. (eds) *Evolution of Crop Plants*, Longman, London, pp. 266–270.

Khawaja, H.I.T. (1985) Cytogenetic studies in the genus *Lathyrus*. PhD thesis, University of London, UK.

Khawaja, H.I.T. (1988) A new interspecific *Lathyrus* hybrid to introduce the yellow flower character into the sweet pea. *Euphytica* 37, 69–75.

Knights, E.J., Southwell, R.J., Schwinghamer, M.W. and Harden, S. (2008) Resistance to *Phytophthora medicaginis* Hansen and Maxwell in wild *Cicer* species and its use in breeding root rot resistant chickpea (*Cicer arietinum* L.) *Australian Journal of Agricultural Research* 59, 383–387.

Knott, D.R. and Dvorak, J. (1976) Alien germplasm as a source of resistance to diseases. *Annual Review of Phytopathology* 14, 211–235.

Kobuyama, T., Shintaku, Y. and Takeda, G. (1991) Hybrid plant of *Phaseolus vulgaris* L. and *P. lunatus* L. obtained by means of embryo rescue and confirmed by restriction endonuclease analysis of rDNA. *Euphytica* 62, 171–180.

Kouadio, D., Echikh, N., Toussaint, A., Pasquet, R.S. and Baudoin, J.P. (2007) Organisation of the gene pool of *Vigna unguiculata* (L.) Walp., crosses between the wild and cultivated forms of cowpea. *Biotechnologie, Agronomie, Societé et Environment* 11, 47–57.

Krishnan, R. and De, D.N. (1968) Cytogenetical studies in *Phaseolus* II. *Phaseolus mungo* × tetraploid *phaseolus* species and the amphidiploid. *Indian Journal of Genetics and Plant Breeding* 28, 23–30.

Kumar, A.S., Reddy, T.P. and Reddy, G.M. (1985) Genetic analysis of certain *in vitro* and *in vivo* parameters in pigeon pea. *Theoretical and Applied Genetics* 70, 151–156.

Kumar, J., Yadav, S.S., Malhotra, R.S., Bharadwaj, C., Imtiaz, M. and Hegde, V. (2010) Chickpea improvement using wild *Cicer* species. In: *5th International Food Legumes Research Conference (IFLRC V) & 7th European Conference on Grain Legumes (AEP VII)*, 26–30 April 2010, Antalya, Turkey.

Kumar, N.P., Pandiyan, M. and Veerabadhiran, P. (2007) Prefertilization barriers in *Vigna radiata* × *Vigna umbillata. Plant Archives* 7, 377–380.

Kumar, S. and Dua, R.P. (2006) Chickpea. In: Dhillon, B.S., Saxena, S., Agrawal, A. and Tyagi, R.K. (eds) *Plant Genetic Resource, Foodgrain Crops*. Narosa Publishing House, New Delhi, India, pp. 302–313.

Kumar, S., Singh, B.B. and Singh, D.P. (2004) Genetics and cytogenetics of mungbean. In: Ali, M. and Kumar, S. (eds) *Advances in Mungbean and Urdbean*. Indian Institute of Pulses Research, Kanpur, India, pp. 40–68.

Kumar, S., Bejiga, G., Ahmed, S., Nakkoul, H. and Sarker, A. (2010) Genetic improvement of grass pea for low neurotoxin (β-ODAP) content. *Food and Chemical Toxicology* 49, 589–600.

Ladizinsky, G. (1979) The origin of lentil and its wild gene pool. *Euphytica* 28, 179–187.

Ladizinsky, G. (1993) Wild lentils. *Critical Review in Plant Science* 12, 169–184.

Ladizinsky, G. (1999) Identification of the lentil's wild genetic stock. *Genetic Resources and Crop Evolution* 46, 115–118.

Ladizinsky, G. and Abbo, S. (1993) Cryptic speciation in *Lens culinaris. Genetic Resources and Crop Evolution* 40, 1–5.

Ladizinsky, G. and Adler, A. (1976a) Genetic relationships among the annual species of *Cicer. Theoretical and Applied Genetics* 48, 197–203.

Ladizinsky, G. and Adler, A. (1976b) The origin of chickpea *Cicer arietinum* L. *Euphytica* 25, 211–217.

Ladizinsky, G., Braun, D., Goshen, D. and Muehlbauer, F.J. (1984) The biological species of the genus *Lens* L. *Botany Gazette* 145, 253–261.

Ladizinsky, G., Cohen, D. and Muehlbauer, F.J. (1985) Hybridization in the genus *Lens* by means of embryo culture. *Theoretical and Applied Genetics* 70, 97–101.

Ladizinsky, G., Pickersgill, B. and Yamamoto, K. (1988) Exploitation of wild relatives of the food legumes. In: Summerfield, R.J. (ed.) *World Crops, Cool Season Food Legumes*, Kluwer Academic Publishers, Dordrecht, The Netherlands, pp. 967–987.

Leonard, M.F., Stephens, L.C. and Summers, W.L. (1987) Effect of maternal genotype on development of *Phaseolus vulgaris* L. × *P. lunatus* L. interspecific hybrid embryos. *Euphytica* 36, 327–332.

Li, L., Yang, Q., Hu, Y., Zhu, L. and Ge, H. (1995) Discovery of parent interaction sterile material of soyabean cultivar and its genetic inference [in Chinese]. *Journal of Anhui Agricultural Sciences* 23, 304–306.

Luo, J.P. and Jia, J.F. (1998) Plant regeneration from callus protoplasts of the forage legume *Astragalus adsurgens* Pall. *Plant Cell Reports* 17, 313–317.

Luo, M.C., Yen, C. and Yang, J.L. (1993) Crossability percentage of bread wheat landraces from Shaanxi and Henan provinces, China with rye. *Euphytica* 67, 1–8.

Lyons, M.E., Dickson, M.H. and Hunter, J.E. (1987) Recurrent selection for resistance to white mold in *Phaseolus* species. *Journal of the American Society of Horticultural Sciences* 112, 149–152.

Machado, M., Tai, W. and Baker, L.R. (1982) Cytogenetic analysis of the interspecific hybrid *Vigna radiata* × *V. umbellata. Journal of Heredity* 73, 205–208.

Mahuku, G., Jara, C., Cajiao, C. and Beebe, S. (2003) Sources of resistance to angular leaf spot (*Phaeoisariopsis griseola*) in common bean core collection, wild *Phaseolus vulgaris* and secondary gene pool. *Euphytica* 130, 303–313.

Malhotra, R.S., Imtiaz, M., Clarke, H.J. and Sandhu, J.S. (2009) Genetic enhancement for cold tolerance in chickpea. In: *International Conference on Grain Legumes – Quality Improvement, Value Addition and Trade (ICGL 2009)*, 14–16 February 2009. Indian Institute of Pulses Research, Kanpur, India.

Mallikarjuna, N. (1999) Ovule and embryo culture to obtain hybrids from interspecific incompatible pollinations in chickpea. *Euphytica* 110, 1–6.

Mallikarjuna, N. and Moss, J.P. (1995) Production of hybrids between *Cajanus platycarpus* and *C. Cajan. Euphytica* 83, 43–46.

Mallikarjuna, N. and Saxena, K.B. (2002) Production of hybrids between *Cajanus acutifolius* and *C. cajan. Euphytica* 124, 107–110.

Mallikarjuna, N. and Saxena, K.B. (2005) A new cytoplasmic male-sterility system derived from cultivated pigeon pea cytoplasm. *Euphytica* 142, 143–148.

Mallikarjuna, N., Jadhav, D. and Reddy, P. (2006) Introgression of *Cajanus platycarpus* genome into cultivated pigeon pea, *C. cajan. Euphytica* 149, 161–167.

Markhart, A.H. (1985) Comparative water relations of *Phaseolus vulgaris* L. and *Phaseolus acutifolius* Gray. *Plant Physiology* 77, 113–117.

Marsden-Jones, M. (1919) Hybrids of *Lathyrus*. *Journal of the Royal Horticultural Society* 45, 92–93.

McElroy, J.B. (1985) Breeding dry beans, *P. vulgaris* L., for common bacterial blight resistance derived from *Phaseolus acutifolius* A. Gray. PhD dissertation (Diss. Abstr. Intl. 46(7), 2192B], Cornell University, Ithaca, New York.

Mejía-Jiménez, A., Muñoz, C., Jacobsen, H.J., Roca, W.M. and Singh, S.P. (1994) Interspecific hybridization between common and tepary beans: increased hybrid embryo growth, fertility, and efficiency of hybridization through recurrent and congruity backcrossing. *Theoretical and Applied Genetics* 88, 324–331.

Mercy, S.T. and Kakar, S.N. (1975) Barrier to interspecific crosses in *Cicer*. *Proceedings of the Indian National Science Academy* 41, 78–82.

Miklas, P.N. and Santiago, J. (1996) Reaction of selected tepary bean to bean golden mosaic virus. *Horticulture Science* 31, 430–432.

Miklas, P.N. and Stavely, J.R. (1998) Incomplete dominance of rust resistance in tepary bean. *Horticulture Science* 33, 143–145.

Miklas, P.N., Beaver, J.S., Grafton, K.F. and Freytag, G.F. (1994a) Registration of TARS VCI-4B multiple disease resistant dry bean germplasm. *Crop Science* 34, 1415.

Miklas, P.N., Zapata, M., Beaver, J.S. and Grafton, K.F. (1994b) Registration of four dry bean germplasm resistant to common bacterial blight, ICB-3, ICB-6, ICB-8, and ICB-10. *Crop Science* 39, 594.

Miklas, P.N., Grafton, K.F., Kelly, J.D., Steadman, J.R. and Silbernagel, M.J. (1998a) Registration of four white mold resistant dry bean germplasm lines, 19365-3, 19365-5, 19365-31, and 92BG-7. *Crop Science* 38, 1728.

Miklas, P.N., Schwartz, H.F., Salgado, M.O., Nina, R. and Beaver, J.S. (1998b) Reaction of selected tepary bean to ashy stem blight and fusarium wilt. *Horticulture Science* 33, 136–139.

Mohan, S.T. (1982) Evaluation of *Phaseolus coccineus* Lam. germplasm for resistance to common bacterial blight of bean. *Turrialba* 32, 489–490.

Muehlbauer, F.J. and McPhee, K.E. (2005) Lentil (*lens culinaris* Medik). In: Singh, R.J. and Jauhar, P.P. (eds) *Genetic Resources, Chromosome Engineering and Crop Improvement, Grain Legumes*. Taylor & Francis, Boca Raton, Florida, pp. 219–230.

Muehlbauer, F.J., Cho, S., Sarker, A., McPhee, K.E., Coyne, C.J., Rajesh, P.N. *et al*. (2006) Application of biotechnology in breeding lentil for resistance to biotic and abiotic stress. *Euphytica* 147, 149–165.

Munoz-Florez, L.C. and Baudoin, J.P. (1994a) Anther culture in some *Phaseolus* species. In: Roca, W.M., Mayer, J.E., Pastor, C.M.A. and Tohme, M.J. (eds) *Proceedings of the International Scientific Meeting of the* Phaseolus *Bean Advanced Biotechnology Research Network*, February 1993, Cali, Colombia. Centro Internacional de Agricultura Tropical (CIAT), Colombia, pp. 205–212.

Munoz-Florez, L.C. and Baudoin, J.P. (1994b) Influence of the cold pretreatment and the carbon source on callus induction from anthers in *Phaseolus*. *Bean Improvement Cooperative Annual Report* (USA) 37, 129–130.

Nagaraj, N.C., Muniyappa, V., Satyan, B.A., Shanmugam, N., Jayarajan, R. and Vidhyasekaran, P. (1981) Resistance source for mung bean yellow mosaic virus. In: *Proceedings of the National Seminar on Disease Resistance in Crop Plants*, pp. 69–72.

Ocampo, B., Conicella, C. and Moss, J.P. (2000) Wide crossing, opportunities and progress. In: Knight, R. (ed.) *Linking Research and Marketing Opportunities for Pulses in the 21st Century*. Kluwer Academic Publishers, Dordrecht, The Netherlands, pp. 411–419.

Ochatt, S.J., Mousset-Declas, C. and Rancillac, M. (2000) Fertile pea plants regenerate from protoplast when calluses have not undergone endoreduplication. *Plant Science* 156, 177–183.

Ockendon, D.J., Currah, L. and Taylor, J.D. (1982) Transfer of resistance to halo-blight (*Pseudomonas phaseolicola*) from *Phaseolus vulgaris* to *P. coccineus*. *Annual Report of the Bean Improvement Cooperation* 25, 84–85.

Osorno, J.M., Beaver, J.S., Ferwerda, F.H. and Miklas, P.N. (2003) Two genes from *Phaseolus coccineus* L. confer resistance to bean golden yellow mosaic virus. *Annual Report of the Bean Improvement Cooperation* 46, 147–148.

Pal, S.S., Sandhu, J.S. and Singh, I. (2005) Exploitation of genetic variability in interspecific cross between *Vigna mungo* × *V. umbellata*. *Indian Journal of Pulses Research* 18, 9–11.

Palmer, J.L., Lawn, R.J. and Atkins, S.W. (2002) An embryo rescue protocol for *Vigna* interspecific hybrids. *Australian Journal of Botany* 50, 331–338.

Pande, K., Raghuvanshi, S.S. and Prakash, D. (1990) Induced high yielding amphiploid of *Vigna radiata* × *V. mungo*. *Cytologia* 55, 249–253.

Pandiyan, M., Ramamoorthi, N., Ganesh S.K., Jebaraj, S., Pagarajan, P. and Balasubramanian, P. (2008) Broadening the genetic base and introgression of MYMV resistance and yield improvement through unexplored genes from wild relatives in mung bean. *Plant Mutation Reports* 2, 33–38.

Park, S.J. and Dhanvantari, B.N. (1987) Transfer of common blight (*Xanthomonas compestris* pv. *phaseoli*) resistance from *Phaseolus coccineus* Lam. to *P. vulgaris* L. through interspecific hybridization. *Canadian Journal of Plant Science* 67, 685–695.

Parker, J.P. and Michaels, T.E. (1986) Simple genetic control of hybrid plant development in interspecific crosses between *P. vulgaris* and *P. acutifolius* A. Gray. *Plant Breeding* 97, 315–323.

Parsons, L.R. and Howe, T.K. (1984) Effect of water stress on the water relations of *Phaseolus vulgaris* and the drought resistant *Phaseolus acutifolius. Plant Physiology* 60, 197–202.

Percy, R.G. (1986) Effects of environment upon ovule abortion in interspecific F_1 hybrids and single species cultivars of cotton. *Crop Science* 26, 938–942.

Peters, J.E., Crocomo, O.J., Sharp, W.R., Paddock, E.F., Tegenkamp, I. and Tegenkamp, T. (1977) Haploid callus cells from anthers of *Phaseolus vulgaris. Phytomorphology* 27, 79–85.

Plucknett, D.L., Smith, N.J.H., Williams, J.T. and Anishetty, N.M. (1987) *Gene Banks and the World's Food.* Princeton University Press, Princeton, New Jersey.

Powers, J.B., Frearson, E.M., Hayward, C., George, D., Evans, P.K., Berry, S.F. *et al.* (1976) Somatic hybridization of *Petunia hybrid* × *P. parodii. Nature* 263, 500–502.

Pratap, A., Priya, R., Nandeesha, P. and Kumar, S. (2009) Haploid embryogenesis in pigeonpea (*Cajanus cajan* L.) through anther culture. *International Conference on Grain Legumes*, 14–16 February, Kanpur, India, pp. 155.

Prescott-Allen, C. and Prescott-Allen, R. (1986) *The First Resource: Wild Species in the North American Economy*. Yale University, New Haven, Connecticut.

Prescott-Allen, C. and Prescott-Allen, R. (1988) *Genes from the Wild: Using Wild Genetic Resources for Food and Raw Materials*. International Institute for Environment and Development, London.

Pundir, R.P.S. and Mengesha, M.H. (1995) Cross compatibility between chickpea and its wild relative, *Cicer echinospermum* Davis. *Euphytica* 83, 241–245.

Pundir, R.P.S. and Singh, R.B. (1985) Gene pools in *Phaseolus* and *Vigna* cultigens. *Euphytica* 34, 303–305.

Rabakoarihanta, A., Mok, D.W.S. and Mok, M.C. (1979) Fertilization and early embryo development in reciprocal interspecific crosses of *Phaseolus. Theoretical Applied Genetics* 54, 55–59.

Rashid, K.A., Smartt, J. and Haq, N. (1988) Hybridization in the genus *Vigna*. In: Shanmugasundaram, S. and Mclean, B.T. (eds) *Mungbean, Proceedings of the Second International Symposium*. Asian Vegetable Research and Development Centre, Shanhua, Taiwan, pp. 205–214.

Rathnaswamy, R., Yolanda, J.L., Kalaimagal, T., Surya Kumar, M. and Sashi Kumar, D. (1999) Cytoplasmic genetic male sterility in pigeon pea. *Indian Journal of Agricultural Sciences* 69, 159–160.

Ravi, J., Singh, J.P. and Minocha, J.L. (1987) Meiotic behaviour of interspecific hybrids of *Vigna radiata* × *V. mungo*. In: *Proceedings of the First Symposium on Crop Improvement*, Tamil Nadu Agricultural University, Coimbatore, India, pp. 58–59.

Reddy, K.R. and Singh, D.P. (1990) The variation and transgressive segregation in the wide and varietal crosses of mung bean. *Madras Agricultural Journal* 77, 12–14.

Reddy, L.J. (1981) Pachytene analyses in *Cajanus cajan*, *Atylosia lineatus* and their hybrids. *Cytologia* 46, 397–412.

Reddy, L.J. and De, D.N. (1983) Cytomorphological studies in *C. cajan* x *A. lineatus. Indian Journal of Genetics and Plant Breeding* 43, 96–103.

Reddy, L.J., Green, J.M. and Sharma, D. (1981) Genetics of *Cajanus cajan* × *Atylosia* spp. In: *Proceedings of the Intrnational Workshop on Pigeonpea*, 15–19 December 1980, ICRISAT, Patancheru, Andhra Pradesh, India, pp. 39–50.

Reddy, M.V., Raju, T.N. and Sheila, V.K. (1996) Phytophthora blight disease in wild pigeonpea. *International Chickpea and Pigeonpea Newsletter* 3, 52–53.

Reed, W. and Lateef, S.S. (1990) Pigeonpea, pest management. In: Nene, Y.L., Hall, S.D. and Sheila, V.K. (eds) *The Pigeonpea*. CAB International, Wallingford, UK, pp. 349–374.

Robertson, L.D. and Abd-El-Moneim, A.M. (1997) Status of *Lathyrus* germplasm held at ICARDA and its use in breeding programs. In: Mathur, P.N., Rao, V.R. and Arora, R.K. (eds) *Lathyrus Genetic Resources Network; Proceedings of IPGRI-ICARDA-ICAR Regional Working Group Meeting*, 8–10 December 1997, New Delhi, India.

Robertson, L.D., Singh, K.B., Erskine, W. and Abd-El-Moneim, A.M. (1996) Useful genetic diversity in germplasm collections of food and forage legumes from West Asia and North Africa. *Genetic Resources and Crop Evolution* 43, 447–460.

Rosas, J.C., Erazo, J.D. and Moncada, J.R. (1991) Tolerancia a la sequía en germoplasma de frijol comun y frijol tepari. *CEIBA* 32, 91–106.
Rozwadowski, K.L., Saxena, P.K. and King, J. (1990) Isolation and culture of *Lens culinaris* Medik. *Plant Cell Tissue and Organ Culture* 15, 175–182.
Sandhu, J.S., Singh, I. and Pal, S.S. (2005) Mash 1008: a new variety of summer urdbean. *Journal of Research of Punjab Agricultural University* 12, 150–155.
Saxena, K.B. (2008) Genetic improvement of pigeonpea – a review. *Tropical Plant Biology* 1, 159–178.
Saxena, K.B. and Kumar, R.V. (2003) Development of a cytoplasmic nuclear male-sterility system in pigeonpea using *C. scarabaeoides* (L.) Thouars. *Indian Journal of Genetics and Plant Breeding* 63, 225–229.
Saxena, K.B. and Sharma, D. (1995) Sources of dwarfism in pigeon pea. *Indian Journal of Pulses Research* 8, 1–6.
Saxena, K.B., Singh, L., Reddy, M.V., Singh, U., Lateef, S.S., Sharma, S.B. *et al.* (1990) Intra-species variation in *Atylosia scarabaeoides* (L.) Benth., a wild relative of pigeonpea [*Cajanus cajan* (L.) Millsp.]. *Euphytica* 49, 185–191.
Saxena, K.B., Ariyanyagam, R.P. and Reddy, L.J. (1992) Genetics of high-selfing trait in pigeon pea. *Euphytica* 59, 125–127.
Saxena, K.B., Rao, A.N., Singh, U. and Ramanandan, P. (1996) Intraspecies variation in *Cajanus platycarpus* for some agronomic traits and crossability. *International Pigeonpea Newsletter* 3, 49–51.
Saxena, K.B., Kumar, R.V. and Rao, P.V. (2002) Pigeon pea nutrition and its improvement. In: *Quality Improvement in Crops*. The Food Products Press, Crop Science, USA, pp. 227–260.
Saxena, K.B., Kumar, R.V., Madhavilatha, K. and Dalvi, V.A. (2006) Commercial pigeonpea hybrids are just a few steps away. *Indian Journal of Pulses Research* 19, 7–16.
Schmit, V. and Baudoin, J.P. (1992) Screening for resistance to *Ascochyta* blight in populations of *Phaseolus coccineus* L. and *P. polyanthus* Greenman. *Field Crops Research* 30, 155–165.
Scott, M.E. and Michaels. T.E. (1992) *Xanthomonas* resistance of *Phaseolus* interspecific cross selections confirmed by field performance. *Hortscience* 27, 348–350.
Shade, R.E., Pratt, R.C. and Pomeroy, M.A. (1987) Development and mortality of the bean weevil, *Acanthoscelides obtectus* (Coleoptera, Bruchidae) on mature seeds of tepary beans, *Phaseolus acutifolius* and common beans, *Phaseolus vulgaris. Environment and Entomology* 16, 1067–1070.
Shahi, V.K., Choudhary, S.C., Kumari, N. and Kumar, H. (2006) Development of *Cajanus platycarpus* × *Cajanus cajan* hybrids through embryo rescue. *Indian Journal of Genetics and Plant Breeding* 66, 212–217.
Shanmungam, A.S., Rathnaswamy, R. and Rangasamy, S.R. (1983) Crossability studies between green gram and black gram. *Current Science* 52, 1018–1020.
Sharma, H.C. (1995) How wide can a wide cross be? *Euphytica* 82, 43–64.
Sharma, HC. (2004) *A Little Help from Wild: Exploiting Wild Relatives of Chickpea for Resistance to Helicoverpa armigera*. ICRISAT, Patancheru, India.
Sharma, J. and Satija, C.K. (1996) *In vitro* hybridization in incompatible crosses of *Vigna* species. *Crop Improvement* 23, 29–32.
Shiela, V.K., Moss, J.P., Gowda, C.L.L. and Rheenen, H.A. (1992) Interspecific hybridization between *Cicer arietinum* and wild *Cicer* species. *International Chickpea Newsletter* 27, 11–13.
Shrivastava, S. and Chawla, H.S. (1993) Effects of seasons and hormones on pre-and post-fertilization barriers of crossability and *in vitro* hybrid development between *Vigna unguiculata* and *V. mungo* crosses. *Biologia Plantarum* 35, 505–512.
Siddique, K.H.M., Loss, S.P., Herwig, S.P. and Wilson, J.M. (1996) Growth, yield and neurotoxin (ODAP) concentration of three *Lathyrus* species in Mediterranean type environments of Western Australia. *Australian Journal of Experimental Agriculture* 36, 209–218.
Sidhu, M.C. (2003) Cytogenetic and isozyme studies in interspecific hybrids of *Vigna radiata* and *V. mungo* with *V. trilobata. Crop Improvement* 30, 140–145.
Silbernagel, M.J. and Hannan, R.M. (1992) Use of plant introductions to develop U.S. bean cultivars. In: Shands, H.L. and Wiesner, L.E. (eds) *Use of Plant Introductions in Cultivar Development. Part 2*. CSSA Special Publication No. 20, CSSA, Madison, Wisconsin, pp. 1–8.
Singh, A.K., Singh, N., Singh, S.P., Singh, N.B. and Smartt, J. (2006) Pigeon pea. In: Dhillon, B.S., Saxena, S., Agrawal, A. and Tyagi, R.K. (eds) *Plant Genetic Resource: Foodgrain Crops*. Narosa Publishing House, New Delhi, India, pp. 323–239.
Singh, B.B. and Dikshit, H.K. (2002) Possibilities and limitations of interspecific hybridization involving green gram (*Phaseolus radiatus*) and black gram (*Phaseolus mungo*). *Indian Journal of Agricultural Sciences* 72, 676–678.

Singh, D.P. (1990) Distant hybridization in genus *Vigna* – a review. *Indian Journal of Genetics and Plant Breeding* 50, 268–276.
Singh, G., Kapoor, S. and Singh, K. (1982) Screening of chickpea for grey mold resistance. *International Chickpea Newsletter* 7, 13–14.
Singh, J., Sidhu, P.S., Verma, M.M., Gosal, S.S. and Singh, J. (1993) Wide cross hybridization in *Cajanus*. *Crop Improvement* 20, 27–30.
Singh, K.B. and Ocampo, B. (1993) Interspecific hybridization in annual *Cicer* species. *Journal of Genetics and Breeding* 47, 199–204.
Singh, K.B. and Ocampo, B. (1997) Exploitation of wild *Cicer* species for yield improvement in chickpea. *Theoretical and Applied Genetics* 95, 418–423.
Singh, K.B. and Reddy, M.V. (1993) Sources of resistance to *Ascochyta* blight in wild *Cicer* species. *Netherlands Journal of Plant Pathology* 99, 163–167.
Singh, K.B. and Weigand, S. (1996) Registration of three chickpea leaf miner resistant lines, ILC 3800, ILC 5901, and ILC 7738. *Crop Science* 36, 472.
Singh, K.B., Malhotra, R.S. and Saxena, M.C. (1990) Sources of tolerance to cold in *Cicer* species. *Crop Science* 30, 1136–1138.
Singh, K.B., Malhotra, R.S., Haldia, H., Knights, E.J. and Verma, M.M. (1994) Current status and future strategy in breeding chickpea for resistance to biotic and abiotic stresses. *Euphytica* 73, 137–149.
Singh, K.B., Malhotra, R.S. and Saxena, M.C. (1995) Additional sources of tolerance to cold in cultivated and wild *Cicer* species. *Crop Science* 35, 1491–1497.
Singh, K.P., Monika, Sareen, P.K. and Kumar, A. (2003) Interspecific hybridization studies in *Vigna radiata* (L.) Wilczek and *Vigna umbellate* L. *National Journal of Plant Improvement* 5, 16–18.
Singh, S., Gumber, R.K., Joshi, N. and Singh, K. (2005) Introgression from wild *Cicer reticulatum* to cultivated chickpea for productivity and disease resistance. *Plant Breeding* 124, 477–480.
Singh, S.P. and Munoz, C.G. (1999) Resistance to common bacterial blight among *Phaseolus* species and common bean improvement. *Crop Science* 39, 80–89.
Singh, S.P., Debouck, D.G. and Roca, W.M. (1997) Successful Interspecific hybridization between *Phaseolus vulgaris* L and *P. costaricensis* Freytag and Debouck. *Annual Report of the Bean Improvement Cooperation* 40, 40–41.
Siriwardhane, D., Egawa, Y. and Tomooka, N. (1991) Cross-compatibility of cultivated adzuki bean (*Vigna angularis*) and rice bean (*V.umbellata*) with their wild relatives. *Plant Breeding* 107, 320–325.
Sirkka, A.T.I., Verugesse, G., Pfeifer, W.H. and Mujeeb-Kazi, A. (1993) Crossability of tetraploid and hexaploid wheats with ryes for primary triticale production. *Euphytica* 65, 203–210.
Smartt, J. (1979) Interspecific hybridization in grain legumes – a review. *Economic Botany* 33, 329–337.
Smartt, J. (1981) Gene pools in *Phaseolus* and *Vigna* cultigens. *Euphytica* 30, 445–459.
Smartt, J. (1985) Evolution of grain legumes. III. Pulses in the genus *Vigna*. *Experimental Agriculture* 21, 87–100.
Smartt, J. (1990) *Grain Legumes: Evolution and Genetic Resources*. Cambridge University Press, Cambridge, UK.
Stalker, H.T. (1980) Utilization of wild species for crop improvement. *Advances in Agronomy* 33, 111–147.
Stoddard, F.L., Balko, C., Erskine, W., Khan, H.R., Link, W. and Sarker, A. (2006) Screening techniques and sources of resistance to abiotic stresses in cool-season food legumes. *Euphytica* 147, 167–186.
Subba Rao, G.V. (1988) Salinity tolerance in pigeonpea (*Cajanus cajan*) and its wild relatives. PhD dissertation, Indian Institute of Technology, Kharagpur, India.
Subba Rao, G.V., Johansen, C., Jana, M.K. and Rao, J.V.D.K. (1990) Physiological basis of differences in salinity tolerance of pigeonpea and its related wild species. *Journal of Plant Physiology* 137, 64–71.
Tanksley, S.D. and McCouch, S.R. (1997) Seed banks and molecular maps, unlocking genetic potential from the wild. *Science* 277, 1063–1066.
Tanksley, S.D. and Nelson, J.C. (1996) Advanced back cross QTL analysis, a method for the simultaneous discovery and transfer of valuable QTLs from unadapted germplasm into elite breeding lines. *Theoretical and Applied Genetics* 92, 191–203.
Tanksley, S.D., Grandillo, S., Fulton, T.M., Zamir, D., Eshed, Y., Petiard, V., Lopez, J. and Beck-Bunn, T. (1996) Advanced back cross QTL analysis in a cross between an elite processing line of tomato and its wild relative *L. pimpinnellifolium*. *Theoretical and Applied Genetics* 92, 213–224.
Thiyagu, K., Jayamani, P. and Nadarajan, N. (2008) Pollen pistil intraction in inter-specific crosses of *Vigna* species. *Cytologia* 73(3), 251–257.

Thomas, C.V., Manshardt, R.M. and Waines, J.G. (1983) Teparies as a source of useful traits for improving common beans. *Desert Plants* 5, 43–48.

Tian, D. and Rose, R.J. (1999) Asymmetric somatic hybridization between the annual legumes *Medicago truncatula* and *Medicago scutellata*. *Plant Cell Reports* 18, 989–996.

Tikka, S.B.S., Panwar, L.D. and Chauhan, R.M. (1997) First report of cytoplasmic genic male sterility in pigeon pea (*Cajanus cajan* L. Millsp) through wide hybridization. *GAU Research Journal* 22, 160–162.

Toker, C., Canci, H. and Yildirim, T. (2007) Evaluation of perennial wild *Cicer* species for drought resistance *Genetic Resources and Crop Evolution* 54, 1781–1786.

Tomooka, N.C., Lairungreang, R., Nakeeraks, P., Egawa, Y. and Thavarasook, C. (1992) Development of bruchid-resistant mungbean line using wild mungbean germplasm in Thailand. *Plant Breeding* 109, 60–66.

Tomooka, N., Kashiwaba, K., Vaughan, D., Ishimoto, M. and Egawa, Y. (2000) The effectiveness of evaluating wild species, searching for sources of resistance to bruchid beetle in the genus *Vigna* subspecies *Caratotropis*. *Euphytica* 115, 27–41.

Trankovskij, D.A. (1962) Interspecific hybridization in the genus *Lathyrus*. *Bulletin of Moscow, Nature and Biology Series* 67, 140–141.

Tullu, A., Buchwaldt, L., Lulsdorf, M., Banniza, S., Barlow, B., Slinkard, A.E. *et al.* (2006) Sources of resistance to anthracnose (*Colletotrichum truncatum*) in wild *Lens* species. *Genetic Resources and Crop Evolution* 53, 111–119.

Tyagi, D.K. and Chawla, H.S. (1999) Effects of seasons and hormones on crossability barriers and *in vitro* hybrid development between *Vigna radiata* and *V. unguiculata*. *Acta Agronomica Hungarica* 47, 147–154.

Upadhyaya, H.D. (2006) Improving pigeonpea with the wild. *SA Trends* January.

van der Maesen, L.L.G. and Pundir, R.P.S. (1984) Availability and use of wild *Cicer* germplasm. *Plant Genetic Resources Newsletter* 57, 19–24.

van der Maesen, L.L.G., Maxted, N., Javad, F., Coles, S. and Davies, A.M.R. (2007) Taxonomy of the genus *Cicer* revisited. In: Yadav, S.S., Redden, R.J., Chen, W. and Sharma, B. (eds) *Chickpea Breeding and Management*, CAB International, Wallingford, UK, pp. 14–46.

van Oss, H., Aron, Y. and Ladizinsky, G. (1997) Chloroplast DNA variation and evolution in the genus *Lens* Mill. *Theoretical and Applied Genetics* 94, 452–457.

Varshney, R.K., Close, T.J., Singh, N.K., Hoisington, D.A. and Cook, D.R. (2009) Orphan legume crops enter the genomics era! *Current Opinion in Plant Biology* 11, 1–9.

Verma, M.M. and Brar, J.S. (1996) Breeding approaches for increasing yield potential of mung bean. In: Asthana, A.N. and Kim, D.H. (eds) *Recent Advances in Mungbean Research*. Indian Society of Pulses Research and Development, Kanpur, India, pp. 102–123.

Verma, M.M., Sandhu, J.S., Brar, H.S. and Brar, J.S. (1990) Crossability studies in different species of *Cicer*. *Crop Improvement* 17, 179–181.

Verma, R.P.S. and Singh, D.P. (1986) Problems and prospects of interspecific hybridization involving green gram and black gram. *Indian Journal of Agricultural Sciences* 56, 535–537.

Verulkar, S.B., Singh, D.P. and Bhattacharya, A.K. (1997) Inheritance of resistance to podfly and pod borer in the interspecific cross of pigeon pea. *Theoretical and Applied Generics* 95, 506–508.

Vessal, S.R., Bagheri, A. and Safarnejad, A. (2002) The possibility of *in vitro* haploid production in chickpea (*Cicer arietinum* L.). *Journal of Science and Technology of Agricultural and Natural Resources* 6, 67–76.

Wallace, D.H. and Wilkinson, R.E. (1965) Breeding for *Fusarium* root rot resistance in beans. *Phytopathology* 55, 1227–1231.

Wilkinson, R.E. (1983) Incorporation of *Phaseolus coccineus* germplasm may facilitate production of high yielding *P. vulgaris* lines. *Annual Report of the Bean Improvement Cooperation* 26, 28–29.

Xiao, J., Grandillo, S., Ahn, S.N., McCouch, S.R., Tanksley, S.D., Li, J. *et al.* (1996) Genes from wild rice to improve yield. *Nature* 384, 223–224.

Yamamoto, K., Fujiware, T. and Blumenreich, L. (1989) Isozymic variation and interspecific crossability in annual species of the genus *Lathyrus* L. In: Kaul, A.K. and Combes, D. (eds) *Lathyrus and Lathyrism*. Third World Medical Research Foundation, New York, pp. 118–121.

Yamini, K.N., Gomathinayagam, P., Devasena, N. and Mohanbabu, R. (2001) Isozyme analysis of interspecific hybrids of *Vigna* spp. *Journal of Soil and Crops* 11, 36–39.

Ye, G., McNiel, D.L. and Hill, G.D. (2002) Breeding for resistance to lentil ascochyta blight. *Plant Breeding* 121, 185–191.

Yerkes, W.D. and Freytag, G.F. (1956) *Phaseolus coccineus* as a source of root-rot resistance for the common bean. *Phytopathology* 46, 32.

Yuko, M., Kato, M., Takamizo, T., Kanbe, M., Inami, S. and Hattori, K. (2006) Iterspecific hybrids between *Medicago sativa* L. and annual *Medicago* containing alfafa weevil resistance. *Plant Cell, Tissue and Organ Culture* 84, 80–89.

Yunus, A.G. (1990) Biosystematics of *Lathyrus* section *Lathyrus* with special reference to the grasspea, *L. sativus* L. PhD thesis, University of Birmingham, UK.

Yunus, A.G. and Jackson, M.T. (1991) The gene pools of the Grasspea (*Lathyrus sativus* L.). *Plant Breeding* 106, 319–328.

7 Polyploidy

S. Safari and J.A Schlueter

7.1 Introduction

Polyploidy is widespread in the plant kingdom, with estimates of 70–100% of angiosperms and upwards of 95% of pteridophytes having a polyploid history (Masterson, 1994; Lockton and Gaut, 2005). The term polyploidy was originally coined in 1916 by Winkler to describe organisms whose genomes have a greater amount of genetic material and chromosomes than their ancestors. Among legumes, peanut, lucerne and soybean (being a diploidized recent polyploidy) are cultivated species. Most of these crop legumes show some evidence for duplication in their evolutionary history and are ancient polyploids or paleopolyploids, while few are more recent polyploids or neopolyploids. The genomes of these species have undergone cyclic rounds of duplication and diploidization, which is a process of allowing major genome rearrangements to revert the genome to a near diploid state (Stebbins, 1966; Blanc and Wolfe, 2004; Schlueter *et al.*, 2004).

7.2 Mechanism of Gene Duplication

Gene duplication can occur by a variety of mechanisms: duplication of regions or segments of chromosomes, tandem duplication, reverse-transcriptase-mediated duplication and whole genome doubling, or polyploidy (Wendel, 2000; Bennetzen, 2002; Schmidt, 2002; Lawton-Rauh, 2003). Regional duplications, often called dispersive processes, can occur through abnormal crossing-over events while tandem duplications are frequently the result of replication slippage or transposon activity (Bennetzen, 2002; Lawton-Rauh, 2003). Single gene or regional duplication is seen in all plant species, while whole genome duplication or polyploidy has probably played the greatest role in the evolution of plant genomes (Lawton-Rauh, 2003).

Gene duplication appears to occur at a higher rate in plants (Mable, 2004), although it is found across eukaryotic lineages. It has been seen as a driving force in the evolution and expansion of eukaryotic genomes (Stebbins, 1966; Ohno, 1970). In plants, most genes are members of gene families, indicating that gene duplication is a widespread phenomenon in the origin and formation of diverse gene functions (Wendel, 2000; Adams and Wendel, 2005). The high incidence of gene duplication in plants could be due to its potential impact on genetic diversity and adaptation (Lawton-Rauh, 2003). Differential patterns of gene silencing following polyploidy may provide the genetic context to facilitate speciation (Werth and Windham, 1991). Gene and genome duplication is also seen as a mechanism for the creation of genetic diversity and as a source of new genes and gene functions, as well as leading to silenced genes or pseudogenes (Ohno, 1970; Pickett and Meeks-Wagner, 1995).

7.3 Polyploidization

Two main paths exist for a polyploid event to occur: either a doubling of unreduced gametes from a single species (autopolyploid) or the hybridization of unreduced gametes or somatic doubling from two different genomes (allopolyploid; Pikaard, 2001; Lawton-Rauh, 2003). The most common mechanism for polyploid formation is the fusion of unreduced gametes containing a diploid rather than haploid chromosome number and subsequence crossing with similar individuals (Pikaard, 2001). In some cases the diploid progenitors of an allopolyploid can be identified, and this type of polyploidy event has been determined, for example, in neopolyploid *Glycine* (Wendel and Cronn, 2002; Parkin *et al.*, 2003; Doyle *et al.*, 2004). However, if the polyploid event is not recent or the species has undergone multiple rounds of duplication and rearrangement, determining whether the event was allo- or autopolyploid can be very complex, as with soybean (Doyle *et al.*, 2003; Straub *et al.*, 2006).

Polyploids often are formed on multiple occasions from the same or similar diploid progenitors, which is often known as recurrent formation (Soltis and Soltis, 1999, 2000). Recurrent origins of a polyploid are seen as 'the rule, not the exception' (Soltis and Soltis, 1999). For example in soybean, it has been seen that at least six hybridization events led to the evolution of polyploidy in *Glycine tabacina* (Doyle *et al.*, 1990). The multiple origins have also been reported for *Glycine tomentella* polyploid (Kollipara *et al.*, 1994). Recurrent formation may account for the large array of genetic diversity found within polyploid species (Soltis and Soltis, 1995). Furthermore, gene flow between genetically different polyploids allows for recombination events and increased genotypes (Soltis and Soltis, 1999). Generation of synthetic amphidiploid from a cross (*Arachis batizocoi* × *Arachis cardenasii* × *Arachis diogoi*; Husted, 1933, 1936; Smartt *et al.*, 1978) clearly suggests that multiple hybridizations resulted in polyploids (i.e. recurrent formation) in groundnut.

7.4 Genome Restructuring and Diploidization (Rearrangement within Genome)

After a polyploidy event duplicated regions begin to diverge from one another at both the sequence and chromosomal levels, either through mutational or epigenetic means such that the polyploid becomes genetically diploidized (Stebbins, 1966; Grant, 1981; Pickett and Meeks-Wagener, 1995). Diploidization is probably a response to the stress or 'genomic shock' experienced by a plant while in a polyploid state (Stebbins, 1966; McClintock, 1984). Allopolyploids have been shown to undergo numerous physical changes, ranging through DNA sequence elimination, heterchromatin expansion and reciprocal chromosome segment translocations and inversions, all thought to have a role in diploidization (Pikaard, 2001). Additionally, diploidization is not simply chromosomal/structural in nature, it also involves the diploidization of gene expression. In other words, RNA content in a diploidizing tetraploid is thought to be reduced to the level of the related diploids (Leipoldt and Schmidtke, 1982). On a genic level, diploidization involves the silencing of one copy or a divergence leading to a change in function of a copy (Pickett and Meeks-Wagner, 1995).

Following polyploidy, there seems to be a genome-wide removal of some but not all of the redundant genomic material. It has been suggested that 'differential gene loss' after a major duplication event may be responsible for much of the differences between closely related plants (Adams and Wendel, 2005). Diploidization at the chromosomal level is caused by additions, deletions, mutations and rearrangements that rapidly inhibit non-homologous pairing of chromosomal tetravalents (Ohno, 1970). The primary effect of diploidization is the switch from tetrasomic to disomic inheritance in meiosis (Wolfe, 2001).

Studies conducted on non-legume crops such as *Brassica* (Song *et al.*, 1995; Lagercrantz and Lydiate, 1996) and *Gossypium* (Cronn *et al.*, 1999; Liu *et al.*, 2000) have suggested that genomic reorganization often occurs rapidly after polyploidy and is extensive in most polyploids (Soltis and Soltis, 1999). Therefore, understanding the process of cyclic duplication and diploidization is key to understanding the role of duplication in many legumes. EST-based studies have found that duplication is likely to have occurred around 54 million years ago across many of the major crop legumes (Blanc and Wolfe, 2004; Schlueter *et al.*, 2004).

7.5 Role of Polypoidy in Improvement of Food Legumes

Over recent decades, polyploidy has been considered important for crop improvement because it enhancs allele doses, allelic diversity, fixed heterozygosity and generates the opportunity for novel phenotypic variation that arises due to duplicated genes acquiring new functions (Udall and Wendel, 2006). In this context, the following text focuses on studies conducted in the cultivated polyploid legume species.

Groundnut

Arachis hypogaea (groundnut) is a member of tribe Aeschynomeneae, subtribe Stylosanthinae, genus *Arachis*. Krapovickas and Gregory (1994) have described this genus as containing 69 diploid and tetraploid species, but recently 11 more species have been described (Valls and Simpson, 2005). The cultivated peanut, *A. hypogaea* is an allotetraploid (2n = 2x = 40) (Kochert *et al.*, 1991; Halward *et al.*, 1992; Lanham *et al.*, 1992; Garcia *et al.*, 1995). *Arachis monticola* (Krapovickas and Gregory, 1994; Valls and Simpson, 2005), *Arachis glabrata*, *Arachis pseudovillosa* and *Arachis nitida* belonging to sections Extranervosae and Rhizomatosae are tetraploid species. It appears that there are similarities between genomes of tetraploids in sections Rhizomatosae and Erectoides and *Arachis* (Stalker, 1985). Along with those species, three aneuploid species (2n=2x=18) (*Arachis decora*, *Arachis palustris* and *Arachis praecox*) are presented in this genus (Lavia, 1998). Polyploidy in these sections is believed to have occurred through independent events (Smartt and Stalker, 1982). *A. hypogaea* probably originated from a single recent polyploidization (Kochert *et al.*, 1996; Young *et al.*, 1996).

The allopolypoid *A. hypogaea* has A and B genomes, which are derived from wild species of *Arachis*. The diploid species *Arachis cardenasii* and *Arachis batizocoi* are reported to have contributed the A and B genomes, respectively, in the evolution of cultivated teraploid species. However, other data (Kochert *et al.*, 1996; Raina and Mukai, 1999) suggest that *Arachis ipaensis* is most likely the B genome donor to *A. hypogaea* (Burow *et al.*, 2001). A genome species can be identified by a cytogenetic difference on a single chromosome (Husted, 1936; Seijo *et al.*, 2004). However, other diploid species not having such a cytogenetic difference have been considered more heterogeneous, usually being deemed to share a B genome (Moretzsohn *et al.*, 2004). Since *Arachis glandulifera* does not show any homology with species having either the A or B genome, the genome of this species has been categorized into a separate class, which is known as the D genome (Stalker, 1997; Robledo and Seijo, 2008). Using RFLP (restriction fragment length polymorphism) markers, 17 diploid species belonging to different sections of *Arachis* and three *A. hypogaea* accessions have been studied in order to determine the ancestral species for the A and B genomes. This suggested that *Arachis duranensis* and *A. ipaensis* contribute the A and B genome, respectively. A unique cross between these two species has resulted in a hybrid, which was followed by a rare spontaneous duplication of chromosomes for generating the cultivated allotetraploid species (Halward *et al.*, 1991; Kochert *et al.*, 1996; Seijo *et al.*, 2004, 2007). However, in contrast to this, *in situ* hybridization techniques used to analyse 13 *A. hypogaea* accessions and 15 wild species have suggested that *Arachis villosa* (A genome) and *A. ipaensis* (B genome) are the progenitors of *A. hypogaea* (Raina and Mukai, 1999; Raina *et al.*, 2001).

Cultivated groundnut is thought to be of monophyletic origin, harbouring relatively little genetic diversity (Burow *et al.*, 2001). Several studies show that, following duplications, cultivated groundnut has been isolated from its wild diploid relatives and natural introgression of alleles from wild species into cultivated species has not been demonstrated (Hopkins *et al.*, 1999). These selective pressures have resulted in a highly conserved genome across varieties (Young *et al.*, 1996). Molecular markers such as RAPDs, AFLPs and RFLPs showed that this isolation led to low nucleotide diversity in groundnut (He and Prakash, 1997; Subramanian *et al.*, 2000; Gimenes *et al.*, 2002; Herselman, 2003; Milla *et al.*, 2005). In addition, being a natural inbreeding species, the breeding process also reduced variation (Isleib and Wynne, 1992; Uphadhyaya *et al.*, 2006). Therefore, development of synthetic amphidiploid in groundnut could help to broaden the genetic base, and useful genes have been introgressed from wild species to cultivated species (Burow *et al.*, 2001). For example, synthetic amphidiploid 'TxAG-6' (Simpson *et al.*, 1993) has been used in introducing root-knot nematode resistance into cultivated groundnut (Burow *et al.*, 1996; Simpson and Starr, 2001).

Lucerne

Medicago sativa (lucerne) is an important perennial food crop of the family Leguminosae, tribe Trifolieae genus *Medicago*. It is an outcrossing autotetraploid (Stanford, 1951), with 2n = 4x = 32 (Armstrong, 1954; Demarly, 1954), allogamous and seed-propagated (Barnes *et al.*, 1988) and is included in the *Medicago sativa* complex along with diploid and tetraploid relatives. Due to the outcrossing nature of lucerne and the buffering capacity of polyploidy, it carries a high level of deleterious recessive alleles (Brouwer and Osborn, 1999). Genetic characterization of lucerne has lagged behind other major crops, due to tetrasomic inheritance and inbreeding depression (McCoy and Bingham, 1988; Mengoni *et al.*, 2000).

Fusion between different ploidy levels of *Medicago* species has occurred through asymmetric hybridization (Kuchuk *et al.*, 1990). Pupilli *et al.* (1992) reported the only symmetric hybrid between different levels of ploidy among *Medicago* species; they fused *M. sativa* (2n = 4x = 32) with *Medicago coerulea* (2n = 2x = 16). Although these species are very similar genetically (Quiros and Bauchan, 1988), they have different ploidy levels. Therefore unreduced gametes are necessary for sexual crosses between them (McCoy and Bingham, 1988). Since *M. coerulea* and *Medicago falcata* belong to the 'sativa–falcata–coerulea' *Medicago* complex, fertilization is possible with *M. sativa* at the same ploidy level (Mariani and Veronesi, 1979). Most genetic maps of lucerne have been constructed in diploids because of the complexity of tetrasomic inheritance (Brummer *et al.*, 1993; Echt *et al.*, 1993; Kiss *et al.*, 1993; Tavoletti *et al.*, 1996; Kalo *et al.*, 2000). However, two genetic maps have been constructed in tetraploid populations (Brouwer and Osborn, 1999; Julier *et al.*, 2003).

Soybean (paleopolyploid nature of the genome)

The north Asian subgenus *soja* has been suggested to be the probable wild progenitor of the cultigen *Glycine max* (L.) Merr. (Doyle *et al.*, 2003). However, the soybean genome has been described as having both allo- and autopolyploid origin. An allopolyploid soybean genome was first hypothesized based on cytogenetic (Singh and Hymowitz, 1985) and molecular studies (Lee and Verma, 1984b; Shoemaker *et al.*, 1996). However, on the basis of the phylogenetic analysis of nuclear genes, its autopolyploid origin has also been hypothesized (Doyle *et al.*, 2003; Straub *et al.*, 2006). Although due to the absence of diploid progenitors or their close relatives the allopolyploid origin of soybean is not supported, a novel cytogenetic approach was used to provide nearly incontrovertible evidence for an allopolyploid origin in soybean (Udall and Wendel, 2006). Fluorescence *in situ* hybridization (FISH) has also distinguished ten chromosome pairs, suggesting that the soybean nucleus contains two distinct, co-resident genomes having two types of centromere, presumably reflecting divergence in its two diploid progenitors (Udall and Wendel, 2006).

Haploid genome studies have suggested that soybean is a diploidized ancient tetraploid (Hadley and Hymowitz, 1973), and the high number of gene families has long supported this hypothesis (Lee and Verma, 1984a; Hightower and Meagher, 1985; Grandbastien *et al.*, 1986; Nielsen *et al.*, 1989; Shoemaker *et al.*, 2002). The genetic map data of soybean reveal multiple nested duplications that appear to reflect an even more ancient round of polyploidy at some point in the ancestry of the genus (Shoemaker *et al.*, 2006). It is suggested that the ancestral 'diploid' genome donors of modern 'allopolyploid' soybean were themselves stabilized paleopolyploids from an earlier round of genome duplication. This nested history of cyclical or episodic polyploidy is the rule rather than the exception for all plant genomes that have been investigated in detail (Udall and Wendel, 2006). Shoemaker *et al.* (1996) compared the relative positions of RFLP probes across nine different mapping populations of soybean and found more than 90% of the probes detected two or more hybridizing genomic fragments, and ~60% detected three or more fragments. By comparing the markers duplicated across different linkage groups, they observed that each chromosome segment is duplicated on average 2.55 times, suggesting that one of the soybean genomes may have undergone additional duplication prior to tetraploidization (Shoemaker *et al.*, 1996; Lee *et al.*, 1999, 2001). A study of 256 duplicated genes identified by EST (expressed sequence tag) sequences showed that soybean has undergone at least two major rounds of duplication at approximately 14.5 and 45 MYA (Blanc and Wolfe, 2004; Schlueter *et al.*, 2004). A phylogenetic approach used by Pfeil *et al.* (2005) determined that the ancient duplication in soybean was shared between soybean and *Medicago*, and probably with all of legumes approximately 50 MYA.

Sequencing of BACs (bacterial artificial chromosomes) anchored by duplicated genes suggests that while the soybean genome is a diploidized paleopolyploid, an astounding amount of sequence is conserved (Schlueter *et al.*, 2006, 2007; Innes *et al.*, 2008). The full genome sequence supports the numerous previous studies suggesting cyclic rounds of duplication. Schumtz *et al.* (2010) found that nearly 70% of the gene space still exists in multiple copy, and hypothesized the most recent duplication event to have occurred 9–13 MYA.

A number of perennial diploid relatives of *Glycine* have been found throughout Australia and Papua New Guinea, and, among these, diploid species have intercrossed and resulted in some allopolyploid taxa (Doyle *et al.*, 2004). Doyle *et al.* (2004) have defined the *tomentella* and *tabacina* complexes, which have been described as allopolyploids found in the wild. These resulted from various combinations of diploid progenitors, which support the view that these polyploids have arisen through multiple origins. Though these species are not considered food legumes, they are important indicators of the propensity for polyploidy formation in wild legumes and for generating variation for soybean improvement.

7.6 Conclusion

We must not forget that most crop legumes are actually ancient polyploids with a major duplication event shared across many genera prior to speciation approximately 54 MYA (Blanc and Wolfe, 2004; Schlueter *et al.*, 2004; Schumtz *et al.*, 2010). Evidence for this duplication event has been found in many legumes for which sequence resources are available. Polyploidy across the legumes – and specifically in the crop legumes – is still being investigated. The Doyle Laboratory is currently working to determine 'cryptic-polyploids' using next-generation sequencing technologies (J.J. Doyle, 2010, personal communication). It is certain that the costs of sequencing will steadily continue to decrease, and that genomes of the so-called 'orphan' legumes will be sequenced allowing for evolutionary studies potentially to identify duplication events. What is evident is that polyploidy has played a significant role in shaping the role of many legumes as crop species.

References

Adams, K.L. and Wendel, J.F. (2005) Novel patterns of gene expression in polyploid plants. *Trends in Genetics* 10, 539–543.

Armstrong, J.M. (1954) Cytological studies in lucerne polyploids. *Candian Journal of Botany* 32, 531–542.

Barnes, D.K., Goplen, B.P. and Baylor, J.E. (1988) Highlights in the USA and Canada. In: Hanson, A.A., Barnes, D.K. and Hill R.R. (eds) *Alfalfa and Alfalfa Improvement. Journal of the American Society of Agronomy* Monograph 29, 1–24.

Bennetzen, J.L. (2002) Mechanisms and rates of genome expansion and contraction in flowering plants. *Genetica* 115, 29–36.

Blanc, G. and Wolfe, K.H. (2004) Widespread paleopolyploidy in model plant species inferred from age distributions of duplicate genes. *Plant Cell* 16, 1667–1678.

Brouwer, D.J. and Osborn, T.C. (1999) A molecular marker linkage map of tetraploid lucerne (*Medicago sativa* L.). *Theoretical and Applied Genetics* 99, 1194–1200.

Brummer, E.C., Bouton, J.H. and Kochert, G. (1993) Development of an RFLP map in diploid lucerne. *Theoretical and Applied Genetics* 86, 329–332.

Burow, M.D., Simpson, C.E., Paterson, A.H. and Starr, J.L. (1996) Identification of peanut (*Arachis hypogaea* L.) RAPD markers diagnostic of root-knot nematode (*Meloidogyne arenaria* (Neal) Chitwood) resistance. *Molecular Breeding* 2, 369–379.

Burow, M.D., Simpson, C.E., Starr, J.L. and Paterson, A.H. (2001) Transmission genetics of chromatin from a synthetic amphidiploid to cultivated peanut (*Arachis hypogaea* L.): Broadening the gene pool of a monophyletic polyploid species. *Genetics* 159, 823–837.

Cronn, R.C., Small, R.L. and Wendel, J.F. (1999) Duplicated genes evolve independently after polyploid formation in cotton. *Proceedings of the National Academy of Sciences U.S.A.* 96, 14406–14411.

Demarly, Y. (1954) Etude de l'hérédité de la bigarrure de la fleur chez la luzerne. *Ann Amélior Plantes* 4, 5–20.

Doyle, J.J., Doyle, J.L., Brown, A.H.D. and Grace, J.P. (1990) Multiple origins of polyploids in the *Glycine tabacina* complex inferred from chloroplast DNA polymorphism. *Proceedings of the National Academy of Sciences U.S.A.* 87, 714–717.

Doyle, J.J., Doyle, J.L. and Harbison, C. (2003) Chloroplast-expressed glutamine synthetase in *Glycine* and related Leguminosae: phylogeny, gene duplication, and ancient polyploidy. *Systemic Botany* 28, 567–577.

Doyle, J.J., Doyle, J.L., Rauscher, J.T. and Borwn, A.H.D. (2004) Evolution of the perennial soybean polyploidy complex (*Glycine* subgenus *Glycine*): a study of contrasts. *Biological Journal of the Linnean Society* 82, 583–597.

Echt, C.S., Kidwell, K.K., Knapp, S.J, Osborn, T.C. and McCoy, T.J. (1993) Linkage mapping in diploid lucerne (*Medicago sativa*). *Genome* 37, 61–71.

Garcia, G.M., Stalker, H.T. and Kochert, G. (1995) Introgression analysis of an interspecific hybrid population in peanuts (*Arachis hypogaea* L.) using RFLP and RAPD markers. *Genome* 38, 166–176.

Gimenes, M.A., Lopes, C.R. and Valls, J.F.M. (2002) Genetic relationships among *Arachis* species based on AFLP. *Genetics and Molecular Biology* 25, 349–353.

Grandbastien, M.A., Berry-Lowe, S., Shirley, B.W. and Meagher, R. (1986) Two soybean ribulose-1,5-bisphosphate carboxylase small subunit genes share extensive homology even in distant flanking sequences. *Plant Molecular Biology* 7, 451–465.

Grant, V. (1981) *Plant Speciation*. Columbia University Press, New York.

Hadley, H.H. and Hymowitz, T. (1973) Speciation and cytogenetics. In: Caldwell, B.E. (ed.) *Soybeans: Improvement, Production, and Uses, 1st edn*. American Society of Agronomy, Madison, Wisconsin, pp. 97–116.

Halward, T.M., Stalker, H.T., LaRue, E.A. and Kochert, G. (1991) Genetic variation detectable with molecular markers among unadapted germplasm resources of cultivated peanut and related wild species. *Genome* 34, 1013–1020.

Halward, T.M., Stalker, H.T., LaRue, E.A. and Kochert, G. (1992) Use of single-primer DNA amplification in genetic studies of peanut (*Arachis* hypogaea L.). *Plant Molecular Biology* 18, 315–325.

He, G. and Prakash, C.S. (1997) Identification of polymorphic DNA markers in cultivated peanut (*Arachis* hypogaea L.). *Euphytica* 97, 143–149.

Herselman, L. (2003) Genetic variation among Southern African cultivated peanut (*A. hypogaea* L.) genotypes as revealed by AFLP analysis. *Euphytica* 133, 319–327.

Hightower, R. and Meagher, R. (1985) Divergence and differential expression of soybean actin genes. *EMBO Journal* 4, 1–8.

Hopkins, M.S., Casa, A.M., Wang, T., Mitchell, S.E., Dean, R., Kochert, G.D. *et al.* (1999) Discovery and characterization of polymorphic simple sequence repeats (SSRs) in peanut. *Crop Science* 39, 1243–1247.

Husted, L. (1933) Cytological studies of the peanut *Arachis*. I. Chromosome number and morphology. *Cytologia* 5, 109–117.

Husted, L. (1936) Cytological studies on the peanut, *Arachis*. II. Chromosome number, morphology and behavior, and their application to the problem of the origin of the cultivated forms. *Cytologia* 7, 396–422.

Innes, R.W., Ameline-Torregrosa, C., Ashfield, T., Cannon, E., Cannon, S.B., Chacko, B. *et al.* (2008) Differential accumulation of retroelements and diversification of NB-LRR disease resistance genes in duplicated regions following polyploidy in the ancestor of soybean. *Plant Physiology* 148, 1740–1759.

Isleib, T.G. and Wynne, J.C. (1992) Use of plant introductions in peanut improvement. In: Shands, H.L. (ed.) *Use of Plant Introductions in Cultivar Development, vol. 2*. Crop Science Society of America, Madison, Wisconsin, pp. 75–116.

Julier, B., Flajoulot, S., Barre, P., Cardinet, G., Santoni, S., Huguet, T. *et al.* (2003) Construction of two genetic linkage maps in cultivated tetraploid lucerne (*Medicago sativa*) using microsatellite and AFLP markers. *BMC Plant Biology* 3, 9.

Kalo, P., Endre, L., Zimanyi, G., Csanadi, G. and Kiss, G.B. (2000) Construction of an improved linkage map of diploid lucerne (*Medicago sativa*). *Theoretical and Applied Genetics* 100, 641–657.

Kiss, G.B., Csanadi, G., Kalman, K., Kalo, P. and Okresz, L. (1993) Construction of a basic genetic map for lucerne using RFLP, RAPD, isozyme and morphological markers. *Molecular and General Genetics* 238, 129–137.

Kochert, G., Halward, T., Branch, W.D. and Simpson, C.E. (1991) RFLP variability in peanut (*Arachis hypogaea* L.) cultivars and wild species. *Molecular and General Genetics* 81, 565–570.

Kochert, G., Stalker, H.T., Gimenes, M.A., Galgaro, M.L., Lopes, C.R. and Moore, K, (1996) RFLP and cytogenetic evidence on the origin and evolution of allotetraploid domesticated peanut, *Arachis hypogaea* (Leguminosae). *American Journal of Botany* 83, 1282–1291.

Kollipara, K.P., Singh, R.J. and Hymowitz, T. (1994) Genomic diversity and multiple origins of tetraploid (2n = 78, 80) *Glycine tomentella*. *Genome* 37, 448–459.

Krapovickas, A. and Gregory, W.C. (1994) Taxonomia del género *Arachis* (Leguminosae). *Bonplandia* 8, 1–186 [in Spanish, with English abstract].

Kuchuk, N.V., Borisynk, N.V. and Gleba, Y.Y. (1990) Isolation and analysis of somatic hybrid cell lines and plants of Medicago. In: *Abstract of the 7th International Congress on Plant Tissue and Cell Culture*, Amsterdam, p. 213.

Lagercrantz, U. and Lydiate, D.J. (1996) Comparative genome mapping in *Brassica*. *Genetics* 144, 1905–1910.

Lanham, P.G., Fennell, S., Moss, J.P. and Powell, W. (1992) Detection of polymorphic loci in *Arachis* germplasm using random amplified polymorphic DNAs. *Genome* 35, 885–889.

Lavia, G.I. (1998) Karyotypes of *Arachis palustris* and *A. praecox* (Section *Arachis*), two species with basic chromosme number x = 9. *Cytologia* 63, 177–181.

Lawton-Rauh, A. (2003) Evolutionary dynamics of duplicated genes in plants. *Molecular Phylogenetics and Evolution* 29, 396–409.

Lee, J.M., Bush A., Specht, J.E. and Shoemaker, R. (1999) Mapping duplicate genes in soybean. *Genome* 42, 829–836.

Lee, J.M., Grant, D., Vallejos, C.E. and Shoemaker, R. (2001) Genome organization in dicots. II. Arabidopsis as a 'bridging species' to resolve genome evolution events among legumes. *Theoretical and Applied Genetics* 103, 765–773.

Lee, J.S. and Verma, D.P.S. (1984a) Chromosomal arrangement of leghemoglobin genes in soybean. *Nucleic Acids Research* 11, 5541–5553.

Lee, J.S., and Verma, D.P.S. (1984b) Structure and chromosomal arrangement of leghemoglobin genes in kidney bean suggest divergence in soybean leghemoglobin gene loci following tetraploidization. *EMBO Journal* 3, 2745–2752.

Leipoldt, M. and Schmidtke, J. (1982) Gene expression in phylogenetically polyploid organisms. In: Dover, G.A. and Flavell, R.B. (eds) *Genome Evolution*. Academic Press, London.

Liu, B., Brubaker, C.L., Mergeai, G., Cronn, R.C. and Wendel, J.F. (2000) Polyploid formation in cotton is not accompanied by rapid genomic changes. *Genome* 44, 321–330.

Lockton, S. and Gaut, B.S. (2005) Plant conserved non-coding sequences and paralogue evolution. *Trends in Genetics* 21, 60–65.

Mable, B.K. (2004) 'Why polyploidy is rarer in animals than plants' myths and mechanisms. *Biological Journal of the Linnean Society* 82, 453–466.

Mariani and Veronesi (1979) Cytological and fertility relationships of different *Medicago* species and cytogenetic behaviour of their hybrids. *Genetics and Agriculture* 33, 245–268.

Masterson, J. (1994) Stomatal size in fossil plants: evidence for polyploidy in majority of angiosperms. *Science* 264, 421–424.

McClintock, B. (1984) The significance of responses of the genome to challenge. *Science* 226, 792–801.

McCoy, T.J. and Bingham, E.T. (1988) Cytology and cytogenetics of lucerne. In: Hanson, A.A., Barnes, D.K. and Hill, R.R. (eds) *Alfalfa and Alfalfa Improvement, American Society of Agronomy* Monograph 29, 737–776.

Mengoni, A., Gori, A. and Bazzicalupo, M. (2000) RAPD and microsatellite (SSR) variation used for assessing genetic relationships among populations of tetraploid lucerne, *Medicago sativa. Plant Breeding* 119, 309–315.

Milla, S.R., Isleib, T.G. and Stalker, H.T. (2005) Taxonomic relationships among *Arachis* sect. *Arachis* species as revealed by AFLP markers. *Genome* 48, 1–11.

Moretzsohn, M.C., Hopkins, M.S., Mitchell, S.E., Kresovich, S., Valls, J.F. and Ferreira, M.E. (2004) Genetic diversity of peanut (*Arachis hypogaea* L.) and its wild relatives based on the analysis of hypervariable regions of the genome. *BMC Plant Biology* 4, 11.

Nielsen, N.C., Dickinson, C., Cho, T.J., Thanh, V.H., Scallon, B.J., Fischer, R.L. *et al.* (1989) Characterization of the glycinin gene family in soybean. *Plant Cell* 1, 313–328.

Ohno, S. (1970) *Evolution by Gene Duplication*. Springer-Verlag, New York.

Parkin, I.A., Sharpe, A.G. and Lydiate, D.J. (2003) Patterns of genome duplication within the *Brassica napus* genome. *Genome* 46, 291–303.

Pfeil, B.E., Schlueter, J.A., Shoemaker, R.C. and Doyle, J.J. (2005) Placing paleopolyploidy in relation to taxon divergence: a phylogenetic analysis in legumes using 39 gene families. *Systematic Biology* 54, 441–454.

Pickett, F.B. and Meeks-Wagner, D.R. (1995) Seeing double: Appreciating genetic redundancy. *Plant Cell* 7, 1347–1356.

Pikaard, C.S. (2001) Genomic change and gene silencing in polyploids. *Trends in Genetics* 17, 675–677.

Pupilli, F., Scarpa, M.G., Damiani, F. and Arcioni, S. (1992) Production of interspecific somatic hybrid plants in the genus *Medicago* through protoplast fusion. *Theoretical and Applied Genetics* 84, 792–797.

Quiros, C.F. and Bauchan, G.R. (1988) The genus *Medicago* and the origin of the *Medicago sativa* complex. In: Hanson, A.A., Barnes, D.K. and Hill, R.R. (eds) *Alfalfa and Alfalfa Improvement*. American Society of Agronomy, Madison, Wisconsin, pp. 93–124.

Raina, S.N., and Mukai, Y. (1999) Genomic *in situ* hybridization in *Arachis* (Fabaceae) identifies the diploid wild progenitors of cultivated (*A. hypogaea*) and related wild (*A. monticola*) peanut. *Proceedings of the National Academy of Sciences U.S.A.* 92, 7719–7723.

Raina, S.N., Rani, V., Kojima, T., Ogihara, Y., Singh, K.P. and Devarumath, R.M. (2001) RAPD and ISSR fingerprints as useful genetic markers for analysis of genetic diversity, varietal identification, and phylogenetic relationships in peanut (*Arachis hypogaea*) cultivars and wild species. *Genome* 44, 763–772.

Robledo, G. and Seijo, G. (2008) Characterization of the *Arachis* (Leguminosae) D genome using fluorescence in situ hybridization (FISH) chromosome markers and total genome DNA hybridization. *Genetics and Molecular Biology* 31, 717–724.

Schlueter, J.A., Dixon, P., Granger, C., Grant, D., Clark, L., Doyle, J.J. *et al.* (2004) Mining EST databases to resolve evolutionary events in major crop species. *Genome* 47, 868–876.

Schlueter, J.A., Scheffler, B.E., Schlueter, S.D. and Shoemaker, R.C. (2006) Sequence conservation of homeologous BACs and expression of homeologous genes in soybean (*Glycine max* L. Merr). *Genetics* 174, 1017–1028.

Schlueter, J.A., Lin, J.-Y., Schlueter, S.D., Vasylenko-Sanders, I.F., Deshpande, S., Yi, J. *et al.* (2007) Genome duplication in soybean and the implications for whole genome sequence assemblies. *BMC Genomics* 8, 330.

Schmidt, R. (2002) Plant genome evolution: lessons from comparative genomics at the DNA level. *Plant Molecular Biology* 48, 21–37.

Schmutz, J., Cannon, S.B., Schlueter, J.A., Ma, J., Hyten, D., Song, Q. *et al.* (2010) Genome sequence of the paleopolyploid soybean (*Glycine max* (L.) Merr.). *Nature* 463, 178–183.

Seijo, J.G., Lavia, G.I., Fernández, A., Krapovickas, A., Ducasse, D. and Moscone, E.A. (2004) Physical mapping of 5S and 18S-25S rRNA genes evidences that *Arachis duranensis* and *A. ipaensis* are the wild diploid species involved in the origin of *A. hypogaea* (Leguminosae). *American Journal of Botany* 91, 1294–1303.

Seijo, J.G., Lavia, G.I., Fernandez, A., Krapovickas, A., Ducasse, D.A., Bertioli, D.J. *et al.* (2007) Genomic relationships between the cultivated peanut (*Arachis hypogaea*, Leguminosae) and its close relatives revealed by double GISH. *American Journal of Botany* 94, 1963–1971.

Shoemaker, R., Polzin, K., Labate, J., Specht, J., Brummer, E.C., Olson, T. *et al.* (1996) Genome duplication in soybean (*Glycine* subgenus *soja*). *Genetics* 144, 329–338.

Shoemaker, R., Keim, P., Vodkin, L., Retzel, E., Clifton, S.W., Waterston, R. *et al.* (2002) A compilation of soybean ESTs: generation and analysis. *Genome* 45, 329–338.

Shoemaker, R.C., Schlueter, J. and Doyle, J.J. (2006) Paleopolyploidy and gene duplication in soybean and other legumes. *Current Opinion in Plant Biology* 9, 104–109.

Simpson, C.E. and Starr, J.L. (2001) Registration of 'COAN' Peanut. *Crop Science* 41, 918–918.

Simpson, C.E., Starr, J.L., Nelson, S.C., Woodard, K.E. and Smith, O.D. (1993) Registration of TxAG-6 and TxAG-7 peanut germplasm. *Crop Science* 33, 1418.

Singh, R.J. and Hymowitz, T. (1985) The genomic relationships among six wild perennial species of the genus *Glycine* subgenus *Glycine* Wild. *Theoretical and Applied Genetics* 71, 221–230.

Smartt, J. and Stalker, H.T. (1982) Speciation and cytogenetics in *Arachis*. In: Pattee, H.E. and Young, C.E. (eds) *Peanut Science and Technology*. American Peanut Research Education Society, Yoakum, Texas, pp. 21–49.

Smartt, J., Gregory, W.C. and Gregory, M.P. (1978) The genomes of *Arachis hypogaea*. 1. Cytogenetic studies of putative genome donors. *Euphytica* 27, 665–675.

Soltis, D.E. and Soltis, P.S. (1995) The dynamic nature of polyploid genomes. *Proceedings of the National Academy of Sciences U.S.A.* 92, 8089–8091.

Soltis, D.E. and Soltis, P.S. (1999) Polyploidy: recurrent formation and genome evolution. *Trends in Ecology and Evolution* 14, 348–352.

Soltis, P.S. and Soltis, D.E. (2000) The role of genetic and genomic attributes in the success of polyploids. *Proceeding of the National Academy of Science U.S.A.* 97, 7051–7057.

Song, K.M., Lu, P., Tang, K.L. and Osborn, T.C. (1995) Rapid genomic changes in synthetic polyploids of *Brassica* and its implications for polyploid evolution. *Proceedings of the National Academy of Sciences U.S.A.* 92, 7719–7723.

Stalker, H.T. (1985) Cytotaxonomy of *Arachis*. In: Moss, J.P. (ed.) *Proceedings of International Workshop on Cytogenetics of Arachis*, 31 October–2 November 1983. ICRISAT, India, pp. 65–79.

Stalker, H.T. and Moss, J.P. (1987) Speciation, cytogenetics and utilization of *Arachis* species. *Advances in Agronomy* 41, 1–40.

Stanford, E.H. (1951) Tetrasomic inheritance in lucerne. *Agronomy Journal* 43, 222–225.

Stebbins, G.L. (1966) Chromosomal variation and evolution. *Science* 152, 1463–1469.

Straub, S.C.K., Pfeil, B.E. and Doyle, J.J. (2006) Testing the polyploid past of soybean using a low-copy nuclear gene – Is *Glycine* (Fabaceae: Papilionoideae) an auto- or allopolyploid? *Molecular Phylogenetics and Evolution* 39, 580–584.

Subramanian, V., Gurtu, S., Nageswara Rao, R.C. and Nigam, S.N. (2000) Identification of DNA polymorphism in cultivated groundnut using random amplified polymorphic DNA (RAPD) assay. *Genome* 43, 656–660.

Tavoletti, S., Veronesi, F. and Osborn, T.C. (1996) RFLP linkage map of an alfalfa meiotic mutant based on an F1 population. *Journal of Heredity* 87, 167–170.

Udall, J.A. and Wendel, J.F. (2006) Polyploidy and crop improvement. *Crop Science* 46(S1), S3–S14.

Upadhyaya, H.D., Gowda, C.L.L., Buhariwalla, H.K. and Crouch, J.H. (2006) Efficient use of crop germplasm resources: identifying useful germplasm for crop improvement through core and mini core subsets and molecular marker approaches. *Plant Genetic Resources* 4, 25–35.

Valls, J.F.M. and Simpson, C.E. (2005) New species of *Arachis* L. (Leguminosae) from Brazil, Paraguay and Bolivia. *Bonplandia (Argentina)* 14, 35–64.

Wendel, J.F. (2000) Genome evolution in polyploids. *Plant Molecular Biology* 42, 225–249.

Wendel, J.F. and Cronn, R.C. (2002) Polyploidy and the evolutionary history of cotton. *Advances in Agronomy* 78, 139–186.

Werth, C.R. and Windham, M.D. (1991) A model for divergent, allopatric speciation of polyploid pteridophytes resulting from silencing of duplicate-gene expression. *American Naturalist* 137, 515–526.

Wolfe, K.H. (2001) Yesterday's polyploids and the mystery of diploidization. *Nature Reviews Genetics* 2, 333–341.

Young, N.D., Weeden, N. and Kochert, G. (1996) Genome mapping in legumes. In: Paterson, A. (ed.) *Genome Mapping in Plants*. Landes, Austin, Texas, pp. 212–227.

8 Cytology and Molecular Cytogenetics

Nobuko Ohmido

8.1 Introduction

Leguminosae is the second most important family after the Poaceae, as they provide sources of food, feed for livestock and raw materials such as oil and protein for industry (Graham and Vance, 2003). There are 700 genera and 20,000 species in the Fabaceae family that comprises the third largest family of flowering plants and displays a striking variety of plant types, ranging from small annual herbs to massive tropical trees. Within the legumes themselves, nodulation occurs in more than 90% of papilionoid genera and just below that percentage of mimosoid genera (Doyle and Luckow, 2003). Among the legumes, the subfamily Papilionoideae contains the majority of pulse crops such as pea (*Pisum sativum*, $2n=14$, 5000 Mb), lucerne (*Medicago sativa*, $2n=16$, 1600 Mb) and soybean (*Glycine max* L. Merr., $2n=40$, 1100 Mb).

For legume chromosome research with large chromosomes, such as *Vicia faba* ($2n=12$) and *Pisum sativum* ($2n=14$), it is now possible to use ordinary karyotyping and/or banding methods for chromosome identification. A comprehensive survey of the molecular and cytogical features of the chromosome complement was provided for *V. faba* based on fluorescence *in situ* hybridization (FISH) and various Giemsa and fluorescence banding patterns (Fuchs *et al.*, 1998a). Physical mapping by FISH plays an important role in collating information from linkage and chromosome maps, as has been demonstrated for pea (Fuchs *et al.*, 1998b). On the other hand, in the case of legumes with small chromosomes, identification of individual chromosomes and their centromeric positions is difficult, especially chromosomes that are condensed after pretreatment with colchicine, 8-hydroxyquinoline or cold water.

The chromosome image analysing system (CHIAS) for small chromosomes makes use of distinct stainability along mitotic prometaphase chromosomes, due to uneven condensation, a feature specific to small plant chromosomes (Fukui and Iijima, 1991). The density profiles at the centre line of both chromatids (midrib line) of prometaphase chromosomes allowed establishment of the first chromosome maps of several legumes with small chromosomes (Yanagisawa *et al.*, 1991; Ito *et al.*, 2000; Sato *et al.*, 2005). In this chapter, cytogenetic and molecular chromosome research into three kinds of legume species is described, and the future of legume research is then discussed.

8.2 High Resolution of Integrated and Genetic Map of *Lotus japonicus* Chromosomes

Lotus japonicus is characterized by a small genome ($2n=2x=12$; genome size per haploid, 472 Mb), relatively short life cycle (2–3 months) and ease of genetic manipulation (e.g. transformation, it being an autogamous diploid plant; Jiang and Gresshoff, 1997; Udvardi *et al.*, 2005). A large-scale sequencing project was initiated for the *L. japonicus* accessions Miyakojima and Gifu, and subsets of genomic sequences are now available (Sato *et al.*, 2001; Nakamura *et al.*, 2002; Asamizu *et al.*, 2003; Kaneko *et al.*, 2003; Kato *et al.*, 2003). Sato and Tabata (2006) have constructed a high-density genetic linkage map of *L. japonicus* and mapped numerous transformation-competent artificial chromosome (TAC) genomic markers (Table 8.1). These are indispensable for Leguminosae studies in various fields, including comparative genomics, gene identification, gene isolation and marker-assisted breeding. Consequently, many microsatellite and simple-sequence repeat (SSR) markers, as well as derived cleaved amplified polymorphic sequences (dCAPS), have been genetically and physically mapped on the *L. japonicus* genome (http://www.kazusa.or.jp/lotus/). Co-dominant markers can be used for map-based cloning of useful protein-coding genes (i.e. transcription factor receptor-like kinase, and transporter and disease resistance genes; Sato *et al.*, 2008).

Map-based cloning requires a dense and precise linkage map of the trait of interest, followed by establishment of the relationship between genetic and physical distances. The identification of individual mitotic prometaphase chromosomes of *L. japonicus* based on condensation patterns (CPs) became feasible, and their chromosome maps were developed (Ito *et al.*, 2000; Hayashi *et al.*, 2001; Pedrosa *et al.*, 2002). However, mitotic prometaphase chromosomes are much smaller than pachytene chromosomes, and thus the resolution for genetic research is limited, probably because the mitotic chromosome length is 4.29–9.64 μm and 1.51–2.67 μm, respectively (Ito *et al.*, 2000; Pedrosa *et al.*, 2002).

We now discuss quantitative pachytene chromosome maps of the six *L. japonicus* chromosomes based on chromosome length, centromeric position, heterochromatin and euchromatin distribution pattern, as well as the position of major repetitive sequences employing FISH and an imaging method using the chromosome image analysis system ver. 3 (CHIAS3) with high-resolution pachytene chromosomes to determine the precise integration between genetic and physical distances in the *L. japonicus* genome (Fig. 8.1).

The image analysis system CHIAS3 (CHIAS III, 2004) was used to analyse the *L. japonicus* chromosomes. A quantitative chromosome idiogram was constructed based on the digitized intensity of the fluorescent signals after counterstaining with DAPI. The original chromosome images for the construction of the idiogram were RGB images, each with 8-bit grey levels. The procedure used to construct the idiogram was as follows: first, the 24-bit RGB images were converted into 8-bit grey images of R, G and B stack images. The chromosome area was delimited based on the DAPI (B) image for each chromomere, and the chromomere indices were established. Midrib lines were drawn along the axis of the chromosome, and the fluorescence intensity of Cy3 (R), FITC (G) and DAPI (B) measured. Next, the average fluorescence profile was computed by measuring the fluorescent intensities of more than three chromosomes from signal-detected images. Finally, the idiogram was constructed based on the average fluorescence profile. The numerical values of the fluorescent intensities of chromomeres were converted into monochrome binary band patterns.

The genomic library of *L. japonicus* was also constructed via TAC, selected on the basis of the sequences of SSR and dCAPS from *L. japonicus* (Sato and Tabata, 2006). TAC clones were selected from the 3-D DNA pools of the TAC libraries by PCR to amplify SSRs. The TAC clones used for FISH mapping are listed in Table 8.1. The 45S ribosomal RNA (rDNA) gene derived from rice and 5S rDNA isolated from *L. japonicus* were employed. The high copy numbers of tandem repeat DNA, LjTR1, LjTR2, LjTR3 and LjTR4, and the retroelements, LjRE1 and LjRE2 with the highest copy numbers, were isolated and cloned from the *L. japonicus* genome (Sato *et al.*, 2008).

Repeated sequences are mapped on the *L. japonicus* genome (Ohmido *et al.*, 2010). LjRE1, a highly repeated retroelement, has long terminal repeats (LTRs) and

Table 8.1. Repetitive sequences, 45S rDNA, 5S rDNA and transformation-competent artificial chromosome (TAC) clones used as probes for fluorescence *in situ* hybridization (FISH). Linkage and physical position data are cited from the *Lotus* genome database (*Lotus japonicus* News, 2011).

Marker	C bne name	Size (bp)	Position (cM)[a]	Mb	Physical location: Chromosome	Physical location: Chromosomal position (%)[b]
LjRE1	Ty-1 Retroelement (copia type)	12,069		Dispersed		
LjRE2	Ty-3 Retroelement (gypsy type)	6840		Centromeres		
LjTR1	Tandem repeat	190		Constitutive hetrochromatin		
LjTR2	Tandem repeat	237		Euchromatin		
LjTR3	Tandem repeat	172		Hetrochromatin		
LjTR4	Tandem repeat	172		Chromosome terminal region		
45S rDNA	Ribosomal RNA gene			2,5 and 6		
5S rDNA	Ribosomal RNA gene			2		
TM0088	LjT15K21		0.0	0.1	1	0.1
TM0063	LjT09L22		4.8	14.8	1	8.7
TM0910	LjT42H23		71.0	87.2	1	98.1
TM0904	LjT33P02		4.0	6.8	2	7.4
TM0153	LjT28L17		10.8	15.5	2	11.4
TM0081	LjT01G01		24.6	24.3	2	–
TM0225	LjT27K02		25.8	25.7	2	–
TM0124	LjT26I01		33.8	34.6	2	57.0
TM0008	LjT10B11		44.2	42.1	2	58.3
TM0021	LjT04I02		60.9	57.9	2	–
TM0031	LjT16N13		68.5	72.3	2	59.9
TM0380	LjT18K09		72.9	80.6	2	94.9
TM0793	LjT23013		0.0	0.005	3	4.2
TM0059	Lj13M14		6.9	7.2	3	5.8
TM0436	LjT13N17		10.5	11.2	3	6.5
TM0111	LjT40002		26.8	27.8	3	17.8
TM0246	LjT34I09		42.0	50.9	3	53.0
TM0217	LjT09C16		74.8	81.9	3	71.7
TM0261	LjT34I09		83.2	88.2	3	90.5
TM0288	LjT36E18		2.0	1.7	4	3.1
TM0131	LjT21G09		21.3	19.4	4	9.1
TM0087	LjT14P20		28.6	31.8	4	34.6
TM0042	LjT10L16		69.2	68.2	4	87.8
TM0089	LjT14E05		0.4	0.7	5	0
TM0048	LjT05P01		27.6	25.7	5	–
TM04148	LjT30P03		44.1	52.2	5	85.5
TM0180	LjT03D07		54.1	61.8	5	95.0
TM0260	LjT47K21		54.9	62.5	5	95.7
TM1383	LjT26K12		1.7	1.2	6	5.7
TM0057	LjT03B03		27.6	32.8	6	48.4
TM1240	LjT33P12		66.6	68.1	6	92.4

[a]Linkage position;
[b]physical position from the end of the short arm of the corresponding chromosome; – the location of the signal shows much variation, with successful detection uncommon.

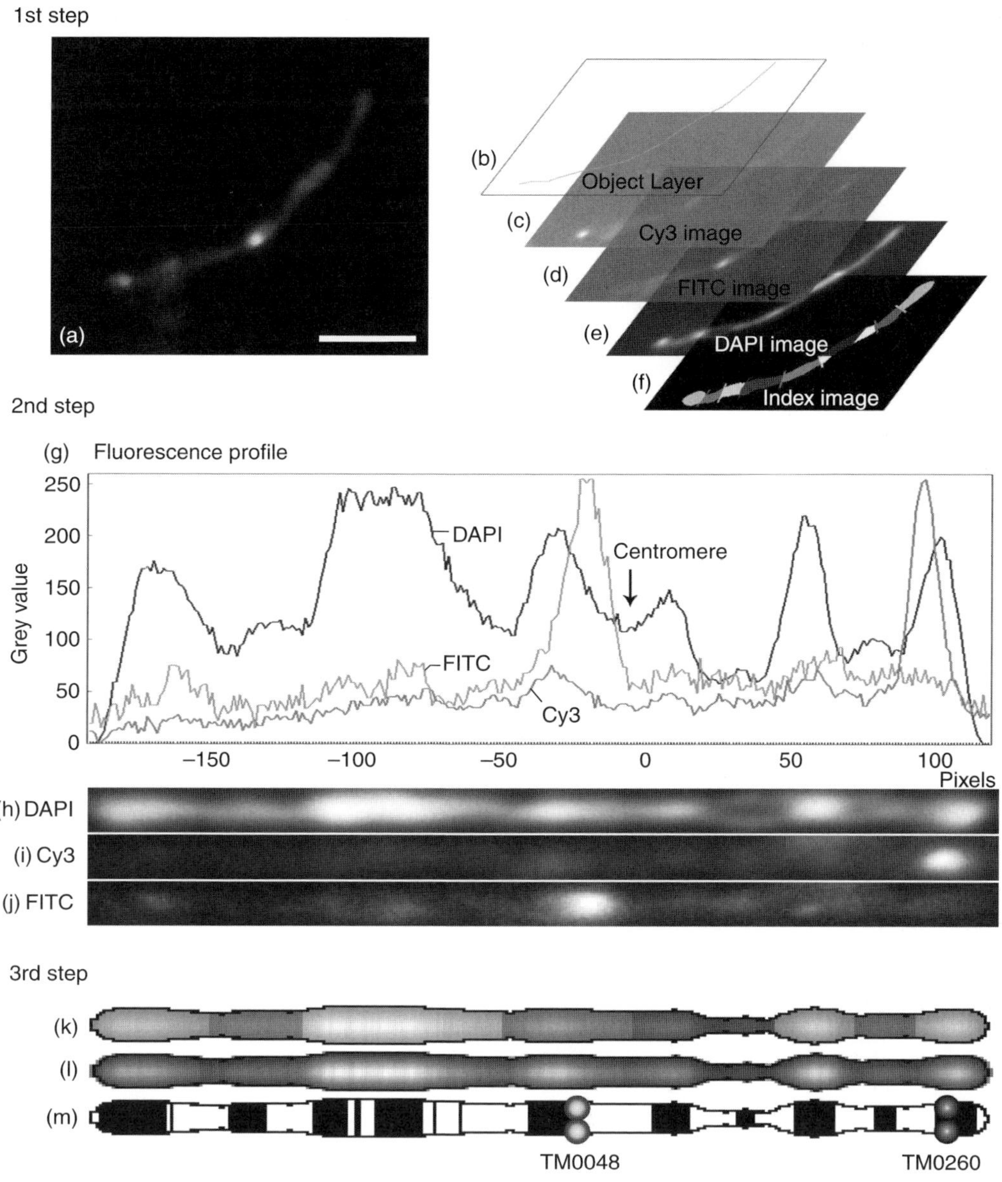

Fig. 8.1. Procedure for development of the quantitative idiogram by CHIAS on *L. japonicus* chromosome 5. A, original RGB image of chromosome 5; B–F, layers of midrib line of the chromosome, Cy3, FITC, DAPI and chromomere-index image; G, fluorescence profiles (FPs) of DAPI, Cy3 and FITC were measured along the midrib line. Each FISH signal was localized into precise chromosomal position; H–J, straightened images of DAPI, Cy3 and FITC, respectively; K–M, idiograms constructed from FP value; K, index idiogram segmented by each chromomere; L, FP image idiogram; M, quantitative pachytene chromosome idiogram with localization of TAC clones.

gag-polymerase genes, and is characterized as a Ty-1 *copia*-type retroelement (Sato *et al.*, 2008). LjRE1 is dispersed throughout euchromatin and heterochromatin regions. The second largest retroelement, LjRE2, characterized by a Ty-3 *gypsy*-type retroelement, was localized on the centromeric regions of *L. japonicus* chromosomes. The fluorescent intensity of the LjRE2 retroelement differed among the six chromosomes; the intensity on chromosome 1 was strong but was weak on chromosome 5. This variation was due to differences in the copy numbers at the pericentromeric regions of each chromosome. The tandem repeat sequence LjTR1 (190 bp unit size) comprised 4.6% of TAC clones in the *L. japonicus* genomic library, which was revealed using end sequences from anchored TACs (Sato *et al.*, 2008). FISH data have shown that LjTR1 was localized at the highly condensed constitutive heterochromatic regions in the *L. japonicus* nucleus and chromosomes (Fig. 8.2). LjTR2 (237 bp unit size) comprised 4.1% of TAC clones and was localized at the decondensed euchromatic regions of all chromosomes (Fig. 8.2). Furthermore, LjTR3 (172 bp unit size) comprised 1.4% of the TAC clones and was localized at specific heterochromatin regions; LjTR4 (172 bp unit size) was localized at the terminal region of all chromosomes, except for the short arms of chromosomes 1 and 2 (data not shown).

An integrated map based on the mitotic chromosome, the pachytene map and linkage map was developed for six individual *L. japonicus* chromosomes using data on the positions of TAC clones and somatic chromosome maps (Ito *et al.*, 2000). The comparison of these maps shows the centromeric position and some interstitial regions, albeit with some recombination distortion (Fig. 8.2). Based on the recombinant frequency, the distance at the terminal regions is apparently evaluated as larger than the physical distance of the chromosomes. The distance between TM0111 and TM0246, including the centromere and heterochromatin, is 2.80 cM/μm, while the terminal region between TM0793 and TM0059 on the short arm is 27.6 cM/μm. These findings suggest that in *L. japonicus*, the recombination frequency at the centromeric region is suppressed by approximately tenfold compared with the terminal region. However, the recombination ratio

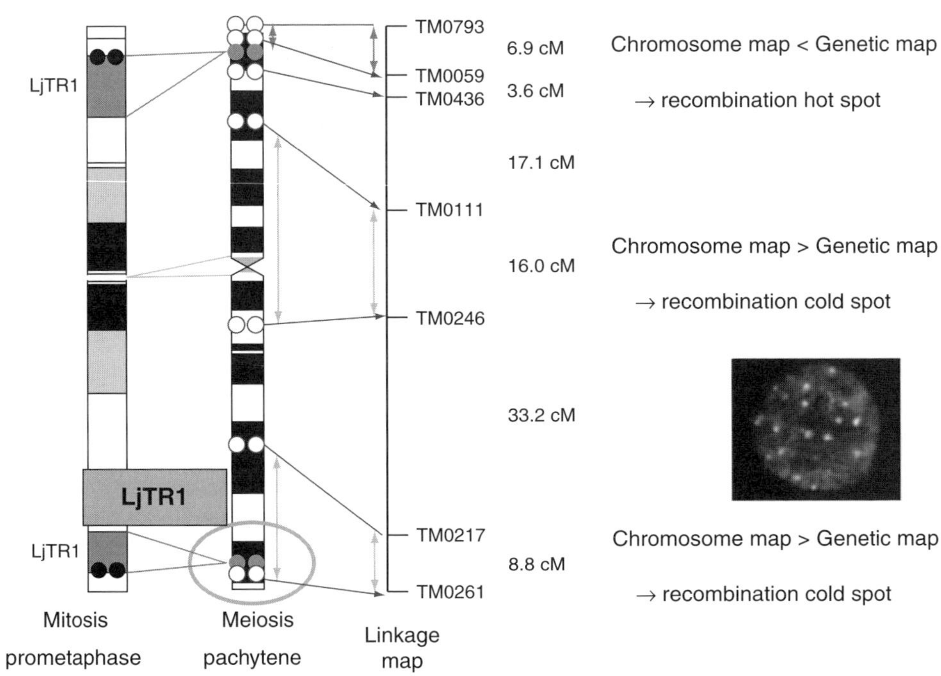

Fig. 8.2. Relationships among cytological features, recombination frequency and the chromosome structure of chromosome 3 by FISH mapping of seven TAC clones in *L. japonicus*. Interphase image represents the FISH mapping of LjTR1 and 45SrDNA.

of the terminal region between TM0217 and TM0261 is similar (2.90 cM/μm) to that of the centromeric region. The large constitutive heterochromatic block comprising LjTE1 found between TM0217 and TM0261 should influence suppression of the recombination frequency on chromosome 3.

The quantification of chromosome density by CHIAS3, *in situ* localization of repetitive sequences and high-resolution mapping of genes and/or markers by FISH are expected to facilitate the analysis of gene density, segment duplication and other chromosome rearrangements and to yield integrated maps for legumes (Ohmido *et al.*, 2010). In particular, probes applicable for *Lotus*, red clover, soybean and other legumes will help in developing a framework for a common genomics of legumes (Ohmido *et al.*, 2007). Molecular cytogenetics may contribute to this goal, for example in the case of rice and tomato (de Jong *et al.*, 1999; Cheng *et al.*, 2001). From the integration of linkage data, chromosome density and physical localization of DNA markers and/or genes, basic research as well as legume breeding will benefit.

8.3 Integrated Chromosome Maps for Red Clover

Red clover has a small genome size (440 Mb), $2n=2x=14$ and its allogamous diploid (Taylor and Chen, 1988). Intra-population genomic heterozygosity in red clover is higher than inter-populations, because it is extremely polymorphic due to its strong self-incompatible fertilization system (Milligan, 1991; Kongkiatngam *et al.*, 1995; Campos-de-Quiroz and Ortega-Klose, 2001). Genomic characteristics have long hampered intensive genetic and genomic analyses in red clover. Recently, Kölliker *et al.* (2003) investigated diverse genetic resources of red clover using amplified fragment length polymorphism (AFLP) markers. In other *Trifolium* species, Isobe *et al.* (2003) reported the first genetic linkage map with RFLP markers and Sato *et al.* (2005) reported 15,000 SSR markers. However, it is not clear whether each linkage group was connected accurately to the corresponding chromosome. In a previous study (Sato *et al.*, 2005), we investigated the consistency between the linkage group and chromosome of red clover strains, HR and R130 using FISH. We developed a red clover chromosome map by the chromosome image analysis system (CHIAS4), which is invaluable for genome comparison.

Red clover karyotyping analysis using metaphase chromosomes was reported (Taylor and Chen, 1988). However, as the seven chromosomes were too highly condensed to identify, the smaller chromosomes were not clear except for NOR chromosome, and the remaining seven chromosomes were similar in size. This karyotype was analysed by microscopic observation of prometaphase chromosomes stained by DAPI. The lengths of the prometaphase chromosomes ranged from 5.1 to 7.4 μm, and uneven condensation patterns that have proved useful in chromosome identification were observed. The resolution of individual chromosomes was better than that found in a previous report (Taylor and Chen, 1988), in which the length of condensed metaphase chromosomes ranged from 1.9 to 2.9 μm, but seven chromosomes could not be definitively distinguished.

Using SSR markers, 26S rDNA, 5S rDNA and BAC clones selected from the 3-D DNA pools of the BAC libraries were used for FISH detection (Sato *et al.*, 2005). Karyotyping of the red clover chromosomes was analysed by mitotic prometaphase chromosomes counterstained with DAPI and 16 BAC clones mapping by FISH. The lengths of the prometaphase chromosomes ranged from 3.3 to 5.6 μm, and uneven CP has proved useful in chromosome identification. The 26S rDNA loci could be detected as the most intensive signals in the nucleolar organizer regions (NORs) of chromosome 1 and as weak signals on the short arms of the internal regions of chromosome 7 (Fig. 8.3a, b). The 5S rDNA loci were detected in the proximal regions on the short arm of chromosome 1 adjacent to NOR, and two minor loci on the short arm of chromosome 2 (Fig. 8.3c).

The cytological map of red clover HR is shown in Fig. 8.3d. Seven microsatellite

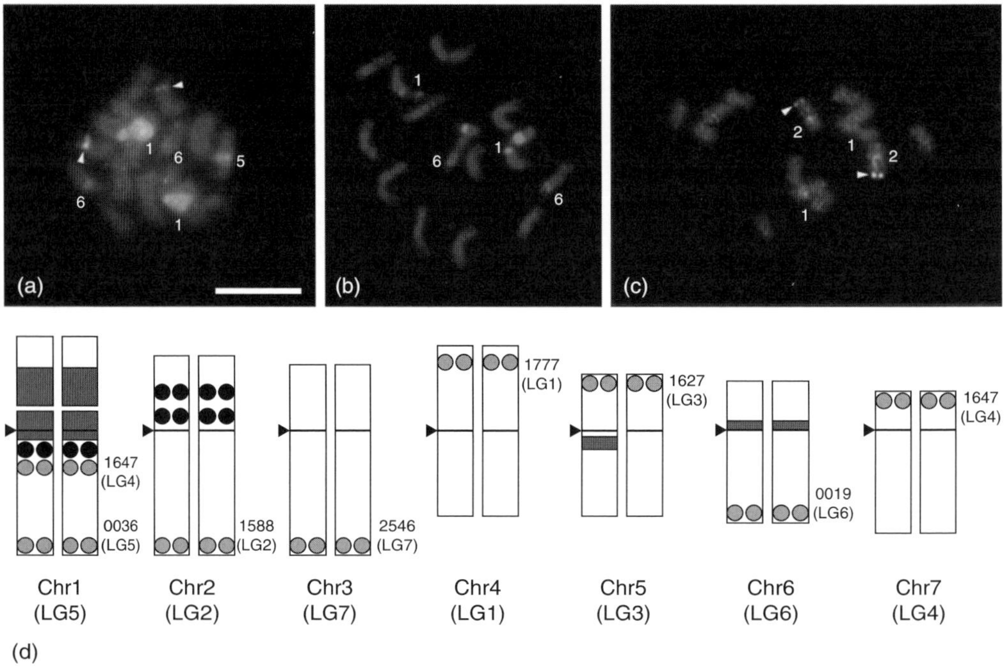

Fig. 8.3. Cytological analysis of red clover. (a) FISH signals for RCS1777 and 28S rDNA on chromosomes of accession HR stained with DAPI. Numbers indicate 28S rDNA loci on chromosomes 1, 5 and 6. Bar=10μm. (b) FISH signals for 28S rDNA on chromosomes 1 and 6 of accession R130. (c) FISH signals for RCS1588 and 5S rDNA (indicated by chromosome numbers 1 and 2) in accession R130. (d) Chromosome map of red clover. Solid light grey circles, loci of seven BACs harbouring linkage group-specific microsatellite markers; dark grey boxes, 28S rDNA loci; solid black circles, 5S rDNA loci. The 28S rDNA locus of chromosome 5 is detected in accession HR but not in accession R130. Arrowheads indicate centromere positions.

markers located close to the end of each linkage group are selected as representatives. BAC clones LG1, LG2, LG3, LG4, LG5, LG6 and LG7 exclusively hybridized on chromosomes 4, 2, 6, 5, 1, 7 and 3, respectively. All signals of BAC clones on seven chromosomes were detected at the portion of each chromosome coinciding with the positions of the respective markers on the corresponding linkage groups. This proves that there is a one-to-one relationship between the seven linkage groups and each chromosome. This study is the first to report on a red clover chromosome map constructed by chromosome mapping. The integration of physical, genetic and quantitative chromosome maps provides valuable information on the genetic data of red clover and should provide further insight into legume genetics.

Six red clover varieties (HR, R130, NS10, H17L, Violetta and M366) in different mapping population were used for polymorphism analysis; 26S rDNA was detected in chromosome 1 in all varieties. A heterozygous 26S rDNA site was detected in HR chromosome 6, but not in other varieties. In chromosome 6 of HR, condensation patterns of homologous chromosomes are different on account of the presence of the 26S rDNA locus. Small signals of 26S rDNA were detected in chromosome 7 of HR, R130, NS10, H17L and Violetta. M366 had one 26S rDNA locus on chromosome 1 only. RCB32E03/RCS6954 was localized on the pericentromeric regions of all chromosomes in all varieties. The differences in the size of the FISH signals was assumed to reflect differences in copy numbers on each chromosome. *Arabidopsis*-type telomere repeats

(TTTAGGG)*n* were localized on the pericentromeric regions of all chromosomes in HR, R130 and NS10 (data not shown). The six red clover varieties from different mapping populations using F_1 populations revealed haplotypes on only specific rDNA gene loci. Specific BAC clones were mapped on the same loci on red clover chromosomes, which should prove the chromosomal co-linearity of even allogamous red clover varieties.

8.4 Chromosome Analysis of Soybean

The haploid soybean (*Glycine max* L. Merr.) genome consists of 1100 Mb packaged into 20 chromosome pairs (Arumuganathan and Earle, 1991) and approximately 40–60% of the DNA is repetitive (Goldberg 1978; Gurley *et al.*, 1979). The mitotic chromosomes are quite small, being only 4–6 µm in length during mitotic prometaphase (Yanagisawa *et al.*, 1991). Previous FISH studies revealed a single 18S-5.8S-28S rDNA locus (Skorupska *et al.*, 1989) and a single 5S rDNA locus (Shi *et al.*, 1996) for *G. max*. One 45S rDNA locus was detected in 17 accessions of 14 diploid species of the genus *Glycine*, including *G. max* and *G. soja* (Singh *et al.*, 2001).

Pachytene chromosomes are much less compact and very useful in molecular cytogenetics. Walling *et al.* (2006) probed chromosomal-level homology in chromosome 19 of soybean. FISH mapping of seven putatively gene-rich BACs from linkage group L (chromosome 19) revealed that most of the genetic map correlates to the highly euchromatic long arm and that there is extensive homeology with another chromosome pair, although the co-linearity of some loci in the genome appears to be conserved.

Soybean represents paleopolyploidy 50 M years ago, when genomes were duplicated and established as a diploid plant. Soybean genome structure is complicated by at least two rounds of polyploidization, called paleopolyploidy. Paleopolyploidy chromosome analyses using the synteny between *Lotus* and soybean have been performed in cytogenetic research and phylogenetic gene analyses. Soybean pachytene chromosomes were mapped using FISH with genomic BAC DNA libraries of soybean selected by common microsatellite markers developed in *L. japonicus*. These results showed two alternately stronger and weaker intensities of fluorescent signals on two different pachytene chromosomes (Fig. 8.4). This represents the presence of the orthologous region of NRF1 (Nod-factor receptor 1) in the genome. The NRF1 gene refers to symbiosis and the genes orders are highly conserved in the two orthologous regions. However, the order of genes in soybean is different in comparison with the orthologous region of *L. japonicus*. It was concluded that internal DNA in the orthologue of soybean had changed, but that genes and mini-satellite markers are conserved beyond the species. Integrated physical, genetic and

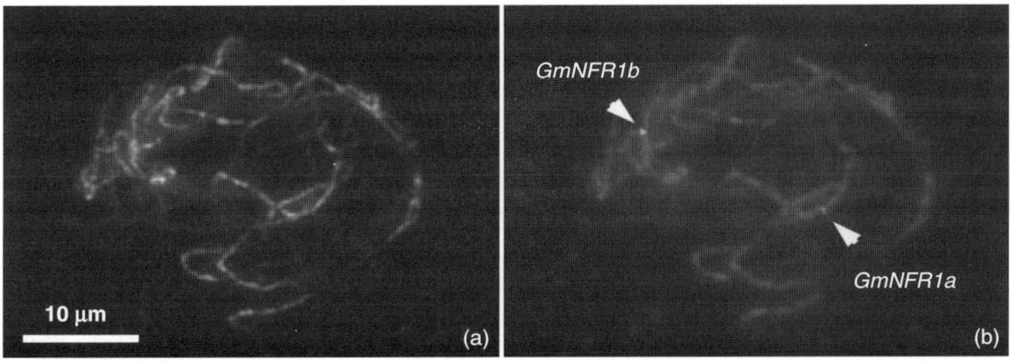

Fig. 8.4. Pachytene chromosomes of soybean using two types of orthologue gene. (a) DAPI image. (b) NFR1 gene sites.

chromosome maps corresponding to the linkage map have been demonstrated. This approach, using common markers derived from the *L. japonicus* genome, would allow the design of chromosome density maps for the complicated soybean paleopolyploidy.

8.5 Conclusions

The quantification of chromosome density by CHIAS, *in situ* localization of repetitive sequences and high-resolution mapping of genes and/or markers by FISH are expected to facilitate the analysis of gene density, segment duplication and other chromosome rearrangements and to yield integrated maps for legumes. Probes especially applicable for *Lotus*, red clover and soybean will help in developing a framework for a common genomics of legumes. Legume sequencing research is in progress (VandenBosch and Stacey, 2003; Schmutz *et al.*, 2010), and molecular cytogenetics may contribute to this goal, as for example in the case of tomato (Szinay *et al.*, 2010). From the integration of linkage data, chromosome density and the physical localization of DNA markers and/or genes, basic research as well as legume breeding will benefit.

Acknowledgements

I sincerely thank Professor Kiichi Fukui (Osaka University) for her excellent support and discussion in conducting this study. I also thank Drs Satoshi Tabata, Shusei Sato and Sachiko Isobe (Kazusa DNA Inst.) for providing the DNA and plant materials and useful discussions. Mr Seiji Kato (Yamanashi Prefectural Agricultural Technology Center), Miss Akiko Ishimaru and Mr Ryohei Kataoka (Kobe University) contributed to research using CHIAS and FISH analysis. This work was supported in part by a grant from Japan Science and Technology: Integration of chromosome maps in allogamous plants, red clover (No. 20580006).

References

Arumuganathan, K. and Earle, E.D. (1991) Nuclear DNA content of some important plant species. *Plant Molecular Reporter* 9, 208–218.

Asamizu, E., Kato, T., Sato, S., Nakamura, Y. and Kaneko, T. (2003) and Satoshi Tabata Structural analysis of a *Lotus japonicus* genome. IV. Sequence features and mapping of seventy-three TAC clones which cover the 7.5 Mb regions of the genome. *DNA Research* 10, 115–122.

Campos-de-Quiroz, H. and Ortega-Klose, F. (2001) Genetic variability among elite red clover (*Trifolium pratense* L.) parents used in Chile as revealed by RAPD markers. *Euphytica* 122, 61–67.

Cheng, Z., Buell, C.R., Wing, R.A., Gu, M. and Jiang, J. (2001) Toward a cytological characterization of the rice genome. *Genome Research* 11, 2133–2141.

CHIAS III (2004) available at http://www2.kobe-u.ac.jp/~ohmido/cl/chiasIII/index.htm (accessed 24 February 2011).

de Jong, J.H., Fransz, P. and Zabel, P. (1999) High resolution FISH in plants – techniques and applications. *Trends in Plant Science* 4, 258–263.

Doyle, J.J. and Luckow, M.A. (2003) The rest of the iceberg. Legume diversity and evolution in a phylogenetic context. *Plant Physiology* 131, 900–910.

Fuchs, J., Strehl, S., Brandes, A., Schweizer, D. and Schubert, I. (1998a) Molecular-cytogenetic characterization of *Vicia faba* genome–heterochromatin differentiation, replication patterns and sequence localization. *Chromosome Research* 6, 219–230.

Fuchs, J., Kuhne, M. and Schubert, I. (1998b) Assignment of linkage groups to pea chromosomes after karyotyping and gene mapping by fluorescent *in situ* hybridization. *Chromosoma* 107, 272–276.

Fukui, K. and Iijima, K. (1991) Somatic chromosome map of rice by imaging methods. *Theoretical and Applied Genetics* 81, 589–596.

Goldberg, R.B. (1978) DNA sequence organization in the soybean plant. *Biochemical Genetics* 16, 45–68.

Graham, P.H. and Vance, C.P. (2003) Legumes, importance and constraints to greater use. *Plant Physiology* 131, 872–877.

Gurley, W.B., Hepburn, A.G. and Key, J.L. (1979) Sequence organization of the soybean genome. *Biochimica et Biophysica Acta* 561, 167–183.
Hayashi, M., Miyahara, A., Sato, S., Kato, T.,Yoshikawa, M., Taketa, M. *et al.* (2001) Construction of a genetic linkage map of the model legume *Lotus japonicus* using an intraspecific F_2 population. *DNA Research* 8, 301–310.
Isobe, S., Klimenko, I., Ivashuta, S., Gau, M. and Kozlov, N.N. (2003) First RFLP linkage map of red clover (*Trifolium pratense* L.) based on cDNA probes and its transferability to other red clover germplasm. *Theoretical and Applied Genetics* 108, 105–112.
Ito, M., Miyamoto, J., Mori, Y., Fujimoto, S., Uchiumi, T., Abe, M. *et al.* (2000) Genome and chromosome dimensions of *Lotus japonicus. Journal of Plant Research* 113, 435–442.
Jiang, Q. and Gresshoff, P.M. (1997) Classical and molecular genetics of the model legume *Lotus japonicus. Molecular Plant Microbe Interact* 10, 59–68.
Kaneko, T., Asamizu, R., Kato, T., Sato, S., Nakamura, Y. and Tabata, S. (2003) Structural analysis of a *Lotus japonicus* genome. III. Sequence features and mapping of sixty-two TAC clones which cover the 6.7 Mb regions of the genome. *DNA Research* 10, 27–33.
Kato, T., Sato, S., Nakamura, Y., Kaneko, T., Asamizu, E. and Tabata, S. (2003) Structural analysis of a *Lotus japonicus* genome. V. Sequence features and mapping of sixty-four TAC clones which cover the 6.4 Mb regions of the genome. *DNA Research* 10, 277–285.
Kölliker, R., Herrmann, D., Boller, B. and Widmer, F. (2003) Swiss Mattenklee landraces, a distinct and diverse genetic resource of red clover (*Trifolium pratense* L.). *Theoretical and Applied Genetics* 107, 306–315.
Kongkiatngam, P., Waterway, M.J., Fortin, M.G. and Coulman, B.E. (1995) Genetic variation within and between two cultivars of red clover (*Trifolium pratense* L.), Comparisons of morphological, isozyme, and RAPD markers. *Euphytica* 84, 237–246.
Lotus japonicus News (2011) available at http://www.kazusa.or.jp/lotus/ (accessed 24 February 2011).
Milligan, B.G. (1991) Chloroplast DNA diversity within and among population of *Trifolium pratense. Current Genetics* 19, 411–416.
Nakamura, Y., Kaneko, T., Asamizu, E., Kato, T., Sato, S. and Tabata, S. (2002) Structural analysis of a *Lotus japonicus* genome. II. Sequence features and mapping of sixty-five TAC clones which cover the 6.5-Mb regions of the genome. *DNA Research* 9, 63–70.
Ohmido, N., Sato, S., Tabata, S. and Fukui, K. (2007) Chromosome maps of legumes. *Chromosome Research* 15, 97–103.
Ohmido, N., Ishimaru, A., Kato, S., Shusei, S., Satoshi, T. and Kiichi F. (2010) Integration of cytogenetic and genetic linkage maps of *Lotus japonicus*, a model plant for legumes. *Chromosome Research* 18, 287–299.
Pedrosa, A., Sandal, N., Stougaard, J., Schweizer, D. and Bachmair, A. (2002) Chromosomal map of the model legume *Lotus japonicus. Genetics* 161, 1661–1672.
Sato, S. and Tabata, S. (2006) *Lotus japonicus* as a platform for legume research. *Current Opinion in Plant Biology* 9, 128–132.
Sato, S., Kaneko, T., Nakamura, Y., Asamizu, E., Kato, T. and Tabata, S. (2001) Structural analysis of a *Lotus japonicus* genome. I. Sequence features and mapping of fifty-six TAC clones which cover the 5.4 Mb regions of the genome. *DNA Research* 8, 311–318.
Sato, S., Isobe S., Asamizu, E., Ohmido, N., Kataoka, R., Nakamura, Y. *et al.* (2005) Comprehensive structural analysis of the genome of red clover (*Trifolium pratense* L.). *DNA Research* 12, 301–364.
Sato, S., Nakamura, Y., Kaneko, T., Asamizu, E., Kato, T., Nakao, M. *et al.* (2008) Genome structure of the legume, *Lotus japonicus. DNA Research* 15, 227–239.
Schmutz, J., Cannon, S.B., Schlueter, J., Ma, J., Mitros, T., Nelson, W. *et al.* (2010) Genome sequence of the palaeopolyploid soybean. *Nature* 463, 178–183.
Shi, L., Zhu, T. and Keim, P. (1996) Ribosomal RNA genes in soybean and common bean, Chromosomal organization, expression, and evolution. *Theoretical and Applied Genetics* 93, 136–141.
Singh, R.J., Kim, H.H. and Hymowitz, T. (2001) Distribution of rDNA loci in the genus *Glycine* Willd. *Theoretical and Applied Genetics* 103, 212–218.
Skorupska, H., Albertsen, M.C., Langholz, K.D. and Palmer, R.G. (1989) Detection of ribosomal RNA genes in soybean, *Glycine max* (L.) Merr., by *in situ* hybridization. *Genome* 32, 1091–1095.
Szinay, D., Bai, Y., Visser, R. and de Jong, H. (2010) FISH applications for genomics and plant breeding strategies in tomato and other solanaceous crops. *Cytogenetic and Genome Research* 129, 199–210.
Taylor, N.L. and Chen, K. (1988) Isolation of trisomics from crosses of diploid, triploid, and tetraploid red clover. *Crop Science* 28, 209–213.

Udvardi, M.K., Tabata, S., Parniske, M. and Stougaard, J. (2005) *Lotus japonicus*, legume research in the fast lane. *Trends in Plant Science* 10, 222–228.

VandenBosch, K. and Stacey, G. (2003) Summaries of legume genomics projects from around the globe. Community resources for crops and models. *Plant Physiology* 131, 840–865.

Walling, J.G., Shoemaker, R., Young, N., Mudge, J. and Jackson, S. (2006) Chromosome-level homeology in paleopolyploid soybean (*Glycine max*) revealed through integration of genetic and chromosome maps. *Genetics* 172, 1893–1900.

Yanagisawa, T., Tano, S., Fukui, K. and Harada, K. (1991) Marker chromosomes commonly observed in the genus *Glycine*. *Theoretical and Applied Genetics* 81, 606–612.

9 Molecular Cytogenetics in Physical Mapping of Genomes and Alien Introgressions

H.K. Chaudhary, V.K. Sood, T. Tayeng, V. Kaila and A. Sood

9.1 Introduction

Legume breeders are usually confined within the primary gene pools in their varietal improvement programmes and have not exploited much in the secondary, tertiary or quaternary gene pools (Singh *et al.*, 2007). Pre-breeding can provide an opportunity to introgress the novel genes in the targeted backgrounds required for generating outstanding recombinants or unique genetic stocks in order to realize the potential of novel genes for resolving the constraints related to breaking the plateau in terms of grain production and enhancement of nutritional value (for details see Chapter 6). When breeders switch on to any wide hybridization endeavour, it becomes very important to keep track of the validity of the wide hybrids and actual retention of the alien chromatin during generation advancement. Such efforts can be made successful by employing the molecular cytogenetics and the methods of *in situ* hybridization that have revolutionized our understanding of the structure, function, organization and evolution of genes and the genome. These methods made it feasible to link the molecular data on DNA sequences with chromosomal and expression information at the tissue, cellular and sub-cellular levels and hence changed the way we apply cytogenetics to agriculture (Schwarzacher and Heslop-Harrison, 2000).

Various versions of molecular cytogenetic approaches that have emerged recently (e.g. genomic *in situ* hybridization (GISH), fluorescence *in situ* hybridization (FISH), multicolour FISH and extended DNA fibre mapping) have excellent applications in various crop improvement programmes. Since the first application in identifying chromosomes (Schwarzacher *et al.*, 1989) and visualizing DNA sequences on plant chromosomes (Yamamoto and Mukai, 1989), GISH and FISH are now the techniques of choice for physical visualization of genomes and chromosomes and the order of chromosome segments, genes and DNA sequences. Many applications and refinements in the technology have opened new vistas for microscopic visualization of DNA manifestation *in situ*, previously confined to gel blot hybridization. Simultaneous detection of multiple targets has become quite easy through multicolour FISH and is now exercised in various cereal plants (e.g. rye (Leitch *et al.*, 1991); wheat (Mukai *et al.*, 1993; Komeda *et al.*, 2007; Chaudhary, 2008, 2009; Chaudhary *et al.*, 2009); barley (Leitch and Heslop-Harrison, 1993); *Aegilops* (Yamamoto and Mukai, 1995); and triticale (Cuadrado and Jouve, 1994). Although the innovative techniques of molecular cytogenetics have been extensively utilized in cereals to physically map whole genomes and the targeted alien introgressions, these tools also exhibit the potential to be employed in food legumes to resolve various fundamental issues related to origin of the species, assessment of variability and physical mapping at the chromosomal level. Exhaustive attempts have been made in this chapter to review the work concerning aspects related to the dynamics of molecular cytogenetic approaches for the resolution of problems in respect of physical mapping of the genomes and alien introgressions in important food legumes.

9.2 Cool Season Food Legumes

Chickpea

As mentioned above, FISH, a modern and powerful molecular cytological technique, has been used by various workers to detect and localize the repetitive DNA sequences of ribosomal DNA (rDNA), which are also known as nucleolus-organizing regions (NORs) as well as in detecting alien introgressed genes. Khattak *et al.* (2007) carried out FISH to detect rDNA sites in chickpea, and they detected rDNA sites on three pairs of chromosomes. Three pairs of rDNA sites were observed in 40 somatic metaphase cells of ten cultivated chickpea varieties; among these three pairs of chromosomes, one pair exhibited both 25S rDNA and 5S rDNA sites, while in the other two pairs the 25S rDNA and 5S rDNA sites were located separately on different pairs of chromosomes. The co-localized site of 5S rDNA appeared with low fluorescent signals as compared with the independent 5S rDNA site. This may have been due to either the lower copies of ribosomal genes or a more divergent sequence than the other 5S rDNA site. Hybridization sites of rDNA probe coding for 18S, 5.8S and 26S genes were detected for the first time by Abbo *et al.* (1994) in chickpea. They also reported three pairs of rDNA sites in cultivated chickpea.

Staginnus *et al.* (1999) performed physical mapping of the repetitive families by FISH on mitotic chromosomes from root tips of cultivated chickpea. The 16 metaphase chromosomes visible in diploid nuclei exhibited large heterochromatic regions with bright DAPI fluorescence around the centromeres, but not in the subtelomeric parts of the chromosomes. Chromosome A carries a secondary constriction corresponding to the NOR region adjacent to a large block of heterochromatin, as reported by Galasso and Pignone (1992). After probing with *CaSat1* repeat family, very strong signals were detected on two chromosome pairs and minor sites were found in distal regions of all other chromosomes. The major signals were visible in the heterochromatin close to the secondary constriction of chromosome A and in the pericentric heterochromatin block of chromosome B. Double-target hybridization revealed a close vicinity of major *CaSat1* sites to the 18S-5.8S-25S rRNA gene clusters on both chromosome pairs: the *CaSat1* signal is located adjacent to the rDNA site of the secondary constriction of chromosome A but does not cover it. On chromosome B, *CaSat1* sequences reside in the distal part of the heterochromatic block next to the rDNA site. *CaSat2* hybridized to the brightly DAPI-stained pericentric heterochromatin blocks of all 16 chromosomes. In double-colour *in situ* hybridization with differentially labelled probes of *CaSat1* and *CaSat2*, the *CaSat2* probe detected sites in close vicinity but clearly separated from the major *CaSat1* sites on chromosomes A and B. The intensity of the hybridization signals found in all metaphases confirms the high abundance of the *CaSat2* family in the chickpea genome. The retrotransposon-like sequences *CaRep1* and *CaRep2* produced uniform hybridization signals along the DAPI-positive heterochromatic blocks in pericentric regions of all chromosomes. However, *CaRep1* elements extended further into the euchromatin, which was weakly stained with DAPI, whereas *CaRep2* repeats were mostly restricted to the heterochromatin. Weak or no signals could be detected at the centromeres and their close vicinity, indicating that this sequence is largely excluded from centromeric regions consisting of *CaSat2* sequences. *CaSat1* elements detected in the heterochromatin of chromosomes A and B under stringent conditions do not interfere with the signals of *CaRep1* or 2 after double-colour hybridization, but reside in the distal areas of the heterochromatic block adjacent to the more proximally located *CaRep1* and 2 elements. The 18S-5.8S-25S rRNA gene clusters at the secondary constriction on chromosome A lack *CaRep1* and *CaRep2* elements.

The FISH technique was also used to probe the physical distribution of *CaEn/Spm* sequences on chickpea chromosomes (Staginnus *et al.*, 2001). Five cloned *En/Spm* fragments from chickpea were used as hybridization probes on metaphase spreads from chickpea root tips, and discrete hybridization signals were detected on at least six of eight chromosome pairs. The loci were observed in the distal parts of the large pericentric heterochromatin regions adjacent to euchromatic regions. Signals were detected on both chromatids on one or both ends of the hetrochromatic block. The largest chromosome pairs, A and B, revealed additional sites in pericentromeric regions within the hetrochromatin. The secondary constriction carrying the NOR region on chromosome A and the central parts of the pericentric heterochromatin did not show hybridization signals, suggesting that the transposon sequences are largely excluded from these chromosomal regions.

Lentil

The chromosomal distribution of the repetitive sequence families, pLc30 and pLc7 was carried out by Galasso *et al.* (2001) through FISH. The hybridization pattern of pLc30 is typical for a satellite DNA family, showing large sequence arrays of varying size distributed along chromosomes. Only chromosome pair number 6 did not show detectable signals after hybridization with the pLc30 probe. Four chromosome pairs (1, 2, 3 and 4) showed signals close to the centromere. There were also signals at interstitial and subtelomeric positions. In contrast, the sequence pLc7 was found at the intercalary position on a single chromosome pair (1) and hence represents a chromosome-specific marker.

Using FISH with pLc30 enabled unambiguous discrimination of all seven *Lens culinaris* ssp. *culinaris* chromosome pairs. FISH with pLc7, pTa71, pTa794 and pLT11 provided additional landmarks for some chromosome arms. Multiple-target FISH was applied on mitotic chromosomes of seven *Lens* taxa using two highly repetitive sequences (pLc30 and pLc7) isolated from the cultivated lentil and the multigene families for the 18S-5.8S-25s (pTa71) and 5S rDNA (pTa794) from wheat simultaneously as probes (Galasso, 2003). The number and location of pLc30 and pLc7 sites on chromosomes varied markedly among the species, whereas the hybridization pattern of 5S rDNA and 18S-5.8S-25S rDNA was less variable. It was also reported that each species showed a typical FISH karyotype, and few differences were observed among accessions belonging to the same species, except for the accessions of *Lens odemensis*. The most similar FISH karyotype to the cultivated lentil is that of *L. c.* subsp. *orientalis*, whereas *Lens nigricans* and *Lens tomentosus* are the two species that elucidated the most divergent FISH patterns compared with all taxa for number and location of pLc30 and 18S–5.8S–25S rDNA sites (Galasso, 2003).

Fernandez *et al.* (2005) performed FISH using the heterologous pTa71 to detect 18S–5.8S–26S rDNA and pTa794 to detect 5S rDNA on chromosome spreads of *L. c.* subsp. *culinaris*. Two digoxigenin hybridization sites corresponding to the 18S–5.8S–26S and 4 rhodamine hybridization sites marking 5S rDNA loci were observed on the metaphase spreads, substantiating previous findings (Abbo *et al.*, 1994; Galasso *et al.*, 2001; Balyan *et al.*, 2002). This indicated that one chromosome pair carried a NOR locus and two chromosome pairs carried 5S loci in this species. The NOR was located in a position close to or on the centromere of metacentric chromosomes. A 5S rDNA locus was located in a proximal position to the centromere of an acrocentric chromosome pair, whereas the other locus was located in a distal position in a submetacentric chromosome pair. When simultaneous FISH analysis of both subspecies of *L. culinaris* at metaphase was performed using pLc451, which encompassed the homologous intergenic spacer (IGS), to detect NOR loci and the C-l NTS to detect 5S rDNA loci as probes, differences in the hybridization patterns were observed. Whereas the 2 digoxigenin IGS hybridization signals for the NOR loci showed a similar signal to the pTa71 probe, the 4 rhodamine C-l hybridization signals for the 5S loci showed different signal intensities to the pTa794 probe. Two more intense rhodamine signals were located in a proximal position of an acrocentric chromosome pair, and two less intense rhodamine signals were located in a distal position in a submetacentric chromosome pair. To further investigate the identity of the major and minor sites of the 5S rDNA, simultaneous FISH, using the long (C-l) and short (C-s) lentil NTS as probes was performed. Two digoxigenin sites for the long NTS (C-l) were observed in a proximal position to the centromere of an acrocentric chromosome pair. Two rhodamine sites for the short NTS (C-s) were detected in a distal position of a submetacentric chromosome pair. FISH results indicated that no appreciable cross-hybridization of the two NTS probes occurred. When *in situ* hybridization analyses of *L. c.* subsp. *orientalis* BG-1688 and *L. odemensis* metaphases were performed using pLc451, C-l, and C-s as probes, the chromosome locations of the NOR and 5S rDNA loci were similar to those seen with *L. c.* subsp. *culinaris*, except for the 5S locus of *L. c.* subsp. *odemensis* hybridized by the short NTS that was located on a metacentric chromosome pair. In the accession ILWL-7 of *L. c.* subsp. *orientalis*, the NOR signal was detected in a distal position of a shorter chromosome, which agreed with the pattern described by Abbo *et al.* (1994) in accession 133 of *orientalis*, when a single 5S signal corresponding to the C-l probe was observed. The probe pLc451 hybridized with *L. culinaris* and *L. odemensis* but did not hybridize with *L. nigricans*. Thus a homologous *nigricans* IGS probe pLn451 was used for this species. In *L. nigricans* metaphases, hybridization signals corresponding to the IGS and long NTS were observed on the short arm of an acrocentric chromosome pair. The physical distance between these rDNA sites was sufficiently large to discriminate NOR from 5S rDNA sites. The NOR signals were located on a distal position on the short arm, whereas the C-1 5S signals were located on a more proximal position on the same arm. The 5S hybridized by the short NTS were located on a distal position of a submetacentric chromosome pair. The two hybridization patterns observed in *L. c.* subsp. *orientalis* agree with different karyotype arrangements described in this subspecies. One of these (BG-16880) is similar to the karyotype observed in the cultivated lentil, whereas the other (ILWL-7) is similar to the karyotype observed in some accessions in which around three-quarters of the satellite was transferred to another chromosome, the metacentric-satellited chromosome became acrocentric and one of the submetacentric chromosomes lengthened (Ladizinsky, 1993; Abbo *et al.*, 1994).

Garden pea

Analysis of genome size variation

Genome size variation is an important issue in the evolutionary and developmental karyology of higher plants. While initial studies were concerned more with genome size differences between species and their ecological and evolutionary interpretation, recent studies are focused on intraspecific genome size variation. *Pisum sativum* L. is one of the species where intraspecific genome size variation, up to 1.29-fold between cultivars, has been reported. Greilhuber and Ebert (1994) used Feulgen cytophotometric analysis to study genome size variation in 25 wild accessions, landraces and cultivars of pea of different geographic origin. Differences between accessions were maximally 1.054-fold in single experiments but proved to be non-reproducible upon repeated measurements. Seedlings of the same accession often differed significantly, up to 1.056-fold, but values from root and shoot tips in one individual were not significantly correlated, indicating the absence of true genome size variation among plants. Upon calibration against *Allium cepa* a 1C value of 4.42 pg was estimated for *P. sativum*. In addition, molecular cytogenetic approaches such as flow cytometry and Feulgen densiometry have been used in *Pisum* spp. to study genome variation in *P. sativum* cultivation and its wild relatives. DAPI and ethidium bromide flow cytometric and Feulgen densiometric analyses of genome size variation in 38 accessions of *P. sativum* and 14 samples of *Pisum elatius*, *Pisum abyssinicum*, *Pisum humile* and *Pisum fulvum* revealed that no genomic size variation existed among *P. sativum* cultivars, whereas *P. abyssinicum* and *P. fulvum* differed from *P. sativum* by about 1.066- and 1.070-fold, respectively. One accession of *P. humile* and two of *P. elatius* differed by 1.089- and 1.12-fold, respectively, from *P. sativum*, while the remainder of the accessions of these texa were homogeneous with cultivated pea (Baranyi and Greilhuber, 1996). In a similar study, Baranyi *et al.* (1996) measured genome size in 25 samples of *P. abyssinicum*, 23 of *P. elatius*, 5 of *P. fulvum* and 22 of *P. humile* using ethidium bromide flow cytometry and Feulgen densiometry. They reported wide variations between samples of *P. abyssinicum*, *P. elatius* and *P. humile*, whereas *P. fulvum* was homogeneous in genome size.

Confirmation of hybrid origin

Wild relatives are used to undertake distant hybridization, which is helpful in transferring environmental plasticity, such as resistance to biotic stresses (aschochyta blight and root rot) and abiotic stresses (drought and extreme temperature). Such traits are present in *P. fulvum* (Ali *et al.*, 1994), which is cross-incompatible with cultivated pea (Conicella and Errico, 1993) as it is clearly the most divergent species of the taxon (Ben-Ze'ev and Zohary, 1973; Hoey *et al.*, 1996). Wroth (1998) suggested use of a wild accession of *P. sativum* as a bridging parent between cultivated pea and *P. fulvum* using the latter as the male parent to produce hybrids of low fertility. However, hybrids were reported without the use of a bridging species by Ochatt *et al.* (2004), who confirmed the hybrid origin of plants obtained from *P. sativum* × *P. fulvum* using flow cytometry as well as GISH (genomic *in situ* hybridization). Flow cytometry revealed intermediate 2C and 4C peaks of hybrids in comparison with the parents. The mitotic index of hybrids was also intermediate between parents. Use of GISH resulted in a clear discrimination of the two parental genomes, using the total genomic DNA probe from *P. fulvum*. The F_1 hybrid exhibited seven chromosomes from *P. sativum* stained yellow and seven from *P. fulvum* fluoresced in red, due to propidium iodide counterstaining. The application of GISH in advanced generations indicated translocation events taking place between two parental genomes.

Identification of chromosomes

Uncertainties remain regarding the unambiguous identification of seven chromosome pairs of *P. sativum* and the assignment of genetic linkage groups to individual chromosome types (Fuchs *et al.*, 1998). Biotin-labelled DNA probes for tandemly repeated sequences were used in *in situ* hybridization experiments as chromosome-specific markers by Simpson *et al.* (1990). Six of the seven chromosome pairs could be marked at single sites in this way. Translocations from a standard karyotype are revealed as chromosomes that have two hybridization sites rather than one. By probing a tester set of reciprocal translocation (or interchange) lines, some markers can be assigned to chromosomes. The method is rapid and simple and, in the absence of well-resolved chromosome bands, provides a mean for clarifying some of the problems in pea cytology. Neumann *et al.* (1998) carried out flow cytometry analysis to discriminate chromosomes by comparing theoretical flow karyotypes with the standard karyotype; while only two chromosomes (5 and 7) were discriminated in the standard karyotype, four chromosomes (3, 5, 6 and 7) could clearly be discriminated in a line containing a stable reciprocal translocation between chromosomes 3 and 6. Neumann *et al.* (2002) used FISH and satellite-repeat Pis TR-B to discriminate all chromosome types based on their signal patterns and morphology. Chromosomes 4 and 7, which were difficult to discriminate due to morphological similarities, were identified since chromosome 4 exhibited three Pis TR-B signals whereas one was on chromosome 7. Chromosome 1 was identified on the basis of the presence of 5S rDNA on the same arm as Pis TR-B.

Samatadze *et al.* (2005) used FISH on pea chromosomes with telomeric repeated sequences for the identification of chromosomes. Chromosomes 2 and 4 always showed less intense signals. The detection of telomeres permitted precise identification of even poorly condensed chromosomes. The translocation lines L-108 ($T_{2\text{-}4s}$) M-10 ($T_{2\text{-}7s}$) were also evaluated by this group through FISH using telomeric repetitive probes pTa71 (45S rDNA) and pTa794 (5S rDNA).

9.3 Warm Season Food Legumes

Common bean

All species of the genus are diploid and most have 22 chromosomes (2n=2x=22). The genome of common bean is one of the smallest in the legume family, at 625 Mbp per haploid genome. Normal mitotic or meiotic chromosomes are very small, metacentric or sub-metacentric. Cytological studies in the Phaseoleae to date have been predominantly of a karyosystematic nature and restricted to chromosome counts and gross karyotype descriptions. The mitotic metaphase chromosomes of the *Phaseolus* species studied cytologically have proved to be barely distinguishable because of their minute size, their homomorphic structure and because of the lack of distinct chromosomal landmarks (Lackey, 1980). Techniques such as Giemsa C-banding, fluorescent banding and Ag-NOR staining (Schweizer and Ambros, 1979; Zheng *et al.*, 1991, 1993) brought some refinements as compared with the observations made by classical procedures. However, detailed karyotype analysis remained as an unsolved problem in *Phaseolus* taxa. Recently, FISH with ribosomal RNA gene probes has been applied to mitotic chromosomes of *Phaseolus vulgaris* (Shi *et al.*, 1996a) and to polytene as well as to mitotic chromosomes of *Phaseolus coccineus* (Nenno *et al.*, 1994; Guerra *et al.*, 1996).

Moscone *et al.* (1999) used FISH followed by DAPI counterstaining for the chromosomal assignment of 5S and 18S–25S rRNA genes in the four cultivated *Phaseolus* species (*P. vulgaris*, *P. coccineus*, *P. acutifolius* and *P. lunatus*). The 18S–25S rRNA gene loci display intraspecific variation, as reflected in differences of signal size and/or number. The numbers of 18S–25S rDNA loci ranged from one pair in *P. lunatus* and *P. acutifolius* var. *latifolius* to seven pairs in *P. vulgaris* cv. Wax, while the numbers of 5S rRNA gene loci ranged from one pair in *P. lunatus* to three pairs in *P. a.* var. *latifolius*. The 5S rRNA gene loci were frequently syntenic to 18S–25S rDNA loci. Exceptions were observed in chromosome pairs 2 and 10 of *P. acutifolius* and chromosome 8 of *P. lunatus*.

Congruency in rRNA gene distribution patterns between *P. vulgaris* and *P. a.* var. *latifolius* (homeologous chromosomes 8) and no congruency between *P. vulgaris* and *P. lunatus* reflects the greater phylogenetic distance. Therefore, on the basis of karyological characters, *P. a.* var. *latifolius* appears somehow closer to *P. vulgaris* and *P. coccineus* by sharing with those species a presumably homeologous chromosome 8, which carries 5S and 18S–25S rRNA gene clusters in its long arm. Finally, *P. lunatus* is unique in possessing predominantly DAPI-negative telomeric heterochromatin and the lowest number of rRNA gene loci, that is, a single 18S–25S rDNA cluster (NOR) on chromosome 1 and a single 5S band on the short arm of chromosome 8. Based on FISH, chromosome morphology and heterochromatin-banding patterns, chromosome 8 in *P. lunatus* is likely to correspond to chromosome 10 of *P. a.* var. *latifolius*. Furthermore, low-copy and single-copy gene-mapping studies should help to establish these, and additional presumptive chromosomal homeology between the cultivated *Phaseolus* species (Vallejos *et al.*, 1992; Nodari *et al.*, 1993).

Pedrosa-Harand *et al.* (2009) used FISH of BAC and a few other genomic clones for the construction of cytogenetic maps of common bean chromosomes 3, 4 and 7. All clones were selected with genetically mapped markers, mostly with single-copy RFLPs, a large subset of BACs from 13 different genomic regions, containing repetitive sequences, as concluded from the regional distribution patterns of multiple FISH signals on chromosomes: pericentromeric, subtelomeric and dispersed. Pericentromeric repeats were present in all 11 chromosome pairs with different intensities, whereas subtelomeric repeats were present in several chromosome ends. The correlation of genetic and physical distance along the three studied chromosomes was obtained for 23 clones. This correlation suggests suppression of recombination around extended pericentromeric regions in a similar way to that previously reported for plant species with larger genomes. These results indicate that a relatively small plant genome may also possess a large proportion of repeats interspersed with single single-copy sequences in regions other than the pericentromeric heterochromatin and consequently, exhibit lower recombination around the pericentromeric fraction of the genome.

Vigna

Two common and effective fluorochromes (Chromomycin A3 (CMA) and DAPI) have been widely used in cytogenetics for karyotype analysis in blackgram (Schweizer, 1976; Alam and Kondo, 1995; Akter and Alam, 2005; Jessy *et al.*, 2005; Mahbub *et al.*, 2007). Alam and Mahbub (2007), while studying the karyotype in two varieties, Barimash-1 and Barimash-3 of *Vigna mungo* using orcein and CMA staining, reported marked differences in karyotype and properties of interphase nuclei and prophase chromosomes, which was not possible using conventional karyotypic techniques. The interphase nucleus of Barimash-1 depicted many prominent dot-like, CMA-positive bands. The prophase chromosomes of this variety had six bright CMA positive bands. Four prominent and many dots like CMA-positive bands were found in the interphase nuclei of Barimash-3. The prophase chromosome of this variety showed five bright CMA-positive bands. The nature of CMA-stained interphase nuclei and prophase chromosomes are beneficial for characterization. In Barimash-1, 16 entirely fluoresced banded chromosomes were found; the remainder did not show any band. In Barimash-3, 19 different CMA positive bands were observed, of which 11 were entirely, 4 were terminal- and 4 were centromeric-banded chromosomes; the karyotypic formula of this variety was 11+4+4+3. The polymorphism of the CMA-positive banding pattern of these two varieties indicates the probable occurrence of minute structural aberration and presence of different heterochromatins. The banded chromosomes were stable and made each karyotype unique.

In green gram (*Vigna radiata*), the detection of 25S and 5S rDNA sites through FISH and active NORs through silver staining was reported for the first time by Khattak *et al.* (2007). They detected four pairs of rDNA sites in 60 somatic metaphase cells of 12 cultivated mungbean varieties. Each 25S rDNA and 5S rDNA had separate sites on two pairs of chromosomes. One of the 5S rDNA pair of chromosomes exhibited very low fluorescent signals sites compared with the same types of site on the other pair of chromosomes. The active NORs were also detected through the silver staining technique, and it was observed that two pairs of chromosomes were active in mung bean for NORs.

Soybean

The cytological study of soybean metaphase chromosomes (2n = 40) is a challenging task due to its small size (1–2 μm) and large number (2n = 40). Moreover, there exists very little morphological diversity (Sen and Vidyabhusan, 1960; Palmer and Kilen, 1987; Clarindo *et al.*, 2007). With the exception of a single acrocentric pair, soybean chromosomes are all metacentric or sub-metacentric, making them difficult to distinguish in routine mitotic preparations. Furthermore, the low mitotic index characteristic of soybean root meristems (Ahmad *et al.*, 1983) means that chromosome preparation for karyotyping is rather inefficient. The first cytological description of domesticated soybean (*Glycine max*) was developed by using pachytene chromosomes numbered 1–20 on the bases of total chromosomes length, arm length ratios and relative proportions of euchromatin and heterochromatin (Singh and Hymowitz, 1988).

In situ hybridization of DNA probes to soybean chromosomes was first reported by Skorupska *et al.* (1989) and later by Griffor *et al.* (1991). Soybean repetitive DNA has been used to develop a cocktail of fluorescent *in situ* hybridization probes that can differentially label mitotic chromosomes in root tip preparations. Genetically anchored BAC clones were used to identify individual chromosomes in metaphase spreads and to complete a FISH-based karyotyping cocktail that permitted simultaneous identification of all 20 chromosome pairs. These karyotyping tools were applied to wild soybean (*Glycine soja* Sieb. and Zucc.), which represents a large gene pool of potentially agronomically valuable traits. Reciprocal chromosome translocations between chromosomes 11 and 13 in two accessions of wild soybean were identified and characterized. The translocation is widespread in *G. soja* accessions and probably accounts for the semi-sterility found in *G. soja* × *G. max* crosses.

Shi *et al.* (1996b) used repetitive DNA sequences and single-copy DNA sequences. PCR-PRINS (PCR-primed *in situ* hybridization) can detect relatively small chromosomal regions that cannot be observed using standard FISH protocols. Both propidium iodide and DAPI are frequently used as counterstains for chromosomal images in FISH and PCR-PRINS; however, PI staining was found to mask some low intensity. Only eight major sites of the repetitive sequence STR 120 AB were detected with PI counterstaining, while more than 20 sites were observed with DAPI counterstaining under the same hybridization condition. Eleven probes from different types of DNA sequences to tag and characterize soybean chromosomes were used. All 40 soybean chromosomes were tagged by FISH, GISH or PCR-PRINS by either positive or negative labelling. Among these, 36 chromosomes were labelled by repetitive DNA probes while eight were tagged by single-copy sequences. In addition, more than ten chromosomes were negatively labelled by repetitive sequences or total genomic DNA.

Apart from identification of chromosomes, molecular cytogenetics has also been used to suggest polyploidy in *G. max*. Two soybean centromere-specific satellite repeat classes in its genome suggest the existence of two sub-genomes (Gill *et al.*, 2009). The ancestor of soybean and the remainder of the genus *Glycine* has been hypothesized as having being formed via a polyploidy event within the last 15 million years (Shoemaker *et al.*, 2006); however, it remains unclear whether this event was allo- or autopolyploid (Kumar and Hymowitz, 1989; Straub *et al.*, 2006).

Lackey (1980) suggested that there have been several rounds of polyploidization and segmental duplication in soybean, on the basis of chromosome number. Shoemaker *et al.* (2006) agreed with this, on the basis of multiple hybridizing RFLP fragments, as did Blanc and Wolfe (2004) and Schlueter *et al.* (2004) on the basis of implicated ESTs.

Faba bean

In *Vicia faba* (2n = 12), five chromosome pairs are acrocentric whereas one pair is metacentric. The faba bean was one of the first plant species to feature reports on: (i) the duration of mitotic cycle stages (Howard and Pelc, 1953); (ii) Giemsa banding (Vosa and Marchi, 1972; Doebel *et al.*, 1973; Schweizer, 1973; Takehisa and Utsumi, 1973); (iii) a map of cold reactive chromosome segments; (iv) restriction endonuclease-mediated banding (Frediani *et al.*, 1987); (v) *in situ* hybridization of rRNA to metaphase chromosomes (Scheuermann and Knaelmann, 1975); (vi) silver staining of NORs and interphase nucleoli (Schubert *et al.*, 1979); (vii) differential staining of sister chromatids (Kihlman and Kronborg, 1975); (viii) lateral A/T asymmetry (Schubert and Rieger, 1979); and (ix) differential histone acetylation along metaphase chromosomes (Houben *et al.*, 1996; Belyaev *et al.*, 1997, 1998). In well-spread metaphases, it is possible to distinguish even acrocentric chromosome pairs, especially after differential staining procedures.

Evolutionary studies

Faba bean is a suitable model crop for the study of evolutionary relationships and functional significance of repetitive elements within the genomes of individual plant species. It represents one of the largest legume genomes (Bennett and Leitch, 1995), having a high proportion (> 85%) of repetitive DNA (Flavell *et al.*, 1974). The most abundant repeat, the *Fok* I element, is present at about 10^7 copies per haploid genome (Kato *et al.*, 1984; Maggini *et al.*, 1991). *Fok* I repeats are arranged in tandem, individual elements being 59 bp long and concentrated at a limited number of genomic loci. Visualization of these loci by *in situ* hybridization on metaphase chromosome revealed several bands, which corresponded with some of the heterochromatic chromosomal regions. The two other families represent dispersed repeats. The *Bam HI* family includes seven classes of repeats 250–1750 bp long that share partial homology. Each of the classes comprises about 3% of the genome (Kato *et al.*, 1985). *Tyl-copia* retrotransposons have been detected in the faba bean genome by PCR amplification using primers derived from conserved regions. The isolated 250 bp fragment was estimated to comprise about 2% of the genome (Pearce *et al.*, 1996). However, if all of these fragments represent parts of full-length copies of *Tyl-copia* elements, this retrotransposon would comprise 40% of the faba bean genome (Pearce *et al.*, 1996). Nouzova *et al.* (1999) localized TIII15 with *Fok* I repeats using a combined PRINS- FISH technique. In this procedure, the *Fok* I repeats were first labelled by fluorescence in the PRINS reaction using sequence-specific primer, and the chromosomes were then subjected to FISH to visualize the TIII15 sequences. Since the labelling of *Fok* I elements produces characteristic bands at defined positions on faba bean chromosomes (Fuchs *et al.*, 1994), it allowed determination of the positions of TIII15 signals on individual chromosome pairs. Twenty-two major hybridization sites were reproducibly detected, some of them located near to NOR, telomeric and centromeric regions. TIII15 signals were present within the heterochromatic regions containing *Fok* I repeats on chromosomes 1, 4 and 6. However, some signals were also associated with heterochromatic regions lacking *Fok* I sequences, as well as with euchromatin.

Physical location of transgenes

The use of FISH for the localization of transgene constructs in plant chromosomes has been described previously (Wang *et al.*, 1995; Moscone *et al.*, 1996; ten Hoopen *et al.*, 1996, 1999; Pedersen *et al.*, 1997; Jakowitsch *et al.*, 1999), but the resolution and reliability of signal detection is not always reproducible. Snowdon *et al.* (2001) described how direct labelling of transgene constructs by PCR with degenerate oligonucleotide primers (Telenius *et al.*, 1992) can also yield FISH probes with optimal probe length and labelling that are highly suitable for physical detection of transgenes. Direct incorporation of 11-FITC-dUTP in the DOP-PCR reaction generated FISH probes of approximately 300–500 bp in length, which gave strong, reproducible signals in transgenic *Vigna faba* and allowed accurate physical location of the transgene with little to no background hybridization. Clean-up of PCR products was not necessary when sheared *V. faba* DNA was added as competitor in probe solutions.

Lathyrus

All species belonging to the genus *Lathyrus* are diploid (2n = 14), but autopolyploid cytotypes of four species are reported to occur as natural populations. In addition to the marked similarities in chromosome number, species are consistently similar in chromosome morphology and karyotype arrangement. In all *Lathyrus* complements, chromosomes are either median or submedian in shape. Divergence and species differentiation on the other hand have resulted in a three- to fourfold increase in chromosome size, which is directly correlated with a fivefold increase in 2C DNA amounts. The total amounts of constitutive heterochromatin and euchromatin differ widely between species, and hence also for their pattern of distribution within complements. It has been established that, during evolution, both heterochromatin and euchromatin have been increased with an increase in 2C DNA (Narayan, 1991). 2C nuclear DNA levels for 24 species of *Lathyrus* were determined using flow cytometry, where a greater than twofold variation was observed, ranging from 10.2 pg in *Lathyrus basalticus* to 24.2 pg in *Lathyrus latifolius*. In general, perennial species have more DNA than annuals. Significant intraspecific variation was observed in five species of *Lathyrus* (from 10.1% in *Lathyrus annuus* to 28% in *Lathyrus tingitanus*). A positive correlation was observed between DNA values obtained by flow cytometry and those previously determined by microdensitometry. Finally, the distribution of DNA amounts in species within section lathyrus appears to be continuous (Nandini *et al.*, 1997). In contrast, Murray *et al.* (1992a) reported constancy in karyotype and genome size of *Lathyrus odoratus* using flow cytometry. Cox *et al.* (1993) generated a telomere-specific probe by PCR and used it to localize chromosome telomeres in *Lathyrus sativus* and nine other unrelated species. The concatenation of the simple monomer 5′ - (TTTAGGG) - 3′ derived from the sequence of *Arabidopsis thaliana* telomeres yielded a stable versatile and reliable probe that gave a signal of high intensity following FISH (Fig. 9.1).

Murray *et al.* (1992b) used rRNA gene probe for *in situ* hybridization and silver staining for identification of secondary constrictions and NORs of *Lathyrus*. Four well-stained NORs at the end of the short arm of two acrocentric pairs and faint staining of centromeres of several other chromosomes were observed on the basis of silver staining. These workers also revealed lightly stained NORs but densely stained centromeres. *L. tingitanus* exhibited silver-positive spots on all chromosomes, and each pair of homologous chromosomes could be distinguished by its silver pattern. In other species (*L. blepharicarpus*, *L. odoratus*, *L. sativus*, *L. cassius* and *L. hirsutus*), NORs were easily identified. Two *in situ* hybridization sites were revealed in *L. blepharicarpus*, *L. cassius* and *L. hirsutus*, which was in agreement with silver staining results. *L. tingitanus* also had a pair of hybridization sites corresponding to silver-positive sites, whereas *L. sativus*, with only three silver-positive sites, showed four sites of *in situ* hybridization. Both *L. sativus* and *L. odoratus* had two *in situ* hybridization sites clearly larger than the other two (Fig. 9.2).

Ali *et al.* (2000) investigated phylogenetic relationships among different *Lathyrus* spp. by studying their DNA content, FISH and DAPI bands. The nuclear DNA content of seven *Lathyrus* spp. ranged from 8.77 pg/2C in *Lathyrus clymenum* to 15.7 pg/2C in *L. tingitanus*. Species belonging to sections aphaca and clymenum showed a lower DNA content. FISH with digoxigenin-labelled 25S rDNA and biotin-labelled 5S rDNA probes revealed one locus of 25S rDNA for all the examined species except *L. sativus*, which has two sites. All 25S rDNA loci were associated with the secondary constriction; no minor loci were observed. Two 5S rDNA loci were observed

Fig. 9.1. Demonstration of telomeres by FISH in *Lathyrus sativus*. Source: Cox *et al.* (1993); reprinted with permission from Oxford University Press, 2010.

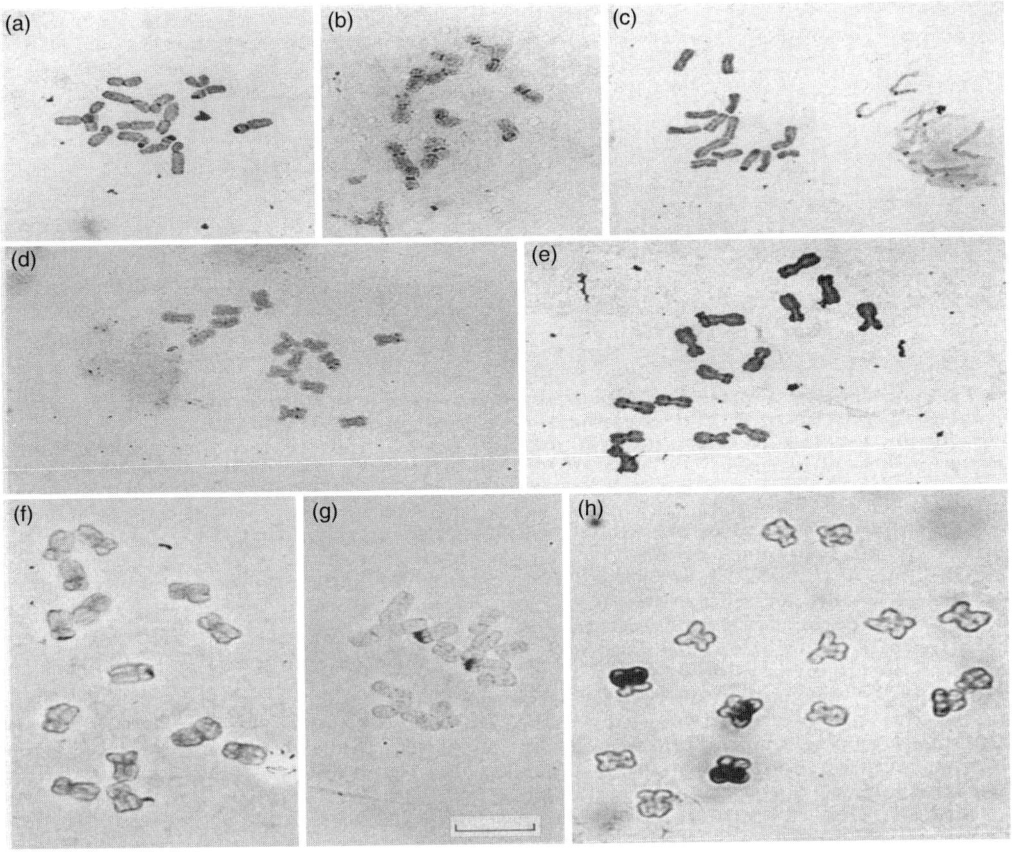

Fig. 9.2. Silver-stained chromosomes of (a and b) *Lathyrus odoratus*, (c) *L. blepharicarpus*, (d) *L. sativus* and (e) *L. tingitanus*; *in-situ* hybridization of the probe pTa71 on the chromosomes of (f) *L. cassius*, (g) *L. blepharicarpus* and (h) *L. sativus*. Scale = 10 μm. Source: Murray *et al.* (1992b); reprinted with permission from Macmillan Publishers Ltd., 2010.

in *L. aphaca*, *L. ochrus*, *L. annuus* and *L. sativus*, and three loci in *L. cicera*, *L. clymenum* and *L. tingitanus*. The DAPI bands were present at the centromeres of all species except for *L. tingitanus*, which showed DAPI-negative centromeres and blocks of DAPI-positive bands at the pericentromeric regions of all chromosomes. Except for *L. ochrus* and *L. clymenum*, all species exhibited some terminal bands, and apart from *L. aphaca*, all showed at least some mostly dot-like interstitial bands. The combination of two-colour FISH for 5S and 25S rDNA loci with DAPI banding on the same metaphases and consideration of arm ratios could distinguish at least three (*L. annuus*, *L. aphaca*), four (*L. cicera*, *L. ochrus*, *L. tingitanus*) and five (*L. sativus*, *L. clymenum*) individual chromosome pairs unambiguously. All data taken together correlate well with the phylogenetic distance of these species. The two species of section clymenum (*L. clymenum*, *L. ochrus*), both with two 5S rDNA loci on the long arm of chromosome 2, are the only ones without terminal heterochromatic bands. *L. aphaca* of section aphaca takes an intermediate position between species of the sections clymenum and lathyrus, differing from section clymenum by the presence of terminal bands, from section lathyrus by a lower DNA content, similar to that of the species belonging to section clymenum, and differs from both in that interstitial DAPI positive bands are absent. *L. tingitanus* apparently takes a peripheral position within section lathyrus, as indicated by unique features such as its high DNA content, the presence of DAPI-negative instead of dot-like DAPI-positive centromeric bands and the presence of strong pericentromeric and only a few terminal and (or) interstitial DAPI bands.

Nandini (1997) utilized FISH with ribosomal probes to confirm that the secondary constrictions in *L. chloroanthus* and *L. chrysanthus* are present on different locations, i.e. six and eight sites, respectively. These results were supported by silver staining, which also failed to localize specific NORs in *Lathyrus*. In another study, FISH was used to investigate the chromosomal distribution of the two sequence families of 45S and 5S ribosomal genes.

The species-specific sequences of *L. sativus* were located around the centromere of chromosome pair IV, where they occupied a very broad region and in a much smaller amount, close to the centromeres in the short arm of pair II. Sequences related to the repeat units isolated from *L. sylvestris* were found, both in this species and *L. latifolius* in all of the chromosome pairs at the terminal and interstitial regions, where they co-localize with the vast majority of DAPI bands. The pattern of hybridization of the satellite DNA sequences investigated, together with that of DAPI bands and ribosomal DNA, allowed each chromosome pair in the three complements studied to be identified unambiguously (Ceccarelli *et al.*, 2010).

9.4 Conclusion

Molecular cytogenetics have not been carried out to any great extent in food legumes because of the small size of the chromosomes, homomorphic structure, lack of distinct chromosome landmarks and low mitotic index as compared with cereal crops such as wheat and rice, which have large chromosome size and high mitotic index. However, FISH for highly repetitive DNA sequences has proved to be a valuable tool in many food legumes for karyotype analysis, and also to elucidate the phylogenetic relationship of a species within a genus or at the family level. FISH also proved to be a powerful tool for the physical location of transgene integration sites in *Vicia faba*. A cytogenetic map of common bean has been prepared by *in situ* hybridization of 35 BACs selected with markers mapping to eight linkage groups using 5S and 45S rDNA and one bacteriophage. An interspecific hybrid between *Pisum sativum* and *P. fulvum* and translocation in an advanced generation of this cross could be identified using GISH. Further efforts are needed to refine the technology for chromosome preparations with high mitotic index and well-condensed metaphase chromosomes, so that the technique can be used efficiently for monitoring alien introgressions in food legume breeding programmes.

References

Abbo, S., Miller, T.E., Reader, S.M., Dunford, R.P. and King, I.P. (1994) Detection of ribosomal DNA sites in lentil and chickpea by fluorescent *in situ* hybridization. *Genome* 37, 713–716.

Ahmad, Q.N., Britten, E.J. and Byth, D.E. (1983) A quantitative method of karyotypic analysis applied to the soybean, *Glycine max*. *Cytologia* 48, 879–892.

Akter, S. and Alam, Sk.S. (2005) Differential fluorescent banding pattern in three varieties of *Cicer arietinum* L. (Fabaceae). *Cytologia* 70, 441–445.

Alam, Sk.S. and Kondo, K. (1995) Differential staining with Orcein, Giemsa, CMA and DAPI for comparative chromosome study of 12 species of Australian *Drosera* (Droseraceae). *American Journal of Botany* 82, 1278–1286.

Alam, Sk.S. and Mahbub, M.N. (2007) Karyotype comparison in two varieties of *Vigna mungo* L. after staining with orcein and CMA. *Bangladesh Journal of Botany* 36, 167–170.

Ali, H.B.M., Meister, A. and Schubert, I. (2000) DNA content, rDNA loci, and DAPI bands reflect the phylogenetic distance between *Lathyrus* species. *Genome* 43, 1027–1032.

Ali, S.M., Sharma, B. and Ambrose, M.J. (1994) Current status and future strategy in breeding pea to improve resistance to biotic and abiotic stresses. *Euphytica* 73, 115–126.

Balyan, H.S., Houben, A. and Ahne, R. (2002) Karyotype analysis and physical mapping of 18S–5.8S–25S and 5S ribosomal RNA loci in species of genus *Lens* Miller (Fabaceae). *Caryologia* 55, 121–128.

Baranyi, M. and Greilhuber, J. (1996) Flow cytometric and Feulgen densitometric analysis of genome size variation in *Pisum. Theoretical and Applied Genetics* 92, 297–307.

Baranyi, M., Greilhuber, J. and Swieciki, W.K. (1996) Genome size in wild *Pisum* species. *Theoretical and Applied Genetics* 93, 717–721.

Belyaev, N.D., Houben, A., Baranczewski, P. and Schubert, I. (1997) Histone H4 acetylation in plant heterochromatin is altered during cell cycle. *Chromosoma* 106, 193–197.

Belyaev, N.D., Houben, A., Baranczewski, P. and Schubert, I. (1998) The acetylation patterns of histones H3 and H4 along plant chromosomes are different. *Chromosome Research* 6, 59–63.

Bennett, M.D. and Leitch, I.J. (1995) Nuclear DNA amounts in angiosperms. *Annals of Botany* 76, 113–176.

Ben-Ze'ev, N. and Zohary, S. (1973) Species relationships in the genus *Pisum* L. *Israel Journal of Botany* 22, 73–91.

Blanc, G. and Wolfe, K.H. (2004) Widespread paleopolyploidy in model plant species inferred from age distributions of duplicate genes. *Plant Cell* 16, 1667–1678.

Ceccarelli, M., Sarri, V., Polizzi, E., Andreozzi, G. and Cionini, P.G. (2010) Characterization, evolution and chromosomal distribution of two satellite DNA sequence families in *Lathyrus* species. *Cytogenetic and Genome Research* 128, 236–244.

Chaudhary, H.K. (2008) Dynamics of doubled haploidy breeding and molecular cytogenetic approaches in bread wheat: Focus on north-west Himalayan regions. *Advances in Chromosome Science* 3, 67–69.

Chaudhary, H.K. (2009) New frontiers in chromosome engineering: Genetic upgradation of bread wheat for varied agroclimatic situations in north-west Himalayas. In: *Proceedings of National Seminar on Designing Crops for the Changing Climate*, 30–31 October 2009, Ranchi, Jharkhand, India, pp. 51–52.

Chaudhary, H.K., Chahota, R.K., Mukai, Y., Jeberson, M.S., Kishore, N. and Kumar, V. (2009) Molecular cytogenetic mapping of the targeted rye chromatin introgressed bread wheat lines associated with drought tolerance and rust resistance suitable for rainfed regions of north- west Himalayas. In: *Proceedings of National Seminar on Designing Crops for the Changing Climate*, 30–31 October 2009, Ranchi, Jharkhand, India, pp. 84.

Clarindo, W.R., Carvalho, C.R. and De Alves, B.M.G. (2007) Mitotic evidence for the tetraploid nature of *Glycine max* provided by high quality karyograms. *Plant Systematics and Evolution* 265, 101–107.

Conicella, C. and Errico, A. (1993) Embryology of ovule abortion in reciprocal crosses between diploids and tetraploids in *Pisum sativum* and *P. fulvum*. *Journal of Genetics and Breeding* 47, 157–162.

Cox, A.V., Bennett, S.T., Parokonny, A.S., Kenton, A., Callimassia, M.A. and Bennett, M.D. (1993) Comparison of plant telomere locations using a PCR- generated synthetic probe. *Annals of Botany* 72, 239–247.

Cuadrado, A. and Jouve, N. (1994) Mapping and organization of highly-repeated DNA sequences by means of simultaneous and sequential FISH and C-banding in 6x Triticale. *Chromosome Research* 2, 331–338.

Doebel, P., Rieger, R. and Michaelis, A. (1973) The Giemsa banding patterns of the standard and four reconstructed karyotypes of *Vicia faba*. *Chromosoma* 43, 409–422.

Fernández, M., Ruiz, M.L., Linares, C., Fominaya, A. and Perez de la Vega, M. (2005) 5S rDNA genome regions of *Lens* species. *Genome* 48, 937–942.

Flavell, R.B., Bennett, M.D., Smith, J.B. and Smith, D.B. (1974) Genome size and the proportion of repeated nucleotide sequence DNA in plants. *Biochemical Genetics* 12, 257–269.

Frediani, M., Mezzanotte, R., Vanni, R., Pignone, D. and Cremonini, R. (1987) The biochemical and cytological characterization of *Vicia faba* DNA by means of *Mbo*I, *Alu*I and *BamHI* restriction endonucleases. *Theoretical and Applied Genetics* 75, 46–50.

Fuchs, J., Pich, U., Meister, A. and Schubert, I. (1994) Differentiation of faba bean heterochromatin by *in situ* hybridization with a repeated *Fok*I sequence. *Chromosome Research* 2, 25–28.

Fuchs, J., Kühne, M. and Schubert, I. (1998) Assignment of linkage groups to pea chromosomes after karyotyping and gene mapping by fluorescent *in situ* hybridization. *Chromosoma* 107, 272–276.

Galasso, I. (2003) Distribution of highly repeated DNA sequences in species of the genus *Lens* Miller. *Genome* 46, 1118–1124.

Galasso, I. and Pignone, D. (1992) Characterization of chickpea chromosomes by banding techniques. *Genetic Resources and Crop Evolution* 39, 115–119.

Galasso, I., Schmidt, T. and Pignone, D. (2001) Identification of *Lens culinaris* ssp. *culinaris* chromosomes by physical mapping of repetitive DNA sequences. *Chromosome Research* 9, 199–209.

Gill, N., Findley, S., Walling, J.G., Hans, C., Ma, J., Doyle, J. *et al.* (2009) Molecular and chromosomal evidence for allopolyploidy in soybean. *Plant Physiology* 151, 1167–1174.

Greilhuber, J. and Ebert, I. (1994) Genome size variation in *Pisum sativum*. *Genome* 37, 646–655.

Griffor, M.C., Vodkin, L.O., Singh, R.J. and Hymowitz, T. (1991) Fluorescent *in situ* hybridization to soybean metaphase chromosomes. *Plant Molecular Biology* 17, 101–109.

Guerra, M., Kenton, A. and Bennett, M.D. (1996) rDNA sites in mitotic and polytene chromosomes of *Vigna unguiculata* (L.) Walp. and *Phaseolus coccineus* L. revealed by *in situ* hybridization. *Annals of Botany* 78, 157–161.

Hoey, B.K., Crowe, K.R., Jones, V.M. and Polans, N.O. (1996) A phylogenetic analysis of *Pisum* based on morphological characters, and allozyme and RAPD markers. *Theoretical and Applied Genetics* 92, 92–100.

Houben, A., Belyaev, N.D., Turner, B.M. and Schubert, I. (1996) Differential immunostaining of plant chromosomes by antibodies recognizing acetylated histone H4 variants. *Chromosome Research* 4, 191–194.

Howard, A. and Pelc, S.R. (1953) Synthesis of deoxyribonucleic acid in normal and irradiated cells and its relation by chromosome breakage. *Heredity* 6, 261–273.

Jakowitsch, J., Papp, I., Moscone, E.A., Van der Winden, J., Matzke, M. and Matzke, A.J.M. (1999) Molecular and cytogenetic characterization of transgene locus that induces silencing and methylation of homologous promoters in *trans*. *Plant Journal* 17, 131–140.

Jessy, N.S., Begum, R., Khatun, M. and Alam, Sk.S. (2005) Differential fluorescent chromosome banding of four species in *Haworthia duval* (Aloaceae). *Cytologia* 70, 435–440.

Kato, Y., Yakura, K. and Tanifuji, S. (1984) Sequence analysis of *Vicia faba* repeated DNA, the *Fok*I repeat element. *Nucleic Acids Research* 13, 6415–6426.

Kato, Y., Yakura, K. and Tanifuji, S. (1985) Repeated DNA sequences found in the large spacer of *Vicia faba* rDNA. *Biochimica et Biophysica Acta* 825, 411–415.

Khattak, G.S.S., Wolny, E. and Saeed, I. (2007) Detection of ribosomal DNA sites in chickpea (*Cicer arietinum* L.) and mungbean (*Vigna radiata* (L).Wilczek) by fluorescence *in situ* hybridization. *Pakistan Journal of Botany* 39, 1511–1515.

Kihlman, B.A. and Kronborg, D. (1975) Sister chromatid exchanges in *Vicia faba*. 1. Demonstration by a modified fluorescent plus Giemsa (FPG) technique. *Chromosoma* 51, 1–10.

Komeda, N., Chaudhary, H.K., Suzuki, G. and Mukai, Y. (2007) Cytological evidence for chromosome elimination in wheat x *Imperata cylindrica* hybrids through fluorescence *in situ* hybridization. *Genes and Genetic Systems* 82, 241–248.

Kumar, P.S. and Hymowitz, T. (1989) Where are the diploid (2n = 2x = 20) genome donors of *Glycine Willd.* (Leguminosae, Papilionoideae)? *Euphytica* 40, 221–226.

Lackey, J.A. (1980) Chromosome numbers in the Phaseoleae (Fabaceae, Faboideae) and their relationship to taxonomy. *American Journal of Botany* 67, 595–602.

Ladizinsky, G. (1993) Wild lentils. *Critical Reviews in Plant Sciences* 12, 169–184.

Leitch, I.J. and Heslop-Harrison, J.S. (1993) Physical mapping of four sites of 5S rDNA sequences and one site of the a-amylase-2 gene in barley (*Hordeum vulgare*). *Genome* 36, 517–523.

Leitch, I.J., Leitch, A.R. and Heslop-Harrison, J.S. (1991) Physical mapping of plant DNA sequences by simultaneous *in situ* hybridization of two differently fluorescent probes. *Genome* 34, 329–333.

Maggini, F., Cremonini, R., Zolfino, C., Tucci, G.F., D'Oxidio, R., Delre, V. *et al.* (1991) Structure and chromosomal localization of DNA sequences related to ribosomal sub repeats in *Vicia faba*. *Chromosoma* 100, 229–234.

Mahbub, M.N., Rubaiyath Bin Rahman, A.N.M. and Alam, Sk.S. (2007) Development of marker chromosomes in three varieties of *Vigna radiata* L. (Fabaceae). *Cytologia* 72, 221–225.

Moscone, E.A., Matzke, M.A. and Matzke, A.J.M. (1996) The use of combined FISH/GISH in conjunction with DAPI counterstaining to identify chromosomes containing transgene inserts in amphidiploids tobacco. *Chromosoma* 105, 231–236.

Moscone, E.A., Klein, F., Lambrou, M., Fuchs, J. and Schweizer, D. (1999) Quantitative karyotyping and dual-color FISH mapping of 5S and 18S–25S rDNA probes in the cultivated *Phaseolus* species (Leguminosae). *Genome* 42, 1224–1233.

Mukai, Y., Nakahara, Y. and Yamamoto, M. (1993) Simultaneous discrimination of the three genomes in hexaploid wheat by multicolour fluorescence *in situ* hybridization using total genomic and highly repeated DNA probes. *Genome* 36, 489–494.

Murray, B.G., Hammett, K.R.W. and Standring, L.S. (1992a) Genomic constancy during the development of *Lathyrus odoratus* cultivars. *Heredity* 68, 321–327.

Murray, B.G., Bennett, M.D. and Hammett, K.R.W. (1992b) Secondary constrictions and NORs of *Lathyrus* investigated by silver staining and *in-situ* hybridization. *Heredity* 68, 473– 478.

Nandini, A.V. (1997) Cytogenetic and interspecific hybridization in *Lathyrus* (L.). PhD thesis, University of Auckland, New Zealand.

Nandini, A.V., Murray, B.G., O'Brien, I.E.W. and Hammett, K.R.W. (1997) Intra- and interspecific variation in genome size in *Lathyrus* (Leguminosae). *Botanical Journal of the Linnean Society* 125, 359–366.

Narayan, R.K.J. (1991) Molecular organisation of the plant genome, Its relation to structure, recombination and evolution of chromosomes. *Journal of Genetics* 70, 43–61.

Nenno, M., Schumann, K. and Nagl, W. (1994) Detection of rRNA and phaseolin genes on polytene chromosomes of *Phaseolus coccineus* by fluorescence *in situ* hybridization after pepsin pre-treatment. *Genome* 37, 1018–1021.

Neumann, P., Lysak, M., Dolezel, J. and Macas, J. (1998) Isolation of chromosomes from *Pisum sativum* L. hairy root cultures and their analysis by flow cytometery. *Plant Science* 137, 205–215.

Neumann, P., Pozarkova, D., Vrana, J., Dolezel, J. and Macas, J. (2002) Chromosome sorting and PCR-based physical mapping in pea (*Pisum sativum* L.). *Chromosome Research* 10, 63–71.

Nodari, R.O., Tsai, S.M., Gilbertson, R.L. and Gepts, P. (1993) Towards an integrated linkage map of common bean. 2. Development of an RFLP-based linkage map. *Theoretical and Applied Genetics* 85, 513–520.

Nouzova, M., Kubalakova, M., Zelova, M.D., Koblizkova, A., Neumann, P., Dolezel, J. *et al.* (1999) Cloning and characterization of new repetitive sequences in faba bean (*Vicia faba* L.) *Annals of Botany* 83, 535–541.

Ochatt, S.J., Benabdelmouna, A., Marget, P., Aubert, G., Moussy, F., Pontécaille, C. *et al.* (2004) Overcoming hybridization barriers between pea and some of its wild relatives. *Euphytica* 137, 353–359.

Palmer, R.G. and Kilen, T.C. (1987) Qualitative genetics and cytogenetics. In: Wilcox, J.R. (ed.) *Soybeans, Improvement, Production and Uses*. American Society of Agronomy, Madison, Wisconsin, pp. 135–197.

Pearce, S.R., Harrison, G., Dongtao, L., Heslop-Harrison, J.S., Kumar, A. and Flavell, A.J. (1996) The *Ty1-copia* group retrotransposons in *Vicia* species, copy number, sequence heterogeneity and chromosomal localisation. *Molecular and General Genetics* 250, 305–315.

Pedersen, C., Zimny, J., Becker, D., Jahne-Gartner, A. and Lorz, H. (1997) Localisation of introduced genes in the chromosomes of transgenic barley, wheat and triticale by fluorescent *in situ* hybridization. *Theoretical and Applied Genetics* 94, 749–757.

Pedrosa-Harand, A., Kami, J. and Gepts, P. (2009) Cytogenetic mapping of common bean chromosomes reveals a less compartmentalized small-genome plant species. *Chromosome Research* 17, 405–417.

Samatadze, T.E., Muravenko, O.V., Bolsheva, N.L., Amosova, A.V., Gostimsky, S.A. and Zelenin, A.V. (2005) Investigation of chromosomes in varieties and translocation lines of pea *Pisum sativum* L. by FISH, Ag-NOR, and differential DAPI staining. *Russian Journal of Genetics* 41, 1381–1388.

Scheuermann, W. and Knaelmann, M. (1975) Localization of ribosomal cistrons in metaphase chromosomes of *Vicia faba* (L.). *Experimental Cell Research* 90, 463–464.

Schlueter, J.A., Dixon, P., Granger, C., Grant, D., Clark, L., Doyle, J.J. *et al.* (2004) Mining EST databases to resolve evolutionary events in major crop species. *Genome* 47, 868–876.

Schubert, I. and Rieger, R. (1979) Asymmetric banding of *Vicia faba* chromosomes after BrdU incorporation. *Chromosoma* 70, 385–391.

Schubert, I., Anastassova-Kristeva, M. and Rieger, R. (1979) Specificity of NOR-staining in *Vicia faba*. *Experimental Cell Research* 120, 433–435.

Schwarzacher, T. and Heslop-Harrison, J.S. (2000) *Practical in-situ Hybridization*. Bios, Oxford, UK, pp. 203–XII.

Schwarzacher, T., Leitch, A.R., Bennett, M.D. and Heslop-Harrison, J.S. (1989) *In situ* localization of parental genomes in a wide hybrid. *Annals of Botany* 64, 315–324.

Schweizer, D. (1973) Differential staining of plant chromosomes with Giemsa. *Chromosoma* 40, 307–320.

Schweizer, D. (1976) Reverse fluorescent chromosome banding with chromomycin and DAPI. *Chromosoma* 58, 307–324.

Schweizer, D. and Ambros, P. (1979) Analysis of nucleolus organizer regions (NORs) in mitotic and polytene chromosomes of *Phaseolus coccineus* by silver staining and Giemsa C-banding. *Plant Systematics and Evolution* 132, 27–51.

Sen, N.K. and Vidyabhusan, R.V. (1960) Tetraploid soybeans. *Euphytica* 9, 317–322.

Shi, L., Zhu, T. and Keim, P. (1996a) Ribosomal RNA genes in soybean and common bean: chromosomal organization, expression, and evolution. *Theoretical and Applied Genetics* 93, 136–141.

Shi, L., Zhu, T., Morgante, M., Rafalski, J.K. and Kelm, P. (1996b) Soybean chromosome painting, a strategy for somatic cytogenetics. *Journal of Heredity* 87, 308–313.

Shoemaker, R.C., Schlueter, J. and Doyle, J.J. (2006) Paleopolyploidy and gene duplication in soybean and other legumes. *Current Opinion in Plant Biology* 9, 104–109.

Simpson, P.R., Newman, M., Davies, D.R., Noel Ellis, T.H., Matthews, P.M. and Lee, D. (1990) Identification of translocations in pea by *in situ* hybridization with chromosome-specific DNA probes. *Genome* 33, 745–749.

Singh, R.J. and Hymowitz, T. (1988) The genomic relationship between *Glycine max* (L) Merr. and *G. soja* Sieb. and Zucc. as revealed by pachytene chromosome analysis. *Theoretical and Applied Genetics* 76, 705–711.

Singh, R.J., Chung, G.H. and Nelson, R.L. (2007) Landmark research in legumes. *Genome* 50, 525–537.

Skorupska, H., Albertsen, M.C., Langholz, K.D. and Palmer, R.G. (1989) Detection of ribosomal RNA genes in soybean, *Glycine max* (L) Merr., by *in situ* hybridization. *Genome* 32, 1091–1095.

Snowdon, R.J., Bottinger, P., Pickardt, P., Kohler, W. and Friedt, W. (2001) Physical localization of transgenes on *Vicia faba* chromosomes. *Chromosome Research* 9, 607– 610.

Staginnus, C., Winter, P., Desel, C., Schmidt, T and Kahl, G. (1999) Molecular structure and chromosomal localization of major repetitive DNA families in the chickpea (*Cicer arietinum* L.) genome. *Plant Molecular Biology* 39, 1037–1050.

Staginnus, C., Huettel, B., Desel, C., Schmidt, T. and Kahl, G. (2001) A PCR- based assay to detect *En/Spm*-like transposon sequences in plants. *Chromosome Research* 9, 591–605.

Straub, S.C.K., Pfeil, B.E. and Doyle, J.J. (2006) Testing the polyploid past of soybean using a low-copy nuclear gene—is *Glycine* (Fabaceae, Papilionoideae) an auto- or allopolyploid? *Molecular Phylogenetics and Evolution* 39, 580–584.

Takehisa, S. and Utsumi, S. (1973) Visualization of metaphase heterochromatin in *Vicia faba* by the denaturation-renaturation Giemsa staining method. *Experimentia* 29, 120–121.

Telenius, H., Pelmear, A.H. and Tumacliffer, A. (1992) Cytogenetic analysis by chromosome painting using DOP-PCR amplified flow sorted chromosomes. *Gene Chromosome Cancer* 4, 257–263.

ten Hoopen, R., Robbins, T.P. and Fransz, P.F. (1996) Localization of T-DNA insertions in *Petunia* by fluorescence *in situ* hybridization: Physical evidence for suppression of recombination. *Plant Cell* 8, 823–830.

ten Hoopen, R., Montijn, B.M., Veuskens, J.I.M., Oud, O.J.L. and Nanninga, N. (1999) The spatial localization of T-DNA insertions in *Petunia* interphase nuclei. Consequences for chromosome organization and transgene insertion sites. *Chromosome Research* 7, 611–623.

Vallejos, C.E., Sakiyama, N.S. and Chase, C.D. (1992) A molecular marker-based linkage map of *Phaseolus vulgaris* L. *Genetics* 131, 733–740.

Vosa, C.G. and Marchi, P. (1972) Quinacrine fluorescence and Giemsa staining in plants. *Nature* 237, 191.

Wang, J., Lewis, M.E., Whallon, J.H. and Sink, K.C. (1995) Chromosomal mapping of T-DNA inserts in transgenic *Petunia* by *in situ* hybridization. *Transgenic Research* 4, 241–246.

Wroth, J.M. (1998) Possible role for wild genotypes of *Pisum* spp. to enhance ascochyta blight resistance in pea. *Australian Journal of Experimental Agriculture* 38, 469–479.

Yamamoto, M. and Mukai, Y. (1989) Application of fluoroscence *in-situ* hybridization to molecular cytogenetics of wheat. *Wheat Information Service* 69, 30–32.

Yamamoto, M. and Mukai, Y. (1995) Physical mapping of ribosomal RNA genes in *Aegilops* and *Triticum*. In, Li, Z.S. and Xin, Z.Y. (eds) *Proceedings of the 8th International Wheat Genetics Symposium*, Beijing, China, pp. 807–811.

Zheng, J., Nakata, M., Uchiyama, H., Morikawa, H. and Tanaka, R. (1991) Giemsa C-banding patterns in several species of *Phaseolus* L. and *Vigna* Savi., Fabaceae. *Cytologia* 56, 459–466.

Zheng, J., Nakata, M., Irifune, K., Tanaka, R. and Morikawa, H. (1993) Fluorescent banding pattern analysis of eight taxa of *Phaseolus* and *Vigna* in relation to their phylogenetic relationships. *Theoretical and Applied Genetics* 87, 38–43.

10 Micropropagation

E. Skrzypek, I. Czyczyło-Mysza and M. Wędzony

10.1 Introduction

Micropropagation is the process of *in vitro* multiplication of the donor plant to produce a large number of true-to-type progeny. The goal is to obtain a large number of healthy plants in a short period at minimal expense. Although this is not easy to achieve, many protocols have been elaborated for food legumes, none of them being universal (Table 10.1).

Micropropagation is based on ability of plant somatic cells to differentiate into whole plants under specific culture conditions. If embryo-like structures emerge from the explant and 'germinate' into plants, the process is termed direct (or primary) somatic embryogenesis. Most often, under *in vitro* conditions, somatic cells first divide into unorganized cell masses called calli, which produce shoots or roots (organogenesis) or embryo-like structures (secondary somatic embryogenesis), capable of developing further into plants. Somatic embryos and young callus tissue may be the object of genetic transformation, or they can be used to initiate cell or protoplast suspension culture, suitable for alternative methods of transformation or *in vitro* mutagenesis. Micropropagation is often used to speed up breeding.

The success of protocols relies on many factors: stock plant care, explant selection and its disinfection, media composition, light, temperature and the length of treatment during subsequent culture phases leading to plants *in vitro*, their *ex vitro* acclimatization and conditions suitable for further growth. Currently, screening for conditions promoting higher regeneration capacity is the main goal of legume culture improvements.

Yield and productivity of many economically important crops have been improved through *in vitro* techniques, including genetic transformation. However, reliable *in vitro* regeneration systems for many genotypes, including those of legumes, are lacking. This chapter reviews the most important recent publications in this area of research. Selected species and some key aspects of protocols are discussed in more detail.

10.2 Soybean (*Glycine max* L. Merrill)

The history of *Glycine max* illustrates well the main problems faced in micropropagation. Barwale *et al.* (1986) succeeded in obtaining fertile plants in 54 soybean genotypes using callus cultures derived from immature embryos. Plant growth regulators had the greatest impact on the process of callus differentiation. The medium, composed of MS basal salts (Murashige and Skoog, 1962) and B5 vitamins (Gamborg *et al.*, 1968),

 Biology and Breeding of Food Legumes
(eds A. Pratap and J. Kumar)

Table 10.1. Examples of successful micropropagation protocols in food legumes.

Species	Explants[a]	Medium[b]	Growth regulators[c]	Reference(s)
Arachis correntina	L	MS	NAA, Kin, TDZ	Mroginski *et al.* (2004)
Arachis glabrata	L	MS	NAA, TDZ	Vidoz *et al.* (2004)
Arachis hypogaea	L	MSB5	BAP, NAA	Chengalrayan *et al.* (1997); Akasaka *et al.* (2000); Tiwari and Tuli (2009)
Arachis pintoi	L, ST	MS	BAP, NAA, PIC	Rey *et al.* (2000); Rey and Mroginski (2006)
Cajanus cajan	CN, L	MS	IBA, TDZ	Singh *et al.* (2003)
Cicer arietinum	CN, EA, N, ST	MSB5	BAP, NAA, Kin	Sarker *et al.* (2005); Naz *et al.* (2007); Rekha and Thiruvengadam (2009)
Glycine max	C, CN, EA, H, IE	KP8, MS, MSB5, MS ½	2,4-D, BA, BAP, GA_3, IBA, NAA, TDZ	Barwale *et al.* (1986); Finer and Nagasawa (1988); Dhir *et al.* (1992); Bailey *et al.* (1993); Walker and Parrott (2001); Tomlin *et al.* (2002); Franklin *et al.* (2004); Hofmann *et al.* (2004); Shan *et al.* (2005); Radhakrishnan *et al.* (2009)
Lathyrus sativus	B, H, ST	MSB5	BAP, IAA, NAA, TDZ	Zambre *et al.* (2002); Ochatt *et al.* (2002)
Lotus corniculatus	R	Rr, MS	BAP	Akashi *et al.* (1998, 2003)
Macrotyloma uniflorum	IC	MS	NAA, Zea, GA_3	Mohamed *et al.* (2005)
Phaseolus acutifolius	B	MS, B5	TDZ, IAA, BAP	Dillen *et al.* (1996)
Phaseolus coccineus	CN, ST	MSB5	BAP, NAA, GA_3	Genga and Allavena (1991); Vaquero *et al.* (1993)
Phaseolus vulgaris	C, CN, EA	MSB5, MS	TDZ, BAP	Cruz de Carvalho *et al.* (2000); Veltcheva *et al.* (2005); Delgado-Sanchez *et al.* (2006)
Phaseolus polyanthus	C, EA	B5	TDZ, IAA	Zambre *et al.* (2001)
Pisum sativum	C, CN, E, EA, H, IE, ST	MS, MSB5, KM	Pic, Zea, NAA, BAP, 2,4-D, TDZ	Griga (1998, 2000, 2002); Griga *et al.* (2007); Franklin *et al.* (2000); Ochatt *et al.* (2000); Zhihui *et al.* (2009)
Vicia faba	E, EA, ST	MS, KM, SH, B5	BAP, 2,4-D, NAA, GA_3, Kin, IBA	Skrzypek (2001); Hamdy and Hattori (2006); Bahgat *et al.* (2009)
Vigna aconitifolia	CN	MS	2,4-D, Kin, BA, GA_3	Choudhary *et al.* (2009)
Vigna mungo	CN, ST	MS, MS ½	TDZ, NAA	Das *et al.* (1998)
Vigna radiata	C, H, L, N	MS	2,4-D, IBA, BA, NAA, Kin	Devi *et al.* (2004); Vidoz *et al.* (2004); Kaviraj *et al.* (2006)
Vigna unguiculata	CN, EA, ST	MSB5, MS	BAP, NAA, IBA	Odutayo *et al.* (2005); Aasim *et al.* (2009); Raveendar *et al.* (2009)

[a]B, vegetative and generative buds; C, cotyledons; CN, cotyledonary nodes; E, epicotyl; EA, embryo axes; H, hypocotyl; IE, immature embryos; L, leaves; N, stem nodes; R, roots; ST, shoot tip.
[b]B5, Gamborg *et al.*'s B5 (1968); KM, Kao and Michayluk (1975); MS, Murashige and Skoog (1962); MSB5, Murashige and Skoog with Gamborg's vitamins (1962); SH, Schenk and Hildebrandt (1972); Rr, Raggio root (Raggio *et al.* 1957); KP8, (Kao, 1977).
[c]BAP, 6-benzylaminopurine; BA, benzylamine; 2,4-D, 2,4-dichlorophenoxyacetic acid; GA_3, gibberellic acid; IAA, indole-3-acetic acid; IBA, indole-3-butyric acid; Kin, kinetin; NAA, 1-naphthaleneacetic acid; Pic, picloram; TDZ, thidiazuron; Zea, zeatin.

was supplemented either by 8 mg/l naphthaleneacetic acid (NAA) or 3 mg/l benzylaminopurine (BAP) and 0.037 mg/l NAA. Either somatic embryogenesis or callusing and organogenesis were achieved. Embryos were converted into plants on the medium supplemented with 0.38 mg/l BAP and 0.04 mg/l indol-3-butyric acid (IBA), while shoot elongation was achieved on media supplemented by 1.13 mg/l BAP, 2 mg/l IBA and 1.73 mg/l GA_3. Rooting media were based on MS salts without growth regulators. The proper sequence of growth regulators in subsequent media is responsible for the success of the procedure, and thus parts of different protocols cannot be combined without careful consideration. The type of explant should be taken into account, since it has a key impact on endogenous phytohormone levels. Here, and in many other leguminous protocols, immature embryos or their parts were used.

The next breakthrough in soybean was reported by Finer and Nagasawa (1988), who elaborated the suspension culture system based on a high level of synthetic auxin analogue 2,4-D in the induction medium. Their protocol was applied for soybean transformation (Finer and McMullen, 1991; Trick and Finer, 1998; Santarem and Finer, 1999) and *in vitro* mutagenesis (Van *et al.*, 2008). Bailey *et al.* (1993) made further improvements to the protocol, testing additional growth regulators, source of carbohydrates and other medium additives. Plant recovery was improved via further modifications (Walker and Parrott, 2001; Tomlin *et al.*, 2002; Schmidt *et al.*, 2005). The latter authors found maltose superior to routinely used sucrose in the conversion rate of embryo to plant. Interestingly, seed pre-treatment with thidiazuron (TDZ) and its addition to the medium in multiple passages enabled longer maintenance of callus tissue without lowering its potential for shoot regeneration (Shan *et al.*, 2005).

Yang *et al.* (2009), working on a large genotype spectrum, found that the addition of 5 mg/l abscisic acid to the regeneration medium beneficial for embryo conversion to plants. The effect was, however, genotype dependent – genotype was reported to influence the protocol's efficiency whenever this aspect was studied (Barwale *et al.*, 1986; Parrott *et al.*, 1989; Dhir *et al.*, 1992; Bailey *et al.*, 1993; Walker and Parrott, 2001; Tomlin *et al.*, 2002; Van *et al.*, 2008). Dan and Reighceri (1998) and Reichert *et al.* (2003) found that the method of utilizing adventitious shoots induced from hypocotyl sections of 7-day-old seedlings was relatively less genotype dependent. Song *et al.* (2010) found six QTL associated with somatic embryogenesis that provided potential for marker-assistant selection of genotypes with higher *in vitro* potential.

10.3 Groundnut (*Arachis hypogaea* L.)

Arachis hypogaea L. cultivars are known to be relatively recalcitrant to plant regeneration. Successful results were achieved via organogenesis (Daimon and Mii, 1991; McKently *et al.*, 1991; Cheng *et al.*, 1992, 1996; Kanyand *et al.*, 1994; Chengalrayan *et al.*, 1995; Akasaka *et al.*, 2000; Tiwari and Tuli, 2009) and somatic embryogenesis (Sellars *et al.*, 1990; Durham and Parrott, 1992; Eapen *et al.*, 1993; Chengalrayan *et al.*, 1994, 1997; Baker *et al.*, 1995; Murthy *et al.*, 1995, Joshi *et al.*, 2003). Similar to soybean, a strong influence of genotype was reported (McKently *et al.*, 1990; Matand and Prakash, 2007).

Growth regulators and the type of explant are the key factors for groundnut regeneration. Thidiazuron (TDZ) is applied most frequently at the start of the culture (Gill and Saxena, 1992; Kanyand *et al.*, 1994; Li *et al.*, 1994; Murthy *et al.*, 1995; Akasaka *et al.*, 2000; Joshi *et al.*, 2003; Matand and Prakash, 2007), while BAP (6-benzylaminopurine) alone or in combination with NAA (1-naphthaleneacetic acid) is also efficient (Chengalrayan *et al.*, 1995; Akasaka *et al.*, 2000; Banerjee *et al.*, 2007). The immature leaflets isolated from young seedlings are most widely used as explants (Cheng *et al.*, 1992; Chengalrayan *et al.*, 1995, 1997, Akasaka *et al.*, 2000; Joshi *et al.*, 2003; Mroginski *et al.*, 2004; Vidoz *et al.*, 2004; Tiwari and Tuli, 2009). However, petioles, mature or immature embryos or their parts and the whole seed were efficient in protocols involving shoot regeneration (Ozias-Akins, 1989; McKently *et al.*, 1990; Cheng *et al.*, 1992; Gill and Saxena, 1992;

Kanyand *et al.*, 1994; Radhakrishnan *et al.*, 2000; Vasanth *et al.*, 2006). Multiple shoots were induced by Radhakrishnan *et al.* (2000) from de-embryonated cotyledons, embryo axes and whole mature seeds on MS medium supplemented with BAP. Significant progress in shoot induction rate was claimed in a report by Akasaka *et al.* (2000). Treatment of 10 mg/l TDZ for 7 days or 1 mg/l TDZ for 21 days was applied to reduce abnormalities in shoot development.

Tiwari and Tuli (2009) obtained excellent results for shoot bud formation (85.1%) and shoot elongation (6.2 shoots/explant) when immature leaflets were pre-incubated for 7 days on a medium containing 3 mg/l BAP and 0.92 mg/l NAA. Li *et al.* (1994) and Tiwari and Tuli (2008) did not observe significant variations in response among cultivated groundnut varieties, similar to the reports of Matand and Prakash (2007). Somatic embryogenesis was induced in leaflets by Narasimhulu and Reddy (1983) and Chengalrayan *et al.* (1995). Globular embryo-like structures appeared on the cut leaf base on MS medium with 20 mg/l 2,4-D. A high frequency of recovery was found after transfer to a medium with 3 mg/l 2,4-D within 20 days, and subsequent culture on that medium with 0.5 mg/l BAP and kinetin (Kin). Micropropagation and *in vitro* conservation of wild *Arachis* species considered as potential sources of novel genes for crop improvement was reviewed by Pacheco *et al.* (2009).

10.4 Phaseolum (*Phaseolus* sp.)

Plant regeneration in *Phaseolus* sp. L. was reviewed by Nagl *et al.* (1997) and Veltcheva *et al.* (2005). Successful regeneration is reported mainly for *Phaseolus vulgaris* L. (Benedicic *et al.*, 1991; Malik and Saxena, 1991; Santalla *et al.*, 1998; Cruz de Calvalho *et al.*, 2000). Regeneration from other *Phaseolus* species was achieved in *Phaseolus coccineus* L. (Rubluo and Kartha, 1985; Angelini and Allavena, 1989; Genga and Allavena, 1991; Malik and Saxena, 1992; Santalla *et al.*, 1998), *Phaseolus acutifolius* (Dillen *et al.*, 1996; Zambre *et al.*, 1998) and *Phaseolus polyanthus* (Zambre *et al.*, 2001).

Organogenesis *via* shoot apex cultures was described by Kartha *et al.*, (1981) and Martins and Sondahl (1984). Cotyledonary nodes and primary leaves were used by McClean and Grafton (1989), Mohamed *et al.* (1992) and Vaquero *et al.* (1993). Axillary meristems or shoot apical meristems (Kartha *et al.*, 1981; Martins and Sondahl, 1984; Rubluo and Kartha, 1985; McClean and Grafton, 1989) were replaced by cotyledons, cotyledonary nodes or the embryonic axis (Mohamed *et al.*, 1992; Santalla *et al.*, 1998).

An enhanced differentiation of somatic embryos in cotyledonary leaf-derived callus but low regeneration frequency has been reported for *P. vulgaris* L. by Mohamed *et al.* (1993). A high frequency of direct shoot formation from intact seedlings has been established by Malik and Saxena (1992) using TDZ and BAP, while seedling-derived thin layers were used to improve regeneration (Cruz de Carvalho *et al.*, 2000). The latter group reported successful development of shoots from bud primordia on a medium with TDZ and $AgNO_3$, with a high rate of development of fertile plants. A protocol based on embryo-axes derived from mature seeds was reported by Delgado-Sanchez *et al.* (2006). All results cited above point to strong genotype dependence and lack of universal protocol for *Phaseolus* species.

10.5 Pea (*Pisum sativum* L.)

Studies reported for *Pisum sativum* L. use various explants: cotyledonary node (Jordan and Hobbs, 1993; Bean *et al.*, 1997; Popiers *et al.*, 1997), immature embryos (Natali and Cavallini, 1987; Tétu *et al.*, 1990; Kosturkova *et al.*, 1997), immature cotyledon (Özcan *et al.*, 1993; Grant *et al.*, 1995), thin layers of nodal explants (Nauerby *et al.*, 1991; Madsen *et al.*, 1998), shoot apices (Griga *et al.*, 1986), and embryonic axis sections (Schroeder *et al.*, 1993; Polowick *et al.*, 2000) as the explants. Regeneration in pea has been achieved by different paths such as somatic embryogenesis (Bencheikh and Gallais, 1996; Griga 1998, 2002), direct and indirect organogenesis (Kartha *et al.*, 1974; Kallak and Koiveer, 1990;

Kosturkova *et al.*, 1997) and protoplast culture (Lehminger-Mertens and Jacobsen, 1989a, b; Boehmer *et al.*, 1995). However, none of the methods above was successful in the routine production of plants.

Hildebrand *at al.* (1963) were the first to describe the development of pea shoots from stem-derived callus. Kartha *et al.* (1974) showed the first successful regeneration using apical meristems. Jacobsen and Kysely (1984) were the first to induce somatic embryogenesis in pea. Plant regeneration via the embryogenic pathway was reported (Kysely *et al.*, 1987). Morphological alterations (in leaflets and tendrils, fasciations, etc.) of a chimeric nature have been observed in plants derived from organogenesis and somatic embryogenesis, often resulting in sterility (Stejskal and Griga, 1992). Ochatt *et al.* (2000) suggested a clear effect of growth regulators used during the *in vitro* stages on the DNA levels of the subsequently regenerated plants. Pniewski *et al.* (2003) observed that a high BAP dose was disadvantageous for long term micropropagation – newly formed shoots were dwarf, vitrified and incapable of forming roots. These observations suggest the application of initially high cytokinin doses for organogenesis induction but subsequently lower concentrations for micropropagation, as postulated earlier (Jackson and Hobbs, 1990). Kysely *et al.* (1987) and Kysely and Jacobsen (1990) found that benzylamine (BA) drastically reduced somatic embryo frequency in pea. Loiseau *et al.* (1995) reported that cytokinins added to an auxin medium reduced embryo conversion. Zhihui *et al.* (2009) showed that shoot development was accomplished when the bud-containing tissues (BCT) were left on MS medium supplemented with 4 mg/l TDZ without subculture prior to transfer onto MS medium supplemented with 0.5 mg/l BA. Tzitzikas *et al.* (2004) initiated BCT on nodal sections isolated from *in vitro*-propagated plants. High cytokinin and very low auxin content appeared to be essential for the initiation of morphogenesis via callus (Malmberg, 1979; Hussey and Gunn, 1984; Rubluo *et al.*, 1984; Natali and Cavallini, 1987; Tétu *et al.*, 1990; Özcan *et al.*, 1992; Kosturkova *et al.*, 1997; Pniewski *et al.*, 2003).

Frequently, *in vitro*-regenerated shoots were rooted directly without any preconditioning phase (Hussey and Gunn, 1984; Griga *et al.*, 1986; Natali and Cavallini, 1987; Nauerby *et al.*, 1991; Özcan *et al.*, 1992; Nadolska-Orczyk *et al.*, 1994; Pniewski *et al.*, 2003). The latter authors introduced the additional step of subculturing on 0.02 mg/l BAP to make the pass from micropropagation to rooting more moderate, and found that half-strength MS with B5 vitamins and 1.0 mg/l NAA the most efficient for rooting. Full-strength MS was generally inappropriate to induce rooting, whereas half-strength MS was recommended (Hussey and Gunn, 1984; Griga *et al.*, 1986; Özcan *et al.*, 1992). Rhisogenesis was proved to be genotype dependent (Nauerby *et al.*, 1991; Nadolska-Orczyk *et al.*, 1994). Madsen *et al.* (1998) showed that the addition of silver nitrate to the medium decreased shoot vitrification but greatly reduced rooting frequency. In pea, the protocols of direct somatic embryogenesis (Griga, 1998) and organogenesis (Pniewski *et al.*, 2003) are relatively well elaborated and thus can be recommended as starting points for new cultivars.

10.6 Cowpea (*Vigna unguiculata* L.)

The regeneration of *Vigna unguiculata* L. via somatic embryogenesis has been achieved by starting the culture with either immature cotyledons (Anand *et al.*, 2001), mature embryonic axes or embryos (Amitha and Reddy, 1996a; Odutayo *et al.*, 2005; Popelka *et al.*, 2006) or young leaves (Muthukumar *et al.*, 1995; Ramakrishnan *et al.*, 2005). The basal medium developed for somatic embryogenesis by Pellegrineschi (1997) was a starting point for media optimization by Machuka *et al.* (2000). Cell suspensions can be obtained from callus (Kulothungan *et al.*, 1995; Anand *et al.*, 2000). The maximum frequency of somatic embryogenesis was obtained when callus was transferred to liquid MS with 0.5 mg/l 2,4-D (Machuka *et al.*, 2000).

In contrast to somatic embryogenesis, numerous protocols were standardized for *in vitro* cowpea organogenesis using hypocotyls, epicotyls and cotyledons (Cheema and Bawa, 1991; Amitha and Reddy, 1996b; Muthukumar *et al.*, 1996; Pellegrineschi, 1997; Brar *et al.*,

1999a; Van Le *et al.*, 2002; Chaudhury *et al.*, 2007; Raveendar *et al.*, 2009). Organogensis was also induced in cultures of shoot meristems (Kartha *et al.*, 1981; Brar *et al.*, 1997; Mao *et al.*, 2006; Aasim *et al.*, 2009) and leaflets (Muthukumar *et al.*, 1995). Pellegrineschi *et al.* (1997) reported regeneration of shoots in the presence of 0.1 mg/l zeatine (ZEA). The variability in methods has involved almost every aspect of the regeneration systems explored, such as optimal explant tissues, basal salt composition, plant growth regulators and sucrose levels (Pellegrineschi, 1997; Popelka *et al.*, 2006). Successful cowpea regeneration was achieved with a wide range of basal media depending on genotype and explant type (Muthukumar *et al.*, 1995; Pellegrineschi, 1997; Brar *et al.*, 1999a). Direct organogenesis was obtained on MS medium containing either BA or BAP (Muthukumar *et al.*, 1995; Pellegrineschi, 1997; Brar *et al.*, 1999a; Mao *et al.*, 2006). It has been indicated that BA plays a key role in shoot formation. A regeneration system successful for 17 cowpea genotypes was reported by Brar *et al.* (1999a). Shoot regeneration from cotyledons was initiated on 1/3 MS with 15–35 mg/l of BA followed by culture on MS with 1.0 mg/l of BA (Machuka *et al.*, 2000).

Apart from BA, successful plant regeneration was also achieved using 2,4-D (Anand *et al.*, 2000; Ramakrishnan *et al.*, 2005), 2,4,5-trichloro-phenoxyacetic acid (2,4,5-T) (Muthukumar *et al.*, 1995), ZEA (Anand *et al.*, 2000) and TDZ (Aasim *et al.*, 2009). Fertile cowpea plants were regenerated from cotyledonary node thin cell layer explants (TCL) by the application of TDZ (Van Le *et al.*, 2002). These authors reported that a 2.20 mg/l TDZ pre-treatment, shoot tip removal and excision of longitudinal TCL at the level of the cotyledonary nodes, with subsequent culture on a MSB5 medium supplemented with 0.20 mg/l IBA and 0.22 mg/l TDZ, were optimal for maximum bud proliferation. On average, 32.5 buds per explant were harvested with an 80% recovery rate, which is far superior to other results reported for cowpea, i.e. 1–11 buds per explant with a survival frequency of 36–55.3% (Muthukumar *et al.*, 1995; Pellegrineschi 1997; Brar *et al.*, 1999b). Brar *et al.* (1999a) showed poor shoot rooting on a hormone-free medium, while Raveendar *et al.* (2009) reported strong root formation on hormone-free MSB5 medium. Supplementing the culture with 1.0 mg/l IAA or 0.05 mg/l NAA significantly enhanced rooting and *ex vitro* plant survival (Machuka *et al.*, 2000). According to Mao *et al.* (2006), IBA had no effect on rooting, whereas results obtained by Aasim *et al.* (2009) showed that IBA had positive effects not only on root induction but also on secondary shoot regeneration. Shoots were easily rooted on MS medium supplemented with 0.5 mg/l IBA (Anand *et al.*, 2001; Aasim *et al.*, 2009). Inconsistent data on optimal protocol for *in vitro* rooting might be due to variability in genotypes used or differences in earlier phases of protocols.

Recently, Raveendar *et al.* (2009) described a rapid and efficient regeneration system via organogenesis for four genotypes of cowpea, where cotyledonary nodes of 3-day-old seedlings appeared suitable for plant regeneration. The seeds were pre-treated with 3 mg/l BAP for 3 days and cultured on MSB5 medium supplemented with 1.49 mg/l BAP for 2–3 weeks. Multiple shoots were then transferred to a medium supplemented with 0.11 mg/l BAP for shoot elongation and rooted on growth regulator-free MSB5 medium. The plantlets were transferred to soil after 12 days, when 90–95% survived – a high percentage.

10.7 Conclusion

Most food legumes are considered difficult to culture *in vitro*, and their regeneration depends to a large extent on genotype and explant type. Many recent advances include explant pre-treatment with growth regulators prior to *in vitro* culture, which enhances induction rate. Effective plant regeneration seems to be the problem in many protocols. Comparison of various culture systems is difficult, since the same protocols were seldom applied to numerous genotypes. While almost every media component was tested in order to improve efficiency, the role of light and temperature was not regularly examined during subsequent culture phases; this might be a field suitable for further optimization of protocols.

References

Aasim, M., Khawar, K.M. and Özcan, S. (2009) *In vitro* micropropagation from plumular apices of Turkish cowpea (*Vigna unguiculata* L.) cultivar Akkiz. *Scientia Horticulturae* 122, 468–471.

Akasaka, Y., Daimon, H. and Mii, M. (2000) Improved plant regeneration from cultured leaf segments in peanut (*Arachis hypogaea* L.) by limited exposure to thidiazuron. *Plant Science* 156, 169–175.

Akashi, R., Hoffmann-Tsay, S.S. and Hoffmann, F. (1998) Selection of a super-growing legume root culture that permits controlled switching between root cloning and direct embryogenesis. *Theoretical and Applied Genetics* 96, 758–764.

Akashi, R., Kawano, T., Hashiguchi, M., Kutsuna, Y., Hoffmann-Tsay, S.S. and Hoffmann, F. (2003) Super roots in *Lotus corniculatus*: A unique tissue culture and regeneration system in a legume species. *Plant and Soil* 255, 27–33.

Amitha, K. and Reddy, T.P. (1996a) Regeneration of plantlets from different explants and callus cultures of cowpea (*Vigna unguiculata* L.). *Phytomorphology* 46, 207–211.

Amitha, K. and Reddy, T.P. (1996b) Induction of somatic embryogenesis and regeneration in cowpea (*Vigna sinensis* L.). *Current Advances in Plant Science* 9, 23–28.

Anand, R.P., Ganapathi, A., Rnbazhagan, V., Vengadesan, G. and Selvaraj, N. (2000) High frequency plant regeneration via somatic embryogenesis in cell suspension cultures of cowpea [*Vigna unguiculata* (L.) Walp.]. *In Vitro Cellular and Developmental Biology – Plant* 36, 475–480.

Anand, R.P., Ganapathi, A., Vengadesan, G., Selvaraj, N., Anbazhagan, V.R. and Kulothungan, S. (2001) Plant regeneration from immature cotyledon-derived callus of *Vigna unguiculata* (L.) Walp. (cowpea). *Current Science* 80, 671–674.

Angelini, R.R. and Allavena, A. (1989) Plant regeneration from immature cotyledon explant cultures of bean (*P. coccineus* L.). *Plant Cell Tissue and Organ Culture* 19, 167–174.

Bahgat, S., Shabban, O.A., El-Shihy, O., Lightfoot, D.A. and El-Shemy, H.A. (2009) Establishment of the regeneration system for *Vicia faba* L. *Current Issues in Molecular Biology* 11, 47–54.

Bailey, M.A., Boerma, H.R. and Parrott, W.A. (1993) Genotype effects on proliferative embryogenesis and plant regeneration of soybean. *In Vitro Cellular and Developmental Biology – Plant* 29, 102–108.

Baker, C.M., Durham, R.E., Austin Burns, J., Parrott, W.A. and Wetzstein, H.Y. (1995) High frequency somatic embryogenesis in peanut (*Arachis hypogaea* L.) using mature, dry seed. *Plant Cell Reports* 15, 38–42.

Banerjee, P., Maity, S., Maiti, S.S. and Banerjee, N. (2007) Influence of genotype on in vitro multiplication potential of *Arachis hypogaea* L. *Acta Botanica Croatica* 66, 15–23.

Barwale, U.B., Kerns, H.R. and Widholm, J.M. (1986) Plant regeneration from callus cultures of several soybean genotypes via embryogenesis and organogenesis. *Planta* 167, 473–481.

Bean, S.J., Gooding, P.S., Mullineaux, P.M. and Davies, D.R. (1997) A simple system for pea transformation. *Plant Cell Reports* 16, 513–519.

Bencheikh, M. and Gallais, A. (1996) Somatic embryogenesis in pea (*Pisum sativum* L. and *Pisum arvense* L.): Diallele analysis and genetic control. *Euphytica* 90, 257–264.

Benedicic, D., Ravnikar, M. and Gogala, N. (1991) The influence of jasmonic acid on the development of *Phaseolus vulgaris* shoot culture. *Acta Horticulturae* 289, 85–86.

Boehmer, P., Meyer, B. and Jacobson, H.J. (1995) Thidiazuron induced high frequency of shoot induction and plant regeneration in protoplast-derived pea callus. *Plant Cell Reports* 15, 26–29.

Brar, M.S., Al-Khayri, J.M., Shamblin, C.E., McNew, R.W., Morelock, T.E. and Anderson, E.J. (1997) In vitro shoot tip multiplication of cowpea *Vigna unguiculata* (L.) Walp. *In Vitro Cellular and Developmental Biology – Plant* 33, 114–118.

Brar, M.S., Al-Khayri, J.M., Morelock, T.E. and Anderson, E.J. (1999a) Genotypic response of cowpea *Vigna unguiculata* (L.) to *in vitro* regeneration from cotyledon explants. *In Vitro Cellular and Developmental Biology – Plant* 35, 8–12.

Brar, M.S., Moore, M.J., Al-Khayri, J.M., Morelock, T.E., and Anderson, E.J. (1999b) Ethylene inhibitors promote *in vitro* regeneration of cowpea (*Vigna unguiculata* L.). *In Vitro Cellular and Developmental Biology – Plant* 35, 222–225.

Chaudhury, D., Madanpotra, S., Jaiwal, R., Saini, R., Kumar, P.A. and Jaiwal, P.K. (2007) *Agrobacterium tumefaciens*-mediated high frequency genetic transformation of an Indian cowpea (*Vigna unguiculata* L. Walp.) cultivar and transmission of transgenes into progeny. *Plant Science* 172, 692–700.

Cheema, H.K. and Bawa, J. (1991) Clonal multiplication via multiple shoots in some legumes (*Vigna unguiculata* and *Cajanus cajan*). *Acta Hortculturae* 289, 93–96.

Cheng, M., Hsi, D.C.H. and Phillips, G.C. (1992) *In vitro* regeneration of Valencia-type peanut (*Arachis hypogaea* L.) from cultured petioles, epicotyl sections and other seedling explants. *Peanut Science* 19, 82–87.

Cheng, M., Jarret, R.L., Li, Z., Xing, A. and Demski, J.W. (1996) Production of fertile transgenic peanut (*Arachis hypogaea* L.) plants using *Agrobacterium tumefaciens*. *Plant Cell Reports* 15, 653–657.

Chengalrayan, K., Sathaye, S.S. and Hazra, S. (1994) Somatic embryogenesis from mature embryo-derived leaflets of peanut (*Arachis hypogaea* L.). *Plant Cell Reports* 13, 578–581.

Chengalrayan, K., Mhaske, V.B. and Hazra, S. (1995) *In vitro* regulation of morphogenesis in peanut (*Arachis hypogaea* L.). *Plant Science* 110, 259–268.

Chengalrayan, K., Mhaske, V.B. and Hazra, S. (1997) High-frequency conversion of peanut somatic embryos. *Plant Cell Reports* 16, 783–786.

Choudhary, K., Singh, M., Rathore, M.S. and Shekhawat, N.S. (2009) Somatic embryogenesis and *in vitro* plant regeneration in moth bean [*Vigna aconitifolia* (Jacq.) Maréchal]: a recalcitrant grain legume. *Plant Biotechnology Reports* 3, 205–211.

Cruz de Carvalho, M.H., Le, B.V., Zuily-Fodil, Y., Thi, A.T.P. and Van, K.T.T. (2000) Efficient whole plant regeneration of common bean (*Phaseolus vulgaris* L.) using thin-cell-layer culture and silver nitrate. *Plant Science* 159, 223–232.

Daimon, H. and Mii, M. (1991) Multiple shoot formation and plantlet regeneration from cotyledonary node in peanut (*Arachis hypogaea* L.). *Japanese Journal of Breeding* 41, 461–466.

Dan, Y. and Reighceri, N.A. (1998) Organogenic regeneration of soybean from hypocotyl explants. *In Vitro Cellular and Developmental Biology – Plant* 34, 14–21.

Das, D.K., Prakash, N.S. and Bhalla-Sarin, N. (1998) An efficient regeneration system of black gram (*Vigna mungo* L.) through organogenesis. *Plant Science* 134, 199–206.

Delgado-Sanchez, P., Saucedo-Ruiz, M., Guzman-Maldonado, S.H., Villordo-Pineda, E., Gonzalez-Chavira, M., Fraire-Velazquez, S. *et al.* (2006) An organogenic plant regeneration system for common bean (*Phaseolus vulgaris* L.). *Plant Science* 170, 822–827.

Devi, P., Radha, P., Sitamahalakshmi, L., Syamala, D. and Kumar, S.M. (2004) Plant regeneration via somatic embryogenesis in mung bean [*Vigna radiata* (L.) Wilczek]. *Scientia Horticulturae* 99, 1–8.

Dhir, S.K., Dhir, S. and Widholm, J.M. (1992) Regeneration of fertile plants from protoplasts of soybean (*Glycine max* L. Merr.): genotypic differences in culture response. *Plant Cell Reports* 11, 285–289.

Dillen, W., De Clercq, J., Van Montagu, M. and Angenon, G. (1996) Plant regeneration from callus in a range of *Phaseolus acutifolius* A. Gray genotypes. *Plant Science* 118, 81–88.

Durham, R.E. and Parrott, W.A. (1992) Repetitive somatic embryogenesis from peanut cultures in liquid medium. *Plant Cell Reports* 11, 122–125.

Eapen, S., George, L. and Rao, P.S. (1993) Plant regeneration through somatic embryogenesis in peanut (*Arachis hypogaea* L.). *Biologia Plantarum* 35, 499–504.

Finer, J.F. and McMullen, D. (1991) Transformation of soybean via particle bombardment of embyrogenic suspension culture tissue. *In Vitro Cellular and Developmental Biology – Plant* 27, 175–182.

Finer, J.J. and Nagasawa, A. (1988) Development of an embryogenic suspension culture of soybean [*Glycine max* (L.) Merrill]. *Plant Cell, Tissue and Organ Culture* 15, 125–136.

Franklin, G., Pius, P.K. and Ignacimuthu, S. (2000) Factors affecting *in vitro* flowering and fruiting of green pea (*Pisum sativum* L.). *Euphytica* 115, 65–73.

Franklin, G., Carpenter, L., Davis, E., Reddy, C.S., Al-Abed, D., Abou Alaiwi, W. *et al.* (2004) Factors influencing regeneration of soybean from mature and immature cotyledons. *Plant Growth Regulation* 43, 73–79.

Gamborg, O.L., Miller, R.A. and Ojima, K. (1968) Nutrient requirements of suspension cultures of soybean cells. *Experimental Cell Research* 50, 151–158.

Genga, A. and Allavena, A. (1991) Factors affecting morphogenesis from immature cotyledons of *Phaseolus coccineus* L. *Plant Cell Tissue and Organ Culture* 27, 189–196.

Gill, R. and Saxena, P.K. (1992) Direct somatic embryogenesis and regeneration of plants from seedlings explants of peanut (*Arachis hypogaea*): promotive role of thidiazuron. *Canadian Journal of Botany* 70, 1186–1192.

Grant, J.E., Cooper, P.A., McAra, A.E. and Frew, T.J. (1995) Transformation of peas (*Pisum sativum* L.) using immature cotyledons. *Plant Cell Reports* 15, 254–258.

Griga, M. (1998) Direct somatic embryogenesis from shoot apical meristems of pea, and thidiazuron-induced high conversion rate of somatic embryos. *Biologia Plantarum* 41, 481–495.

Griga, M. (2000) Morphological alteration in sterile mutant of *Pisum sativum* obtained via somatic embryogenesis. *Biologia Plantarum* 43, 161–165.

Griga, M. (2002) Morphology and anatomy of *Pisum sativum* somatic embryos. *Biologia Plantarum* 45, 173–182.

Griga, M., Tejklova, E., Novak, F.J. and Kubalakova, M. (1986) *In vitro* clonal propagation of *Pisum sativum* L. *Plant Cell Tissue and Organ Culture* 6, 95–104.

Griga, M., Horáček, J. and Klenotičová, H. (2007) Protein patterns associated with *Pisum sativum* somatic embryogenesis. *Biologia Plantarum* 51, 201–211.

Hamdy, A.M.A. and Hattori, K. (2006) Regeneration of (*Vicia faba* L.) cultivars from mature seeds cotyledons. *Asian Journal of Plant Sciences* 5, 623–629.

Hildebrand, A.C., Wilmar, J.C., Johons, H. and Riker, A.J. (1963) Growth of edible chlorophyllous plant tissues *in vitro*. *American Journal of Botany* 50, 248–254.

Hofmann, N., Nelson, R.L. and Korban, S.S. (2004) Influence of media components and pH on somatic embryo induction in three genotypes of soybean. *Plant Cell, Tissue and Organ Culture* 77, 157–163.

Hussey, G. and Gunn, H.V. (1984) Plant production in pea (*Pisum sativum* L. cvs. Puget and Upton) from long-term callus with superficial meristems. *Plant Science Letters* 3, 143–148.

Jackson, J.A. and Hobbs, S.L.A. (1990) Rapid multiple shoot production from cotyledonary node explants of pea (*Pisum sativum* L.). *In Vitro Cellular and Developmental Biology* 26, 835–838.

Jacobsen, H.J. and Kysely, W. (1984) Induction of somatic embryos in pea, *Pisum sativum* L. *Plant Cell Tissue and Organ Culture* 3, 319–324.

Jordan, M.C. and Hobbs, L.A. (1993) Evaluation of a cotyledonary node regeneration system for *Agrobacterium*-mediated transformation of pea (*Pisum sativum* L.). *In Vitro Cellular and Developmental Biology* 29, 77–82.

Joshi, M.V., Sahasrabudhe, N.A. and Hazra, S. (2003) Responses of peanut somatic embryos to thidiazuron. *Biologia Plantarum* 46, 187–192.

Kallak, H. and Koiveer, A. (1990) Induction of morphogenesis in meristems of different cultivars of *Pisum sativum* L. *Plant Science* 67, 221–226.

Kanyand, M., Dessai, A.P. and Prakash, C.S. (1994) Thidiazuron promotes high frequency regeneration of peanut (*Arachis hypogaea*) plants *in vitro*. *Plant Cell Reports* 14, 1–5.

Kao, K.N. (1977) Chromosomal behaviour in somatic hybrids of Soybean – *Nicotiana glauca*. *Molecular and General Genetics* 150, 225–230.

Kao, K.N. and Michayluk, M.R. (1975) Nutritional requirements for growth of *Vicia hajastana* cells and protoplasts at a very low population density in liquid media. *Planta* 126, 105–110.

Kartha, K.K., Gamborg, O.L. and Constabel, F. (1974) Regeneration of pea *Pisum sativum* plants from shoot apical meristems. *Zeitschrift für Pflanzenphysiologie* 72, 172–176.

Kartha, K.K., Pahl, K., Leung, N.L. and Mroginski, L.A. (1981) Plant regeneration from meristems of grain legumes soybean, cowpea, peanut, chickpea and bean. *Canadian Journal of Botany* 59, 1671–1679.

Kaviraj, C.P., Kiran, G., Venugopal, R.B., Kishor, P.B.K. and Rao, S. (2006) Somatic embryogenesis and plant regeneration from cotyledonary explants of green gram [*Vigna radiata* (L.) Wilczek.] – a recalcitrant grain legume. *In Vitro Cellular and Developmental Biology – Plant* 42, 134–138.

Kosturkova, G., Mahandjiev, A., Dobreva, I. and Tzvetkova, V. (1997) Regeneration systems for immature embryos of Bulgarian pea genotypes. *Plant Cell, Tissue and Organ Culture* 18, 139–142.

Kulothungan, S., Ganapathi, A., Shajahan, A. and Kathiravan, K. (1995). Somatic embryogenesis in cell suspension culture of cowpea [*Vigna unguiculata* (L.) Walp.]. *Israel Journal of Plant Science* 43, 385–390.

Kysely, W. and Jacobsen, H.J. (1990) Somatic embryogenesis from pea embryos and shoot apices. *Plant Cell, Tissue and Organ Culture* 20, 7–14.

Kysely, W., Myers, J.R., Lazzeri, P.A., Collins, G.B. and Jacobsen, H.J. (1987) Plant regeneration via somatic embryogenesis in pea (*Pisum sativum* L.). *Plant Cell Reports* 6, 305–308.

Lehminger-Mertens, R. and Jacobsen, H.J. (1989a) Plant regeneration from pea protoplasts via somatic embryogenesis. *Plant Cell Reports* 8, 379–382.

Lehminger-Mertens, R. and Jacobsen, H.J. (1989b) Protoplast regeneration and organogenesis from pea protoplasts. *In Vitro Cellular and Developmental Biology – Plant* 25, 571–574.

Li, Z., Jarret, R.L., Pittman, R.N. and Demski, J.W. (1994) Shoot organogenesis from cultured seed explants of peanut (*Arachis hypogaea* L.) using thidiazuron. *In Vitro Cellular and Developmental Biology – Plant* 30, 187–191.

Loiseau, J., Marche, C. and Le Deunff, Y. (1995) Effect of auxins, cytokinins, carbohydrates, and amino acids on somatic embryogenesis induction from shoot apices of pea. *Plant Cell, Tissue and Organ Culture* 41, 267–275.

Machuka, J., Adesoye, A. and Obembe, O.O. (2000) Regeneration and genetic transformation in cowpea. In: Fatokun, C.A., Tarowali, S.A., Singh, B.B., Kormana, P.M. and Tamo, M. (eds) Challenges and opportunities for enhancing sustainable cowpea production. In: *Proceedings of the World Cowpea Conference III*, 2000, International Institute of Tropical Agriculture (IITA), Ibadan, Nigeria, pp. 185–196.

Madsen, M.H., Nauerby, B., Frederiksen, C.G. and Wyndaele, R. (1998) Regeneration of pea (*Pisum sativum* L.) by the thin cell layer nodal system: influence of explant culture media on rooting and plantlet formation. *Acta Agriculturae Scandinavica, Section B – Soil and Plant Science* 48, 58–64.

Malik, K.A. and Saxena, P.K. (1991) Regeneration in *Phaseolus vulgaris* L. promotion role of N6-benzylaminopurine in cultures from juvenile leaves. *Planta* 184, 148–150.

Malik, K.A. and Saxena, P.K. (1992) Regeneration in *Phaseolus vulgaris* L.: High-frequency induction of direct shoot formation in intact seedlings by N6-benzylaminopurine and thidiazuron. *Planta* 186, 384–389.

Malmberg, R.L. (1979) Regeneration of whole plants from callus culture of diverse genetic lines of *Pisum sativum* L. *Planta* 146, 243–244.

Mao, J.Q., Zaidi, M.A., Aranson, J.T. and Altosaar, I. (2006) *In vitro* regeneration of *Vigna unguiculata* (L.) Walp. cv. Black eye cowpea via shoot organogenesis. *Plant Cell, Tissue and Organ Culture* 87, 121–125.

Martins, I.S. and Sondahl, M.R. (1984) Early stages of somatic embryo differentiation from callus cells of bean (*Phaseolus vulgaris* L.) grown liquid medium. *Journal of Plant Physiology* 117, 97–103.

Matand, K. and Prakash, C.S. (2007) Evaluation of peanut genotypes for *in vitro* plant regeneration using thidiazuron. *Journal of Biotechnology* 130, 202–207.

McClean, P. and Grafton, K.F. (1989) Regeneration of dry bean (*Phaseolus vulgaris* L.) via organogenesis. *Plant Science* 60, 117–122.

McKently, A.H., Moore, G.A. and Gardner, F.P. (1990) *In vitro* plant regeneration of peanut from seed explants. *Crop Science* 30, 192–196.

McKently, A.H., Moore, G.A. and Gardner, F.P. (1991) Regeneration of peanut and perennial peanut from cultured leaf tissue. *Crop Science* 31, 833–837.

Mohamed, M.F., Read, P.E. and Coyne, D.P. (1992) Plant regeneration from in vitro culture of embryonic axis explants in common and terapy beans. *Journal of the American Society for Horticultural Science* 117, 332–336.

Mohamed, M.F., Coyne, D.P. and Read, P.E. (1993) Shoot organogenesis in callus induced from pedicel explants of common bean (*Phaseolus vulgaris* L.). *Journal of the American Society for Horticultural Science* 118, 158–162.

Mohamed, S.V., Sung, J.M., Jeng, T.L. and Wang, C.S. (2005) Optimization of somatic embryogenesis in suspension cultures of horsegram [*Macrotyloma uniflorum* (Lam.) Verdc.] A hardy grain legume. *Scientia Horticulturae* 106, 427–439.

Mroginski, E., Rey, H.Y. and Gonzalez, A.M. (2004) Thidiazuron promotes *in vitro* plant regeneration of *Arachis correntina* (*Leguminosae*) via organogenesis. *Journal of Plant Growth Regulation* 23, 129–134.

Murashige, T. and Skoog, F. (1962) A revised medium for rapid growth and bioassays with tobacco tissue cultures. *Physiologia Plantarum* 15, 473–497.

Murthy, B.N.S., Murch, S.J. and Saxena, P.K. (1995) Thidiazuron induced somatic embryogenesis in intact seedlings of peanut (*Arachis hypogaea*): endogenous growth regulator levels and significance of cotyledons. *Physiologia Plantarum* 94, 268–276.

Muthukumar, B., Mariamma, M. and Gnanam, A. (1995) Regeneration of plants from primary leaves of cowpea. *Plant Cell, Tissue and Organ Culture* 42, 153–155.

Muthukumar, B., Mariamma, M., Veluthamb, K. and Gnanam, A. (1996) Genetic transformation of cotyledon explants of cowpea (*Vigna unguiculata* L. Walp.) using *Agrobacterium tumefaciens. Plant Cell Reports* 15, 980–985.

Nadolska-Orczyk, A., Milkowska, L. and Orczyk, W. (1994) Two ways of plant regeneration from immature cotyledons of pea. *Acta Societatis Botanicorum Poloniae* 63, 153–157.

Nagl, W., Ignacimuthu, S. and Becker, J. (1997) Genetic engineering and regeneration of *Phaseolus* and *Vigna*. State of the art and new attempts. *Plant Physiology* 150, 625–644.

Narasimhulu, S.B. and Reddy, G.M. (1983) Plantlet regeneration from different callus cultures of *Arachis hypogaea* L. *Plant Science Letters* 31, 157–163.

Natali, L. and Cavallini, A. (1987) Regeneration of pea (*Pisum sativum* L.) plantlets by *in vitro* culture of immature embryos. *Plant Breeding* 99, 172–176.

Nauerby, B., Madsen, M., Christiansen, J. and Wyndaele, R. (1991) A rapid and efficient regeneration system for pea (*Pisum sativum*), suitable for transformation. *Plant Cell Reports* 9, 676–679.

Naz, S., Ali, A, Siddique, F.A. and Iqbal, J. (2007) Multiple shoot formation from different explants of chick pea (*Cicer arietinum* l.). *Pakistan Journal of Botany* 39, 2067–2073.

Ochatt, S.J., Mousset-Declas, C. and Rancillac, M. (2000) Fertile pea plants regenerate from protoplasts when calluses have not undergone endoreduplication. *Plant Science* 156, 177–183.

Ochatt, S.J., Muneaux, E., Machado, C., Jacas, L. and Pontécaille, C. (2002) The hyperhydricity of *in vitro* regenerants of grass pea (*Lathyrus sativus* L.) is linked with an abnormal DNA content. *Journal of Plant Physiology* 159, 1021–1028.

Odutayo, O.I., Akinirinusi, F.B., Odunbososye, I., and Oso, R.T. (2005) Multiple shoot induction from embryo derived callus cultures of cowpea (*Vigna unguiculata* L.). Walp. *African Journal of Biotechnology* 4, 1214–1216.

Özcan, S., Barghchi, M., Firek, S. and Draper J. (1992) High frequency adventitious shoot regeneration from immature cotyledons of pea (*Pisum sativum* L.). *Plant Cell Reports* 11, 44–47.

Özcan, S., Barghchi, M., Firek, S. and Draper, J. (1993) Efficient adventitious shoot regeneration and somatic embryogenesis in pea. *Plant Cell, Tissue and Organ Culture* 34, 271–277.

Ozias-Akins, P. (1989) Plant regeneration from immature embryos of peanut. *Plant Cell Reports* 8, 217–218.

Pacheco, G., Gagliardi, R.F., Valls, J.F.M. and Mansur, E. (2009) Micropropagation and *in vitro* conservation of wild *Arachis* species. *Plant Cell, Tissue and Organ Culture* 99, 239–249.

Parrott, W.A., Williams, E.G., Hildebrand, D.F. and Collins, G.B. (1989) Effect of genotype on somatic embyrogenesis from immature cotyledons of soybean. *Plant Cell, Tissue and Organ Culture* 16, 15–21.

Pellegrineschi, A. (1997) *In vitro* plant regeneration via organogenesis of cowpea (*Vigna unguiculata* (L.) Walp). *Plant Cell Reports* 17, 89–95.

Pniewski, T., Wachowiak, J., Kapusta, J. and Legocki, A.B. (2003) Organogenesis and long-term micropropagation of Polish pea cultivars. *Acta Societatis Botanicorum Poloniae* 72, 295–302.

Polowick, P.L., Quandt, J. and Mahon J.D. (2000) The ability of pea transformation technology to transfer genes into peas adapted to western Canadian growing conditions. *Plant Science* 153, 161–170.

Popelka, J.C., Gollasch, S., Moore, A., Molvig, L. and Huggins, T.J.V. (2006) Genetic transformation of cowpea and stable transmission of the transgenes to progeny. *Plant Cell Reports* 25, 304–312.

Popiers, D., Flandre, F. and Sangwan-Norreel, B.S. (1997) Intensification of the regeneration of the pea (*Pisum sativum* L.) by thidiazuron, via formation of organogenic stem structures. *Canadian Journal of Botany* 75, 492–500.

Radhakrishnan, R., Ramachandran, A. and Kumari, B.D.R. (2009) Rooting and shooting: dual function of thidiazuron in *in vitro* regeneration of soybean (*Glycine max* L). *Acta Physiologiae Plantarum* 31, 1213–1217.

Radhakrishnan, T., Murthy, T.G.K., Chandran, K. and Bandyopadhyay, A. (2000) Micropropagation in peanut (*Arachis hypogaea* L.). *Biologia Plantarum* 43, 447–450.

Raggio, M., Raggio, N. and Torrey, J.G. (1957) The nodulation of isolated leguminous roots. *American Journal of Botany* 44, 325–334.

Ramakrishnan, K., Gnanam, R., Sivakumar, P. and Manickam, A. (2005) *In vitro* somatic embryogenesis from cell suspension cultures of cowpea [*Vigna unguiculata* (L) Walp]. *Plant Cell Reports* 24, 449–461.

Raveendar, S., Premkumar, A., Sasikumar, S., Ignacimuthu, S. and Agastian, P. (2009) Development of a rapid, highly efficient system of organogenesis in cowpea *Vigna unguiculata* (L.) Walp. *South African Journal of Botany* 75, 17–21.

Reichert, N.A., Young, M.M. and Woods, A.L. (2003) Adventitious organogenic regeneration from soybean genotypes representing nine maturity groups. *Plant Cell, Tissue and Organ Culture* 75, 273–277.

Rekha, K.T. and Thiruvengadam, M. (2009) An efficient micropropagation of chickpea (*Cicer arietinum* L.). *Philippine Agricultural Scientist* 3, 320–326.

Rey, H.Y. and Mroginski, L.A. (2006) Somatic embryogenesis and plant regeneration in diploid and triploid *Arachis pintoi. Biologia Plantarum* 50, 152–155.

Rey, H.Y., Scocchi, A.M., Gonzalez, A.M. and Mroginski, L.A. (2000) Plant regeneration in *Arachis pintoi* (Leguminosae) through leaf culture. *Plant Cell Reports* 19, 856–862.

Rubluo, A. and Kartha, K.K. (1985) *In vitro* culture of shoot apical meristems of various *Phaseolus* species and cultivars. *Plant Physiology* 119, 425–433.

Rubluo, A., Kartha, K.K., Mroginski, L.A. and Dyck, J. (1984) Plant regeneration from pea leaflets cultured *in vitro* and genetic stability of regenerants. *Journal of Plant Physiology* 117, 119–130.

Santalla, M., Power, J.B. and Davey, M.R. (1998) Efficient *in vitro* shoot regeneration responses of *Phaseolus vulgaris* and *P. coccineus*. *Euphytica* 102, 195–202.

Santarem, E.R. and Finer, J.J. (1999) Transformation of soybean (*Glycine max* (L.) Merrill) using proliferative embryogenic tissue maintained on semi-solid medium. *In Vitro Cellular and Developmental Biology – Plant* 35, 451–455.

Sarker, R.H., Ferdous, T. and Hoque, M.I. (2005) *In vitro* direct regeneration of three indigenous chickpea (*Cicer arietinum* L.) varieties of Bangladesh. *Plant Tissue Culutre and Biotechnology* 15, 135–144.

Schenk, R.U. and Hildebrandt, A.C. (1972) Medium and techniques for induction and growth of monocotyledonous and dicotyledonous plant cell cultures. *Canadian Journal of Botany*, 50, 199–204.

Schmidt, M.A., Tucker, D.M., Cahoon, E.B. and Parrott, W.A. (2005) Towards normalization of soybean somatic embryo maturation. *Plant Cell Reports* 24, 383–391.

Schroeder, H.E., Schotz, A.H., Wardley-Richardson, T., Spencer, D. and Higgins, T.J.V. (1993) Transformation and regeneration of two cultivars of pea (*Pisum sativum* L.). *Plant Physiology* 101, 751–757.

Sellars, R.M., Southward, G.M. and Phillips, G.C. (1990) Adventitious somatic embryogenesis from cultured immature zygotic embryos of peanut and soybeans. *Crop Science* 30, 408–414.

Shan, Z., Raemakers, K., Tzitzikas, E.N., Ma, Z. and Visser, R.G.F. (2005) Development of a highly efficient, repetitive system of organogenesis in soybean (*Glycine max* (L.) Merr). *Plant Cell Reports* 24, 507–512.

Singh, N.D., Sahoo, L., Sarin, N.B. and Jaiwal, P.K. (2003) The effect of TDZ on organogenesis and somatic embryogenesis in pigeonpea (*Cajanus cajan* L. Millsp). *Plant Science* 164, 341–347.

Skrzypek, E. (2001) Optimisation of the regeneration abilities of field bean (*Vicia faba* ssp. *minor*) in *in vitro* culture. *Biological Bulletin of Poznań* 38, 87–95.

Song, X., Han, Y., Teng, W., Sun, G. and Li, W. (2010) Identification of QTL underlying somatic embryogenesis capacity of immature embryos in soybean (*Glycine max* (L.) Merr.). *Plant Cell Reports* 29, 125–131.

Stejskal, J. and Griga, M. (1992) Somatic embryogenesis and plant regeneration in *Pisum sativum* L. *Biologia Plantarum* 34, 15–22.

Tétu, T., Sangwan, R.S. and Sangwan-Norreel, B.S. (1990) Direct somatic embryogenesis and organogenesis in cultured immature zygotic embryos of *Pisum sativum* L. *Plant Physiology* 137, 102–109.

Tiwari, S. and Tuli, R. (2008) Factors promoting efficient *in vitro* regeneration from deembryonated cotyledon explants of *Arachis hypogaea* L. *Plant Cell Tissue and Organ Culture* 92, 15–24.

Tiwari, S. and Tuli, R. (2009) Multiple shoot regeneration in seed-derived immature leaflet explants of peanut (*Arachis hypogaea* L.). *Scientia Horticulturae* 121, 223–227.

Tomlin, E.S., Branch, S.R., Chamberlain, D., Gabe, H., Wright, M.S. and Stewart Jr., C.N. (2002) Screening of soybean, *Glycine max* (L.) Merrill, lines for somatic embryo induction and maturation capability from immature cotyledons. *In Vitro Cellular and Developmental Biology – Plant* 38, 543–548.

Trick, H.N. and Finer, J.J. (1998) Sonication-assisted *Agrobacterium*-mediated transformation of soybean [*Glycine max* (L.) Merrill] embryogenic suspension culture tissue. *Plant Cell Reports* 17, 482–488.

Tzitzikas, E.N., Bergervoet, M., Raemakers, K., Vincken, J.P., van Lammeren, A. and Visser, R.G.F. (2004) Regeneration of pea (*Pisum sativum* L.) by a cyclic organogenic system. *Plant Cell Reports* 23, 453–460.

Van, K., Jang, H., Young-Eun, J. and Lee Suk-Ha, J. (2008) Regeneration of plants from EMS-treated immature embryo cultures in soybean [*Glycine max* (L.) Merr.]. *Crop Science and Biotechnology* 11, 119–126.

Van Le, B.U.I., De Carvalho, M.H.C., Zuily-Fodil, Y., Thi, A.T.P. and Van, K.T.T. (2002) Direct whole plant regeneration of cowpea [*Vigna unguiculata* (L.) Walp] from cotyledonary node thin cell layer explants. *Journal of Plant Physiology* 159, 1255–1258.

Vaquero, F., Robles, C. and Ruiz, A. (1993) A method for long-term micropropagation of *Phaseolus coccineus* L. *Plant Cell Reports* 12, 395–398.

Vasanth, K., Lakshmiprabha, A. and Jayabalan, N. (2006) Amino acids enhancing plant regeneration from cotyledon and embryonal axis of peanut (*Arachis hypogaea* L.). *Indian Journal of Crop Science* 1, 79–83.

Veltcheva, M., Svetleva, D., Petkova, S. and Perl, A. (2005) *In vitro* regeneration and genetic transformation of common bean (*Phaseolus vulgaris* L.) – problems and progress. *Scientia Horticulturae* 107, 2–10.

Vidoz, M.L., Rey, H.Y., Gonzalez, A.M. and Mroginski, L.A. (2004) Somatic embryogenesis and plant regeneration through leaf culture in *Arachis glabrata* (Leguminosae). *Acta Physiologiae Plantarum* 26, 59–66.

Walker, D.R. and Parrott, W.A. (2001) Effect of polyethylene glycol and sugar alcohols on soybean somatic embryo germination and conversion. *Plant Cell, Tissue and Organ Culture* 64, 55–62.

Yang, C., Zhao, T., Yu, D. and Gai, J. (2009) Somatic embryogenesis and plant regeneration in Chinese soybean (*Glycine max* (L.) Merr.) – impacts of mannitol, abscisic acid, and explant age. *In Vitro Cellular and Developmental Biology – Plant* 45, 180–188.

Zambre, M.A., De Clercq, J., Vranova, E., Van Montagu, M., Angenon, G. and Dillen, W. (1998) Plant regeneration from embryo-derived callus in *Phaseolus vulgaris* L. (common bean) and *P. acutifolius* A. Gray (Tepary Bean). *Plant Cell Reports* 17, 626–630.

Zambre, M.A., Geerts, P., Maquet, A., Van Montagu, M., Dillen, W. and Angenon, G. (2001) Regeneration of fertile plants from callus in *Phaseolus polyanthus* Greenman (year bean). *Annals of Botany* 88, 371–377.

Zambre, M., Chowdhury, B., Kuo, Y.H., Van Montagu, M., Angenon, G. and Lambein, F. (2002) Prolific regeneration of fertile plants from green nodular callus induced from meristematic tissues in *Lathyrus sativus* L. (grass pea). *Plant Science* 163, 1107–1112.

Zhihui, S., Tzitzikas, M., Raemakers, K., Zhengqiang, M. and Visser, R. (2009) Effect of TDZ on plant regeneration from mature seeds in pea (*Pisum sativum*). *In Vitro Cellular and Developmental Biology – Plant* 45, 776–782.

11 Androgenesis and Doubled-Haploid Production in Food Legumes

M.M. Lulsdorf, J.S Croser and S. Ochatt

11.1 Introduction

In conventional breeding programmes, more than four segregating generations are needed to reach a level of near-homozygosity that allows the selection of traits of interest to begin. In contrast, doubled-haploid (DH) technology produces complete homozygosity in one generation (Palmer and Keller, 2005; Forster *et al.*, 2007). The use of molecular markers as a selection tool in breeding programmes becomes easier because it depends on homozygous populations. Using DHs improves selection efficiency since fewer populations must be screened in order to cover a wide spectrum of recombinants (Forster *et al.*, 2007). Haploid cells, prior to doubling, are also ideal targets for genetic manipulation (Kumlehn, 2009; Resch *et al.*, 2009), benefitting legumes such as chickpea because of low intraspecific variability.

Haploids have the same chromosome complement as the gametes of the species. They may be obtained by chromosome elimination via wide crosses (Kasha and Kao, 1970; Devaux and Kasha, 2009); parthenogenesis and apomixis (Germanà, 2006); culture of female gametes (gynogenesis) (Tulecke, 1964; Bohanec, 2009); or androgenesis from anthers or isolated microspores (Nitsch and Nitsch, 1969; Wedzony *et al.*, 2009). A new approach to haploid development was suggested by Ravi and Chan (2010) using mutants with CENH3 centromeres that have specific affinity towards spindle microtubules. Chromosomes from the mutant parent of *Arabidopsis thaliana* (L. Heynh.) were selectively eliminated and either male- or female-derived haploids produced. Within the *Fabaceae*, anther or microspore culture are commonly used, while reports on the other techniques are few (Reddy and Reddy, 1996; Mallikarjuna *et al.*, 2005). Grain legumes are well known for their recalcitrance to most *in vitro* approaches, and doubled-haploidy is no exception (Croser *et al.*, 2006; Germanà, 2006; Skrzypek *et al.*, 2008; Ochatt *et al.*, 2009). However, in the last 5 years, significant advances have been made with dry pea, chickpea, grass pea and also the model legume species, *Medicago truncatula* Gaertn., all through androgenesis (Grewal *et al.*, 2009; Ochatt *et al.*, 2009).

The rationale behind the use of androgenesis is the developmental shift from the gametophytic to the sporophytic pathway, inducing sustained cell divisions and cell differentiation, respectively leading to production of shoots or of embryos, either directly, or via a callus phase (Maluszynski *et al.*, 2003). The various aspects of androgenesis are discussed in the literature; for example, the triggers for embryo development (Pauls *et al.*, 2006; Segui-Simarro and Nuez, 2008a); the different types and effects of stresses

(Touraev *et al.*, 1997; Shariatpanahi *et al.*, 2006); the role of hormones (Feng *et al.*, 2006; Yang *et al.*, 1997; Zur *et al.*, 2008); and chromosome doubling (Segui-Simarro and Nuez, 2008b). This chapter provides a review of the current status of androgenesis and DH production in the different food legumes and outlines some strategies to overcome this recalcitrance.

11.2 Food Legume Species

The first paper on production of haploid pea callus was published by Gupta *et al.* in 1972 and on soybean by Tang *et al.* (1973) but, after over 30 years, haploid protocols are still not routinely used in any food legume breeding programme. However, recent progress in pea (Ochatt *et al.*, 2009) and chickpea (Grewal *et al.*, 2009) androgenesis, through combination of various stresses (cold, electroporation, centrifugation and osmotic shock), suggests that this recalcitrance can be overcome in the *Fabaceae*. Table 11.1 lists the androgenesis studies conducted during over the years in different food legume species, and these are discussed below in detail.

Field pea (*Pisum sativum* L. subsp. *sativum* var. *arvense*)

Since the onset of genetic studies with plants, pea (2n = 14) has been a preferred species for study. In terms of haploid development, this was also true with the first report on haploid callus induction of anthers (Gupta *et al.*, 1972), and with the recovery of a few haploid pea plants (Gupta 1975), although these results could not be reproduced subsequently (Table 11.1). Using cold treatment for 72 h, Gosal and Bajaj (1988) obtained 0.34% embryoid formation. Croser and Lulsdorf (2004) tested cold or heat stress for the induction of microspores resulting in symmetrical microspore nuclei division. Recent research underlined the difficulty of producing confirmed haploids, with most results stopping short of the recovery of pea plants (Croser *et al.*, 2005, 2007; Sidhu and Davies, 2005), irrespective of the use of male organs (anthers) or reduced gametophytes (microspores) as starting material. Recently, a small number of plants were recovered from isolated microspores of a few field pea genotypes (Ochatt *et al.*, 2009). Thus, five plants were obtained through organogenesis from microspore-derived calli (one from cv. Victor and four from cv. Frisson), and three more plants were produced via embryogenesis from the microspores of cv. CDC April (Table 11.1).

Chickpea (*Cicer arietinum* L.)

Khan and Ghosh (1983) were the first to report *in vitro* androgenesis in chickpea (2n = 14). Three pollen embryoids were regenerated from calli, but plants were not obtained. Altaf and Ahmad (1986) used a cold pre-treatment of buds at 4°C for 3–7 days and centrifugation for 45 min at 1000 RPM, resulting in callus development from the anthers. However, shoots could not be obtained and the ploidy status of the callus cells was not determined. Bajaj and Gosal (1987) induced callus from anthers cold-treated for 3 days, on MS medium with various hormones; a few multicellular embryoids were obtained. Later, Huda *et al.* (2001) found that cold treatment of anthers and a B5 (Gamborg *et al.*, 1968) medium with either 2,4-D or NAA was suitable for induction of androgenesis. After callus induction, a few embryos and shoots developed, but ploidy level was not determined. Mature embryos were obtained by Vessal *et al.* (2002), using cold treatment of buds for 7–10 days, followed by anther culture on MS medium with 1 mg/l 2,4-D and 0.2 mg/l kinetin. Embryos were regenerated from haploid callus on a modified Blaydes' (1966) medium with 0.5 mg/l kinetin and 10% sucrose. Callus growth consisted of cells with haploid to polyploid chromosome numbers. Similarly, Croser *et al.* (2005) used isolated microspore culture and a modified MS medium to obtain androgenesis in three chickpea cultivars (Table 11.1).

The first confirmed haploid plants from anther culture were reported by Grewal *et al.* (2009) for cv. CDC Xena (*kabuli*) and cv. Sonali (*desi*) (Table 11.1). Induction required a four-step stress treatment consisting of: (i) a 72 h cold treatment of buds; (ii) centrifugation (168 g)

Table 11.1. Overview of target explants, stresses and media used for induction of androgenesis in food legume species.

Target explants[a]	Stress sequence	Medium[b]	Reference(s)
Pea			
A	–	I: White + 2,4-D + coconut milk	Gupta *et al.* (1972)
A	–	I: White + NAA + coconut milk	Gupta (1975)
A	Cold for 72 h (A)	I: Various MS-based media	Gosal and Bajaj (1988)
M	Cold or heat (buds)	I: Various semi-solid media with 2,4-D; S: hormone-free medium	Croser and Lulsdorf (2004)
M	Cold (buds)	I: Modified ML6 + 1 mg/l NAA + 15% fructose maltose, or 9% sucrose	Croser *et al.* (2005, 2007)
A	Cold for 72 h (A)	I: B5 + 2 mg/l Dicamba + 300 mg/l casein hydrolysate + 9% sucrose; S: ELS on L2 + 1 mg/lBAP + 2% sucrose	Sidhu and Davies (2005)
M	a) Cold > 48 h (buds) b) Electroporation	I: Liquid stationary culture on NLN or HSO, 1 month S: Same media but semi-solid	Ochatt *et al.* (2009)
Chickpea			
A		I: MS + 2 mg/l2,4-D + 10% coconut milk; S: as I but + 500 mg/l acalbumin hydrolysate	Khan and Ghosh (1983)
A	a) Cold 72–168 h (buds) b) Centrifugation at 1000 RPM for 45 min at 4°C (buds)	I: MS or B5 + 2.21 mg/l 2,4-D + 0.225 mg/l BAP	Altaf and Ahmad (1986)
A	Cold 72 h (A)	I: MS + 4 mg/l IAA + 2 mg/l Kin	Bajaj and Gosal (1987)
A	Cold 72–168 h (A)	I: cv. Nabin on B5 + 2 mg/l 2,4-D + 2 mg/l BAP; I: cv. ICCL83105 on B5 + 2 mg/l NAA + 2 mg/l BAP ; S: B5 + 0.5 mg/l IAA + 1 mg/l BAP + 0.5 mg/l Kin	Huda *et al.* (2001)
A	Cold 168–240 h (buds)	I: MS + 1 mg/l 2,4-D + 0.2 mg/l Kin S: Modified Blaydes + 0.5 mg/l Kin + 10% sucrose	Vessal *et al.* (2002)
M	Cv. Narayen 32.5°C for 16 h Cv. Sona 48 h cold (buds) Cv. Rupali none	I: Modified MS + 1 mg/l 2,4-D + 0.25 mg/l Pic + 0.1 mg/l BAP + 9% sucrose	Croser *et al.* (2005)
A	a) Cold 72 h (buds) b) Centrifugation of 168 g for 10 min (anthers) c) Electroporation with 625 V/cm, 25 µF and 25 Ω (A) d) High osmotic liquid medium for 4 days (A)	I: RM-IK + 4 mg/l IAA + 0.4 mg/l Kin + 17% sucrose S1: Modified L2 + 1 mg/l Pic + 0.40 mg/l 2iP + 4% sucrose + 5% maltose; S2: Modified L2 + 4 mg/l IAA + 1 mg/l ZR + 5 mg/l GA_3 + 1 mg/l ABA; S3: Modified MS + 0.01 mg/l NAA + 0.1 mg/l BA + 4.5% sucrose + 4.5 % maltose	Grewal *et al.* (2009)

Continued

Table 11.1. Continued.

Target explants[a]	Stress sequence	Medium[b]	Reference(s)
Lentil			
A, M	Heat or cold not effective	I: ML6 + 2 mg/l 2,4,5-T + 1 mg/l BAP + 6% sucrose	Keller and Ferrie (2002)
M	Cold 96 h	I: Modified R&D + 1mg/l 2,4-D + 1 mg/l NAA + 1 mg/l Kin 10% sucrose	Croser and Lulsdorf (2004)
Soybean			
A		I: Miller's + 20 mg/l NAA + 1 mg/l Kin	Ivers *et al.* (1974)
A		I: B5 + 2 mg/l 2,4-D + 12% sucrose	Yin *et al.* (1982)
A		I: Modified B5 + 2 mg/l 2,4-D + 2 mg/l BAP + 0.5 mg/l Kin + 12% sucrose	Jian *et al.* (1986)
	Cold 120–192 h + 2 mg/l 2,4-D (buds)	I: Enriched B5 + 0.5 - 1.0 mg/l NAA + 0.1- 0.5 mg/l zeatin	Liu and Zhao (1986)
A	Cold 4–8 days; 37°C for 24 h (buds)	I: B5 'long' + 2 mg/l 2,4-D + 0.5 mg/l BA + 9% sucrose + 0.3% agarose	Zhuang *et al.* (1991)
A	Cold 72–120 h (buds)	I: Modified MS and B5 + 2 mg/l 2,4-D + 12% sucrose S: B5 + 0.5 mg/l NAA + 1 mg/l Kin + 1% sucrose S: Modified MS + 0.5 mg/l IBA + 0.5 mg/l BAP, 0.5 mg/l Kin, O.5 mg/l zeatin + 5% sucrose + 1% maltose	Ye *et al.* (1994)
A	Cold 96 and 192 h or heat (37°C)	I: B5 'long' + 2 mg/l 2,4-D + 0.5 mg/l BAP + 9–12% sucrose + 0.35% agarose	Hu *et al.* (1996)
A	Cold 0–10 days (buds)	I: B5 'long'+ 2 mg/l 2,4-D + 0.5 mg/l BAP + 9% sucrose + 0.8% agarose	Kaltchuk-Santos *et al.* (1997)
A	Cold 24–48 h (buds)	I: B5 or B5 'long' + YS amino acids + 2 mg/l 2,4-D + 0.5 mg/l BAP + 9% sucrose + 0.3% phytagel	Cardoso *et al.* (2004)
A	Cold 12 h (buds)	I: B5 'long' + YSaa + 2 mg/l 2,4-D + 0.5 mg l^{-1}BAP + 9% sucrose + 0.25% phytagel; S: as above but 1 mg/l 2,4-D + 1 mg/l BAP; S: MSO: MS salts + B5 vitamins + 3% sucrose + 0.25% phytagel; S: MSO + 1% sucrose	de Moraes *et al.* (2004)
A	Cold 0–10 days (buds)	I: B5 'long'+ YSaa + 2 mg/l 2,4-D + 0.5 mg/l BAP + 9% sucrose + 0.8% agarose; S: B5 + 1 mg/l 2,4-D + 3 mg/l BAP + 3% sucrose	Rodrigues *et al.* (2004a, b)
A	Cold 3–5 days (buds)	I: B_5DBIG + 2 mg/l 2,4-D + 0.5 mg/l IBA + 100 mg/l myo-inositol + 360 mg/l L-glutamine + 9% sucrose + 0.7% agar S: MS + 0.4 mg/l NAA + 0.4 mg/l BAP + 2% sucrose + 0.8% agar	Tiwari *et al.* (2004)
A		I: B5 'long'+ YSaa + 2 mg/l 2,4-D + 0.5 mg/l BAP + 9% sucrose + 0.25% phytagel	Rodrigues *et al.* (2005a)
M		I: Modified PTA-15	Rodrigues *et al.* (2006)
A	Cold 24–48 h (buds)	I: B5 and B5 'long'+ YS amino acids + 2 mg/l 2,4-D + 0.5 mg/l BAP + 9% sucrose + 0.3% phytagel or modified PTA-15	Cardoso *et al.* (2007)

Common bean			
A		I: B5 + 2 mg/l 2,4-D + 0.2 mg/l Kin + 2% sucrose	Haddon and Northcote (1976)
A		I: 67V + 1 mg/l 2,4-D (or 1mg/l NAA + 2 mg/l IAA + 0.2 mg/l Kin) + 0.2% casein hydrolysate + 2% sucrose	Peters *et al.* (1977)
A		I: B5 + 2 mg/l 2,4-D + 1 mg/l Kin	Tai and Cheng (1990)
A	Cold 0–48 h (buds)	I: MS + 2 mg/l 2,4-D + 2 mg/l Kin + 0.2% casein hydrolysate + 1.25–5.0% sucrose or maltose	Muñoz-Florez *et al.* (1992); Muñoz *et al.* (1993); Muñoz and Baudoin (1994, 2001–2002),
Lupin			
A		I: MS + 1 mg/l 2,4-D + 1 mg/l BAP + 1^{-1} mg/l IAA or 1.5 mg/l+ 0.2 mg/l NAA or 2 mg l^{-1}BAP + 0.2 mg/l NAA	Sator (1985)
M		I: NNB5 + 1 mg/l NAA + 0.5 mg/l 2,4-D + 1 mg/l Kin + 0.5 mg/l BAP + 5% sucrose or maltose + 0.8 g/l L-proline + 0.1 g/ l L-serine S: S&H + 0.09 mg/l GA_3 + 5% sucrose or maltose + 0.8% agar	Ormerod and Caligari (1994)
M	a) 6, 22 or 30°C for 24 h, 48 h and 72 h (buds) b) Centrifugation for 15 min at 130 × *g* then 2 × 5 min at 100 × *g* (M)	I: N1 = NN basal medium + 36.7 mg/l NaFeEDTA + 13% sucrose + 30 mg/l glutathione + 0.8 g/l glutamine + 0.1 g/l serine S1: N2 = N1 but with 0.1 mg/l IAA + 0.01 mg/l zeatin + 6% sucrose S2: N3 = N2 + 10% coconut milk	Campos-Andrada *et al.* (2001)
A and M	a) Cold 72 h + heat 24 h (buds) b) Centrifugation 10 min at 2000 × g then 2 × 5 min at 2000 × g (M)	I: Medium B = KM salts & vitamins + 0.3 M mannitol + 166 mg/l $CaCl_2$ $2H_2O$ + 40 mg/l FeEDTA; S: Medium B + 2% sucrose + 0.4% PEG + 2% coconut water + 250–500 mg/l casein hydrolysate	Bayliss *et al.* (2004)
A	Cold or heat not effective	I: NN macro- + B5 micro-elements + 0.5 mg/l 2,4-D + 1 mg/l NAA + 1 mg/l Kin + 0.5 mg/l BAP + 5% sucrose or maltose + 0.8 g/l L-proline + 0.1 g/l L-serine; S: MS + 0.5 mg/l NAA + 1 mg/l BAP + 0.25 mg/l GA_3 + 5% sucrose or maltose + 0.6% agar	Skrzypek *et al.* (2008)
Cowpea			
A		Modified MS + 0.5 mg/l NAA + 0.1 mg/l Kin	Ladeinde and Bliss (1977)
		I: MS + 1 mg/l NAA + 2 mg/l BAP + 3% sucrose for cvs. Tvu91 and Tvu1987; Cv. Pipo same except for 6% sucrose S: Cv. Tvu 91 using MS + 0.25 mg/l NAA + 0.25 mg/l IAA + 0.5 mg/l 2-iP + 2% sucrose; Cv. Tvu 1987 MS + 0.05 mg/l BA + 6% sucrose; Cv. Pipo as above but + 0.1 mg/l BAP + 6% sucrose	Mix and Wang (1988)

Continued

Table 11.1. Continued.

Target explants[a]	Stress sequence	Medium[b]	Reference(s)
Mung bean			
A		I: MS + 0.5 mg/l IAA + 1 mg/l Kin; S: MS + 1 mg/l BAP + 0.5 mg/l IBA	Arya and Chandra (1989)
		I: MS + 2 mg/l IAA + 2 mg/l 2,4-D + 2 mg/l Kin + 0.7% agar	Bajaj and Singh (1980)
Urd bean			
A	Cold for 72 h (A)	I: MS + 2 mg/l 2,4-D + 0.2 mg/l Kin+ 8% sucrose + 70 ml l^{-1} coconut water	Gosal and Bajaj (1988)
Pigeon pea			
A		I: MS + 2 mg/l 2,4-D + 0.2 mg/l Kin + 8% sucrose + 200 mg/lpotato extract + 0.8% agar ; S: MS + 1 mg/l 2,4-D	Gosal and Bajaj (1979)
A		MS + 4 mg/l IAA + 2 mg/l Kin	Bajaj *et al.* (1980)
A	Cold 5–7 days	I: Modified MS + 2 mg/l 2,4-D + 0.2 mg/l Kin; S: as above + 1% agar	Sudhakar *et al.* (1986)
A		I: MS + 1.5 mg/l IAA + 0.5 mg/l Kin + 0.8% agar	Fougat *et al.* (1992)
M	Cold 3–7 days (buds)	I: ½ MS macro + NN micro-elements + vitamins + 0.1 mg/l NAA + 0.1 mg/l BAP + 2% sucrose + 2% glucose; S1: MS + 0.5 mg/l BAP; S2: MS + 2 mg/l NAA + 0.1 mg/l Kin	Kaur and Bhalla (1998)
A		I: B5 + 1.75 mg/lIAA + 2.25 mg/l BAP + 0.22 mg/l Kin + 1.73 mg/l GA_3	Narasimham (1999)
A		I: MS + 2 mg/l 2,4-D + 0.5 mg/l Kin; S: MS + 2 mg/l BAP	Vishukumar *et al.* (2000)

[a]A, anthers; M, microspores.

[b]I, induction; S, subculture; ELS, embryo-like structures; Base media, B5 (Gamborg *et al.*, 1968); HSO (Ochatt *et al.*, 2009); L2 (Phillips and Collins, 1979); ML6 (Kumar *et al.*, 1988); NLN (White, 1963; Lichter, 1981, 1982); MS (Murashige and Skoog, 1962); RM-IK modified HSO (Ochatt *et al.*, 2009); ML6 (Kumar *et al.*, 1988); R&D (Rao and De, 1987); Enriched B5 (modified B5 by Kao, 1982); B5 'long' (modified B5 by Zhuang *et al.*, 1991; Hu *et al.*, 1996; Carolina Biological Supply Co., Burlington, North Carolina); B_5DBIG (modified B5 by Tiwari *et al.*, 2004); Miller's (Miller, 1963); MSO (de Moraes *et al.*, 2004); modified PTA-15 (Skinner and Liang, 1996); YS amino acids (Yeung and Sussex, 1979); 67 V (Veliky and Martin, 1970); KM (Kao and Michayluk, 1975); NNB5 NN macro-B5 micro-elements (Nitsch and Nitsch, 1969); S&H (Schenk and Hildebrandt, 1972); N_6 (Chu, 1978).

of anthers in medium RM-IK-17 (modified HSO) (Ochatt *et al.*, 2009) for 10 min followed by; (iii) electroporation of anthers in the same medium, using 625 V/cm. The final stress treatment was (iv) a 4-day high osmotic medium (563 mmol, RM-IK17/HSO) prior to transfer of anthers onto modified Phillips and Collins (1979) embryo development medium and then maturation medium containing different hormones. Plants were regenerated on a modified MS medium with a low amount of BAP (0.10 mg/l) and NAA (0.01 mg mg/l). Flow cytometry and chromosome counts showed that callus cells were initially haploid but ploidy levels increased with age, resulting in spontaneously doubled haploid embryos and plants.

Lentil (*Lens culinaris* Medik. ssp. culinaris)

Lentil (2n = 14) is the least explored species in terms of haploid technology; calli with a few pro-embryos were obtained but no plants regenerated (Keller and Ferrie, 2002). In another study, buds from cvs CDC Crimson and CDC Robin were cold-treated for 96 h prior to microspore extraction, resulting in multinucleate microspores, but no embryos were regenerated (Croser and Lulsdorf, 2004).

Soybean (*Glycine max* L. Merr.)

Over the past 30 years, there has been an intensive research effort from both the private and public sectors into the cell biology and biotechnology of soybean (2n = 40). However, no routine protocol has been established for haploid or DH plant regeneration, and no DH lines of soybean are currently available (Rodrigues *et al.*, 2004a; Croser *et al.*, 2006).

Initial reports demonstrated induction of callus from anthers (Tang *et al.*, 1973; Ivers *et al.*, 1974; Liu and Zhao, 1986), shoot organogenesis (Yin *et al.*, 1982; Jian *et al.*, 1986) and embryo-like structures (ELS) from anther-derived callus (Zhuang *et al.*, 1991; Hu *et al.*, 1996; Kaltchuk-Santos *et al.*, 1997). In a few cases, a small number of plants were regenerated, but the haploid origin of the plants was uncertain (Yin *et al.*, 1982; Jian *et al.*, 1986; Hu *et al.*, 1996; Zhao *et al.*, 1998; de Moraes *et al.*, 2004; Rodrigues *et al.*, 2004a; Tiwari *et al.*, 2004). A haploid chromosome number (n = 20) was confirmed in a single plant (de Moraes *et al.*, 2004). Detailed cytological studies of soybean anthers were carried out *in vivo* (Kaltchuk-Santos *et al.*, 1993; da Silva Lauxen *et al.*, 2003) and *in vitro* (Yin *et al.*, 1982; Kaltchuk-Santos *et al.*, 1997; Cardoso *et al.*, 2004) describing cellular events related to the androgenic pathway, such as the symmetrical mitotic division of microspores and formation of multinucleate and multicellular pollen grains. Yin *et al.* (1982) reported multinucleate grains after 15–20 days *in vitro*. Kaltchuk-Santos *et al.* (1997) were the first to show that these grains were not present at dissection, but started to appear during *in vitro* incubation, reaching an overall frequency of 0.3% by four weeks of culture.

There is no general consensus regarding the most appropriate microspore developmental stage for induction of androgenesis in soybean. Yin *et al.* (1982) and Ye *et al.* (1994) found that the early- to mid-uninucleate stage was best for induction. Later reports suggested the mid- to late uninucleate and early binucleate stage of pollen development as appropriate (Kaltchuk-Santos *et al.*, 1997; da Silva Lauxen *et al.*, 2003; Cardoso *et al.*, 2004). This could be due to the propensity of soybean to have varying developmental stages within the same bud, thereby making it difficult to establish the original pollen source.

There has been little consensus on the effect of pre-treatment stress on androgenesis from soybean. To date, authors have focused on testing temperature stress applied to the buds prior to, or directly after, anther or microspore isolation and culture (Liu and Zhao, 1986; Zhuang *et al.*, 1991; Rodrigues *et al.*, 2005b). Hu *et al.* (1996) recommended the use of sonication to improve sterilization of buds prior to anther isolation. Sonication is now showing potential under testing in our laboratories as an effective elicitation stress in a range of species (Ochatt and Croser, unpublished results).

For most species, androgenesis requires an auxin, a cytokinin or a combination of both in the medium (Smýkal, 2000), with soybean most likely requiring both (Table 11.1). In general, B5 medium with 16 organic compounds ('B5 long') (Zhuang *et al.*, 1991) and with Yeung's amino acids (Yeung and Sussex, 1979) is appropriate for anther culture. De Moraes *et al.* (2004) obtained one confirmed haploid plant (2n=20), following induction of embryogenic calli from anthers on this basal medium supplemented with 2.0 mg/l 2,4-D, 0.5 mg/l BAP, 9% sucrose and 0.25% phytagel. This result further confirms the finding of Hu *et al.* (1996) that 2,4-D is essential for soybean microspore callus induction, although Rodrigues *et al.* (2004b) noted that this growth regulator favours morphogenic response from sporophytic tissue.

Cardoso *et al.* (2004) showed that a high percentage of soybean microspores doubled their chromosome number within the first ten days of culture, suggesting spontaneous doubling may be at a rate high enough to avoid the requirement for an artificial doubling step. However, it also makes determination of the androgenic origin of regenerated plants more difficult. Rodrigues *et al.* (2004a) confirmed that soybean androgenic and somatic ELS were induced simultaneously under the same culture conditions. The presence of both heterozygous and homozygous ELS within the same culture (but not within the same anther) confirmed that somatic embryogenesis and androgenesis were promoted under identical conditions. Zhuang *et al.* (1991) demonstrated that calli derived from anthers in the first three months of culture were mainly of anther somatic tissue in origin. If this initial callus was removed upon transfer of anthers to fresh medium, four weeks later a few newly grown calli developed embryoids that were more likely of haploid origin. Another strategy to overcome somatic embryogenesis is to culture isolated microspores that are free of the somatic anther tissue. This technique has been applied widely in other species, but rarely in soybean (Liu and Zhao, 1986; Rodrigues *et al.*, 2006).

While genotypic effects have been recognized in soybean, there is little discussion of the effect of donor plant growth conditions, which can have a profound effect on embryogenic response. Soybean protocols use anthers collected from the field (Zhuang *et al.*, 1991; Kaltchuk-Santos *et al.*, 1997; da Silva Lauxen *et al.*, 2003; Cardoso *et al.*, 2004; de Moraes *et al.*, 2004; Rodrigues *et al.*, 2004b) in contrast to most other species, where donor plants are grown under controlled conditions.

Common bean (*Phaseolus vulgaris* L.)

Given the first report of bean (2n=22) anther culture (Haddon and Northcote, 1976), little progress has been made in this species in 34 years (Table 11.1). However, the androgenic origin of callus cells could not be determined in the first study because DNA analysis showed only diploid to polyploid chromosome levels. In contrast, Peters *et al.* (1977) reported near equal amounts of haploid and diploid callus cells with fewer than 3% of cells showing polyploidy. Tai and Cheng (1990) cultured anthers of common bean on B5 medium with 2 mg/l 2,4-D and 1 mg/l kinetin. Bean callus growth was the poorest among the four legume species tested. Origin of the callus cells is unknown since ploidy levels were not determined. Muñoz and co-workers (Muñoz and Baudoin, 1994, 2001/2002; Muñoz, *et al.*, 1992, 1993) conducted a more detailed study into bean anther culture (Table 11.1). In 1992, these authors reported that the early to mid-uninucleate microspore stage was the most responsive to androgenesis induction and that a larger size of Petri dish (55 mm diameter) resulted in more callus growth than smaller ones (35 mm). A few modifications to the MS base medium (Veliky and Martin, 1970) were also tried for better callus growth. The medium for anther induction was modified to MS macro- and micro-nutrients, B5 vitamins, 2 g/l casein hydrolysate, 2.5% sucrose and 2 mg/l each of 2,4-D and kinetin (Muñoz and Baudoin, 2001/2002). Cold pretreatment of anthers did not have a beneficial effect. Callus cells during the early growth stages were predominantly haploid but, with age, ploidy levels increased, thus indicating spontaneous doubling of chromosomes.

Lupin (*Lupinus* spp.)

To date, there has been no confirmed report on haploid embryo or plantlet regeneration from any of the four grain lupin species. Sator (1985) first obtained callus production following anther culture of *Lupinus luteus* and *Lupinus angustifolius*. Ormerod and Caligari (1994) produced cotyledonary-stage embryos from microspores that were released from cultured anthers of *Lupinus albus*, but no plants were regenerated. Campos-Andrada *et al.* (2001) demonstrated *in vivo* pollen dimorphism in pearl lupin. Culture of the isolated microspores led to symmetrical division and procallus formation. Bayliss *et al.* (2004) reported isolated microspore-derived pro-embryos in *L. albus* and *L. angustifolius* and, most recently, Skryzpek *et al.* (2008) achieved callus induction from microspores released from anthers of *L. albus*, *L. angustifolius* and *L. luteus*. A feature of these studies was the spontaneous release of microspores into the surrounding medium after anther dehiscence during culture, similar to that seen in *Nicotiana tabacum* L. Bayliss *et al.* (2004) compared this natural dehiscence with a mechanical microspore isolation system. All reports agree that the uninucleate and/or early binucleate microspore stage is optimal in lupin.

Bayliss *et al.* (2004) obtained haploid pro-embryos from isolated microspores in *L. albus* and *L. angustifolius* but found further embryo development to be restricted by the failure of the outer exine layer to rupture. Pro-embryos were induced from microspores that were mechanically isolated from buds stored at 4°C for 72 h and then cultured for 24 h at 32°C (Kao and Michayluk, 1975). The mechanical isolation method included a 10 min centrifugation step at 2000 × *g*, more vigorous than that used as a stress treatment for enhancing androgenesis in chickpea (Grewal *et al.*, 2009). After the 24 h heat and starvation treatment, microspores were transferred to modified KM medium. This transfer resulted in an osmotic stress treatment, similar in nature to that described for haploid plant production in other legumes by Grewal *et al.* (2009) and Ochatt *et al.* (2009).

It appears that the best androgenic response, observed by Bayliss *et al.* (2004), came after a rigorous stress treatment of cold, heat, centrifugation, starvation and osmotic stress, thus providing further evidence of the efficacy of combining stress agents for induction of androgenesis from the grain legumes. In contrast, Skrzypek *et al.* (2008) reported cold and heat pre-treatment either did not improve, or was inhibitory, to callus induction from anthers of *L. albus*, *L. angustifolius* and *L. luteus*. This report compared field- with glasshouse-grown donor material, observing that androgenic response was higher in the field-grown plants. The results of Skrzypek *et al.* (2008) contrasted with those of Bayliss *et al.* (2004) with regard to the pollen wall limiting further androgenic development. However, no cytological evidence was presented to support this observation. If the outer exine limits embryo development from microspores, electrostimulation may assist in overcoming this issue as one of its effects is to 'loosen' the cell wall (Cole, 1968; Neumann and Rosenheck, 1973).

Other food lgumes

Research on haploid development of cowpea (2n = 22) and other food legumes is sparse (Table 11.1), although a few reports (e.g. Ladeinde and Bliss, 1977; Arya and Chandra, 1989) on production of callus are available. Mix and Wang (1988) were the only authors to report haploid plant production in cowpea. Donor plants were grown at 30°C/22°C (day/night) with 30–40% humidity. Flower bud length was 2–4 mm, with anther colour being whitish-green and containing uninucleated microspores. Upon culture, such anthers provided a callus from which 38 shoots were regenerated, with five of them confirmed as haploid.

In mung bean (*Vigna radiata*; 2n = 22), Bajaj and Singh (1980) obtained callus and immature embryos from three genotypes. Although callus cells were initially predominantly haploid, large variations in chromosome complements were observed over time. No mature embryos or haploid plants were recovered. For urd bean (*Vigna mungo*; 2n = 22) there is only a single report, by Gosal and Bajaj (1988), where regeneration of haploid plants was achieved at a low frequency.

Gosal and Bajaj (1979) cultured anthers of pigeon pea (*Cajanus cajan* (L.) Millsp.; 2n = 22) but obtained only callus. The following year, Bajaj and co-workers (1980) encased anthers in small droplets using a MS medium with 4 mg/l IAA and 2 mg/l kinetin. They obtained embryoids and callus with haploid to mixoploid (8–28) chromosome numbers. In some other studies a modified MS medium with 2,4-D or B5 medium was used in combination with IAA or kinetin, but all resulting callus cells were of diploid origin (Sudhakar *et al.*, 1986; Narasimham, 1999; Vishukumar *et al.*, 2000). Fougat *et al.* (1992) reported initially haploid callus cells with mixoploidy occurring after several sub-cultures. Kaur and Bhalla (1998) were the first to achieve haploid pigeon pea plants by using a modified MS medium with 0.1 mg/l of each NAA and BAP in combination with 2% sucrose and glucose. Shoots were rooted on semi-solid MS medium with 2 mg/l NAA and 0.1 mg/l kinetin.

11.3 Strategies for Developing Doubled-Haploid Technology for Legumes

Anther versus microspore culture

Isolated microspore culture has the advantage of producing plants from haploid sources, whereas anther culture regenerates can be of either sporophytic or gametophytic origin (Table 11.1). However, anther culture seems to be the more promising method for induction of androgenesis in legumes, partly due to the low number of donor plants required, the relative ease of use and also because of the nutritive environment that the anthers provide for the microspores. The anther wall acts as a filter, and the slow uptake or diffusion of nutrients from the medium to the microspores could provide a starvation environment until the anther wall degrades (Aruga and Nakajima, 1985; Kyo and Harada, 1986). After 10 days of culture, accumulation of large amounts of asparagine and glutamine might cause embryo formation in anthers (Aruga and Nakajima, 1985). Another function of the anther wall could be the protection of pollen from inhibitory factors in the medium (Aslam *et al.*, 1990).

The disadvantage of anther culture is that the anthers consist not only of haploid cells but also of diploid sporophytic tissue of maternal origin. This is especially important for determining the origin of the callus cells, since spontaneous doubling during early phases is quite common (Gupta, 1975; Peters *et al.*, 1977; Grewal *et al.*, 2009). Many researchers fall prey to the fallacy that the larger the callus volume induced, the better for androgenesis. In fact, the callus phase should be kept short and the amount of callus low due to increasing ploidy levels with increasing number of cell divisions (Haddon and Northcote 1976; Grewal *et al.* 2009).

Donor plants, genotype, bud size and microspore stage

High-quality donor plants grown in a controlled environment, with little or no stress, are a prerequisite for an androgenetic response. One exception is soybean, where field-grown plants are routinely used (Zhuang *et al.*, 1991; Cardoso *et al.*, 2004). Legumes generally require high light intensity (> 600 μmol/m/s) and good light quality. Bud size and microspore stage are also closely related and usually easy to determine, with some exceptions. It is generally agreed that the developmental window of embryogenic competence lies between the mid-unicellular and mid-bicellular stage, although this varies between species (Smýkal, 2000). Uninucleate microspores with their high auxin content (Feng *et al.*, 2006) are also a target for androgenesis in legumes.

As with most other species, for grain legumes the genotype is of paramount importance (Jain *et al.*, 1996/97; Maluszynski *et al.*, 2003; Germanà, 2006). In their work with various legume species and genotypes, Ochatt *et al.* (2009) tested ten field pea genotypes and only three (cvs Victor, Frisson and CDC April) permitted the recovery of haploid plants from the cultured microspores. This is particularly surprising when considering that among the genotypes tested were included three single loci EMS-mutants of Frisson (P64, P79 and P90). These mutants are capable of proliferating as callus and differentiating shoots and early-stage embryos, but failed to regenerate any plants. Likewise, with *Medicago truncatula*, haploid plants could be recovered from isolated microspores of genotype A17, but not from two of its nodulation and mycorrhizogenesis mutants (TRV25 and TR122) (Ochatt *et al.*, 2009). In *Lathyrus* species, out of ten genotypes, only one white-seeded cultivar (LB) and one coloured-seeded cultivar (L3) produced haploid plants. It is also noteworthy that none of the elicitation treatments applied to such microspores could modify this trend.

From a cytological viewpoint, the window of androgenetic response from microspores is narrow for many species (Maluszynski *et al.*, 2003). The arrest of the first asymmetric mitotic division of microspores is required to initiate embryogenesis (Jain *et al.*, 1996/1997), i.e. the precise stage of microsporogenesis when the symmetrical division starts yielding two identical cells. In pea, it was consistently found for all genotypes studied that uninucleate microspores were best for initiation of haploid cultures (Gupta, 1975; Croser *et al.*, 2006; Ochatt *et al.*, 2009). This corresponds to a flower bud length of 6–7 mm and anther size of 1 mm (Croser *et al.* 2006; Ochatt *et al.* 2009). Ochatt *et al.* (2009) established the kinetics of microsporogenesis during flower bud growth in pea. In lupin and chickpea, uninucleate microspores also provided the best responses (Skrzypek *et al.*, 2008; Grewal *et al.*, 2009).

Stress treatments

Prior to 2009, legume androgenesis protocols used mostly temperature (heat or cold) as stress pre-treatments (Table 11.1), although at least 16 other stresses had been used for the induction of androgenesis in other species (Shariatpanahi *et al.*, 2006). The application of different stresses might be the way to overcome the recalcitrance of legumes, probably mediated through increases in hormone levels in stressed anthers. Since the use of electroporation for induction of asparagus anthers (Delaitre *et al.*, 2001), this technique has proved useful in particular for pea, grass pea (Ochatt *et al.*, 2009) and chickpea (Grewal *et al.*, 2009). Combining several stress-inducing factors, one on top of the other, is the way forward to trigger the switch of isolated microspores from the gametophytic to the androgenetic developmental pathway in species as recalcitrant as the temperate legumes. Thus, in field pea, the key to success was to superimpose a cold treatment of flower buds with electrostimulation and an osmotic shock. In chickpea, Grewal *et al.* (2009) found that adding a centrifugation step for anthers (at 168 × *g* for 15 min) to these factors was also beneficial. In recent work (Ochatt *et al.*, unpublished), it was determined that sonication of anthers (30 s, 38 Hz), prior to their culture, may further increase their androgenic potential when added to the other stress agents used.

Temperature

The effect of a cold storage period on anthers and flower buds prior to culture has been studied for many species, including legumes (Jain *et al.*, 1996/97; Touraev *et al.*, 1997; Delaitre *et al.*, 2001; Lionneton *et al.* 2001; Maluszynski *et al.*, 2003). For pea (Croser *et al.*, 2006), chickpea (Croser *et al.*, 2006; Grewal *et al.*, 2009) and lupin (Skrzypek *et al.*, 2008), cold storage of flower buds was needed to foster microspore division. In an early study, anthers of the field pea cv. Bonneville and the breeding lines T163 and P88 were subjected to a 72 h cold pre-treatment whereby callus and heart-shaped-stage embryos were obtained even if plants were not recovered (Gosal and Bajaj, 1988).

High and low temperatures with increasing lengths of time were tested on flower buds of field pea prior to their culture (Ochatt *et al.*, 2009). It was apparent that high temperatures were detrimental to microspore viability, even when delivered for just a few hours (Fig. 11.1). In contrast, cold storage was always beneficial, even for periods as long as one month. Buds can be kept in cold storage before or after surface disinfection and for several weeks without any detrimental effect on

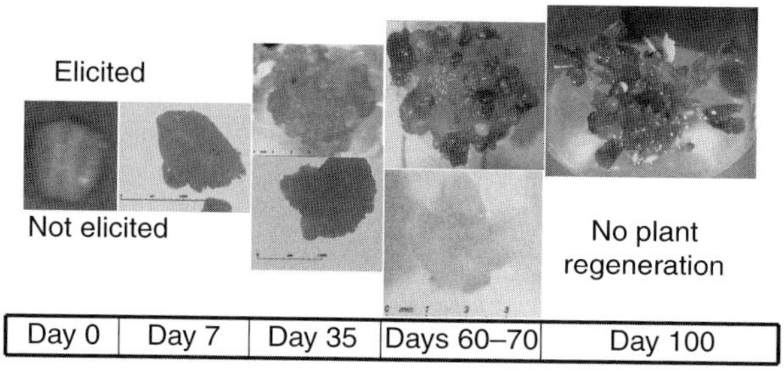

Fig. 11.1. Elicitation of anthers (cold shock, followed by electroporation, centrifugation and sonication) prior to their culture; osmotic shock during culture induces faster growth, somatic embryo formation and, ultimately, haploid plant regeneration, as shown here for field pea.

the subsequent viability of cultured anthers or the division competence of cultured microspores. Ochatt *et al.* (2009) cold-stored flower buds individually rather than on their stems, as reported by Croser *et al.* (2005) for pea and Grewal *et al.* (2009) for chickpea.

Centrifugation

Shariatpanahi *et al.* (2006) mentioned centrifugal treatment as one of the neglected stresses. A centrifugal force of about 10,000 × *g* was used by Tanaka (1973) on tobacco anthers. After cold treatment of buds, Altaf and Ahmad (1986) used centrifugation as additional stress treatment for induction of androgenesis in chickpea. However, plants were not regenerated and ploidy level of callus cells was not determined. In contrast, Grewal *et al.* (2009) effectively used centrifugation of chickpea anthers after cold treatment of buds prior to electroporation and high osmotic shock treatment of anthers (Table 11.1). Centrifugation was also successful for induction of lupin microspores (Campos-Andrada *et al.*, 2001; Bayliss *et al.*, 2004).

Electro-stimulation

When a cell is exposed to an electric field, pores are formed through an enhancement of its trans-membrane potential (Cole, 1968; Neumann and Rosenheck, 1973). This formation depends on the cell radius, the electric field strength delivered, the angle between the normal vector of the membrane and the direction of the electric field applied (Chang, 1992). The application of an electroporation treatment has been known to improve division and initial proliferation competence of protoplasts (Rech *et al.*, 1987) and callus cultures (Rathore and Goldsworthy, 1985). The effect of electroporation on the androgenetic competence of isolated microspores and intact anthers was assessed for pea (Ochatt *et al.*, 2009) and chickpea (Grewal *et al.* 2009). In these studies, differences in pulse duration only marginally affected the viability of the electro-manipulated microspores. This suggests that the field strengths and durations examined are still well below the threshold values required for a significant and irreversible dielectric breakdown of cell membranes. For isolated pea microspores, either square or exponential wave electric fields could be applied, with little difference in viability. For intact anthers, an electric field using exponential waves (i.e. with electricity delivered by discharging capacitors) was preferred to avoid detrimental effects on anther viability. Microspores are surrounded by a thick cell wall that confers a strong physical barrier to electricity and thus may hold the membrane integrity for longer (Saunders *et al.*, 1992). Voltage application must be long enough to give the pores time to form and reseal in order to avoid cell death. In intact anthers, all diploid cells will be more strongly affected by electricity and, if killed, may release substances into the medium that may negatively affect microspore growth and proliferation. The electrical parameters fostering the proliferation of undifferentiated tissues from the cultured microspores differed from those inducing somatic embryogenesis (Ochatt *et al.*, 2009). Electrical parameters necessary for induction of embryos from cultured anthers and microspores are likely not only to be species specific but also genotype specific.

Osmotic pressure of the medium

The eliciting effects of osmotic pressure on androgenesis have been known for a long time, first in the Brassicaceae (Lichter, 1982) and other species (Delaitre *et al.*, 2001) and more recently in legumes. A consistent effect of osmotic pressure modifications in the medium was observed in isolated microspores and in cultured anthers of pea (Croser *et al.*, 2006; Ochatt *et al.*, 2009) and chickpea (Croser *et al.*, 2005; Grewal *et al.*, 2009). Ochatt *et al.* (2009) found that the osmotic stress needed to foster androgenesis from isolated microspores was stronger than reported by Croser *et al.* (2005) for both pea and chickpea. Ochatt *et al.* (2009) obtained the best responses with a 7-day osmotic stress treatment (17% w/v sucrose) followed by transfer to a medium with 10% (w/v) sucrose, which is in line with previous observations made with isolated microspore culture in *Brassica juncea* (L.) Czern. and confirms the positive effect of a changing medium osmolarity at the onset of embryogenesis, as previously observed with cell suspensions of pea and other grain legumes (Ochatt *et al.*, 2009). In their work, Ochatt *et al.* (2009) tested several osmotic pressure regimes during early culture of isolated microspores and compared sucrose with mannitol as an osmoticum. The results obtained demonstrated that sucrose yielded a better response than mannitol. In addition, a large difference between the osmolarity of the initial medium (at 17% sucrose) against that used for subsequent culture (10%) was required to support microspore viability and subsequently trigger their sustained division.

Culture conditions

There is no clear consensus in the literature on culture conditions required for DH of grain legumes (reviewed by Croser *et al.*, 2006). Light conditions ranged from culture in darkness and a photoperiodic light regime to constant illumination, with different effects depending on species and genotypes. The same applies to the temperature during culture.

Differences were reported for microspore culture in terms of the initial plating density required. Ochatt *et al.* (2009) identified the optimum density as 2×10^5 microspores/ml of medium for pea, with lower densities not responding and higher ones resulting in culture and cell oxidation and growth arrest.

Culture medium composition is important (Table 11.1). While various authors reported the effects of medium composition on androgenesis, in particular concerning the content of growth regulators added, most have used various modifications of the MS formula. Ochatt *et al.* (2009) compared three different basal media: NLN medium (Lichter, 1981, 1982) (originally devised for *Brassica* microspore culture), LMJ medium (as used for protoplast culture in pea by Ochatt *et al.* (2000)) and HSO (purpose-prepared for isolated pea microspores). They found that medium composition, although important, would not be crucial for responses, as all three media supported reproducible and comparable responses in the absence of any treatment of microspores but following cold storage of the donor flower buds. Alternatively, some genotypes remained recalcitrant irrespective of the basal medium, treatment or culture conditions employed, thereby indicating that the genotype is the main parameter governing androgenetic capacity in legume species.

Plant regeneration

Plant regeneration is still the Achilles' heel of androgenesis as in many other legume protocols, probably requiring a multiple step approach for induction of androgenesis, embryo development and maturation, plant conversion and rooting. Androgenesis induction often takes place in high-osmotic media (e.g. 17% sucrose; Ochatt *et al.*, 2009), however, embryos retained in these types of media often fail to grow (George and Rao, 1982). Embryos should be regenerated as soon as possible, especially due to the negative effects of many hormones on plant regeneration and rooting in legumes. Hormone-free or low hormone-containing media seem to be best suited for this purpose. If rooting cannot be achieved, progress has been made in grafting (Gurusamy *et al.*, 2010) or *in vitro* flowering of many legume species, which could be used for the generation of DH populations (Ochatt *et al.*, 2002).

Ontogeny and ploidy levels

The commonly used methods for confirmation of haploid origin are chromosome counting, flow cytometry, cytological tracking of embryogenesis directly from individual microspores or the use of heterozygous starting material followed by molecular or morphological confirmation. Too many publications completely omit this step (Table 11.1), but it is vital in the case of anther culture since anthers consist of both haploid and diploid tissues. Spontaneous chromosome doubling is commonplace during the regeneration stages of many species (Jain *et al.*, 1996/1997; Maluszynski *et al.*, 2003). Recently, this has also been confirmed in field pea (Ochatt *et al.*, 2009) and chickpea (Grewal *et al.*, 2009). Furthermore, many researchers reported increasing ploidy levels with increasing age of callus cultures (Gupta, 1975; Haddon and Northcote, 1976; Grewal *et al.*, 2009), making the ontogeny of embryos even more difficult to report. Isolated microspores divided (tracked with DAPI-stained microspores observed under UV) and subsequently proliferated on a solid medium with 2,4-D and, for cv. Highlight, cotyledonary-stage embryos were produced and one plant was regenerated (Croser and Lulsdorf, 2004). This plant was determined to be diploid and, although being unable to root, it could set seed *in vitro*. This plant probably underwent spontaneous chromosome doubling during early regeneration stages.

Anthers should be routinely checked during the induction phase for microspore development either via DAPI (Widholm, 1972) or FDA staining techniques (Dunwell, 1985). Flow cytometric techniques also offer a reliable way of determining ploidy level, and nowadays require only small amounts of tissue (Ochatt, 2008).

11.4 Conclusion

A fundamental understanding of the molecular and biochemical basis for plant gametophyte to sporophyte transition and morphogenesis remains elusive. Research directed toward this aim has predominantly been undertaken using responsive species from the *Brassicaceae*, *Poaceae* and *Solanaceae*. The absence of a robust haploid production system for androgenesis in the model species *Arabidopsis thaliana* has been a constraint on attempts to elucidate these processes. Without the benefit of this knowledge, the current empirical efforts to adapt DH production techniques to recalcitrant species of the *Fabaceae* will continue to be time consuming and difficult.

At this point, anther culture seems to be the most promising method for induction of androgenesis. However, this is coupled with problems of determining whether the induced calli originate from gametophytic or sporophytic tissue. The goal needs to be to keep the callus phase short, the amount of callus produced low and to regenerate embryos or shoots as soon as possible, with the possible exception of soybean. Combining different stresses seems to be the pathway to androgenesis in legumes, especially a combination of cold and other stresses such as electroporation, sonication, centrifugation and a short, high-osmotic medium period. However, even under the 'best circumstances' plant regeneration remains difficult and, currently, the numbers of DH plants produced remain too low for use in breeding programmes.

References

Altaf, N. and Ahmad, M.S. (1986) Plant regeneration and propagation of chickpea (*Cicer arietinum* L.) through tissue culture techniques. In: I.A.E. Agency (ed.) *Nuclear Techniques and in vitro Culture for Plant Improvement*. Vienna, Austria, pp. 407–417.

Aruga, K. and Nakajima, T. (1985) Role of anther on pollen embryogenesis in anther culture of *Nicotiana tabacum* L. *Japanese Journal of Breeding* 35, 390–397.

Arya, I.D. and Chandra, N. (1989) Organogenesis in anther-derived callus culture of cowpea (*Vigna unguiculata* (L.) Walp). *Current Science* 58, 257–259.

Aslam F.N., MacDonald, M.V., Loudon, P.T. and Ingram, D.S. (1990) Rapid-cycling *Brassica* species. Inbreeding and selection of *Brassica napus* for anther culture ability and an assessment of its potential for microspore culture. *Annals of Botany* 66, 331–339.

Bajaj, Y.P.S. and Gosal, S.S. (1987) Pollen embryogenesis and chromosomal variation in cultured anthers of chickpea. *International Chickpea Newsletter* 17, 12–13.

Bajaj, Y.P.S. and Singh, H. (1980) *In vitro* androgenesis in mung bean (*Phaseolus aureus* L.). *Indian Journal of Experimental Biology* 18, 1316–1318.

Bajaj, Y.P.S., Singh, H. and Gosal, S.S. (1980) Haploid embryogenesis in anther cultures of pigeonpea. *Theoretical and Applied Genetics* 58, 157–159.

Bayliss, K.L., Wroth, J.M. and Cowling, W.A. (2004) Pro-embryos of *Lupinus* spp. produced from isolated microspore culture. *Australian Journal of Agricultural Research* 55, 589–593.

Blakeslee, A.F., Belling, J., Farnham, M.E. and Bergner, A.D. (1922) A haploid mutant in the Jimson weed, *Datura stramonium*. *Science* 55, 646–647.

Blaydes, D.F. (1966) Interaction of kinetin and various inhibitors in the growth of soybean tissue. *Physiologia Plantarum* 19, 748–753.

Bohanec, B. (2009) Doubled haploids via gynogenesis. In: Touraev, A., Forster, B.P. and Jain, S.M. (eds) *Advances in Haploid Production in Higher Plants*. Springer, Berlin, pp. 35–46.

Campos-Andrada, da Paz M., Filomena-Carneiro, M. and Mota M. (2001) Isolated microspore culture in *Lupinus mutabilis* Sweet. *Agronomia Lusitana* 49, 119–135.

Cardoso, M.B., Kaltchuk-Santos, E., de Mundstock, E.C and Bodanese-Zanettini, M.H. (2004) Initial segmentation patterns of microspores and pollen viability in soybean cultured anthers: Indication of chromosome doubling. *Brazilian Archives of Biology and Technology* 47, 703–712.

Cardoso, M.B., Bodanese-Zanettini, M.H., de Mundstock E.C. and Kaltchuk-Santos, E. (2007) Evaluation of gelling agents on anther culture: response of two soybean cultivars. *Brazilian Archives of Biology and Technology* 50, 933–939.

Chang, D.C. (1992) Design of protocols for electroporation and electrofusion: selection of electrical parameters. In: Chang, D.C., Chassy, B.M., Saunders, J.A. and Sowers, A.E. (eds) *Guide to Electroporation and Electrofusion*. Academic Press, San Diego, California, pp. 429–455.

Chu, C.C. (1978) The N_6 medium and its applications to anther culture of cereal crops. In: *Proceedings of the Symposium on Plant Tissue Culture*, Science Press, Beijing, pp. 43–50.

Cole, K.S. (1968) *Membranes, Ions and Impulses: a Chapter of Classical Biophysics*. University of California Press, Berkeley, California.

Croser, J.S. and Lulsdorf, M.M. (2004) Progress towards haploid division in chickpea (*Cicer arietinum* L.), field pea (*Pisum sativum* L.) and lentil (*Lens culinaris* Medik.) using isolated microspore culture. In: *European Grain Legume Conference*, AEP, Dijon, France, pp. 189.

Croser, J., Lulsdorf, M.M., Davies, P., Wilson, J., Sidhu, P., Grewal, R. *et al.* (2005) Haploid embryogenesis from chickpea and field pea – progress towards a routine protocol. In: *Proceedings of the Australian Branch of the International Association for Plant Tissue Culture and Biotechnology – Contributing to a Sustainable Future*, 21–24 September, Perth, Australia, pp. 71–82.

Croser, J.S., Lulsdorf, M.M., Davies, P.A., Clarke, H.J., Bayliss, K.L., Mallikarjuna, N. *et al.* (2006) Toward doubled haploid production in the *Fabaceae*: progress, constraints, and opportunities. *Critical Reviews in Plant Sciences* 25, 139–157.

Croser, J.S., Clarke, H.J., Lulsdorf, M.M., Mallikarjuna, N., Usher, K., Edwards, K. *et al.* (2007) Haploid and hybrids – new biotechnology tools for chickpea. In: *6th European Conference on Grain Legumes: Integrating Legume Biology for Sustainable Agriculture*, November, Lisbon, Portugal, pp. 52.

da Silva Lauxen, M., Kaltchuk-Santos, E., Hu, C.Y., Callegari-Jacques, S.M. and Bodanese-Zanettini, M.H. (2003) Association between floral bud size and developmental stage in soybean microspores. *Brazilian Archives of Biology and Technology* 46, 515–520.

Delaitre, C., Ochatt, S. and Deleury, E. (2001) Electroporation modulates the embryogenic responses of asparagus (*Asparagus officinalis* L.) microspores. *Protoplasma* 216, 39–46.

de Moraes, A.P., Bonadese-Zanettini, M.H., Callegari-Jacques, S.M. and Kaltchuk-Santos, E. (2004) Effect of temperature shock on soybean microspore embryogenesis. *Brazilian Archives of Biology and Technology* 47, 537–544.

Devaux, P. and Kasha, K.J. (2009) Overview of barley doubled haploid production. In: Touraev, A., Forster, B.P. and Jain, S.M. (eds) *Advances in Haploid Production in Higher Plants*. Springer Scientific, Berlin, pp. 47–63.

Dunwell, J.M. (1985) Haploid cell cultures. In: Dixon, R.A. (ed.) *Plant Cell Culture: a Practical Approach*. OUP, Oxford, UK, pp. 21–36.

Feng, X.L., Ni, W.M., Elge, S., Mueller-Roeber, B. and Xu, Z.H. (2006) Auxin flow in anther filaments is critical for pollen grain development through regulating pollen mitosis. *Plant Molecular Biology* 61, 215–226.

Forster, B.P., Heberle-Bors, E., Kasha, K.J. and Touraev, A. (2007) The resurgence of haploids in higher plants. *Trends in Plant Science* 12, 368–375.

Fougat, R.S., Pathak, A.R. and Bharodia, P.S. (1992) Regeneration of haploid callus from anthers of pigeonpea. *Gujarat Agricultural University Research Journal* 17, 151–152.

Gamborg, O.L., Miller, R.A. and Ojima, K. (1968) Nutrient requirements of suspension cultures of soybean root cells. *Experimental Cell Research* 50, 150–158.

George, L. and Rao, P.S. (1982) *In vitro* induction of pollen embryos and plantlets in *Brassica juncea* through anther culture. *Plant Science Letters* 26, 111–116.

Germanà, M.A. (2006) Doubled haploid production in fruit crops. *Plant Cell, Tissue and Organ Culture* 86, 131–146.

Gosal, S.S. and Bajaj, Y.P.S. (1979) Establishment of callus-tissue cultures, and the induction of organogenesis in some grain-legumes. *Crop Improvement* 6, 154–160.

Gosal, S.S. and Bajaj, Y.P.S. (1988) Pollen embryogenesis and chromosomal variation in anther of three food legumes – *Cicer arietinum*, *Pisum sativum* and *Vigna mungo*. *Sabrao Journal* 20, 51–58.

Grewal, R.K., Lulsdorf, M., Croser, J., Ochatt, S., Vandenberg, A. and Warkentin, T.D. (2009) Doubled-haploid production in chickpea (*Cicer arietinum* L.): role of stress treatments. *Plant Cell Reports* 28, 1289–1299.

Gupta, S. (1975) Morphogenetic response of haploid callus tissue of *Pisum sativum* (var. B22). *Indian Agriculture* 19, 11–21.

Gupta, S., Ghosal, K.K. and Gadgil, V.N. (1972) Haploid tissue culture of *Triticum aestivum* var. Sonalika and *Pisum sativum* var. B22. *Indian Agriculture* 16, 277–278.

Gurusamy, V., Bett, K.E. and Vandenberg, A. (2010) Grafting as a tool in common bean breeding. *Canadian Journal of Plant Sciences* 90, 299–304.

Haddon, L. and Northcote, D.H. (1976) The effect of growth conditions and origin of tissue on the ploidy and morphogenic potential of tissue cultures of bean (*Phaseolus vulgaris* L.). *Journal of Experimental Botany* 27, 1031–1051.

Hu, C.Y., Yin, G.C. and Bodanese Zanettini, M.H. (1996) Haploid of soybean. In: Jain, S.M., Sopory, S.K. and Veilleux, R.E. (eds) *In vitro Haploid Production in Higher Plants*. Kluwer Academic Publisher, Dordrecht, The Netherlands, pp. 377–395.

Huda, S., Islam, R., Bari, M.A. and Asaduzzaman, M. (2001) Anther culture of chickpea. *International Chickpea and Pigeonpea Newsletter* 8, 24–26.

Ivers, D.R., Palmer, R.G. and Fehr, W.R. (1974) Anther culture in soybeans. *Crop Science* 14, 891–893.

Jain, S.M., Sopory, S.K. and Veilleux, R.E. (1996/1997) *In vitro Haploid Production in Higher Plants*. Kluwer Academic Publishers, Dordrecht, The Netherlands.

Jian, Y.Y., Liu, D.P., Luo, X.M. and Zhao, G.L. (1986) Studies on induction of pollen plants in *Glycine max* (L.) Merr. *Journal of Agricultural Sciences* (China) 2, 26–30.

Kaltchuk-Santos, E., Bodanese-Zanettini, M.H. and Mundstock, E. (1993) Pollen dimorphism in soybean. *Protoplasma* 174, 74-78.

Kaltchuk-Santos, E., Mariath, J.E.A., Mundstock, E., Hu, C.Y. and Bodanese-Zanettini, M.H. (1997) Cytological analysis of early microspore divisions and embryo formation in cultured soybean anthers. *Plant Cell, Tissue and Organ Culture* 49, 107–115.

Kao, K.N. (1982) Plant protoplast fusion and isolation of heterokaryocytes. In: Wettel, L.R. and Constabel, F. (eds) *Plant Tissue Culture Methods* (2nd revised edn), National Research Council of Canada, Prairie Regional Laboratory, Saskatoon, Canada, pp. 49–56.

Kao, K.N. and Michayluk, M.R. (1975) Nutritional requirements for growth of *Vicia hajastana* cells and protoplasts at a very low population density in liquid media. *Planta* 126, 105–110.

Kasha, K.J. and Kao, K.N. (1970) High frequency haploid production in barley (*Hordeum vulgare* L.). *Nature* 225, 874–875.

Kaur, P. and Bhalla, J.K. (1998) Regeneration of haploid plants from microspore culture of pigeonpea (*Cajanus cajan* L.). *Indian Journal of Experimental Biology* 36, 736–738.

Keller, W.A. and Ferrie, A.M.R. (2002) Development of haploid production technology applicable for genetic improvement of lentils. *ADF project* No. 98000297.

Khan, S.K. and Ghosh, P.D. (1983) *In vitro* induction of androgenesis and organogenesis in *Cicer arietinum* L. *Current Science* 52, 891–893.

Kumar, A.S., Gamborg, O.L. and Nabors, M.W. (1988) Plant regeneration from cell suspension cultures of *Vigna aconitifolia*. *Plant Cell Reports* 7, 138–141.

Kumlehn, J. (2009) Embryogenic pollen culture: a promising target for genetic transformation. In: Touraev, A., Forster, B.P. and Jain, S.M. (eds) *Advances in Doubled Haploid Technology in Higher Plants*. Springer Scientific, Berlin, pp. 295–305.

Kyo, M. and Harada, H. (1986) Control of the developmental pathway of tobacco pollen *in vitro*. *Planta* 168, 427–432.

Ladeinde, T.A.O. and Bliss, F.A. (1977) A preliminary study on the production of plantlets from anthers of cowpea. *Tropical Grain Legume Bulletin* 8, 13.

Lichter, R. (1981) Anther culture of *Brassica napus* in a liquid culture medium. *Zeitschrift fuer Pflanzenphysiologie* 103, 229–237.

Lichter, R. (1982) Induction of haploid plants from isolated pollen of *Brassica napus*. *Zeitschrift fuer Pflanzenphysiologie* 105, 427–434.

Lionneton, E., Beuret, W., Delaitre, C., Ochatt, S. and Rancillac, M. (2001) Improved microspore culture and doubled-haploid plant regeneration in the brown condiment mustard (*Brassica juncea*). *Plant Cell Reports* 20, 126–130.

Liu, D.P. and Zhao, G.L. (1986) Callus formation from pollen and culture *in vitro* of soybean. *Soybean Science* 5, 17–20.

Mallikarjuna, N., Jadhav, D., Clarke, H., Coyne, C. and Muehlbauer, F. (2005) Induction of androgenesis as a consequence of wide crossing in chickpea. *ejournal.icrisat.org* 1, 1–3.

Maluszynski, M., Kasha, K.J, Forster, B.P. and Szarejko, I. (2003) *Doubled Haploid Production in Crop Plants*. Kluwer Academic Publishers, Dordrecht, The Netherlands.

Miller, C.O. (1963) Kinetin and kinetin-like compounds. *Moderne Methoden der Pflanzenanalyse* 6, 194–202.

Mix, G. and Wang, H.M. (1988) *In vitro* Erzeugung von haploiden Cowpea-Pflanzen (*Vigna unguiculata* L.). *Landbauforschung Völkenrode* 4, 305–309.

Muñoz, L.C. and Baudoin, J.P. (1994) Influence of the cold pretreatment and the carbon source on callus induction from anthers in *Phaseolus*. *Annual Report of the Bean Improvement Cooperative* 37, 129–130.

Muñoz, L.C. and Baudoin, J.P. (2001/2002) Improvement of *in vitro* induction of androgenesis in *Phaseolus* beans (*P. vulgaris* L. and *P. coccineus* L.). *Acta Agronomica* 51, 81–87.

Muñoz, L.C., Baudoin, J.P. and Schmit, V. (1992) Finding out an efficient technique for inducing callus from *Phaseolus* microspores. *Annual Report of the Bean Improvement Cooperative* 35, 217–218.

Muñoz, L.C., Baudoin, J.P. and Bradfer, C. (1993) *In vitro* induction of androgenesis in *Phaseolus*. *Annual Report of the Bean Improvement Cooperative* 36, 18–19.

Murashige, T. and Skoog, F. (1962) A revised medium for rapid growth and bioassays with tobacco tissue cultures. *Physiologia Plantarum* 15, 473–497.

Narasimham, N. (1999) Induction of androgenesis in pigeonpea – possible reason for low response of pollen to tissue culture conditions. In: Plant tissue culture and biotechnology: emerging trends, *Proceedings of the Symposium*, 29–31 January, Hyderabad, India, pp. 232–239.

Neumann, E. and Rosenheck, K. (1973) An alternate explanation for the permeability changes induced by electrical impulses in vesicular membranes. *Journal of Membrane Biology* 14, 193–196.

Nitsch, J.P. and Nitsch, C. (1969) Haploid plants from pollen grains. *Science* 163, 85–87.

Ochatt, S.J. (2008) Flow cytometry in plant breeding. *Cytometry* A 73, 581–598.

Ochatt, S.J., Mousset-Declas, C. and Rancillac, M. (2000) Fertile pea plants regenerate from protoplasts when calluses have not undergone endoreduplication. *Plant Science* 156, 177–183.

Ochatt, S.J., Sangwan, R.S., Marget, P., Assoumou Ndong, Y., Rancillac, M. and Perney, P. (2002) New approaches towards shortening of generation cycles for faster breeding of protein legumes. *Plant Breeding* 121, 436–440.

Ochatt, S.J., Pech, C., Grewal, R., Conreux, C., Lulsdorf, M.M. and Jacas, L. (2009) Abiotic stress enhances androgenesis from isolated microspores of some legume species (*Fabaceae*). *Journal of Plant Physiology* 166, 1314–1328.

Ormerod, A.J. and Caligari, P.D.S. (1994) Anther and microspore culture of *Lupinus albus* in liquid culture medium. *Plant Cell, Tissue and Organ Culture* 36, 227–236.

Palmer, C.E. and Keller, W.A. (2005) Overview of haploidy. In: Palmer, C.E., Keller, W.A. and Kasha, K.J. (eds) *Haploids in Crop Improvement*. II. Springer-Verlag Scientific, Berlin, pp. 3–9.

Pauls, K.P., Chan, J., Woronuk, G., Schulze, D. and Brazolot, J. (2006) When microspores decide to become embryos – cellular and molecular changes. *Canadian Journal of Botany* 84, 668–678.

Peters, J.E., Crocomo, O.J., Sharp, W.R., Paddock, E.F., Tegenkamp, I. and Tengenkamp, T. (1977) Haploid callus cells from anthers of *Phaseolus vulgaris*. *Phytomorphology* 27, 79–85.

Phillips, G.C. and. Collins, G.B. (1979) *In vitro* tissue culture of selected legumes and plant regeneration from callus culture of red clover. *Crop Science* 19, 59–64.

Rao, P.V.L. and De, D.N. (1987) Haploid plants from *in vitro* anther culture of the leguminous tree, *Peltophorum pterocarpum* (DC) K. Hayne (Copper pod). *Plant Cell, Tissue and Organ Culture* 11, 167–177.

Rathore, K.S. and Goldsworthy, A. (1985) Electrical control of shoot regeneration in plant tissue culture. *Bio/Technology* 3, 1107–1109.

Ravi, M. and Chan, S.W.L. (2010) Haploid plants produced through seeds by centromere-mediated genome elimination. *Nature* 464, 615–618.

Rech, E.L., Ochatt, S.J., Chand, P.K., Power, J.B. and Davey, M.R. (1987) Electro-enhancement of division of plant protoplast-derived cells. *Protoplasma* 141, 169–176.

Reddy, V.D. and Reddy, G.M. (1996) *In vivo* production of haploids in chickpea (*Cicer arietinum* L.). *Journal of Genetics and Breeding* 51, 29–32.

Resch, T., Ankele, E., Badur, R., Reiss, B., Heberle-Bors, E. and Touraev, A. (2009) Immature pollen as target for gene targeting. In: Touraev, A., Forster, B.P. and Jain, S.M. (eds) *Advances in Haploid Production in Higher Plants*. Springer Scientific, Berlin, pp. 307–317.

Rodrigues, L.R., Terra, T.F., Bered, F. and Bodanese-Zanettini, M.H. (2004a) Origin of embryo-like structures in soybean anther culture investigated using SSR markers. *Plant Cell, Tissue and Organ Culture* 77, 287–289.

Rodrigues, L.R., Forte, B.C., Oliveira, J.M.S., Mariath, J.E.A. and Bodanese-Zanettini, M.H. (2004b) Effects of light conditions and 2,4-D concentration in soybean anther culture. *Plant Growth Regulation* 44, 125–131.

Rodrigues, L.R., Oliveira, J.M.S., Mariath, J.E.A. and Bodanese-Zanettini, M.H. (2005a) Histology of embryogenic responses in soybean anther culture. *Plant Cell, Tissue and Organ Culture* 80, 129–137.

Rodrigues, L.R., Oliveira, J.M.S., Mariath, J.E.A., Iranco, L.B. and Bodanese-Zanettini, M.H. (2005b) Anther culture and cold treatment of floral buds increased symmetrical and extra-nuclei frequencies in soybean pollen grains. *Plant Cell, Tissue and Organ Culture* 81, 101–104.

Rodrigues, L.R., Forte, B.C. and Bodanese-Zanettini, M.H. (2006) Isolation and culture of soybean (*Glycine max* L. Merrill) microspores and pollen grains. *Brazilian Archives of Biology and Technology* 49, 537–545.

Sator, C. (1985) Regeneration von Lupinenpflanzen aus Antheren [Regeneration of lupin plants from anthers]. *Landbauforschung Volkenrode* 35, 5–7.

Saunders, J.A., Matthews, B.F. and Van Wert, S.L. (1992) Pollen electrotransformation for gene transfer in plants. In: Chang, D.C., Chassy, B.M., Saunders, J.A. and Sowers, A.E. (eds) *Guide to Electroporation and Electrofusion*. Academic Press, San Diego, California, pp. 227–247.

Schenk, R.U. and Hildebrandt, A.C. (1972) Medium and techniques for induction and growth of monocotyledonous and dicotyledonous plant cell cultures. *Canadian Journal of Botany* 50, 199–204.

Segui-Simarro, J.M. and Nuez, F. (2008a) How microspores transform into haploid embryos: changes associated with embryogenesis induction and microspore-derived embryogenesis. *Physiologia Plantarum* 134, 1–12.

Segui-Simarro, J.M. and Nuez, F. (2008b) Pathways to doubled haploidy: chromosome doubling during androgenesis. *Cytogenetics and Plant Breeding* 120, 358–369.

Shariatpanahi, M.E., Bal, U., Heberle-Bors, E. and Touraev, A. (2006) Stresses applied for the re-programming of plant microspores towards *in vitro* embryogenesis. *Physiologia Plantarum* 127, 519–534.

Sidhu, P. and Davies, P. (2005) Pea anther culture: callus initiation and production of haploid plants. In: Bennett, I.J., Bunn, E., Clarke, H. and McComb, J.A. (eds) Contributing to a sustainable future, *Proceedings of the Australian Branch of the IAPTC&B*, Perth, Australia, pp. 180–186.

Skinner, D.Z. and Liang, G.H. (1996) Haploidy in alfalfa. In: Jain, S.M., Sopory, S.K. and Veilleux, R.E. (eds) *In vitro Haploid Production in Higher Plants*. Kluwer Academic Publishers, Dordrecht, The Netherlands, pp. 365–375.

Skrzypek, E., Czyczylo-Mysza, I., Marcinska, I. and Wedzony, M. (2008) Prospects of androgenetic induction in *Lupinus* spp. *Plant Cell, Tissue and Organ Culture* 94, 131–137.

Smýkal, P. (2000) Pollen embryogenesis – the stress mediated switch from gametophytic to sporophytic development. Current status and future prospects. *Biologia Plantarum* 43, 481–489.

Sudhakar, Y., Singh, I.S. and Singh, C.P. (1986) Anther culture of pigeonpea: physiology of callus formation and ploidy analysis. *Indian Journal of Plant Physiology* 29, 67–70.

Tai, K.S. and Cheng, S.H. (1990) Tissue culture of cultivated *Leguminosae*. *Journal of the Agricultural Association of China* 149, 42–52.

Tanaka, M. (1973) The effect of centrifugal treatment on the emergence of plantlet from cultured anther of tobacco. *Japanese Journal of Breeding* 23, 171–174.

Tang, W.T., Ling, T.S. and Chang, C.H. (1973) Effect of kinetin and auxin on callus formation of anther culture of soybean. *Journal of the Agricultural Association of China* 83, 1–7.

Tiwari, S., Shanker, P. and Tripathi, M. (2004) Effects of genotype and culture medium on *in vitro* androgenesis in soybean (*Glycine max* Merr.). *Indian Journal of Biotechnology* 3, 441–444.

Touraev, A., Vicente, O. and Heberle-Bors, E. (1997) Initiation of microspore embryogenesis by stress. *Trends in Plant Science* 2, 297–302.

Tulecke, W. (1964) A haploid tissue culture from the female gametophyte of *Ginkgo biloba* L. *Nature* 203, 94–95.

Veliky, I.A. and Martin, S.M. (1970) A fermenter for plant cell suspension cultures. *Canadian Journal of Microbiology* 16, 223–226.

Vessal, S.R., Bagheri, A. and Safarnejad, A. (2002) The possibility of *in vitro* haploid production in chickpea (*Cicer arietinum* L.). *Journal of Science and Technology in Agriculture and Natural Resources* 6, 67–76.

Vishukumar, U., Patil, M.S. and Nayak, S.N. (2000) Anther culture studies in pigeonpea. *Karnataka Journal of Agricultural Sciences* 13, 16–19.

Wedzony, M. Forster, B.P., Zur, I., Golemiec, E., Szechynska-Hebda, M., Dubas, E. *et al.* (2009) Progress in doubled haploid technology in higher plants. In: Touraev, A., Forster, B.P. and Jain, S.M. (eds) *Advances in Haploid Production in Higher Plants*. Springer Sciences, New York, pp. 1–33.

White, P.R. (1963) *The Cultivation of Animal and Plant Cells*. Ronald Press, New York, pp. 57–63.

Widholm, J.M. (1972) The use of fluorescein diacetate and phenosafranine for determining the viability of cultured cells. *Stain Technology* 47, 189–194.

Yang, H., Wei, Z. and Xu, Z. (1997) Effects of specific expression of iaaL gene in tobacco tapetum on pollen embryogenesis. *Science in China* (C) 40, 384–391.

Ye, X.G., Fu, Y.Q. and Wang, L.Z. (1994) Study on several problems of soybean anther culture. *Soybean Science* 13, 193–199.

Yeung, E.C. and Sussex, I.M. (1979) Embryogeny of *Phaseolus coccineus*: the suspensor and the growth of the embryo-proper *in vitro*. *Zeitschrift für Pflanzenphysiologie* 91, 423–433.

Yin, G.C., Zhu, Z.Y., Xu, Z., Chen, L., Li, X.Z. and Bi, F.Y. (1982) Studies on induction of pollen plant and their androgenesis in *Glycine max* (L.). *Soybean Science* 1, 69–76.

Zhao, G., Liu, Y. and Li, J. (1998) Germination of embryo in soybean anther culture. *Chinese Science Bulletin* 43, 1991–1995.

Zhuang, X.J., Hu, C.Y., Chen, Y. and Yin, G.C. (1991) Embroids from soybean anther culture. *Soybean Genetics Newsletter* 18, 265.

Zur, I., Dubas, E., Golemiec, E., Szechynska-Hebda, M., Janowiak, F. and Wedzony, M. (2008) Stress-induced changes important for effective androgenic induction in isolated microspore culture of triticale (× *Triticosecale* Wittm.). *Plant Cell, Tissue and Organ Culture* 94, 319–328.

12 Genetic Transformation

G. Angenon and T.T. Thu

12.1 Introduction

The majority of the economically important grain legumes are subject to a number of biotic (fungi, bacteria, insects, viruses, nematodes and weeds) and abiotic (drought, salinity, waterlogging, cold) stresses, which limit the productivity and quality of these crops considerably, especially in tropical and subtropical countries (Dita *et al.*, 2006). Conventional grain legume breeding has a long history and has made available a large number of improved varieties; however suitable solutions for all the above-mentioned problems have not yet been provided, particularly because of the absence of desirable characteristics in the (primary) gene pool. In this regard, genetic transformation can be considered a complementary tool in breeding strategies, as it can overcome the limitations imposed by sexual compatibility. In addition, transformation technology, together with the rapidly expanding sets of genomics data for several leguminous plants (Varshney *et al.*, 2009), may unravel biological processes through a molecular genetics approach, thus generating knowledge that can be applied for innovative breeding strategies. Finally, because of their high protein content, transgenic leguminous plants can be attractive hosts for novel applications in the field of 'molecular farming', for example for the production of vaccines or antibodies (Boothe *et al.*, 2010).

Although numerous applications of transgene technology have been or are being developed in the grain legumes, the majority of these species remain difficult to transform. This may seem surprising, given the huge commercial success of transgenic soybean plants. Also, the first reports on creating transgenic legumes appeared only a few years after the pioneering work on transformation of easily regenerable plants such as tobacco: for instance, the reports on transgenic *Vigna aconitifolia* (Köhler *et al.*, 1987) and soybean plants (Hinchee *et al.*, 1988; McCabe *et al.*, 1988). Since then, most important food legumes have been added to the list of transformable species and a large number of studies were conducted focusing on improvements of all aspects of DNA transfer, regeneration and selection of transgenic plants. Although the list of publications on this subject is quite long, unfortunately, it is difficult to point out really routine and easily applicable protocols for any of the grain legumes. Almost all grain legumes should still be considered recalcitrant to transformation, the main bottleneck being the limited regeneration capacity. Indeed, an efficient system for gene transformation in plants comprises various factors, but at least high regeneration capacity and efficient delivery of transgenes to a large number of cells from the target explants and effective selectable markers have to be considered as essential and crucial factors (Karami *et al.*, 2009).

12.2 Regeneration

In general, the Fabaceae species are difficult to regenerate *in vitro*, and display high genotype specificity for regeneration; grain legumes have generally less regeneration potential compared with the forage legumes (Somers *et al.*, 2003; Svetleva *et al.*, 2003). Embryogenic calli have been shown to be suitable explants for transformation of various species, including model and forage legumes (Chabaud *et al.*, 1996; Trinh *et al.*, 1998; Wigdorovitz *et al.*, 1999). Embryogenesis has been tested with several grain legume species, for instance pigeon pea (George and Eapen, 1994; Mohan and Krishnamurthy, 2002), chickpea (Kumar *et al.*, 1994; Murthy *et al.*, 1996), soybean (Bailey *et al.*, 1993) and pea (Griga, 1998). However, using embryogenic calli for gene transformation gave low efficiencies for many grain legume species, except for soybean. Regeneration via embryogenesis remains a major method for obtaining transgenic soybean plants, using both particle bombardment and *Agrobacterium*-mediated gene transfer (Ko *et al.*, 2003; Kita *et al.*, 2007). Also, transformation of peanut can be achieved via somatic embryogenesis (Ozias-Akins *et al.*, 1993).

Another pathway for legume regeneration avoiding the low-regenerable callus phase is through direct organogenesis. In legume transformation protocols, direct organogenesis has been obtained from a variety of explants, including intact shoot tips, meristems, cotyledons, cotyledonary nodes and embryo axes derived from either germinating mature seeds or immature seeds, in addition to leaf discs, stems, complete immature seeds, etc. (for more detail see Somers *et al.*, 2003; Eapen, 2008). Embryonic axes and cotyledonary nodes from germinated seeds have been used most widely as explants for gene transfer and regeneration. Efficient plant regeneration often requires cytokinins at relatively low concentrations to stimulate multiple shoot formation at the target location of the explants (Thu *et al.*, 2003; Popelka *et al.*, 2006) and is sometimes enhanced by pre-treatment with cytokinins during seed germination (Mohamed *et al.*, 1992; Thu *et al.*, 2003). One disadvantage of these direct organogenesis systems is that the obtained shoots are often of multicellular origin, which may prevent strict selection for transgenic shoots and may lead to high numbers of 'escapes' (non-transgenic plants that survive selection) (Popelka *et al.*, 2006; Solleti *et al.*, 2008; Patil *et al.*, 2009). Some authors have indicated that pre-existing meristems, which are abundant in explants from mature seeds, can produce chimeric transgenic plants and, to avoid this, immature seeds should be used as in the case of mung bean transformation (Muruganantham *et al.*, 2007). Nevertheless, mature seeds remain the preferred source of explants, not only because they can be stored and are easily available, but also because the problem of chimeric transgenic plants appeared to be minor in several optimized protocols (Popelka *et al.*, 2006; Rech *et al.*, 2008). Immature embryos show high regeneration potential, but are also highly sensitive to co-cultivation conditions and, therefore, the efficiency of transformation may be low (Thu *et al.*, 2003).

As mentioned above, shoot regeneration from callus is often difficult to obtain in grain legumes and is probably more genotype dependent than direct organogenesis. On the other hand such systems allow for strict selection of transgenic callus and shoots, thus avoiding the problems of chimerism and escapes. Accordingly, several highly reliable legume transformation methods are based on shoot regeneration from transgenic callus, for example in the case of pea (Schroeder *et al.*, 1993; Grant *et al.*, 1995) and *Phaseolus acutifolius* (De Clercq *et al.*, 2002; Zambre *et al.*, 2005).

To regenerate shoots, various phytohormones have been supplemented to the regeneration media, but the majority of the protocols for grain legumes use the cytokinin benzyl aminopurine. Thidiazuron (TDZ) has been reported to induce shoot organogenesis in several recalcitrant woody plants (Murthy *et al.*, 1998). To increase the regeneration rate and consequently acquire a high efficiency of transformation, TDZ has been supplied to the shoot induction media of many legume species, for example pea (Richter *et al.*, 2006), pigeon pea (Singh *et al.*, 2003), chickpea (Ignacimuthu and Prakash, 2006), bean (Zambre *et al.*, 1998) and *Vicia faba* (Hanafy *et al.*, 2005). The positive effect of TDZ on plant regeneration that has been observed depends on the applied concentration. In the case of pigeon pea regeneration, the continuous presence of TDZ at concentrations of 0.05–1.0 μM induced multiple shoots, but at higher concentrations (10.0, 20.0 μM) direct somatic embryogenesis was obtained (Singh *et al.*, 2003). TDZ, especially at high concentrations, has also been observed to have negative effects on shoot formation, for example in *Phaseolus angularis* (Mohamed *et al.*, 2006) and cowpea (Popelka *et al.*, 2006).

12.3 Transformation

Many different methods have been developed to deliver transgenes into plants, but only *Agrobacterium*-mediated transformation and particle bombardment (biolistics) have been extensively used to create transgenic plants in major grain legumes (see Popelka *et al.*, 2004; Eapen, 2008). *Agrobacterium*-mediated transformation is the most widely used transformation technology for plants in general, as well as for legumes (Eapen, 2008), partly because it often gives rise to simple transgene integration patterns, which is desirable for correct and stable transgene expression. Particle bombardment, on the other hand, is expected to be less genotype dependent because, in contrast to *Agrobacterium*-mediated transformation, it does not depend on the interaction between two living organisms.

Agrobacterium-mediated transformation

Agrobacterium tumefaciens and its close relative *Agrobacterium rhizogenes* are bacteria that genetically colonize host plants: they have the unique capacity to transfer a set of genes, the T-DNA genes, to wounded plant cells. The finding that the T-DNA genes are dispensable for the transfer process and can be replaced by any gene(s) of interest allowed for the development of *Agrobacterium* as a versatile tool for plant transformation three decades ago. Based on a detailed knowledge of the *A. tumefaciens*–plant cell interaction and of the T-DNA transfer process (Zupan *et al.*, 2000; Tzfira and Citovsky, 2006), *Agrobacterium* has subsequently been used as a vector for transformation of nearly every plant species of interest (and even non-plant species, primarily a large number of fungi; Lacroix *et al.*, 2006).

Widely used *Agrobacterium* strains (Hellens *et al.*, 2000) such as LBA4404, EHA101, EHA105, AGL1, C58C1Rif^R (pMP90), C58C1Rif^R (pGV2260) and KYRT1, have been reported to infect a wide range of legume species. EHA101, EHA105 and AGL1 contain *vir* genes from the oncogenic strains A281, whereas KYRT1 is derived from the oncogenic strain Chry5. As both A281 and Chry5 are supervirulent on several plant species, including legumes (Hood *et al.*, 1986, 1987; Torisky *et al.*, 1997), the derived strains are often considered specifically useful for legume transformation. Several publications focus on comparison of the transformation efficiency between different *Agrobacterium* strains. Among these, Nadolska-Orczyk and Orczyk (2000) have reported a significantly better effect of strain EHA105 on transformation of pea compared with LBA4404 or C58C1Rif^R (pMP90). However, these *Agrobacterium* strains gave a different effect on mung bean transformation, for which EHA105 was not an optimal choice (Jaiwal *et al.*, 2001). Solleti *et al.* (2008) used LBA4404, C58C1Rif^R (pGV2260), AGL1 and EHA105 strains for cowpea transformation and noted that EHA105 gave the highest efficiency (76%), followed by LBA4404 (64%), AGL1 (61%) and C58C1Rif^R (pGV2260) (23%), however additional copies of *vir*G, *vir*C and *vir*B genes in LBA4404 were able to enhance the efficiency, up to 100%. Comparing the effects of C58C1Rif^R (pMP90), C58C1Rif^R (pGV2260) and EHA101 on callus transformation of tepary bean (*Phaseolus acutifolius*), De Clercq *et al.* (2002) indicated that among these, EHA101 was the least efficient. For pea transformation, KYRT1 was found to be threefold more efficient than AGL1 (Grant *et al.*, 2003). From the above examples it is clear that no general conclusions can be drawn regarding which *Agrobacterium* strain is most efficient, and that careful comparison of different strains is advisable for each species.

Injuries to explants before infection are recommended in nearly all published protocols. Wounding not only provides an entry point for bacteria but also activates the release of phenolic substances critical for *Agrobacterium vir* gene induction (Bolton *et al.*, 1986; Zupan *et al.*, 2000). In general, plant tissues are injured by scalpels or needles but additional enforcement of wounding can be obtained by vacuum infiltration or sonication (sonication-assisted *Agrobacterium*-mediated transformation – SAAT). Enhanced efficiencies of transformation using the SAAT method have been observed with soybean and chickpea (Santarem *et al.*, 1998; Pathak and Hamzah, 2008). The combination of sonication and vacuum infiltration has been successfully applied for bean transformation (Liu *et al.*, 2005). The negative side of strong wounding is that the wounding may result in extensive enzymatic browning and cell death, and disrupt tissue organization such that *de novo* shoot production cannot occur near the wounded surfaces (Wright *et al.*, 1986).

Supplementation of the *vir* gene inducer acetosyringone (AS) to assist the gene transfer process can be found in many publications concerning legume transformation, although its presence is not always considered as absolutely necessary. For instance, addition of AS to the bacterial re-suspension medium as well as co-cultivation medium resulted in a non-significant increase in transformation frequency of mung bean (Sonia *et al.*, 2007), and transgenic pigeon pea can be obtained without using AS (Kumar *et al.*, 2004; Surekha *et al.*, 2005). Transgenic chickpea can be obtained when using AS (Chakraborti *et al.*, 2009) as well as without AS (Sarmah *et al.*, 2004). However, Polowick *et al.* (2004) claimed that no transgenic plants from chickpea were recovered after co-cultivation without AS. A positive effect on *P. acutifolius* transformation was observed when AS was used at concentrations of 20–200 μM, but a higher concentration (2000 μM) proved inhibitory (De Clercq *et al.*, 2002).

Other parameters affecting the transformation efficiency are the temperature and light conditions during bacterial infection and co-culture. The effect of temperature on *Agrobacterium*-mediated gene transfer was first described in detail with tobacco and *P. acutifolius* (Dillen *et al.*, 1997). The transformation was carried out at temperatures between 15 and 29°C and the authors reported that, irrespective of the *Agrobacterium* strain used, the transfer of the transgene (*uidA*) was optimal at 22°C. A similar effect on stable transformation was subsequently found in several leguminous as well as non-leguminous species (e.g. Sunilkumar and Rathore, 2001; Dang and Wei, 2007). Also, the light conditions affect transgene transfer from *Agrobacterium* to plant cells, as has been found in *P. acutifolius* where continuous light or a 16 h light/8 h dark photoperiod drastically enhanced T-DNA transfer compared with co-cultivation in the dark (Zambre *et al.*, 2003).

Particle bombardment

Among the direct gene transfer techniques, particle bombardment is by far the most popular. This technology has been applied to different legumes including groundnut (Ozias-Akins *et al.*, 1993), pigeon pea (Thu *et al.*, 2003), chickpea (Husnain *et al.*, 1997), cowpea (Ikea *et al.*, 2003; Ivo *et al.*, 2008), lentil (Gulati *et al.*, 2002), soybean and common bean (Rech *et al.*, 2008).

One disadvantage of this technique is that it sometimes results in complex transgene integration patterns, thus enhancing the likelihood of transgene silencing (Travella *et al.*, 2005; Yang *et al.*, 2005). An example of this phenomenon in legumes is a study concerning transformation with isoflavone biosynthetic genes in soybean (Zernova *et al.*, 2009). The transgenic lines carried multiple transgene inserts and, although the lines were transformed with sense constructs aiming at overexpression of isoflavone biosynthetic enzymes, the transgenic lines actually contained lower levels of isoflavones, suggesting co-suppression of the homologous soybean genes (Zernova *et al.*, 2009). In this regard, an appealing technique is the use of recombinase-mediated DNA cassette exchange (RMCE) as applied by Li *et al.* (2009) in soybean. This allows the introduction of a single copy of a transgene at a defined, previously characterized position in the genome (Li *et al.*, 2009), thus reducing position and silencing effects and ensuring correct expression of the transgene.

An interesting feature of the particle bombardment technique is that it can be used for introduction of genes in the plastid genome, in addition to generating nuclear transformation events. Plastid transformation has several attractive features, including: (i) potentially high expression levels; (ii) transgene integration at defined positions through homologous recombination; (iii) the absence of gene silencing phenomena; and (iv) the lack of transgene transmission via pollen (Bock, 2007). Plastid transformation in soybean was first reported by Dufourmantel *et al.* (2004), and has subsequently been used to obtain high-level expression of Cry1Ab protein and 4-hydroxyphenylpyruvate dioxygenase, conferring strong insecticidal activity and herbicide tolerance, respectively (Dufourmantel *et al.*, 2005, 2007).

12.4 Selection

Irrespective of the gene transfer method used, the number of cells that stably integrate and express introduced transgenes is small. Therefore, selectable marker genes are needed to distinguish these cells efficiently from a large excess of untransformed cells. The classical antibiotic and herbicide resistance genes (Miki and McHugh, 2004) have been widely used for selection of genetically transformed legumes, notably the neomycin phosphotransferase gene (*nptII*, conferring resistance to antibiotics such as kanamycin, geneticin and paromomycin); the hygromycin phosphotransferase gene (*hpt*, conferring resistance to the antibiotic hygromycin B); the herbicide resistance genes *bar* and *pat* (encoding phosphinothricin acetyl transferase and conferring resistance to bialaphos, phosphinothricin or glufosinate ammonium); genes encoding herbicide-insensitive 5-enolpyruvylshikimate-3-phosphate synthase (EPSPS, providing resistance to the herbicide glyphosate); and genes encoding herbicide-insensitive acetolactate synthase (ALS, providing resistance to several classes of herbicides, including imidazolinones and sulfonylureas).

Production of chimeric transformants and escapes of non-transgenic materials that survive selection are problems that have often been described in legume transformation, for example in soybean (Hinchee *et al.*, 1988), pigeon pea (Thu *et al.*, 2003), mung bean (Muruganantham *et al.*, 2007; Saini and Jaiwal, 2007) and cowpea (Popelka *et al.*, 2006). As mentioned before, these phenomena are linked to the mode of regeneration, but also the choice of the selective agent and its concentration can be important. For example, many legume tissue cultures show a high tolerance for kanamycin; however, using geneticin instead of kanamycin for selection in conjunction with the *nptII* selectable marker prevented the escape of non-transgenic transformants in various cases (Zambre *et al.*, 2005; Popelka *et al.*, 2006). The herbicide imazapyr appears to be an efficient selection agent when using apical meristems as the target for particle bombardment-mediated transformation, and the mutant *als* as selectable marker gene (Rech *et al.*, 2008). This has been ascribed to the fact that imazapyr, in contrast to many other selective agents, is capable of translocating and concentrating in the apical meristem of the explant. In general, precise optimization of the concentration of the selective agent used is often necessary and may substantially improve the transformation efficiency (Zambre *et al.*, 2005).

In addition to antibiotic and herbicide resistance genes, other selectable markers have been used successfully for legume transformation, including phosphomannose isomerase (Patil *et al.*, 2009) and desensitized aspartate kinase, providing resistance to toxic levels of lysine and threonine (Tewari-Singh *et al.*, 2004). Although selectable marker genes are indispensable in nearly all current plant transformation protocols, they are of little use once transgenic plants have been obtained. On the contrary, their continued presence may pose certain problems and hence it may be desirable to remove the marker genes. Although the most commonly used selectable markers are safe from a human health and environmental perspective and have been approved by regulatory agencies (Ramessar *et al.*, 2007), a considerable proportion of the public remains concerned about the widespread use of antibiotic and herbicide resistance genes in particular. In addition, more scientifically grounded reasons may dictate marker gene removal; indeed, some marker genes or their regulatory elements may have pleiotropic effects (e.g. Miki *et al.*, 2009). Moreover, removal of the selectable marker gene from a transgenic plant allows for retransformation with the same selectable marker system; this strategy has, for example, been used to introduce consecutively two genes involved in fatty acid biosynthesis in soybean (Eckert *et al.*, 2006).

Two main strategies for marker gene removal are available: the co-transformation strategy and the use of site-specific recombinase systems such as Cre-*lox*, R-*RS* or FLT-*FRT* (Darbani *et al.*, 2007). In the co-transformation strategy, the marker gene and the gene(s) of interest are present on two different plasmids or two different T-DNAs. Plant cells selected for the presence of the marker gene are often found to be co-transformed with the unselected gene of interest. If the marker gene and the gene of interest are integrated at different loci, they can segregate independently and marker-free progeny can be obtained. This strategy has, for example, been adopted for the production of marker-free transgenic soybean (Sato *et al.*, 2004; Behrens *et al.*, 2007) and chickpea (Acharjee *et al.*, 2010) plants.

In the site-specific recombination strategy, the selectable marker gene is flanked by recombinase recognition sites in direct repeat, allowing excision of the marker gene after the transformation and selection procedures by the cognate site-specific recombinase enzyme. Various ways of providing the recombinase gene have been developed, including re-transformation with a recombinase construct, crossing with a recombinase-expressing plant or viral delivery of the recombinase (Darbani *et al.*, 2007). However, the most versatile systems are 'auto-excision' vectors that contain on a single vector the marker and the recombinase genes flanked by recombinase recognition sites, and the gene(s) of interest outside of the recognition sites. In auto-excision methods, the recombinase gene should not be expressed until after the selection stage. This can be achieved by placing the recombinase gene under control of a chemically inducible (Zuo *et al.*, 2001), heat-inducible (Zhang *et al.*, 2003) or developmentally regulated promoter (Verweire *et al.*, 2007). An example of the latter approach is the production of marker-free transgenic soybean plants using the *cre* recombinase gene under the control of an embryo-specific promoter (Li *et al.*, 2007). Recombinase-mediated excision of marker genes has the additional advantage that it may convert complex transgene loci to less complex or single-copy integrations (Verweire *et al.*, 2007). To date, marker removal strategies have only been used to a limited extent in legume transformation; however, it is likely that this situation will rapidly change, especially for transgenic plants destined for commercial release.

12.5 Applications of Genetic Transformation

Methods of reproducibly obtaining large numbers of transgenic plants are not yet available for the majority of the legume species, and further improvement of existing transformation protocols is certainly needed. This has, however, not impeded the application of transgene technology for legume crop improvement, as most clearly testified by the herbicide-tolerant transgenic soybean varieties that are commercially grown on more than 69 million hectares worldwide (James, 2009). Other transgenic food legumes have not yet been commercialized, although a large number of transgenic strategies and prototypes have been developed and are being tested in laboratory, greenhouse or field tests. Table 12.1 gives an overview of recent examples of the application of transformation technology for food legume improvement.

Table 12.1. Recent examples of food legumes improved through genetic engineering.

Legume species	Introduced gene(s)	Purpose	Reference
Arachis hypogaea	*cry1EC*	Insect resistance	Tiwari *et al.* (2008)
	Rice chitinase and alfalfa glucanase	Fungal resistance	Chenault *et al.* (2005)
	Oxalate oxidase	Fungal resistance	Livingstone *et al.* (2005)
	Coat protein of peanut stripe virus	Viral resistance	Higgins *et al.* (2004)
	AtDREB1A	Drought tolerance	Bhatnagar-Mathur *et al.* (2007)
	Ara h 2 silencing construct	Allergen elimination	Chu *et al.* (2008)
	Ara h 2 silencing construct	Allergen elimination	Dodo *et al.* (2008)
	Haemagglutinin gene of rinderpest virus	Oral vaccine	Khandelwal *et al.* (2003)
Cajanus cajan	Synthetic *cryIE-C gene*	Insect resistance	Surekha *et al.* (2005)
	cryIA(b)	Insect resistance	Sharma *et al.* (2006)
	Chitinase gene (*chit30*)	Fungal resistance	Kumar *et al.* (2004)
	Feedback-insensitive *DHDPS*	Nutritional quality improvement	Thu *et al.* (2007)
	Haemagglutinin-neuraminidase (*HN*) gene of Peste des petits ruminants virus (PPRV)	Oral vaccine	Prasad *et al.* (2004)
	Haemagglutinin gene (*H*) of rinderpest virus	Oral vaccine	Satyavathi *et al.* (2003)
Cicer arietinum	α-amylase inhibitor gene	Insect resistance	Sarmah *et al.* (2004)
	α-amylase inhibitor gene	Insect resistance	Ignacimuthu and Prakash (2006)
	cry1Ac	Insect resistance	Sanyal *et al.* (2005)
	cryIAc	Insect resistance	Indurker *et al.* (2007)
	Modified *cry2Aa*	Insect resistance	Acharjee *et al.* (2010)
	Agglutinin gene (*ASAL*)	Insect resistance	Chakraborti *et al.* (2009)
	Mutant *P5CS*	Drought tolerance	Bhatnagar-Mathur *et al.* (2009)
Glycine max	*cryIA(c)* and *Pinellia ternata* agglutinin (*pta*) genes	Insect resistance	Dang and Wei (2007)
	cry1Ab	Insect resistance	Dufourmantel *et al.* (2005)
	Coat protein of soybean mosaic virus	Virus resistance	Furutani *et al.* (2006)
	Inverted repeat of coat protein of soybean dwarf virus	Virus resistance	Tougou *et al.* (2006)
	Oxalate decarboxylase	Fungal resistance	Cunha *et al.* (2010)
	RNAi construct targeting cyst nematode *MSP* gene	Nematode resistance	Steeves *et al.* (2006)
	4-hydroxyphenylpyruvate dioxygenase	Weed control	Dufourmantel *et al.* (2007)
	Dicamba monooxygenase	Weed control	Behrens *et al.* (2007)
	Mutated anthranilate synthase	Nutritional quality improvement	Ishimoto *et al.* (2010)
	SLC1	Increased oil content	Rao and Hildebrand (2009)
	Ribozyme terminated fatty acid desaturase and thioesterase	Modified seed oil composition	Buhr *et al.* (2002)
	Borago officinalis fatty acid Δ^6 desaturase	Modified seed oil composition	Sato *et al.* (2004)

Continued

Table 12.1. Continued.

Legume species	Introduced gene(s)	Purpose	Reference
	Fatty acid Δ^6 desaturase and Δ^{15} desaturase	Modified seed oil composition	Eckert *et al.* (2006)
	Fatty acid Δ^6 desaturase, fatty acid elongase and fatty acid Δ^5 desaturase	Modified seed oil composition	Chen *et al.* (2006)
	Gly m Bd 30K	Allergen elimination	Herman *et al.* (2003)
	Heat-labile toxin (*LT*) B subunit	Oral vaccine	Moravec *et al.* (2007)
Lens culinaris	Mutant acetolactate synthase gene	Weed control	Gulati *et al.* (2002)
Phaseolus acutifolius	Arcelins	Insect resistance	Zambre *et al.* (2005)
P. vulgaris	Inverted repeat of *AC1* gene of bean golden mosaic virus	Virus resistance	Bonfim *et al.* (2007)
	bar gene	Weed control	Aragão *et al.* (2002)
	lea gene	Salt and drought tolerance	Liu *et al.* (2005)
Pisum sativum	Polygalacturonase-inhibiting protein (PGIP) and stilbene synthase	Fungal resistance	Richter *et al.* (2006)
	α-galactosidase	Nutritional quality improvement	Polowick *et al.* (2009)
	Amino acid permease *VfAAP1*	Increased protein content	Rolletschek *et al.* (2005)
	Rabbit haemorrhagic disease virus *VP60*	Oral vaccine	Mikschofsky *et al.* (2009)
Vicia faba	*SFA8* gene, *lysC*	Nutritional quality improvement	Hanafy *et al.* (2005)
V. narbonensis	Bacterial phosphoenolpyruvate carboxylase	Increased seed protein content	Rolletschek *et al.* (2004)
Vigna angularis	Mutated anthranilate synthase	Nutritional quality improvement	Hanafy *et al.* (2006)
	Δ^6-fatty-acid desaturase gene	Modified seed oil composition	Chen *et al.* (2005)
	α-amylase inhibitor	Insect resistance	Nishizawa *et al.* (2007)
V. radiata	α-amylase inhibitor	Insect resistance	Sonia *et al.* (2007)
V. unguiculata	*bar*	Weed control	Popelka *et al.* (2006)

Pests and diseases are major constraints for food legume production (Dita *et al.*, 2006) and have thus received a lot of attention from plant biotechnologists. Insect resistance is one of the main traits introduced in leguminous crops, mostly through expression of the *cry* genes of *Bacillus thuringiensis*, but also through lectin and α-amylase inhibitor genes (see Table 12.1). Knowledge of pathogen life cycles and plant–pathogen interactions led to development of strategies to counteract fungal and viral infections. Resistance against *Sclerotinia* has, for example, been obtained by the expression of oxalate-degrading enzymes (Livingstone *et al.*, 2005; Cunha *et al.*, 2010). Other strategies seeking fungal resistance are the expression of chitinases and glucanases (Kumar *et al.*, 2004; Chenault *et al.*, 2005). Virus resistance has been obtained in grain legumes through expression of viral proteins, mostly the coat protein. Although resistance is sometimes correlated to high-level accumulation of the viral protein (e.g. Furutani *et al.*, 2006), more often it appears to be due to induction of RNA silencing, i.e. the sequence-specific degradation of transgene derived and viral RNA (e.g. Higgins *et al.*, 2004). Thus, exploitation of the RNA-silencing mechanism, by the introduction of inverted repeats of viral sequences, appears to be the most promising technique towards obtaining virus resistance (Tougou *et al.*, 2006; Bonfim *et al.*, 2007). RNA silencing may perhaps also be exploited to obtain nematode resistance (Steeves *et al.*, 2006). In addition, herbicide-tolerant varieties have been developed for several legume crops, opening the way to new weed control strategies. Promising results with regard to abiotic stress tolerance, especially in improving drought tolerance, have already been obtained (see Table 12.1).

Nutritional quality improvement is another important area of research, mainly from the viewpoint of increasing the level of the essential amino acids methionine, lysine and tyrosine (e.g. Thu *et al.*, 2007; Ishimoto *et al.*, 2010). Also, fatty acid metabolism has been manipulated, which resulted for example in soybean with reduced levels of saturated and polyunsaturated fatty acids and a concomitant significant increase in those of oleic acid (Buhr *et al.*, 2002) and long-chain polyunsaturated fatty acids (Chen *et al.*, 2006; Eckert *et al.*, 2006). Furthermore, transgenic soybean and peanut plants have been bred from which the major seed allergens have been eliminated, resulting in a significant decrease in binding of IgEs from allergic patients to extracts of these transgenic seeds (Herman *et al.*, 2003; Chu *et al.*, 2008; Dodo *et al.*, 2008).

The field of 'molecular farming', i.e. the utilization of plant systems as a platform for the production of biopharmaceuticals such as vaccines and antibodies, has strongly progressed during the last decade (Ma *et al.*, 2005; Kaiser, 2008). Seeds of grain legumes are particularly interesting in this regard, because of their large size and their capacity to accumulate large amounts of protein in a stable form. The production of edible vaccines in the seeds of soybean, pea, pigeon pea and groundnut has been reported (Khandelwal *et al.*, 2003; Satyavathi *et al.*, 2003; Prasad *et al.*, 2004; Moravec *et al.*, 2007; Mikschofsky *et al.*, 2009). To achieve the required high expression levels of proteins in seeds, many factors need to be taken into account, including appropriate promoters, leader sequences and 3' non-coding elements, optimized codon usage, choice of the subcellular compartment, etc. (Streatfield, 2007; Boothe *et al.*, 2010). Vectors incorporating several of these factors have been developed to produce vaccines and other biologically active proteins in seeds of legumes and other dicotyledonous hosts (De Jaeger *et al.*, 2002).

12.6 Conclusions

The examples mentioned above clearly illustrate the wide range of applications of transgene technology in grain legume improvement. Obviously, our knowledge on legume biology will further increase through research on genetics and genomics of legume plants, the regulation of their metabolic pathways and their interactions with the environment, as provided through several legume projects (Harrison, 2000; VandenBosch and Stacey, 2003). This in turn will allow the development of novel biotechnological crop improvement strategies. To date, only herbicide-tolerant soybean is cultivated on a large scale, largely due to the heavy regulatory process accompanying commercialization of transgenic plants and the low public acceptance of this technology in some parts of the world. Nevertheless, many transgenic legume varieties are moving beyond laboratory experiments, examples being the successful field tests of bean golden mosaic virus-resistant beans (Aragão and Faria, 2009); protection of peas from pea weevil (Morton *et al.*, 2000); a new class of transgenic herbicide-tolerant soybean that showed complete resistance to the herbicide dicamba in field trials (Behrens *et al.*, 2007); nutritional quality improvement observed in feeding trials with tryptophan-enriched soybean seeds (Ishimoto *et al.*, 2010); and immune responses detected in cattle orally immunized with haemagglutinin protein of rinderpest virus expressed in transgenic peanut (Khandelwal *et al.*, 2003). We can therefore be confident of seeing new transgenic varieties coming on to the market in the years to come, albeit most probably at a slow pace.

References

Acharjee, S., Sarmah, B.K., Kumar, P.A., Olsen, K., Mahon, R., Moar, W.J. *et al.* (2010) Transgenic chickpeas (*Cicer arietinum* L.) expressing a sequence-modified *cry2Aa* gene. *Plant Science* 178, 333–339.

Aragão, F.J.L. and Faria, J.C. (2009) First transgenic geminivirus-resistant plant in the field. *Nature Biotechnology* 27, 1086–1088.

Aragão, F.J.L., Vianna, G.R., Albino, M.M.C. and Rech, E.L. (2002) Transgenic dry bean tolerant to the herbicide glufosinate ammonium. *Crop Science* 42, 1298–1302.

Bailey, M. A., Boerma, H.R. and Parrott, W. A. (1993) Genotype-specific optimization of plant regeneration from somatic embryos of soybean. *Plant Science* 93, 117–120.

Behrens, M.R., Mutlu, N., Chakraborty, S., Dumitru, R., Jiang, W.Z., LaVallee, B.J. *et al.* (2007) Dicamba resistance: enlarging and preserving biotechnology-based weed management strategies. *Science* 316, 1185–1188.

Bhatnagar-Mathur, P., Devi, M.J., Reddy, D.S., Lavanya, M., Vadez, V., Serraj, R. *et al.* (2007) Stress-inducible expression of *At* DREB1A in transgenic peanut (*Arachis hypogaea* L.) increases transpiration efficiency under water-limiting conditions. *Plant Cell Reports* 26, 2071–2082.

Bhatnagar-Mathur, P., Vadez, V., Devi, M.J., Lavanya, M., Vani, G. and Sharma, K.K. (2009) Genetic engineering of chickpea (*Cicer arietinum* L.) with the P5CSF129A gene for osmoregulation with implications on drought tolerance. *Molecular Breeding* 23, 591–606.

Bock, R. (2007) Plastid biotechnology: prospects for herbicide and insect resistance, metabolic engineering and molecular farming. *Current Opinion in Biotechnology* 18, 100–106.

Bolton, G.W., Nester, E.W. and Gordon, M.P. (1986) Plant phenolic compounds induce expression of the *Agrobacterium tumefaciens* loci needed for virulence. *Science* 232, 983–985.

Bonfim, K., Faria, J.C., Nogueira, E.O.P.L., Mendes, E.A. and Aragão, F.J.L. (2007) RNAi-mediated resistance to *bean golden mosaic virus* in genetically engineered common bean (*Phaseolus vulgaris*). *Molecular Plant Microbe Interactions* 20, 717–726.

Boothe, J., Nykiforuk, C., Shen, Y., Zaplachinski, S., Szarka, S., Kuhlman, P. *et al.* (2010) Seed-based expression systems for plant molecular farming. *Plant Biotechnology Journal* 8, 588–606.

Buhr, T., Sato, S., Ebrahim, F., Xing, A., Zhou, Y., Mathiesen, M. *et al.* (2002) Ribozyme termination of RNA transcripts down-regulate seed fatty acid genes in transgenic soybean. *The Plant Journal* 30, 155–163.

Chabaud, M., Larsonneau, C., Marmouget, C. and Huguet, T. (1996) Transformation of barrel medic (*Medicago truncatula* Gaertn.) by *Agrobacterium tumefaciens* and regeneration *via* somatic embryogenesis of transgenic plants with the *MtENOD12* nodulin promoter fused to the *gus* reporter gene. *Plant Cell Reports* 15, 305–310.

Chakraborti, D., Sarkar, A., Mondal, H.A. and Das, S. (2009) Tissue specific expression of potent insecticidal, *Allium sativum* leaf agglutinin (ASAL) in important pulse crop, chickpea (*Cicer arietinum* L.) to resist the phloem feeding *Aphis craccivora*. *Transgenic Research* 18, 529–544.

Chen, R., Tsuda, S., Matsui, K., Fukuchi-Mizutani, M., Ochiai, M., Shimizu, S. *et al.* (2005) Production of γ-linolenic acid in *Lotus japonicus* and *Vigna angularis* by expression of the Δ6-fatty-acid desaturase gene isolated from *Mortierella alpine*. *Plant Science* 169, 599–605.

Chen, R., Matsui, K., Ogawa, M., Oe, M., Ochiai, M., Kawashima, H. *et al.* (2006) Expression of Δ6, Δ5 desaturase and GLELO elongase genes from *Mortierella alpina* for production of arachidonic acid in soybean [*Glycine max* (L.) Merrill] seeds. *Plant Science* 170, 399–406.

Chenault, K.D., Melouk, H.A. and Payton, M.E. (2005) Field reaction to Sclerotinia blight among transgenic peanut lines containing antifungal genes. *Crop Science* 45, 511–515.

Chu, Y., Faustinelli, P., Ramos, M.L., Hajduch, M., Stevenson, S., Thelen, J.J. *et al.* (2008) Reduction of IgE binding and nonpromotion of *Aspergillus flavus* fungal growth by simultaneously silencing Ara h 2 and Ara h 6 in peanut. *Journal of Agricultural and Food Chemistry* 56, 11225–11233.

Cunha, W.G., Tinoco, M.L.P., Pancotti, H.L., Ribeiro, R.E. and Aragão, F.J.L. (2010) High resistance to *Sclerotinia sclerotiorum* in transgenic soybean plants transformed to express an oxalate decarboxylase gene. *Plant Pathology* 59, 654–660.

Dang, W. and Wei, Z. (2007) An optimized *Agrobacterium*-mediated transformation for soybean for expression of binary insect resistance genes. *Plant Science* 173, 381–389.

Darbani, B., Eimanifar, A., Stewart, C.N. Jr and Camargo, W.N. (2007) Methods to produce marker-free transgenic plants. *Biotechnology Journal* 2, 83–90.

De Clercq, J., Zambre, M., Van Montagu, M., Dillen, W. and Angenon, G. (2002) An optimized *Agrobacterium*-mediated transformation procedure for *Phaseolus acutifolius* A. Gray. *Plant Cell Reports* 21, 333–340.

De Jaeger, G., Scheffer, S., Jacobs, A., Zambre, M., Zobell, O., Goossens, A. *et al.* (2002) Boosting heterologous protein production in transgenic dicotyledonous seeds using *Phaseolus vulgaris* regulatory sequences. *Nature Biotechnology* 20, 1265–1268.

Dillen, W., De Clercq, J., Kapila, J., Zambre, M., Van Montagu, M. and Angenon, G. (1997) The effect of temperature on *Agrobacterium tumefaciens*-mediated gene transfer to plants. *The Plant Journal* 12, 1459–1463.

Dita, M.A., Rispail, N., Prats, E., Rubiales, D. and Singh, K.B. (2006) Biotechnology approaches to overcome biotic and abiotic stress constraints in legumes. *Euphytica* 147, 1–24.

Dodo, H.W., Konan, K.N., Chen, F.C., Egnin, M. and Viquez, O.M. (2008) Alleviating peanut allergy using genetic engineering: the silencing of the immunodominant allergen Ara h 2 leads to its significant reduction and a decrease in peanut allergenicity. *Plant Biotechnology Journal* 6, 135–145.

Dufourmantel, N., Pelissier, B., Garçon, F., Peltier, G., Ferullo, J.M. and Tissot, G. (2004) Generation of fertile transplastomic soybean. *Plant Molecular Biology* 55, 479–489.

Dufourmantel, N., Tissot, G., Goutorbe, F., Garçon, F., Muhr, C., Jansens, S. *et al.* (2005) Generation and analysis of soybean plastid transformants expressing *Bacillus thuringiensis* Cry1Ab protoxin. *Plant Molecular Biology* 58, 659–668.

Dufourmantel, N., Dubald, M., Matringe, M., Canard, H., Garçon, F., Job, C. *et al.* (2007) Generation and characterization of soybean and marker-free tobacco plastid transformants over-expressing a bacterial 4-hydroxyphenylpyruvate dioxygenase which provides strong herbicide tolerance. *Plant Biotechnology Journal* 5, 118–133.

Eapen, S. (2008) Advances in development of transgenic pulse crops. *Biotechnology Advances* 26, 162–168.

Eckert, H., La Vallee, B., Schweiger, B.J., Kinney, A.J., Cahoon, E.B. and Clemente, T. (2006) Co-expression of the borage Δ^6 desaturase and the Arabidopsis Δ^{15} desaturase results in high accumulation of stearidonic acid in the seeds of transgenic soybean. *Planta* 224, 1050–1057.

Furutani, N., Hidaka, S., Kosaka, Y., Shizukawa, Y. and Kanematsu, S. (2006) Coat protein gene-mediated resistance to soybean mosaic virus in transgenic soybean. *Breeding Science* 56, 119–124.

George, L. and Eapen, S. (1994) Organogenesis and embryogenesis from diverse explants in pigeonpea (*Cajanus cajan* L.). *Plant Cell Reports* 13, 417–420.

Grant, J.E., Cooper, P.A., McAra, A.E. and Frew, T.J. (1995) Transformation of peas (*Pisum sativum* L.) using immature cotyledons. *Plant Cell Reports* 15, 254–258.

Grant, J.E., Thomson, L.M.J., Pither-Joyce, M.D., Dale, T.M. and Cooper P.A. (2003) Influence of *Agrobacterium tumefaciens* strain on the production of transgenic peas (*Pisum sativum* L.). *Plant Cell Reports* 21, 1207–1210.

Griga, M. (1998) Direct somatic embryogenesis from shoot apical meristems of pea, and thidiazuron-induced high conversion rate of somatic embryos. *Biologia Plantarum* 41, 481–495.

Gulati, A., Schryer, P. and McHughen, A. (2002) Production of fertile transgenic lentil (*Lens culinaris* Medik) plants using particle bombardment. *In Vitro Cellular & Developmental Biology – Plant* 38, 316–324.

Hanafy, M., Pickardt, T., Kiesecker, H. and Jacobsen, H.J. (2005) *Agrobacterium*-mediated transformation of faba bean (*Vicia faba* L) using embryo axes. *Euphytica* 142, 227–326.

Hanafy, M.S., Rahman, S.M., Khalafalla, M.M., El-Shemy, H.A., Nakamoto, Y., Ishimoto, M. *et al.* (2006) Accumulation of free tryptophan in azuki bean (*Vigna angularis*) induced by expression of a gene (OASA1D) for a modified α-subunit of rice anthranilate synthase. *Plant Science* 171, 670–676.

Harrison, M.J. (2000) Molecular genetics of model legumes. *Trends in Plant Science* 5, 414–415.

Hellens, R., Mullineaux, P. and Klee, H. (2000) A guide to *Agrobacterium* binary Ti vectors. *Trends in Plant Science* 5, 446–451.

Herman, E.M., Helm, R.M., Jung, R. and Kinney, A.J. (2003) Genetic modification removes an immunodominant allergen from soybean. *Plant Physiology* 132, 36–43.

Higgins, C.M., Hall, R.M., Mitter, N., Cruickshank, A. and Dietzgen, R.G. (2004) Peanut stripe potyvirus resistance in peanut (*Arachis hypogaea* L.) plants carrying viral coat protein gene sequences. *Transgenic Research* 13, 59–67.

Hinchee, M.A.W., Connor-Ward, D.V., Newell, C.A., McDonnell, R.E., Sato, S.J., Gasser, C.S. *et al.* (1988) Production of transgenic soybean plants using *Agrobacterium*-mediated DNA transfer. *Bio/Technology* 6, 915–921.

Hood, E.E., Helmer, G.L., Fraley, R.T. and Chilton, M.D. (1986) The hypervirulence of *Agrobacterium tumefaciens* A281 is encoded in a region of pTiBo542 outside of T-DNA. *Journal of Bacteriology* 168, 1291–1301.

Hood, E.E., Fraley, R.T. and Chilton, M.D. (1987) Virulence of *Agrobacterium tumefaciens* strain A281 on legumes. *Plant Physiology* 83, 529–534.
Husnain, T., Malik T., Riazuddin, S. and Gordon, M.P. (1997) Studies on the expression of marker genes in chickpea. *Plant Cell, Tissue and Organ Culture* 49, 7–16.
Ignacimuthu, S. and Prakash, S. (2006) *Agrobacterium*-mediated transformation of chickpea with α-amylase inhibitor gene for insect resistance. *Journal of Biosciences* 31, 339–345.
Ikea, J., Ingelbrecht, I., Uwaifo, A. and Thottappilly, G. (2003) Stable gene transformation in cowpea (*Vigna unguiculata* L. Walp.) using particle gun method. *African Journal of Biotechnology* 2, 211–218.
Indurker, S., Misra, H.S. and Eapen, S. (2007) Genetic transformation of chickpea (*Cicer arietinum* L.) with insecticidal crystal protein gene using particle gun bombardment. *Plant Cell Reports* 26, 755–763.
Ishimoto, M., Rahman, S.M., Hanafy, M.S., Khalafalla, M.M., El-Shemy, H.A., Nakamoto, Y. *et al.* (2010) Evaluation of amino acid content and nutritional quality of transgenic soybean seeds with high-level tryptophan accumulation. *Molecular Breeding* 25, 313–326.
Ivo, N.L., Nascimento, C.P., Vieira, L.S., Campos, F.A.P. and Aragão, F.J.L. (2008) Biolistic-mediated genetic transformation of cowpea (*Vigna unguiculata*) and stable Mendelian inheritance of transgenes. *Plant Cell Reports* 27, 1475–1483.
Jaiwal, P.K., Kumari, R., Ignacimuthu, S., Potrykus, I. and Sautter, C. (2001) *Agrobacterium tumefaciens*-mediated genetic transformation of mungbean (*Vigna radiata* L. Wilczek) – a recalcitrant grain legume. *Plant Science* 161, 239–247.
James, C. (2009) Global status of commercialized biotech/GM crops: 2009. *ISAAA Brief* No. 41, ISAAA, Ithaca, New York.
Kaiser, J. (2008) Is the drought over for pharming? *Science* 320, 473–475.
Karami, O., Esna-Ashari, M., Kurdistani, G.K. and Aghavaisi, B. (2009) *Agrobacterium*-mediated genetic transformation of plants: the role of host. *Biologia Plantarum* 53, 201–212.
Khandelwal, A., Lakshmi Sita, G. and Shaila, M.S (2003) Oral immunization of cattle with hemagglutinin protein of rinderpest virus expressed in transgenic peanut induces specific immune responses. *Vaccine* 21, 3282–3289.
Kita,Y., Nishizawa, K., Takahashi, M., Kitayama, M. and Ishimoto, M. (2007) Genetic improvement of the somatic embryogenesis and regeneration in soybean and transformation of the improved breeding lines. *Plant Cell Reports* 26, 439–447.
Ko, T-S., Lee, S., Krasnyanski, S. and Korban, S.S. (2003) Two critical factors are required for efficient transformation of multiple soybean cultivars: *Agrobacterium* strain and orientation of immature cotyledonary explant. *Theoretical and Applied Genetics* 107, 439–447.
Köhler, F., Golz, C., Eapen, S., Kohn, H. and Schieder, O. (1987) Stable transformation of moth bean *Vigna aconitifolia* via direct gene transfer. *Plant Cell Reports* 6, 313–317.
Kumar, S.M., Kumar, B.K., Sharma, K.K. and Devi, P. (2004) Genetic transformation of pigeonpea with rice chitinase gene. *Plant Breeding* 123, 485–489.
Kumar,V.D., Kirti, P.B., Sachan, J.K.S. and Chopra, V.L. (1994) Plant regeneration via somatic embryogenesis in chickpea (*Cicer arietinum* L.). *Plant Cell Reports* 13, 468–472.
Lacroix, B., Tzfira, T., Vainstein, A. and Citovsky, V. (2006) A case of promiscuity: *Agrobacterium*'s endless hunt for new partners. *Trends in Genetics* 22, 29–37.
Li, Z., Xing, A., Moon, B.P., Burgoyne, S.A., Guida, A.D., Liang, H. *et al.* (2007) A Cre/*loxP*-mediated self-activating gene excision system to produce marker gene free transgenic soybean plants. *Plant Molecular Biology* 65, 329–341.
Li, Z., Xing, A., Moon, B.P., McCardell, R.P., Mills, K. and Falco, S.C. (2009) Site-specific integration of transgenes in soybean via recombinase-mediated DNA cassette exchange. *Plant Physiology* 151, 1087–1095.
Liu, Z., Park, B.J., Kanno, A. and Kameya, T. (2005) The novel use of a combination of sonication and vacuum infiltration in *Agrobacterium*-mediated transformation of kidney bean (*Phaseolus vulgaris* L.) with *lea* gene. *Molecular Breeding* 16, 189–197.
Livingstone, D.M., Hampton, J.L., Phipps, P.M. and Grabau, E.A. (2005) Enhancing resistance to *Sclerotinia minor* in peanut by expressing a barley oxalate oxidase gene. *Plant Physiology* 137, 1354–1362.
Ma, J.K.C., Chikwamba, R., Sparrow, P., Fischer, R., Mahoney, R. and Twyman, R.M. (2005) Plant-derived pharmaceuticals – the way forward. *Trends in Plant Science* 10, 580–585.
McCabe, D.E., Swain, W.F., Martinell, B.J. and Christou, P. (1988) Stable transformation of soybean (*Glycine max*) by particle acceleration. *Bio/Technology* 6, 923–926.
Miki, B. and McHugh, S. (2004) Selectable marker genes in transgenic plants: applications, alternatives and biosafety. *Journal of Biotechnology* 107, 193–232.

Miki, B., Abdeen, A., Manabe, Y. and MacDonald, P. (2009) Selectable marker genes and unintended changes to the plant transcriptome. *Plant Biotechnology Journal* 7, 211–218.

Mikschofsky, H., Schirrmeier, H., Keil, G.M., Lange, B., Polowick, P.L., Keller, W. *et al.* (2009) Pea-derived vaccines demonstrate high immunogenicity and protection in rabbits against rabbit haemorrhagic disease virus. *Plant Biotechnology Journal* 7, 537–549.

Mohamed, M.F., Read, P.E. and Coyne, D.P. (1992) Dark preconditioning, CPPU & thidiazuron promote shoot organogenesis on seedling node explants of common and faba beans. *Journal of the American Society for Horticultural Science* 117, 668–672.

Mohamed, S.V., Sung, J., Jeng, T. and Wang, C. (2006) Organogenesis of *Phaseolus angularis* L.: high efficiency of adventitious shoot regeneration from etiolated seedlings in the presence of N6-benzylaminopurine and thidiazuron. *Plant Cell, Tissue and Organ Culture* 86, 187–199.

Mohan, M.L. and Krishnamurthy, K.V. (2002) Somatic embryogenesis and plant regeneration in pigeonpea. *Biologia Plantarum* 45, 19–25.

Moravec, T., Schmidt, M.A., Eliot, M., Herman, E.M. and Woodford-Thomas, T. (2007) Production of *Escherichia coli* heat labile toxin (LT) B subunit in soybean seed and analysis of its immunogenicity as an oral vaccine. *Vaccine* 25, 1647–1657.

Morton, R.L., Schroeder, H.E., Bateman, K.S., Chrispeels, M.J., Armstrong, E. and Higgins, T.J.V. (2000) Bean α-amylase inhibitor 1 in transgenic peas (*Pisum sativum*) provides complete protection from pea weevil (*Bruchus pisorum*) under field conditions. *Proceedings of the National Academy of Sciences of the U.S.A.* 97, 3820–3825.

Murthy, B.N.S., Victor, J., Singh, R.P., Fletcher, R.A. and Saxena, P.K. (1996) *In vitro* regeneration of chickpea (*Cicer arientinum* L.): stimulation of direct organogenesis and somatic embryogenesis by thidiazuron. *Plant Growth Regulation* 19, 233–240.

Murthy, B.N.S., Murch, S.J. and Saxena, P.K. (1998) Thidiazuron: a potent regulator of *in vitro* morphogenesis. *In Vitro Cellular and Developmental Biology – Plant* 34, 267–275.

Muruganantham, M., Amutha, S., Selvaraj, N., Vengadesan, G. and Ganapathi, A. (2007) Efficient *Agrobacterium*-mediated transformation of *Vigna mungo* using immature cotyledonary-node explants and phosphinothricin as the selection agent. *In Vitro Cellular and Developmental Biology – Plant* 43, 550–557.

Nadolska-Orczyk, A. and Orczyk, W. (2000) Study of the factors influencing *Agrobacterium*-mediated transformation of pea (*Pisum sativum* L.). *Molecular Breeding* 6, 185–194.

Nishizawa, K., Teraishi, M., Utsumi, S. and Ishimoto, M. (2007) Assessment of the importance of α- amylase inhibitor-2 in bruchid resistance of wild common bean. *Theoretical and Applied Genetics* 114, 755–764.

Ozias-Akins, P., Schnall, J.A., Anderson, W.F., Singsit, C., Clemente, T.E., Adang, M.J. *et al.* (1993) Regeneration of transgenic peanut plants from stably transformed embryogenic callus. *Plant Science* 93, 185–194.

Pathak, M.R. and Hamzah, R.Y. (2008) An effective method of sonication-assisted *Agrobacterium*-mediated transformation of chickpeas. *Plant Cell, Tissue and Organ Culture* 93, 65–71.

Patil, G., Deokar, A., Jain, P.K., Thengane, R.J. and Srinivasan, R. (2009) Development of a phosphomannose isomerase-based *Agrobacterium*-mediated transformation system for chickpea (*Cicer arietinum* L.). *Plant Cell Reports* 28, 1669–1676.

Polowick, P.L., Baliski, D.S. and Mahon, J.D. (2004) *Agrobacterium tumefaciens*-mediated transformation of chickpea (*Cicer arietinum* L.): gene integration, expression and inheritance. *Plant Cell Reports* 23, 485–491.

Polowick, P.L., Baliski, D.S., Bock, C., Ray, H. and Georges, F. (2009) Over-expression of α-galactosidase in pea seeds to reduce raffinose oligosaccharide content. *Botany* 87, 526–532.

Popelka, J.C., Terryn, N. and Higgins, T.J.V. (2004) Gene technology for grain legumes: can it contribute to the food challenge in developing countries? *Plant Science* 167, 195–206.

Popelka, J.C., Gollasch, S., Moore, A., Molvig, L. and Higgins, T.J.V. (2006) Genetic transformation of cowpea (*Vigna unguiculata* L.) and stable transmission of the transgenes to progeny. *Plant Cell Reports* 25, 304–312.

Prasad,V., Satyavathi, V.V., Valli, S.K.M., Khandelwal, A., Shaila, M.S. and Lakshmi Sita, G. (2004) Expression of biologically active hemagglutinin-neuraminidase protein of *Peste des petits ruminants* virus in transgenic pigeonpea [*Cajanus cajan* (L) Millsp.]. *Plant Science* 166, 199–205.

Ramessar, K., Peremarti, A., Gomez-Galera, S., Naqvi, S., Moralejo, M., Munoz, P. *et al.* (2007) Biosafety and risk assessment framework for selectable marker genes in transgenic crop plants: a case of the science not supporting the politics. *Transgenic Research* 16, 261–280.

Rao, S.S. and Hildebrand, D. (2009) Changes in oil content of transgenic soybeans expressing the yeast *SLC1* gene. *Lipids* 44, 945–951.

Rech, E.L., Vianna, G.R. and Aragão, F.J.L. (2008) High-efficiency transformation by biolistics of soybean, common bean and cotton transgenic plants. *Nature Protocols* 3, 410–418.

Richter, A., de Kathen, A., de Lorenzo, G., Briviba, K., Hain, R., Ramsay, G. *et al.* (2006) Transgenic peas (*Pisum sativum*) expressing polygalacturonase inhibiting protein from raspberry (*Rubus idaeus*) and stilbene synthase from grape (*Vitis vinifera*). *Plant Cell Reports* 25, 1166–1173.

Rolletschek, H., Borisjuk, L., Radchuk, R., Miranda, M., Heim, U., Wobus, U. *et al.* (2004) Seed-specific expression of a bacterial phosphoenolpyruvate carboxylase in *Vicia narbonensis* increases protein content and improves carbon economy. *Plant Biotechnology Journal* 2, 211–219.

Rolletschek, H., Hosein, F., Miranda, M., Heim, U., Götz, K.P., Schlereth, A. *et al.* (2005) Ectopic expression of an amino acid transporter (VFAAP1) in seeds of *Vicia narbonensis* and pea increases storage proteins. *Plant Physiology* 137, 1236–1249.

Saini, R. and Jaiwal, P.K. (2007) *Agrobacterium tumefaciens*-mediated transformation of blackgram: an assessment of factors influencing the efficiency of *uidA* gene transfer. *Biologia Plantarum* 51, 69–74.

Santarem, E.R., Trick, H.N., Essig, J.S. and Finer, J.J. (1998) Sonication-assisted *Agrobacterium*-mediated transformation of soybean immature cotyledons: optimization of transient expression. *Plant Cell Reports* 17, 752–759.

Sanyal, I., Singh, A.K., Kaushik, M. and Amla, D.V. (2005) *Agrobacterium*-mediated transformation of chickpea (*Cicer arietinum* L.) with *Bacillus thuringiensis cry1Ac* gene for resistance against pod borer insect *Helicoverpa armigera*. *Plant Science* 168, 1135–1146.

Sarmah, B.K., Moore, A., Tate, W., Molvig, L., Morton, R.L., Rees, R.P. *et al.* (2004) Transgenic chickpea seeds expressing high levels of a bean α-amylase inhibitor. *Molecular Breeding* 14, 73–82.

Sato, S., Xing, A., Ye, X., Schweiger, B., Kinney, A., Graef, G. *et al.* (2004) Production of γ-linolenic acid and stearidonic acid in seeds of marker-free transgenic soybean. *Crop Science* 44, 646–652.

Satyavathi, V., Prasad, V., Khandelwal, A., Shaila, M.S. and Lakshmi Sita, G. (2003) Expression of hemagglutinin protein of Rinderpest virus in transgenic pigeonpea (*Cajanus cajan* (L.) Millsp.) plants. *Plant Cell Reports* 21, 651–658.

Schroeder, H.E., Scholtz, A.H., Wardley-Richardson, T., Spencer, D. and Higgins, T.J.V. (1993) Transformation and regeneration of two cultivars of pea (*Pisum sativum* L.). *Plant Physiology* 101, 751–757.

Sharma, K.K., Lavanya, M. and Anjaiah, V. (2006) *Agrobacterium*-mediated production of transgenic pigeonpea (*Cajanus cajan* L Millsp) expressing the synthetic Bt cryIAb gene. *In Vitro Cellular and Developmental Biology – Plant* 42, 165–173.

Singh, N.D., Sahoo, L., Sarin, N.B. and Jaiwal, P.K. (2003) The effect of TDZ on organogenesis and somatic embryogenesis in pigeonpea (*Cajanus cajan* L. Millsp). *Plant Science* 164, 341–347.

Solleti, S.K., Bakshi, S. and Sahoo, L. (2008) Additional virulence genes in conjunction with efficient selection scheme, and compatible culture regime enhance recovery of stable transgenic plants in cowpea via *Agrobacterium tumefaciens*-mediated transformation. *Journal of Biotechnology* 135, 97–104.

Somers, D.A., Samac, D.A. and Olhoft, P.M. (2003) Recent advances in legume transformation. *Plant Physiology* 131, 892–899.

Sonia, R., Saini, R., Singh, R.P. and Jaiwal, P.K. (2007) *Agrobacterium tumefaciens* mediated transfer of *Phaseolus vulgaris* α-amylase inhibitor-1 gene into mungbean *Vigna radiata* (L.) Wilczek using *bar* as selectable marker. *Plant Cell Reports* 26, 187–198.

Steeves, R.M., Todd, T.C., Essig, J.S. and Trick, H.N. (2006) Transgenic soybeans expressing siRNAs specific to a major sperm protein gene suppress *Heterodera glycines* reproduction. *Functional Plant Biology* 33, 991–999.

Streatfield, S.J. (2007) Approaches to achieve high-level heterologous protein production in plants. *Plant Biotechnology Journal* 5, 2–15.

Sunilkumar, G. and Rathore, K.S. (2001) Transgenic cotton: factors influencing *Agrobacterium*-mediated transformation and regeneration. *Molecular Breeding* 8, 37–52.

Surekha, C., Beena, M.R., Arundhati, A., Singh, P.K., Tuli, R., Dutta-Gupta, A. *et al.* (2005) *Agrobacterium*-mediated genetic transformation of pigeon pea (*Cajanus cajan* (L.) Millsp.) using embryonal segments and development of transgenic plants for resistance against *Spodoptera*. *Plant Science* 169, 1074–1080.

Svetleva, D., Velcheva, M. and Bhowmik, G. (2003) Biotechnology as a useful tool in common bean (*Phaseolus vulgaris* L.) improvement. *Euphytica* 131, 189–200.

Tewari-Singh, N., Sen, J., Kiesecker, H., Reddy, V.S., Jacobsen, H.J. and Guha-Mukherjee, S. (2004) Use of a herbicide or lysine plus threonine for non-antibiotic selection of transgenic chickpea. *Plant Cell Reports* 22, 576–583.

Thu, T.T., Mai, T.T.X., Dewaele, E., Farsi, S., Tadesse, Y., Angenon, G. *et al.* (2003) *In vitro* regeneration and transformation of pigeonpea [*Cajanus cajan* (L.) Millsp]. *Molecular Breeding* 11, 159–168.

Thu, T.T., Dewaele, E., Trung, L.Q., Claeys, M., Jacobs, M. and Angenon, G. (2007) Increasing lysine levels in pigeonpea (*Cajanus cajan* (L.) Millsp) seeds through genetic engineering. *Plant Cell Tissue and Organ Culture* 91, 135–143.

Tiwari, S., Mishra, D.K., Singh, A., Singh, P.K. and Tuli, R. (2008) Expression of a synthetic *cry1EC* gene for resistance against *Spodoptera litura* in transgenic peanut (*Arachis hypogaea* L.). *Plant Cell Reports* 27, 1017–1025.

Torisky, R.S., Kovacs, L., Avdiushko, S., Newman, J.D., Hunt, A.G. and Collins, G.B. (1997) Development of a binary vector system for plant transformation based on the supervirulent *Agrobacterium tumefaciens* strain Chry5. *Plant Cell Reports* 17, 102–108.

Tougou, M., Furutani, N., Yamagishi, N., Shizukawa, Y., Takahata, Y. and Hidaka, S. (2006) Development of resistant transgenic soybeans with inverted repeat-coat protein genes of soybean dwarf virus. *Plant Cell Reports* 25, 1213–1218.

Travella, S., Ross, S.M., Harden, J., Everett, C., Snape, J.W. and Harwood, W.A. (2005) A comparison of transgenic barley lines produced by particle bombardment and *Agrobacterium*-mediated techniques. *Plant Cell Reports* 23, 780–789.

Trinh, T.H., Ratet, P., Kondorosi, E., Durand, P., Kamaté, K., Bauer, P. *et al.* (1998) Rapid and efficient transformation of diploid *Medicago truncatula* and *Medicago sativa* ssp. *falcata* lines improved in somatic embryogenesis. *Plant Cell Reports* 17, 345–355.

Tzfira, T. and Citovsky, V. (2006) *Agrobacterium*-mediated genetic transformation of plants: biology and biotechnology. *Current Opinion in Biotechnology* 17, 147–154.

VandenBosch, K.A. and Stacey, G. (2003) Summaries of legume genomics projects from around the globe. Community resources for crops and models. *Plant Physiology* 131, 840–865.

Varshney, R.K., Close, T.J., Singh, N.K., Hoisington, D.A. and Cook, D.R. (2009) Orphan legume crops enter the genomics era! *Current Opinion in Plant Biology* 12, 202–210.

Verweire, D., Verleyen, K., De Buck, S., Claeys, M. and Angenon, G. (2007) Marker-free transgenic plants through genetically programmed auto-excision. *Plant Physiology* 145, 1220–1231.

Wigdorovitz, A., Carrillo, C., Dus Santos, M.J., Trono, K., Peralta, A., Gomez, M.C. *et al.* (1999) Induction of a protective antibody response to foot and mouth disease virus in mice following oral or parenteral immunization with alfalfa transgenic plants expressing the viral structural protein VP1. *Virology* 255, 347–353.

Wright, M.S., Koehler, S.M., Hinchee, M.A. and Carnes, M.G. (1986) Plant regeneration by organogenesis in *Glycine max*. *Plant Cell Reports* 5, 150–154.

Yang, G., Lee, Y.H., Jiang, Y., Kumpatla, S.P. and Hall, T.C. (2005) Organization, not duplication, triggers silencing in a complex transgene locus in rice. *Plant Molecular Biology* 58, 351–366.

Zambre, M.A., De Clercq, J., Vranová, E., Van Montagu, M., Angenon, G. and Dillen, W. (1998) Plant regeneration from embryo-derived callus in *Phaseolus vulgaris L.* (common bean) and *P. acutifolius* A. Gray (tepary bean). *Plant Cell Reports* 17, 626–630.

Zambre, M., Terryn, N., De Clercq, J., De Buck, S., Dillen, W., Van Montagu, M. *et al.* (2003) Light strongly promotes gene transfer from *Agrobacterium tumefaciens* to plant cells. *Planta* 216, 580–586.

Zambre, M., Goossens, A., Cardona, C., Van Montagu, M., Terryn, N. and Angenon, G. (2005) A reproducible genetic transformation system for cultivated *Phaseolus acutifolius* (tepary bean) and its use to assess the role of arcelins in resistance to the Mexican bean weevil. *Theoretical and Applied Genetics* 110, 914–924.

Zernova, O.V., Lygin, A.V., Widholm, J.M. and Lozovaya, V.V. (2009) Modification of isoflavones in soybean seeds via expression of multiple phenolic biosynthetic genes. *Plant Physiology and Biochemistry* 47, 769–777.

Zhang, W., Subbarao, S., Addae, P., Shen, A., Armstrong, C., Peschke, V. *et al.* (2003) Cre/*lox*-mediated marker gene excision in transgenic maize (*Zea mays* L.) plants. *Theoretical and Applied Genetics* 107, 1157–1168.

Zuo, J., Niu, Q.W., Moller, S.G. and Chua, N.H. (2001) Chemical-regulated, site-specific DNA excision in transgenic plants. *Nature Biotechnology* 19, 157–161.

Zupan, J., Muth, T.R., Draper, O. and Zambryski, P. (2000) The transfer of DNA from *Agrobacterium tumefaciens* into plants: a feast of fundamental insights. *The Plant Journal* 23, 11–28.

13 Male Sterility and Hybrid Production Technology

R.G. Palmer, J. Gai, V.A. Dalvi and M.J. Suso

13.1 Introduction

Sexual reproduction in angiosperms is a complex process that includes a portion of sporophytic (vegetative) generation and all of the gametophytic (sexual) generation. For normal sexual reproduction, coordination of both female and male reproductive ontogenies must occur. An abnormality anywhere in this process may lead to sterility. Classification of sterility into various categories has been reported (Gottschalk and Kaul, 1974; Johns *et al.*, 1981; Horner and Palmer, 1995). This chapter will focus on genetic male sterility: nuclear and cytoplasmically inherited mechanisms that have been used to produce hybrids in pulses.

Hybrid vigour or heterosis is the superior performance of the heterozygous hybrid. High-parent heterosis is the superior performance of the hybrid over both parents; while mid-parent heterosis is the superior performance of the hybrid over the mid-parent value of the two parents. Heterosis has been exploited in many cross- and often cross-pollinated crops, but the flower structure and small size of flowers of many leguminous crops make manual cross-pollination in the production of commercial quantities of hybrid seed not economically feasible. There are five components that are crucial for the successful development of hybrid food legumes (Palmer *et al.*, 2001; Perez-Prat and Van Lookeren Campagne, 2002):

- parental combinations that produce heterosis levels superior to the best pure-line cultivars;
- a stable male–sterile, female–fertile system;
- a selection system to obtain 100% female (pod parent) parents that set seed normally and can be harvested mechanically;
- an efficient pollen transfer mechanism from pollen parent to pod parent; and
- an economical level of seed increase for the seedsman and growers that ultimately benefits the consumer.

A number of studies for hybrid production in food legumes have been conducted; however, the above five components are lacking in most of them, making hybrid research an uphill task in pulses. This chapter discusses the efforts made in various food legume crops for developing hybrid varieties.

13.2 Adzuki Bean

Adzuki bean, *Vigna angularis* [(Willd.) Ohwi and Ohashi], is a self-pollinating plant grown mainly in the Far East. Male sterility is known in adzuki bean but it was not determined whether this is genetic or cytoplasmic genetic (Nakashima *et al.*, 1980). Using hand-pollination, six fertile hybrid combinations were generated that showed

 Biology and Breeding of Food Legumes (eds A. Pratap and J. Kumar)

seed yield heterosis over the best parent from −11 to +21% (Kunkaew *et al.*, 2006).

13.3 Chickpea

Cultivated chickpea is an autogamous crop with less than 1% outcrossing, even though cleistogamous flowers are visited by bumblebees and honeybees (Tayyar *et al.*, 1996). Toker *et al.* (2006) reported outcrossing rates ranging from 0.0 to 1.25%. Rubio *et al.* (2010), using three independent microsatellite markers and a multilocus approach for mating system estimation, reported that the open flower mutant exhibited increased outcrossing rates (5.9%) under field conditions; this value was much higher than previously reported. Male sterility and the production of hybrids have not been reported in chickpea; this higher level of outcrossing might be used to develop chickpea cultivars with increased levels of heterozygosity. The discovery of the open flower trait (Pundir and Reddy, 1998), a potential gene for increasing outcrossing in chickpea populations, could allow the possibility of exploitation of hybrid vigour in developing synthetics or hybrid varieties.

13.4 Common Bean

Male sterility systems

Male sterile plants in *Phaseolus* were observed by Singh *et al.* (1980), while CMS (cytoplasmic male sterility) was confirmed as being maternally inherited (Bassett and Shuh, 1982). Five sterile cytoplasms have been identified that differ in their mitochondrial DNA restriction patterns, but are functional with the same set of maintainers (Bannerot, 1989). It has been shown that CMS in *Phaseolus vulgaris* is caused by a unique mitochondrial DNA sequence that occurs on a subgenomic molecule (containing only a portion of the DNA sequence) and maintained by autonomous replication (Mackenzie, 1991). This DNA sequence occurs in all studied accessions of *Phaseolus* and induces male sterility only when present in high copy number (Arrieta-Montiel *et al.*, 2001). Male fertility can be restored when the subgenomic molecule is spontaneously lost (Mackenzie and Chase, 1990) or when the molecule is eliminated or reduced to low levels by a nuclear 'reversion' allele, *Fr* (Mackenzie and Bassett, 1987; Janska *et al.*, 1998).

The cultivar 'Sprite' was identified as a nuclear maintainer genotype (CMS-Sprite), and fertility was observed to be restored by a single dominant nuclear gene (Mackenzie and Bassett, 1987). Most of the *P. vulgaris* genotypes tested, however, are poor maintainers, resulting in partial male sterility. Fortunately, several good maintainers were found among different growth habit types. Dominant restorers have been found in *P. vulgaris*, as well as in *P. vulgaris* × *Phasaseolus coccineus* crosses (Bannerot, 1989).

Hybrid development

The flowers of common bean are frequently visited by pollinating insects, which increases the level of cross-pollination in a normally self-pollinated plant (Andersson and de Vicente, 2010). Outcrossing rates are usually less than 1% (Tucker and Harding, 1975), though there are reports of 6–10% (Antunes *et al.*, 1973) and 0–85% (Wells *et al.*, 1988). The latter report considered six white-seeded beans with two planting dates at Irvine, California, and the authors concluded that there was considerable genetic and environmental variation for outcrossing. However, attempts to manually cross-pollinate field-grown plants have resulted in limited success. Most cross-pollinations are made on plants grown in a growth chamber, greenhouse or shelter house (Bliss, 1980).

Most heterosis studies in *Phaseolus* have been conducted on hand-emasculation and -pollination to produce F_1 seed. In addition, the difficulty in producing adequate numbers of hybrid seed for large-scale agronomic performance tests has led to limited progress in *Phaseolus* improvement. Gutierrez and Singh (1985) studied heterosis and inbreeding depression in 13 parental combinations produced by hand-emasculation and -pollination. Mid-parent heterosis values were given and six crosses showed positive heterosis (28–47%) for seed yield, but none of the F_1 hybrids yielded significantly higher than the highest-yielding parental line (Gutierrez and Singh, 1985). Some crosses that did not have either non-significant or negative heterotic values for seed yield showed positive effects of inbreeding, i.e. the F_2 outperformed the corresponding F_1 hybrids.

A total of 72 F_1 combinations of three *Phaseolus* plant growth habits resulted from all possible cross combinations, including reciprocals. Heterosis for yield above the high parent was observed for 20 crosses at location 1, while 4 crosses at location 2 were above the high parent (Nienhuis and Singh, 1986). These results are in agreement with previous results showing that F_1 heterosis is greater in *Phaseolus* crosses between, rather than within, growth habit types.

13.5 Cowpea

Cowpea is highly self-pollinated in most environments, the result of a cleistogamous flower structure and simultaneous pollen shed and stigma receptivity (Ehlers and Hall, 1997). The flowers open early in the morning and close before noon on the same day. Honeybees and bumblebees are attracted mainly by the extra floral nectaries on its petioles and leaflets; insects large enough to manipulate the floral mechanism would be required to transport pollen from the male parent to the male-sterile plants. e.g. bumblebees. Outcrossing rates of up to 15% are known (Duke, 1981). Nuclear male sterility has been reported in cowpea (Sen and Bhowal, 1962; Rachie *et al.*, 1975), but has not been utilized in hybrid seed production. Hybrid cultivars are not likely to become available in cowpea, even though substantial hybrid vigour has been shown (Hall *et al.*, 1997).

13.6 Faba Bean

The evolution of *Vicia* crops since their domestication has been driven by the selection towards selfing (Rick, 1988). Hybrids in autogamous crops such as *Vicia* species were considered impractical because of the strict self-pollination mechanisms that discourage cross-pollination. Thus hybrids in *Vicia* have received very little attention, except in *Vicia faba*. In faba bean the level of allogamy ranges from 4 to 84%, with a mean of around 30–60% and with large genotypic and environmental variation (Link, 1990; Link *et al.*, 1994; Suso and Moreno, 1999; Suso *et al.*, 2001; Gasim *et al.*, 2004). The importance of heterosis in faba bean is evident (Link, 2006). Hybrid varieties in faba bean offer great potential because of high heterosis for yield and stability in yield performance (Stelling *et al.*, 1994; Link *et al.*, 1996). Heterosis also is expressed in traits like seedling biomass, plant height, winter survival, tillering ability and autofertility (Link *et al.*, 2010).

Male sterility systems

No comprehensive studies on male sterility have been conducted to date on *Vicia* species, except for *Vicia faba*. Two reviews described and assessed *Vicia faba* male-sterile systems (Picard *et al.*, 1982; Bond, 1989). Articles presenting results of faba bean research (Duc, 1997) and CMS studies (Link *et al.*, 1997) summarize available information about male sterility and its consequences on breeding. Little can be added to these reviews, with the exception of the more recent studies of Vaupel (2000). New perspectives for the exploitation of heterosis-mediated yield and resistance to biotic and abiotic stresses are based mainly on the development of synthetics (Link *et al.*, 2010).

The first report on male sterility was that of Bond *et al.* (1964). The recessive form *(ms 1)* of genetic male sterility was observed to occur spontaneously at the Plant Breeding Institute (PBI), Cambridge, UK. Additionally, Duc *et al.* (1985a) at INRA, Rennes, France induced a dominant genetic male sterile, *Ms-d*, by mutagenesis with EMS, which was proposed to be used in improving outcrossing and gene randomization in both natural and breeding populations.

The interest of breeders was focused on CMS systems (Duc, 1997). First, two CMS systems were described in faba bean, the first of these discovered at PBI, Cambridge, and known as CMS 447 (Bond *et al.*, 1966). The second was discovered by Berthelem at INRA, Rennes, known as CMS 350 (Picard *et al.*, 1982). These were independently discovered in natural populations and are different in the sense that they do not accept the same restorer lines. Faba bean lines Ad23 and G58 maintain male sterility in CMS 447 and CMS 350, respectively (Berthelem and Le Guen, 1967). Male fertility is restored by the gene *Rf*1 in CMS 447 and by *Rf*3 in CMS 350 (Bond, 1989). Duc *et al.* (1985b) obtained 421 and 417 CMS cytoplasms from the mutagenesis of 447 CMS cytoplasm.

A major problem with both 447 and 350 CMS systems is large fluctuations in male-sterile expression in backcross generations or while multiplying female lines, this barrier preventing their successful exploitation in commercial hybrid seed production (Berthelem and Le Guen, 1974). Thus, the approach of Link *et al.* (1997) was to search for a CMS system based on the interaction of cytoplasm with a restorer allele *Rf* and a maintainer allele *rf* at one specific nuclear locus that showed stable and homogeneous expression of male sterility. Accordingly, two new CMS systems, CMS 199 and CMS 297, were identified. However, these CMS systems also were unstable to different degrees, and spontaneous reversion to fertility occurred similarly to the CMS 447 and CMS 350 systems. The CMS 199 and CMS 297 systems have been used extensively for experimental production of minor × major hybrid cultivars (Vaupel, 2000), but CMS instability prevents the use for hybrid seed production on a commercial scale.

Potential causes of this instability have been analysed. Electron microscopy and molecular studies of the CMS 447 cytoplasm have detected spherical virus-like particles, 73nm in diameter, which are linked to male sterility (Edwardson *et al.*, 1976). These particles were shown to contain high-molecular weight double-stranded RNA and an endogenous RNA-dependent RNA-polymerase (Scalla *et al.*, 1981; Lefebvre *et al.*, 1990; Pfeiffer, 1998).

Hybrid development

In CMS systems, pollen transport needs insects and cross-pollination is essential for seed set. *V. faba* is vulnerable to the effects of poor pollination, as some experiments have shown poor pollination in a CMS line leading to higher fertility levels in the progeny (Bond, 1989). Thus, good pollination is necessary for limiting the development of fertility in CMS lines. The diversity, density and behaviour of pollinator fauna vary geographically. Research in France has shown that the most frequent pollinators are among the genus *Bombus* (*Bombus terrestris* L. and *Bombus lucorum* L.) and honeybees (*Apis mellifera* L.) that often behave as nectar robbers. However, in Spain, the pollination fauna is largely composed of solitary bees, mainly *Eucera* (*Eucera numida* Lep.), which behaves as a positive pollinator and is present at high density and frequency (Pierre *et al.*, 1996, 1999). In the UK, solitary bees (*Anthophora plumipes*) were observed to visit flowers more efficiently and in greater numbers than bumblebees and honeybees (Bond and Kirby, 1999, 2001). *Vicia faba* plants exhibit spectacular variation in flower phenology, design and display, and much of the functional basis of this diversity is associated with levels of cross-pollination (Suso *et al.*, 2005; Suso and Maalouf, 2010). Further experimentation is necessary to determine whether the floral variation can be effectively utilized for the development of exclusively cross-pollinated crops, and for use in hybrid breeding programmes.

Although good hybrid combinations have been found, none of the many published CMS systems are employed in practical breeding, mostly due to instability and spontaneous reversion to pollen fertility. Cytoplasm and nuclear genes are maintained in the work germplasm collection of G. Duc and W. Link (Duc and Link, Spain, 2010, personal communication).

Although heterosis is fully realized in hybrid cultivars and partly so in synthetics, important improvements in synthetic populations have decreased the interest for hybrids in faba bean. Breeders continue to improve inbreds, to define the best parental combinations and to develop synthetic populations. Faba bean varieties currently commercialized are mainly population varieties obtained by mass selection or synthetic varieties (Link *et al.*, 2010). Compared with 10 years ago, breeders now have new approaches for the exploitation of heterosis based on the development of synthetics. They have new tools, such as hypervariable DNA markers and improved bioinformatic models for estimating the mating system. Multilocus likelihood-based estimation of outcrossing (Ritland, 2002), in combination with multivariate regression analysis, enables the plant breeder to identify floral traits related to outcrossing. Such traits provide the basis for developing heterotic varieties within which heterozygosity is maintained, due to floral behaviour rather than the use of male sterility.

13.7 Mung Bean

Mung bean (*Vigna radiata*) is an important source of protein in South-east Asian countries. Mung bean hybrids have been produced by manual cross-pollination for agronomic performance tests of F_1 plants. Fifteen hybrid combinations were evaluated by Khattak *et al.* (2002), of which 11 showed lower seed yield while four combinations had positive seed yield, with the highest combination of 27% heterosis. Xin *et al.* (2003) reported maximum heterosis for grain yield of 10% for 34 parental combinations. Heterosis for seed yield was determined for four parental combinations and ranged from 52 to 96%, although these data were from plants grown in pots (Soehendi and Srinives, 2005). In general, the major limiting factor in mung bean is the lack of a sterility system that could be used to produce large quantities of hybrid seed for agronomic performance tests.

13.8 Pigeon Pea

Male sterility systems

Nuclear male sterility and genetic male sterility (GMS) systems were reported during the 1980s in pigeon pea. The first GMS-based hybrid in this crop, ICPH 8, was released at ICRISAT (International Crops Research Institute for the Semi-Arid Tropics) in 1993 (ICRISAT, 1993) and, following this, many national institutes in India started to develop GMS-bred hybrids. The performance of these hybrids was good, but large-scale seed production was the major bottleneck for their popularization. The range of heterosis observed in GMS-based hybrids was extremely encouraging (Table 13.1), and provided impetus in the search for CMS systems in pigeon pea.

There have been many reports of CMS in pigeon pea (Mallikarjuna and Saxena, 2005; Chauhan *et al.*, 2008) and a number of CMS lines are available from different cytoplasm sources (Table 13.2). Various new germplasms are being tested for restoration/maintenance reactions to the CMS lines. Due to their stable nature, more emphasis is given to development of hybrids in the A_4 CMS system.

Three CMS lines were tested with seven restorers (testers) for fertility restoration (Table 13.3.); only four restorers gave complete fertility restoration. The CMS line from A_1 cytoplasm source is more sensitive to the environment; that from A_2 cytoplasm is less sensitive and can be used for hybrid development in specific environmental conditions; the A_4 CMS system is the most stable and should be given preference for the development of hybrids.

Table 13.1. Heterosis (%) in selected genetic male sterility-based hybrids of pigeon pea.

Hybrid	Year released	Days to maturity	Grain yield (kg/ha)	Superiority over control (%)
ICPH 9	1991	125	1780	41 over UPAS 120
PPH 4	1994	137	1930	14 over UPAS 120
CoH 1	1994	117	1210	32 over Vamban 1
CoH 2	1997	120–130	1050	35 over Co 5
AKPH 4104	1997	130–140	NA	64 over UPAS 120
AKPH 2022	1998	180–200	NA	35 over BDN 2

NA, data not available. Source: Saxena *et al.* (2006).

Table 13.2. Different sources of male sterility systems in pigeon pea.

Cytoplasm	Source (wild species)	Reference(s)
A_1	*Cajanua sericeus* (Benth.ex Bak.) van der Maesen comb. nov.	Ariyanayagam *et al.* (1995); Saxena *et al.* (1996)
A_2	*C. scarabaeoides* (L.) Thou. var. *pedunculatus*	Tikka *et al.* (1997); Saxena and Kumar (2003)
A_3	*C. volubilis* (Blanco) Blanco	Wanjari *et al.* (1999)
A_4	*C. cajanifolius* (Hains) van der Maesen comb. nov.	Dalvi *et al.* (2008); Saxena (2009)
A_5	*C. cajan* (L.) Millsp.	Mallikarjuna and Saxena (2002); Saxena (2008)
A_6	*C. lineatus* (W. & A.) van der Maesen comb. nov.	Saxena (unpublished)
A_7	*C. platycarpus* (Benth.) van der Maesen comb. nov.	Mallikarjuna *et al.* (2006)

Table 13.3. Fertility restoration in F_1 hybrids of pigeon pea averaged over three locations during the 2005 rainy season.

	ICPA[1] 2067		ICPA 2052		ICPA 2039	
Tester	Total plants	Fertility restoration (%)	Total plants	Fertility restoration (%)	Total plants	Fertility restoration (%)
ICPL 129-3	105	100	102	0	100	0
Nirmal 2	105	100	151	0	118	66
BWR 23	93	100	132	78	141	72
BSMR 736	76	71	144	63	143	100
BSMR 175	105	35	132	0	125	79
BDN 2	115	70	138	75	133	84
BSMR 853	96	83	116	61	146	77
Mean (across testers)		80 ± 0.50		40 ± 0.80		68 ± 1.00

[1]International Conference on Precision Agriculture.

Pollen transfer and stigma receptivity of CMS lines

For any male sterility system to be commercially effective, the essential factor is large-scale seed production in isolation. Natural outcrossing with insect-mediated pollen transfer is highly successful in pigeon pea. Furthermore, the long duration of stigma receptivity aids in higher seed production. Luo *et al.* (2009) studied the duration of stigma receptivity in two CMS lines with A_4 cytoplasm, both showing ~120h of effective stigma receptivity. Such a long duration of stigma receptivity was sufficient with high levels of insect populations for good seed production. Dalvi and Saxena (2009) studied stigma receptivity in early-duration CMS line ICPA 2039 and found that the stigma was receptive for a longer period. Both studies indicated that pigeon pea has sufficiently long stigma receptivity to possess high levels of natural outcrossing, which may facilitate hybrid seed production on a large scale.

Hybrid development

At ICRISAT various CMS lines with different cytoplasms have been developed, but CMS with A_4 cytoplasm has been used extensively for the production of experimental hybrids, in which a large range of heterosis was observed (Saxena *et al.*, 2010). The best hybrid, ICPH 2671, was released for commercial cultivation in 2007, and yielded 36% more than the best control (Saxena *et al.*, 2010). This hybrid was also tested in China for its productivity.

13.9 Soybean

The genus *Glycine* consists of two subgenera, *Glycine* (perennials) and *Soja* (annuals); the perennials consist of 22 recognized species and the annuals 2 species, *Glycine max* L. Merr. (cultigen) and *Glycine soja* Sieb. & Zucc. (wild species and progenitor of *G. max*) (Hymowitz, 2004). Natural cross-pollination is usually less than 1% in the highly self-pollinated annual *G. max*, altough it may also reach up to 2–3% (Palmer *et al.*, 2004). The perennial species were reported to show up to 60% outcrossing for *Glycine argyrea* and *Glycine clandestina* (Brown *et al.*, 1986; Schoen and Brown, 1991). In a study of seed set in chasmogamous and cleistogamous flowers of *G. clandestina*, Hempel (2004) found that pollination limitation in chasmogamous flowers was an important factor limiting seed production.

Male sterility systems

Genetic mutations affecting microsporogenesis and microgametogenesis in soybean have generated male-sterile and female-fertile lines. A detailed list of genes controlling sterility and their corresponding phenotype has been given by Palmer *et al.* (2004).

Sterility mutations are sometimes linked to other morphological characteristics. Thus, nuclear male-sterile, female-fertile plants can be identified by selecting for another trait. Examples of these traits include: (i) seed size differential (Carter *et al.*, 1984); and (ii) linkage between genes controlling the green cotyledon trait and the *Ms5* locus (Burton and Carter, 1983), as well as the *W1* flower colour locus and the *Ms6* locus (Lewers and Palmer, 1997; Lewers *et al.*, 1998a, b). Stine and Eby (2002) identified male-sterile, female-fertile soybean plants by using the linkage of the nuclear Midwest Oilseed (MWO) male-sterile, female-fertile trait with a chemical resistance locus. All nuclear male sterility mutations in soybean are stable, except for the partial male-sterile (*msp*) mutant (Palmer and Hymowitz, 2004) and the *ms8* mutation (Frasch *et al.*, 2011). The stability of CMS lines, restorers and maintainers has been well documented in China (Wang *et al.*, 2009). In hybrid seed production fields, female rows will be segregated for the nuclear male sterility (*ms*) mutation. Thus, a method to identify male-sterile, female-fertile plants is necessary. A workable CMS system with appropriate maintainers and restorers is a prerequisite for commercialization of a hybrid (Wang *et al.*, 2009).

The identification of cytoplasmic-nuclear male-sterile lines along with their maintainers and restorers has been achieved by intraspecific (*G. max* × *G. max*) and interspecific (*G. max* × *G. soja*) hybridizations (Davis, 1987; Sun *et al.*, 1994, 1997; Gai *et al.*, 1995; Zhang and Dai, 1997; Ding *et al.*, 1998; Zhao *et al.*, 1998; Bai and Gai, 2003; Zhao and Gai, 2006). CMS systems have been identified several times in soybean (Table 13.4). Restorers of all these lines have been found (Dr. Junyi Gai, China, 2009, personal communication). These systems are being exploited extensively for commercial soybean hybrid production in China, and world's first commercial soybean hybrid was released here in 2003, using a CMS system with nuclear restoration.

Hybrid development

Interest in hybrid soybean developed after the identification of the first male-sterile, female-fertile mutant (Brim and Young, 1971). Its use in recurrent selection breeding programmes (Brim and Stuber, 1973; Lewers and Palmer, 1997) increased the awareness of its potential in the production of commercial hybrid soybean. Several components are crucial for the successful development of hybrid soybean (Palmer *et al.*, 2001).

Heterosis studies have shown that levels above the better parent are possible (Brim and Cockerham, 1961; Nelson and Bernard, 1984; Cerna *et al.*, 1997; Manjarrez-Sandoval *et al.*, 1997; Lewers, 1998a, b; Sun *et al.*, 1999;

Table 13.4. Cytoplasmic-nuclear male-sterile soybean lines (modified from Palmer *et al.*, 2004).

Designation	Source of cytoplasm	Source of nuclear gene(s)	Nuclear male-sterile gene	Reference(s)
OA	167	O35	Recessive	Sun *et al.* (1994, 1997)
NJCMS1A	N8855	N2899	Two dominant	Gai *et al.* (1995); Ding *et al.* (1998)
NJCMS2A	N8855	N1628	Two dominant	Bai and Gai (2003)
NJCMS3A	N21566	N21249	One pair of genes	Zhao and Gai (2006)
Fu CMS1A	ZD8319	SG01	Dominant gene	Li *et al.* (1995); Xu *et al.* (1999)
Fu CMS2A	ZD8319	JX03	Incomplete dominant gene	Li *et al.* (1995); Xu *et al.* (1999)
Fu CMS3A	ZD8319	PI004	Six genes	Li *et al.* (1995); Xu *et al.* (1999)
ZA	ZD8319	YB	Recessive	Zhao *et al.* (1998)
YA	OA	YB	Recessive	Zhao *et al.* (1998)
W931A	Zhongyu 89B	W206	Recessive	Zhang and Dai (1997); Zhang *et al.* (1999a)
W936A	Zhongyu 89B	W203	Recessive	Zhang *et al.* (1999a)
W933A	Zhongyu 89B	W207	Recessive	Zhang *et al.* (1999a)
W945A	Zhongyu 89B	W210	Not reported	Zhang *et al.* (1999b)
W948A	Zhongyu 89B	W212	Not reported	Zhang *et al.* (1999b)

Burton and Brownie, 2006; Ortiz-Perez *et al.*, 2007; Perez *et al.*, 2009a, b; Yang and Gai, 2009a, b). In some cases, the better hybrids yielded 10–20% more than the better parent (Palmer *et al.*, 2001). Many of the studies in hybrid soybean have been conducted in single rows with spaced plants, conditions that are different from those in commercial fields. In other studies, where more hybrid seed was available, yield tests were done in replicated plots in several environments (Table 13.5).

Upon obtaining a stable male sterility system, it is necessary to transfer the pollen from the male parent to the female parent. In soybean, manual cross-pollination to produce large quantities of hybrid seed is difficult and time consuming. The small size of the soybean flower, the low success rate and the low number of seeds obtained per hybrid pod contribute to the difficulty in manually producing large quantities of hybrid seed (Fehr, 1991). Even though soybean is a self-pollinated species, soybean flowers possess most of the floral characteristics of entomophilous plants (Erickson, 1975; Erickson and Garment, 1979; Horner *et al.*, 2003). Insect-mediated cross-pollination of male-sterile soybean plants may facilitate the production of hybrid seed (Nelson and Bernard, 1984; Ortiz-Perez *et al.*, 2007; Zhao *et al.*, 2009). Pollinator insects such as honeybees (*Apis melliphera*) and alfalfa leaf cutter bee (*Megachile rotundata* F.) are attracted to soybean flowers and can be used in hybrid soybean production. In addition, some wild native bees, primarily from the families Megachilidae, Halictidae, Anthophoridae and Andrenidae, could be efficient pollinators (Ortiz-Perez *et al.*, 2007). An extensive study of seed production from a CMS line (JLCMS82A) under field conditions showed that the most effective arrangement for hybrid seed production was a 1 female:1 male parent row (Zhao *et al.*, 2009). In a recent field inspection by J. Wei and J. Gai of seed increased-fields of male-sterile lines and hybrid seed production fields isolated among the hills in Shanxi Province, China, pod-set of 70–100% of the maintainer and restorer lines was observed with natural insect pollination in a 2 female:1 male parent row condition (J. Wei and J. Gai, China, 2010, personal communication). This implies that it is possible to solve the hybrid seed production problem under natural insect pollination conditions. However, further studies are needed before large-scale production fields become a reality.

Table 13.5. Grain yield heterosis of soybean measured in replicated bordered row plots in more than one environment.[a]

Reference(s)	Population (*n*)	HPH[b] (%)	MPH[c] (%)	Environments (locations* years)
Brim and Cockerham (1961)	Single cross (2)	20	28	4
Hillsman and Carter (1981)	Single cross (8)	6	13	2
Nelson and Bernard (1984)	Single cross (27)	3	8	2–4
Loiselle *et al.* (1990)	Single cross (55)	–	11	3
Gizlice *et al.* (1993)	Single-cross (10)	3	9	4
Lewers *et al.* (1998a, b)	Single cross (18)	–4 to 2	2–8	4–6
Manjarrez-Sandoval *et al.* (1997)	Single cross (24)	3	7	2
Burton and Brownie (2006)	Single cross (2)	5–16	–	11
Ortiz-Perez *et al.* (2007)	Single cross (9)	–66 to 17	–59 to 37	6
	Three-way cross (8)	–25 to –5	–14 to 16	6
	BC_1F_1 (8)	–16 to 42	–7 to 42	6
Perez *et al.* (2009a)	Single cross (12)	–23 to 1	–29 to 32	4–6
Perez *et al.* (2009b)	Single cross (3)	–41 to 11	–34 to 15	2
	Three-way cross (8)	–31 to –5	–30 to 16	2
	Four-way cross (8)	–44 to –26	–35 to –16	2
	Five-way cross (6)	–38 to 1	–30 to 3	2
	BC_1F_1(7)	–34 to 22	–29 to 22	2
	BC_2F_1 (6)	–21 to –8	–14 to –2	2
	BC_3F_1 (6)	–22 to 3	–22 to 10	2
Yang and Gai (2009a, b)	Single cross (28)	–5 to 77	–1 to 81	3

[a]Modified from Palmer *et al.* (2001);
[b]Heterosis expressed as a percentage of the mid-parent;
[c]Heterosis expressed as a percentage of the high-parent.

13.10 Conclusions

Food legumes, in general, have not benefitted from male sterility systems that are widely used in maize, sorghum, rice, onion, tomato, etc. to produce hybrids. To date, hybrid pigeon pea is the only success story in pulses (Stakstad, 2007); the research by ICRISAT and its collaborators with sterility systems and agronomic performance studies was the catalyst that ensured the success of hybrid pigeon pea. Soybean hybrid research has been a key focus of Chinese scientists, and a number of CMS systems with appropriate maintainers and restorers are now available. However, the major limitation is pollen movement from male parents to female parents.

Faba bean and common bean have CMS systems, but the prerequisite of stable CMS systems with restorer and maintainer genotypes have not been identified. Efforts to develop novel CMS systems for release as non-GMO or non-transgenic germplasm seem successful (Sandhu *et al.*, 2007). If this methodology or other technology becomes viable, food legumes would be a major beneficiary of this science.

Acknowledgements

This is a joint contribution from the Iowa Agriculture and Home Economics Experiment Station, Ames, Iowa, USA, Project No. 4403 and the USDA Agricultural Research Service, Corn Insects and Crop Genetics Research Unit, and was supported by the Hatch Act and the State of Iowa. The mention of a trademark or proprietary product does not constitute a guarantee or warranty of the product by Iowa State University or the USDA, and the use of the name by Iowa State University or the USDA does not imply its approval to the exclusion of other products that may be suitable. M.J. Suso gratefully acknowledges the support of the AGL2005-07497-CO2-02 project. V.A. Dalvi gratefully acknowledges the input by Dr. K.B. Saxena, Principal Scientist, ICRISAT, India.

References

Andersson, M.S. and de Vicente, M.C. (2010) Cowpea. In: Andersson, M.S. and de Vicente, M.C. (eds) *Gene Flow Between Crops and their Wild Relatives*. Johns Hopkins University Press, Baltimore, Maryland, pp. 223–241.

Antunes, I.F., da Costa, J.G.C. and Oliveira, E.H. (1973) Natural hybridization in *Phaseolus vulgaris* in Pelotas. R.S. Brasil. *Annual Report of the Bean Improvement Cooperative* 16, 61–62.

Ariyanayagam, R.P., Rao, A.N. and Zaveri, P.P. (1995) Cytoplasmic male-sterility in interspecific matings of *Cajanus*. *Crop Science* 35, 981–985.

Arrieta-Montiel, M., Lyznik, A., Woloszynska, M., Janska, H., Tohme, J. and Mackenzie, S. (2001) Tracing evolutionary and developmental implications of mitochondrial stoichiometric shifting in the common bean. *Genetics* 158, 851–864.

Bai, Y.N. and Gai, J.Y. (2003) Development of soybean cytoplasmic-nuclear male sterile line NJCMS2A and restorability of its male fertility. *Scientia Agricultura Sinica* 36, 740–745.

Bannerot, H. (1989) The potential of hybrid bean. In: Beeke, S. (ed.) *Current Topics in Breeding of Common Bean*. Centro Internacional de Agricultura Tropical, Cali, Columbia, pp. 111–134.

Bassett, M.J. and Shuh, D.M. (1982) Cytoplasmic male sterility in common bean. *Journal of the American Society for Horticultural Science* 107, 791–793.

Bernard, R.L. and Cremeens, C.R. (1975) Inheritance of the Eldorado male-sterile trait. *Soybean Genetics Newsletter* 2, 37–39.

Berthelem, P. and Le Guen, J. (1967) *Rapport d'Activite 1967*. INRA, Rennes, France, pp. 39–44.

Berthelem, P. and Le Guen, J. (1974) *Rapport d'Activite 1971–74*. INRA, Rennes, France, pp. 119–163.

Bliss, F.A. (1980) Common bean. In: Fehr, W.R. and Hadley, H.H. (eds) *Hybridization of Crop Plants*. American Society of Agronomy–Crop Science Society of America, Madison, Wisconsin, pp. 283–284.

Boerma, H.R. and Cooper, R.L. (1978) Increased female fertility associated with the *ms1* locus in soybeans. *Crop Science* 18, 344–346.

Bond, D.A. (1989) A short review of research on male sterility and prospects for F_1 hybrid varieties in field beans (*Vicia faba* L.). *Euphytica* 41, 87–90.

Bond, D.A. and Kirby, E.J.M. (1999) *Anthophora plumipes* (Hymenoptera:Anthophoridae) as a pollinator of broad bean (*Vicia faba* major). *Journal of Apicultural Research* 38, 199–203.

Bond, D.A. and Kirby, E.J.M. (2001) Further observations of *Anthophora plumipes* visiting autumn-sown broad bean (*Vicia faba* major) in the United Kingdom. *Journal of Apicultural Research* 40, 113–114.

Bond, D.A., Drayner, J.M., Fyfe, J.L. and Toynbee-Clarke, G. (1964) A male sterile bean (*V. faba* L.) inherited as a Mendellian recessive character. *Journal of Agricultural Science Cambridge* 63, 223–234.

Bond, D.A., Fyfe, J.L. and Toynhe-Clarke, G. (1966) Male sterility with a cytoplasmic type of inheritance in field beans. *Journal of Agricultural Science Cambridge* 66, 359–367.

Brim, C.A. and Cockerham, C.C. (1961) Inheritance of quantitative characters in soybeans. *Crop Science* 1, 187–190.

Brim, C.A. and Stuber, C.W. (1973) Application of genetic male sterility to recurrent selection schemes in soybeans. *Crop Science* 13, 528–530.

Brim, C.A. and Young, M.F. (1971) Inheritance of a male-sterile character in soybeans. *Crop Science* 11, 564–566.

Brown, A.H.D., Grant, J.E. and Pullen, R. (1986) Outcrossing and paternity in *Glycine argyrea* by paired fruit analysis. *Biological Journal of the Linnean Society* 29, 283–294.

Burton, J.W. and Brownie, C. (2006) Heterosis and inbreeding depression in two soybean single crosses. *Crop Science* 46, 2643–2648.

Burton, J.W. and Carter, T.E., Jr (1983) A method for production of experimental quantities of hybrid soybean seed. *Crop Science* 23, 388–390.

Buss, G.R. (1983) Inheritance of a male-sterile mutant from irradiated Essex soybeans. *Soybean Genetics Newsletter* 10, 104–108.

Carter, T.E., Jr., Burton, J.W. and Huie, E.B., Jr. (1984) Mechanical separation of seed from male-sterile and fertile plants by seed size. *Soybean Genetics Newsletter* 11, 146–149.

Cerna, F.J., Cianzio, S.R., Rafalski, A., Tingey, S. and Dyer, D. (1997). Relationship between seed yield heterosis and molecular heterozygosity in soybean. *Theoretical and Applied Genetics* 95, 460–467.

Cervantes-Martinez, I.G., Xu, M., Zhang, L., Huang, Z., Kato, K.K., Horner, H.T. *et al.* (2005) Molecular mapping of male-sterile loci *ms2* and *ms9* in soybean. *Crop Science* 47, 374–379.

Chaudhari, H.K. and Davis, W.H. (1977) A new male-sterile strain in Wabash soybeans. *Journal of Heredity* 68, 266–267.

Chauhan, R.M., Panwar, L.D., Patel, P.T. and Tikka, S.B.S. (2008) Identification of heterotic combination of CMS lines and restorers of pigeonpea. *Journal of Food Legumes* 21, 25–27.

Dalvi, V.A. and Saxena, K.B. (2009) Stigma receptivity in pigeonpea (*Cajanus cajan*). *Indian Journal of Genetics* 69, 247–249.

Dalvi, V.A., Saxena, K.B. and Madrap, I.A. (2008) Fertility restoration in cytoplasmic-nuclear male-sterile lines derived from three wild relatives of pigeonpea. *Journal of Heredity* 99, 671–673.

Davis, W.H. (1987) Process for forming seeds capable of growing hybrid soybean plants. *United States Patent* 4, 648–204.

Delannay, X. and Palmer, R.G. (1982) Genetics and cytology of the *ms4* male-sterile soybean. *Journal of Heredity* 73, 219–223.

Ding, D., Cui, Z. and Gai, J. (1998) Development of cytological features of the cytoplasmic-nuclear male-sterile soybean line NJCMS1A. *Soybean Genetics Newsletter* 25, 34–35.

Duc, G. (1997) Faba bean (*Vicia faba* L.). *Field Crops Research* 53, 99–109.

Duc, G., Le Guen, J., Picard, J. and Berthelem, P. (1985a) Proposed use for a newly-discovered dominant male ~sterile allele for breeding purposes in *Vicia faba* L. *Fabis* 12, 8–10.

Duc, G., Picard, J., Le Guen, J. and Berthelem, P. (1985b) Note on the appearance of a new nuclecytoplasmic male sterility in *Vicia faba* appeared after mutagenesis. *Agronomie* 5, 851–854.

Duke, J.A. (1981) *Handbook of Legumes of World Economic Importance*. Plenum Press. New York.

Edwardson, J.R., Bond, D.A. and Christie, R.G. (1976) Cytoplasmic sterility factors in *Vicia faba* L. *Genetics* 82, 443–449.

Ehlers, J.D. and Hall, A.E. (1997) Cowpea (*Vigna unguiculata* L. Walp.). *Field Crops Research* 53, 187–204.

Erickson, E.H. (1975) Variability of floral characteristics influences honeybee visitation to soybean blossoms. *Crop Science* 15, 767–771.

Erickson, E.H. and Garment, M.B. (1979) Soya-bean flowers: nectary ultrastructure, nectar guides, and orientation on the flower by foraging honeybees. *Journal of Apicultural Research* 18, 3–11.

Fehr, W.R. (1991) *Principles of Cultivar Development. Theory and Technique*. Macmillan Publishing Company, Ames, Iowa.

Frasch, R., Weigand, C., Perez, P.T., Palmer, R.G. and Sandhu, D. (2011) Molecular mapping of two environmentally sensitive male-sterile mutants in soybean. *Journal of Heredity* 102, 11–16.

Gai, J., Cui, Z., Ji, D., Ren, A. and Ding, D. (1995) A report on the nuclear cytoplasmic male sterility from a cross between two soybean cultivars. *Soybean Genetics Newsletter* 22, 55–58.

Gasim, S., Abel, S. and Link, W. (2004) Extent, variation and breeding impact of natural cross-fertilization in German winter faba beans using hilum colour as marker. *Euphytica* 136, 193–200.

Graybosch, R.A. and Palmer, R.G. (1987) Analysis of a male-sterile character in soybeans. *Journal of Heredity* 78, 66–70.

Gutierrez, J.A. and Singh, S.P. (1985) Heterosis and inbreeding depression in dry bush beans. *Canadian Journal of Plant Science* 65, 243–249.

Hall, A.E., Singh, B.B. and Ehlers, J.B. (1997) Cowpea breeding. *Plant Breeding Reviews* 15, 215–274.

Hempel, K.A. (2004) Advantages of chasmogamy and cleistogamy in a perennial *Glycine clandestina* Wendl. (Fabaceae). PhD thesis, Australian National University, Canberra, Australia.

Horner, H.T., Healy, R.A., Cervantes-Martinez, T. and Palmer, R.G. (2003) Floral nectary fine structure and development in *Glycine max* L. (Fabaceae). *International Journal of Plant Science* 164, 675–690.

Hymowitz, T. (2004) Speciation and cytogenetics. In: Specht, J.E. and Boerma, H.R. (eds) *Soybean: Improvement, Production, and Uses*, 3rd edn. Monograph 16. American Society of Agronomy, Madison, Wisconsin, pp. 97–136.

ICRISAT (1993) Pigeonpea hybrid ICPH 8 (ICPH 82008). *Plant Material Description no. 40*. Patancheru 502 344, International Crops Research Institute for the Semi-Arid Tropics, Andra Pradesh, India.

Ilarslan, H., Horner, H.T. and Palmer, R.G. (1999) Genetics and cytology of a new male-sterile, female-fertile soybean [*Glycine max* (L.) Merr.] mutant. *Crop Science* 39, 58–64.

Janska, H., Sarria, R., Woloszynska, M., Arrieta-Montiel, M. and Mackenzie, S.A. (1998) Stoichiometric shifts in the common bean mitochondrial genome leading to male sterility and spontaneous reversion to fertility. *Plant Cell* 10, 1163–1180.

Khattak, G.S.S., Haq, M.A., Marwat, E.U.T., Ashraf, M. and Srinives, P. (2002) Heterosis for seed yield and yield components in mungbean (*Vigna radiata* (L.) Wilczek). *ScienceAsia* 28, 345–350.

Kunkaew, W., Julsrigival, S., Senthong, C. and Karladee, D. (2006) Estimation of heterosis and combining ability in azuki bean under highland growing conditions in Thailand. *Chiang Mai University Journal* 5, 162–168.

Lefebvre, A., Scalla, R. and Pfeiffer, P. (1990) The double-stranded RNA associated with the '447' cytoplasmic male sterility in *Vicia faba* is packaged together with its replicase in cytoplasmic membranous vesicles. *Plant Molecular Biology* 14, 477–490.

Lewers, K.S. and Palmer, R.G. (1997) Recurrent selection in soybean. *Plant Breeding Reviews* 16, 275–313.

Lewers, K.S., St. Martin, S.K., Hedges, B.R. and Palmer, R.G. (1998a) Effects of the Dt_2 and *S* alleles on agronomic traits of F1 hybrid soybean. *Crop Science* 38, 1137–1142.

Lewers, K.S., St. Martin, S.K., Hedges, B.R. and Palmer, R.G. (1998b) Testcross evaluation of soybean germplasm. *Crop Science* 38, 1143–1149.

Li, L., Yang, Q., Hu, Y., Zhu, L. and Ge, H. (1995) Discovery of parent interaction sterile material of soybean cultivars and its genetic inference. *Journal of Anhui Agricultural Sciences* 23, 304–306.

Link, W. (1990) Autofertility and rate of cross-fertilization: crucial characters for breeding synthetic varieties in faba bean (*Vicia faba* L.). *Theoretical and Applied Genetics* 79, 713–717.

Link, W. (2006) Methods and objectives in faba bean breeding. In: Avila, C., Cubero, J.I., Moreno, M.T., Suso, M.J. and Torres, A.M. (eds) *International Workshop on Faba Bean Breeding and Agronomy*. Junta de Andalucía, Córdoba, Spain, pp. 35–40.

Link, W., Ederer, W., Metz, P. and Buiel, H. (1994) Genotypic and environmental variation for degree of cross-fertilization in faba bean. *Crop Science* 34, 960–964.

Link, W., Schill, B., Barbera, A.C. and Cubero, J.I. (1996) Comparison of intra- and inter-pool crosses in faba beans (*Vicia faba* L.). I. Hybrid performance and heterosis in Mediterranean and German environments. *Plant Breeding* 115, 352–360.

Link, W., Ederer, W., Gumberm, R.K. and Melchinger, A.E. (1997) Detection and characterization of two new CMS systems in faba bean. *Plant Breeding* 116, 158–162.

Link, W., Balko, C. and Stoddard, F.L. (2010) Winter hardiness in faba bean: Physiology and breeding. *Field Crops Research* 115, 287–296.

Luo, R.H., Dalvi, V.A., Li, Y.R. and Saxena, K.B. (2009) A study on stigma receptivity of cytoplasmic-nuclear male-sterile lines of pigeonpea, *Cajanus cajan* (L.) Millsp. *Journal of Plant Breeding and Crop Science* 1, 254–257.

Mackenzie, S.A. (1991) Identification of a sterility-inducing cytoplasm in a fertile accession line of *Phaseolus vulgaris* L. *Genetics* 127, 411–416.

Mackenzie, S.A. and Bassett, M.J. (1987) Genetics of restoration in cytoplasmic male sterile *Phaseolus vulgaris* L. *Theoretical and Applied Genetics* 74, 642–645.

Mackenzie, S.A. and Chase, C.D. (1990) Fertility restoration is associated with a loss of a portion of the mitochondrial genome in a cytoplasmic-male sterility common bean. *Plant Cell* 2, 905–912.

Mallikarjuna, N. and Saxena, K.B. (2002) Production of hybrids between *Cajanus acutifolius and C. cajan*. *Euphytica* 124, 107–110.

Mallikarjuna, N. and Saxena, K.B. (2005) A new cytoplasmic nuclear male-sterility system derived from cultivated pigeonpea cytoplasm. *Euphytica* 142, 143–149.

Mallikarjuna, N., Jadhav, D. and Reddy, P. (2006) Introgression of *Cajanus platycarpus* genome into cultivated pigeonpea genome. *Euphytica* 149, 161–167.

Manjarrez-Sandoval, P., Carter, T.E. Jr., Webb, D.M. and Burton, J.W. (1997) Heterosis in soybean and its prediction by genetic similarity measures. *Crop Science* 37, 1443–1452.

Nakashima, H., Tsuda, C., Murata, K. and Narikawa, T. (1980) Histological features and inheritance of male sterile adzuki bean [*Vigna angularis* (Willd.) Ohwi & Ohashi]. *Japanese Journal of Breeding* 30, 241–245.

Nelson, R.L. and Bernard, R.L. (1984) Production and performance of hybrid soybeans. *Crop Science* 24, 549–553.

Nienhuis, J. and Singh, S.P. (1986) Combining ability analysis and relationships among yield, yield components and architectural traits in dry beans. *Crop Science* 26, 21–27.

Ortiz-Perez, E., Cianzio, S.R., Wiley, H., Horner, H.T., Davis, W.H. and Palmer, R.G. (2007) Insect-mediated cross-pollination in soybean [*Glycine max* (L.) Merr.]: I. Agronomic performance. *Field Crops Research* 101, 259–268.

Palmer, R.G. (2000) Genetics of four male-sterile, female-fertile soybean mutants. *Crop Science* 40, 78–83.

Palmer, R.G. and Hymowitz, T. (2004) Soybean genetics and breeding. In: Wrigley, C., Corke, H. and Walker, C. (eds) *Encylopedia of Grain Sciences*. Academic Press, London, pp. 136–146.

Palmer, R.G. and Skorupska, H. (1990) Registration of a male-sterile genetic stock (T295H) of soybean. *Crop Science* 30, 241.

Palmer, R.G., Winger, C.L. and Albertsen, M.C. (1978) Four independent mutations at the *ms1* locus in soybeans. *Crop Science* 18, 727–729.

Palmer, R.G., Winger, C.L. and Muir, P.S. (1980) Genetics and cytology of the *ms3* male-sterile soybean. *Journal of Heredity* 71, 343–348.

Palmer, R.G., Gai, J., Sun, H. and Burton, J.W. (2001) Production and evaluation of hybrid soybean. *Plant Breeding Reviews* 21, 263–307.

Palmer, R.G., Ortiz-Perez, E., Cervantes-Martinez, I.G., Wiley, H., Hanlin, S.J., Healy, R.A. *et al.* (2003) *Hybrid Soybean – Current Status and Future Outlook: 33rd Soybean Seed Research Conference*. American Seed Trade Association, Seed Expo 2003 (available on CD-ROM).

Palmer, R.G., Pfeiffer, T.W., Buss, G.R. and Kilen, T.C. (2004) Qualitative genetics. In: Specht, J.E. and Boerma, H.R. (eds) *Soybean, Improvement, Production, and Uses*, 3rd edn. Monograph 16. American Society of Agronomy, Madision, Wisconsin, pp. 137–233.

Perez, P.T., Cianzio, S.R. and Palmer, R.G. (2009a) Evaluation of soybean [*Glycine max* (l.) Merr.] F_1 hybrids. *Journal of Crop Improvement* 23, 1–18.

Perez, P.T., Cianzio, S.R., Ortiz-Perez, E. and Palmer, R.G. (2009b) Agronomic performance of soybean hybrids from single, three-way, four-way, and five-way crosses, and backcross populations. *Journal of Crop Improvement* 23, 95–118.

Perez-Prat, E. and Van Lookeren Campagne, M.M. (2002) Hybrid seed production and the challenge of propagating male-sterile plants. *Trends in Plant Science* 7, 199–203.

Pfeiffer, P. (1998) Nucleotide sequence, genetic organization and expression strategy of the double-stranded RNA associated with the '447' cytoplasmic male sterility trait in *Vicia faba*. *Journal of General Virology* 79, 2349–2358.

Picard, J., Berthelem, P., Duc, G. and Le Guen, J. (1982) Male sterility in *Vicia faba*. Future prospects for hybrid cultivars. In: Hawtin, G. and Webb, C. (eds) *Faba Bean Improvement*. ICARDA, Aleppo, Syria, pp. 53–69.

Pierre, J., Le Guen, J., Pham Delegue, M.H., Mesquida, J., Marilleau, R. and Morin, G. (1996) Comparative study of nectar secretion and attractivity to bees of two lines of spring-type faba bean (*Vicia faba* L. var. equina Steudel). *Apidologie* 27, 65–75.

Pierre, J., Suso, M.J., Moreno, M.T., Esnault, R. and Le Guen, J. (1999) Diversite et efficacite de l'entomofaune pollinisatrice (Hymenoptera: Apidae) de la feverole (Vicia faba L.) sur deux sites, en France et en Espagne. *Annales de la Société Entomolgique de France* 35 (suppl.), 312–318.

Pundir, R.P.S. and Reddy, G.V. (1998) Two new traits: open flower and small leaf in chickpea (*Cicer arietinum* L.). *Euphytica* 102, 357–361.

Rachie, K.O., Rawal, K., Franckowiak, J.D. and Akinpelu, N.A. (1975) Two out-crossing mechanisms in cowpea [*Vigna unguiculata* (L.) Walp.]. *Euphytica* 24, 159–163.

Rick, C.M. (1988) Evolution of mating systems in cultivated plants. In: Gottlieb, L.D. and Jain, S. (eds) *Plant Evolutionary Biology*. Chapman and Hall, London, pp. 133–147.

Ritland, K. (2002) Extensions of models for the estimation of mating systems using independent loci. *Heredity* 88, 221–228.

Rubio, J., Fernandez-Romero, M.D., Millán, T., Gil, J. and Suso, M.J. (2010) Outcrossing rate and genetic structure on an open-flowering population of *Cicer arietinum* based on microsatellite markers. *5th International Food Legumes Research Conference (IFLRC V) & 7th European Conference on Grain Legumes (AEP VII)*, Book of Abstracts, Antalya, Turkey, pp. 220.

Sandhu, A.P.S., Abdelmoor, R.V. and Mackenzie, S.A. (2007) Transgenic induction of mitochondrial rearrangements for cytoplasmic male sterility in crop plants. *Proceedings of the National Academy of Sciences U.S.A.* 104, 1766–1770.

Saxena, K.B. (2008) Genetic improvement of pigeonpea – a review. *Tropical Plant Biology* 1, 159–178.

Saxena, K.B. (2009) Evolution of hybrid breeding technology in pigeonpea. In: Ali, M. and Kumar, S. (eds) *Milestones in Food Legume Research*. Indian Institute of Pulses Research, Kanpur, India, pp. 82–114

Saxena, K.B. and Kumar, R.V. (2003) Development of a cytoplasmic nuclear male-sterility system in pigeonpea using *C. scarabaeoides* (L.) Thouars. *Indian Journal of Genetics and Plant Breeding* 63, 225–229.

Saxena, K.B., Singh, L., Kumar, R. and Rao, A.N. (1996) Development of CMS System in pigeonpea at ICRISAT Asia Center. *Procedural Working Group on CMS in Pigeonpea*, 9–10 May 1996, ICRISAT, Patancheru, Andra Pradesh, India, pp. 32–50.

Saxena, K.B., Kumar, R.V., Madhavi Latha, K. and Dalvi, V.A. (2006) Commercial pigeonpea hybrids are just a few steps away. *Indian Journal of Pulses Research* 19, 7–16.

Saxena, K.B., Ravikoti, V.K., Dalvi, V.A., Pandey, L.B. and Gaddikeri, G. (2010) Development of cytoplasmic-nuclear male sterility, its inheritance, and potential use in hybrid pigeonpea breeding. *Journal of Heredity* 101, 497–503.

Scalla. R., Duc, G., Rigaud, J., Lefebvre, A. and Meignoz, R. (1981) RNA-containing intracellular particles in cytoplasmic male-sterile faba bean (*Vicia faba* L.). *Plant Science Letters* 22, 269–277.

Schoen, D.J. and Brown, A.H.D. (1991) Whole- and part-flower self-pollination in *Glycine clandestina* and *G. argyrea* and the evolution of autogamy. *Evolution* 45, 1651–1664.

Sen, N.K. and Bhowal, J.G.A. (1962) A male-sterile mutant cowpea. *Journal of Heredity* 53, 44–46.

Singh, S.P., White, J.W. and Gutierrez, J.A. (1980) Male sterility in dry beans. *Annual Report of the Bean Improvement Cooperative* 23, 55–57.

Skorupska, H. and Palmer, R.G. (1989) Genetics and cytology of the *ms6* male-sterile soybean. *Journal of Heredity* 80, 304–310.

Skorupska, H.T. and Palmer, R.G. (1990) Additional sterile mutations in soybean *Glycine max* (L.) Merr. *Journal of Heredity* 81, 296–300.

Soehendi, R. and Srinives, P. (2005) Significance of heterosis and heterobeltiosis in an F1 hybrid of mungbean (*Vigna radiata* L. Wilczek) for hybrid seed production. *SABRAO Journal of Breeding and Genetics* 37, 97–105.

Stakstad, E. (2007) The plant breeder and the pea. *Science* 316, 196–197.

Stelling, D., Ebmeyer, E. and Link, W. (1994) Yield stability in faba bean, *Vicia faba* L. 2. Effects of heterozygosity and heterogeneity. *Plant Breeding* 112, 30–39.

Stelly, D.M. and Palmer, R.G. (1980) A partially male-sterile mutant line of soybeans, *Glycine max* (L.) Merr.: Inheritance. *Euphytica* 29, 295–303.

Stine, H.H. and Eby, W.H. (2002) Hybrid soybeans and methods of production. *International Patent Application* WO 02/007504 A3.

Sun, H., Zhao, L. and Huang, M. (1994) Studies on cytoplasmic-nuclear male sterile soybean. *Chinese Science Bulletin* 39, 175–176.

Sun, H., Zhao, L. and Huang, M. (1997) Cytoplasmic-nuclear male-sterile soybean line from interspecific crosses between *G. max* and *G. soja*. *World Soybean Research Conference V*. Kasetsart University Press, Bangkok, pp. 99–102.

Sun, H., Zhao, L., Li, J. and Wang, S. (1999) The investigation of heterosis and pollen transfer in soybean. In: Kauffman, H.E. (ed.) *World Soybean Research Conference VI*. Superior Printing, Champaign, Illinois, p. 489.

Suso, M.J. and Maalouf, F. (2010) Direct and correlated responses to upward and downward selection for outcrossing in *Vicia faba*. *Field Crops Research* 116, 116–126.

Suso, M.J. and Moreno, M.T. (1999) Variation in out crossing rate and genetic structure on six culitvars of *Vicia faba* L. as affected by geographic location and year. *Plant Breeding* 118, 347–350.

Suso, M.J., Pierre, J., Moreno, M.T., Esnault, R. and Le Guen, J. (2001) Variation in outcrossing levels in faba bean cultivars: role of ecological factors. *Journal of Agricultural Science Cambridge* 136, 399–405.

Suso, M.J., Harder, L.D., Moreno, M.T. and Maalouf, F. (2005) New strategies for increasing heterozygosity in crops: *Vicia faba* mating system as a study case. *Euphytica* 143, 51–65.

Tayyar, R.I., Federici, C.V. and Waines, J.G. (1996) Natural outcrossing in chickpea (*Cicer arietinum* L.). *Crop Science* 36, 203–205.

Tikka, S.B.S., Parmar, L.D. and Chauhan, R.M. (1997) First record of cytoplasmic-genic male-sterility system in pigeonpea [*Cajanus cajan* (L.) Millsp.] through wide hybridization. *Gujarat Agricultural University Research Journal* 22, 160–162.

Toker, C., Canci, H. and Ceylan, F.O. (2006) Estimation of outcrossing rate in chickpea (*Cicer arietinum* L.) sown in autumn. *Euphytica* 151, 201–205.

Tucker, C.L. and Harding, J. (1975) Outcrossing in common bean *Phaseolus vulgaris* (L.). *Journal of the American Society for Horticultural Science* 100, 283–285.

Vaupel, J.C. (2000) New CMS-systems of the production of minor × major-hybrid cultivars *Vicia faba* L.: genetic analysis and line development. *Grain Legumes* 30, 7.

Wang, S., Sun, H., Zhao, L., Wang, Y., Peng, B., Fan, X. *et al.* (2009) Progress and problem analysis on soybean male sterility and heterosis exploitation in China. *Soybean Science* 28, 1089–1096.

Wanjari, K.B., Patil, A.N., Manapure, P., Manjaya, J.G. and Manish, P. (1999) Cytoplasmic male-sterility in pigeonpea with cytoplasm from *Cajanus volubilis*. *Annals of Plant Physiology* 13, 170–174.

Wells, W.C., Isom, W.H. and Waines, J.G. (1988) Outcrossing rates of six common bean lines. *Crop Science* 28, 177–178.

Xu, Z., Li, L., Qiu, L., Chang, R., Wang, M., Li, Z. *et al.* (1999) Selection of three lines and localization of the restorer genes in soybean using SSR markers. *Scientia Agricultura Sinica* 32, 32–38.

Zhang, L. and Dai, O. (1997) Selection of cytoplasm-nuclear male-sterile soybean line W931A. *Scientia Agricultura Sinica* 30, 90–91.

Zhang, L., Dai, O., Huang, Z. and Li, J. (1999a) Selection of soybean male-sterile line of nucleo-cytoplasmic interaction and its fertility. *Scientia Agricultura Sinica* 32, 34–38.

Zhang, L., Dai, O. and Zhang, L. (1999b) Breeding of soybean male-sterile line of nucleo-cytoplasmic interaction. *Soybean Science* 18, 327–330.

Zhao, L., Sun, H. and Huang, M. (1998) The development and preliminary studies on cytoplasmic male sterile soybean line ZA. *Soybean Science* 17, 268–270.

Zhao, L., Sun, H., Peng, B., Li, J., Wang, S., Li, M. *et al.* (2009) Pollinator effects on genotypically distinct soybean cytoplasmic male sterile lines. *Crop Science* 49, 2080–2086.

Zhao, T.J. and Gai, J.Y. (2006) Discovery of new male-sterile cytoplasm sources and development of a new cytoplasmic nuclear male-sterile line NJCMS 3A in soybean. *Euphytica* 152, 387–396.

14 Mutagenesis

K.H. Oldach

14.1 Introduction

Breeding programmes rely on genetically diverse germplasm that can be explored to select for desirable traits. Much of the naturally occurring genetic diversity is captured in germplasm collections from which crop species and their relatives can be sourced. In cases such as chickpea (*Cicer arietinum*) where the genetic variation of the species is low due to its monophyletic origin from *Cicer reticulatum* (Ladizinsky and Adler, 1976), genetic diversity can be enriched by hybridization with other related species or by induced mutations.

Even if the genetic variation of a species is relatively large, desirable traits from an end-user point of view (humans) are often quite different from what is beneficial for a plant in its natural and highly competitive environment. For example, pathogen-deterring toxins are undesirable for human purposes if they affect taste or represent a risk to health. Therefore, additional genetic diversity is desirable and can be found in mutant populations that carry variations of specific traits that are not present in naturally occurring germplasm. Identifying a desirable trait requires selection methods that are 'simple' in the sense of being able to assess large numbers of plants at a reasonable cost to find the desirable mutation. Once identified, the mutant trait can be introgressed through crosses into elite breeding lines.

Physical, chemical or biological means are used to induce changes to or remove genes that lead to functional alterations or disruptions/eliminations of gene functions, and thus traits. The earliest report on intentional mutagenesis in plants is for barley and maize 80 years ago (Stadler, 1930). As a breeding tool, mutagenesis became very popular from the 1950s onwards, when a large range of crop and ornamental plant species were treated, predominantly by irradiation, to increase trait variation.

14.2 Mutant Varieties

There are currently (June 2010) 3084 released mutant varieties listed in the mutant variety database established by the Food and Agriculture Organization of the United Nations (FAO) and the International Atomic Energy Agency (IAEA) in Vienna, Austria (IAEA, 2011). The database contains information on released mutant varieties across 216 plant species. Between 1950 and 2010, 446 mutant varieties from 21 food legume species have officially been released, with the number of released mutant varieties in cereals being even higher, 1490. Information on the most common traits modified in the different food legume species is summarized in Table 14.1.

Table 14.1. Mutant varieties released (1959–2009) in different legumes.

Crop	Varieties (*n*)	Mutant trait	Mutagen used
Arachis hypogea (groundnut, peanut)	69	Higher yield, fungal resistance[a], early/uniform maturity, seed size, oil content, pod size/number, dwarfism, habit (erect, lodging resistant), branching type, seed dormancy, drought tolerance, leaf morphology (colour or size), tolerance to acidic soils, miscellaneous[b]	Gamma or X-rays, beta rays, laser treatment of seeds, EMS
Cajanus cajan (pigeon pea)	5	Seed size, early/uniform maturity, high yield, drought tolerance, seed dormancy	Neutron rays, gamma rays, EMS
Cicer arietinum (chickpea)	18	Fungal resistance, yield, early/uniform maturity, erect habit, seed size, branching type, miscellaneous	Gamma or X-rays, EMS, neutron rays
Glycine max (soybean)	154	Yield, early/uniform maturity, resistance to biotic and abiotic stresses, high protein, dwarfism, lodging resistance, oil content, ranch type, seed colour, seed size, pod size/number, vigour, late maturity, hypernodulation, summer type, adaptability, miscellaneous	Gamma or X-rays, EMS, neutron rays, EI, laser, spontaneous, DES, DMS, NEU, NMH, NMU, undefined chemical mutagen
Lens culinaris (lentil)	7	Fungal resistance, yield, early/uniform maturity, miscellaneous	Gamma or X-rays, neutron rays, EMS, Datura seed extract
Lathyrus sativus (grass pea)	2	Seed yield, low content of neurotoxin BOAA, seed colour. Disease, insect and drought resistance	Gamma or X-rays, NEU
Lupinus albus (white lupin)	13	Low alkaloid content, fungal resistance, early/uniform maturity, insect resistance, lodging resistance, high protein, yield, miscellaneous	EI, gamma or X-rays, undefined chemical mutagen, NMU, DMS, EMS, MNH, NEU
Lupinus angustifolius (blue lupin)	2	Early/uniform maturity, non-branching	NMH, EI
Lupinus consentini (sandplain lupin)	1	Early/uniform maturity, low alkaloid content, flower colour, non-shattering seeds and flowers	Undefined
Lupinus luteus (yellow lupin)	3	Fusarium resistance, early/uniform maturity, yield	Gamma or X-rays, undefined chemical mutagen
Medicago sativa (lucerne)	2	Fungal resistance	Magnetic field free space, undefined
Onobrychis vicifolia (sainfoin)	2	Yield, branching	Undefined chemical mutagen, NMU
Phaseolus coccineus (scarlet runner bean)	1	Dwarfism, seed size, suitable for mechanical harvesting	X-rays
Phaseolus vulgaris (common bean)	56	Early/uniform maturity, seed colour, fungal resistance, flower colour, virus resistance, habit (bush type), yield, high protein, cooking quality (reduced cooking time), miscellaneous	Gamma or X-rays, EMS, EI, NMU
Pisum sativum (pea)	33	Lodging resistance, yield, early/uniform maturity, dwarfism, resistance to seed shedding, seed size, tendrils instead of leaflets, suitable for processing, seed shape/colour/smoothness, high protein, mechanical harvesting, late maturity, miscellaneous	Gamma or X-rays, undefined chemical mutagen, EI, DES, NEU, spontaneous mutation

Continued

Table 14.1. Continued.

Crop	Varieties (*n*)	Mutant trait	Mutagen used
Vicia faba (faba bean)	19	Early/uniform maturity, plant architecture, yield, dwarfism (6), high protein, disease resistance, lodging resistance, wilt disease	Gamma or X-rays, NMU, NEU, EMS, DES, DMS, EI
Vicia sativa (vetch)	3	Leaf shape/size, vigour, branching, dwarfism	EMS, DES
Vigna aconitifolia (moth bean)	1	Early/uniform maturity, dwarf, yield	Gamma rays and EMS
Vigna mungo (black gram)	8	Yield, early/uniform maturity, seed size, fungal resistance, miscellaneous	Gamma or X-rays
Vigna radiata (mung bean)	35	Yield, fungal resistance, early/uniform maturity, virus resistance, seed size, non-shattering pods, pod number, dwarfism, plant architecture, miscellaneous	Gamma or X-rays, EMS
Vigna unguiculata (cowpea)	12	Yield, early/uniform maturity, viral resistance, fungal resistance, bacterial resistance, strong vegetative growth, cowpea aphid resistance, miscellaneous	Gamma or X-rays, DMS

BOAA, beta-N-oxalylamino-L-alanine; DES, diethyl sulfate; DMS, dimethyl sulphate; EI, ethyleneimine; EMS, ethyl methanesulfonate; ENH, N-ethyl-N-nitrosourea ENU; MNH, N-methyl-N-nitrosourea MNU; NEU, N-nitroso-N-ethylurethane; NMH, N-nitroso-N-methylurea; NMU, N-nitroso-N-methylurethane.
[a]Resistance to fungi including *Ascochyta*, *Fusarium*, powdery mildew, *Cercospora* sp., *Colletotrichum lindemuthianum*, *Sclerotinia sclerotiorum*, rusts and rots.
[b]Miscellaneous refers to traits that appear in only one variety per species, e.g. content of Vitamins A or C, or taste.

This list also includes those new varieties that were developed through crosses of elite lines with existing mutant varieties.

All mutant varieties among the different food legume species are selections for higher yield, disease resistance (predominantly against fungi) and earliness or uniformity in maturity. In addition, species-specific traits can be identified in the list of modified traits. For example, in common bean (*Phaseolus vulgaris*) a strong emphasis is on seed colour, which has been mutated in 22 of 56 released varieties.

In general, the frequency of a specific mutation within a legume species is determined by several factors:

1. Species-specific selection takes place due to the economical importance of an individual trait such as seed colour, which is a major criterion for marketability in common bean but not in faba bean.
2. The preference for a specific mutation can lead to its frequent use in crosses within the breeding programme and make it a 'common' trait. This is the case for the three mutant pea varieties Wasata, Sum and Hamil, which carry the 'afila-type' mutation with leaflets being changed into tendrils that was originally induced by gamma irradiation to generate the variety Wasata.
3. If different mutations within the same gene lead to the same phenotype, respective mutation is more frequent. This has been observed with tolerance to the herbicide sulfonylurea, which is achieved if one of six possible amino acids is changed due to mutations in the *ALS* gene (Tranel and Wright, 2002; Whaley *et al.*, 2007), whereas tolerance to the herbicide glyphosate relies on mutation in one specific amino acid of the EPSP gene (Powles and Preston, 2006), which is less likely.

14.3 Generating Mutations with Physical or Chemical Mutagens

Physical mutagen (irradiation) or chemical treatments of seeds are generally used to generate induced mutations. Radiation was not only the first mutagen known in plants

(Stadler, 1930), but also stands out for its simplicity of application. It is applied on dry seeds that can be stored until needed and it requires less handling than any chemical seed treatment. For this reason seed irradiation by gamma or X-rays has been the most commonly used method of mutagenesis, with about two-thirds of released mutant food legume varieties developed using irradiation. Less common radiation treatments include beta and neutron rays, and the application of laser treatment for the development of five varieties of groundnut and soybean. A range of chemical mutagens was employed in the development of about one quarter of the reported varieties (Table 14.1).

Ionizing radiation and chemical mutagens differ in the type of mutations they cause. Chemical mutagens mostly induce point mutations, whereas gamma or X-rays tend to produce larger chromosomal abnormalities such as chromosome breakages, which lead to translocations or deletions of chromosomes or chromosome segments carrying the affected genes.

Not all of the effects that are caused by a mutagen will be inherited by the next generation; when seeds are treated with a mutagen, damage occurs to the DNA of different cell types. Only the damage that has occurred in genetically effective germline cells will be stably inherited, whereas DNA damage in somatic cells affects only the current generation. The number of germline cells (genetically effective cell number, GECN) in the seed varies between species and is considered to range from 1 to 10 (Redei, 1975), with two germline cells estimated in soybean seeds (Carroll *et al.*, 1985), three in *Medicago truncatula* (Le Signor *et al.*, 2009) and six in *Lotus japonicus* (Tadege *et al.*, 2009).

Physical mutagenesis

For physical mutagenesis, an irradiation source such as the radioactive isotope ^{60}Cobalt is required. Irradiation sources can be accessed in nuclear research centres or quarantine services that use radiation to sterilize imported goods. This is a common procedure in countries with strict quarantine regulations such as Australia. Although dose information at which released mutant varieties of different species have been produced is publicly available (e.g. IAEA, 2011), a kill-curve specific to the seed batch to be used should be established, as genotypic differences, seed quality and moisture content impact on the mutation rate. For most food legumes, radiation dose varies between 100 Gy and 250 Gy. Establishing a kill-curve for a food legume species could comprise six batches of seeds treated with either 50, 100, 150, 200, 250 or 300 Gy. Sowing 100 seeds for each treatment and an untreated control batch and observing the growth over two to four weeks will give an indication of the most efficient dose or doses to be applied to generate the larger actual mutant population.

In the past – and still today – doses that lead to 50% lethality (LD_{50}) were often chosen. It can be argued that an LD_{50} is quite arbitrary and might lead to such a high number of (mostly deleterious) mutations in every plant that desirable mutations are either lost or overlooked due to either plant mortality or poor agronomic performance in generations following the mutagenesis. Therefore, a mutation rate targeting a lower LD (e.g. LD_{20}) with a survival rate of 80% might be more suitable for mutation breeding in selfing plant species. Maluszynski *et al.* (2009) also suggest that the final doses for mutagenic treatment should be rather low if the aim is to add new traits to an already high-quality genetic background, such as varieties or elite breeding lines. They conclude that the doses with an LD_{50} generally applied in the mutation breeding programmes of the 1960s and 1970s were too high and thus did not lead to the success expected with this technology.

The mutagen dose used should be a compromise between mutation load and the chance to find desirable mutations, and this greatly depends on the feasibility of cost-effective, high-throughput selection. For traits with simple phenotypic selection criteria, such as early maturity, screening of larger mutant populations that originate from a lower mutagen dose is feasible. On the other hand, screening the same population for complex phenotypic traits such as seed protein quality would not be feasible.

Chemical mutagenesis

The chemical mutagenesis of seeds is slightly more involved than irradiation, and extra care must be taken for health protection during the procedure as most mutagens are highly carcinogenic. Material and safety data sheets (MSDS) for the specific chemical mutagen chosen should be carefully read and the agent should be appropriately inactivated before disposal. The most commonly used chemical mutagen is EMS (ethyl methanesulfonate), an alkylating agent that can be inactivated by adjusting the solutions (treatment and wash solutions) to a final concentration of 10% sodium thiosulfate ($Na_2S_2O_3$) and 1% sodium hydroxide (NaOH) (Johnson *et al.*, 2007), with incubation for 24h at room temperature. A clear advantage of the point mutations created by chemical mutagens is their potential to generate not only loss-of-function but also gain-of-function phenotypes if the mutation leads to a modified protein activity or affinity, as in tolerances to the herbicides glyphosate (Bradshaw *et al.*, 1997) or sulfonylurea shown in the legume *Medicago truncatula* (Oldach *et al.*, 2008).

The protocol we use for mutagenesis of seeds with EMS involves the following steps:

1. Imbibe seeds in reverse osmosis (RO) or distilled water for 12h.
2. Decant RO water and rinse once with RO water.
3. Directly add EMS at desired final concentration; mix EMS well and shake occasionally over the next 12h.
4. Add sodium thiosulfate to a final concentration of 10% and incubate for 30min with treated seeds to inactivate EMS.
5. Decant solution into waste container and add NaOH (final conc. 1%) and incubate for 24h before disposal.
6. Wash seeds with tap water and treat as in step 5.
7. Repeat previous step 8–10 times.
8. Drain seeds well and transfer to tray covered with filter paper.
9. Sow seeds directly or maintain at 4–6°C for up to 2 days; any longer will make machine-sowing difficult (seed damage).

The concentration of the mutagen, the length of treatment and the temperature at which the experiment is carried out all affect the efficiency of mutagenesis. As chemical mutagens are very reactive, it is important to use fresh batches of the chemical or chemicals that have been appropriately stored. To determine the treatment dose, the above procedure is first applied to sets of 100 seeds (all from the same batch) to establish a kill-curve. To calculate the inhibitory effect of the EMS treatment on seedling growth, a negative control is required, comprising another set of 100 seeds treated as described in steps 1 to 9, but without EMS (no step 3). Each set of 100 seeds is then planted in soil or simply on to filter paper and monitored over the next few weeks, for potential germination inhibition or growth reduction compared with controls. An example of such a kill-curve is shown in Fig. 14.1 for faba bean.

In faba bean treated seeds germinate at all EMS concentrations, possibly due to the large seed reserves in this species. However, variation in seedling growth was noticeable at EMS concentrations of 0.08, 0.16% and 0.32% (trays 4 to 6 in Fig. 14.1). The mutagenic effect of EMS varies between the legume species; for example, a kill-curve experiment in the smaller-seeded lentil species is expressed as germination inhibition, rather than the growth seen in young faba bean seedlings (Fig. 14.2).

Fig. 14.1. EMS kill-curve for faba bean. Increasing concentrations of EMS were applied to faba bean seeds cv. Nura, each tray containing 100 seeds. Tray 1, control (EMS 0%); 2, 0.02% (1.9mM); 3, 0.04% (3.9mM); 4, 0.08% (7.8mM); 5, 0.16% (15.6mM); 6, 0.32% (31.1mM).

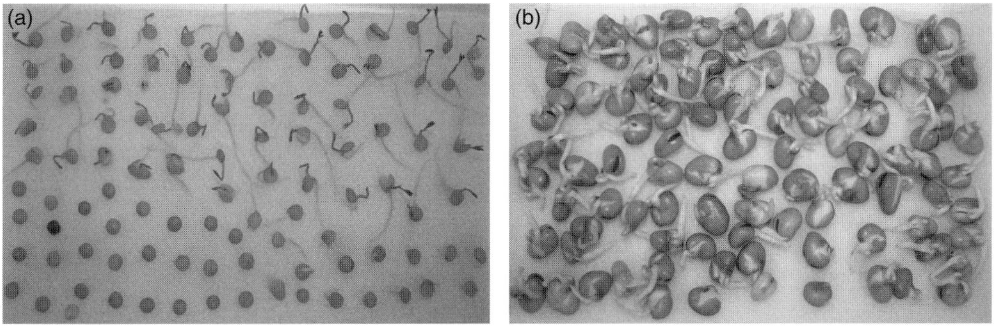

Fig. 14.2. Comparison of EMS kill-curve in lentil and faba bean; 100 seeds each of lentil (A) and faba bean (B) shown 8 days after EMS seed treatment at 0.16%. In lentil, the germination rate was reduced from 100% in the control treatment (not shown) to 73% (A), whereas in faba bean it was comparable to controls, at 97% (B). Although nearly all faba bean seeds germinated, growth reduction occurred later in nearly all seedlings (tray 5, Fig. 14.1).

Based on the kill-curve seen in the faba bean example above (Fig.14.1), an EMS concentration of 0.10% was chosen for a bulk experiment to generate a faba bean mutant population.

14.4 Development of Mutant Populations and Selection

Mutant populations are best generated using a genotype that is well characterized and has a range of desirable agronomical characteristics, e.g. varieties or advanced breeding lines. Familiar agronomical characteristics, e.g. time to flowering, habit, yield, disease resistance, etc. serve as a reference to potential mutants detected in mutagenized populations. The untreated seeds are referred to as the M_0 generation. Once the M_0 seeds have been treated with a mutagen they are referred to as the M_1 generation, which carries heterozygous mutations comparable to the heterozygous status of an F_1 generation in a bi-parental cross; the M_1 plants produce the seeds of the M_2 generation, and so forth. In the M_2 generation, homozygous mutant plants appear and recessive mutations can be identified phenotypically, but the population is still segregating. The fact that plant species carry between one and ten germline cells (Redei, 1975) can lead to independent mutations in each M_1 plant, and thus lead to M_2 plants that carry different mutations although originating from the same chimeric M_1 plant.

Determination of the chance or probability to identify desirable new traits depends on the size of the M_1, but also on subsequent generations, the mutagenesis protocol (e.g. mutagen concentration, mutagen exposure time, etc.), the seed production per plant and the selection methodology. It has been observed that a relationship exists between the ploidy level in species and the tolerable mutation density. For example, in *Medicago truncatula* (2n = 2x), mutation frequency in EMS-induced populations varied from 1/400 kb to 1/485 kb (Porceddu *et al.*, 2008; Le Signor *et al.*, 2009), while in *Glycine max* (2n = 4x) it varied from 1/140 to 1/550 kb (Cooper *et al.*, 2008). This positive correlation can be explained by the gene redundancy that exists in polyploid species. A deleterious mutation in a gene in one subgenome can be complemented by a functional version of the same gene in another subgenome, the homoeologous gene. Consequently, amongst food legumes, soybean and faba bean should be more tolerant to a higher mutation frequency in the genome, but not necessarily to a higher mutagen dose, as the somatic effects of a mutagen also determine the maximum dose. Common sizes of mutant populations range between 1000 and 8000 in crop species. The chimeric M_1 plants are more appropriately referred to as M_1 families, due to the GECN usually being >1, e.g. the GECN in *Lotus japonicus* is 6 (Tadege *et al.*, 2009).

In the case where a species with GECN = 3 has been mutagenized, the three germline cells carry independent mutations and are heterozygous for each mutation in the M_1 plant. If only one seed is being taken from the M_1 plant to develop the M_2 generation (single-seed descent approach), the chance that a specific M_2 plant carries the mutation (homozygous or heterozygous) in a germline cell is only 1 in 4 (1 × aa, 2 × Aa, 9 × AA; Fig. 14.3). If the phenotype of this mutation is recessive, it has only a 1 in 12 chance of being phenotypically expressed in M_2 and thus is unlikely to be found in this hypothetical single-seed descent M_2 population.

To calculate the number of M_2 plants that give a realistic chance of capturing most of the mutations that are present in the M_1 families, Le Signor *et al.* (2009) recommend using the following equation:

$$\text{Number of mutations recovered in } M_2 = \text{number of } M_1 \text{ plants x GECN} \times p$$

where p is the probability at which at least one mutation is recovered in M_2, and depends on the GECN. For example, if GECN = 1, then $p = 0.75$ (1:2:1 segregation ratio in a diploid species), if only one seed is being collected from an M_1 plant.

For most plant species the GECN has not been reported, and a specific recommendation for an M_2 population size that carries all the mutations that have been generated in the M_1 families is not possible. As a general rule, the M_2 population should be several times (e.g. 5–10) larger than the M_1 population to ensure that recessive mutations are visible in homozygous plants, even if the plant species has more than one or two genetically efficient cell lines. In grain legumes, it is suggested that between 5000 and 10,000 M_1 seeds are treated to obtain at least 50,000 M_2 plants that will be phenotyped (Maluszynski *et al.*, 2009). A typical process of isolating desirable mutants in a breeding programme for grain legumes is shown in Figure 14.4 (modified from Maluszynski *et al.*, 2009).

The breeding process described above (Fig. 14.4) has been successfully used for mutation breeding of a range of mutant varieties in groundnut (*Arachis hypogea*), black gram (*Vigna mungo*), mung bean (*Vigna radiata*), pigeon pea (*Cajanus cajan*) and soybean (*Glycine max*) at the Bhabha Atomic Research Centre, Bombay, India (Maluszynski *et al.*, 2009). It is interesting to note that, although most mutations have a deleterious effect on plant performance, improved yield components such as pod number per plant, pod size, seed number per pod and seed weight were reportedly easily found. A modified yield component does generally not lead to improved yield in the mutant plant, but yield can be improved significantly if crossed with a second mutant plant that carries another modified yield component (Maluszynski *et al.*, 2009).

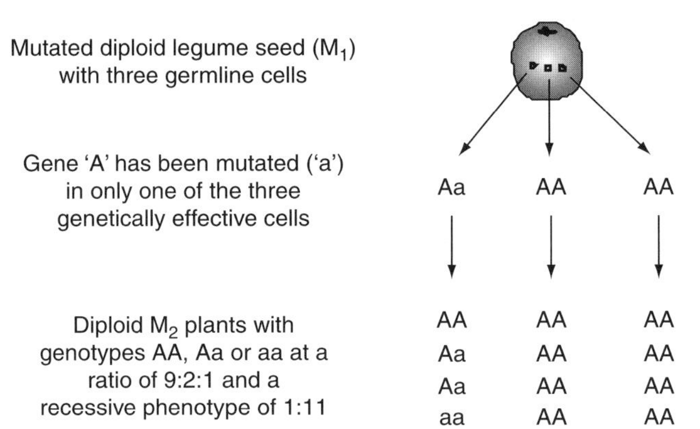

Fig. 14.3. Example of the effects of GECN and sampling on the possibility of finding recessive mutations.

M_1 5 -10,000 seeds are mutagenised
- M_1 plants are grown under optimal glasshouse or field conditions to secure M_2 progeny
- Subset of pods or all pods from each M_1 plant are harvested

M_2 M_2 plants are grown in progeny rows, (>50,000)
- Phenotypic assessment of M_2 plants and identification of interesting mutants

M_3 Seeds from selected mutants are grown to confirm phenotype and assess segregation ratio
- Selection of single plants for next generation analysis

M_4 Mutants are assessed for agronomic traits in comparison to parents and check varieties
- Selection of single plants for next generation analysis

M_x Further generations follow the usual breeding process

Fig. 14.4. Isolation of desirable mutants in a breeding programme for grain legumes.

Crosses between selected mutant plants and the original parental line lead to a reduction in the number of undesirable mutations that might impact negatively on the mutant's agronomical performance. The process can be accelerated if crosses can be made independent of season in a glasshouse. Each backcross has the potential to considerably reduce the number of undesirable mutations, so that after four backcrosses about 97% of undesirable mutations have been removed. The presence of the desirable mutation needs to be monitored and selected at each generation by phenotypic analysis.

14.5 Mutant Populations and their Use for Gene Function Analysis

The benefit of mutagenesis in the context of breeding for improved varieties is obvious. However, mutagenesis and mutant populations are not only useful in generating new traits in well-adapted germplasm, but have also become an invaluable tool in gene discovery and functional analysis of genes.

The development of genetic resources in legumes is expanding rapidly. Genomes of the two model legume species, *Medicago truncatula* (Johnson *et al.*, 2007) and *Lotus japonicus* (*Lotus japonicus* News, 2011) are nearly fully sequenced (Sato *et al.*, 2010), and the soybean genome sequence has just been completed (Schmutz *et al.*, 2010). The latest sequencing technologies have already drastically accelerated genome sequencing by cost reduction and high throughput, and new developments continue this trend (Edwards and Batley, 2010). In the mid-term, it is expected that genomes of the larger grain legumes will also be sequenced as a result of international efforts (Sato *et al.*, 2010).

Availability of the genome sequence allows the linkage of gene sequence information with gene function or, in other words, linking genes with specific plant traits. Mutant populations can play an important role in achieving this objective, as they carry mutated versions of potentially every gene present in the genome. Comparing mutant and non-mutated versions of a target gene and aligning the two versions to the corresponding plant phenotypes (mutant versus wild type) allows potential gene functions to be inferred. Mutant populations that are used for such reverse genetics approaches are called Targeting Induced Local Lesions in Genomics (TILLING) populations. These are similar to aforementioned populations used for mutation breeding, but TILLING populations are generated to carry a maximum number of mutations, so that fewer plants have to be analysed molecularly to find the mutation in the target gene.

Development of a TILLING mutant population

- A maximum mutagen dose (EMS is common) should be used to generate a high density of point mutations: 2000–10,000 M_1 plants treated at LD_{50} to saturate the genome with mutations.
- M_2 seeds are harvested separately from each M_1 plant; as M_1 plants are chimeric, pods can be genetically different and should be kept separate for analysis.
- One M_2 seed per M_1 family (or one M_2 seed from different pods per M_1 family) is sown.
- Fresh leaf material is harvested from each seedling for DNA extraction.
- M_3 seeds are harvested separately from each M_2 plant and stored for later plant analysis.
- DNA is extracted from each individual M_2 plant and subsampled into pools comprising DNA of several (often eight) M_2 plants.

Identification of mutations in a gene of interest

- PCR amplification performed using gene-specific primers and pooled DNA as template.
- PCR products of each DNA pool are heat-denatured and allowed to re-anneal.
- DNA strands in pools that contain PCR products from mutant and wild-type plants form heteroduplexes, due to mismatches at sites where a mutation has occurred.
- Mismatching sites are cleaved by single-strand specific nucleases, such as the enzyme *Cel* 1.
- Nuclease-treated PCR products are separated on denaturing polyacrylamide gels.
- Smaller fragments on the PAGE indicate the presence of mutant DNA in the pool of eight.
- Individual pool members are assessed by PCR and gene sequencing.
- Stored M_3 seeds of mutant plants are sown and plants phenotypically assessed.

Table 14.2. Reported TILLING populations in legumes (modified from Tadege *et al.*, 2009).

Species	Reference(s)
Arachis hypogea (groundnut, peanut)	Ramos *et al.* (2009)
Cicer arietinum (chickpea)	Muehlbauer and Rajesh (2008); Cooper *et al.* (2008)
Glycine max (soybean)	Horst *et al.* (2007)
Lotus japonicus (birdsfoot trefoil)	Perry *et al.* (2003, 2009); Le Signor *et al.* (2009)
Medicago truncatula (barrel medic)	Porceddu *et al.* (2008)
Phaseolus vulgaris (common bean)	Porch *et al.* (2009)
Pisum sativum (pea)	Dalmais *et al.* (2008)

The TILLING approach has been applied in numerous non-leguminous species (Arabidopsis, barley, Canola, maize, rice and sorghum) and in a range of legume species (groundnut, chickpea, *Lotus japonicus*, *Medicago truncatula*, common bean and pea (Table 14.2).

The investment over nearly two decades into genetic resource building of the model legumes *L. japonicus* (Handberg and Stougaard, 1992) and *M. truncatula* (Cook, 1999) has greatly facilitated the search for genes that control agronomically important traits in the crop legumes (Varshney *et al.*, 2009; Young and Udvardi, 2009; Sato *et al.*, 2010). Genes with similar sequences to genes of known function in *Lotus* or *Medicago* can easily be isolated from grain legumes. Verification of the function of these sequence-related genes can be carried out by either TILLING or transgenesis. There are several advantages in the use of TILLING over the use of transgenesis to verify the function of a gene:

- A TILLING population represents a long-lasting resource that can be used to find gene variations in every gene of the genome whereas, in the case of transgenesis, the isolated gene has to be cloned into specific expression vectors and transferred into the crop legume.
- Transgenesis requires established transformation protocols for each species, and corresponding facilities for the culture and gene transfer technology.

- Mutant genes in a mutant plant are under their endogenous gene regulation, that is, gene-specific promoters control the expression in contrast to transgenic gene expression that is under less specific regulation, often constitutive. Transgenic phenotypes can misrepresent a gene's function if the transgene expression varies from the temporal and spatial expression pattern of the endogenous gene.
- In the case of over-expression, the transgene and the endogenous gene are both present in the same plant, which could mask the effect of the transgene.
- Mutant plants can be assessed under realistic environments in the field and do not require approval by gene technology regulatory authorities in most countries, except in the USA, where mutagenized plants are considered as being genetically modified.

Transgenesis, however, has other advantages over mutagenesis, particularly if traits are targeted that are either very rare or impossible. For example, lethal mutations are impossible to recover in mutant populations, but the expression of a lethal gene version can be studied in transgenic plants if the gene is expressed in a tissue or during a developmental stage that does not affect the viability of the plant.

Importantly, it is the ability to introduce genes outside of the natural gene pool that gives genetic engineering great potential for modern plant breeding. Examples of transgenic food legume are the herbicide-tolerant soybean varieties. An impressive 77% of soybean production worldwide is in varieties that carry bacterial genes that mediate tolerance to the herbicides glyphosate or glufosinate ammonium. In the USA, 93% of the cultivated soybean crop carries these transgenic traits.

14.6 Conclusions

Estimates of the frequency of spontaneous mutations in plants suggest about 7×10^{-9} base substitutions per DNA site and generation. Taking into account the genome sizes, the offspring carries numerous DNA sequence differences compared with their parents. For example, in homozygous and selfing pea (*Pisum sativum*) with a genome size of about 5×10^9 base pairs, seven new mutations can be expected on average in the next generation.

This mutation rate, together with diverse selection pressures in different environments (natural or man-made), drives the genetic diversity of the germplasm. Mutagenesis is an easy-to-use tool for increasing genetic diversity, either for direct breeding purposes, to introduce trait variation within adapted germplasm or to better understand gene function by employing a TILLING approach. Mutation rates, calculated on the basis of measured mutation rate and estimated genome size, suggest that several thousand induced mutations are present in a mutant plant. The wide range of traits being developed with the help of mutagenesis provides evidence of the commercial value of mutation breeding; using mutagenesis, new traits are still being developed that might further be developed into commercial varieties. Recently, Campion *et al.* (2009) described the discovery of a common bean (*Phaseolus vulgaris*) mutant line with a seed in which the phytic acid concentration was reduced by 90%. The mutant line was found by screening for levels of free phosphate content in seeds of around 1000 M_2 families from a larger EMS-mutant population. The low phytic acid line is expected to increase the bioavailability of important micronutrients, such as iron and zinc, that are usually deficient in plant-based diets in developing countries (Campion *et al.*, 2009).

The trend in recent years towards development of TILLING populations in major crop species such as the cereals has led to the development of valuable mutant populations in major grain legumes such as chickpea, common bean, pea, groundnut and soybean. TILLING populations are a versatile instrument, and the mutant lines can be utilized for two different genetics approaches:

- forward genetics approaches, where an interesting phenotype is known and the corresponding gene is to be identified via methods such as positional or map-based cloning; and

- reverse genetics approaches, where a gene sequence is known and its function analysed by identifying corresponding mutant plants by using the aforementioned screening approach of a TILLING population.

The function of genes previously characterized in model legumes or even in non-legume species can be quickly verified in grain legume species to further elucidate gene function or to use the candidate gene for molecular breeding in a crop legume. Candidate genes with a validated function in crop species represent perfect molecular markers that can be used for marker-assisted selection of the desirable trait, either in a breeding programme or for genetic engineering using transgenesis.

The versatility of mutant populations has secured their role in plant and legume research over the last 60 years. It is a technology with the ability to merge traditionally separated disciplines, the rather applied area of plant breeding and the more fundamentally oriented area of functional genomics. Progress in grain legumes will be supported by findings in other crop or model plant species. Advances in sequencing technologies will facilitate the investigation of traits that are specific to grain legumes, such as quality traits that cannot be addressed with the current model legumes. Trait variation through mutation will remain a means to harvest the enormous potential of food legumes in providing a sustainable protein source.

References

Bradshaw, L.D., Padgette, S.R., Kimball, S.L. and Wells, B.H. (1997) Perspectives on glyphosate resistance. *Weed Technology* 11, 189–198.

Campion, B., Sparvoli, F., Doria, E., Tagliabue, G., Galasso, I., Fileppi, M. *et al.* (2009) Isolation and characterisation of an *lpa* (low phytic acid) mutant in common bean (*Phaseolus vulgaris* L.). *Theoretical and Applied Genetics* 118, 1211–1221.

Carroll, B.J., McNeil, D.L. and Gresshoff, P.M. (1985) A supernodulation and nitrate-tolerant symbiotic (Nts) soybean mutant. *Plant Physiology* 78, 34–40.

Cook, D.R. (1999) *Medicago truncatula* – a model in the making! *Current Opinion in Plant Biology* 2, 301–304.

Cooper, J.L., Till, B.J., Laport, R.G., Darlow, M.C., Kleffner, J.M., Jamai, A. *et al.* (2008) TILLING to detect induced mutations in soybean. *BMC Plant Biology* 8, 9–18.

Dalmais, M., Schmidt, J., Le Signor, C., Moussy, F., Burstin, J., Savois, V. *et al.* (2008) UTILLdb, a *Pisum sativum* in silico forward and reverse genetics tool. *Genome Biology* 9, R43.

Edwards, D. and Batley, J. (2010) Plant genome sequencing: applications for crop improvement. *Plant Biotechnology Journal* 8, 2–9.

Handberg, K. and Stougaard, J. (1992) *Lotus japonicus*, an autogamous, diploid legume species for classical and molecular genetics. *Plant Journal* 2, 487–496.

Horst, I., Welham, T., Kelly, S., Kaneko, T., Sato, S., Tabata, S. *et al.* (2007) TILLING mutants of *Lotus japonicus* reveal that nitrogen assimilation and fixation can occur in the absence of nodule-enhanced sucrose synthase. *Plant Physiology* 144, 806–820.

IAEA (2011) Available at http://mvgs.iaea.org/ (accessed 24 February 2011).

Johnson, S, Grunwald, D, Driever, W and Mullins, M. (2007) *Genetic Methods: Chemical Mutagenesis.* Available at http://zfin.org/zf_info/zfbook/chapt7/7.10.html (accessed 30 October 2010).

Ladizinsky, G. and Adler, A. (1976) Origin of chickpea *Cicer arietinum* L. *Euphytica* 25, 211–217.

Le Signor, C., Savois, V., Aubert, G., Verdier, J., Nicolas, M., Pagny, G. *et al.* (2009) Optimizing TILLING populations for reverse genetics in *Medicago truncatula. Plant Biotechnology Journal* 7, 430–441.

Lotus japonicus News (2011) Available at http://www.kazusa.or.jp/lotus/ (accessed 24 February 2011).

Maluszynski, M., Szarejko, I., Bhatia, C.R., Nichterlein, K. and Lagoda, P.J.L. (2009) Methodologies for generating variability. In: Ceccarelli, S., Guimar, E.P. and Weltzien, E. (eds) *Plant Breeding and Farmer Participation*. Food and Agriculture Organization (FAO), Rome.

Muehlbauer, F.J. and Rajesh, P.N. (2008) Chickpea, a common source of protein and starch in the semi-arid tropics. In: Moore P.H. and Ming, R. (eds) *Genomics of Tropical Crop Plants*. Springer, New York, pp. 171–186.

Oldach, K.H., Peck, D.M., Cheong, J., Williams, K.J. and Nair, R.M. (2008) Identification of a chemically induced point mutation mediating herbicide tolerance in annual medics (*Medicago* spp.). *Annals of Botany* 101, 997–1005.

Perry, J.A., Wang, T.L., Welham, T.J., Gardner, S., Pike, J.M., Yoshida, S. *et al.* (2003) A TILLING reverse genetics tool and a web-accessible collection of mutants of the legume *Lotus japonicus*. *Plant Physiology* 131, 866–871.

Perry, J., Brachmann, A., Welham, T., Binder, A., Charpentier, M., Groth, M. *et al.* (2009) TILLING in *Lotus japonicus* identified large allelic series for symbiosis genes and revealed a bias in functionally defective ethyl methanesulfonate alleles toward glycine replacements. *Plant Physiology* 151, 1281–1291.

Porceddu, A., Panara, F., Calderini, O., Molinari, L., Taviani, P., Lanfaloni, L. *et al.* (2008) An Italian functional genomic resource for *Medicago truncatula*. *BMC Research Notes* 1, 12.

Porch, T.G., Blair, M.W., Lariguet, P., Galeano, C., Pankhurst, C.E. and Broughton, W.J. (2009) Generation of a mutant population for TILLING common bean genotype BAT 93. *Journal of the American Society for Horticultural Science* 134, 348–355.

Powles, S.B. and Preston, C. (2006) Evolved glyphosate resistance in plants: Biochemical and genetic basis of resistance. *Weed Technology* 20, 282–289.

Ramos, M.L., Huntley, J.J., Maleki, S.J. and Ozias-Akins, P. (2009) Identification and characterization of a hypoallergenic ortholog of Ara h 2.01. *Plant Molecular Biology* 69, 325–335.

Redei, G.P. (1975) Induction of auxotrophic mutations in plants. In: Ledoux, L. (ed.) *Genetic Manipulations with Plant Material*. Plenum Press, New York, pp. 329–350.

Sato, S., Isobe, S. and Tabata, S. (2010) Structural analyses of the genomes in legumes. *Current Opinion in Plant Biology* 13, 146–152.

Schmutz, J., Cannon, S.B., Schlueter, J., Ma, J.X., Mitros, T., Nelson, W. *et al.* (2010) Genome sequence of the palaeopolyploid soybean. *Nature* 463, 178–183.

Stadler, L.J. (1930) Some genetic effects of X-rays in plants. *Journal of Heredity* 2, 3–20.

Tadege, M., Wang, T.L., Wen, J.Q., Ratet, P. and Mysore, K.S. (2009) Mutagenesis and beyond! Tools for understanding legume biology. *Plant Physiology* 151, 978–984.

Tranel, P.J. and Wright, T.R. (2002) Resistance of weeds to ALS-inhibiting herbicides: what have we learned? *Weed Science* 50, 700–712.

Varshney, R.K., Close, T.J., Singh, N.K., Hoisington, D.A. and Cook, D.R. (2009) Orphan legume crops enter the genomics era! *Current Opinion in Plant Biology* 12, 202–210.

Whaley, C.M., Wilson, H.P. and Westwood, J.H. (2007) A new mutation in plant ALS confers resistance to five classes of ALS-inhibiting herbicides. *Weed Science* 55, 83–90.

Young, N.D. and Udvardi, M. (2009) Translating *Medicago truncatula* genomics to crop legumes. *Current Opinion in Plant Biology* 12, 193–201.

15 Breeding for Biotic Stresses

**Ashwani K. Basandrai, Daisy Basandrai,
P. Duraimurugan and T. Srinivasan**

15.1 Introduction

Food legumes are an important source of human food and animal feed and are in great demand, particularly in countries where vegetarian diets dominate the food habits. However, pulses are preferred less by farmers due to their inherent low yields and vulnerability to various biotic and abiotic stresses, leading to low productivity. Biotic stresses, for example diseases, insect pests and plant-parasitic nematodes, exact a heavy toll on crop productivity and cause yield instability in food legumes. Some diseases may even force farmers to abandon cultivation of certain food legumes, thereby threatening the sustainability of whole crop production systems. For example, drastic reductions in areas of chickpea cultivation have occurred in northern India due to ascochyta blight and pod borer (Kaur *et al.*, 2008; Pande *et al.*, 2008; Sarker *et al.*, 2008). The full genetic potential of these crops is seldom realized, due to the cultivation of susceptible cultivars and improper crop management practices employed in coping with these biotic stress factors. The pulse deficit of 2–3 million tonnes per annum in countries like India must be curtailed by minimizing production losses through suitable management options easily extendable equally to the small and marginal pulse growers in the developing world and large farmers of the Western world.

In this context, host plant resistance is the most sought-after solution, being an economical, durable, environmentally safe and acceptable means to manage these biotic stresses. Emphasis in breeding programmes has been to incorporate disease, insect pest and nematode resistance genes in addition to improving yield and quality. Most recently, biotechnological tools have been adopted to tag molecular markers with resistance genes to enhance breeding efficiency through marker-assisted selection. Mutation breeding and transgenic technology offers the opportunity for genetic enhancement where genes for resistance are not available in nature. This chapter discusse the progress of and prospects through genetic enhancement in combating biotic stresses in important food legumes.

15.2 Biotic Stresses and Extent of Losses in Food Legumes

Food legumes are cultivated on 23 million ha worldwide, accounting for over 18% of the total arable area, but only 8% of the total grain production. There is a large disparity between yields of cereals and legumes. Losses due to biotic stresses, i.e. diseases, insect pests and plant-parasitic nematodes are the most serious; Kaur *et al.* (2008)

reported 10–15% production losses in food legumes due to diseases. Avoidable losses due to insect pests at current production levels of 60.45 million t would equate to nearly 18.14 million t (at an average loss of 30%), valued at nearly US$10 billion (Sharma *et al.*, 2008). Sasser and Freckman (1987) estimated that worldwide average yield losses caused by plant-parasitic nematodes were 13.7, 13.2, 15.1 and 10.9% in chickpea, pigeon pea, cowpea and field bean, respectively. The major biotic stresses associated with food legumes are listed in Table 15.1.

15.3 Mechanisms of Resistance against Major Biotic Stresses

In nature, many defence mechanisms operate in plants for their protection against pathogens, insect pests and nematodes. The mechanisms of resistance against major biotic stresses in food legumes are detailed below.

Mechanisms of resistance against pathogens

Activation of resistance genes following infection leads to the biosynthesis and accumulation of phytoalexins and secondary metabolites toxic to pathogens, thus making plants more resistant to attack. Plants also accumulate a novel class of proteins called pathogenesis-related proteins (PR proteins) in response to pathogen attack (Sarker *et al.*, 2008). Srivastava (2009) reported that phenolic acids such as chlorogenic, coumaric, caffeic, ferulic and protocathuic acids present in plant roots play an important role in imparting resistance against *Fusarium oxysporum* f.sp. *ciceri* in chickpea; cholorogenic acid levels in the roots of resistant varieties were 400–1500 mg/kg, compared with 150–200 mg/kg in susceptible varieties. The leaves of wilt-resistant plants also had high cholorogenic and coumaric or ferulic acids compared with susceptible plants. Root exudates from the susceptible chickpea cultivar, JG 62 stimulated mycelial growth and conidial and chlamydospore germination of the *Fusarium* wilt fungus, while root exudates from the resistant cultivar, CPS 1 inhibited germination and growth (Satyaprasad and Rama Rao, 1983).

In the chickpea–*Ascochyta rabiei* interaction system, genotypic response to pathogens is controlled by a diverse set of morphological, anatomical, biochemical and genetic characters. The presence of glandular hair along with non-glandular hair reduces the susceptibility of the genotype, probably acting as a mechanical barrier to the entry of the pathogen. Leaf and stem cuticle thickness, palisade tissue thickness, palisade index and epidermal cell wall thickness are closely associated with resistance to ascochyta blight in chickpea. Phenolic components and certain enzymes (peroxidase and β-1,3 glucanase) are linked to wilt and *Ascochyta* resistance in chickpea (Singh *et al.*, 2003). An increased number of malic acid-secreting glandular hairs was found to be linked to ascochyta blight resisance. Furthermore, seven biochemical characters (total phenol, ortho-dihyroxyphenol, flavanols, lignin, silica, epicuticular waxes and copper content) were found to be closely associated with resistance to ascochyta blight in chickpea (Hafiz, 1952). Some resistance mechanisms in foliar diseases may be determined by plant growth habits, e.g. varieties with erect, non-bushy and a non-spreading canopy may provide better aeration and light and thus reduce relative humidity, which may help in the management of diseases like ascochyta blight and botrytis grey mould in chickpea.

The fungitoxic isoflavonoid phytoalexin, cajanol, was identified as the main antifungal compound against fusarium wilt (*Fusarium udum*) of pigeon pea (Marley and Hillocks, 1993). In this study, the concentration of cajanol 15 days after inoculation was 329.4 μg/g in the resistant cultivar compared with 88.6 μg/g in the susceptible one. Crude extract from the resistant plants sampled 24 h after inoculation contained 34.8 μg ml of cajanol, and the LD_{50} value of cajanol for spore germination was 35 μg/ml. The cajanol content of fungus-infected ICP 9145 totally inhibited conidial germination of *F. udum* after 10 days of inoculation.

The viral coat protein (CP) gene is the first and one of the most widely used

Table 15.1. Major biotic stresses in food legumes.

Crop	Diseases	Insects	Parasitic nematodes
Chickpea (*Cicer arietinum*)	Fusarium wilt (*Fusarium oxysporum* f.sp. *ciceri*); ascochyta blight (*Ascochyta rabiei*); botrytis grey mould (*Botrytis cinerea*); dry root rot (*Rhizoctonia bataticola*)	Pod borers (*Helicoverpa armigera, H. punctigera*); cutworm (*Agrotis ipsilon*); leaf miner (*Liriomyza cicerina*); bruchids (*Callosobruchus chinensis*)	Root-knot nematodes (*Meloidogyne incognita, M. javanica*); root lesion nematode (*Pratylenchus* spp.); reniform nematode (*Rotylenchulus reniformis*); *Tylenchorhynchus* spp.
Lentil (*Lens culinaris*)	Rust (*Uromyces vicia fabae*); fusarium wilt (*Fusarium oxysporum* f.sp. *lentis*); ascochyta blight (*Ascochyta lentis*); stemphylium blight (*Stemphylium botryosum*)	Spiny pod borer (*Etiella zinckenella*); aphids (*Aphis craccivora*); bruchids (*Callosobruchus* spp.)	Root-knot nematodes (*Meloidogyne incognita, M. javanica*); reniform nematode (*Rotylenchulus reniformis*); *Helicotylenchus* spp.
Mung bean (*Vigna radiata*) and urd bean (*Vigna mungo*)	Cercospora leaf spot (*Cercospora cruenta, C. canescens*); powdery mildew (*Erysiphe polygon, Sphaerotheca fuligenea*); mung bean yellow mosaic virus	Spotted pod borer (*Maruca vitrata*); whitefly (*Bemisia tabaci*); aphids (*Aphis craccivora*); thrips (*Caliothrips indicus, Megalurothrips usitatus*); stem fly (*Ophiomyia phaseoli*); bruchids (*Callosobruchus chinensis*)	Root-knot nematodes (*Meloidogyne incognita, M. javanica*); cyst nematode (*Heterodera cajani*)
Pigeon pea (*Cajanus cajan*)	Fusarium wilt (*Fusarium udum*); phytophthora blight (*Phytophthora drechsleri* f.sp. *cajani*); alternaria blight (*Alternaria alternata, A. tennissima*); sterility mosaic (pigeon pea sterility mosaic virus)	Pod borers (*Helicoverpa armigera, H. punctigera*); spotted pod borer (*Maruca vitrata*); pod fly (*Melanagromyza obtusa*); bruchids (*Callosobruchus chinensis*)	Cyst nematode (*Heterodera cajani*); root-knot nematodes (*Meloidogyne incognita, M. javanica*); reniform nematode (*Rotylenchulus reniformis*); *Hoplolaimus* spp.
Cowpea (*Vigna unguiculata*)	Cercospora leaf spot (*Cercospora canescens, Pseudocercospora cruenta*); cowpea mosaic virus	Spotted pod borer (*Maruca vitrata*)	Root-knot nematodes (*Meloidogyne* spp.); cyst nematode (*Heterodera cajani*); reniform nematode (*Rotylenchulus reniformis*)
Pea (*Pisum sativum*)	Powdery mildew (*Erysiphe pisi*); downy mildew (*Peronospora viciae*); rust (*Uromyces vicia fabae*), white rot (*Sclerotinia sclerotiorum*)	Spiny pod borer (*Etiella zinckenella*); stem fly (*Ophiomyia phaseoli*); pod borers (*Helicoverpa armigera, H. punctigera*)	Root-knot nematodes (*Meloidogyne incognita, M. javanica*); reniform nematode (*Rotylenchulus reniformis*)

genes to have been used to confer pathogen-derived resistance (PDR) against plant viruses. PDR can be categorized as trangene-encoded protein-mediated resistance and trangene RNA-mediated resistance. An enhanced level of resistance to pea enation mosaic virus (PEMV) was obtained by introducing its coat protein gene (PEMV-CP).

The key anti-fungal proteins are chitinases and β-1,3 glucanase. Chitinase catalyses the hydrolysis of chitin, whereas glucanase hydrolyses β-1,3 glucan, both enzymes inhibiting fungal growth through the breakdown of cell components. Among the four classes of hydrolases, class I hydrolases, localized in the plant vacuole, showed anti-fungal properties. Class I glucanases, in combination with chitinases, showed a very strong growth inhibition of many parasitic fungi. Plant ribosomal-inactivating proteins (RIPs) inhibit protein synthesis in target cells by a specific modification of 28 S rRNA. RIPs do not inhibit the protein synthesis machinery of plants, but inhibit fungal ribosomes, and a strong synergy is observed when RIPs are combined with chitinases or glucanases. The gene encoding the polygalacturonase inhibitor has been cloned and characterized, and is now being used in fungal disease resistance programmes. High level of phenols and low concentrations of carbohydrates are linked to resistance to cercospora leaf spot in mung bean.

Mechanisms of resistance against insect pests

Multiple types of resistance (antixenosis, antibiosis, tolerance and escape) have been reported against *Helicoverpa armigera* in chickpea. Oviposition non-preference is one of the major components of resistance to this pest in both chickpea (Cowgill and Lateef, 1996) and pigeon pea (Sharma *et al.*, 2001; Kumari *et al.*, 2006). The acid exudates (pH 1.3) with a high concentration of malic acid secreted from the glandular hairs on the leaves, stems and pods have been suggested as a marker for resistance (Rembold, 1981). The genotypes resistant to *H. armigera* accumulated more oxalic acid on the leaves than susceptible genotypes. Oxalic acid showed significant growth inhibition on *H. armigera* larvae when included in a semi-artificial diet. Inhibition of larval growth by oxalic acid was not caused by anti-feedant effects but was more likely attributable to antibiosis (Yoshida *et al.*, 1995). High percentages of cellulose, hemicellulose and lignin in the pod husk and high percentages of crude fibre and non-reducing sugars and a low percentage of starch in chickpea seeds have been found associated with resistance against *H. armigera* (Chhabra *et al.*, 1990). Pupae of *H. armigera* reared on genotypes ICC 506 and ICCV 7 weighed less than those reared on ICC 37 (Cowgill and Lateef, 1996). Cultivars resistant to cutworm (*Agrotis ipsilon*) attack in chickpea have been shown to have well-developed secondary xylem at the base of the stem.

All four known mechanisms of resistance are reported in pigeon pea against *H. armigera*. The presence of dense glandular hairs, the concentration of tannin-like substances beneath the outer epidermis and the thickness of the fibrous cell layer above the inner epidermis may influence oviposition preference, in both pigeon pea and its wild relatives. Venugopal Rao *et al.* (1991) observed that *H. armigera* females laid more eggs on ICPL 270, while ICPL 332, ICPL 84060 and LRG 30 were less preferred.

Antibiosis effects are expressed in terms of weight and size of insects, sex ratio and proportion of insects entering into diapause. An experiment conducted at ICRISAT (International Crops Research Institute for the Semi-Arid Tropics), Patancheru showed that *H. armigera* females did not lay eggs on the wild relatives *Cajanus platycarpus*, *Cajanus scarabaeoides* and *Cajanus sericeus*, whereas egg laying was seen in cultivated pigeonpea. Pigeon pea pods produce chemicals from glandular trichomes that act as a phagostimulant to *Helicoverpa* larvae. Pods of *C. scarabaeoides* have a dense shoot and non-glandular trichomes that act as physical barriers to feeding by the young larvae (Green *et al.*, 2002). Guercetrin, isoguercetrin, guercetrin-3-methyl ether (feeding stimulants) and genistein (anti-feedant) are present only in *C. scarabaeoides*, and two other flavonoids have been identified in pod surface extracts (Sharma *et al.*, 2001). Guaiene and beta-caryophyllene, which are present in cultivated pigeon pea but absent in *C. scarabaeoides*, act as attractants to *Helicoverpa* adults. Romeis *et al.* (1999) identified different types of trichomes in pigeon pea and its wild relatives. Shanower *et al.* (1997) suggested that increasing the density of non-glandular trichomes on pigeon pea pods could reduce damage and losses due to pod-feeding insect pests. Sharma *et al.* (2009) studied the morphological and biochemical components associated with expression of resistance to *H. armigera* in wild relatives of pigeon pea. Among the wild relatives, oviposition non-preference was observed in *Cajanus scarabaeoides*. Accessions of *Rhyncosia aurea*, *C. scarabaeoides*, *C. sericeus*, *C. acutifolius* and *Flemingia bracteata* showed high levels of resistance, with the non-glandular trichomes (types C and D) on calyces and pods being associated with resistance.

Resistance was also associated with low levels of sugars and high concentrations of tannins and polyphenols. Therefore, accessions of wild relatives of pigeon pea with non-glandular trichomes (types C and D), or low densities of glandular trichomes (type A), and high levels of polyphenols and tannins, may be used in wide hybridization to develop pigeon pea cultivars with resistance to *H. armigera*.

Both ovipositional non-preference and antibiosis have been suggested as modes of resistance for pigeon pea pod fly, *Melanogromyza obtusa* (Reed and Lateef, 1990). Several plant characters have been implicated in pod fly ovipositional preference, including pod trichomes, concentrations of tannin-like substances beneath the outer epidermis and thickness of the fibrous cell layer above the inner epidermis (Sithanantham *et al.*, 1981). Lal and Yadava (1994) observed that resistant pigeon pea selections had fewer pod fly eggs than susceptible selections, indicating that their ovipositional non-preference may be an important character in pod fly resistance. Pod and seed size also showed a direct relationship with pod fly resistance (Durairaj, 1999). Lal *et al.* (1988) reported that the pods and seeds of most of the pod fly-resistant types were small.

A high density of trichomes on stems and leaves, purplish stem of small diameter and small, unifoliate leaves all contribute towards the biophysical basis of resistance to stem fly (*Ophiomyia phaseoli*) in mung bean (Talekar *et al.*, 1988). Antibiosis appeared to be an important component of resistance to stemfly in mung bean. Insects feeding on the stems of resistant accessions had a longer larval period than those feeding on susceptible lines. In many cases, larvae feeding on resistant accessions had tenfold more tannins than that of the susceptible control (Talekar, 1983). Talekar *et al.* (1988) found that, in the smaller unifoliate and first two trifoliate leaves, a higher trichome density on leaf surface and stem, smaller diameter of stem and shorter internode between first and second nodes were associated with resistance of mung bean to stem fly.

Analysis of the stem cortex of V 4281 and one accession of *Vigna glabrescens* (a close relative of mung bean, which is highly resistant to agromyzids) showed 0.51 and 0.50 mg phenolic compounds per gram of cortex tissue (dry weight basis), respectively, compared with the susceptible V 2184, with 0.06 mg/g phenolic compounds. Durairaj and Sakthivel (2007) reported higher phenol levels as the biochemical basis for resistance in mung bean and urd bean against stem fly. In mung bean and urd bean, high levels of biochemicals such as phenols, amino acids and non-reducing sugars are responsible for imparting resistance against whitefly (*Bemisia tabaci*) and green jassid (*Empoasca* spp.; Chhabra *et al.*, 1981, 1993). Chandra *et al.* (1992) reported that the variation in resistance to aphid (*Aphis craccivora*) infestation was correlated mainly with the colour of genotype foliage; they observed that *A. craccivora* is attracted towards the colour yellow. Chhabra *et al.* (1994) reported low concentrations of free amino acids, total phenols, total minerals, total sugars, non-reducing sugars, calcium and potassium, and high levels of total carbohydrates, as the mechanisms responsible for resistance against bean thrips, *Megalurothrips distalis*.

Cowpea resistance to the flower and pod borer (*Maruca testulalis*) is governed by biochemical, anatomical and/or morphological factors. These factors may act either in tandem or independently to confer resistance, depending on the genotype or the insect species (Oghiakhe *et al.*, 1992). Macfoy *et al.* (1983) studied ovipositional preferences of *M. testulalis* and reported that TVu 946 was least preferred, due to nutritional and antibiotic factors. Factors such as nutritional composition (primary and secondary metabolites), trichomes and pod position and angle are more important in cowpea pod resistance to *M. testulalis* than its anatomy (Jackai and Oghiakhe, 1989). Among other factors, pod trichome density is important in reducing damage to cowpea pods by the larvae of *M. testulalis*. However, pod wall toughness had no effect on resistance (Oghiakhe *et al.*, 1992).

Mechanisms of resistance against plant-parasitic nematodes

Peroxidase plays an important role in the resistance mechanism of plants. It is a key enzyme required for lignin synthesis, as well as other trapezoids involved in phytoalexin production. Peroxidase catalyses several reactions, including those involved in the metabolism of phenols and indoles. IC 4928 to IC 4848 (25 genotypes) were screened on the basis of increase in peroxidase activity. IC 4941, IC 4942 and IC 4944 were reported to be tolerant against *Meloidogyne incognita* (Siddiqui and Hussain, 1992). Chickpea cv. K 850 was inoculated with 1000–2500 J2 *M. incognita*; after 60 days, biochemical analysis revealed that there were increases of 10–18% and 26–54% in total protein and amino acid concentrations, respectively, which were found to be greater in the stem and at higher levels of infection. An increase in the protein content of chickpea was dependent on the level of infection by root-knot nematodes (Upadhyay and Banerjee, 1986). Cai *et al.* (1997) reported that the gene conferring resistance to the beet cyst nematode (*Heterodera schachtii*) encoded an LRR-containing protein, which led to the developmental arrest of the nematode and breakdown of the feeding structure. A similar mechanism of resistance to *Heterodera cajani* has been noted in the two accessions of *Cajanus platycarpus*.

15.4 Genetics of Resistance

Resistance to a particular disease may be under either monogenic or polygenic control in different genetic backgrounds, e.g. resistance to ascochyta blight in chickpea and lentil may be controlled by major as well as minor genes. Resistance to aphids and weevils in pea and lentil is under polygenic control. The degree of dominance will determine whether only one or both parents need to be resistant. Knowledge of the number of genes and their interaction enables the breeder to calculate the population size needed to achieve the desired probability of obtaining resistant plants in segregating generations. Such information on mode of inheritance of resistance to major biotic stresses in different food legumes is given in Table 15.2.

15.5 Screening for Resistance

Screening for resistance to diseases

Wilt

Wilt caused by *Fusarium* species constitutes an important problem in food legumes. Screening of test materials is generally done in plots and glasshouse conditions; in the glasshouse it is done according to the methods suggested by Nene and Haware (1980) and Pande *et al.* (2006). Field screening is done in sick plots.

Root rot

Root rot testing can be done with the addition of inocula in soil in either glass- or greenhouses. The use of associated traits and an estimation of the concentration of root exudates have been performed in different studies (Dua *et al.*, 2002).

Powdery mildew

Screening for resistance to *Erysiphe polygoni* is done in the field under natural infestation or by the use of infector rows planted after every five rows of the test material. This procedure restricts the screening to the cool and

Table 15.2. Inheritance of resistance to major biotic stresses in major food legumes.

Crop	Biotic stresses	Mode of inheritance	Reference
Diseases			
Chickpea	Fusarium wilt	Two genes, one each for early and late wilting	Gumber *et al.* (1995)
		Two recessive and one dominant genes at three different loci	Kumar (1998)
		Single recessive gene linked to race 1	Tullu *et al.* (1998)
	Ascochyta blight	Monogenic recessive for race 3	Singh and Reddy (1989)
		Five dominant genes with inter-allelic interactions	Dey and Singh (1993)
Pigeon pea	Fusarium wilt	Monogenic dominant	Singh *et al.* (1988b)
	Sterility mosaic	Digenic recessive complementary	Singh *et al.* (1991)
		Two independent non-allelic genes, at least three multiple alleles, at each locus	Srinivas *et al.* (1997)
	Phytopthora blight	Monogenic dominant	Sharma *et al.* (1982)
	Alternaria blight	Monogenic recessive	Singh *et al.* (1988b)
Lentil	Fusarium wilt	Monogenic dominant	Eujayl *et al.* (1998)
		Five independent segregating genes	Kamboj *et al.* (1990)
	Rust	Duplicate dominant gene	Lal *et al.* (1996)
		Monogenic dominant	Kumar *et al.* (1997)
	Ascochyta blight	Monogenic dominant	Ford *et al.* (1999)
Field pea	Fusarium wilt	Monogenic recessive	Marx and Providenti (1979)
	Ascochyta blight	Monogenic dominant	Darby *et al.* (1985)
	Powdery mildew	Monogenic recessive	Timmerman-Vaughan *et al.* (1994)
Mung bean	MYMV	Digenic recessive	Sandhu *et al.* (1985)
		Digenic with dominant and recessive epistasis	–
Insect pests			
Pigeon pea	Pod fly (*Melanagromyza obtusa*);	Two recessive genes	Verulkar *et al.* (1997)
	pod borer (*Helicoverpa armigera*)	Single dominant gene	Verulkar *et al.* (1997)
Cowpea	Aphid (*Aphis craccivora*);	Monogenic dominant	Githiri *et al.* (1996)
	Weevil (*Callosobruchus maculatus*)	Additive dominance and maternal components	Githiri *et al.* (1996)
Plant-parasitic nematodes			
Cowpea	Root-knot nematode (*Meloidogyne incognita*)	Recessive and governed by a single gene	Ehlers *et al.* (2000)
Soybean	*Meloidogyne javanica*;	Quantitative inheritance	Luzzi *et al.* (1995)
	Heterodera glycines	Quantitative inheritance (estimated broad sense heritability 0.48–0.81)	Mansur *et al.* (1993)

dry seasons, when powdery mildew disease is prevalent. In a glasshouse screening trial, plants were inoculated with *E. polygoni* at the seedling stage by dusting with conidia from infected leaves (Sokhi *et al.*, 1979). Scoring can be done on a scale of 0–5, as suggested by Munjal *et al.* (1964); the results compared closely with field screening of the same strains.

Ascochyta blight

Efficient inoculation techniques for use in both the greenhouse and field have been standardized. Inoculating plants grown in pots, bags or trays and covering them with polythene or cloth bags or cages for 24–48 h results in satisfactory levels of infection, temperatures suitable for infection ranging from 15 to 25°C. The presence of a moisture film on the leaf surface is also essential for infection. In the field, inoculation by spreading of disease debris or spraying a spore suspension over the plants followed by sprinkler irrigation results in high and uniform disease levels (Reddy *et al.*, 1984). Rating scales for the scoring of disease severity have been standardized (Reddy and Singh, 1984). Pande *et al.* (2010) reported field, cloth chamber, cut twig and detached leaf techniques for screening resistance against ascochyta blight. In the field-screening technique (followed at hot-spot locations, i.e. Dhaulkaun, Hisar and Ludhiana), test material was planted in rows 3–5 m spaced 40 cm apart in replicated trials. The indicator-cum-infector rows of the highly susceptible varieties L550/ ILC1929 were planted every 2/4/8 rows. At flowering in February, material was artificially inoculated by a spore suspension of *Ascochyta rabiei* at the level 4×10^5 ml/l. Adequate moisture and relative humidity > 85% was maintained by running the perfo-spray from 10.00 to 16.00 h daily at 1 h intervals. Signs of disease began to appear 7–8 days after inoculation, and 100% mortality was observed in both susceptible material and controls after 15 days. Final observations were recorded after 21 days, before maturing of the crop.

The cloth chamber technique is very quick, reliable, economical and useful for utilizing resistant donors in the same season for crossing, backcrossing and shortlisting of large germplasm collections, evaluation of research material against different phenotypes and identification/detection of new pathotypes. Results of screening are available within 14 days of inoculation. The cut twig technique is used for testing precious materials, where the whole plant cannot be risked, vis-à-vis the same plant can be used for testing against other pathotypes, diseases and agronomic traits. Materials identified as resistant can be used in crosses/backcrosses in the same season (Pande *et al.*, 2010). The detached leaf technique is quicker than the cut twig technique and can be used for interspecific hybridization and backcrossing resistance breeding programmes.

Rust

Screening for rust disease can be done by planting test material surrounded on all sides by a susceptible variety. Pal *et al.* (1979) suggested using one spreader row for every two rows of the test material. Artificial inoculations can be done by spraying an urdo-/ aeciosporic suspension from infected plants to induce maximum infection. After spraying the inoculum in the evening, the plots may be irrigated to maintain humidity.

Leaf spot

Artificial inoculation to evaluate resistance to leaf spot was at first difficult due to the poor sporulation of inocula on the artificial medium. This problem was overcome by growing the pathogen on a carrot leaf juice/oatmeal agar medium, where it sporulated abundantly (Mew *et al.*, 1975), permitting inoculation by spraying of spore suspension. The infector row technique was adopted using the highly susceptible mung bean variety, Kopergaon as an infector-cum-spreader row for every two rows of a test entry, grown in 3–5 m lengths. To increase disease pressure, the inoculum is obtained by macerating the infected leaves and incubating either in a moist chamber or on moist blotting paper in Petri dishes at 25°C for 48 h to allow the pathogen to sporulate profusely. The field is properly irrigated prior to spraying of the inoculum. To maintain high relative humidity in the screening nursery, a sprinkler-spray system is used at 1 h intervals during daytime for 2–3 weeks. For recording the results, a 1–9-point rating scale has been devised (Singh and Naimuddin, 2009).

Phytopthora stem blight

Nene *et al.* (1981) reported field, pot and greenhouse techniques in screening for resistance against PSB in pigeon pea. The field technique involved rubbing of inocula at the base of the stem of one-month-old plants individually and providing light irrigation. In the pot technique, the drench inoculation method is followed in which 5–10-days-old seedlings are inoculated by pouring 100ml inoculum around their base. A 1–9-point rating scale is followed to measure the severity of disease.

Botrytis grey mould

Growth room, cut twig techniques and field inoculations have been used for screening against botrytis grey mould (Pande *et al.*, 2007a). In the growth room technique, test entries are raised in polyethylene bags filled with sandy loam soil. The 25-day-old plants are moved to a growth room maintained at 22–24°C. The plants are then spray-inoculated with a spore suspension of *Botrytis cinerea* in water (10^4/ml) and covered with wet polyethylene, and given 16h light/8h darkness. Data are recorded after 6 days of inoculation on a 1–9-point scale. In the cut twig method, the cut twigs are first placed in test tubes filled with water and, after inoculation, placed in polyethylene covers as in the above technique. The results are available after 6 days. A large-scale unique and reliable controlled-environment screening facility for botrytis grey mould resistance has been established at the International Crops Research Institute for the Semi-Arid Tropics (ICRISAT), Patancheru, India (Pande *et al.*, 2007a).

Mung bean yellow mosaic virus

For screening against MYMV, the infector- or spreader-row method has been recommended by various workers. The reaction was evaluated by determining the percentage of plants/foliage affected and assigning different categories, based on a rating scale by Singh *et al.* (1988a). A laboratory screening procedure described by Nene (1972) involves rearing whitefly in cages, acquisition feeding on susceptible source plants and then transferring equal numbers of viruliferous whitefly to test plants under controlled conditions. With regard to screening for vector resistance, use of the modified sampler-split cage has been recommended for obtaining a reliable estimate of the insect population (Chhabra *et al.*, 1979). An infector-row technique for field screening of mung bean and urd bean genotypes against MYMV has been reported by Singh and Naimuddin (2009). The success of this technique depends on the availability of a susceptible control that can be used as an infector-cum-spreader every two rows of test genotype. In India, the urd bean (*Vigna mungo*) genotype, Barabanki Local was identified as an infector-cum-spreader and has been used for many years. However, a change in the disease reaction led to the identification of the mungbean genotype Kopergaon/Palampur 93/Kullu 4 as a better option for this method. In order to adopt a simple and uniform scoring method, a 1–9-point scale for MYMV reaction has been developed, taking both disease incidence and severity into account.

Bean common mosaic virus

A procedure for evaluating resistance to BCMV was described by Drijfhout (1978), in which inoculum was sprayed with a suspension of BCMV particles mixed with carborundum.

Screening for resistance to insect pests

The infester rows technique has been reported for screening against the pod borer, *Helicoverpa armigera*. Planting infester rows of a susceptible cultivar along the field borders or at regular intervals in the field can be used to increase *H. armigera* infestation in the test material; there should be sufficient time for the insect to multiply on the infester rows and then to infest the test material. The infester rows may be planted 20–30 days earlier than the test material, or an early-flowering crop or cultivar can also be planted in infester rows along with the test material so that flowering in the infester rows occurs 20–25 days earlier than the test material. Removal of infester rows after infestation of the test material can be done so that it does not compete with the test material (Sharma *et al*., 1988; Smith *et al*., 1994). The artificial infestation technique is also used for screening against *H. armigera*. Several artificial diets have been developed to rear *Helicoverpa* in the laboratory, the most widely used has been described by Armes *et al*. (1992).

For artificial infestation, the crop is raised in the field without the application of insecticide, except for control of non-target insects with selective insecticides. The crop is infested with neonate larvae at either the seedling (ten larvae/plant in chickpea), flowering (ten neonate larvae/plant) or podding stage (five third-instar larvae/plant or inflorescence). After one week, observations are recorded on the number of surviving larvae and larval weights. At maturity, the samples are combined from all the infested plants/inflorescences in each plot/genotypes and data recorded on the number of pods in the infested plant/inflorescence, the number of pods damaged and grain yield in the infested and non-infested plants/inflorescences. The resistant genotypes are selected in comparison with the resistant control based on larval survival, larval weight, pod set, pod damage and grain yield.

Jackai (1982) developed a method for screening resistance in a large number of collections of cowpea to the legume pod borer, *Maruca testulalis*. Several damage parameters were measured, including those to stem, flowers, pods and seeds. The stem and pod damage measurements provided the assessment of resistance to the borer at the initial stage. At a later stage, when the number of cultivars has been reduced considerably, larval counts in flowers and seed damage measurements can be included. Verulkar *et al*. (1997) developed a technique for the screening of breeding materials against the pod fly, *Melanagromyza obtusa* in pigeon pea sown under field conditions. The reproductive phase of the off-season crop coincided with the peak of pest population in April, when > 90% of pod damage was observed in Pant A-3, the susceptible cultivar. All pods of individual plants were examined for the presence of the typical pinhead exit hole, a marker of susceptibility, and percentage pod damage at maturity was recorded. The test entries were then graded on a susceptibility rating scale of 1–9.

Screening for resistance to nematodes

Sharma *et al*. (1994) identified the sources of resistance in cool season legumes (chickpea, faba bean and pea) for cyst (*Heterodera* spp.), root-knot (*Meloidogyne* spp.) and stem (*Ditylenchus dipsaci*) nematodes. Based on the number of cysts on roots, root-knot nematodes induced gall index and stem nematodes affected reproduction in shoot tissue. Ali and Ahmad (2000) screened chickpea breeding lines in a field heavily infested with nematode species (*Meloidogyne javanica*, *Pratylenchus thornei*, and *Rotylenchulus reniformis*). The initial inoculation was 1, 2 and 5 juveniles/g of soil of *M. javanica*, *P. thornei* and *R. reniformis*, respectively. Two-month-old chickpea plants were uprooted carefully, the roots washed and stained with acid fuchsin and lactophenol and visually observed. Observations were made on galls on a scale of 1–5 for root-knot nematode, and the lesions counted on a 1–10 scale for root-lesion nematode. For reniform nematodes, infestation was recorded by counting the exposed females on the root system on a 1–10 scale. Sharma *et al*. (1991) reported a greenhouse technique for screening pigeon pea resistance to *Heterodera cajani*. The effects of different infestation levels on the ratings were not significant, although the use of higher inoculum density (16–27 eggs and juveniles/cm^3 soil) was effective in reducing variability.

15.6 Breeding for Resistance

Conventional breeding methods

The first step in resistance breeding programmes is the collection of natural variability, followed by discovery of the sources of resistance. The next step is to incorporate the resistant gene(s) from the donor parent(s) using various methods including induced mutations, where the susceptible alleles are altered by the use of mutagens. Selection of resistance to pests and diseases is relatively easy, but certain host plant genotypes may show different reactions to races/biotypes. It is desirable to test host genotypes against a wide range of variants of a pathogen/pest before selection is made. This could be done by using variants separately or in known composition. The use of individual variants is important when studying the genetics of the host–pathogen/pest relationship, but is not important for practical breeding tests. The mixture of variants can be maintained either individually on a range of host genotypes or on culture media. The breeding material can also be grown at several locations (sites) where different variants of pathogen/pest are expected to occur (Singh and Singh, 2005).

The primary, secondary, and tertiary gene pools of the food legumes represent potential genetic diversity that may eventually be exploited in cultivated types to overcome biotic stresses. Resistant genotypes can be developed by backcrossing or the use of other appropriate breeding methods. In some cases, the bulk pedigree method could be useful. When two independent recessive genes control resistance, it is relatively easy to transfer these to agronomically desirable types. Most sources of resistance to soil-borne fungi in chickpea and pigeon pea show low levels of resistance or tolerance. Such partial resistance is presumably governed by two or more genes and is assumed to be similar to horizontal resistance. There are practical difficulties in incorporating this type of resistance into germplasm with the desired agronomic traits. The strategy has been to breed for a low level of host–pathogen coexistence that is stable, environmentally balanced and economically useful. Successful use of such varieties requires excellent management skills that simultaneously reduce disease severity and inoculum reproduction. In many cases a combination of two methods, such as bulk pedigree and backcross-pedigree, is generally applied. When multiple disease resistance is needed, it is difficult to accumulate enough polygenes to provide a good level of resistance to all diseases, if the genes governing resistance are inherited independently. Attempts to incorporate polygenes for resistance to two diseases may result in the loss of resistance to one disease as selection occurs for the second disease. Therefore, gene pyramiding is required for development of multiple resistance. The biochemical and genetic parameters of phenolic content offer an alternative method of evaluating the breeding material (Ali *et al.*, 1994). Halila and Harrabi (1990) reported 'shuttle' screening in chickpea to combine *Ascochyta rabiei*, *Fusarium oxysporum*, *Verticillium albo-atrum* and other *Fusarium* resistance through breeding.

Among wild species, *Cicer bijugum* was found resistant to the Italian isolate of fusarium wilt (Infantino *et al.*, 1996). Dey *et al.* (1993) reported that *C. pinnatifidum* and *C. judaicum* were the most resistant to ascochyta blight. An accession of *C. echinospermum*, ICWC 35/S1, was found resistant to ascochyta blight by Singh *et al.* (1991). Wild relatives of lentils, *Lens nigricans* ssp. *ervoides*, *L. odomensis* and *L. culnaris* ssp. *orientalis*, were found to be valuable sources of resistance for vascular wilt and *Ascochyta rabiei*. Ali *et al.* (1994) reported the wild relatives of pea, *Pisum fulvum* and *P. humile*, as a valuable source of resistance to rust. Wild relatives of *Vigna*, i.e. *V. radiata* var. *sublobata* and *V. mungo* var. *sylvestris*, were found resistant to MYMV. Although many reports on successful transfer of single gene resistance are available and much of the literature reports the identification of resistance and production of interspecific hybrids, rarely has the actual release of a new cultivar and its use by farmers occurred. Conventional crossing has been successful in producing interspecific hybrids in *Lens*, *Cicer* and *Pisum*, and those hybrids are being evaluated for desired recombinants. *In vitro* culture of hybrid embryos has been successful in overcoming barriers to wider crosses in *Lens*. The successful transfer of genes from wide sources to cultivated types can be assisted by repeated backcrossing and selection designed to eliminate undesired traits while transferring genes of interest.

Mutation breeding

Isolation of micromutations or polygenic mutations for higher yield, coupled with some other desirable attributes like disease and pest resistance, has been reported in chickpea (Kharkwal *et al.*, 2008). Mutagenic treatments for inducing mutation for a specific trait often result in alteration of several traits. Such changes may be due either to the pleiotropic effects of a single mutant allele or to simultaneous mutations in other loci (Kharkwal *et al.*, 2008). Some crop varieties improved for yield or morphological traits through induced mutations exhibited improved tolerance to biotic and abiotic stresses, and these were therefore used as donors in the breeding programme for disease and insect pest resistance. Chickpea mutant varieties Pusa 408, Pusa 413, Pusa 417 and Pusa 547 with resistance to ascochyta blight and fusarium wilt have been released in India. Similarly, varieties CM-72, CM-88, NIFA-95 and CM 1918 were released in Pakistan in the high-yielding mung bean mutant variety MUM-2 resistant to MYMV (mung bean yellow mosaic virus), cercospora leaf spot, leaf crinkle, bacterial blight and macrophomina blight. The radiation-induced mutant cultivar CAZRI Moth-1 of moth bean (*Vigna aconitifolia*) was resistant to YMV (yellow mosaic virus) disease (Kharkwal *et al.*, 2008). The variety NM-92, developed at NIAB (the National Institute of Agricultural Botany) showed durable resistance to YMV and cercospora leaf spot; this variety occupies about 51% of the cultivated area under mung bean (Kharkwal *et al.*, 2008). The variety TARM-1, resistant to powdery mildew and YMV, is the first of its kind to be released for rabi/rice fallow cultivation in India (Kharkwal *et al.*, 2008). Mutant cultivars with improved insect resistance to aphid in cowpea include ICV-11 and ICV-12 (Kharkwal *et al.*, 2008).

Transgenic approach

Transgenic plants resistant to pod borer are being researched globally in chickpea and pigeon pea, using the *Bt* crystal protein gene from a soil bacterium. In chickpea, few reports are available on genetic transformation. Transformed callus was obtained in chickpea using wild strains of *Agrobacterium*, and transformed chickpea plants possessing the *Cry 1Ac* construct for resistance to *Helicoverpa armigera* have been reported (Sarker *et al.*, 2008). Sharma *et al.* (2006) developed an efficient method of producing transgenic pigeon pea plant by incorporating the *cry1Ab* gene of *Bacillus thuringiensis* through *Agrobacterium tumefaciens*-mediated genetic transformation, based on the direct regeneration of adventitious shoot buds in the axillary bud region of *in vitro*-germinating seedlings. The tissue with potential to produce adventitious shoot buds could be used as an explant and for co-cultivation with *A. tumefaciens* carrying the synthetic *cry1Ab* on a binary vector and driven by a CaMV 35S promoter. PCR analysis of the progenies from independent transformants followed gene inheritance in a Mendelian ratio, and 65% of the transformants showed the presence of single-copy inserts of the introduced genes. The transcripts of the introduced genes were normally transcribed and resulted in the expression of *Cry1Ab* protein in the tested T_2 generation plants.

In pigeon pea, transformed callus and plantlets possessing foreign genes have been reported from various institutes in India. Trangenics have been developed in chickpea at ICRISAT for *Cry 1Ab* and trypsin inhibitor (SbTI) genes (Sarker *et al.*, 2008). Among the *Bt* toxins, *Cry1Ac* is known to be the most effective against *Helicoverpa* larvae, followed by *Cry1Aa*, *Cry2Aa* and *Cry2Ab*. Sanyal *et al.* (2005) transformed chickpea with *Cry1Ac* gene using *Agrobacterium* and successfully generated several transgenic chickpea lines. When such plants were challenged in bioassays, most *H. armigera* larvae ceased feeding on transgenic chickpea leaves after 2 days, and showed high levels of mortality and reduced weight gain compared with insects fed on conventional leaves. Acharjee *et al.* (2010b) reported for the first time on the production of transgenic chickpea with a sequence-modified *cry2Aa* gene; the new *Bt* chickpea can be used to complement the existing lines carrying the *cry1Ac* gene.

Transgenics have also been produced in lentil for bean golden mosaic virus resistance (Aragao *et al.*, 1998; Sarker *et al.*, 2008) using the particle bombardment method. A gene encoding a multi-domain proteinase inhibitor precursor was expressed in transgenic pea, resulting in higher mortality of *H. armigera* larvae. The tobacco proteinase inhibitor (PI) has shown enhanced resistance against *H. armigera* in transgenic pea (Chairity *et al.*, 1999). The alfa-amylase inhibitor from the common bean (*Phaseolus vulgaris*) inhibits alfa-amylases in the mid-gut of coleopteran insects and storage pests of the genera *Callosobruchus* and *Bruchus*, and blocks larval development (Ishimoto *et al.*, 1996). Arora *et al.* (2005) reported that pigeon pea and garlic lectins resulted in reduced pupation and adult emergence of *H. armigera*. Moreover, garlic lectin had an adverse effect on larval and pupal weight, but did not affect the duration of larval and pupal development. Lectins from garlic and pigeon pea could therefore potentially be deployed in transgenic plants in combination with *Bt* genes to increase the level of plant resistance to *H. armigera*.

15.7 Major Achievements

Chickpea

Diseases

Resistant sources to ascochyta blight have been identified and used in breeding programmes (Malhotra *et al.*, 2003; Pande *et al.*, 2005, 2007b; Basandrai *et al.*, 2008). Lines with moderate resistance to this disease have continuously been delivered to national programmes from ICARDA (International Center for Agricultural Research In the Dry Areas) and ICRISAT (Malhotra *et al.*, 2003; Pande *et al.*, 2005; Basandrai *et al.*, 2008; Kaur *et al.*, 2008; Sarker *et al.*, 2008). Pande *et al.* (2006) reported three accessions as being moderately resistant to ascochyta blight, but to date resistance sources have not been identified against pathotypes III and IV, as identified in Syria (Sarker *et al.*, 2008).

Interspecific crosses have been investigated in attempts to introduce alien genes, and promising resistant lines were identified. Gene pyramiding in lines has resulted in higher resistance (Kaur *et al.*, 2008; Gaur *et al.*, 2010). Fourteen genotypes and some accessions of wild *Cicer* species (*C. judaicum*, *C. reticulatum*, *C. echinospermum* and *C. pinnatifidum*) have shown resistance to botrytis grey mould (BGM; Basandrai *et al.*, 2006; Pande *et al.*, 2006; Pande *et al.*, 2007a; Basandrai *et al.*, 2008). Some lines derived through interspecific hybridization have shown a high level of resistance to BGM (Kaur *et al.*, 2008). Achievement of resistance has also been attempted through gene pyramiding, and some ICRISAT lines have shown promise against the disease. Of 428 Australian advanced breeding lines evaluated under controlled growth room conditions at ICRISAT, and in field conditions at Ishurdi and Jessure in Bangladesh, 99 moderately resistant lines were identified (Pande *et al.*, 2005). Recently, Pande *et al.* (2006) identified BGM-resistant genotypes among the minicore collections at ICRISAT.

The varieties JG 315, Avrodhi, DCP 92-3, JG 74, BG 372 and KWR 108 were reported as being resistant against fusarium wilt (Chaudhary, 2009). Pande *et al.* (2006) reported high levels of resistance to this disease, where 21 accessions were free from the disease and 25 were resistant.

A number of varieties moderately resistant to dry root rot have been identified (Pande *et al.* 2006; Kaur *et al.*, 2008). Pande *et al.* (2006) reported six accessions with moderate resistance to dry root rot among 211 accessions in the *desi* chickpea mini-core collection. Under natural epiphytotic conditions, lines GL 84102, GL 88223, GLK 88114 and GF 89-75 showed moderate resistance to stem rot.

The wild *Cicer* species *C. judaicum*, *C. reticulatum*, *C. pinnatifidum* and *C. yamashitae* are reported as being tolerant to stem rot (Kaur *et al.*, 2008). Viral stunt disease caused by chickpea chlorotic dwarf mono gemini virus and chickpea luteovirus is common in the Indian subcontinent. Four varieties with improved resistance and 17 resistance sources have been identified (Kaur *et al.*, 2008). Genotypes ICC 11284 and ICC 13441 showed combined resistance against ascochyta blight and BGM, and dry root rot and fusarium wilt, respectively (Anonymous 2010a). Eleven accessions showed combined resistance against BGM and FW (Pande *et al.*, 2006). Some lines of chickpea having multiple disease resistance are listed in Table 15.3.

Table 15.3. Chickpea genotypes with multiple disease resistance.

Genotypes	Diseases
ICV 12237, ICC12269	Fusarium wilt, dry root rot, black root rot
ICC 1069	Fusarium wilt, ascochyta blight, botrytis grey mould
ICC 1046	Fusarium wilt, dry root rot, stunt
ICC 858, 959, 4918, 8933, 9001	Fusarium wilt, sclerotinia stem rot

Insect pests

A large amount of germplasm, including cultigens and wilds was evaluated for resistance to pod borer, and low to moderate levels of resistance were reported, with line ILL 506 possessing a good level of resistance (Pratap *et al.*, 2002; Sharma *et al.*, 2003). Few resistance sources for leaf miner have been identified at ICARDA (Malhotra *et al.*, 1996).

Nematodes

A moderate level of resistance to the root-knot nematode (*Meloidogyne* spp.) was reported in germplasm, and a high level of resistance was identified against the cyst nematode (*Heterodera* spp.) in wild relatives. Progress has been made in the introgression of resistance gene(s) to cultigens.

Lentil

Sources of resistance to fusarium wilt have been identified through rigorous screening in a wilt-sick plot at ICARDA, Tel Hadya and at various locations throughout India. Thirty-four stable sources of resistance were identified at ICARDA and were included in the international breeding programme (Sarker *et al.*, 2004). Eight varieties with moderate resistance to wilt have been released in India. Rust-resistant sources have been reported in India, Bangladesh and Ethiopia (Sarker *et al.*, 2008). In India, 49 resistant sources have been identified (Mishra *et al.*, 2005). Resistant sources were used in breeding programmes following backcross breeding, and the bulk pedigree method has led to the development of resistant varieties in Bangladesh, Ethiopia, India and Morocco (Tikoo *et al.*, 2005; Sarker *et al.*, 2008). A high level of resistance to ascochyta blight has been identified among cultivated germplasm and wild relatives in Australia, Canada, India, New Zealand, Pakistan and India (Basandrai *et al.*, 2000; Sarker *et al.*, 2008). The stemphylium blight-resistant cultivars 'Barimasur-4', 'Barimasur-5' and 'Barimasur-6' have been developed in Bangladesh (Chen *et al.*, 2008).

Cowpea

Scientists from IITA (the International Institute for Tropical Agriculture), Nigeria have developed varieties by incorporating resistance genes in the variety Ife Brown, which was used as recurrent parent. These new varieties are resistant to all the major pests except *Maruca* and pod bugs, and are now being used as donors as well as parents in breeding programmes. Sources of resistance against cowpea mosaic virus, cercospora leaf spot and anthracnose have been identified in India (Basandrai *et al.*, 2004; Mishra *et al.*, 2008).

Mung bean

Sources of resistance to cercospora leaf spot have been identified at AVRDC (the World Vegetable Center) and in Taiwan, Bangladesh, India, Pakistan and the Philippines, and to MYMV in Bangladesh, India, Pakistan and Sri Lanka (Sarker *et al.*, 2008). In the Indian mung bean breeding programme, resistance to MYMV is an important component and 41 MYMV resistant varieties have been released to date (Kaur *et al.*, 2008; Anonymous 2010b). In addition to this, some commercially released varieties have been found to be resistant to powdery mildew, macrophomina blight and leaf crinkle virus (Kaur *et al.*, 2008; Anonymous, 2010b). AVRDC accessions V 4281, V 2396 and V 3495 were resistant to agromyzids, while accessions, V 2709 and V 2802 were resistant to bruchids (Sarker *et al.*, 2008). Resistance to MYMV and bruchids was introgressed through wide crosses (*V. radiata* × *V. radiata* var. *sublobata*). Useful disease resistance genes were also identified from amphidiploids of mungbean × ricebean crosses (Dar *et al.*, 1991). Mutation breeding has led to the development of the large-seeded, high-yielding and MYMV-resistant lines NM-51 and NM-54, which were released in Pakistan. Mutagenic treatment has been used for generating variability for resistance against a number of diseases, i.e. leaf spot and MYMV.

Black gram

Varieties of black gram resistant to yellow mosaic virus, powdery mildew, root rot and macrophomina blight have been released for cultivation in India (Anonymous, 2010b). Sources of resistance to MYMV have been identified and used in breeding programmes to develop resistant varieties (Basandrai *et al.*, 1999, 2003). The wild relatives of *Vigna* (*V. trilobata* (syn. *Phaseolous trilobus* Wall), *V. umbellata* (Thunb) Ohwhi & Ohashi (syn., *P. calcaratus* Roxb.) and the wild tetraploid species, *V. glabrescenes* are highly resistant to YMV. Germplasm lines/cultivars resistant to powdery mildew have been reported in India (Basandrai *et al.*, 2003; Kaur *et al.*, 2008), whereas varieties showing a high level of resistance have been released in Bangladesh (Afzal *et al.*, 2002). Combined resistance to anthracnose, cercospora leaf spot and MYMV and to anthracnose, cercospora leaf spot, powdery mildew and MYMV (TEU 95-1) have been reported (Basandrai *et al.*, 1999, 2003).

Pigeon pea

Diseases

Systematic and intensive work has been undertaken in India on the identification of resistant sources and the development of wilt-resistant varieties of pigeon pea by the Indian Council of Agricultural Research (ICAR) and ICRISAT. Wilt-resistant/-tolerant cultivars are available for medium- and long-duration groups (Vishwa Dhar *et al.*, 2005; Anonymous, 2010c). In the long-duration group, varieties with high wilt resistance are not available; however, a few tolerant varieties have been released (Anonymous, 2010c). For sterility mosaic virus, sources with a high level of resistance have been reported (Kaur *et al.*, 2008; Sarker *et al.*, 2008), and high-yielding resistant varieties have been released in India. For phytophthora stem blight, 13 resistance sources have been identified (Kaur *et al.*, 2008; Sarker *et al.*, 2008). In India, the phytophthora stem blight-resistant varieties Jawahar (JKM 7), Narendera Arhar 1, MAL 13 and PA 291 (Anonymous, 2010c) have been released for cultivation.

Insect pests

Lines with a moderate level of resistance to *Maruca vitrata* (ICPL 4, ICPL 93015 and Pusa 6), pod fly (PDA 88-2E, PDA 92-1E, ICPL 5036, ICP 8102-5, Mukta, Malviya Vikalap) and pod borer (UPAS 120, JA 4, GT 100, AKT 8811, Abhaya, Co 6, GTH 1 and pa 29) have been identified (Sarker *et al.*, 2008; Anonymous, 2010c).

15.8 Conclusion

Food legumes are prone to attack by several plant pathogens, insect pests and plant-parasitic nematodes, resulting in huge economic losses globally. The conventional approaches of resistance breeding have provided several improved varieties of food legumes with resistance to important biotic stresses. There is no substitute for these approaches, and these will continue to be the mainstay in the future. However, efforts are needed on improving the effectiveness of these approaches by further refining screening methods for resistance to stresses and identifying new sources of resistance genes in both cultivated and wild species. There is a need to use diverse sources of resistance in breeding programmes and to develop cultivars with resistance to multiple stress factors.

Mutagenesis has the potential for creating the desired variability, including resistance to stresses, and thus should find a role in resistance breeding. Wild species are valuable sources of resistance genes, and concerted efforts are needed towards their exploitation.

Marker-assisted selection, particularly MABC (marker-assisted backcross) breeding, has a greater role to play in resistance breeding, especially when the direct assessment of the phenotype is difficult and a large number of resistance genes are to be combined. Transgenic technology has already proved its worth in many crops, including some legumes such as soybean, *Phaseolus* and groundnut. However, in pulses such as pea, lentil, chickpea and pigeon pea, gene transfer methods are yet to be perfected and transgenic varieties to be developed having resistance to biotic stresses like wilt, rust, powdery mildew and pod borer. There is an urgent need for the isolation, characterization and cloning of disease-, insect pest- and nematode-resistant genes from other plants and microbes. Finally, it can be concluded that the support of biotechnology approaches to conventional breeding methods would lead to rapid progress in the development of improved cultivars of food legumes with resistance to biotic stresses.

References

Acharjee, S., Sarmaha, B.K., AnandaKumar, P., Olsenc, K., Mahonc, R., Moard, W.J. *et al.* (2010) Transgenic chickpeas (*Cicer arietinum* L.) expressing a sequence-modified *cry2* Aagene. *Plant Science* 178, 333–339.

Afzal, M.A., Bakr, M.A., Rahman, M.M. and Luna, N.K. (2002) Registration of 'Barimash 2' blackgram. *Crop Science* 42, 985.

Ali, S.S. and Ahmad, R. (2000) Screening of chickpea germplasm against nematode. *International Chickpea and Pigeonpea Newsletter* 7, 8.

Ali, S.M., Sharma, B., Ambrose, M.J., Muehlbauer, F.J. and Kaiser, W.J. (1994) Current status and future strategy in breeding pea to improve resistance to biotic and abiotic stress. Expanding the production and use of cool season food legumes. In: *Proceedings of the 2nd International Food Legumes Research Conference on Pea, Lentil, Fababean, Chickpea and Grasspea*, 12–16 April, Cairo, pp. 540–558.

Anonymous (2010a) *Project Coordinator's Report 2009–10 of All-India Coordinated Research Project on Chickpea.* Indian Institute of Pulses Research, Kanpur, India.

Anonymous (2010b) *Project Coordinator's Report of All-India Coordinated Research Project on MULLaRP.* Indian Institute of Pulses Research, Kanpur, India.

Anonymous (2010c). *Project Coordinator's Report of All-India Coordinated Research Project on Pigeonpea.* Indian Institute of Pulses Research, Kanpur, India.

Aragao, F.J.L., Riberiro, S.G., Barros, L.M.G., Brasileiro, A.C.M., Maxwell, D.P., Rech, E.L. *et al.* (1998) Trangenic bean (*Phaseolus vulgaris* L.) engineered to express viral antisense RNA shows delayed and attenuated symptoms to bean golden mosaic Gemini virus. *Molecular Breeding* 4, 491–499.

Armes, N.J., Bond, G.S. and Cooters, R.J. (1992, The laboratory culture and development of *Helicoverpa armigera. Natural Resources Institute Bulletin* 57, Natural Resources Institute, Chatham, UK.

Arora, R., Sharma, K.K., Sharma, H.C. and Dreissche, E. van. (2005) Biological activity of lectins from grain legumes and garlic against the legume pod borer, *Helicoverpa armigera. Journal of SAT* 1, 3.

Basandrai, A.K., Gartan, S.L., Basandrai, D. and Kalia, V. (1999) Black gram (*Phaseolus mungo*) germplasm evaluation against different diseases. *Indian Journal of Agricultural Sciences* 69, 506–508.

Basandrai, D., Basandrai, A.K. and Kalia, V. (2000) Evaluation of lentil germplasm against rust (*Uromyces viciae fabae*) and *Ascochyta* blight. *Indian Journal of Agricultural Sciences* 70, 804–805.

Basandrai, D., Basandrai, A.K. and Singh, I. (2003) Multiple disease resistance against anthracnose, leaf spot, powdery mildew and mung bean yellow mosaic virus in blackgram (*Vigna mungo). Journal of Mycology and Plant Pathology* 33, 56–58.

Basandrai, D., Thakur, H.L., Basandrai, A.K. and Kumar, S. (2004) Genetic divergence among cowpea (*Vigna unguiculata* L. Walp.) genotypes and their reaction to important diseases. *Indian Journal of Arid Legumes* 1, 92–95.

Basandrai, A.K., MacLeod, W.J., Siddique, K., Pande, S. and Payne, P. (2006) Evaluation of wild chickpea germplasm against Australian isolates of *Botrytis cinerea* the causal organism of *Botrytis* grey mould of chickpea. *BGM Newsletter* November.

Basandrai, A.K., Pande, S. and Basandrai, D. (2008) Recent advances in host–parasite interaction, chickpea-*ascochyta rabiei* and *Botrytis cinerea* systems. In: Setia, R.C., Nayyar, H. and Setia, N. (eds) *Crop Improvement, Strategies and Applications*. IK International, New Delhi, India, pp. 265–301.

Cai, D., Kleine, M., Kifle, S., Harloff, H., Sandal, N.N., Marcioer, K.A. *et al.* (1997) Positional cloning of a gene for nematode resistance in sugar beet. *Science* 275, 832–834.

Chairity, J.A., Anderson, M.A., Brittisnich, D.J., Whitecross, M. and Higgins, T.J.V. (1999) Transgenic tobacco and pea expressing proteinase inhibitor from *Nicotiana alata* have increased insect resistance. *Molecular Breeding* 5, 357–365.

Chandra, S., Ahamd, R. and Asthana, A.N. (1992). Field tolerance to aphid in lentil. *LENS Newsletter* 19, 45.

Chaudhary, R.G. (2009) Chickpea diseases. In: *25 Years of Research at IIPR*. Indian Institute of Pulses Research, Kanpur, India, pp. 85–115.

Chen, W., Basandrai, D., Banniza, S., Bayaa, B., Buchwaldt, L., Davidson, J. *et al.* (2008) Lentil diseases and their management. In: Erskine, W., Muehlbauer, F.J., Sarker, A. and Sharma, B. (eds) *The Lentil: Botany, Production and Uses*. CAB International, Wallingford, UK, pp. 262–281.

Chhabra, K.S., Kooner, B.S. and Singh, G. (1979) Field resistance of certain cultivars of mungbean (*V. radiata*) to whitefly, *Bemisia tabaci* G. and yellow mosaic. *Journal of Research of Punjab Agricultural University* 16, 385–388.

Chhabra, K.S., Kooner, B.S., Saxena, A.K. and Sharma, A.K. (1981) Effect of biochemical components on the incidence of insect pest complex and yellow mosaic virus in mungbean. *Crop Improvement* 8, 56–59.

Chhabra, K.S., Kooner, B.S., Sharma, A.K. and Saxena, A.K. (1990) Sources of resistance in chickpea, role of biochemical components in the incidence of gram pod bore (*Helicoverpa armigera* Hb.). *Indian Journal of Entomology* 52, 423–430.

Chhabra, K.S., Kooner, B.S., Sharma, A.K. and Saxena, A.K. (1993) Screening of black gram genotypes against whitefly, jassids and yellow mosaic virus, role of pohytochemicals in resistance. *Indian Journal of Pulses Research* 6, 76–81.

Chhabra, K.S., Kooner, B.S., Saxena, A.K. and Sharma, A.K. (1994) Role of allomones in varietal resitance in summer mungbean against thrips (*Megalurothrips distalis* Karny). In: *Proceedings of International Symposium on Pulses Research*, 2–6 April 1994, New Delhi, India, p. 161.

Cowgill, S.E. and Lateef, S.S. (1996) Identification of antibiotic and antixenotic resistance to *Helicoverpa armigera* (Lepidoptera, Noctuidae) in chickpea. *Journal of Economic Entomology* 89, 224–229.

Dar, G.M., Verma, M.M., Gosal, S.S. and Brar, J.S. (1991) Characterization of some interspecific hybrids and amphiploids in *Vigna*. In: Sharma, B. and Mehra, R.B. (eds) *Golden Jubilee Celebration Symposium on Grain Legumes*, 9–11 February 1991, IARI, New Delhi, India, pp. 73–78.

Darby, P., Lewis, B.G. and Matthews, P. (1985) Inheritance and expression of resistance to *Ascochyta pisi*. In: Hebblethwaite, P.D., Heath, M.C. and Dawkins, T.C.K. (eds) *The Pea Crop*. Butterworth, London, pp. 231–236.

Dev, S.K. and Singh, G. (1993) Resistance to aschochyta blight in chickpeas – genetic basis. *Euphytica* 68, 147–153.

Dey, S.K., Singh, G., Gosal, S.S. and Verma, M.M. (1993) Tissue culture response and resistance to *Ascochyta* blight in some interspecific crosses in chickpea. *Annals of Biology – Ludhiana* 9, 235–238.

Drijfhout, E. (1978) Genetic interaction between *Phaseolus vulgaris* and bean common mosaic virus with implications for strain identification and breeding for resistance. Agricultural Research Report 872, Wageningen, The Netherlands.

Dua, R.P., Chaturvedi, S.K. and Gupta, S. (2002) Breeding for disease resistance in pulse crops. Pulses for sustainable agriculture and nutritional security. In: Ali, M., Chaturvedi, S.K. and Gurha, S.N. (eds) *Proceedings of National Symposium*, Indian Institute of Pulses Research, Kanpur, India, pp. 52–60.

Durairaj, C. (1999) Influence of pigeonpea pod and seed characters on pigeonpea pod fly infestation. *Madras Agriculture Journal* 86, 594–596.

Durairaj, C. and Sakthivel, P. (2007) Sources of resistance against stem fly *Ophiomyia phaseoli* Tryon (Diptera, Agromyzidae) in mungbean and urdbean. In: *National Symposium on Legumes for Ecological Sustainability, Emerging Challenges and Opportunities*, 3–5 November 2007, Indian Institute of Pulses Research, Kanpur, India, p. 144.

Ehlers, J.D., Matthews, W.C. Jr, Hall, A.E. and Roberts, P.A. (2000) Inheritance of a broad-based form of root-knot nematode resistance in cowpea. *Crop Science* 40, 611–618.

Eujayl, I., Erskine, W., Bayaa, B., Baum, M. and Pehu, E. (1998) *Fusarium* vascular wilt in lentil, inheritance and identification of DNA markers for resistance. *Plant Breeding* 117, 497–499.

Ford, R., Pang, E.C.K. and Taylor, P.W.J. (1999) Genetics of resistance to *Ascochyta* blight (*Ascochyta lentis*) of lentil and identification of closely linked RAPD markers. *Theoretical and Applied Genetics* 98, 93–98.

Gaur, P., Pande, S., Khan, T., Tripathi, S., Sharma, M., Sandhu, J.S. *et al.* (2010) Multiple disease resistance in chickpea. In: *4th Workshop on* Medicago truncatula and *5th International Congress on Legume Genetics and Genomics Pathogens, Pests & Symbionts*, 2–8 July 2010, Washington State University, Washington.

Githiri, S.M., Ampong-Nyarko, K., Osir, E.D. and Kimani, P.M. (1996) Genetics of resistance to *Aphis craccivora* in cowpea. *Euphytica* 89, 371–376.

Green, P.W.C., Stevenson, P.C., Simmonds, M.S.J. and Sharma, H.C. (2002) Can larvae of the pod borer, *Helicoverpa armigera* (Lepidoptera, Noctuidae), select between wild and cultivated pigeon pea [*Cajanus* sp. (Fabaceae)]. *Bulletin of Entomological Research* 92, 45–51.

Gumber, R.K., Kumar, J. and Haware, M.P. (1995) Inheritance of resistance in *Fusarium* wilt in chickpea. *Plant Breeding* 114, 272–279.

Hafiz, A. (1952) Basis of resistance in gram to Mycosphaerella blight. *Phytopathology* 42, 422–424.

Halila, M.H. and Harrabi, M.M. (1990) Breeding for dual resistance to *Ascochyta* and wilt diseases in chickpea. *Options Méditerranéennes – Série Séminaires* 9, 163–166.

Infantino, A., Porta-Puglia, A. and Singh, K.B. (1996) Screening wild *Cicer* species for resistance to *Fusarium* wilt. *Plant Disease* 80, 42–44.

Ishimoto, M., Sato, T., Chrispeels, M.J. and Kitamura, K. (1996) Bruchid resistance of transgenic azuki bean expressing seed alpha-amylase inhibitor of common bean. *Entomology Experimentation and Application* 79, 309–315.

Jackai, L.E. (1982) A field screening technique for resistance of cowpea (*Vigna unguiculata*) to the pod-borer, *Maruca testulalis* (Geyer) (Lepidoptera, Pyralidae). *Bulletin of Entomological Research* 72, 145–156.

Jackai, L.E.N. and Oghiakhe, S. (1989) Pod wall trichomes and resistance of two wild cowpea *Vigna vexillata* accessions to *Maruca testulalis* (Geyer) (Lepidoptera, Pyralidae) and *Clavigralla tomentosicollis* Stal. (Heliptera, Coreidae). *Bulletin of Entomological Research* 79, 595–605.

Kamboj, R.K., Pandey, R.K. and Caube, H.S. (1990) Inheritance of resistance to *Fusarium* wilt in Indian germplasm of *Lens culinaris* Medic. *Lens* 50, 113–117.

Kaur, L., Tripathi, H.S., Vishwa, D., Reddy, M.V., Singh, G. and Kharkwal, M.C. (2008) Management of diseases in food legumes. In: Kharkwal, M.C. (ed.) *Proceedings of the 4th International Food Legumes Research Conference vol. 1*, 18–22 October 2005, New Delhi, India, pp. 608–637.

Kharkwal, M.C., Gopalakrishna, T., Pawar, S.E. and Ahsanul Haq, M. (2008) Mutation breeding for improvement of food legumes. In: Kharkwal, M.C. (ed.) *Proceedings of the 4th International Food Legumes Research Conference vol. 1*, 18–22 October 2005, New Delhi, India, pp. 194–221.

Kumar, S. (1998) Inheritance of resistance to *Fusarium* wilt (race 2) in chickpea. *Plant Breeding* 117, 139–142.

Kumar, V., Singh, B.M. and Singh, S. (1997) Genetics of lentil resistance to rust. *Lens Newsletter* 24, 23–25.

Kumari, D.A., Reddy, D.J. and Sharma, H.C. (2006) Antixenosis mechanism of resistance in pigeonpea to the pod borer, *Helicoverpa armigera*. *Journal of Applied Entomology*, 130, 10–14.

Lal, C., Sharma, S.K. and Chahota, R.K. (1996) Inheritance of rust resistance in lentil. *Indian Journal of Genetics and Plant Breeding* 56, 350–351.

Lal, S.S. and Yadava, C.P. (1994) Oviposition response of pod fly (*Melanoaromyza obtusa*) on resistant pigeonpea (*Cajanus cajan*) selections. *Indian Journal of Agricultural Science* 64, 658–660.

Lal, S.S., Yadav, C.P. and Sachan, J.N. (1988) Studies on some aspects of oviposition and damage of podfly in relation to host plant phenology. *Indian Journal of Pulses Research* 1, 83–88.

Luzzi, B.M, Tamulonis J.P., Hussy, R.S. and Boerma, H.R. (1995) Inheritance of resistance to the Javanese root-knot nematode in soybean. *Crop Science* 35(5), 1240–1243.

Macfoy, C.A., Dabrowski, Z.T. and Okech, S. (1983) Studies on the legume pod borer, *Maruca testulalis* (Geyer). VI. Cowpea resistance to oviposition and larval feeding. *Insect Science and its Application* 4, 147–152.

Malhotra, R.S., Singh, K.B., Van Rheenen, H.A. and Pala, M. (1996) Genetic improvement and agronomic management of chickpea with emphasis on the Mediterranean region. In: Saxena, N.P., Saxena, M.C., Johnsen, C., Virmani, S.M. and Harris, H. (eds) *Adaptation of Chickpea in the West Asia and North Africa Region*. ICRISAT, Patancheru, India and ICARDA, Aleppo, Syria, pp. 217–232.

Malhotra, R.S., Baum, M., Udupa, S.M., Bayaa, B., Kabbabe, S. and Khalaf, G. (2003) *Ascochyta* blight resistance in chickpea with emphasis on the Mediterranean region. In: Sharma, R.N., Srivastava, G.K., Rathore, A.L., Sharma, M.L. and Khan, M.A. (eds) *Proceedings of the International Chickpea Conference on Chickpea Research for the Millennium*, 20–22 January, Indira Gandhi Agricultural University, Raipur, India, pp. 108–117.

Mansur L.M., Carriquiry, A.L. and Rao, A.P.A. (1993) Generation mean analysis of resistance to race 3 of soybean cyst nematode. *Crop Science* 33(6), 1249–1253.

Marley, P.S. and Hillocks, R.J. (1993) The role of phytoalexins in resistance to *Fusarium* wilt in pigeon pea (*Cajanus cajan*). *Plant Pathology* 42, 212–218.

Marx, G.A. and Providenti, R. (1979) Linkage relations of mo. *Pisum Newsletter* 11, 28–29.

Mew, I.C., Wang, T.C. and Mew, T.W. (1975) Inoculum production and evaluation of mungbean varieties for resistance to *Cercospora canescens. Plant Disease Reporter* 59, 397–401.

Mishra, S.K., Singh, B.B., Sarker, A., Basandrai, D. and Basandrai, A.K. (2005) Slow rusting and its potential donors for resistance in lentil (*Lens culinaris*). *Indian Journal of Genetics and Plant Breeding* 65, 319–320.

Mishra, S.K., Sharma, B., Tyagi, M.C., Singh, B.B., Basandrai, D., Basandrai, A.K. *et al.* (2008) Screening of cowpea for biotic and biotic stress. *Indian Journal of Genetics and Plant Breeding* 68, 446–448.

Munjal, R.L., Chenulu, V.V. and Hora, T.S. (1964) Assessment of losses due to powdery mildew (*Erysiphe polygoni* D.C.) in pea. *Indian Phytopathology* 16, 268–270.

Nene, Y.L. (1972) A survey of viral diseases of pulses crops in U.P. Research bulletin no. 4, G.B. Pant University of Agriculture and technology, Pantnagar, India.

Nene, Y.L. and Haware, M.P. (1980) Screening chickpeas for resistance to wilt. *Plant Diseases* 64, 379–380.

Nene, Y.L., Kannaiyan, J. and Reddy, M.V. (1981) Pigeonpea diseases, resistance screening techniques. Bull 9, ICRISAT, Hyderabad, India, pp. 15.

Oghiakhe, S., Jackai, L.E.N., Makanjuola, W.A. and Hodgson, C.J. (1992) Morphology, distribution and the role of trichomes in cowpea (*Vigna unguiculata*) resistance to the legume pod borer, *Maruca testulalis* (Lepidoptera, Pyralidae). *Bulletin of Entomological Research* 82, 499–506.

Pal, A.B., Sohi, H.S. and Rawal, R.D. (1979) Studies on inheritance of resistance to rust (*Uromyces fabae* (Pers) de Berry) on peas. *SABRAO Journal* 11, 101–103.

Pande, S., Siddique, K.H.M., Kishore, G.K., Baya, B., Gaur, P.M., Gowda, C.L.L. *et al.* (2005) *Ascochyta* blight of chickpea: Biology, pathogenicity, and disease management. *Australian Journal of Agricultural Research* 56, 317–322.

Pande, S., Krishna Kishore, G., Upadhyay, H.D. and Narayana Rao, J. (2006) Identification of sources of multiple disease resistance in mini-core collection of chickpea. *Plant Disease* 90, 1214–1218.

Pande, S., Galloway, J., Gaur, P.M., Siddique, K.H.M., Tripathi, H.S.P., Taylor, M.W. *et al.* (2007a) Botrytis grey mould of chickpea, a review of biology, epidemiology, and disease management. *Australian Journal of Agricultural Research* 2006 57, 1137–1150.

Pande, S., Gaur, P.M., Sharma, M., Rao, J.N., Rao, B.V. and Krishna Kishore, G. (2007b) Identification of single and multiple disease resistance in desi chickpea genotypes to *Ascochyta* blight, Botrytis gray mold and *Fusarium* wilt. *SAT eJournal* 3, 36–38 (available at www.ejournal.icrisat.org).

Pande, S., Sharma, M. and Rao, J.N. (2008) Etiology, biology, and management of diseases of food legumes. In: Kharkwal, M.C. (ed.) *Proceedings of the 4th International Food Legumes Research Conference vol. 2*, 18–22 October 2005, New Delhi, India, pp. 363–377.

Pande, S., Sharma, M.M., Gaur, P.M. and Gowda, C.L.L. (2010) Host plant resistance to *Ascochyta* blight of chickpea. *Information Bulletin* 82, ICRISAT, Patancheru, India, 40 pp.

Partap, A., Basandrai, D., Mehta, P.K. and Basandrai, A.K. (2002) Evaluation of chickpea (*Cicer arietinum* L.) genotypes against pod borer. *Indian Journal of Genetics and Plant Breeding* 62, 131–134.

Reddy, M.V. and Singh, K.B. (1984) Evaluation of world collection of chickpea germplasm accession of resistance to *Ascochyta* blight. *Crop Science* 32, 1079–1080.

Reddy, M.V., Singh, K.B. and Nene, Y.L. (1984) Screening techniques for *Ascochyta* blight of chickpea. In: Saxena, M.C. and Singh, K.B. (eds) *Ascochyta Blight and Winter Sowing of Chickpeas*, Martinus Nijhoff/ Dr. W. Junk Publishers, The Hague, The Netherlands, pp. 45–54.

Reed, W. and Lateef, S.S. (1990) Pigeonpea, pest management. In: Nene, Y.L., Hall, S.D. and Sheila, V. (eds) *The Pigeonpea*. CAB International, Wallingford, UK, pp. 349–374.

Rembold, H. (1981) Malic acid in chickpea exudate – a marker for *Heliothis* resistance. *International Chickpea Newsletter* 4, 18–19.

Romeis, J., Shanowet, T.G. and Peter, A.J. (1999) Trichomes on pigeonpea [*Cajanus cajan* (L.) Mill. sp.] and two wild *Cajanus* spp. *Crop Science* 39, 564–569.

Sandhu, T.S., Brar, J.S., Sandhu, S.S. and Verma, M.M. (1985) Inheritance of resistance to mungbean yellow mosaic virus in greengram. *Journal of Research Punjab Agricultural University* 22, 607–611.

Sanyal, I., Singh, A.K., Kaushik, M.A. and Amla, D.V. (2005) *Agrobacterium*-mediated transformation of chickpea (*Cicer arietinum* L.) with *Bacillus thuringiensis Cry1Ac* gene for resistance against pod borer insect *Helicoverpa armigera*. *Plant Science* 168, 1135–1146.

Sarker, A., Bayaa, B., El-Hassan, H. and Erskine, W. (2004) New sources of resistance to *Fusarium* wilt in lentil (*Lens culinaris* Medikus subsp. *culinaris*). *Journal of Lentil Research* 1, 30–33.
Sarker, A., Singh, B.B., Malhotra, R.S., Gupta, S., Khalil, S., Dixit, G.P. *et al.* (2008) Insect pest management in food legumes, the future strategies. In: Kharkwal, M.C. (ed) *Proceedings of the 4th International Food Legumes Research Conference vol. 1*, 18–22 October 2005, New Delhi, India, pp. 588–607.
Sasser, J.N. and Freckman, D.W. (1987) A world perspective on nematology, the role of the society. In: Veech, J.A. and Dickson, D.W. (eds) *Vistas on Nematology, a Commemoration of the Twenty-Fifth Anniversary of the Society of Nematalogists.* Society of Nematologists, Hyattsville, Maryland, pp. 7–14.
Satyaprasad, K. and Rama Rao, P. (1983) Effect on chickpea root exudates on *Fusarium oxysporum* f.sp. *ciceri. Indian Phytopathology* 36, 77–81.
Shanower, T.G., Yoshida, M. and Peter, A.J. (1997) Survival, growth, fecundity and behaviour of *Helicoverpa armigera* (Lepidoptera, Noctuidae) on pigeonpea and two wild *Cajanus* species. *Journal of Economic Entomology* 90, 837–841.
Sharma, D., Kannaiyan, J. and Reddy, L.J. (1982) Inheritance of resistance to blight in pigeonpea. *Plant Diseases* 66, 22–25.
Sharma, H.C., Vidyasagar, P. and Leuschner, K. (1988) Field screening for resistance to sorghum midge (Diptera, Cecidomyiidae). *Journal of Economic Entomology* 81, 327–334.
Sharma, H.C., Stevenson, P.C., Simmonds, M.S.J. and Green, P.W.C. (2001) Identification of *Helicoverpa armigera* (Hubner) feeding stimulants and the location of their production in the pod surface of pigeonpea [*Cajanus cajan* (L.) Millsp.]. In: Final technical report, Competitive Research Facility Project [R 7029 (C)], ICRISAT, Andra Pradesh, India, pp. 85.
Sharma, H.C., Clement, S.L., Ridsdill-Smith, T.J., Ranga Rao, G.V., El Bouhssini, M., Ujagir, R. *et al.* (2008) Insect pest management in food legumes, the future strategies. In: Kharkwal, M.C. (ed.) *Proceedings of the 4th International Food Legumes Research Conference vol. 1*, 18–22 October 2005, New Delhi, India, pp. 522–544.
Sharma, H.C., Sujana, G. and Manohar Rao, D. (2009) Morphological and chemical components of resistance to pod borer, *Helicoverpa armigera* in wild relatives of pigeonpea. *Arthropod Plant Interaction* 3, 151–161.
Sharma, K.K., Lavanya, M. and Anjaiah, V. (2006) *Agrobacterium*-mediated production of transgenic pigeonpea (*Cajanus cajan* L.Millsp.) expressing the synthetic Bt *cry1Ab* gene. *In Vitro Cellular and Developmental Biology – Plant* 42, 165–173.
Sharma, R.N., Srivastava, G., Rathore, A.L., Sharma, M.L. and Khan, M.A. (2003) Chickpea research for the millennium. In: *Proceedings of the International Chickpea Conference*, 20–22 January 2003, Indira Gandhi Agricultural University, Raipur, India.
Sharma, S.B. and Ashokkumar, P. (1991) A greenhouse technique to screen pigeonpea for resistance to *Heterodera cajani. Annals of Applied Biology* 118, 351–356.
Sharma, S.K., Sikora, R.A., Greco, N., Vito, M.D. and Caubel, G. (1994) Screening techniques and sources of resistance to nematodes in cool season food legumes. *Euphytica* 129, 109–117.
Siddiqui, Z.A. and Hussain, S.I. (1992) Response of 20 chickpea cultivars to *Meloidogyne incognita* race 3. *Nematologia Mediterranea* 20, 33–36.
Singh, D.P. and Singh, A. (2005) *Disease and Insect Resistance in Plants.* Oxford and IBH Publishing Co. Pvt. Ltd., New Delhi, India, pp. 417.
Singh, D.P., Sharma, B.L., Verma, R.P.S. and Reddy, K.R. (1991) Evaluation of greengram germplasm. *ND Journal of Agricultural Research* 6, 211–213.
Singh, G., Kapoor, S. and Singh, K. (1988a) Multiple disease resistance in mungbean with special emphasis on MYMV. In: Shanmugasundaram, S. (ed.) *Proceedings of the 2nd International Symposium on Mungbean*, AVRDC, Shanhua, Taiwan, pp. 290–296.
Singh, I.P., Singh, D.P. and Singh, B.B. (1991) Inheritance of resistance to sterility mosaic in pigeonpea. In: *Proceedings of Symposium on Grain Legumes*, 9–11 February 1991. Indian Society of Genetics and Plant Breeding, Indian Agricultural Research Institute, New Delhi, India, pp. 291–293.
Singh, K.B. and Reddy, M.V. (1989) Genetics of resistance to *Ascochyta* blight in four chickpea lines. *Crop Science* 29, 657–659.
Singh, R.A. and Naimuddin (2009) Mungbean and urdbean diseases. In: *25 Years of Research at IIPR.* Indian Institute of Pulses Research, Kanpur, India, pp. 85–115.
Singh, R., Sindhu, A.H. and Singal, R. (2003) Biochemical basis of resistance in chickpea (*Cicer arietinum* L.) against *Fusarium* wilt. *Acta Phytopathologica et Entomologica Hungarica* 38, 13–19.
Singh, U.P., Singh, P. and Singh, R.M. (1988b) Inheritance of field resistance to *Alternaria* blight in pigeonpea. *International Pigeonpea Newsletter* 7, 4–5.

Sithanantham, S., Lateef, S.S. and Reed, W. (1981) Pod fly susceptibility in pigeonpea, some aspects of oviposition preference. In: *Proceedings of International Workshop on Pigeonpea, vol. 2*, ICRISAT, Patancheru, Andhra Pradesh, India, pp. 329–335.

Smith, C.M., Khan, Z.R. and Pathak, M.D. (1994) *Techniques for Evaluating Insect Resistance in Crop Plants*. CRC Press, Boca Raton, Florida, pp. 320.

Sokhi, S.S., Jhooty, J.S. and Bains, S.S. (1979) Resistance in pea against powdery mildew. *Indian Phytopathology* 32, 571–574.

Srinivas, T., Reddy, M.V., Jain, K.C. and Reddy, M.S.S. (1997) Inheritance of resistance to two isolates of sterility mosaic pathogen in pigeonpea (*Cajanus cajan* (L.) Millisp.). *Euphytica* 97, 45–52.

Srivastava, R.P. (2009) *25 Years of Research at IIPR*. Indian Institute of Pulses Research, Kanpur, India, pp. 166–170.

Talekar, N.S. (1983) Mungbean – Agromyzid fly resistance characterization. *Annual Plant Resistance to Insects Newsletter* 9, 75.

Talekar, N.S., Yang, H.L. and Lee, Y.H. (1988) Morphological and physiological traits associated with agromyzid (Diptera, Agromyzidae) resistance in mungbean. *Journal of Economic Entomology* 82, 1352–1358.

Tikoo, J.L., Sharma, B., Mishra, S.K. and Dikshit, H.K. (2005) Lentil (*Lens culinaris*) in India, present status and future perspectives. *Indian Journal of Agricultural Sciences* 75, 539–562.

Timmerman-Vaughan G.M., Frew T.J. and Weeden, N.F. (1994) Linkage analysis of *er1*, a recessive *Pisum sativum* gene for resistance to powdery mildew fungus (*Erysiphe pisi* D.C). *Theoretical and Applied Genetics* 88, 1050–1055.

Tullu, A., Muehlbauer, F.J., Simer, C.J., Meyer, M.S., Kumar, J., Kaiser, W.J. *et al.* (1998) Inheritance and linkage of a gene resistant to race 4 of *Fusarium* wilt and RAPD markers in chickpea. *Euphytica* 102, 227–232.

Upadhyay, K.D. and Banerjee, H. (1986) Some chemical changes in chickpea plant infected with root-knot nematode, *Meloidogyne javanica*. *Indian Journal of Nematology* 16, 286–288.

Venugopal Rao, N., Tirumala Rao, K. and Reddy, A.S. (1991) Ovipositional and larval development sites of gram caterpillar (*Helicoverpa armigera*) in pigeonpea. *Indian Journal of Agricultural Sciences* 61, 608–609.

Verulkar, S.B., Singh, D.P. and Bhattaacharya, A.K. (1997) Inheritance of resistance to podfly and pod borer in the interspecific cross of pigeonpea. *Theoretical and Applied Genetics* 95, 506–508.

Vishwa, D., Reddy, M.V. and Chaudhary, R.G. (2005) Major diseases of pigeonpea and their management. In: Ali, M. and Kumar, S. (eds) *Advances in Pigeonpea Research*. Indian Institute of Pulses Research, Kanpur, India, pp. 229–261.

Yoshida, M., Cowgill, S.E. and Wightman, J.A. (1995) Mechanism of resistance to *Helicoverpa armigera* (Lepidoptera, Noctuidae) in chickpea: Role of oxalic acid in leaf exudate as an antibiotic factor. *Journal of Economic Entomology* 88, 1783–1786.

16 Breeding for Abiotic Stresses

C. Toker and N. Mutlu

16.1 Introduction

Food legumes are divided into two groups according to their eco-geographic distributions in the world and climatic requirements, such as cool season food legumes and warm or tropical season food legumes (Hall, 2001; Toker and Yadav, 2010). The genera *Cicer* L., *Lathyrus* L., *Lens* Mill., *Lupinus* L., *Pisum* L. and *Vicia* L. are referred to as cool season food legumes (Singh and Saxena, 1993; Muehlbauer and Kaiser, 1994). On the other hand, the genera *Arachis* L., *Cajanus* L., *Glycine* Willd., *Phaseolus* L., *Vigna* Savi and some minor food legumes are referred to as warm season food legumes (Clarke *et al.*, 2008). A summary of these two groups is given in Table 16.1.

The yield of cool season food legumes increased slightly from 1961 to 2008 (FAOSTAT, 2008), while the increase in yield of warm season food legumes (except soybean) has been even less, despite increased efforts to improve these crops (ISI, 2010). Although food legumes have high yield potential (Table 16.2), their yields globally are low and unstable, mainly due to biotic and abiotic stresses (FAOSTAT, 2008). On a global basis, annual yield losses due to biotic and abiotic stresses in food legumes are estimated to be close to current production, since their yield potential is three or four times higher than the average global yield (Table 16.2).

The most common abiotic stresses affecting production of food legumes are drought accompanied by heat and cold (Troedson *et al.*, 1990; Singh and Saxena, 1993; Muehlbauer and Kaiser, 1994; Burton, 1997; Dracup *et al.*, 1998; Singh and Matsui, 2002; Materne *et al.*, 2007; Toker *et al.*, 2007a; Toker and Yadav, 2010). Other abiotic stresses specific to some regions of the world are salinity, waterlogging, soil alkalinity and acidity, and nutrient deficiencies and toxicities (Ryan, 1997; Siddique *et al.*, 2000; Toker *et al.*, 2007a). This chapter aims to review current knowledge of the main abiotic global constraints facing important food legume production. It also summarizes selection criteria and available genetic resources for stress resistance under abiotic stress conditions.

16.2 Drought

As a meteorological term and environmental event, drought is defined as a water stress due to lack or insufficiency of rainfall and/or inadequate irrigation. Drought stress is affected by several climatic, edaphic and agronomic factors, and involves three main parameters: timing, duration and intensity (Serraj *et al.*, 2003). Drought stress depends not only on rainfall and its distribution, but also on evaporation, soil water-holding capacity and crop water requirements (Toker *et al.*, 2007a).

Table 16.1. Important characteristics of cool and warm season food legumes.

Important characteristics	Cool season food legumes	Warm season food legumes
Germination shape	Generally hypogeal[a]	Generally epigeal[b]
Germination minimum (°C)	4	10–12
Optimum temperature (°C)	15–25	25–35
Vernalization response	Quantitative	No
Low temperature response	Cold tolerant	Cold susceptible
High temperature response	Generally susceptible	Generally tolerant
Photoperiodic response	Quantitative long day and neutral	Short day and neutral
Irrigation	Generally rainfed	Generally supplemental

[a]Lupines germinate epigeally;
[b]Pigeon pea, runner bean and adzuki bean germinate hypogeally.

Types of drought

The major cool season food legume-growing areas in the world are in the arid and semi-arid zones (Singh and Saxena, 1993), while the majority of the warm season food legumes are grown as rainfed crops in the tropics (van der Maesen and Somaadmadja, 1992). Among warm season legumes, beans are predominantly grown in the rainy season in the tropics when rainfall distribution is bimodal. Beans are normally irrigated when they face drought, owing to erratic rainfall distribution under rainfed conditions (Thung, 1991). However, drought has equal importance to soil fertility problems, since approximately 60% of production suffers from serious drought conditions (White and Singh, 1991).

Cowpea is widely grown in the semi-arid tropics where drought is a major limiting factor of production. It is often subjected to drought stress at both the seedling and terminal growth stages, due to irregular distribution of rainfall at the beginning and towards the end of the rainy season (Singh and Matsui, 2002). With deep and extensive roots (Reddy, 1990), pigeon pea is a rainfed crop well adapted to drought-prone environments (Lawn and Troedson, 1990). Pigeon pea is mainly sown as a rainy-season crop in India (Troedson *et al.*, 1990), where about 90% of world production occurs (FAOSTAT, 2008),

Table 16.2. Comparative yield and yield potential analysis of food legumes (1961–2008).

Food legumes	Crop	Yield (kg/ha)				Reference
		1961	2008	Increase	Potential	
Cool season	Chickpea	649	760	111	5000	Singh (1997)
	Lentil	528	944	416	3000	Erskine and Saxena (1993)
	Lupine	580	1280	700	5000	Huyghe (1997)
	Pea	973	1658	685	5000	Cousin (1997)
	Faba bean	896	1484	588	5000	Duc (1997)
Warm season	Pigeon pea	817	844	27	5000	Chauhan (1990)
	Bean	493	727	234	5000	Graham and Ranalli (1997)
	Cowpea	361	456	95	4000	Ehlers and Hall (1997)
	Soybean	1127	2384	1257	7000	–

and is grown through to maturity in the subsequent dry season on stored soil water (Lawn and Troedson, 1990). Thus, the crop is exposed to intermittent or transient drought during the vegetative stage, followed by terminal drought during most of its reproductive stage (Troedson *et al.*, 1990). Water is often the primary limiting factor in soybean production, and therefore is an important management concern (Pendleton and Hartwig, 1973). In soybean, yield is reduced more by drought at the pod-filling stage than at the flowering stage (Mederski *et al.*, 1973).

The first step in breeding for resistance to drought in cool season food legumes is to determine the type of drought. Food legumes are generally are subjected to: (i) terminal drought, increasing towards the generative stage, due to the depletion of soil moisture; and/or (ii) intermittent or transient (unpredictable) drought, caused by a break in rainfall followed by insufficient rains at the vegetative stage (Singh and Saxena, 1993; Materne *et al.*, 2007; Toker *et al.*, 2007a).

Effects of drought

The growing season of food legumes may be shortened by drought, affecting the production of yield components, i.e. total biomass, pod number, seed number, seed weight and quality, and seed yield per plant (Lawn and Troedson, 1990; Materne *et al.*, 2007; Toker *et al.*, 2007a; Charlson *et al.*, 2009; Khan *et al.*, 2010).

Chickpea and lentil are known as drought-resistant genera; in contrast, pea and faba bean are known as drought-sensitive (Toker and Yadav, 2010). Although drought resistance is relatively higher in chickpea than in lentil, field pea and faba bean (Siddique *et al.*, 2000), seed yield losses due to drought range from 30 to 100% (Saxena *et al.*, 1993a; Leport *et al.*, 1999; Canci and Toker, 2009a) depending on genotype and the type of drought experienced in the target environment. Benjamin and Nielsen (2006) reported that chickpea was superior to pea for dryland crop production in semi-arid climates due to an adaptive root distribution. Yield losses in lentil due to drought can range from 6 to 60% in rainfed environments (Saxena *et al.*, 1993a; Materne *et al.*, 2007). Pea is subjected to drought in some parts of the world (Saxena, 1993).

Drought has several effects, including the prevention of nitrogen fixation and reducing the total biomass in pea (Cousin, 1997). Drought and high-temperature stresses caused yield losses of 21–54% in India, Syria and New Zealand (Saxena *et al.*, 1993a, b). Faba bean is known as a drought-susceptible species among the cool season food legumes (Bond *et al.*, 1994), especially during its flowering period (Duc, 1997). However, Link *et al.* (1999) and Ricciardi *et al.* (2001) reported that there is a genotypic variation for drought tolerance in faba bean, especially in North African and Latin American genotypes (Link *et al.*, 1999). In drought conditions, dry matter yield in faba bean, pea and chickpea was reduced to 36.4, 23.9 and 14.5%, respectively (Amede *et al.*, 2003). Although narrow-leafed lupine is one of the most drought-resistant species among the cultivated lupines (Cowling *et al.*, 1998; Palta *et al.*, 2004), considerable yield reduction was reported in narrow-leafed lupine. More than 50% yield reduction was reported in lupines, including *Lupine albus*, *L. angustifolius*, *L. pilosus* and *L. atlanticus* in rainfed plots compared with the irrigated (Dracup *et al.*, 1998).

Among the warm season food legumes, ranking of the crops in increasing order of drought resistance was soybean, followed by black gram, green gram, groundnut, bambara nut, lablab and cowpea (Singh *et al.*, 1999). However, Likoswe and Lawn (2008) reported that total dry matter per plant ranked in the order cowpea > soybean > pigeon pea when water was withheld. At the Centro Internacional de Agricultura Tropical (CIAT), the yield of beans was reduced from 16 to 94% when subjected to drought (White and Singh, 1991). Similarly, yield reduction in drought and irrigated plots of beans was approximately 40% and 80% for drought-tolerant and -susceptible genotypes, respectively (White and Izquiero, 1991). In 2000, Singh and Matsui (2002) found that drought-tolerant varieties of cowpea had significantly higher grain yields than -susceptible varieties in the field at Minjibir and Zinder (Niger Republic), where rainfall is normally low.

Yield reduction in the drought-tolerant varieties at Minjibir ranged from 8 to 69% (Singh and Matsui, 2002). In soybean, Sincik *et al.* (2008) demonstrated a 45% seed yield reduction in non-irrigated specimens when compared with fully irrigated conditions. When pigeon pea genotypes were grown in irrigated and non-irrigated (drought) conditions at the International Crops Research Institute for the Semi-Arid Tropics (ICRISAT), yield reduction in 26 genotypes was found to be 67% by Subbarao *et al.* (2000). As a result of drought accompanied by high temperature, food legume yields decline according to the time and occurrence of drought (Table 16.3).

Drought resistance mechanisms

A range of different mechanisms utilized by food legumes in adapting to drought conditions has been suggested in chickpea (Toker *et al.*, 2007a), lentil (Materne *et al.*, 2007), lupines (Dracup *et al.*, 1998), pea (Wery *et al.*, 1993), faba bean (Khan *et al.*, 2010), pigeon pea (Singh *et al.*, 1990), beans (White and Singh; 1991) and soybean (Charlson *et al.*, 2009). Common drought resistance mechanisms include: (i) escape; (ii) avoidance; and (iii) tolerance (Ludlow and Muchow, 1990; Subbarao *et al.*, 1995; Turner *et al.*, 2001). Escape, which can be engineered by early sowing and earliness (days to first flowering, 50% flowering and maturity) is the most important mechanism in avoiding the onset of drought under terminal conditions (Toker *et al.*, 2007a). Although short-duration varieties maturing before the onset of severe terminal drought have proved to be successful in increasing yield under drought-prone conditions (Kumar *et al.*, 1996), these hold no advantage in intermittent or unpredictable drought conditions. Maximum yield depends upon water availability, and varieties need to be matched with the longest growing period (Toker *et al.*, 2007a). Avoidance can be achieved by maintaining water uptake and reducing water loss through roots and leaves. Tolerance is the ability of cells to metabolize at low leaf water status.

Table 16.3. Yield reduction in food legumes due to drought.

Crop	Yield reduction (%)	Reference(s)
Chickpea	30–100	Leport *et al.* (1999); Canci and Toker (2009a)
Lentil	6–66	Saxena *et al.* (1993a, b)
Lupine	> 50	Dracup *et al.* (1998)
Pea	21–54	Saxena *et al.* (1993a, b)
Faba bean	36	Amede *et al.* (2003)
Pigeon pea	67	Subbarao *et al.* (2000)
Bean	16–94	White and Singh (1991); Singh *et al.* (2008)
Cowpea	8–69	Singh and Matsui (2002)
Soybean	45	Sincik *et al.* (2008)

Inheritance of drought resistance

Drought resistance mechanisms can also be categorized as: (i) morphological; (ii) physiological; or (iii) molecular (Toker and Yadav, 2010). Morphological and physiological characters related to drought resistance were reported to follow different types of inheritance patterns (monogenic and polygenic) and gene actions (additive and non-additive). A single recessive gene controlling for earliness was reported in chickpea (Kumar and van Rheenen, 2000). Both the early-flowering trait and photoperiodic response, being simply inherited, may easily be introduced into late-flowering genetic backgrounds (Kumar and Abbo, 2001). Hovav *et al.* (2003) reported that genetic correlations between time to flowering and seed weight were positive and relatively high, suggesting that for certain genetic backgrounds it might be difficult to breed early-flowering cultivars without compromising seed weight. Growth vigour in chickpea is controlled by two genes with duplicate dominant epistasis (Sabaghpour *et al.*, 2003).

Root length density and root dry weight among recombinant inbred lines derived from ICC 4958 and Annigeri appeared to be under polygenic control. The broad-sense heritability for root length and root dry weight was estimated to be 0.23 and 0.27, respectively (Serraj *et al.*, 2004). Hegde (2010) found that duplicate dominant genes with cumulative but unequal effect govern flowering time in chickpea. A genotype with two dominant alleles in homozygous or heterozygous conditions at both loci (*Efl1, Efl2*) controls late flowering. A genotype with a dominant allele in homozygous or heterozygous condition at one of the loci and a homozygous recessive allele at the other (*Efl1,efl2*) controls earliness, and a genotype with homozygous recessive alleles at both loci (*efl1,efl2*) is responsible for super-earliness (< 25 days). As a drought-escape mechanism, early flowering in lentil is governed by single recessive gene (*sn*), and transgressive segregants for early flowering in F_2 populations are based on the interaction between major and minor genes for earliness (Sarker *et al.*, 1999). Earliness in flowering for *Pisum* is controlled by a single recessive gene (Murfet, 1975). In faba bean, Toker (2004) estimated 97% broad-sense heritability for days to flowering and maturity. Abdelmula *et al.* (1999) found that the heritability of drought tolerance was 48% in F_1 hybrids and 70% in parents. In white lupine, broad-sense heritability ranged from 91 to 97% for flowering date in dwarf, determinate and dwarf-determinate genotypes (LeSech and Huyghe, 1991).

Although pigeon pea is considered a drought-tolerant crop due to its deep root systems and indeterminate growth habit, it often suffers from drought in semi-arid tropics and is subjected to drought under rainfed conditions. Heritability for days to flowering was estimated both as medium (50–75%) and high (> 75%), with mainly additive and additive plus non-additive gene action (for details see Saxena and Sharma, 1990). Leaf pubescence density is an important component for the adaptation of soybean to a drought-prone environment (Du *et al.*, 2009), and the gene *P* (glabrous) having monogenic inheritance is epistatic to this, controlling hair density (*Pd*) (Bernard and Singh, 1969). In bean, Bouwkamp and Summers (1982) found that the combined resistance to drought and heat was controlled by a single dominant gene in the accession PI 297079, and by two epistatic genes in the accession PI 151062 in controlled conditions. In cowpea, Mai-Kodomi *et al.* (1999) reported simple inheritance of drought tolerance using a box-screening method, and they identified two types of shoot drought tolerance: (i) type 1 plants stayed green for a long time after withholding of water, and the whole plant died under continuing dry conditions; (ii) type 2 plants remained alive for a much longer period, but the whole plant did not die with continuing dry conditions (Singh, 2002). Drought tolerance of both types 1 (*Rds1*) and 2 (*Rds2*) is inherited as monogenic dominant traits (Singh and Matsui, 2002).

Cattivelli *et al.* (2008) have recently summarized some genes conferring drought tolerance. Using marker-assisted selection (MAS), these genes will play a crucial role in selecting for resistance to drought in food legumes.

Assessment of resistance to drought

Fischer and Maurer (1978) proposed a drought susceptibility index (*S*) to evaluate for adaptation of advanced lines to semi-arid environments, defined as

$$S = (1 - Y/YP)/D$$

where, *Y* is the yield under drought stress, *YP* is the yield under non-stress conditions and *D* is drought intensity; $D = 1 - X/XP$, where *X* and *XP* represent the mean yield of all varieties under stress and non-stress conditions, respectively; and *D* ranges from 0 to 1.

A drought response index (DRI) was proposed by Bidinger *et al.* (1987) to describe the response of individual genotypes to drought conditions, and fitted a multiple regression of stressed grain yield on unstressed grain yield and days to flowering:

$$Y0 = a - bF + cYi \text{ and DRI} = (Y0 - \hat{Y}0)/\text{standard error of } \hat{Y}0$$

where, *Y*0 is yield under drought, $\hat{Y}0$ is regression estimate of yield under drought, *Yi* represents yield potential and *F* is the days to flowering.

Baker (1994) chose to approximate the response to stress as a linear function of increasing stress. Productivity (Yj) at any level of stress (Sj) can be represented by a linear regression equation such as;

$$Yj = M - TSj$$

where M is the maximum productivity in the absence of stress, T represents a measure of tolerance and Sj is a quantitative measure of the level of stress.

High tolerance, represented by low values of T, will result in low-level changes in productivity with changes in stress level. A low-level stress (Sl) and high-level stress (Sh), $T(Sl - Sh)$, will decrease productivity. Abebe *et al.* (1998) explained some indices to select drought-resistant beans: arithmetic mean (AM) and geometric mean (GM), response to drought (RD) and percentage reduction (PR):

$$AM = (Ys + Yns)/2;$$
$$GM = \sqrt{Ys \times Yns};$$
$$RD = (Yns - Ys)/W;$$
$$PR = 100\,[1 - (Ys/Yns)]$$

where Ys is mean seed yield of a cultivar under stress environments over 3 years and Yns is mean seed yield in non-stress environments. W was calculated as the difference (mm) in total seasonal rainfall between the stress and non-stress environments.

The yield (Y) of a crop was modelled by a generalized equation (Hay and Porter, 2006; Khan *et al.*, 2010):

$$Y = Q \times I \times \varepsilon \times HI$$

where Q is the received water, I is the fraction of that input that is intercepted or absorbed by the crop, ε is the efficiency (water use efficiency at the crop level (WUE) or transpiration efficiency at the leaf level (TE)) and HI is the harvest index.

For water-limited crops, therefore, $Q \times I$ is the total amount of water transpired.

Screening and selection for drought resistance

Although there are many screening and selection techniques for drought resistance in food legumes, few techniques have been successful under field conditions: (i) line-source sprinkler irrigation systems (Saxena *et al.*, 1993a); (ii) root trait characteristics (root length, root density, root biomass, root length density (Serraj *et al.*, 2004) and the 'root-box pin board' method (Singh and Matsui, 2002); (iii) delayed sowing strategy (Singh *et al.*, 1997); with this technique, test materials can be evaluated together with drought and heat stresses under field conditions (Canci and Toker, 2009a); (iv) comparison of lines in non-stressed and stress conditions as the defined formula (Silim and Saxena, 1993; Toker and Cagirgan, 1998); and (v) rain-out shelter tunnels (Abdelmula *et al.*, 1999; Amede *et al.*, 1999; Link *et al.*, 1999). For large-scale screening these methods can be useful, but are labour and time intensive.

In addition to field screening techniques, Khan *et al.* (2010) have recently explained some useful traits for selection of drought-resistant genotypes such as water use and transpiration efficiency, relative water content, stomatal conductance, leaf (canopy) temperature, carbon isotope discrimination, leaf cuticle characteristics, osmotic potential, oxidative response and specific leaf area. The use of carbon isotope discrimination ($\Delta^{13}C$) in screening is described for some food legumes (Stoddard *et al.*, 2006; Khan *et al.*, 2007, 2010), but it incurs high costs per sample and thus other, cheaper, methods may be preferred. The delayed leaf senescence (DLS) trait has the potential to enhance the drought adaptation of cowpea in dry areas (Hall *et al.*, 2002), while delayed canopy wilting (DCW) is used to select for resistance to drought in soybean (Charlson *et al.*, 2009). Recovery ability after wilting (RAW) has been proposed in chickpea (Toker *et al.*, 2007b), leaf pubescence density (LPD) is an important component for the adaptation of soybean to drought-prone environments (Du *et al.*, 2009), while chlorophyll content is another trait used for evaluating drought resistance in water-stressed plants (Nayyar *et al.*, 2005). Alterations in levels, distribution and timing of plant growth regulators (abscisic acid, brassinosteroids, jasmonates, phosphatidic and salicylic acids) protect plants from drought effect when externally applied or internally produced (Davies,

1995). Ricciardi *et al.* (2001) showed that leaf water potential and stomatal resistance measurements in faba bean were useful in describing simulated water stress, but were not suitable for discriminating genotypes with tolerance to water stress. However, screening under controlled conditions can allow the rapid and uniform evaluation of test genotypes (Grzesiak *et al.*, 1996); this method should also be non-destructive, accurate and capable of processing many samples.

Sources of drought resistance/tolerance

Using the above methods, Toker and Yadav (2010) have recently selected and identified sources of tolerance or resistance to drought in cool season food legumes. In chickpea, ICC 4958, ICC 8261 and FLIP 87-59C are the most popular drought-tolerant germplasm lines (Saxena *et al.*, 1993b; Singh *et al.*, 1996; Kashiwagi *et al.*, 2005). In lentil, ILL 2914, ILL 2915, ILL 3124, ILL 3397 and ILL 3399 were selected for earliness and early maturity (Erskine and Witcombe, 1984). ILL 784 and ILL 1861 have high yield in drought conditions (Toker and Yadav, 2010). Faba bean and pea are referred to as drought-susceptible genera among cool season food legumes (Toker and Yadav, 2010), faba bean being more sensitive to drought than pea (McDonald and Paulsen, 1997). ILB 938/2 is one of the most successful drought-tolerant faba beans (Khan *et al.*, 2007, 2010). Although some cultivars and accessions of lentil, pea and faba bean were reported to be drought tolerant (Stoddard *et al.*, 2006; Toker and Yadav, 2010), they should be considered as winter-sown crops since they had the highest cold tolerance level among cool season food legumes (Clarke *et al.*, 2008; Toker and Yadav, 2010). Narrow-leafed lupine is one of the most drought-resistant species among the cultivated lupine species (Cowling *et al.*, 1998; Palta *et al.*, 2004).

Late-maturing pigeon pea genotypes are more suitable to intermittent drought conditions, while early-maturing genotypes are likely to be more productive in terminal and severe drought conditions (Troedson *et al.*, 1990). Therefore, selection should be based on the nature of the drought. Medium-duration genotypes such as BDN 5, ICPL 8340, ICP 3233, PBN/A 53 and ICP 4865 were classified as drought tolerant (Singh *et al.*, 1990). 'Essex' soybean was identified as tolerant and a wild soybean PI 407155 (*Glycine soja* Sieb. & Zucc.) as more tolerant to dehydration stress in a greenhouse screen (Chen *et al.*, 2006). The Bean International Yield Trial was carried out by CIAT at several locations, with V8025 and BAT477 having the highest yield over 11 locations under drought conditions (White and Singh, 1991). SEA 5 and SEA 13 were developed as drought-tolerant lines at CIAT (Sing *et al.*, 2001). In the CIAT bean project in 2004, RAB 650 and SEA 23 were two lines from the breeding programme found to have outstanding adaptation to drought (Hillocks *et al.*, 2006). CO46348 is a drought-tolerant, rust-resistant and high-yielding germplasm line (Brick *et al.*, 2008). Genetic variability for drought tolerance was found to be narrow in *Phaseolus vulgaris* but the tepary bean, *Phaseolus acutifolius* was superior for drought tolerance (Hillocks *et al.*, 2006). Crosses with tepary bean have been recovered at CIAT using 'embryo rescue' techniques (White and Singh, 1991). R01-416F and R01-581F soybean germplasm lines have been improved for yield and nitrogen fixation under drought stress (Chen *et al.*, 2007). In cowpea, 'Mouride', 'Melakh' and 'Ein El Gazal' have substantial resistance to vegetative-stage drought (Cisse *et al.*, 1995, 1997; Elawad and Hall, 2002), since 'California Blackeye No. 5' ('CB5') is one of the parents of 'Ein El Gazal' (Elawad and Hall, 2002). Singh and Matsui (2002) reported certain drought-tolerant cowpea lines: 'Type 1' and 'Type 2'.

16.3 Temperature

Low temperature (cold)

Types of cold stress

According to Wery *et al.* (1993), cold-related stress can be defined as heat (high temperature), chilling (low positive temperature) or freezing (negative temperature). Chilling and freezing stresses are commonly known as 'cold'. The following temperatures are considered an approximate threshold for explaining cold-related stresses in cool season food legumes (Wery *et al.*, 1993; Toker *et al.*, 2007a). A daily minimal temperature below 0°C without snow cover is referred to as 'freezing'. A daily average temperature between 0°C and 10°C is called 'chilling', but temperatures of 12°C represent the threshold to distinguish chilling-sensitive and chilling-resistant plants (Wery *et al.*, 1993) – Toker *et al.* (2007a) defined 'chilling as a condition between 0°C and 12°C'. A daily maximal temperature above 25°C is known as 'heat', which could be equivalent to 30°C at the level of a non-transpiring canopy (Wery *et al.*, 1993).

In general, freezing stress is an important yield reducer, from severe to moderate, and is common during vegetative growth in Asia, North Africa, Europe and in the western hemisphere (Johansen *et al.*, 1994; Slinkard *et al.*, 1994). Chickpea, pea and faba bean encounter chilling stress at the reproductive stage (Clarke *et al.*, 2008), while they face freezing stress at the vegetative stage when they are sown in autumn or early spring (Materne *et al.*, 2007; Toker *et al.*, 2007a; Saeed *et al.*, 2010; Toker and Yadav, 2010). Winter hardiness of lentil is similar to that of faba bean and greater than that of pea and chickpea (Murray *et al.*, 1988). Lentil survived exposure to air temperatures of −26.8°C in January at Haymana, Ankara, Turkey with snow cover for 47 days (Erskine *et al.*, 1981). Lentil was exposed to freezing stress in western Canada and the highlands of western Asia at the vegetative stage in spring and at the generative stage in late summer and early autumn (Ali *et al.*, 1991). Faba bean has the highest cold-tolerance level among cool season food legumes (Duc, 1997), with a minimum air temperature of −25°C being reported to have permitted the survival of faba bean under field conditions in Ankara, Turkey (Murray *et al.*, 1988). 'Cote d'Or' can survive −22°C when overwintering (Duc, 1997). Air temperatures of −23°C or below are considered to be lethal for pea (Murray *et al.*, 1988). Breeders have developed winter forage peas because of their good resistance to freezing (Cousin, 1997). Lupines are not only injured by freezing stress during the early vegetative stage when sown in autumn, but they are also damaged by chilling temperatures during the early reproductive stage in late winter or early spring (Dracup *et al.*, 1998).

Although warm season food legumes are not cold tolerant (freezing), some are subjected to low temperature (chilling) during germination (Clarke *et al.*, 2008). In controlled conditions, plant height, node numbers and dry mass of shoots and leaves of pigeonpea are increased by increasing the temperature from 16°C to 32°C (McPherson *et al.*, 1985). Conversely, vegetative growth was slow at temperatures below 20°C and temperatures between −2°C and −3°C caused defoliation (Troedson *et al.*, 1990). Temperatures below 15°C adversely affect growth and development in bean (Singh, 1991). Low temperatures (< 10°C) prevailing at over 2500 m above sea level in South America had detrimental effects on bean, especially bush bean, but climbing bean showed considerable cold tolerance at all stages of growth (Singh, 1991). Early-planted soybean frequently encounters cold soil conditions (Unander *et al.*, 1986); also, soybean is sensitive to chilling temperatures (≈15°C) at flowering time (Takahashi and Asanuma, 1996). Cowpea is also sensitive to chilling temperatures (Hall *et al.*, 2002); the rate of emergence was reported to be slower and the extent of maximal emergence less under chilling (15°C) compared with more favourable (28°C) temperatures. The threshold soil temperature where cowpea exhibits incomplete emergence is quite high, at about 19°C (Ismail *et al.*, 1997).

Effects of cold stress

In chickpea, mean daily temperatures below 15°C cause flower and pod abortion in some parts of India and Australia (Savithri *et al.*, 1980; Srinivasan *et al.*, 1999; Clarke *et al.*, 2004). Similarly, faba bean and pea faced low temperatures, both of freezing and chilling, during the reproductive stage causing stem collapse, flower shedding and pod abortion (Clarke *et al.*, 2008; C. Toker, unpublished data). These effects reduce growth rate and increase chlorosis and necrosis in older leaves at the whole-plant level. Meiosis is adversely affected by cold (Blum, 1988). Reproductive organs are very sensitive to cold, resulting in sterile flowers (Savithri *et al.*, 1980; Clarke *et al.*, 2004). Delayed germination and emergence take place at low temperatures (Auld *et al.*, 1988; Toker *et al.*, 2007a).

Temperatures as low as −3°C were found to be lethal for pigeon pea (Troedson *et al.*, 1990). Chilling stress in beans, soybean and cowpea causes poor germination, poor vigour, pollen and seed production at the reproductive stage and also delayed maturity, resulting in reduced seed yield and physical quality (Singh, 1991).

Cold resistance mechanisms

Three resistance mechanisms are reported in regard to freezing and chilling stresses: (i) escape; (iii) avoidance; and (iii) tolerance (Wery *et al.*, 1993; Toker *et al.*, 2007a).

Inheritance of cold tolerance

Malhotra and Singh (1991) reported that cold tolerance at the vegetative stage, controlled by at least five genes in chickpea with both additive and non-additive gene effects, was dominant over susceptibility, and suggested that selection for cold tolerance would be more effective if dominance and epistatic effects were reduced after selfing generations. Narrow-sense heritability for cold tolerance was estimated at 87.9% (Malhotra and Saxena, 1993). Clarke *et al.* (2008) underlined that there were no published data on the genetics of tolerance to chilling at the reproductive stage in chickpea, despite the fact that molecular markers were linked to chilling tolerance and susceptibility in some varieties. Nevertheless, these markers are absent from marker-assisted selection (MAS) in other chill-tolerant chickpeas (Clarke *et al.*, 2008). In lentil, winter hardiness is determined by several genes, and heritability was estimated at 32–71% by Ali and Johnson (2000) and at 16–91% by Kahraman *et al.* (2004a). Kahraman *et al.* (2004b) also found four QTL markers for winter hardiness. Cold tolerance in lupines is highly and additively inherited (Huyghe, 1997). Winter hardiness in pea is governed by dominant and additive genes (Auld *et al.*, 1983). Link *et al.* (2010) reported that frost tolerance in faba bean increased after hardening, and heritability was estimated at 89%; these workers also identified three QTLs for frost tolerance.

In soybean, Takahashi *et al.* (2005) concluded that the dominant *T* allele might also be useful towards further improvement in chilling tolerance. A cowpea line with chilling tolerance was found and it was hypothesized that such tolerance is due to two independent and additive factors (Ismail *et al.*, 1997). A dehydrin protein related to chilling tolerance was found to be controlled by a single nuclear gene (Ismail and Hall, 2002).

Screening and selection for cold tolerance

Some reliable screening and selection techniques for cold tolerance in food legumes have been reported (Malhotra and Saxena, 1993). Singh *et al.* (1989) proposed a screening and selection technique for cold tolerance in chickpea, which in turn has been clubbed with screening for resistance to ascochyta blight. The technique involves (Toker *et al.*, 2007a): (i) early sowing (in October) of test materials; (ii) using at least one known cold-susceptible (ILC 533) and cold-tolerant accession (ILC 8617 is cold tolerant and ascochyta blight resistant); (iii) using at least one known ascochyta blight-susceptible but cold-tolerant accession (ILC 8262); (iv) inoculation with *Ascochyta*-infected crop debris prior to flowering and ensuring proper moisture provision; and (v) evaluating the test materials for resistance to ascochyta blight and tolerance to cold using a visual scale scored from 1 to 9 (Toker and Canci, 2003). This technique can be useful for a large number of test materials and could easily be adopted for cold and chilling tolerance in other food legumes.

Sources of cold tolerance

The best sources for cold tolerance in chickpea are ILC 8262 (Singh *et al.*, 1992) and ILC 8617 (Singh, 1997), with rosette-type and dark green leaves in the seedling stage plus late flowering. Srinivasan *et al.* (1999)

reported ICCV 88502 and ICCV 88503 to be cold-tolerant genotypes during reproductive growth. Using pollen-selection techniques, Clarke *et al.* (2004) developed two chilling-tolerant genotypes, 'Rupali' and 'Sonali'. Additional cold-tolerant sources include ICCV 88506, ICC 8923, ICCV 88510 and ICCV 88516 (Clarke *et al.*, 2008). Noffsinger and van Santen (2005) reported that the French white lupine cultivar 'Lucky' had sufficient cold tolerance to be selected for later evaluation and breeding. Malhotra and Saxena (1993) documented some winter-hardy pea genotypes. Winter forage peas have been used as parental material due to their good tolerance to freezing in breeding programmes (Clarke *et al.*, 2008). Available winter hardiness cultivars in lentil include 'Kafkas' (Aydogan *et al.*, 2007) and 'Ozbek' (Aydogan *et al.*, 2008), which survived at −29°C in Sivas, Turkey. 'Morton' is another winter-hardy lentil cultivar (Muehlbauer and McPhee, 2007). Link *et al.* (2010) reported that Cote d'Or and BPL 4628 were frost-tolerant faba bean genotypes. Faba bean line F7-(Cor1 × BPL)-95 and Karl are good sources for cold tolerance (Arbaoui *et al.*, 2008). The lines Cote d'Or (−16°C), Hiverna (−15°C), ILB3187, ILB2999, ILB14 and ILB345 (−14°C) were reported to be frost tolerant (Link *et al.*, 2010).

The most cold-tolerant soybean genotypes are related to the Swedish cultivar 'Fiskeby V' (Hume and Jackson, 1981). In common bean, cultivars/lines that germinated best and most rapidly at a constant 8°C were 'Volare', 'Great Northern (G.N.) Tara', 'G.N. Belneb # 1', 'G.N. Spinel' and 'San Cristobal' (Zaiter *et al.*, 1994). Also, 68823, 69345 and AC Polaris were found to be promising for developing cultivars that can germinate under cool temperatures (< 10°C) (Nleya *et al.*, 2005). Rodino *et al.* (2007) found that the commercial cultivars of runner bean (*Phaseolus coccineus* L.) Painted Lady Bi-color, Scarlet Emperor, the Rwanda cultivar NI-15c and the Spanish cultivars PHA-0013, PHA-0133, PHA-0311, PHA-0664 and PHA-1025 exhibited the best performance under cold conditions. In cowpea, the genotype 'UCR 1393-2-11' was identified as being chill tolerant (Ismail and Hall, 2002).

High temperature (heat)

Types of heat stress

According to the interaction of time and temperature, two types of heat stress were defined: (i) heat shock (lethal temperatures from a few minutes to a few hours); and (ii) moderate heat (higher than optimum temperatures during the growing season) (Blum, 1988; Toker *et al.*, 2007a). Heat or high temperature stress is common in major food legume-growing areas around the world, and occurs together with drought in many environments (McDonald and Paulsen, 1997). Interaction of these stresses often coincides with the phase of reproductive development in legumes.

Effects of heat

In general, heat stress accompanied by drought has negative effects on production, especially during gamete development, flowering and podding in food legumes (Malhotra and Saxena, 1993; Clarke *et al.*, 2008). Heat stress conditions reduced the duration of flowering and pod filling, caused withering and burning of lower leaves, desiccation of poorly developed plants, stunting of flowers and pod abortion, and reduced root nodulation and nitrogen (N) fixation, resulting in large yield losses (Saxena *et al.*, 1988; van Rheenen *et al.*, 1997). Flowers are the organs most sensitive to heat (Wery *et al.*, 1993; Toker and Canci, 2008).

Sources of heat tolerance

Heat tolerance in cool season food legumes has not attracted much attention from researchers, due to the difficulty in distinguishing heat stress from drought stress in the field (Malhotra and Saxena, 1993). The significant additive effects observed indicate that gain from selection for improved heat tolerance in common bean should be possible for both traits (Shonnard and Gepts, 1994). A method for breeding cowpea with heat tolerance during reproductive development has been developed (Hall, 2004), and was used to breed 'California Blackeye No. 27' ('CB27') (Ehlers *et al.*, 2000). 'CB27' is both tolerant to heat during reproductive development and heat resistant in that it produces more grain yield than other cowpea cultivars in hot field environments (Ismail and Hall, 1998). Pollen parameters would be more useful than those based on vegetative organs for screening soybean genotype tolerance to high temperature (Salem *et al.*, 2007).

16.4 Nutrient Toxicity

Salinity

Types of salinity

The total area of salt-affected soil worldwide, either by salinity (397 million ha) or sodicity (434 million ha), is estimated at over 800 million ha (Turkan and Demiral, 2009; Munns, 2010), representing over 6% of the world's total land area (Munns, 2005). Saline soils have a high concentration of soluble salts, and a soil is saline when the electrical conductivity (EC) of saturated soil extract is ≥ 4 dS/m, while a soil is sodic when the ESP (exchangeable sodium percentage) is ≥ 15 Ds/m (Munns, 2005). Salt-affected soils can also be divided into the following groups: saline (dominantly Na_2SO_4 and NaCl, seldom $NaNO_3$); alkaline (mainly $NaCO_3$ and $NaHCO_3$, seldom Na_2SiO_3 and $NaHSiO_3$); gypsifer (mainly $CaSO_4$ and seldom $CaCl_2$); magnesium (magnesium ions) and acid sulphate ($Al_2(SO_4)_3$; and $Fe_2(SO_4)_3$ (Szabolcs, 1994).

Effects of salinity

Farmers generally do not consider growing food legumes in salt-affected soils, since they are relatively salt-sensitive compared with cereal crops (Saxena *et al.*, 1993a). The deleterious effects of salinity on plant growth are associated with: (i) water stress; (ii) nutrient ion imbalance; (iii) salt stress due to specific ion effects; and (iv) a combination of these (Ashraf and Harris, 2004). All these factors cause adverse pleiotropic effects on plant growth and development at the physiological, biochemical, molecular and whole-plant levels (Toker *et al.*, 2007a). The salinity effect on bacterial activity with respect to nitrogen fixation is one hypothesis that may explain salt sensitivity in legumes (Pessarakli *et al.*, 1989; Materne *et al.*, 2007; Toker *et al.*, 2007a).

Inheritance of salinity resistance

Salinity resistance in plants may be controlled by the actions of several to many genes, and is also influenced by various environmental factors as it is by various physiological and agronomic characteristics (Foolad, 2004). Therefore, interactions between genotype and environment need to be considered in identifying salt-resistant genotypes for breeding programmes (Flowers *et al.*, 2009). The most salt-tolerant species have high internal salt concentrations (Gorham *et al.*, 1985), which suggests that this is at least as important as the ability to restrict accumulation (Toker *et al.*, 2007a).

Screening and selection for salinity resistance

Despite screening methods in the field for selection of salt-tolerant food legumes, its routine use in breeding programmes seems to be very limited (Saxena *et al.*, 1993a), due to the complex nature of salinity (Flowers *et al.*, 2009). The following characteristics have been used in screening for resistance to salinity: germination percentage, radicle length, shoot length, nodulation, leaf necrosis, salinity susceptibility index (based on biomass yield under saline and non-saline conditions), plant biomass, number of pods per plant and grain yield (Flowers *et al.*, 2009). Several criteria have been used to assess salinity tolerance, including cell survival, germination, dry matter accumulation, leaf death and senescence, ion concentrations (ratio Na^+/K^+ or K^+/Na^+), leaf necrosis, osmoregulation and yield. In conclusion, no single selection criterion is there for salinity tolerance (Toker *et al.*, 2007a). The characteristics used for assessing salinity resistance should be correlated with grain yield, because the ultimate criterion for salinity resistance is grain yield under saline conditions.

Sources of salinity tolerance

There is a wide variation of salinity resistance in food legumes (van Hoorn *et al.*,

2001; Stoddard *et al.*, 2006; Vadez *et al.*, 2007). Yoshida (2002) listed genes useful for enhancement of plant cell salt tolerance. DNA microarray technology is likely to become a powerful tool for this purpose.

16.5 Nutrient Deficiency and Toxicity

Deficiencies of some elements in agricultural soils reduce yield and adversely affect nitrogen fixation in legumes. For example, nitrogen (N) and phosphorus (P) deficiencies in chickpea have been reported to cause worldwide yield losses of 709,000 and 653,000 t per year, respectively. Similarly, yield losses caused by micronutrient deficiencies have been estimated at about 360,000 t/year (Ryan, 1997). In most legume-growing soils, N and P are at either low or medium levels, whereas K is sufficiently available to support growth, but the element is deficient in some soils (Srinivasarao *et al.*, 2003). Calcium (Ca) and Magnesium (Mg) are generally deficient in acid soils (pH < 5.5). Sulphur (S) deficiency has been reported on light-textured soils in India, and the application of S at 20 kg/ha is recommended (Srinivasarao *et al.*, 2003). S deficiency is also seen in calcareous soils with a pH of 8.0 or higher (Toker *et al.*, 2011).

Iron-induced deficiency (FeDC) has been reported in a wide range of legume crops such as chickpea, lentil, lupine, pea, bean and soybean (Wallace, 1960; Erskine *et al.*, 1993; Toker *et al.*, 2010). Studies to determine genetic models for resistance to FeDC in bean showed that resistance may be controlled by either two major gene pairs (Coyne *et al.*, 1982) or one or two major genes (Zaiter *et al.*, 1992). Severe FeDC causes significant yield reduction in dry beans grown on highly calcareous soils (Zaiter *et al.*, 1992). Lime-induced FeDC is common in the Mediterranean area, and represents a major constraint for the majority of legumes (Zaiter and Ghalayini, 1994). A large variability in response to Fe deficiency among either legume species or cultivars has been reported (Ellsworth *et al.*, 1997; Zribi and Gharsalli, 2002; Mahmoudi *et al.*, 2005).

Zinc (Zn)-deficient soils are common throughout the world in both tropical and temperate areas, but are most widespread in India, Pakistan, Iran, China and Turkey (Alloway, 2009). Plant species exhibit differential response to Zn deficiency. Of the legume species, the relative sensitivity of common bean to Zn deficiency is high, and of soybean medium (Alloway, 2009). Lentil, chickpea and pea were found to be more sensitive to Zn deficiency than oilseeds and cereals (Tiwari and Dwivedi, 1990). Differential Zn efficiency was reported among navy bean genotypes (Jolley and Brown, 1991; Moraghan and Grafton, 1999). Zn deficiency is known to delay pod maturity in bean (Blaylock, 1995).

Boron (B) is needed for maintenance of the nodule cell wall and membrane structure, in both pea with indeterminate nodules (Bolanos *et al.*, 1994) and in bean with determinate nodules (Bonilla *et al.*, 1997). Boron does not seem to be required by rhizobia, but is essential for the establishment of effective legume symbioses. Hence, B is required for rhizobial infection and the nodule invasion process (Bolanos *et al.*, 1996).

Boron toxicity is a worldwide problem that significantly limits crop yield in agricultural areas of Australia, North Africa and West Asia. It is characterized by alkaline and saline soils. together with low rainfall and very scarce leaching. Boron-rich soils also occur as a consequence of over-fertilization and/or irrigation, with water containing high levels of B (Nable *et al.*, 1997). Boron toxicity exerts different effects on very diverse processes in vascular plants, such as altered metabolism, reduced root cell division, lower leaf chlorophyll content and photosynthetic rates, and decreased lignin and suberin levels (Nable *et al.*, 1997; Reid, 2007). Plants exposed to high B levels display reduced growth of shoots and roots (Nable *et al.*, 1990). Interestingly, it has been reported that increased B and Ca supplies enhance crop salt tolerance and improve yield production in saline soils (El-Hamdaoui *et al.*, 2003a, b), which could be useful for agriculture. Several B-tolerance genes from lupine and *Arabidopsis* that encode transcription factors or ribosomal proteins conferred tolerance to high B levels in yeast (Nozawa *et al.*, 2006; Reid, 2007).

16.6 Waterlogging Tolerance

Waterlogging is a major problem and causes considerable losses in food legumes (Jackson, 2010). It reduces germination, seedling emergence, root and shoot growth and plant density by up to 80%, besides causing seedling diseases. Yield losses due to waterlogging may reach almost 100% (Siddique, 2000). Management practices to reduce the effects of waterlogging include paddock selection, sowing time, seeding rate and drainage (Jackson, 2010). Genetic variation to waterlogging tolerance in food legumes deserves attention (Toker *et al.*, 2007a). Among cool season food legumes, faba bean is more tolerant to waterlogging than lentil, pea and chickpea (Siddique, 2000).

16.7 Role of Wild Species

Since the agronomically desirable genes have been either spontaneously or artificially induced by mutations (Toker, 2009), selection processes for these characteristics during domestication have resulted in a drastic narrowing of the genetic variation of cultivated crop species (Tanksley and McCouch, 1997). Wild species of cultivated food legumes have useful alleles carrying promising traits (Toker and Yadav, 2010). Wild species that are easily crossable with cultivated species appear to be important gene sources (Hamdi and Erskine, 1996; Hamdi *et al.*, 1996; Pantalone *et al.*, 1997; Singh *et al.*, 1998; Bayuelo-Jimenez *et al.*, 2002; Toker, 2005; Ceylan *et al.*, 2006; Chen *et al.*, 2006; Hillocks *et al.*, 2006; Toker *et al.*, 2007a, b; Canci and Toker, 2009b).

16.8 Breeding Methodologies

Like other plant species, the breeding process in food legumes consists of four stages: (i) creating variation with hybridizations and induced mutations; (ii) selection in early generation; (iii) evaluation of selected lines; and (iv) release of varieties. In this chapter, the means by which plant breeders can select and evaluate their materials and available genetic resources have been discussed, along with the nature of the genetics of characteristics. Most characteristics conferring resistance to drought, heat, cold, salinity and nutrient deficiency or toxicity are quantitatively inherited. In addition, most of the germplasm resources selected for resistance to abiotic stresses are still some distance away from being grown directly in farmers' fields due to lack of resistance to multiple stresses and environment–genotype interactions. Resistance to abiotic and important biotic stresses at the target area should be pyramided using breeding methodologies, i.e. pedigree, single-seed descent (SSD), bulk populations, recurrent selection and modifications of these methods (Ranalli and Cubero, 1997). In mutated populations, SSD is also suggested for pyramiding genes conferring resistance to biotic and abiotic stresses (Toker *et al.*, 2007c). Marker-assisted recurrent selection (MARS) should be used for pyramiding promising genes.

16.9 Conclusions

The world's largest food legume collections at CIAT, ICARDA, ICRISAT, IITA and other national gene banks have been screened for resistance to various abiotic and biotic stresses. Despite new food legume varieties being released by national programmes and major international centres, primitive landraces are still grown over much of the crop area in developing countries. The desirable genes for resistance to abiotic stresses in the target environments should be pyramided with biotic stresses. Identification and utilization of wild species representing potential resistance in primary and secondary gene pools of food legumes should be enhanced in medium- and short-term breeding programmes. Thus, interspecific crosses should be made for the introgression of important alleles from wild species of food legumes to cultivated species. Mutagenesis under bottleneck conditions facilitates an increase the genetic variability for resistance to abiotic stresses in food legumes. Transgenic legumes offer a great opportunity, but genes can flow from transgenics to wild relatives, resulting in environmental pollution (Toker *et al.*, 2006) when transgenics are grown in the areas where wild relatives exist. Although the markers or genes identified for resistance to abiotic stresses using molecular approaches are useful for future strategies, they are absent from MARS in whole-breeding programmes. New QTLs for MARS appear to be suitable for whole genotypes.

References

Abdelmula, A.A., Link, W., von Kittlitz, E. and Stelling, D. (1999) Heterosis and inheritance of drought tolerance in faba bean, *Vicia faba* L. *Plant Breeding* 118, 485–490.

Abebe, A., Bricka, M.A. and Kirkby, R.A. (1998) Comparison of selection indices to identify productive dry bean lines under diverse environmental conditions. *Field Crops Research* 58, 15–23.

Ali, A. and Johnson, D.L. (2000) Heritability estimates of winter hardiness in lentil under natural and controlled conditions. *Plant Breeding* 119, 283–285.

Ali, A., Johnson, D.L. and Stushnoff, C. (1991) Screening lentil for cold hardiness under controlled conditions. *Journal of Agricultural Science* 133, 313–319.

Alloway, B.J. (2009) Soil factors associated with zinc deficiency in crops and humans. *Environmental Geochemistry and Health* 31, 537–548.

Amede, T., von Kittlitz, E. and Schubert, S. (1999) Differential drought responses of faba bean (*Vicia faba* L.) inbred lines. *Journal of Agronomy and Crop Science* 183, 35–45.

Amede, T., Schubert, S. and Starh, K. (2003) Mechanisms of drought resistance in grain legumes. I. Osmotic adjustment. *Ethiopian Journal of Science* 26, 37–46.

Arbaoui, M., Balko, C. and Link, W. (2008) Study of faba bean (*Vicia faba* L.) winter-hardiness and development of screening methods. *Field Crop Research* 106, 60–67.

Ashraf, M. and Harris, P.J.C. (2004) Potential biochemical indicators of salinity tolerance in plants. *Plant Science* 166, 3–16.

Auld, D.L., Adams, K.J., Swensen, J.B. and Murray, G.A. (1983) Diallel analyses of winterhardiness in peas. *Crop Science* 23, 763–766.

Auld, D.L., Bettis, B.L., Crock, J.L. and Kephart, K.D. (1988) Planting date and temperature effects on germination, emergence, and seed yield of chickpea. *Agronomy Journal* 80, 909–914.

Aydogan, A., Sarker, A., Aydin, N., Küsmenoglu, I., Karagöz, A. and Erskine, W. (2007) Registration of 'Kafkas' lentil. *Journal of Plant Registrations* 1, 44–45.

Aydogan, A., Sarker, A., Aydin, N., Küsmenoglu, I., Karagöz, A. and Erskine, W. (2008) Registration of 'Ozbek' lentil. *Journal of Plant Registrations* 2, 16.

Baker, R.J. (1994) Breeding methods and selection indices for improved tolerance to biotic and abiotic stresses in cool season food legumes. In: Muehlbauer, F.J. and Kaiser, W.J. (eds) *Expanding the Production and Use of Cool Season Food Legumes*. Kluwer Academic, Dordrecht, The Netherlands, pp. 429–438.

Bayuelo-Jimenez, J.S., Debouck, D.G. and Lynch, J.P. (2002) Salinity tolerance in *Phaseolus* species during early vegetative growth. *Crop Science* 42, 2184–2192.

Benjamin, J.G. and Nielsen, D.C. (2006) Water deficit effects on root distribution of soybean, field pea and chickpea. *Field Crops Research* 97, 248–253.

Bernard, R.L. and Singh, B.B. (1969) Inheritance of pubescence type in soybean: glabrous, curly, dense, and puberulent. *Crop Science* 9, 192–197.

Bidinger, F.R., Mahalakshmi, V. and Rao, G.D.P. (1987) Assessment of drought resistance in pearl millet [*Pennisetum americanum* (L.) Leeke]. II. Estimation of genotype response to stress. *Australian Journal of Agricultural Research* 38, 49–59.

Blaylock, A.D. (1995) Navy bean yield and maturity response to nitrogen and zinc. *Journal of Plant Nutrition* 18, 163–178.

Blum, A. (1988) *Plant Breeding for Stress Environments*. CRC Press, Boca Raton, Florida.

Bolanos, L., Esteban, E., de Lorenzo, C., Fernandez–Pascual, M., de Felipe, M.R., Garate, A. *et al.* (1994) Essentiality of boron for symbiotic dinitrogen fixation in pea (*Pisum sativum*) – *Rhizobium* nodules. *Plant Physiology* 104, 85–90.

Bolanos L., Brewin, N.J. and Bonilla, I. (1996) Effects of boron on *Rhizobium*–legume cell surface interactions and nodule development. *Plant Physiology* 110, 1249–1256.

Bond, D.A., Jellis, G.J., Rowland, G.G., Guen, J., Robertson, L.D., Khalil, S.A. *et al.* (1994) Present status and future strategy in breeding faba bean (*Vicia faba* L.) for resistance to biotic and abiotic stresses. *Euphytica* 73, 151–166.

Bonilla, I., Mergold-Villasenor, C., Campos, M.E., Sanchez, N., Perez, H., Lopez, L. *et al.* (1997) The aberrant cell walls of boron-deficient bean root nodules have no covalently bound hydroxyprolin-proline-rich proteins. *Plant Physiology* 115, 1329–1340.

Bouwkamp, J.C. and Summers, W.L. (1982) Inheritance resistance to temperature–drought stress in the snap bean. *Journal of Heredity* 73, 385–386.

Brick, M.A., Ogg, J.B., Singh, S.P., Schwartz, H.F., Johnson, J.J. and Pastor-Corrales, M.A. (2008) Registration of drought-tolerant, rust-resistant, high-yielding pinto bean germplasm line CO46348. *Journal of Plant Registrations* 2, 120–124.

Burton, J.W. (1997) Soyabean (*Glycine max* (L.) Merr.). *Field Crops Research* 53, 171–186.

Canci, H. and Toker, C. (2009a) Evaluation of yield criteria for drought and heat resistance in chickpea (*Cicer arietinum* L.). *Journal of Agronomy and Crop Science* 195, 47–54.

Canci, H. and Toker, C. (2009b) Evaluation of annual wild *Cicer* species for drought and heat resistance under field conditions. *Genetic Resources and Crop Evolution* 56, 1–6.

Cattivelli, L., Rizza, F., Badeck, F.W., Mazzucotelli, E., Mastrangelo, A.M., Francia, E. *et al.* (2008) Drought tolerance improvement in crop plants: An integrated view from breeding to genomics. *Field Crops Research* 105, 1–14.

Ceylan, F.O., Canci, H., Ertoy, N. and Toker, C. (2006) Comparison of *Vicia* species for cold tolerance to pea and lupin species. *International Workshop on Faba Bean Breeding and Agronomy*, 25–27 October, Cordoba, Spain, pp. 113–116.

Charlson, D.V., Bhatnagar, S., King, C.A., Ray, J.D., Sneller, C.H., Carter, T.E. Jr *et al.* (2009) Polygenic inheritance of canopy wilting in soybean [*Glycine max* (L.) Merr.]. *Theoretical and Applied Genetics* 119, 587–594.

Chauhan, Y.S. (1990) Pigeonpea: optimum agronomic management. In: Nene, Y.L., Hall, S.D. and Sheila, V.K. (eds) *The Pigeonpea*. CAB International, Wallingford, UK, pp. 257–278.

Chen, P., Sneller, C.H., Purcell, L.C., Sinclair, T.R., King, C.A. and Ishibashi, T. (2007) Registration of soybean germplasm lines R01-416F and R01-581F for improved yield and nitrogen fixation under drought stress. *Journal of Plant Registrations* 1, 166–167.

Chen, Y., Chen, P. and de los Reyes, B.G. (2006) Differential responses of the cultivated and wild species of soybean to dehydration stress. *Crop Science* 46, 2041–2046.

Cisse, N., Ndiaye, M., Thiaw, S. and Hall, A.E. (1995) Registration of 'Mouride' cowpea. *Crop Science* 35, 1215–1216.

Cisse, N., Ndiaye, M., Thiaw, S. and Hall, A.E. (1997) Registration of 'Melakh' cowpea. *Crop Science* 37, 1978.

Clarke, H.J., Khan, T.N. and Siddique, K.H.M. (2004) Pollen selection for chilling tolerance at hybridization leads to improved chickpea cultivars. *Euphytica* 139, 65–74.

Clarke, H., Khan, T., Croser, J., White, P., Singh, S., Lulsdorf, M. *et al.* (2008) Temperature tolerance in food legumes. In: Kharkwal, M.C. (ed.) *Food Legumes for Nutritional Security and Sustainable Agriculture*. Indian Journal of Genetics and Plant Breeding, IARI, New Delhi, India, pp. 678–710.

Cousin, R. (1997) Peas (*Pisum sativum* L.). *Field Crops Research* 53, 111–130.

Cowling, W.A., Huyghe, C. and Swiecicki, W. (1998) Lupin breeding. In: Gladstones, J.S., Atkins, C.A. and Hamblin, J. (eds) *Lupins as Crop Plants: Biology, Production and Utilization*. CAB International, Wallingford, UK, pp. 93–120.

Coyne, D.P., Korban, S.S., Knudsen, D. and Clark, R.B. (1982) Inheritance of iron deficiency in crosses of dry beans (*Phaseolus vulgaris* L.). *Journal of Plant Nutrition* 5, 575–585.

Davies, J. (1995) *Plant Hormones: Physiology, Biochemistry and Molecular Biology*. Kluwer Academic, Dordrecht, The Netherlands.

Dracup, M., Turner, N.C., Thang, C., Reader, M. and Palta, J. (1998) Responses to abiotic stresses. In: Gladstones, J.S., Atkins, C. and Hamblin, J. (eds) *Lupins as Crop Plants: Biology, Production and Utilization*. CAB International, Wallingford, UK, pp. 227–262.

Du, W.J., Fu, S.X. and Yu, D.Y. (2009) Genetic analysis for the leaf pubescence density and water status traits in soybean [*Glycine max* (L.) Merr.]. *Plant Breeding* 128, 259–265.

Duc, G. (1997) Faba bean (*Vicia faba* L.). *Field Crops Research* 53, 99–109.

Ehlers, J.D. and Hall, A.E. (1997) Cowpea (*Vigna unguiculata* L. Walp.). *Field Crops Research* 53, 187–204.

Ehlers, J.D., Hall, A.E., Patel, P.N., Roberts, P.A. and Matthews, W.C. (2000) Registration of 'California Blackeye 27' cowpea. *Crop Science* 40, 854–855.

Elawad, H.O.A. and Hall, A.E. (2002) Registration of 'Ein El Gazal' cowpea. *Crop Science* 42, 1745–1746.

El-Hamdaoui, A., Redondo-Nieto, M., Rivilla, R., Bonilla, I. and Bolanos, L. (2003a) Effects of boron and calcium nutrition on the establishment of the *Rhizobium leguminosarum*-pea (*Pisum sativum*) symbiosis and nodule development under salt stress. *Plant, Cell and Environment* 26, 1003–1011.

El-Hamdaoui, A., Redondo-Nieto, M., Torralba, B., Rivilla, R., Bonilla, I. and Bolanos, L. (2003b) Influence of boron and calcium on the tolerance to salinity of nitrogen-fixing pea plants. *Plant and Soil* 251, 93–103.

Ellsworth, J.W., Jolley, V.D., Nuland, D.S. and Blaylock, A.D. (1997) Screening for resistance to iron deficiency chlorosis in dry bean using iron reduction capacity. *Journal of Plant Nutrition* 20, 1489–1502.

Erskine, W. and Saxena, M.C. (1993) Problems and prospects of stress resistance breeding in lentil. In: Singh, K.B. and Saxena, M.C. (eds) *Breeding for Stress Tolerance in Cool-Season Food Legumes*, Wiley-Sayce Publication, Chichester, UK, pp. 51–62.

Erskine, W. and Witcombe, J.R. (1984) *Lentil Germplasm Catalog*. ICARDA, Aleppo, Syria.

Erskine, W., Meyveci, K. and Izgin, N. (1981) Screening of world lentil collection for cold tolerance. *Lens Newsletter* 8, 5–8.

Erskine, W., Saxena, N.P. and Saxena, M.C. (1993) Iron-deficiency in lentil – yield loss and geographic distribution in a germplasm collection. *Plant and Soil* 151, 249–254.

FAOSTAT (2008) Food and Agriculture Organization Statistics (available at http://faostat.fao.org/site/567/default.aspx#ancor).

Fischer, R.A. and Maurer, R. (1978) Drought resistance in spring wheat cultivars. I. Grain yield responses. *Australian Journal of Agricultural Research* 29, 897–912.

Flowers, T.J., Gaur, P.M., Gowda, C.L.L., Krishnamurthy, L., Samineni, S., Siddique, K.H.M. *et al.* (2009) Salt sensitivity in chickpea. *Plant, Cell and Environment* 33(4), 490–509.

Foolad, M.R. (2004) Recent advances in genetics of salt tolerance in tomato. *Plant Cell Tissue Organ Culture* 76, 101–119.

Gorham, J., Wyn Jones, R.G. and McDonnel, E. (1985) Some mechanisms of salt tolerance in crop plants. *Plant and Soil* 89, 15–40.

Graham, P.H. and Ranalli, P. (1997) Common bean (*Phaseolus vulgaris* L.). *Field Crops Research* 53, 131–146.

Grzesiak, S., Filek, W., Pienkowski, S. and Niziol, B. (1996) Screening for drought resistance: Evaluation of drought susceptibility index of legume plants under natural growth conditions. *Journal of Agronomy and Crop Science* 177, 237–244.

Hall, A.E. (2001) *Crop Responses to Environment*. CRC Press, Boca Raton, Florida.

Hall, A.E. (2004) Breeding for adaptation to drought and heat in cowpea. *European Journal of Agronomy* 21, 447–454.

Hall, A.E., Ismail, A.M., Ehlers, J.D., Marfo, K.O., Cisse, N. *et al.* (2002) Breeding cowpea for tolerance to temperature extremes and adaptation to drought. In: Fatokun, C.A., Tarawali, S.A., Singh, B.B., Kormawa, P.M. and Tamò, M. (eds) *Challenges and Opportunities for Enhancing Sustainable Cowpea Production*, Institute of Tropical Agriculture (IITA), Ibadan, Nigeria, pp. 14–21.

Hamdi, A. and Erskine, W. (1996) Reaction of wild species of the genus *Lens* to drought. *Euphytica* 91, 173–179.

Hamdi, A., Kusmenoglu, I. and Erskine, W. (1996) Sources of winter hardiness in wild lentil. *Genetic Resources and Crop Evolution* 43, 63–67.

Hay, R.K.M. and Porter, J.R. (2006) *The Physiology of Crop Yield*. Blackwell Publishing, Oxford, UK.

Hegde, V.S. (2010) Genetics of flowering time in chickpea in a semi-arid environment. *Plant Breeding* 129, 683–687.

Hillocks, R.J., Madata, C.S., Chirwa, R., Minja, E.M. and Msolla, S. (2006) *Phaseolus* bean improvement in Tanzania, 1959–2005. *Euphytica* 150, 215–231.

Hovav, R., Upadhyaya, K.C., Beharav, A. and Abbo, S. (2003) Major flowering time gene and polygene effects on chickpea seed weight. *Plant Breeding* 122, 539–541.

Hume, D.J. and Jackson, A.K.H. (1981) Pod formation in soybeans at low temperatures. *Crop Science* 21, 933–937.

Huyghe, C. (1997) White lupin (*Lupinus albus* L.). *Field Crops Research* 53, 147–160.

ISI (2010) ISI Web of Knowledge. Available at http://apps.isiknowledge.com/WOS_GeneralSearch_input.do?product=WOS&search_mode=GeneralSearch&SID=V1pKjFEiH9d5@fpa658&preferencesSaved=&highlighted_tab=WOS (accessed July 2010).

Ismail, A.M. and Hall, A.E. (1998) Positive and potential negative effects of heat-tolerance genes in cowpea lines. *Crop Science* 38, 381–390.

Ismail, A.M. and Hall, A.E. (2002) Variation in traits associated with chilling tolerance during emergence in cowpea germplasm. *Field Crops Research* 77, 99–113.

Ismail, A.M., Hall, A.E. and Close, T.J. (1997) Chilling tolerance during emergence of cowpea associated with a dehydrin and slow electrolyte leakage. *Crop Science* 37, 1270–1277.

Jackson, M.B. (2010) The impact of flooding stress on plants and crops. In: Blum, A. (ed.) *Plant Stress* (available at http://www.plantstress.com/Articles/index.asp).

Johansen, C., Baldev, B., Brouwer, J.B., Erskine, W., Jermyn, W.A., Li-Juan, L. *et al.* (1994) Biotic and abiotic stresses constraining productivity of cool season food legumes in Asia, Africa and Oceania. In: Muehlbauer, F.J. and Kaiser, W.J. (eds) *Expanding the Production and Use of Cool Season Food Legumes*, Kluwer Academic Publishers, Dordrecht, The Netherlands, pp. 175–194.

Jolley, V.D. and Brown, J.C. (1991) Factors in iron-stress response mechanism enhanced by Zn deficiency in Sanilac, but not Saginaw navy bean. *Journal of Plant Nutrition* 14, 257–265.

Kahraman, A., Kusmenoglu, I., Aydin, N., Aydogan, A., Erskine, W. and Muehlbauer, F.J. (2004a) Genetics of winter hardiness in 10 lentil recombinant inbred line populations. *Crop Science* 44, 5–12.

Kahraman, A., Kusmenoglu, I., Aydin, N., Aydogan, A., Erskine, W. and Muehlbauer F.J. (2004b) QTL mapping of winter hardiness genes in lentil. *Crop Science* 44, 13–22.

Kashiwagi, J., Krishnamurthy, L., Upadhyaya, H.D., Krishna, H.S., Vandez, C.V. and Serraj, R. (2005) Genetic variability of drought-avoidance root traits in the minicore germplasm collection of chickpea (*Cicer arietinum* L.). *Euphytica* 146, 213–222.

Khan, H.R., Link, W., Hocking, T.J. Stoddard, F.L. (2007) Evaluation of physiological traits for improving drought tolerance in faba bean (*Vicia faba* L.). *Plant and Soil* 292, 205–217.

Khan, H.R., Paull, J.G., Siddique, K.H.M. and Stoddard, F.L. (2010) Faba bean breeding for drought-affected environments: A physiological and agronomic perspective. *Field Crops Research* 115, 279–286.

Kumar, J. and Abbo, S. (2001) Genetics of flowering time in chickpea and its bearing on productivity in semi-arid environments. *Advances in Agronomy* 72, 107–138.

Kumar, J. and van Rheenen, H.A. (2000) A major gene for time of flowering in chickpea. *Journal of Heredity* 91, 67–68.

Kumar, J., Sethi, S.C., Johansen, C., Kelly, T.G., Rahman, M.M. and van Rheenen, H.A. (1996) Potential of short-duration chickpea varieties. *Journal of Dryland Agricultural Research and Development* 11, 28–32.

Lawn, R.J. and Troedson, R.J. (1990) Pigeonpea: physiology of yield formation. In: Nene, Y.L., Hall, S.D. and Sheila, V.K. (eds) *The Pigeonpea*. CAB International, Wallingford, UK, pp. 179–208.

Leport, L., Turner, N.C., French, R.J., Barr, M.D., Duda, R., Davies, S.L. *et al.* (1999) Physiological response of chickpea genotypes to terminal drought in a Mediterranean-type environment. *European Journal of Agronomy* 11, 279–291.

LeSech, L. and Huyghe, C. (1991) Diallel analysis on white lupin: breeding consequences. *Agronomie* 11, 719–726.

Likoswe, A.A. and Lawn, R.J. (2008) Response to terminal water deficit stress of cowpea, pigeonpea and soybean in pure stand and in competition. *Australian Journal of Agricultural Research* 59, 27–37.

Link, W., Abdelmula, A.A., von Kittlitz, E., Bruns, S., Riemer, H. and Stelling, D. (1999) Genotypic variation for drought tolerance in *Vicia faba*. *Plant Breeding* 118, 477–483.

Link, W., Balko, C. and Stoddard, F.L. (2010) Winter hardiness in faba bean: Physiology and breeding. *Field Crops Research* 115, 287–296.

Ludlow, M.M. and Muchow, R.C. (1990) A critical evaluation of traits for improving crop yields in water-limited environments. *Advances in Agronomy* 43, 107–153.

Mahmoudi, H., Ksouri, R., Gharsalli, M. and Lachaal, M. (2005) Differences in responses to iron deficiency between two legumes: lentil (*Lens culinaris*) and chickpea (*Cicer arietinum*), *Journal of Plant Physiology* 162, 1237–1245.

Mai-Kodomi, Y., Singh, B.B., Myers, O. Jr, Yopp, J.H., Gibson, J.P. and Terao, T. (1999) Inheritance of drought tolerance in cowpea. *Indian Journal of Genetics and Plant Breeding* 59, 317–323.

Malhotra, R.S. and Saxena, M.C. (1993) Screening for cold and heat tolerance in cool-season food legumes. In: Singh, K.B. and Saxena, M.C. (eds) *Breeding for Tolerance in Cool Season Food Legumes*. Wiley, Chichester, UK, pp. 227–244.

Malhotra, R.S. and Singh, K.B. (1991) Gene action for cold tolerance in chickpea. *Theoretical and Applied Genetics* 82, 598–601.

Materne, M., McNeil, D., Hobson, K. and Ford R. (2007) Abiotic stresses. In: Yadav, S.S., McNeil, D. and Stevenson, P.C. (eds) *Lentil: An Ancient Crop for Modern Times*. Springer, Dordrecht, The Netherlands, pp. 315–329.

McDonald, G.K. and Paulsen, G.M. (1997) High temperature effects on photosynthesis and water relations of grain legumes. *Plant and Soil* 196, 47–58.

McPherson, H.G., Warrington, I.J. and Turnbull H.L. (1985) The effect of temperature and daylength on the rate of development of pigeonpea. *Annals of Botany* 56, 597–611.

Mederski, H.J., Jeffers, D.L. and Peters, D.B. (1973) Water and water relations. In: Caldwell (ed.) *Soybeans: Improvement, Production, and Uses*. American Society of Agronomy, Inc., Publishers, Madison, Wisconsin, pp. 239–266.

Moraghan, J.T. and Grafton, K. (1999) Seed zinc concentration and the zinc efficiency trait in navy bean. *Soil Science Society of America Journal* 63, 918–922.

Muehlbauer, F.J. and Kaiser, W.J. (1994) *Expanding the Production and Use of Cool Season Food Legumes*. Kluwer Academic Publishers, Dordrecht, The Netherlands.

Muehlbauer, F.J. and McPhee, K.E. (2007). Registration of 'Morton' winter-hardy lentil. *Crop Science* 47, 438–439.

Munns, R. (2005) Genes and salt tolerance: bringing them together. *New Phytologist* 167, 645–663.

Munns, R. (2010) The impact of salinity stress. In: Blum, A. (ed.) *Plant Stress* (available at: http://www.plantstress.com/Articles/index.asp).

Murfet, I.C. (1975) Flowering in *Pisum*: multiple alleles at the *lf* locus. *Heredity* 35, 85–98.

Murray, G.A., Eser, D., Gusta, L.V. and Eteve, G. (1988) Winter hardiness in pea, lentil, faba bean and chickpea. In: Summerfield, R.J. (ed.) *World Crops: Cool Season Food Legumes*. Kluwer Academic Publishers, Dordrecht, The Netherlands, pp. 831–843.

Nable, R.O., Cartwright, B. and Lance, R.C. (1990) Genotypic differences in boron accumulation in barley: relative susceptibilities to boron deficiency and toxicity. In: El Bassam, N., Dambroth, M. and Loughman, B. (eds) *Genetic Aspects of Plant Mineral Nutrition*. Kluwer Academic Publishers, Dordrecht, The Netherlands, pp. 243–251.

Nable, R.O., Banuelos, G.S. and Paull, J.G. (1997) Boron toxicity. *Plant and Soil* 193, 181–198.

Nayyar, H., Kaur, S., Smita, Singh, K.J., Dhir, K.K. and Bains, T. (2005) Water stress induced injury to reproductive phase in chickpea: evaluation of stress sensitivity in wild and cultivated species in relation to abscisic acid and polyamines. *Journal of Agronomy and Crop Science* 191, 450–457.

Nleya, T., Ball, R.A. and Vandenberg, A. (2005) Germination of common bean under constant and alternating cool temperatures. *Canadian Journal of Plant Science* 85, 577–585.

Noffsinger, S.L. and van Santen, E. (2005) Evaluation of *Lupinus albus* L. germplasm for the south eastern USA. *Crop Science* 45, 1941–1950.

Nozawa, A., Miwa, K., Kobayashi, M. and Fujiwara, T. (2006) Isolation of Arabidopsis thaliana cDNAs that confer yeast boric acid tolerance. *Bioscience, Biotechnology, and Biochemistry* 70, 1724–1730.

Palta, J.A., Turner, N.C. and French, R.J. (2004) The yield performance of lupin genotypes under terminal drought in a Mediterranean-type environment. *Australian Journal of Agricultural Research* 55, 449–459.

Pantalone, V.R., Kenworthy, W.J., Slaughter, L.H. and James, B.R. (1997) Chloride tolerance in soybean and perennial *Glycine* accessions. *Euphytica* 97, 235–239.

Pendleton, J.W. and Hartwig, E.E. (1973) Management. In: Caldwell (ed.) *Soybeans: Improvement, Production, and Uses*. American Society of Agronomy, Inc., Publishers, Madison, Wisconsin, pp. 211–237.

Pessarakli, M., Huber, J.T. and Tucker, T.C. (1989) Protein synthesis in green beans under salt stress with two nitrogen sources. *Journal of Plant Nutrition* 12, 1261–1377.

Ranalli, P. and Cubero, J.I. (1997) Bases for genetic improvement of grain legumes. *Field Crops Research* 53, 69–82.

Reddy, L.J. (1990) Pigeonpea: morphology. In: Nene, Y.L., Hall, S.D. and Sheila, V.K. (eds) *The Pigeonpea*. CAB International, Wallingford, UK, pp. 47–87.

Reid, R. (2007) Update on boron toxicity and tolerance in plants. In: Xu, F. *et al*. (eds) *Advances in Plant and Animal Boron Nutrition*. Springer, Dordrecht, The Netherlands, pp. 83–90.

Ricciardi, L., Polignano, G.B. and De Giovanni, C. (2001) Genotypic response of faba bean to water stress. *Euphytica* 118, 39–46.

Rodino, A.P., Lema, M., Perez-Barbeito, M., Santalla, M. and De Ron, A.M. (2007) Assessment of runner bean (*Phaseolus coccineus* L.) germplasm for tolerance to low temperature during early seedling growth. *Euphytica* 155, 63–70.

Ryan, J.G. (1997) A global perspective on pigeonpea and chickpea sustainable production systems: Present status and future potential. In: Asthana, A.N. and Ali, M. (eds) *Recent Advances in Pulses Research*. Indian Society of Pulses Research and Development, Kanpur, India, pp. 1–31.

Sabaghpour, S.H., Kumar, J. and Rao, T.N. (2003) Inheritance of growth vigour and its association with other characters in chickpea. *Plant Breeding* 122, 542–544.

Saeed, A., Darvishzadeh, R., Hovsepyan, H. and Asatryan, A. (2010). Tolerance to freezing stress in *Cicer* accessions under controlled and field conditions *African Journal of Biotechnology* 9, 2618–2626.

Salem, M.A., Kakani, V.G., Koti, S. and Reddy, K.R. (2007) Pollen-based screening of soybean genotypes for high temperatures. *Crop Science* 47, 219–231.

Sarker, A., Erskine, W., Sharma, B. and Tyagi, M.C. (1999) Inheritance and linkage relationships of days to flower and morphological loci in lentil (*Lens culinaris* Medikus subsp. *culinaris*). *Journal of Heredity* 90, 270–275.

Savithri, K.S., Ganapathy, P.S. and Sinha, S.K. (1980) Sensitivity to low temperature in pollen germination and fruit-set in *Cicer arietinum* L. *Journal of Experimental Botany* 31, 475–481.

Saxena, K.B. and Sharma, D. (1990) Pigeonpea: genetics. In: Nene, Y.L., Hall, S.D. and Sheila, V.K. (eds) *The Pigeonpea*. CAB International, Wallingford, UK, pp. 137–157.

Saxena, M.C. (1993) The challenge of developing biotic and abiotic stress resistance in cool-season food legumes. In: Singh, K.B. and Saxena, M.C. (eds) *Breeding for Tolerance in Cool Season Food Legumes*. John Wiley & Sons, Chichester, UK, pp. 3–14.

Saxena, M.C., Saxena, N.P. and Mohamed, A.K. (1988) High temperature stress. In: Summerfield, R.J. (ed.) *World Crops: Cool Season Food Legumes*. Kluwer Academic, Dordrecht, The Netherlands, pp. 845–856.

Saxena, N.P., Johansen, C., Saxena, M.C. and Silim, S.N. (1993a) Selection for drought and salinity tolerance in cool-season food legumes. In: Singh, K.B. and Saxena, M.C. (eds) *Breeding for Tolerance in Cool Season Food Legumes*. John Wiley & Sons, Chichester, UK, pp. 245–270.

Saxena, N.P., Krishnamurthy, L. and Johansen, C. (1993b) Registration of drought-resistant chickpea germplasm. *Crop Science* 33, 1424.

Serraj, R., Bidinger, F.R., Chauhan, Y.S., Seetharama, N., Nigam, S.N. and Saxena, N.P. (2003) Management of drought in ICRISAT cereal and legume mandate crops. In: Kijne, J.W., Barker, R. and Molden, D. (eds) *Water Productivity in Agriculture: Limits and Opportunities for Improvement*. CAB International, Wallingford, UK, pp. 127–144.

Serraj, R., Krishnamurthy, L., Kashiwagi, J., Kumar, J., Chandra, S. and Crouch, J.H. (2004) Variation in root traits of chickpea (*Cicer arietinum* L.) grown under terminal drought. *Field Crops Research* 88, 115–127.

Shonnard, G.C. and Gepts, P. (1994) Genetics of heat tolerance during reproductive development in common bean. *Crop Science* 34, 1168–1175.

Siddique, K.H.M. (2000) Understanding growth stresses of cool season pulses. In: O'Connell, L. (ed.) *Shared Solutions Cropping Manual*, Australian Grain, Berekua, Brisbane, pp. 12–16.

Siddique, K.H.M., Brinsmead, R.B., Knight, R., Knights, E.J., Paull, J.G. and Rose, I.A. (2000) Adaptation of chickpea (*Cicer arietinum* L.) and faba bean (*Vicia faba* L.) to Australia. In: Knight, R. (ed.) *Linking Research and Marketing Opportunities for Pulses in the 21st Century*. Kluwer Academic, Dordrecht, The Netherlands, pp. 289–303.

Silim, S.N. and Saxena, M.C. (1993) Adaptation of spring-sown chickpea to the Mediterranean basin. I. Response to moisture supply. *Field Crops Research* 34, 121–136.

Sincik, M., Candogan, B.N., Demirtas, C., Buyukcangaz, H., Yazgan, S. and Göksoy, A.T. (2008) Deficit irrigation of soya bean [*Glycine max* (L.) Merr.] in a sub-humid climate. *Journal of Agronomy and Crop Science* 194, 200–205.

Singh, B.B. (2002) Recent genetic studies in cowpea. In: Fatokun, C.A., Tarawali, S.A., Singh, B.B., Kormawa, P.M. and Tamò, M. (eds) *Challenges and Opportunities for Enhancing Sustainable Cowpea Production*. IITA, Ibadan, Nigeria, pp. 3–13.

Singh, B.B. and Matsui, T. (2002) Cowpea varieties for drought tolerance. In: Fatokun, C.A., Tarawali, S.A., Singh, B.B., Kormawa, P.M. and Tamò, M. (eds) *Challenges and Opportunities for Enhancing Sustainable Cowpea Production*. IITA, Ibadan, Nigeria, pp. 287–300.

Singh, B.B., Mai-Kodomi, Y. and Terao, T. (1999) Relative drought tolerance of major rainfed crops of the semi-arid tropics. *Indian Journal of Genetics* 59, 1–8.

Singh, B.B., Singh, S.P., Sarker, A. and Chauhan, Y. (2008) Genetics and breeding for drought tolerance in food legumes. In: Kharkwal, M.C. (ed.) *Food Legumes for Nutritional Security and Sustainable Agriculture*. The Indian Journal of Genetics and Plant Breeding, IARI, New Delhi, India, pp. 725–737.

Singh, K.B. (1997) Chickpea (*Cicer arietinum* L.). *Field Crops Research* 53, 161–170.

Singh, K.B. and Saxena, M.C. (1993) *Breeding for Stress Tolerance in Cool-Season Food Legumes*, John Wiley & Sons, Chichester, UK.

Singh, K.B., Malhotra, R.S. and Saxena, M.C. (1989) Chickpea evaluation for cold tolerance under field conditions. *Crop Science* 29, 282–285.

Singh, K.B., Malhotra, R.S. and Saxena M.C. (1992) Registration of ILC 8262, a cold-tolerant germplasm line of chickpea. *Crop Science* 32: 508.

Singh, K.B., Omar, M., Saxena, M.C. and Johansen, C. (1996) Registration of FLIP 87-59C, a drought-tolerant chickpea germplasm line. *Crop Science* 36, 472.

Singh, K.B., Omar, M., Saxena, M.C. and Johansen, C. (1997) Screening for drought resistance in spring chickpea in the Mediterranean region. *Journal of Agronomy and Crop Science* 178, 227–235.

Singh, K.B., Ocampo, B. and Robertson, L.D. (1998) Diversity for abiotic and biotic stress resistance in the wild annual *Cicer* species. *Genetic Resources and Crop Evolution* 45, 9–17.

Singh, L., Gupta, S.C. and Faris, D.G. (1990) Pigeonpea: breeding. In: Nene, Y.L., Hall, S.D. and Sheila, V.K. (eds) *The Pigeonpea*. CAB International, Wallingford, UK, pp. 375–399.

Singh, S.P. (1991) Bean genetics. In: van Schoonhoven, A. and Voyest, O. (eds) *Common Beans: Research for Crop Improvement*. CAB International, Wallingford, UK, pp. 199–286.

Singh, S.P., Teran, H. and Gutierrez, J.A. (2001) Registration of SEA 5 & SEA 13 drought tolerant dry bean germplasm. *Crop Science* 41, 276–277.

Slinkard, A.E., Bascur, G. and Hernandez-Bravo, G. (1994) Biotic and abiotic stresses of cool season food legumes in the western hemisphere. In: Muehlbauer, F.J. and Kaiser, W.J. (eds) *Expanding the Production and Use of Cool Season Food Legumes*, Kluwer Academic Publishers, Dordrecht, The Netherlands, pp. 195–203.

Srinivasan, A., Saxena, N.P. and Johansen, C. (1999) Cold tolerance during early reproductive growth of chickpea (*Cicer arietinum* L.): Genetic variation in gamete development and function. *Field Crops Research* 60, 209–222.

Srinivasarao, C.H., Ganeshamurthy, A.N. and Ali, M. (2003) *Nutritional Constraints in Pulse Production*. Indian Institute of Pulses Research, Kanpur, India.

Stoddard, F.L., Balko, C., Erskine, W., Khan, H.R., Link, W. and Sarker, A. (2006) Screening techniques and sources of resistance to abiotic stresses in cool-season food legumes. *Euphytica* 147, 167–186.

Subbarao, G.V., Johansen, C., Slinkard, A.E., Nageswara Rao, R.C., Saxena, N.P. and Chauhan, Y.S. (1995) Strategies for improving drought resistance in grain legumes. *Critical Review in Plant Science* 14, 469–523.

Subbarao, G.V., Chauhan, Y.S. and Johansen, C. (2000) Patterns of osmotic adjustment in pigeonpea – its importance as a mechanism of drought resistance. *European Journal of Agronomy* 12, 239–249.

Szabolcs, I. (1994) Soils and salinization. In: Pessarakli, M. (ed.) *Handbook of Plant and Crop Stress*. Marcel Dekker, New York, pp. 3–11.

Takahashi, R. and Asanuma, S. (1996) Association of *T* gene with chilling tolerance in soybean. *Crop Science* 36, 559–562.

Takahashi, R., Benitez, E.R., Funatsuki, H. and Ohnishi, S. (2005) Soybean maturity and pubescence color genes improve chilling tolerance. *Crop Science* 45, 1387–1393.

Tanksley, S.D. and McCouch, S.R. (1997) Seed banks and molecular maps: unlocking genetic potential from the wild. *Science* 277, 1063–1066.

Thung, M. (1991) Bean agronomy in monoculture. In: van Schoonhoven, A. and Voyset, O. (eds) *Common Beans: Research for Crop Improvement*. CAB International, Wallingford, UK, pp. 737–816.

Tiwari, K.N. and Dwivedi, B.S. (1990) Response of eight winter crops to zinc fertilizer on a Typic Ustochrept soil. *Journal of Agricultural Science (Cambridge)* 115, 383–387.

Toker, C. (2004) Estimates of broad-sense heritability for seed yield and yield criteria in faba bean (*Vicia faba* L.). *Hereditas* 140, 222–225.

Toker, C. (2005) Preliminary screening and selection for cold tolerance in annual wild *Cicer* species. *Genetic Resources and Crop Evolution* 52, 1–5.

Toker, C. (2009) A note on the evolution of kabuli chickpeas as shown by induced mutations in *Cicer reticulatum* Ladizinsky. *Genetic Resources and Crop Evolution* 56, 7–12.

Toker, C. and Cagirgan, M.I. (1998) Assessment of response to drought stress of chickpea (*Cicer arietinum* L.) lines under rainfed conditions. *Turkish Journal of Agriculture and Forestry* 22, 615–621.

Toker, C. and Canci, H. (2003) Selection of chickpea (*Cicer arietinum* L.) genotypes for resistance to ascochyta blight [*Ascochyta rabiei* (Pass.) Labr.], yield and yield criteria. *Turkish Journal of Agriculture and Forestry* 27, 277–283.

Toker, C. and Canci, H. (2008) Selection for drought and heat resistance in chickpea (*Cicer arietinum* L.) under terminal drought conditions. In: Kharkwal, M.C. (ed) *Food Legumes for Nutritional Security and Sustainable Agriculture*. The Indian Journal of Genetics and Plant Breeding, IARI, New Delhi, India, pp. 511–517.

Toker, C. and Yadav, S.S. (2010) Legume cultivars for stress environments. In: Yadav, S.S., McNeil, D.L., Redden, R. and Patil, S.A. (eds) *Climate Change and Management of Cool Season Grain Legume Crops*. Springer, Dordrecht, The Netherlands, pp. 351–376.

Toker, C., Canci, H. and Ceylan, F.O. (2006) Estimation of outcrossing rate in chickpea (*Cicer arietinum* L.) sown in autumn. *Euphytica* 151, 201–205.

Toker, C., Lluch, C., Tejera N.A. and Siddique, K.H.M. (2007a) Abiotic stresses. In: Yadav, S.S., Redden, R., Chen, W. and Sharma, B. (eds) *Chickpea Breeding and Management*. CAB International, Wallingford, UK, pp. 474–496.

Toker, C., Canci, H. and Yildirim, T. (2007b) Evalution of perennial wild *Cicer* species for drought resistance. *Genetic Resources and Crop Evolution* 54, 1781–1786.

Toker, C., Yadav, S.S. and Solanki, I.S. (2007c) Mutation breeding. In: Yadav, S.S., McNeil, D. and Stevenson, P.C. (eds) *Lentil: An Ancient Crop for Modern Times*. Springer, Dordrecht, The Netherlands, pp. 209–224.

Toker, C., Yildirim, T., Canci, H., Inci, E.N. and Ceylan, F.O. (2010) Inheritance of resistance to iron deficiency chlorosis in chickpea (*Cicer arietinum* L.). *Journal of Plant Nutrition* 33, 1366–1373.

Toker, C., Canci, H. and Siddique K.H.M. (2011) Nutrient disorders. In: Chen, W., Sharma, H.C. and Muehlbauer, F.J. (eds) *Compendium of Chickpea and Lentil Diseases and Pests*. APS Publishers, New York.

Troedson, R.J., Wallis, E.S. and Singh, L. (1990) Pigeonpea: adaptation. In: Nene, Y.L., Hall, S.D. and Sheila, V.K. (eds) *The Pigeonpea*. CAB International, Wallingford, UK, pp. 159–177.

Turkan, I. and Demiral, T. (2009) Recent developments in understanding salinity tolerance. *Environmental and Experimental Botany* 67, 2–9.

Turner, N.C., Wright, G.C. and Siddique, K.H.M. (2001) Adaptation of grain legumes (pulses) to water-limited environments. *Advances in Agronomy* 71, 193–231.

Unander, D.W., Orf, J.H. and Lambert, J.W. (1986) Early season cold tolerance in soybean. *Crop Science* 26, 676–680.

Vadez, V., Krishnamurthy, L., Serraj, R., Gaur, P.M., Upadhyaya, H.D., Hoisington, D.A. *et al.* (2007) Large variation in salinity tolerance in chickpea is explained by differences in sensitivity at the reproductive stage. *Field Crops Research* 104, 123–129.

van der Maesen, L.J.G. and Somaatmadja, S. (1992) *Plant Resources of South-East Asia. No 1. Pulses*. Prosea Fondation, Bogor, Pudoc-DLO, Wageningen, The Netherlands.

van Hoorn, J.W., Katerji, N., Hamdy, A. and Mastrorilli, M. (2001) Effect of salinity on yield and nitrogen uptake of four grain legumes and on biological nitrogen contribution from the soil. *Agricultural Water Management* 51, 87–98.

van Rheenen, H.H., Singh, O. and Saxena, N.P. (1997) Using evaluation techniques for photoperiod and thermo-insensitivity in pulses improvement. In: Asthana, A.N. and Ali, M. (eds) *Recent Advantages in Pulses Research*. Indian Society of Pulses Research and Development, IIPR, Kanpur, India, pp. 443–458.

Wallace, A. (1960) Iron chlorosis in horticultural plants: A review. *Proceedings of the American Society for Horticultural Science* 75, 819–841.

Wery, J., Turc, O. and Lecoeur, J. (1993) Mechanism of resistance to cold, heat and drought in cool-season legumes, with special reference to chickpea and pea. In: Singh, K.B. and Saxena, M.C. (eds) *Breeding for Tolerance in Cool Season Food Legumes*, John Wiley & Sons, Chichester, UK, pp. 271–291.

White, J.W. and Izquiero, J. (1991) Physiology of yield potential and stress tolerance. In: van Schoonhoven, A. and Voyest, O. (eds) *Common Beans: Research for Crop Improvement*. CAB International, Wallingford, UK, pp. 287–382.

White, J.W. and Singh, S.P. (1991) Breeding for adaptation to drought. In: van Schoonhoven, A. and Voyest, O. (eds) *Common Beans: Research for Crop Improvement*. CAB International, Wallingford, UK, pp. 501–560.

Yoshida, K. (2002) Plant biotechnology – genetic engineering to enhance plant salt tolerance. *Journal of Bioscience and Bioengineering* 94, 585–590.

Zaiter, H.Z. and Ghalayini, A. (1994) Iron deficiency in lentils in the Mediterranean region and its control through resistant genotypes and nutrient application. *Journal of Plant Nutrition* 17, 945–952.

Zaiter, H.Z., Coyne, D.P., Clark, R.B., Lindgren, D.T., Nordquist, P.T., Stroup, W.W. *et al.* (1992) Leaf chlorosis and seed yield of dry beans grown on high-pH calcareous soil following foliar iron sprays. *HortScience* 27, 983–985.

Zaiter, H., Baydoun, E. and Sayyedhallak, M. (1994) Genotypic variation in the germination of common bean in response to cold temperature stress. *Plant and Soil* 163, 95–101.

Zribi, K. and Gharsalli, M. (2002) Effect of bicarbonate on growth and iron nutrition of pea. *Journal of Plant Nutrition* 25, 2143–2149.

17 Legume Improvement in Acidic and Less Fertile Soils

C.R. Spehar, E.A. Pereira and L.A.C. Souza

17.1 Introduction

Large areas of the world, including the savannah lands of South America, Africa, southern Asia and Oceania, pose setbacks for agricultural development mainly due to acidic and less fertile soils. However, the Brazilian savannah lands have recently emerged as a frontier for agricultural development and food production (Goedert *et al.*, 2008) and are the best example of advances leading to competitive agriculture. Economical activities in these areas were reduced until the early 1970s, due to environmental problems. These include soil acidity with an excess of toxic aluminium (Al), reduced levels of calcium (Ca) and magnesium (Mg) (Goedert, 1986; Sanchez *et al.*, 2003) and peculiar weather conditions, including alternate wet and dry seasons with prolonged dry spells in the rainy period. These factors considerably affect crops at the reproductive phase and reduce economic yield. During the mid- and late 20th century, few options were available for farming in these areas, which relied on native grassland ranching, followed by upland rice pasture cultivation on partly limed and fertilized soils prepared for grazing, and charcoal and wood production (Spehar, 1998). It was more like extractive subsistence than economic activity to sustain a limited population and avoid poverty and isolation (Spehar, 2006).

However, today the savannah's participation in supplying food and raw materials to meet the country's needs and for export has grown enormously. In these areas soybean, as a pioneer crop in a relatively short period of time, has contributed more than 50% of total Brazilian soybean production, with productivity higher (2900–3000 kg/ha in the states of Mato Grosso and Goiás) than the national average (2782 kg/ha). The importance of other legumes (phaseolus bean, *Vigna*, pea, chickpea, pigeon pea, lentil and groundnut, and a vast number of other tropical and subtropical legumes) has also been well demonstrated in these regions (Spehar, 2006). Research on key legume crops, challenges and the impact on rural development and food security are presented and analysed under the perspective of rural development. The converging forces of events have promoted a 'silent revolution' in less favourable tropical environments (Spehar, 2008). This chapter covers advances and changing paradigms emphasizing the role of legumes, and points at the achievements and development of, and prospects for, of legume cultivation in marginal lands, especially in the savannahs of Brazil, which present unique environmental conditions.

17.2 Environmental Conditions

The important environmental factors, as understood from experience of the savannah of Brazil, can be applied to similar conditions worldwide (Spehar, 2006). The main environmental conditions, including climate (day length, temperature and rainfall) and soils (biology, chemistry and physics), are described here with respective implications for various legume crops.

Climate

The climatic environment of the Brazilian savannah is characterized by alternating rainy and dry seasons, each lasting for about six months. Annual rainfall varies between 1200 and 1700 mm for the entire savannah region, depending on the influence of other biomes, such as the semi-arid Caatinga or the Amazon rain forest. The minimum rainfall (from 1 mm to < 20 mm) occurs from May to September. Evapotranspiration at the savannah core is 1280 mm, while the mean temperature is 21.3°C, with a range of 17.0–27.0°C. In general, temperatures in these regions year-round are favourable for crop growth and development (Spehar, 1995; Souza, 2001), and annual precipitation is sufficient to supply the demands of grain legume, cereal, fibre, forage and tree crops. In more favoured areas, two to three rainfed annual crops can be grown in the rainy season, one of these being a legume (Spehar, 2009).

The cropping sequence involves legumes, grown either in spring/summer (the rainy season) or autumn/winter (the dry season). In general, erratic dry spells may occur during the spring/summer season and may affect the performance of annuals at the reproductive phase. Yield reduction is more severe if they do not develop roots long enough to tap the water from lower layers of soil (Spehar and Souza, 1999), and thus commercial crops such as soybean should be sown at different periods to mitigate this effect.

Soils

Most savannah soils are acidic and devoid of nutrients. Their chemistry varies, having intermediate to low cation-exchange capacity (CEC). In this environment, nitrogen (N) is a limiting nutrient causing severe deficiency in many non-leguminous crops. Therefore, legume adaptation plays an important role in association with selected N_2-fixing bacteria. The predominant Brazilian savannah soils are comparable to those of certain Australian savannahs, being deep, well drained and with low natural fertility and remarkable acidity (Goedert, 1986). They are classified in the following main groups: oxisols, inseptsols, podzols, entisols, alfisols, quartz sands (quartzarenic neosols), laterites and gleys (Prado, 1993). These soils have several physical and chemical properties that are unsuitable for the growth and development of crops (Table 17.1).

In Australia, there are savannah areas whose soils are prone to increase salinity, if not properly managed, taking into consideration the replacement of indigenous vegetation by sole crops (Williams *et al.*, 1997), whereas in the Brazilian savannahs there could be soil compaction problems under soybean monocrop in conventional tillage (Spehar, 1998). In general, chemical analyses

Table 17.1. Physical and chemical characteristics of typical Brazilian savannah oxisol.

	Physical				Chemical				
Area	Sand (g/dm^3)	Silt	Clay	OM[a]	pH	Al ($cmol_{c+}/kg$)	Ca + Mg	P (mg/dm^3)	K
Virgin	340	190	450	2.0	4.7	1.9	0.4	0.9	16
Amended[b]					5.6	0.0	3.4	8.0	50

[a]Organic matter;
[b]Amendments: 4.0 t/ha lime, 240 kg/ha P_2O_5 and 100 kg/ha K_2O in the form of either single or triple superphosphate, and potassium chloride.

Table 17.2. Aluminium saturation and calcium content in savannah subsoil (21–50 cm).

Aluminium (%)		Calcium	
Saturation	Distribution	Content ($cmol_{c+}$)	Distribution (%)
> 40	42	< 0.4	86
40–10	28	0.4–4.0	13
< 10	30	> 4.0	1

Source: Spehar, 2006 (adapted from Cochrane and Azevedo, 1988).

of savannah soils demonstrate very low CEC and organic matter (OM), which are key factors to achieving best agricultural practices in the savannah (Resck *et al.*, 1991). In addition to this, high Al saturation and Ca deficiency are compounding problems found in large areas of the savannah (Table 17.2).

17.3 Targets for Agricultural Improvement

Original contributions to tropical agriculture, valuable in similar savannah environments, are a result of the development of cultivars tolerant to acidic soils; lime and gypsum amendments, and continuous breeding efforts for soybean and other legumes to develop genotypes for low-latitude commercial cultivation, associated with the selection of efficient N_2-fixing bacteria (Spehar, 1998, 2006; Oliveira *et al.*, 2008). This has been the case for soybean and field bean, and could be extended to pea, chickpea, lentil, cowpea, groundnut and pigeon pea, besides other innovative annual and perennial crops (Spehar, 2009).

Crop improvement

Crop improvement has made significant advances in changing the agriculture scenario for tropical savannah lands. Improved varieties of major and potential food legume crops, such as soybean, cowpea, field bean, pigeon pea and perennial legumes, have become available and have received major attention for marketing to the savannahs (Spehar, 2006). These crops, besides providing an advantage of cost reduction in N application, have contributed valuable products to local markets as well as for export.

The limitations, such as soil acidity, high Al levels and moisture stress, have emphasized the need to increase knowledge of biological processes and the genetics and breeding of food legumes. This has set a target of tailoring plants to the specificities of savannah environments, while achieving high yield and quality of final products (Spehar, 2009). Traditional legume-breeding programmes, such as selecting for complex heritable characters, have been supplemented by the tool of molecular biology for genetic gain. These have proved useful in assessing genetic variability and the mode of inheritance, thus assisting in selection for complex traits in legumes. Assisted selection is used in conjunction with induced DNA changes to generate abiotic stress-, pest- and disease-resistant genotypes, and to address the non-biotic problems typical of the savannahs (Spehar, 2009).

Crop diversification

The introduction and selection of potentially adaptable crops has been a major achievement in the savannah lands. Extension agents, the private sector and government agencies have contributed immensely towards crop diversification (Spehar, 2009). Efforts in regard to those crops that are in demand either locally or for the international market – and on new crops for which the savannah is still the fringe area – have been made in regard to investigation of new technology. Soybean, which is exotic to the savannah, was little affected initially by epidemics, and no serious pests and diseases occurred. At this stage, it seemed that the challenge of adapting soybean to cultivation in a modified environment was easy, similar to how other crops had been introduced. However, monoculture over an expanding acreage led to outbreaks of pests and diseases, with increased costs for plant protection. This has also been the case for field bean and other improved crops. Consequently, despite yield surpluses, the compounding effects of epidemics and persistent weeds have caused losses to farmers (Spehar, 2004; Spehar and Pereira, 2006).

The most promising solution to these problems has been crop diversification by the introduction of less exploited species. Actual and potential crops of both global and native savannah species have been studied as alternatives for biological and economical solutions (Spehar, 2009). Their participation in diversified systems comprises a whole sequence of activities, starting with germplasm introduction and enhancement. Selection and evaluation in representative locations within the savannah has resulted in release of new cultivars.

Soil amendments

Chemical analyses of soils have demonstrated that they have very low CEC which, in turn, is pH dependent, favouring anion sorption with a preference for phosphate. There are reduced negative charges relating to pH, indicating the need for liming to increase its value, while supplying Ca and Mg. Calcium sulphate (gypsum), a by-product of the phosphate fertilizer industry, has been utilized to amend Ca-deficient subsoils and to supply sulphur (S) (Ritchey *et al.*, 1980). One question that needed to be addressed was the amount of lime necessary to neutralize Al and to supply Ca and Mg. Local experimentation allowed determination of the response curve for lime and the changes in pH caused by its addition. It has also been found that excess lime causes trace element deficiencies by increasing pH above 6.5 (Spehar, 1994a). Similarly, experiments with P, S and minor elements have defined their levels of requirements for economical yields in these acid soils (Goedert, 1986).

The OM content directly affects the level of negative charge necessary to retain cations, its CEC being 50 times higher than that of savannah clay particles, and so any increase in OM has an impact on cation sorption, mainly K, Ca, Mg and micronutrients (Mascarenhas *et al.*, 1997). Another positive aspect of increasing OM in savannah soils concerns soil particle structuring and increased water retention. For these reasons, OM is relevant to achieving best agricultural practice in the savannah (Resck *et al.*, 1991); the ongoing task is to maintain and appropriately increase its content in these low-fertility soils. From this perspective, appropriate intervention has been critical in changing the whole agricultural scenario. i.e. development of cropping sequences while taking into account the C:N ratio. Once soil constraints and the climate conditions of the savannahs are dealt with, the more predictable climatic conditions in the savannah help to explain the recent outstanding crop performance, while breeding for tolerance to acidity and efficient utilization of nutrients also contribute (Spehar and Souza, 1999, 2006).

17.4 Food Legume Crop Adaptation to Acidic, Low-fertility Soils

The typical characteristics of savannah lands, with limiting soils and restricted moisture conditions, require special provisions for adaptation to legumes (Spehar and Souza, 1996). Advances in production will now be discussed for those food legumes that are adaptable to commercial production systems and which therefore have the potential for widening the prospects of their inclusion in production systems in less favourable environments.

Soybean

Having been restricted to temperate and subtropical areas until the early 1970s, soybean has now been adapted commercially to the low-latitude/acidic soils of the savannahs. In the past, farmers profited from its cultivation only when prices on the international market were attractive. Combined with incentives and technological advances, this crop was established over a range of small to large agricultural concerns from the 1970s to the 1990s. Once incentives disappeared, the option chosen was to develop the processing industry, directed at food and feed production (Spehar, 2006). Various uses for soybeans have sustained high demand and, consequently, its cultivation. Soybean cultivation at low latitudes was triggered by opportunities created through genetic improvement. For example, identification of late-flowering sources under short-day conditions has proved to be beneficial, and the gene sources for delayed bloom have been discovered, allowing sufficient growth before flowering with a direct and positive impact on the reproductive phase (Kiihl and Garcia, 1989). Thus cultivars possessing these traits have proved superior to traditional photosensitive cultivars when they were grown side by side in savannahs. This was a major achievement, in the form of extension of soybean cultivation to all tropical and equatorial regions (Spehar, 1995; Spehar and Souza, 1999).

A variety of development programmes aimed at high yield, absence of lodging, suitable plant architecture and canopy, and resistance to soil factors have been followed (Kiihl and Garcia, 1989). The crossing of late-flowering varieties, originating from the Philippines and the Americas and adapted to southern regions, such as Bragg, Lee, Hale7, Davis, Hardee, Hood, Lee, Hampton, Bossier and Bienville, has resulted in the development of genotypes that have shown resistance to low levels of Al and Ca (Spehar and Souza 2006). Alhough the crosses between these sources and cultivated varieties resulted in a low frequency of suitable agronomic recombinants in the F_2 generation (St. Martin *et al.*, 2009), repeated backcrossing with locally adapted commercial varieties increased the chances of gain from selection (Spehar, 1994b) and also led to pyramiding of favourable genes. However, this has also narrowed the genetic base, exposing the crop to pests and diseases, with limited additional genetic gain for yield (Hiromoto and Vello, 1986).

Segregating populations of the above crosses has been handled by the modified pedigree method (Spehar, 1994b, 1995). In this procedure no selection is practised in early generations, and bulks obtained at F_4–F_5 were sent to savannah areas for selection *in loco*. The near-homozygous progenies performed well in comparison with controls for the maturity group. This also followed indirect selection for soil constraints, as the best performers across uniform trials showed yield stability, probably due to the deeper roots that could seek water and nutrients from subsoils during dry spells (Spehar and Souza, 2006). The field trials used as tools to increase selection efficiency resulted in identification of genotypes attaining Al resistance genes accumulated during crop adaptation (Souza, 2001; Spehar and Souza, 2006). In addition, differential response for P in the savannah genotypes was demonstrated by the analysis of acid phosphatase activity in leaves, proving useful in further adaptation of the crop in the tropics (Raposo *et al.*, 2004).

In other savannah areas of the world, the breeding methodology described here can also be applied, aiming at development of early-maturing soybean suitable for water-scarce production systems. It is expected that the Brazilian varieties having resistance/tolerance to soil problems and the 'leaf angle–open canopy' character might be used as resistance sources in breeding programmes for acidic/less fertile soils. The 'leaf angle–open canopy' character changes canopy shape, avoiding self-shading and assuring light penetration for lower leaves, allowing the most efficient use of radiation (Spehar, 2009).

The genetic improvements in soybean have led to genetic gain for yield and yield-related traits, and so the improved cultivars adopted by farmers have resulted in increased productivity (from 1.06 t/ha to 2.66 t/ha, 1960–2004). It is not only genetic superiority that has led to these gains but also nitrogen fixation, suitable crop husbandry, and soil and pest management (Spehar, 2006).

Phaseolus

The genus *Phaseolus* has two major cultivated bean species, *Phaseolus vulgaris* (field bean) and *Phaseolus lunatus* (lima bean), in addition to less exploited, more wild-related species such as *Phaseolus coccineus* and *Phaseolus accutifolius*. In Brazil, as in many parts of the tropics, low productivity of *Phaeolus* beans is often associated with subsistence farming under suboptimal soil fertility. In the savannahs, however, there has been a relatively constant increase in yield and production of field bean over the last 30 years, and now it has been established as a commercial crop in the dry season under irrigation. It is now possible to harvest 3–4 t/ha of beans in 100 days (Spehar, 2006). Besides the availability of new cultivars available to farmers, suitable soil and plant management complements the technology for commercial production. For breeding of *P. vulgaris*, germplasm has been used from both within the species and related species, such as *P. coccineus*, to obtain recombinants that are resistant to pests and diseases and have high yield, ideal plant architecture and early maturity. However, repeated use of a few parents such as Carioca, Cornell 49-242, Jamapa, Tlalnepantla 64, Tara and Veranic 2 of Meso-American origin for development of improved lines in common bean has restricted possibilities for new recombinants (Alzate-Marin *et al.*, 2003).

Early-generation testing is used in the selection of desirable recombinants across gene pools for yield improvement, ideotype, physiological efficiency and combining ability (Kelly *et al.*, 1998). Interestingly, Carioca, which is a probable mutant retrieved from a farmer's field and released in 1969 (Almeida *et al.*, 1971), covers the largest area of Brazilian dry bean production. This has also been used for further selection, by exploring variability from accumulated mutations (Santos *et al.*, 2002) and natural hybridizations. However, there is a further need to breed genotypes for specific traits including growth habit, days to maturity, seed quality and disease resistance. In addition to this, selection for acidity and low mineral nutrient, in association with drought tolerance, should be taken into consideration in a range of environments where beans are grown (Beebe *et al.*, 2008). For this purpose, the considerable genetic diversity available within the species can be used in breeding programmes (Baudoin, 1988). Since its spread into the Americas and other parts of the world, it has been cultivated over a range of soil fertility, and it is likely that adaptive genes for these environmental constraints can be found. Research investment is needed for its insertion in commercial production, including nitrogen fixation (Ormeño-Orrillo *et al.*, 2006).

Vigna

The species belonging to genus *Vigna* are mostly tropical and subtropical, on the basis of their centres of origin as well as where they have been domesticated. Most of them have become valuable food sources, becoming an integral part of the local diet (Lush and Evans, 1981). The most widespread is *Vigna unguiculata* (L.) Walp, or cowpea, of African origin. Other valuable species include *Vigna mungo* and *Vigna radiata*, originally from the Indian subcontinent, while *Vigna angularis* (adzuki bean) originated in eastern Asia. In Brazil, *V. unguiculata* was probably introduced by Africans during the colonial era. Associated with culinary traditions, it has become an important protein source mainly in northern and north-eastern parts of the country (Lopes *et al.*, 2001). Considering the poor quality of soil in most areas of commercial production, it is likely that the species holds resistance to acidic soils and low nutrient availability. Genotypic differences under partial liming in ultisols and ferralsols at levels of 1.6–2.5 t/ha lime indicate cowpea as a potential grain legume crop on acidic soils (Edwards *et al.*, 1981). There is great potential for *Vigna* species in the acid soils of the tropics. Further studies, however, are needed to enlarge germplasm collections and to exploit genetic diversity for root growth in less fertile acidic subsoils. In addition to this, it is necessary to tailor those plants having upright and indeterminate growth habit (Bezerra *et al.*, 2001). Pioneer work on interspecific hybridizations has pointed to the possibilities of transferring genes conditioning resistance to pests and other limiting factors (Chen *et al.*, 1989). It is likely that germplasm originating and introduced in the poorer soils of Africa, Asia and the Americas, where *Vigna* species have been cultivated for a long time, contains such genes. Through appropriate manipulation, these can be incorporated and utilized to improve genotypes possessing adaptability characters for the commercial cultivation in countries most in need of bean production (Fery, 2002).

Semi-arid and Mediterranean pulses

Guar (*Cyanopsis tetragonoloba*) and other similar legumes from the temperate and subtropical semi-arid zones are grown on a rather small scale. These crops evolved in the absence of soil acidity, being more exposed to excess salts and moisture stress. Therefore, efforts have been made in adapting them to previously amended acidic soils in tropical environments. Root exudates, containing some Al-immobilizing organic acids, have also been studied in these species, aiming at improving crops (Hocking, 2001). The low levels of the key acids in these crops suggest limited genetic gain. Selection can be guided, however, by screening germplasm from a broad range of environments: gene manipulation within the genome of acidic soil-adapted legumes, followed by transfer to these pulses. The impact of attaining adaptive genes is great, broadening cultivation to supply growing global demand.

The migration of pea crops from southern Brazil to the savannah lands illustrates the possibilities for its commercial production in soils previously acidic and devoid of nutrients (Leonel and Giordano, 1984; Reis *et al.*, 1989). Once the ploughed layer of soil has been improved by liming and fertilization, pea shows adaptability by giving higher yields (Reis *et al.*, 1989; Marouelli *et al.*, 1991). Breeding strategies for pea in the savannahs have been based on transferring genes controlling different leaf types and tendrils into local selected varieties and genes controlling adaptability to acid subsoils that may be present in southern cultivars (Reis *et al.*, 1989). Powdery mildew is an important disease in these areas (Café Filho *et al.*, 1987), and further studies are needed to identify resistance sources for this disease in acid soils by screening of germplasm.

In Brazil, attempts have been made to acquire cultivars of chickpea for commercial production in improved savannah acid soils (Nascimento *et al.*, 1998). The promising yields achieved in experiment trials on acid subsoils strongly indicate that chickpea can be included in agricultural production systems, representing an economic opportunity for local farmers. Considering the extension of the savannahs, it is expected that in the near future Brazil may initiate the export of either *kabuli* or *desi* types to Europe, Asia and North Africa (Spehar, 2009). In the meantime, advances in breeding will contribute to enhanced resistance to acidity, leading to stable yields. Preliminary work (Spehar, unpublished results) indicates that exploiting variability for plant type and growth habit leads to profitable yields.

Lentil has shown possibilities for commercial cropping in those savannah areas located on plateaux of 800–1000 m altitude, at 15° latitude. In these areas, the original acid soils have been improved and lentil cultivars have been selected for commercial production (Nascimento and Giordano, 1993). In regard to this, Argentinean germplasm has shown promising results.

Of the lupines, *Lupine luteus*, *L. album* and *L. angustifolius* have been introduced and grown in subtropical areas, and genotypes adapted to acid soils have been identified. Here, crops are grown for fodder production in the winter and also for soil protection (Calegari *et al.*, 2008). The use of their grains is limited by the presence of alkaloids, but breeding efforts are under way aimed at reduced the content of alkaloids to promote their commercial cultivation. The Andean *Lupinus mutabilis* indicates adaptability in preliminary studies (Spehar, unpublished results). However, for all species, germplasm needs to be introduced and evaluated to increase the chances of acquiring genotypes resistant to soil constraints. Faba or broad beans also hold the same promise as lupines in the savannah or tropical acid soil environments, although there has been little experimentation on faba bean under acid soils. It has been observed that genotypes that originated from marginal areas have shown drought tolerance (Link *et al.*, 2008). In Latin American highlands, this crop has prospered since its introduction by the Spanish, centuries ago.

Pigeon pea

In north-east Brazil, semi-perennial types of pigeon pea are grown as isolated shrubs, reaching 3–4 m in height. In the savannah, experiments have been conducted in growing early varieties of pigeon pea in rows 50 cm apart, at densities similar to those of soybean (10–12 plants/m). The aim is to produce grains and forage for intensive dairy production (Spehar, 2004). The studies to date suggest that this crop has the potential to adapt in production systems as it has a slow rate of crop residue decomposition (Carvalho *et al.*, 1996), a narrow C:N ratio and the residue provides good soil cover, which are desirable traits in savannah regions. In sowings at the end of the rainy season, pigeon pea shows reduced biomass production but is suitable as soil cover, preventing weed infestation both during growth and after cutting and placing on the soil (Spehar, 2004). This has positive implications in organic farming, owing to its capacity to control weeds. The species has the ability to increase available P from soils containing the element in its insoluble form. The production of organic acids in roots is the key factor for synergistic effects on production systems (Ae *et al.*, 1995). Tolerance to soil acidity has accumulated in selected genotypes. Besides adaptability to savannah soils and the subtropics, genotypes have been selected for grains and forage, being grown either as sole crop or in combination (Spehar, 2004). In late-maturing cultivars, management trials for either forage or soil cover have also been conducted at 140–180 days after emergence. In forage production the stems are cut at 20–40 cm to allow regrowth. When the purpose is to grow pigeon pea in sequence with other crops, plants should be cut 5–10 cm from the ground, to avoid sprouting.

Groundnut

The genus *Arachis*, originating from South America, which includes groundnut (*Arachis hypogea*) and forage groundnut (*A. silvestris*), is of interest in agriculture worldwide. The high biomass-yielding ability of *A. silvestris* enables the species to be used in integrated crop–livestock systems. Understanding the diversity in the genus is necessary to its adaptation into less favourable environments. Therefore, 96 accessions of *A. hypogaea*, including 36 wild types, have been studied using SSR (simple sequence repeat) markers, where it was found that the Brazilian groundnut germplasm collection has considerable levels of genetic diversity (Moretzsohn *et al.*, 2004). Both *A. hypogea* and *A. silvestris* have varieties with large biomass production that can be utilized as forage and soil cover during the dry season in the savannah, besides nut production. Management of annual species is easier than for the perennial *Arachis pintoi*, although this species has shown tolerance to acidity and reduced available nutrients (Argel and Pizzarro, 1992). This character can be better exploited in future studies to extend adaptation of the genus *Arachis* to the acidic tropical soils of the world.

17.5 Legume Seed Production for the Savannahs

The success in adapting the legume crops for the savannahs depends on seed production and supply to farmers. In soybean, genetic quality has been maintained by standard procedures and checked with molecular markers, becoming a reference for other self-pollinating legumes (Schuster *et al.*, 2004). In the savannahs, the plants are commonly exposed to high moisture levels and temperatures at maturity which favour the development of diseases, leading to seed deterioration. Organized seed production and maintenance of genetic purity have been the key to reaching competitive yields in soybean. This, however, does not always apply to other crops such as common bean and cowpea. Farmers of different production systems differ with regard to the usage of technology, subsistence farmers preferably saving part of their grain as seeds for the following crop, thereby carrying pests and weeds from the previous crop

(Santos *et al.*, 2002). It is expected that as seed supplies increase due to affordable price in both public and private sectors, farmers will prefer to purchase certified seeds.

17.6 Nitrogen Fixation in Less Favourable Soil Conditions

The symbiosis of leguminous species with nitrogen-fixing (N_2) bacteria suppresses the need for N fertilization, reducing costs and contributing to the environment. Through research and experimentation, it has been possible to identify and select for efficient strains of *Bradyrhizobium* spp. and *Rhizobium* spp., of which there are several species adapted to acidic soil conditions (Table 17.3). For this purpose, knowledge of soil biology and chemical properties is necessary, given the peculiarities of acidic soils (Goedert, 1986; Hungria and Vargas, 2000).

The genus *Bradyrhizobium* is specific to soybean, and inoculants containing soil–host-adapted strains have become available to farmers. The most popular vehicle being peat, inoculants are easy to prepare on a commercial scale at low cost, yielding a considerably advantageous cost:benefit ratio (Spehar, 2004). Alternatively, liquid inoculants have been developed but require more care in handling during sowing to maintain population density. Under conditions of high temperature, when unexpected rain may cause sowing delay, peat is still the most practical vehicle for inoculation. In soybean crops alone, the inoculation technique in the Brazilian savannah is responsible for saving the equivalent of 10.5 million t of urea (worth about US$2.1 billion), and also brings environmental gain by preventing wastage and contamination of the water table. Identifying efficient bacterial strains for many legume species has been a long-term target in research, which may become easily accessible to farmers and eventually represent routine practice in cultivation. There are examples of success in soybean (Mendes *et al.*, 2008), groundnut (Giardini *et al.*, 1984), pea (Azevedo *et al.*, 2002), dry bean and chickpea (Voss *et al.*, 1998). In soybean, advances have been made (Kaschuk *et al.*, 2009) but in dry bean, chickpea and lentil N_2 fixation still awaits further studies on soil conditions and host nutritional status (Voss *et al.*, 1998).

Efficient biological nitrogen fixation contributes to consolidation of the participation of legumes in tropical agricultural systems. The soils in these environments are subject to fluctuations in available N, water movement and exposure to high temperatures in the dry season. High temperatures may be compounded by dry spells, causing nodulation failure and affecting rhizobial growth and survival. These stresses may also contribute to undesirable changes causing plasmid deletions, genomic rearrangements and reduced diversity. Acidic conditions limit cultivation of legumes originating from semi-arid, alkaline soils, and hence liming is necessary for commercial crop production. However, the inoculation of such soils with acidic soil, stress-tolerant strains has avoided the need for liming and has increased grain yields for common bean and soybean in Brazil.

The savannah soils of Brazil have antibiotic-producing fungi that are antagonistic to nitrogenfixing bacteria, thus preventing nodulation (Coelho and Drozdowics, 1978). The first experimentation with locally selected genotypes showed that, for initial cultivations, there was little efficiency in N_2-fixation from inoculants. Soybean plants showed nitrogen shortage, mainly after grass pasture, rice or maize, when yields were lower than the potential of selected cultivars, and hence there is a need to identify the causes of deficiency. In a classical experiment, locally selected soybean genotypes having sufficient growth and development suffered from N deficiency, even when seeds were inoculated (Damirgi and Johnson, 1966). In addition, *Streptomyces* spp. represent 75–94% of the population of microorganisms in savannah virgin soils (Coelho and Drozdowics, 1978). The full adaptation of soybean came about when *Bradyrhizobium* strains tolerating 80–160 ug/ml streptomycin were selected.

Advances in soybean have yielded interesting results. In more recent times, efficient and promising strains have been selected, using simple procedures. Screening takes place under acid medium-favouring antibiotic production by pH-specific fungi. Antiserum reaction is measured to allow the selection of efficient N-fixing strains. In fact, the whole process takes into account the population dynamics. Several generations are produced in a short time, leading to antibiotic-resistant mutants that recombine and multiply (Hungria and Vargas, 2000). The aim is to trial these in soybean or other legumes under different soil conditions and to select for antiserum reaction.

The effects of soil acidity have been studied at several locations, irrespective of microorganism interactions causing inhibition of nitrogen-fixing bacteria, for example with groundnut in the acid soils of Africa. Strains have been recovered from soils that had received different lime doses, with pH levels of 5.0 and 6.5. Even though soil acidity negatively affected plant development, nodulation was enhanced, favouring the indigenous populations of Bradyrhizobium (Rossum *et al.*, 1994). The types of associations for respective hosts and N_2-fixing species for legumes are listed in Table 17.3. This list is far from comprehensive, but it serves to illustrate the relations among hosts and microsymbionts, and the data illustrate how bacterial species relate to hosts and vice versa.

It has been demonstrated that associations of legumes with N_2-fixing bacteria are particular and very complex. The process can be described as co-evolution, in the sense that existing strains in centres of origin and dispersion evolve to associate

Table 17.3. Host and N_2-fixing bacteria in regard to pH range, strain response to environmental factors and efficiency (after Fernandes and Reis, 2008).

Host	N_2-fixing genus/species/ biovariety	pH range	Response factors	Efficiency
Glycine max	*Bradyrhizobium japonicum, B. elkanii, B. liaoningense*	5.5–6.5	Al, low moisture, high temperature	High
Phaseolus vulgaris, P. coccineus, P. accutifolius	*Rhizobium etli bv phaseoli, R. gallicum, R. tropici, R. leguminosarum bv phaseoli, R. giardinii, R. gallicum*	5.0–6.8	Al, low moisture, high and low temperatures	Low–medium
Phaseolus lunatus	*Bradyrhizobium yuanmingense*	5.0–7.0	Al, low moisture, high temperature	Medium
Vigna unguiculata, V. radiata, V. mungo, V. angularis	*Azorhizobium, Burkholderia, Bradyrhizobium, Mesorhizobium, R. Sinorhizobium*	5.5–6.8	Al, low moisture, high temperature	Medium–high
Pisum sativum, Lens culinaris	*Rhizobium leguminosarum*	5.6–6.9	Acidity, high moisture	Medium
Cicer arietinum	*Mesorhizobium ciceri*	5.6–6.9	Acidity, high moisture	Low–medium
Arachis hypogaea	*Bradyrhizobium* spp.	5.3–6.6	Al, low moisture, high temperature	Medium–high
Cyamopsis tetragonolobus	*Bradyrhizobium* spp.	5.8–6.9	Temperature, low moisture	Medium–high
Cajanus cajan, Prosopis spp.	*Rhizobium* spp.	5.5–6.5	Acidity, temperature	Medium–high
Leucaena spp.	*Rhizobium* spp.	5.2–6.5	Acidity, temperature, moisture	High
Mimosa spp.	*Burkholderia phymatum*	4.6–6.5	Acidity, temperature, moisture	Medium–high

with hosts when they are introduced into different environments. This has been the case for exotic crops in new environments, such as soybean in the savannahs. From its introduction into the tropics of Brazil, Africa and Asia, soybean has developed an association with native bacteria that were already colonizing soils and associated with other local and introduced legumes. The ability to infect with different strains or species depends on the population size, effectiveness and survival of indigenous or introduced bacteria in the field. Indigenous strains have proved more effective than introduced. Alternatively, the relationship between rhizobial cell counts and promising soybean responses may be used to indicate under which conditions inoculation is beneficial to farmers (Sanging *et al.*, 1996).

The present limitations in efficiency for introduced legumes into less favourable environments are known, as in the case of chickpea, pea and lentil grown in the savannahs. However, this can be overcome, in a similar way to that achieved with soybean in the same environments. Moreover, the understanding of N_2-fixing bacterial associations with native species that have evolved in the acidic, less fertile soils will be of use in agricultural systems. Many less exploited annual and perennial legumes remain to be studied for their associations and benefits from efficiency with N input in the tropics.

17.7 Conclusions

From the above, it is evident that it is possible to adapt originally non-tolerant pulse species to acid savannah soils, such as the examples given of soybean and *Phaseolus*. Their contribution to production systems and the supply of food must be considered under the prospect of development and food security. The introduction and selection of promising genotypes, indispensable for commercial cropping, will also contribute to changes in the genetic constitution of pulse crops for the acidic and less fertile savannah soils, leading to the development of more promising and stable cultivars. In the long term, it is expected that mutants will lead to enhanced crop productivity in acid soils (Spehar and Souza, 2006), leading eventually to the savannahs becoming major producers of pulses in addition to the other legumes. Strategies for the inclusion of legume crops should be taken with regard to the scope of production efficiency. Cropping sequences and combinations, including the production of agro-fuel and other raw materials, become feasible when improvements are monitored by their output and synergistic effects. The approaches are to be based on internal food supply and revenue from exports, generating employment and demand and creating local markets. This will trigger the production chains in acidic and less fertile soils, taking the savannah lands of Brazil as a model.

References

Ae, N., Otani, T. and Arihara, J. (1995) Effects of specific compounds exuded from roots on phosphorus nutrition. In: Johansen, C., Lee, K.K., Sharma, K.K., Subbarao, G.V. and Kueneman, E. (eds) *Genetic Manipulation of Crop Plants to Enhance Integrated Nutrient Management in Cropping Systems 1. Phosphorus*. FAO/IAEA/ICRISAT, Patancheru, India, pp. 117–128.

Almeida, L.D.A.L., Leitão Filho, H.F. and Miyasaka, S. (1971) Características do feijão Carioca, um novo cultivar. *Bragantia* 30, 33–38.

Alzate-Marin, A.L., Costa, M.R., Sartorato, A., Del Peloso, M.J., Barros, E.G. and Moreira, M.A. (2003) Genetic variability and pedigree analysis of Brazilian common bean elite genotypes. *Scientia Agricola* 60, 283–290.

Argel, P.J. and Pizzaro E.A. (1992) Germplasm case study: *Arachis pintoi*. In: *Pastures for the Tropical Lowlands: CIAT's Contribution*. Centro Internacional de Agricultura Tropical, Cali, Colombia, pp. 57–73.

Azevedo, W.R., Faquin, V., Moreira, F.M.S., Oliveira Júnior, A.C. and Lisboa, C.C. (2002) Boron effects on nodule development in pea plants cultivated in lowland soils. *Pesquisa Agropecuária Brasileira* 37, 1137–1143.

Baudoin, J.P. (1988) Genetic resources, domestication and evolution of lima bean, *Phaseolus lunatus*. In: Gepts, P. (ed.). *Genetic Resources of Phaseolus Bean*. Kluwer Academic Publishers, Amsterdam, The Netherlands, pp. 393–407.

Beebe, S.E., Rao, I.M., Cajiao, C. and Grajales, M. (2008) Selection for drought resistance in common bean also improves yield in phosphorus limited and favorable environments. *Crop Science* 48, 582–592.

Bezerra, A.E.C., Anunciação Filho, C.J., Freire Filho, F.R. and Ribeiro, V.Q. (2001) Interrelation among characters of upright cowpea plants with determined growth. *Pesquisa Agropecuária Brasileira* 36, 137–142.

Café Filho, A.C., Menezes, L.W., Giordano, L.B. and Reifschneider, F.J.B. (1987) Controle químico do oídio da ervilha. *Horticultura Brasileira, Brasília* 5, 44.

Calegari, A., Hargrove, W.L., Rheinheimer, D.S., Ralisch, R., Tessier, D., Tourdonnet, S. *et al.* (2008) Impact of long-term no-tillage and cropping system management on soil organic carbon in an oxisol: A model for sustainability. *Agronomy Journal* 100, 1013–1019.

Carvalho, A., Correia, J.R., Blancaneaux, P., Freitas, L.R.S., Menezes, H., Pereira, J. *et al.* (1996) Caracterização de espécies de adubos verdes para o cultivo de milho em latossolo vermelho-escuro originalmente sob cerrado. In: *Proceedings of First International Symposium on Tropical Savannas: Biodiversity and Sustainable Production of Food and Fibbers.* EMBRAPA-CPAC, Planaltina, Brasília, Brazil, pp. 384–388.

Chen, H.K., Mok, M.C., Shanmugasundaram, S. and Mok, D.W.S. (1989) Interspecific hybridization between *Vigna radiata* (L.) Wilczek and *V. glabrescens*. *Theoretical and Applied Genetics* 78, 641–647.

Coelho, R.R.R. and Drozdowics, A. (1978) The occurrence of actinomycetes in a cerrado soil in Brazil. *Revue de Ecologie et Biologie du Sol* 15, 459–473.

Damirgi, S.M. and Johnson, H.W. (1966) Effect of soil actinomycetes on strains of *Rhizobium japonicum*. *Agronomy Journal* 58, 223–234.

Edwards, D.G., Kang, B.T. and Danso, K.A. (1981) Differential response of six cowpea (*Vigna unguiculata* (L.) Walp) cultivars to liming in an ultisol. *Plant and Soil* 59, 61–73.

Fernandes, P.I. Jr and Reis, V.M. (2008) *Algumas Limitações à Fixação Simbiótica de Nitrogêncio em Leguminosas.* Agrobiologia, Embrapa, Seropédica, RJ, Brazil, pp. 33. (Documento 252).

Fery, R. (2002) New opportunities in *Vigna*. In: Janick, J. and Whimpkey, A. (eds) *Trends in New Crops and New Uses*. ASHS Press, Alexandria, Virginia, pp. 424–428.

Giardini, A.R., Lopes, E.S. and Neptune, A.M.L. (1984) Pré-seleção de estirpes de *Rhizobium* sp. *para amendoim. Bragantia* 43, 389–396.

Goedert, W.J. (1986) *Solos dos Cerrados – Tecnologias e Estratégias de Manejo.* Embrapa/Nobel, Brasília, Brazil, 422 pp.

Goedert, W.J., Wagner, E. and Barcelos, A.O. (2008) Savanas tropicais: dimensão histórico e perspectivas. In: Faleiro, F.G. and Farias Neto, A.L. (eds) *Savanas: Desafios e Estratégias Para o Equilíbrio Entre Sociedade, Agronegócio e Recursos Naturais*. Embrapa Cerrados, Planaltina, DF, Brazil, pp. 49–80.

Hiromoto, D.H. and Vello, N. (1986) The genetic base of Brazilian soybean [(*Glycine max* L. (Merrill)] cultivars. *Brazilian Journal of Genetics* 9, 295–306.

Hocking, P.J. (2001) Organic acids exuded from roots in phosphorus uptake and aluminum tolerance of plants in acid soils. *Advances in Agronomy* 74, 63–97.

Hungria, M. and Vargas, M.A.T. (2000) Environmental factors affecting N_2 fixation in grain legumes in the tropics, with an emphasis on Brazil. *Euphytica* 65, 151–164.

Kaschuk, G., Hungria, M., Leffelaar, P.A., Giller, K.E. and Kuyper, T.W. (2009) Differences in photosynthetic behaviour and leaf senescence of soybean (*Glycine max* [L.] Merrill) dependent on N_2 fixation or nitrate supply. *Plant Biology* 12, 60–69.

Kelly, J.D., Kolkman J.M. and Schneider, K. (1998) Breeding for yield in dry bean (*Phaseolus vulgaris* L.). *Euphytica* 102, 343–356.

Kiihl, R.A.S. and Garcia, A. (1989) The use of the long-juvenile trait in breeding soybean cultivars. In: *Conferencia Mundial de Investigacion en Soya*, 4., AASOJA, Buenos Aires, Actas, Buenos Aires, pp. 994–1000.

Leonel, L.A.K. and Giordano, L.B. (1984) A cultura da ervilha visando a produção de grãos secos para reihidratação no Mato Grosso do Sul. *Horticultura Brasileira, Brasília* 2, 38.

Link, W., Abdelmula, A.A., Von Kittlitz, E., Bruns, S. and Sterling, H.R.D. (2008) Genotypic variation for drought tolerance in *Vicia faba*. *Plant Breeding* 118, 477–484.

Lopes, A.C.A., Freire Filho, F.R., Silva, R.B.Q., Campos, F.C. and Rocha, M.M. (2001) Variabilidade e correlações entre caracteres agronômicos em caupi (*Vigna unguiculata*). *Pesquisa Agropecuária Brasileira* 36, 515–520.

Lush, V.M. and Evans, L.T. (1981) The domestication and improvement of cowpeas (*Vigna unguiculata* (L.) Walp.). *Euphytica* 30, 579–587.

Marouelli, W.A., Giordano, L.B., Oliveira, C.A. and Carrijo, O.A. (1991) Desenvolvimento, produção e qualidade de ervilha sob diferentes tensões de água no solo. *Pesquisa Agropecuária Brasileira, Brasília* 26, 1041–1047.

Mascarenhas, H.A.A., Ito, M.F., Tanaka, M.A., Tanaka, R.T., Ambrosano, G.M.B. and Muraoka, T. (1997) Efeito da adubação potássica no cancro da haste da soja. *Summa Phytopathologica* 23, 21–221.

Mendes, I.M.C., Reis Junior, F.B., Hungria, M., Sousa, D.M.G. and Campo, R.J. (2008) Adubação nitrogenada suplementar tardia em soja cultivada em latossolos do Cerrado. *Pesquisa Agropecuáriia Brasileira* 43, 1053–1060.

Moretzsohn, M.C., Hopkins, M.S., Mitchell, S.E., Kresovich, S., Valls, J.F.M. and Ferreira, M.E. (2004) Genetic diversity of *Arachis hypogaea* and its wild relatives based on the analysis of hypervariable regions of the genome. *BMC Plant Biology* 4 (http://www.biomedcentral.com/1471-2229/4/1).

Nascimento, W.M. and Giordano, L.B. (1993) Viabilidade de produção de lentilha no Brasil. *Horticultura Brasileira, Brasília* 11, 51–52.

Nascimento, W.M., Pessoa, H.B.S.V. and Giordano, L.B. (1998) *Cultivo do Grão-de-bico (Cicer arietinum)*. EMBRAPA/CNPH, Brasília, Brazil.

Oliveira, C.A., Sá, N.M.H., Gomes, E.A., Marriel, I.E., Scotti, M.R., Guimarães, C.T. *et al.* (2008) Assessment of the mycorrhizal community in the rhizosphere of maize (*Zea mays* L.) genotypes contrasting for phosphorus efficiency in the acid savannas of Brazil using denaturing gradient gel electrophoresis (DGGE). *Applied Soil Ecology* 41, 249–258.

Ormeño-Orrillo, E., Vinuesa, P., Zúñiga-Dávila, D. and Martínez-Romero, E. (2006) Molecular diversity of native bradyrhizobia isolated from Lima bean (*Phaseolus lunatus* L.) in Peru. *Systematic and Applied Microbiology* 2, 253–262.

Prado, H. (1993) *Manual de Classificação de Solos do Brasil*. FUNEP, Jaboticabal, 218 pp.

Raposo, R.W.C., Takashi, M., Basso, L.C., Lavres, J. and Franzini, V.I. (2004) Acid phosphatase activity and leaf phosphorus content in soybean cultivars. *Scientia Agrícola* 61, 439–445.

Reis, N.V.B., Oliveira, C.A.S. and Giordano, L.B. (1989) Graus-dias e época de plantio para produção de grãos secos de ervilha. *Horticultura Brasileira, Brasília* 7, 12–14.

Resck, D.V.J., Pereira, J. and Silva, J.E. (1991) *Dinâmica da Matéria Orgânica na Região dos Cerrados*. EMBRAPA/CPAC, Planaltina, DF, Brazil, 22 pp.

Ritchey, K.D., Sousa, D.M.G., Lobato, E. and Correia, O. (1980) Calcium leaching to increase rooting depth in a Brazilian savannah Oxisol. *Agronomy Journal* 72, 40–44.

Rossum, D., Muyotcha, A., Hoop, B.M., Verseveld, H.W., Stouthamer, A.H. and Boogerd, F.C. (1994) Soil acidity in relation to groundnut – *Bradyrhizobium* symbiotic performance. *Plant and Soil* 163, 165–175.

Sanchez, P.A., Palm, C.A. and Buol, S.W. (2003). Fertility capability soil classification: a tool to help assess soil quality in the tropics. *Geoderma* 114, 157–185.

Sanging, N., Abaidou, R., Dashiell, K., Carsky, R.J. and Okogu, A. (1996). Persistence and effectiveness of rhizobia nodulating promiscuous soybeans in moist savanna zones of Nigeria. *Applied Soil Ecology* 3, 215–224.

Santos, P.S.J., Abreu, A.F.B. and Ramalho, M.A.P. (2002) Seleção de inhas puras no feijão "Carioca". *Ciência Agrotécnica, Edição Especial* 62, 1492–1498.

Schuster, I., Queiroz, W.T., Teixeira, A.I., Barros, E.G. and Moreira, M.A. (2004) Determination of genetic purity of soybean seeds with the aid of microsatellite molecular markers. *Pesquisa Agropecuária Brasileira* 39, 249–253.

Souza, L.A.C. (2001) Soybean genotypes reaction to aluminium in hydroponics and soil. *Pesquisa Agropecuária Brasileira* 36, 1255–1260.

Spehar, C.R. (1994a) Seed quality of soya bean based on mineral composition of seeds of 45 varieties grown in a Brazilian Savanna acid soil. *Euphytica* 76, 127–132.

Spehar, C.R. (1994b) Breeding soybeans to the low latitudes of Brazilian Cerrados (Savannahs). *Pesquisa Agropecuária Brasileira, Brasilia* 29, 1.167–1.180.

Spehar, C.R. (1995) Impact of strategic genes in soybean on agricultural development in the Brazilian tropical savannahs. *Field Crops Research* 41, 141–146.

Spehar, C.R. (1998) Production systems in the savannas of Brazil: Key factors to sustainability. In: Lal, R. (ed.) *Soil Quality and Agricultural Sustainability*, Ann Arbor Press, Chelsea, Michigan, pp. 301–318.

Spehar, C.R. (2004) Manejo cultural no Plantio Direto. Módulo 11. Associação Brasileira de Educação Agrícola Superior, Universidade de Brasília Faculdade de Agronomia e Medicina Veterinária. Brasília, Brazil, 145 pp.

Spehar, C.R. (2006) Conquering the Brazilian savannah and sonsolidation of agriculture. In. Paterniani, E. (ed.) *Ciência, Agricultura e Sociedade*. Embrapa Informação Tecnológica, Brasília, pp. 195–226.

Spehar, C.R. (2008) Grain, fiber and fruit production in the Cerrado development. In: Faleiro, F.G. and Farias Neto, A.L. (eds) *Savanas: Desafios e Estratégias Para o Equilíbrio Entre Sociedade, Agronegócio e Recursos Naturais*. Embrapa Cerrados, Planaltina, DF, Brazil, pp. 477–504.

Spehar, C.R. (2009) Challenges and prospects to realize diversified agriculture in the tropics. In: *World Congress on Conservation Agriculture, 4: Innovations for Improving Efficiency, Equity and Environment*. OSIDC, New Delhi, pp. 223–229.

Spehar, C.R. and Pereira, R.J.C. (2006) Co-existing with the weeds. In: *Proceedings of the Brazilian Congress of Weed Science: Co-existing with the Weeds*. Brasília, DF, Brazil, pp. v–xx.

Spehar, C.R. and Souza, P.I.M. (1996) Sustainable cropping systems in the Brazilian Cerrados: Identification of analogous land for agro-technology transfer in the savannah zones of developing world. *Integrated Crop Management* 1, 1–25.

Spehar, C.R. and Souza, L.A.C. (1999) Selecting soybean [*Glycine max* (L) Merrill)] tolerant to low-calcium stress in short term hydroponics experiment. *Euphytica* 106, 35–38.

Spehar, C.R. and Souza, L.A.C. (2006) Selection for aluminium tolerance in tropical soybeans. *Pesquisa Agropecuária Tropical* 36, 1–6.

St. Martin, S.K., Xieb, F., Zhangb, H., Zhangb, W. and Songb, X. (2009) Epistasis for quantitative traits in crosses between soybean lines from China and the United States. *Crop Science* 49, 20–28.

Voss, M., Parra, M.S. and Campos, A.D. (1998) Yield of common bean and chickpea as a function of soluble phosphate applied in pits or in the seeding row. *Revista Brasileira de Ciência do Solo* 22, 166–168.

Williams, J., Bui, E.N., Gardner, E.A., Littleboy, M. and Probert, M.E. (1997) Tree clearing and dry land salinity hazard in the upper burdekin catchment of North Queensland. *Australian Journal of Soil Research* 35, 785–802.

18 Molecular Breeding Approach in Managing Abiotic Stresses

M. Ishitani, J. Rane, S. Bebee, M. Sankaran, M. Blair and I.M. Rao

18.1 Introduction

Over the past two decades, persistent efforts have been made to develop alternative approaches for accelerating crop improvement, including marker-assisted selection (MAS) and the transgenic approach (Varshney *et al.*, 2009a). Several studies have shown that desired genes introduced in crop plants can significantly improve tolerance to various abiotic stresses such as drought and salinity (Mittler and Blumwald, 2010). On the other hand, molecular markers are emerging as a promising tool to complement conventional plant breeding, as they are generally not environmentally regulated, are unaffected by the conditions in which the plants are grown and are detectable at all stages of plant growth. In addition, there is now a focus on environmentally induced and developmentally regulated genomic variation (ED-genomic variation or ED-genetic variation) that can be found in both coding and non-coding sequences, and is often non-Mendelian in its inheritance pattern (Li, 2009). The implications of such variations for the improvement of crop plants have not been investigated in detail. Modern molecular biology tools, such as DNA array and next-generation sequencing technologies (Varshney *et al.*, 2009b; Capra *et al.*, 2010; Metzker, 2010) are expected to reveal more a comprehensive evaluation of ED-genomic variation. Genetic resources and gene discovery for trait of interest are the major prerequisites for developing and applying molecular breeding approaches in crop improvement (Glaszmann *et al.*, 2010). Molecular breeding depends on proof of concept related, to the association of discovered genes with desired traits and its relevance to crop productivity. Discovery of candidate genes is determined by genetic resources and the phenotypic analysis that can complement genomic analysis (Ishitani *et al.*, 2004; Blair and Ishitani, 2009). The aim of this chapter is to focus on efforts made to discover genes associated with abiotic stresses and molecular breeding approaches for improvement of grain legumes, with a special emphasis on common bean.

18.2 Genetic Resources for Improving Abiotic Stress Tolerance

There are huge legume germplasm collections worldwide (see Chapter 23 for details). Characterization of genetic diversity within these collections is a necessary prelude to their efficient use (Alonso-Blanco *et al.*, 2009).

For example, the wild common bean represents four gene pools, two of which are amply represented in *Phaseolus vulgaris*, the cultivated bean, and a third for which there is evidence of incipient domestication (Islam *et al.*, 2004; Blair *et al.*, 2009). The two major gene pools in turn have been subdivided into as many as seven independent races (Blair *et al.*, 2006, 2007, 2009; Kwak and Gepts, 2009). Sister species of common bean in the secondary gene pool (see Chapter 23) that can readily be crossed with *P. vulgaris* include both domesticated species (*Phaseolus dumosus* and *Phaseolus coccineus*) and truly wild forms (Chacón *et al.*, 2005). *Phaseolus acutifolius* (the tepary bean) is a fourth domesticated species but can be crossed only with great difficulty, and crosses with *Phaseolus lunatus* (the lima bean) produce only sterile F_1 plants (Fig. 18.1). Recent technological advances in DNA sequencing and high-throughput genotyping with precise and focused phenotyping are now redefining the scope of germplasm characterization and its efficient use for crop improvement through molecular breeding programmes (Ganal *et al.*, 2009; Tester and Langridge, 2010).

18.3 Development of Molecular Breeding Tools for Improvement

Investigations with model legumes

Model legume plants for investigations at the molecular level and advances in genomic tools such as genome sequencing have greatly enriched our knowledge of plant genes (Cannon *et al.*, 2009; Edwards and Batley, 2010). *Medicago truncatula* (Cook, 1999; Young and Udvardi, 2009) and *Lotus japonicus* (Udvardi *et al.*, 2005; Sato and Tabata, 2006) emerged as model species in accelerating the study of legume biology, because these species are rich in a range of powerful molecular, genetic and genomic tools. A comprehensive detail of studies on genetics and genomics aspects in these model legume plants is discussed separately in Chapter 22. For abiotic stress analysis, an expression database for the roots of *Medicago truncatula* under salt stress has been also developed (Li *et al.*, 2009), which is intended to help in selecting gene markers to improve abiotic stress resistance in legumes. Owing to phylogenetic relationships

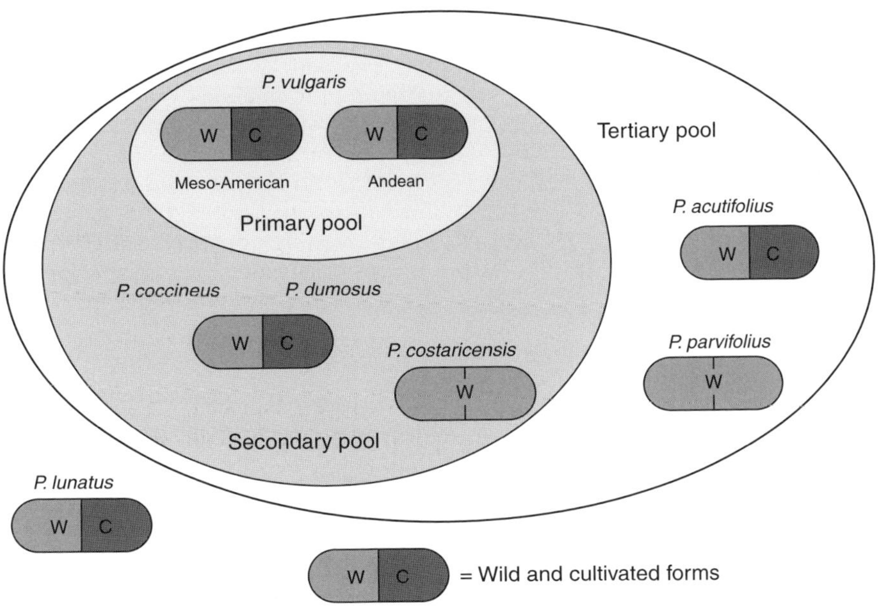

Fig. 18.1. Species involved in different gene pools of common bean (*Phaseolus*).

within the legume family, *Medicago* and *Lotus* genomics are now facilitating research in extending the application of genome knowledge for the improvement of related food and feed legumes (Young and Udvardi, 2009). Several studies have shown conserved synteny among the cool season legumes, particularly between *M. truncatula* and lucerne (Kaló *et al.*, 2004, Chandran *et al.*, 2008) and pea (Macas *et al.*, 2007), as well as between the major papilionoid clades (Gonzales *et al.*, 2005; Bertioli *et al.*, 2009). Thus the conservation of genome structure and function between legume species has facilitated cross-species transfer of genetic markers (Fredslund *et al.*, 2006; Gupta and Prasad, 2009) and microarray chips (Frickey *et al.*, 2008).

The accumulation of nucleotide sequences for agricultural species now allows us to perform genome-wide comparative analyses of model organisms, with the goal of discovering key genes involved in phenotypic characteristics (Sato and Tabata 2006; Neale and Ingvarsson, 2008; Itoh and Watanabe, 2009). This will also accelerate the molecular elucidation of cellular systems related to agronomically important traits. Several attempts have been made to compare the model plant genome with that of the cultivated crop. For example, soybean with a published complete draft genome sequence (Cannon *et al.*, 2009; Schmutz *et al.*, 2010) shares a common ancestor with other domesticated bean species, which will allow us to collate knowledge on traits observed and mapped in beans and their relatives.

The genome sequence is also an essential framework for vast new areas of experimental information such as tissue-specific expression and whole-genome association data (Kumar *et al.*, 2010; Rafalski, 2010). At present, the genome sequencing of *Phaseolus vulgaris* is in progress at the Joint Genome Institute of the US Department of Energy (www.jgi.doe.gov/sequencing/statusreporter/psr.php?projected=400630). An international consortium called Phaseomics (*Phaseolus* genomics) was formed to establish the necessary framework of knowledge and materials for the advancement of genomic studies on bean. The principal goal of this consortium is to increase the genetic resources and tools available for the crop, especially large inserts and cDNA libraries, genomic sequences, expressed sequence tags (ESTs) and genetic markers. An additional long-term goal is the more rational use of the large germplasm collections held for the crop at the CIAT (36,000 accessions) and at national germplasm repositories in the USA (USDA), Brazil (Brazilian Agricultural Research Corporation) and Mexico (National Institute for Forest and Agricultural Research; www.phaseolus.net/). The genome of *M. trunculata* also allows analysis of genes expressed in common bean (*Phaseolus vulgaris* L.) seedlings with the help of bioinformatics (Melotto *et al.*, 2005). These efforts will ultimately assist in the generation of new common bean varieties that are not only suitable for, but also desired by, local farmer and consumer communities.

Advances in sequencing and genotyping technologies

The Sanger method (Sanger *et al.*, 1977), based on the dideoxynucleotide sequencing of DNA, has undergone several transformations from a small-scale industry to a large-scale production enterprise and, in the process, the cost per reaction for DNA sequencing has fallen substantially. These advancements in sequencing have led to the development of sequence-based markers, including simple sequence repeats (SSRs) and single nucleotide polymorphisms (SNPs) (Ganal *et al.*, 2009; Park *et al.*, 2009). The reducing cost of DNA sequencing and increasing availability of large sequence data sets permit the mining of data for large numbers of SSRs and SNPs. These can then be used in genetic linkage analysis and trait mapping, diversity analysis, association studies and MAS (Duran *et al.*, 2009; Rafalski, 2010).

A detailed description of automated methods for the discovery of molecular markers and new technologies for high-throughput and low-cost molecular marker genotyping has been discussed by Appleby *et al.* (2009). Importantly, the combination of high-throughput genotyping with precise and focused phenotyping will facilitate efforts to associate molecular markers with agronomic traits (Tester and Langridge, 2010). The extensive reviews published illustrate in detail the possible application of recent molecular approaches, including MAS, gene pyramiding, marker-aided recurrent selection (MARS), genome-wide or genome selection and analysis of complex traits, through a combination of high-throughput phenotyping and genotyping (Ishitani *et al.*, 2004; Appleby *et al.*, 2009; Salekdeh *et al.*, 2009).

18.4 Gene Discoveries for the Improvement of Abiotic Stress Tolerance

Alhough lagging far behind the progress made in cereals, legume genomics is picking up pace mainly due to reduced sequencing costs (Raju *et al.*, 2010). This is evident from the genomic information generated so far for 12 genera of legumes, as illustrated in Table 18.1. In the future, this information will be highly useful for identifying the trait–gene association and gene function relevant to agronomic traits, including tolerance to abiotic stresses (Hirayama and Shinozaki, 2010). Furthermore, several attempts have been made to investigate the association of genes with key abiotic stresses (Rafalski, 2010).

Recently, efforts have been made to detect and validate single-feature polymorphisms in cowpea (*Vigna unguiculata* L. Walp) using a soybean genome array (Das *et al.*, 2008), to sequence and analyse the gene-rich space of cowpea (Timko *et al.*, 2008) and to identify ESTs from a wild *Arachis* species for gene discovery and marker development (Proite *et al.*, 2007). A comprehensive resource of drought- and salinity-responsive ESTs for gene discovery and marker development in chickpea (*Cicer arietinum* L.) has now been published (Varshney *et al.*, 2009c). Ramirez *et al.* (2005) have provided an initial platform for the functional genomics of the common bean by the identification of almost 8000 unique genes assembled from more than 20,000 ESTs sequenced from various plant organs. These sequences enrich the collection of ESTs in this important crop and provide new understanding of bean metabolism, development and adaptation to stress. Roughly 3400–4900 ESTs were sequenced from each of five cDNA libraries for different bean tissues, and these identified 2226 contigs (with two or more ESTs each), which were classified into 15 functional subgroups. Of these contigs, 36% represented sequences of unknown function or had no homology to previously identified proteins in the UniProt database (Schneider *et al.*, 2005); another 34% corresponded to genes involved in C and N metabolism. These subgroup percentages are similar to those noted for nodules of *M. truncatula* (Györgyey *et al.*, 2000) and *L. japonicus* (Colebatch *et al.*, 2004) and proteoid roots of white lupin, *Lupinus albus* (Uhde-Stone

Table 18.1. Summary of gene bank genomic records of Leguminaceae.[a]

Crop	EST (*n*)	Protein sequences (*n*)	Unigene sequences (*n*)
Arachis	87,170	1,171	11,909
Cicer	35,731	1,265	–
Glycine max	1,429,050	36,119	33,001
Lotus corniculatus	38,765	254	57
Lupinus	–	14	–
Medicago sativa	13,358	1,896	–
Medicago truncatulata	30,102	13,412	18,801
Phaseolus	21,079	169	–
Phaseolus vulgaris	1,07,890	3,155	6,686
Pisum sativum	21,507	3,799	–
Vicia	60,225	1,597	–
Vigna	18	20	–

EST, expressed sequence tag. [a] Accessed 19 May 2010 at the National Center for Biotechnology Information (NCBI, www.ncbi.nlm.nih.gov).

et al., 2003). The third most abundant common bean functional subgroup composed of contigs is involved in signal transduction (8.7%). Transcripts involved in signal transduction also comprised a large proportion of the ESTs noted in *M. truncatula* and *L. japonicus* (Colebatch *et al.*, 2004). Some 5.7% of bean contigs corresponded to genes implicated in biotic/abiotic stress.

The availability of extensive genomic data for cowpea represents a significant step forward in legume research (Timko *et al.*, 2008; Muchero *et al.*, 2009). Not only does the gene space sequence enable detailed analysis of gene structure, gene family organization and phylogenetic relationships within cowpea, but it also facilitates the characterization of syntenic relationships with other cultivated and model legumes, and will help in determining patterns of chromosomal evolution in the Leguminosae (Muchero *et al.*, 2009). The micro- and macro-syntenic relationships detected between cowpea and other cultivated and model legumes should simplify the identification of the informative markers for MAS and map-based gene isolation necessary for cowpea improvement (Muchero *et al.*, 2009).

Expressed sequence tags were produced from four cDNA libraries of RNA extracted from leaves and roots of *Arachis stenosperma* (a wild species). Randomly selected cDNA clones were sequenced to generate 8785 ESTs, of which 6264 (71.3%) had high quality, with 3500 clusters: 963 contigs and 2537 singlets; only 55.9% matched homologous sequences of known genes (Proite *et al.*, 2007). ESTs were classified into 23 different categories according to putative protein functions. Numerous sequences related to disease resistance, drought tolerance and human health were identified. Two hundred and six microsatellites were found, with markers having been developed for 188 of these. The microsatellite profile was analysed and compared to other transcribed and genomic sequence data. This is, to date, the first report on the analysis of transcriptomes of a wild relative of the groundnut. The ESTs produced in this study are a valuable resource for gene discovery, the characterization of new wild alleles and for marker development (Proite *et al.*, 2007). The various efforts made to discover genes associated with drought, salinity, phosphorus (P) deficiency and tolerance to aluminium (Al) toxicity in legumes are now discussed.

Tolerance to drought

Tolerance to soil moisture deficit during crop growth is a very complex mechanism operating at the cellular, organ and whole-plant levels. The physiological, biochemical and molecular basis of plant tolerance to drought has been reviewed by Shao *et al.* (2009). Since drought tolerance is highly influenced by edaphic and environmental factors, the conventional breeding approach with empirical selection focused mainly on grain yield under drought has led to limited progress. Other major constraints in breeding for drought tolerance in plants are lack of appropriate traits for phenotyping drought tolerance and the screening techniques used for evaluating genotypes on a large scale, particularly under field conditions (Raynolds and Tuberosa, 2008; Salekdeh *et al.*, 2009). Although it is known that drought-adaptive traits are complex and multigenic, understanding of their physiological and genetic basis is incomplete, making specific genetic targets rare. The progress made so far in improving productivity of crop plants under drought conditions and opportunities for adopting molecular breeding approaches have been reviewed in detail by Ashraf (2010).

In general, tolerance to drought comprises drought escape, drought avoidance and/or dehydration tolerance, which are ultimately measured by the reproductive success of the species (Araus *et al.* 2002; Salekdeh *et al.*, 2009). For grain crops, measurement is similar but determined by yield per unit area of land. A conceptual framework for phenotyping for drought, largely based on Passiora's equation, can be highly useful in developing phenotyping protocols to complement the genomic approach and conventional breeding through MAS (Salekdeh *et al.*, 2009). The genomic approach, however, is largely determined by the understanding of the mechanisms of drought tolerance at the molecular level (Seki *et al.*, 2007; Shinozaki and Yamaguchi-Shinozaki, 2007). Therefore, some reports in legumes have indicated the isolation of homologues genes reported in model plants (Kwak *et al.*, 2008; Chen *et al.*, 2009; Zhou *et al.*, 2010). Among the grain legumes, much of the progress in gene discovery has been achieved in soybean. The identified soybean candidate genes shown in Table 18.2 are usually tested for their ability to enhance drought tolerance in *Arabidopsis* before pursuing their engineering into soybean. Recently, an attempt was made to identify genes associated with drought tolerance in chickpea (*Cicer aerietinum*), with a focus on the difference between tolerant and intolerant genotypes (Jain and Chattopadhyay, 2010). Alhough not conclusive, the genes identified on the basis of differential expression patterns corroborate the results from other, similar studies on other plants.

Tolerance to phosphorus deficiency

Phosphorus is required for essential metabolic processes including energy transfer, signal transduction, biosynthesis of macromolecules, photosynthesis and respiration (Ramaekers *et al.*, 2010; White and Brown, 2010). Plants have evolved morphological, physiological and molecular adaptations to cope with P deficiency (Yang and Finnegan, 2010). Morphological responses involve the modification of root architecture, principally by decreasing primary root growth and increasing lateral root and root hair formation (Pérez-Torres *et al.*, 2008). The physiological and biochemical responses include: (i) modifications of carbon metabolism to bypass P requiring the steps, synthesis and secretion of acid phosphatases (AP); (ii) the exudation of organic acids; and (iii) the enhanced expression of high-affinity phosphate transporters (Liu *et al.*, 2008; Hurley *et al.*, 2010). The main focus of P stress research in legumes has been in white lupine (*Lupinus albus*) and common bean (*P. vulgaris*) and, to a lesser extent, in *M. truncatula* and soybean (*Glycine max*) (Vance, 2001; Graham and Vance, 2003; Tesfaye *et al.*, 2007; Yang and Finnegan, 2010).

Transcriptional profile in response to P deficiency

There are around 25,000 partially sequenced cDNA inserts or ESTs derived from P-starved tissues of four legume species (barrel medic, soybean, common bean and white lupine) currently available in the public domain (http://compbio.dfci.harvard.edu/tgi/). Microarray and macroarray analysis of P stress in plants shows increased transcript abundance of genes with homology to Pi transporters, organic acid synthesis, purple acid phosphatase, multi-drug and toxin efflux (MATE), transcription factors, signalling and defence (Misson *et al.*, 2005; Valdés-López and Hernández, 2008). From the *MYB* transcription factors *PvPHR1* and *PvmiR399*, Valdés-López *et al.* (2008) identified a microRNA essential for P deficiency signalling in common bean roots.

Two different approaches, namely macroarray and suppressive subtractive cDNA library analyses, have been used to decipher global gene expression in response to P deficiency in common bean (Hernández *et al.*, 2007; Tian *et al.*, 2007). Suppressive

Table 18.2. Genes associated with drought tolerance identified in soybean.

Gene	Transformed plant	Associated trait	Reference
GmDREB3	Arabidopsis	Drought tolerance	Chen *et al.* (2009)
bZIP TFs	*Arabidopsis*	Drought and high salt tolerance	Liao *et al.* (2008)
GMCHI	Tobacco	Drought	Cheng *et al.* (2009)
GmNAC		Drought	Tran *et al.* (2009)
GmUBC2	*Arabidopsis*	Ubiquitination, drought, ion homeostasis, osmolyte synthesis and oxidative stress responses	Zhou *et al.* (2010)

subtractive cDNA library analyses were investigated to identify those genes involved mainly in the modification of carbon metabolism and photosynthesis, to bypass the steps requiring P and P transport (Tian *et al.*, 2007). High-density macroarrays, printed with ESTs derived from a P-deficient bean root cDNA library (Ramírez *et al.*, 2005), identified 126 P-responsive genes representing different functional categories such as secondary metabolism, regulation/signal transduction; genes encoding proteins that participate in intracellular and extracellular transport were also identified (Valdés-López and Hernández, 2008). Both these analyses refer to induced genes that mediate P remobilization and stress/defence processes. In agreement, the non-biased metabolic profile of bean roots revealed that stress-related metabolites such as polyols and proline are accumulated in P-deficient treatment (Hernández *et al.*, 2007).

Transcription factors are key global regulators of gene expression and are known to play critical roles in many biological processes, including the regulation of plant responses to numerous biotic and abiotic stresses (Libault *et al.*, 2009). Phosphorus stress-responsive transcription factors belong to several families, including *MYB*, *SCARECROW*, *APETALA2* domain, homeobox, WRKY and zinc fingers. A recent bioinformatic analysis of legume gene indices (*Medicago*, *Glycine*, *Phaseolus* and *Lupinus*) queried for genes overrepresented in P-stressed tissue revealed the annotation of several putative transcription factor genes, including the *WRKY*, *MYB* and zinc finger gene families (Graham *et al.*, 2006). Furthermore, leaves and root tissues showed non-overlapping sets of transcription factor genes (Wu *et al.*, 2003); in addition, distinct sets of genes that may function as early- and late-responsive genes during P stress were observed (Misson *et al.*, 2005; Hernández *et al.*, 2009).

In the case of common bean, the TF transcript profile revealed 17 genes differentially expressed in P-starved roots (Valdés-López *et al.*, 2008). Of these, four TF genes were induced: three belonging to the *MYB* family and one to the *TIFY* (previously *ZIM*) family. The TC2883 *MYB* TF (DFCI/Common Bean Gene Index v.1.0), induced twofold in P-stressed bean roots, is 63% homologous to *AtPHR1* (*PvPHR1*). Through BLATX analysis in the DFCI/Common Bean Gene Index, bean gene orthologues of *Arabidopsis* genes from the PHR1 signalling pathway (*ure1*) such as *PvSIZ1* (SUMO E3 ligase; TC2445), *PvPHO2* (ubiquitin E2 conjugase; TC1095) and *Pv4* (CV536419) have been identified (Valdés-López *et al.*, 2008). The possible role of *PvPHR1* and other regulatory proteins in P starvation signalling remains to be demonstrated in bean. Hernández *et al.* (2007) reported *TIFY* family TF (TC1670) that is induced twofold in response to P starvation in bean roots.

Genes related to root morphology in response to P stress adaptation

White lupine is characterized by its extreme tolerance to P deficiency that is correlated with a highly coordinated modification of root physiology and biochemistry, resulting in the development of proteoid (cluster) roots (Yan *et al.*, 2002; Uhde-Stone *et al.*, 2003, 2005). Unlike typical lateral roots, proteoid roots develop laterals that emerge from every xylem pole within the axis, accompanied by extensive root hair growth, resulting in more than a 100-fold increase in surface area.

Modified root architecture in response to P stress has been well characterized in common bean. Bean responses to P deficiency include the modification of root growth axis and gravitropism, and the formation of shallower and adventitious roots, which facilitate exploration for P resources in the topsoil (Ho *et al.*, 2004; Rubio *et al.*, 2004). Some QTLs for root architecture showed a good correlation with P acquisition, something that strengthens the importance of root architecture in bean P deficiency adaptation (Yan *et al.*, 2004; Ochoa *et al.*, 2006; Cichya *et al.*, 2009). Several P deficiency adaptation responses are regulated at the transcriptional level by a highly coordinated gene expression programme (Franco-Zorrilla *et al.*, 2004). Phosphorus deficiency stress is known to stimulate ethylene production in many plant species (Tesfaye *et al.*, 2007), including bean and lupine (Gilbert *et al.*, 2000). Stimulation of ethylene production results in an increase in root hair density and length (López -Bucio *et al.*, 2003;

Zhang *et al.*, 2003), and it is noteworthy that bean, lupine and barrel medic plants exposed to P stress have increased density and length of root hairs. Some genes are suggested to be involved in root hair development (Zhang *et al.*, 2003). Graham *et al.* (2006) reported that a key gene in ethylene biosynthesis, 1-aminocyclopropane-1-carboxylicacid oxidase, is overrepresented in the ESTs derived from P-stressed roots of lupine, bean and *Medicago*. These results taken together indicate that ethylene production and/or plant responsiveness to ethylene plays a role in root adaptation to P deficiency. The role of cytokinins in root growth and P deficiency stress remains to be resolved. Traditionally, cytokinins are thought to be negative regulators of root growth while having positive effects on shoot growth (Aloni *et al.*, 2006).

Great variability exists for root traits in common bean (Ochoa *et al.*, 2006; Lynch, 2007), and matching the root system to the environment will be a particular research challenge for the future. Different root systems will be required for different soil environments: shallow roots to maximize P acquisition in a P-poor soil (Ho *et al.*, 2004, Nord and Lynch, 2009); deep roots for water acquisition under drought; greater numbers of root tips for calcium absorption; and exudation of organic acids as a defence mechanism against aluminum toxicity or as a mechanism of P acquisition (Nord and Lynch, 2009).

Tolerance to salinity

More than 800 million ha of land, accounting for more than 6% of the world's total land area, are salt affected (FAO, 2008; Munns and Tester, 2008). Extensive reviews concerning salt tolerance research exist for other crops while much of the research on salt tolerance in legumes is focused on soybean. The mechanisms of salt tolerance have been extensively reviewed (Munns and Tester, 2008). Studies have revealed three salt tolerance mechanisms in soybean: (i) maintenance of ion homeostasis; (ii) restoration of oxidative balance; and (iii) other metabolic and structural adaptations. The differential salt tolerance capability exhibited by different germplasms is determined by their efficiency in operating and coordinating these systems.

Maintenance of ion homeostasis and the possible roles of ion transporters

In contrast to reports of salt-induced mortality due to excess Na^+ (Gao *et al.*, 2007), damage due to excess chlorine (Cl) has been often highlighted in soybean (Abel and MacKenzie, 1964; Abel, 1969), yet whether Na^+ or Cl^- plays the more critical role in NaCl-induced mortality in soybean remains unknown. On the other hand, reports suggest that salt-tolerant soybean germplasm accumulates less Na^+ in leaves (An *et al.*, 2002; Essa, 2002; Luo *et al.*, 2005a, b; Li *et al.*, 2006). These studies clearly indicate the significance of ion homeostasis in salt tolerance in legumes such as soybean. Structural changes in plant tissues such as xylem parenchymal cells often accompany changes in processes related to ion homoeostasis in response to salt stress (Munns and Tester, 2008). In addition, several genes associated with ion transporters and their possible role in salt tolerance mechanisms have been reported (Table 18.3).

Previous reports have highlighted the key role of ATPase in Na^+–K^+ exchange in soybean roots (Durand and Lacan, 1994; Lacan and Durand, 1995, 1996; Yu *et al.*, 2005). These findings indicate that salt-tolerant soybean varieties have a stronger energetic system for the operation of transporters in tonoplasts than salt-sensitive varieties. Several attempts have been made to identify the genes associated with ion homoestasis in salt-affected soybean, including H^+-PPiase (*GmVP1*) and the subunit C of vacuolarly located H^+-ATPase (*GmVHA-C*) (Apse *et al.*, 1999; Apse and Blumwald, 2007), *GmNHX1* and *GmNHX2* (Li *et al.*, 2006; Sun *et al.*, 2006). In *Arabidopsis thaliana*, the Na^+/H^+ antiporter, SOS1, located in the plasma membrane, limits the loading of Na^+ into the xylem sap and enhances Na^+ efflux at the root tip (Shi *et al.*, 2002). The gene encoding a soybean *SOS1* homologue (*GmSOS1*) was found to accumulate in roots upon NaCl challenge (Phang *et al.*, 2008); however, its

Table 18.3. Ion transporter genes related to salt tolerance in soybean.

Gene	Putative gene product	Reference
GmAKT1	Inward-rectifying K^+ channel	Tsai (2003)
GmCLC1	Vacuolar CLC chloride channel	Li *et al.* (2006)
GmNHX1	Vacuolar Na^+/H^+ antiporter	Li *et al.* (2006); Sun *et al.* (2006)
GmSOS1	Na^+/H^+ antiporter	Phang *et al.* (2008)
GmCNGC	Cyclic nucleotide-gated cation channel	Phang *et al.* (2008)
GmGLR3	Glutamate receptor	Phang *et al.* (2008)
GmNKCC	$Na^+/K^+/Cl^-$ co-transporter	Phang *et al.* (2008)
GmCAX1	Cation/proton exchanger	Luo *et al.* (2005a)
GmPHD	Alfin1-type PHD finger protein	Wei *et al.* (2009)

function in salt tolerance remains to be elucidated. Another plasma membrane-localized cation/H^+ antiporter in soybean (*GmCAX1*) was studied in detail (Luo *et al.*, 2005a). The evidence from these studies suggests that soybean possesses the components of Na^+ transportation systems similar to those of other higher plants, but more studies are required to delineate the function of each component in detail.

Taking a lead from discoveries in other crops, attempts have been made to characterize a chickpea (*Cicer arietinum* L.) NAC family gene, *CarNAC5*, which is both developmentally and stress regulated (Peng *et al.*, 2009). A successful attempt at cloning the *PvP5CS* gene, associated with proline synthesis, from common bean (*P. vulgaris*) may be useful in the future in dealing with abiotic stresses, including salt tolerance, in this important legume crop (Chen *et al.*, 2008).

Accumulation of osmoprotectants

It is evident from several studies that when soybean is exposed to salt stress, it accumulates major osmoprotectants such as glycinebetaine (Agboma *et al.*, 1997), trigonelline (TRG) (Cho *et al.*, 1999; Malencic *et al.*, 2003; Chen and Wood, 2004), pinitol (Dittrich and Brandl, 1987; Streeter *et al.*, 2001) and proline (Moftah and Michel, 1987; Krackhard and Guerrier, 1995; Guo and Weng, 2004). QTL analysis also revealed that TRG accumulation is a polygenic trait that is subject to environmental factors (Cho *et al.*, 2002). Two unique microsatellite markers (SSR) on LG-J (Satt285) and LG-C2 (Satt079) were found to be closely associated with foliar TRG accumulation; these two loci explained 28.1% of the total variation for the accumulation of TRG in foliar tissues. Genetic variation in the accumulation of pinitol, a major component of all plant components of soybean, has been reported (Dittrich and Brandl, 1987; Streeter *et al.*, 2001). Other reports have demonstrated the enhancement of pinitol accumulation in legumes under water stress (Ford, 1984) and in soybean under cold stress (Guo and Oosterhuis, 1995).

Recent genomic studies have revealed that the overexpression of two TFIIIA-related TF genes (*MtZPT2-1* and *MtZPT2-2*) in *M. truncatula* led to an increase in the size of the plant root system under salt stress (de Lorenzo *et al.*, 2007). *GmWRKY54* and *GmDREB2* (Chen *et al.*, 2007; Zhou *et al.*, 2008) and *GmUBC2* (Zhou *et al.*, 2010) increased tolerance to both salt and drought when overexpressed in *Arabidopsis*.

18.5 QTL and Molecular Marker Development

A number of minor genes (polygenes) with additive effects in their expression control tolerance of crop plants to many abiotic stresses (Mohammadi *et al.*, 2005; Thi Lang and Chi Buu, 2008). The loci on chromosomes housing such genes are referred to as quantitative trait loci (QTL). QTL mapping allows assessment of the locations, numbers and magnitude of phenotypic effects and the pattern of gene action (Kumar *et al.*, 2010). The role of polygenes in controlling a trait has been widely assessed by traditional means, but the use of DNA markers and QTL mapping has enabled the unravelling of complex traits (Humphreys and Humphreys, 2005). In the following section, molecular markers identified for selected abiotic stresses in legumes have been highlighted.

Drought

Das *et al.* (2008) reported the detection and validation of single-feature polymorphisms (SFPs) in cowpea using a readily available soybean (*G. max*) genome array. Robustified projection pursuit (RPP) was used for statistical analysis, with RNA as a surrogate for DNA. Using a 15% outlying score cut-off, 1058 potential SFPs were enumerated between two parents of a recombinant inbred line (RIL) population segregating for several important traits, including drought tolerance. It was concluded that the Affymetrix soybean genome array is a satisfactory platform for identification of some thousands of SFPs for cowpea (Das *et al.*, 2008). This study provides an example of the extension of genomic resources from a well-supported species to an orphan crop. Presumably, other legume systems will be similarly tractable to SFP marker development using existing legume array resources.

Phosphorus deficiency

Genetic variability with contrasting degrees of root architecture responses to P-limiting conditions is known for a wide range of plant species (Rubio *et al.*, 2001; Chevalier *et al.*, 2003; Yan *et al.*, 2004); Lynch (2007) noted a direct correlation between plant productivity and root architecture. Therefore, P stress tolerance and adaptation have been analysed in common through the QTL analysis approach (Beebe *et al.*, 2006; Ochoa *et al.*, 2006). Ochoa *et al.* (2006) identified 19 QTL for adventitious root formation under P-stress and P-sufficient conditions. Because Pi availability is expected to be greater in topsoil compared with subsoil, selection for root trait QTL markers associated with adventitious rooting and topsoil foraging may enhance P acquisition. Furthermore, a mapping population between 'G19833' (a landrace of the Andean gene pool with superior growth and yield under P-stress conditions) and 'DOR 364' (a genotype of the Meso-American gene pool with low P accumulation efficiency in P-limiting conditions) has been developed for QTL analysis. Using this bi-parental mapping population, 26 more QTLs associated with basal root development and greater P acquisition efficiency under P-stress conditions have recently been identified though the composite interval mapping approach (Beebe *et al.*, 2006). The selection and development of P-efficient legume plants using QTL markers would be beneficial to low-input agriculture and enhance environmentally friendly cropping in intensively cultivated systems. Tesfaye *et al.* (2007) discuss in more detail the genomics of phosphate stress in legumes.

Salt tolerance

Salt tolerance is an inheritable quantitative trait, but the phenotypes may be dominated by a few major loci. For example, a cross was made between soybean genotypes having contrasting capabilities of Cl^- accumulation in the aerial part (Abel and MacKenzie, 1964; Abel, 1969; Pantalone *et al.*, 1997). The inheritance pattern studied in F_2 and BC_1 generations suggests that Cl^- accumulation in the aerial part of soybean is controlled by a single locus with exclusion (*Ncl*) being dominant and a locus with inclusion (*ncl*) being recessive (Abel, 1969). Since the occurrence of salt-induced necrosis was associated with high Cl^- content in the aerial part, it was implied that *Ncl* is a major locus determining the salt tolerance capability of soybean. Another study showed that the mode of inheritance of salt-induced necrosis on leaves (salt sensitivity) is controlled by a major locus, with a non-necrotic (salt-tolerant) allele showing dominance over a necrotic (salt-sensitive) allele. Salt tolerance in soybean has been recently reviewed by Phang *et al.* (2008).

QTL analysis was carried out using a F_2 population derived from a cross between salt-tolerant 'S-100' and salt-sensitive 'Tokyo' (Lee *et al.*, 2004). The scoring for the salt tolerance phenotype on the basis of degree of salt-induced chlorosis in leaves, a major QTL for salt tolerance, was identified on the linkage group LGN (between SSR markers Sat091 and Satt237) and explained 41–79% of phenotypic variation for salt tolerance under field, greenhouse and combined environmental conditions (Lee *et al.*, 2004). Tracking of this QTL in the descendent US cultivars revealed that 'S-100' is the source of the salt tolerance allele in the cultivar 'Lee' (the Cl^- excluder in Abel's studies described above). It is believed that this QTL is the *Ncl* locus (Lee *et al.*, 2004). However, another report suggests that salt tolerance in soybean is controlled by minor genes (Luo *et al.*, 2005b). The discrepancy among different studies is probably due to differences in genetic background of parental germplasms and the variation in scoring parameters. Hence, it is essential to develop efficient phenotyping protocols for the successful detection of efficient and robust molecular markers.

Aluminium toxicity

Lee and Foy (1986) found that Al stress reduced root extract organic acid concentration to a greater extent in Al-sensitive than in Al-resistant genotypes, while Miyasaka *et al.* (1991) demonstrated that other Al-resistant genotypes of common bean exuded eight times more citrate than susceptible genotypes during prolonged exposure. Rangel *et al.* (2005) reported that genotype 'G19833' had a higher level of Al resistance compared with 'DOR364' in common bean. Inheritance of Al resistance is likely to be genetically simpler than that of acid soil tolerance (Kochian *et al.*, 2004). While root traits in the presence of Al are controlled by many genes in common bean, QTL for root morphological characteristics identified under the stresses of Al and low P were found in five regions of the common bean genome (López-Marín *et al.*, 2009). These studies therefore indicate that a common mechanism is probably responsible for root response under Al conditions or low P stresses in both common bean and soybean (Liao *et al.*, 2006).

Rangel *et al.* (2007) predicted an Al exclusion mechanism in common bean, based on its accumulation in the root tips of an Al-resistant genotype ICA Quimbaya decreasing over the duration of Al exposure, while in the sensitive genotype 'VAX 1', accumulation increased. Furthermore, studies by these authors have shown that this exclusion mechanism is based on differential citrate, malate, oxalate or succinate exudation (Rangel *et al.*, 2009), findings that classify common bean as a type II response according to Ma *et al.* (2001) and indicate that gene expression response occurs on exposure to Al. In conclusion, the inheritance of Al resistance in common bean is quantitative and related to the differential response of root architecture traits to exposure to Al, and some common QTL exist for resistance to Al and tolerance to low P stress.

Given the importance of roots in confronting abiotic stress, one would expect that root vigour should favour better yield, which may not necessarily be true. Breeders may instead need to breed for more efficient roots that use the same biomass to best advantage, for example, through longer root hairs, thinner roots and greater specific root length, greater organic acid exudation, etc. (Liao *et al.*, 2004; Yan *et al.*, 2004; Beebe *et al.*, 2006; Nord and Lynch, 2009) while maintaining or improving partitioning to grain. An efficient phenotyping procedure for roots will greatly facilitate our efforts to understand the mechanisms of tolerance to several abiotic stresses and will open new avenues for crop improvement.

18.6 The Transgenic Approach

The transgenic approach has been shown to be a powerful tool in aiding the understanding and manipulation of the responses of plants to stresses, particularly when accompanied by phenotyping procedures that include precise physiological and biochemical investigation of transgenic plants under stress conditions (Halpin, 2005). However, in general, there are few success stories in regard to genetic transformation in legumes compared with those in cereals such as rice. The biolistics transformation method and *Agrobacterium*-mediated transformation exist for model legumes, soybean and cowpea. Details on the development of transformation protocols for the production of transgenic plants are reviewed by Popelka *et al.* (2004). Routine transformation protocols are limited in most grain legumes, this low success having been attributed to poor regeneration ability (especially via the callus) and a lack of compatible gene delivery methods, although some success has been achieved in soybean. Worldwide, soybean is the only transgenic grain legume under cultivation in nearly 63% of the total area under transgenics. The development of *Agrobacterium tumefaciens* as a vector for legume transformation was an important breakthrough. Both micro-particle bombardment (Gulati *et al.*, 2002; Li *et al.*, 2004) and *A. tumefaciens* (De Clercq *et al.*, 2002; Li *et al.*, 2004) have been used for DNA delivery to either embryogenic or organogenic cultures.

Although transgenic plants are yet to be examined for salt tolerance in the field, recent genetic advances suggest that there are good prospects for the development of transgenic legumes with enhanced salt tolerance in the near future (Sharma and Lavanya, 2002). The increase in tolerance to aluminium and cyanamide toxicity in transgenic lucerne (Morphew *et al.*, 2004) and soybean (Zhang *et al.*, 2005) demonstrates the potential of this approach in legumes. Morphew *et al.* (2004) reported the overexpression of malate dehydrogenase that conferred aluminium tolerance in transgenic lucerne plants, while Zhang *et al.* (2005) described the development of cyanamide-tolerant transgenic soybean by overexpression of the fungal cyanamide hydratase (*Cah*) gene. Overexpression of the transcription factor *Alfin 1* improved salt tolerance in *Medicago sativa* (Winicov, 2000), while transgenic *M. truncatula* plants that had accumulated proline (Verdoy *et al.*, 2006), due to overexpression of pyrroline-5-carboxylate synthase (P5CS), displayed enhanced osmotolerance. Xue *et al.* (2007) reported that transgenic soybean plants overexpressing *NTR1* (encoding a jasmonic acid carboxyl methyl transferase that produces methyl jasmonate) were more tolerant to water deficit stress than their wild-type counterparts. The genetic transformation of *Vigna mungo* for value addition of an agronomic trait, wherein the gene of interest (glyoxalase I) is driven by a novel constitutive Cestrum yellow leaf-curling viral promoter, has been transferred for alleviating salt stress (Bhomkar *et al.*, 2008).

18.7 Future Perspectives

Bearing in mind the strong influence of genotype–environmental interactions in food legumes, the future strategy for improvement in legumes should be based on the scientific data amassed on abiotic stress tolerance mechanisms that exist for these crops at the cost of biomass spent on maintenance of structural and functional attributes of plants under harsh environments. This occurs in terms of biomass partitioning to roots and pods or internal carbon and nitrogen resources to metabolites for osmotic substances (for drought and salt) and organic acids (for Al and P). Hence, the research challenge for the future is to explore genomic advances, with the aim of striking a balance between internal spending of carbon resources and the benefits gained from such investments in terms of fraction of biomass that can be efficiently translocated to grains.

Briefly, mechanisms at the molecular level that can stimulate investment in osmoprotectants and organic acids only when necessary will reduce the metabolic cost. On the other hand, biomass investment to produce shallow but finer roots in stress environments will be of immense importance for improving legume productivity. This can be achieved through a systematic and precise approach to plant phenotyping through an understanding of plants' architectural and metabolic strategy in coping with abiotic stresses under natural environments, with a particular focus on uncultivated species or wild relatives. This may give rise to gene-based markers for improving specific traits that can contribute to abiotic stress tolerance. Figure 18.2 presents a future strategy scheme for exploring the use of genomics in achieving these objectives, and the associated opportunities.

Recent advances in plant genomics have tremendous potential to explore genetic variation and to manipulate the genetic potential of crop plants, and to make them more productive even under harsh environments. For many of the legumes, lack of efficient transformation protocol is the major constraint to validation of scientific leads originating from bioinformatics and molecular studies in other crops. This gap needs to be filled through enhanced efforts and increased investment in research. Research on phenomics must be accelerated to keep pace with scientific leads in terms of gene discovery that are emerging from molecular laboratories across the world. A concerted effort is required toward a better understanding of gene functions through global networks integrating phenomics laboratories and multi-location evaluation sites focusing on target population environments for legume crops.

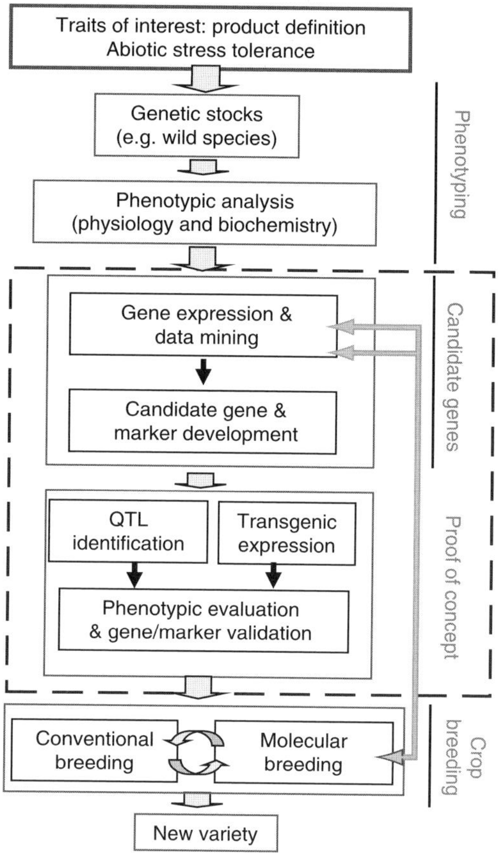

Fig. 18.2. A strategy for achieving objectives and the opportunities for developing improved genotypes through the application of genomics.

References

Abel, G.H. (1969) Inheritance of the capacity for chloride inclusion and chloride exclusion by soybeans. *Crop Science* 9, 697–698.

Abel, G.H. and MacKenzie, A.J. (1964) Salt tolerance of soybean varieties (*Glycine max* L. Merrill) during germination and later growth. *Crop Science* 4, 157–161.

Agboma, P.C., Sinclair, T.R., Jokinen, K., Peltonen–Sainio, P. and Pehu, E. (1997) An evaluation of the effect of exogenous glycinebetaine on the growth and yield of soybean: timing of application, watering regimes and cultivars. *Field Crops Research* 54, 51–64.

Aloni, R., Aloni, E., Langhans, M. and Ullrich, C.I. (2006) Role of cytokinin and auxin in shaping root architecture: regulating vascular differentiation, lateral root initiation, root apical dominance and root ravitropism. *Annals of Botany* 97, 883–893.

Alonso-Blanco, C., Aarts, M.G., Bentsink, L., Keurentjes, J.J., Reymond, M., Vreugdenhil, D. *et al.* (2009): What has natural variation taught us about plant development, physiology, and adaptation? *Plant Cell* 21, 1877–1896.

An, P., Inanaga, S., Cohen, Y., Kafkafi, U. and Sugimoto, Y. (2002) Salt tolerance in two soybean cultivars. *Journal of Plant Nutrition* 25, 407–423.

Appleby, N., Edwards, D. and Batley, J. (2009) New technologies for ultra-high throughput genotyping in plants. *Methods in Molecular Biology* 513, 19–39.

Apse, M.P. and Blumwald, E. (2007) Na^{+} transport in plants. *FEBS Letters* 581, 2247–2254.

Apse, M.P., Aharon, G.S., Snedden, W.A. and Blumwald, E. (1999) Salt tolerance conferred by over expression of a vacuolar Na^{+}/H^{+} antiporter in *Arabidopsis. Science* 285, 1256–1258.

Araus, J.L., Slafer, G.A., Reynolds, M.P. and Royo, C. (2002) Plant breeding and drought in C3 cereals: what should we breed for? *Annals of Botany* 89, 925–940.

Ashraf, M. (2010) Inducing drought tolerance in plants: recent advances. *Biotechnology Advances* 28, 169–183.

Beebe, S.E., Rojas, M., Yan, X., Blair, M.W., Pedraza, F., Muñoz, F. *et al.* (2006) Quantitative trait loci for root architecture traits correlated with phosphorus acquisition in common bean. *Crop Science* 46, 413–423.

Bertioli, D.J., Moretzsohn, M.C., Madsen, L.H., Sandal, N., Leal-Bertioli, S.C., Guimarães, P.M. *et al.* (2009) An analysis of synteny of *Arachis* with *Lotus* and *Medicago* sheds new light on the structure, stability and evolution of legume genomes. *BMC Genomics* 10, 45.

Bhomkar, P., Upadhyay, C.P., Saxena, M., Muthusamy, A., Prakash, N.S., Pooggin, M. *et al.* (2008) Salt stress alleviation in transgenic *Vigna mungo* L. Hepper (blackgram) by overexpression of the glyoxalase I gene using a novel Cestrum yellow leaf curling virus (CmYLCV) promoter. *Molecular Breeding* 22, 169–181.

Blair, M.W. and Ishitani, M. (2009) Common bean: a model food legume for international agriculture. *Grain Legumes* 53, 8–9.

Blair, M.W., Iriarte, G. and Beebe, S. (2006) QTL analysis of yield traits in an advanced backcross population derived from a cultivated Andean x wild common bean (*Phaseolus vulgaris* L.) cross. *Theoretical and Applied Genetics* 112, 1149–1163.

Blair, M.W., Diaz, J.M., Hidalgo, R., Diaz, L.M. and Duque, M.C. (2007) Microsatellite characterization of Andean races of common bean (*Phaseolus vulgaris* L.). *Theoretical and Applied Genetics* 116, 29–43.

Blair, M.W., Torres, M.M., Giraldo, M.C. and Pedraza, F. (2009) Development and diversity of Andean-derived, gene-based microsatellites for common bean (*Phaseolus vulgaris* L.). *BMC Plant Biology* 9, 100.

Cannon, S.B., May, G.D. and Jackson, S.A. (2009) Three sequenced legume genomes and many crop species: rich opportunities for translational genomics. *Plant Physiology* 151, 970–977.

Capra, J.A., Carbone, L., Riesenfeld, S.J. and Wall, J.D. (2010) Genomics through the lens of next-generation sequencing. *Genome Biology* 11, 306.

Chacón, S. M.I., Pickersgill, B. and Debouck, D.G. (2005) Domestication patterns in common bean (*Phaseolus vulgaris* L.) and the origin of the Mesoamerican and Andean cultivated races. *Theoretical and Applied Genetics* 110, 432–444.

Chandran, D., Sharopova, N., VandenBosch, K.A., Garvin, D.F. and Samac, D.A. (2008) Physiological and molecular characterization of aluminum resistance *Medicago truncatula. BMC Plant Biology* 8, 89.

Chen, J.-B., Wang, S.-M., Jing, R.-L. and Mao, X.-G. (2008) Cloning the *PvP5CS* gene from common bean (*Phaseolus vulgaris*) and its expression patterns under abiotic stresses. *Journal of Plant Physiology* 166, 12–19.

Chen, M., Wang, Q.Y., Cheng, X.G., Xu, Z.S., Li, L.C., Ye, X.G. *et al.* (2007) *GmDREB2*, a soybean DRE-binding transcription factor, conferred drought and high-salt tolerance in transgenic plants. *Biochemical and Biophysical Research Communications* 353, 299–305.

Chen, M., Xu, Z., Xia, L., Li, L., Cheng, X., Dong, J. *et al.* (2009) Cold-induced modulation and functional analyses of the DRE-binding transcription factor gene, *GmDREB3*, in soybean (*Glycine max* L.). *Journal of Experimental Botany* 60, 121–135.

Chen, X. and Wood, A.J. (2004). Purification and characterization of S-adenosyl-L-methionine nicotinic acid-N-methyltransferase from leaves of Glycine max. *Biologia Plantarum* 48, 531–535.

Cheng, L., Huan, S., Sheng, Y., Hua, X., Shu, Q., Song, S. *et al.* (2009) *GMCHI*, cloned from soybean [*Glycine max* (L.) Meer.] enhances survival in transgenic *Arabidopsis* under abiotic stress. *Plant Cell Report* 28, 145–153.

Chevalier, F., Pata, M., Nacry, P., Douma, P. and Rossignol, M. (2003) Effects of phosphate availability on the root system architecture: large-scale analysis of the natural variation between *Arabidopsis* accessions. *Plant, Cell and Environment* 26, 1839–1850.

Cho, Y., Lightfoot, D.A. and Wood, A.J. (1999) Trigonelline concentrations in salt stressed leaves of cultivated Glycine max. *Phytochemistry* 52, 1235–1238.

Cho, Y., Njiti, V.N., Chen, X., Triwatayakorn, K., Kassem, M.A., Meksem, K. *et al.* (2002) Quantitative trait loci associated with foliar trigonelline accumulation in *Glycine Max* L. *Journal of Biomedicine and Biotechnology* 2, 151–157.

Cichya, K.A., Blair, M.W., Galeano Mendoza, C.H., Snapp, S.S. and Kelly, J.D. (2009) QTL analysis of root architecture traits and low phosphorus tolerance in an Andean bean population. *Crop Science* 49, 59–68.

Colebatch, G., Desbrosses, G., Ott, T., Krusell, L., Montanari, O., Kloska, S. *et al.* (2004) Global changes in transcription orchestrate metabolic differentiation during symbiotic nitrogen fixation in *Lotus japonicus*. *Plant Journal* 39, 487–512.

Cook, D.R. (1999) *Medicago truncatula*—a model in the making! *Current Opinion in Plant Biology* 2, 301–304.

Das, S., Bhat, P.R., Sudhakar, C., Ehlers, J.D., Wanamaker, S., Roberts, P.A. *et al.* (2008) Detection and validation of single feature polymorphisms in cowpea (*Vigna unguiculata* L. Walp) using a soybean genome array. *BMC Genomics* 9, 107.

De Clercq, J., Zambre, M., VanMontagu, M., Dillen, W. and Angenon, G. (2002) An optimized *Agrobacterium*-mediated transformation procedure for *Phaseolus acutifolius* A. Gray. *Plant Cell Reports* 21, 333–340.

De Lorenzo, L., Merchan, F., Blanchet, S., Megias, M., Frugier, F., Crespi, M. *et al.* (2007) Differential expression of the TFIIIA regulatory pathway in response to salt stress between *Medicago truncatula* genotypes. *Plant Physiology* 145, 1521–1532.

Dittrich, P. and Brandl, A. (1987) A revision of the pathway of D-pinitol formation in Leguminosae. *Phytochemistry* 26, 1925–1926.

Duran, C., Edwards, D. and Batley, J. (2009) Genetic maps and the use of synteny. *Methods in Molecular Biology* 513, 41–55.

Durand, M. and Lacan, D. (1994) Sodium partitioning within the shoot of soybean. *Physiologia Plantarum* 91, 65–71.

Edwards, D. and Batley, J. (2010) Plant genome sequencing: applications for crop improvement. *Plant Biotechnology Journal* 8, 2–9.

Essa, T.A. (2002) Effect of salinity stress on growth and nutrient composition of three soybean (*Glycine max* L. Merrill) cultivars. *Journal of Agronomy and Crop Science* 188, 86–93.

FAO (2008) FAO Land and Plant Nutrition Management Service. http://www.fao.org/

Ford, C.W. (1984) Accumulation of low molecular weight solutes in water-stressed tropical legumes. *Phytochemistry* 23, 1007–1015.

Franco-Zorrilla, J.M., González, E., Bustos, R., Linhares, F., Leyva, A. and Paz–Ares, J. (2004) The transcriptional control of plant responses to phosphate limitation. *Journal of Experimental Botany* 55, 285–293.

Fredslund, J., Madsen, L.H., Hougaard, B.K., Nielsen, A.M., Bertioli, D., Sandal, N. *et al.* (2006) A general pipeline for the development of anchor markers for comparative genomics in plants. *BMC Genomics* 7, 207.

Frickey, T., Benedito, V.A., Udvardi, M. and Weiller, G. (2008) AffyTrees: facilitating comparative analysis of Affymetrix plant microarray chips. *Plant Physiology* 146, 377–386.

Ganal, M.W., Altmann, T. and Röder, M.S. (2009) SNP identification in crop plants. *Current Opinion in Plant Biology* 12, 211–217.

Gao, J.P., Chao, D.Y. and Lin, H.X. (2007) Understanding abiotic stress tolerance mechanisms: recent studies on stress response in rice. *Journal of Integrative Plant Biology* 49, 742–750.

Gilbert, G.A., Knight, J.D., Vance, C.P. and Allan, D.L. (2000) Proteoid root development of phosphorus deficient lupin is mimicked by auxin and phosphonate. *Annals of Botany* 85, 921–928.

Glaszmann, J.C., Kilian, B., Upadhyaya, H.D. and Varshney, R.K. (2010) Accessing genetic diversity for crop improvement. *Current Opinion in Plant Biology* 13, 167–173.

Gonzales, M.D., Archuleta, E., Farmer, A., Gajendran, K., Grant, D., Shoemaker, R. *et al.* (2005) The Legume Information System (LIS): an integrated information resource for comparative legume biology. *Nucleic Acids Research* 33, D660–D665.

Graham, M.A., Ramirez, M., Valdes-López, O., Lara, M., Tesfaye, M., Vance, C.P. *et al.* (2006) Identification of candidate phosphorus stress induced genes in *Phaseolus vulgaris* through clustering analysis across several plant species. *Functional Plant Biology* 33, 789–797.

Graham, P.H. and Vance, C.P. (2003) Legumes: importance and constraints to greater use. *Plant Physiology* 131, 872–877.

Gulati, A., Schryer, P. and McHughen, A. (2002) Production of fertile transgenic lentil (*Lens culinaris* Medik) plants using particle bombardment. *In Vitro Cellular and Developmental Biology – Plant* 38, 316–324.

Guo, B. and Weng, Y. (2004) Salt tolerance mechanism and molecular markers of genes associated with salt tolerance in soybean. *Chinese Bulletin of Botany* 21, 113–120.

Guo, C. and Oosterhuis, D.M. (1995) Pinitol occurrence in soybean plants as affected by temperature and plant growth regulators. *Journal of Experimental Botany* 46, 249–253.

Gupta, S. and Prasad, M. (2009) Development and characterization of genic SSR markers in *Medicago truncatula* and their transferability in leguminous and non-leguminous species. *Genome* 52, 761–771.

Györgyey, J., Vaubert, D., Jiménez-Zurdo, J.I., Charon, C., Troussard, L., Kondorosi, A. *et al.* (2000) Analysis of *Medicago truncatula* nodule expressed sequence tags. *Molecular Plant–Microbe Interactions* 13, 62–71.

Halpin, C. (2005) Gene stacking in transgenic plants – the challenge for 21st century plant biotechnology. *Plant Biotechnology Journal* 3, 141–155.

Hernández, G., Ramírez, M., Valdés-López, O., Tesfaye, M., Graham, M.A., Czechowski, T. *et al.* (2007) Phosphorus stress in common bean: root transcript and metabolic responses. *Plant Physiology* 144, 752–767.

Hernández, G., Valdés-López, O., Ramírez, M., Goffard, N., Weiller, G., Aparicio-Fabre, R. *et al.* (2009) Global changes in the transcript and metabolic profiles during symbiotic nitrogen fixation in phosphorus-stressed common bean plants. *Plant Physiology* 151, 1221–1238.

Hirayama, T. and Shinozaki, K. (2010) Research on plant abiotic stress responses in the post-genome era: past, present and future. *Plant Journal* 61, 1041–1052.

Ho, M.D., McCannon, B.C. and Lynch, J.P. (2004) Optimization modeling of plant root architecture for water and phosphorus acquisition. *Journal of Theoretical Biology* 226, 331–340.

Humphreys, M.O. and Humphreys, M.W. (2005) Breeding for stress resistance: general principles. In: Ashraf, M. and Harris, P.J.C. (eds) *Abiotic Stresses: Plant Resistance through Breeding and Molecular Approaches*. Food Products Press, an imprint of The Haworth Press, Inc., Binghamton, NY, pp. 19–46.

Hurley, B.A., Tran, H.T., Marty, N.J., Park, J., Snedden, W.A., Mullen, R.T. *et al.* (2010) The dual-targeted purple acid phosphatase isozyme AtPAP26 is essential for efficient acclimation of *Arabidopsis* to nutritional phosphate deprivation. *Plant Physiology* 153, 1112–1122.

Ishitani, M., Rao, I., Wenzl, P., Beebe, S. and Tohme, T. (2004) Integration of genomics approach with traditional breeding towards improving abiotic stress adaptation: drought and aluminum toxicity as case studies. *Field Crops Research* 90, 35–45.

Islam, F.M., Beebe, S., Muñoz, M., Tohme, J., Redden, R.J. and Basford, K.E. (2004) Using molecular markers to assess the effect of introgression on quantitative attributes of common bean in the Andean gene pool. *Theoretical and Applied Genetics* 108, 243–252.

Itoh, M. and Watanabe, H. (2009) CGAS: comparative genomic analysis server. *Bioinformatics*. 25, 958–959.

Jain, D. and Chattopadhyay, D. (2010) Analysis of gene expression in response to water deficit of chickpea (*Cicer arietinum* L.) varieties differing in drought tolerance. *BMC Plant Biology* 10, 24.

Kaló, P., Seres, A., Taylor, S.A., Jakab, J., Kevei, Z., Kereszt, A. *et al.* (2004) Comparative mapping between *Medicago sativa* and *Pisum sativum*. *Molecular Genetics and Genomics* 272, 235–246.

Kochian, L.V., Hoekenga, O.A. and Piñeros, M.A. (2004) How do crop plants tolerate acid soils? Mechanisms of aluminum tolerance and phosphorus deficiency. *Annual Review of Plant Biology* 55, 459–493.

Krackhard, M. and Guerrier, G. (1995) Effect of osmotic and ionic stresses on proline and organic acid contents during inhibition and germination of soybean seeds. *Journal of Plant Physiology* 1146, 725–730.

Kumar, G.R., Sakthivel, K., Sundaram, R.M., Neeraja, C.N., Balachandran, S.M., Rani, N.S. *et al.* (2010) Allele mining in crops: prospects and potentials. *Biotechnology Advances* 28, 451–461.

Kwak, M. and Gepts, P. (2009) Structure of genetic diversity in the two major gene pools of common bean (*Phaseolus vulgaris* L., Fabaceae). *Theoretical and Applied Genetics* 118, 979–992.

Kwak, M., Velasco, D. and Gepts, P. (2008) Mapping homologous sequences for determinacy and photoperiod sensitivity in common bean (*Phaseolus vulgaris*). *Journal of Heredity* 99, 283–291.

Lacan, D. and Durand, M. (1995) Na^+ and K^+ transport in excised soybean roots. *Physiologia Plantarum* 93, 132–138.

Lacan, D. and Durand, M. (1996) Na^+–K^+ exchange at the xylem/symplast boundary. *Plant Physiology* 110, 705–711.

Lee, E.H. and Foy, C.D. (1986) Aluminum tolerance of two snapbean cultivars related to organic acid content evaluated by high-performance liquid chromatography. *Journal of Plant Nutrition* 9, 1481–1498.

Lee, G.J., Boerma, H.R., Villagarcia, M.R., Zhou, X., Carter, T.E. Jr, Li, Z. *et al.* (2004) A major QTL conditioning salt tolerance in S-100 soybean and descendent cultivars. *Theoretical and Applied Genetics* 109, 1610–1619.

Li, D., Su, Z., Dong, J. and Wang, T. (2009) An expression database for roots of the model legume *Medicago truncatula* under salt stress. *BMC Genomics* 10, 517.

Li, H.Y., Zhu, Y.M., Chen, Q., Conner, R.L., Ding, X.D., Li, J. *et al.* (2004) Production of transgenic soybean plants with two anti-fungal protein genes via *Agrobacterium* and particle bombardment. *Biology Plantarum* 48, 367–374.

Li, W.Y.F., Wong, F.L., Tsai, S.N., Phang, T.H., Shao, G. and Lam, H.M. (2006) Tonoplast-located *GmCLC1* and *GmNHX1* from soybean enhance NaCl tolerance in transgenic bright yellow (BY)-2 cells. *Plant, Cell and Environment* 29, 1122–1137.

Li, X.Q. (2009) Developmental and environmental variation in genomes. *Heredity* 102, 323–329.

Liao, H., Yan, X., Rubio, G., Beebe, S.E., Blair, M.W. and Lynch, J.P. (2004) Genetic mapping of basal root gravitropism and phosphorus acquisition efficiency in common bean. *Functional Plant Biology* 31, 959–970.

Liao, H., Wan, H., ShaV. J., Wang, X., Yan, X. and Kochian, L.V. (2006) Phosphorus and aluminum interactions in soybean in relation to aluminum tolerance. Exudation of specific organic acids from divergent regions of the intact root system. *Plant Physiology* 141, 674–684.

Liao, Y., Zou, H.F., Wei, W., Hao, Y.J., Tian, A.G., Huang, J. *et al.* (2008) Soybean *GmbZIP44*, *GmbZIP62* and *GmbZIP78* genes function as negative regulator of ABA signaling and confer salt and freezing tolerance in transgenic *Arabidopsis. Planta* 228, 225–240.

Libault, M., Joshi, T., Benedito, V.A., Xu, D., Udvardi, M.K. and Stacey, G. (2009) Legume transcription factor genes: what makes legumes so special? *Plant Physiology* 151, 991–1001.

Liu, J., Versaw, W.K., Pumplin, N., Gomez, S.K., Blaylock, L.A. and Harrison, M.J. (2008) Closely related members of the *Medicago truncatula PHT1* phosphate transporter gene family encode phosphate transporters with distinct biochemical activities. *Journal of Biological Chemistry* 283, 24673–24681.

López-Bucio, J., Cruz-Ramírez, A. and Herrera-Estrella, L. (2003) The role of nutrient availability in regulating root architecture. *Current Opinion in Plant Biology* 6, 280–287.

López-Marín, H.D., Rao, I.M. and Blair, M.W. (2009) Quantitative trait loci for root morphology traits under aluminum stress in common bean (*Phaseolus vulgaris* L.). *Theoretical and Applied Genetics* 119, 449–458.

Luo, G.Z., Wang, H.W., Huang, J., Tian, A.G., Wang, Y.J., Zhang J.S. *et al.* (2005a) A putative plasma membrane cation/proton antiporter from soybean confers salt tolerance in *Arabidopsis. Plant Molecular Biology* 59, 809–820.

Luo, Q., Yu, B. and Liu, Y. (2005b) Differential sensitivity to chloride and sodium ions in seedlings of *Glycine max* and *G. soja* under NaCl stress. *Journal of Plant Physiology* 162, 1003–1012.

Lynch, J.P. (2007) Roots of the second green revolution. *Australian Journal of Botany* 55, 493–512.

Ma, J.F., Ryan, P.R. and Delhaize, E. (2001) Aluminium tolerance in plants and the complexing role of organic acids. *Trends in Plant Science* 6, 273–278.

Macas, J., Neumann, P. and Navrátilová, A. (2007) Repetitive DNA in the pea (*Pisum sativum* L.) genome: comprehensive characterization using 454 sequencing and comparison to soybean and *Medicago truncatula. BMC Genomics* 8, 427.

Malencic, D., Popovic, M. and Miladinovic, J. (2003) Stress tolerance parameters in different genotypes of soybean. *Biologia Plantarum 46*, 141–143.

Melotto, M., Monteiro-Vitorello, C.B., Bruschi, A.G. and Camargo, L.E.A. (2005) Comparative bioinformatic analysis of genes expressed in common bean (*Phaseolus vulgaris* L.) seedlings. *Genome* 48, 562–570.

Metzker, M.L. (2010) Sequencing technologies – the next generation. *Nature Review Genetics* 11, 31–46.

Misson, J., Raghothama, K.G., Jain, A., Jouhet, J., Block, M.A., Bligny, R. *et al.* (2005) A genome-wide transcriptional analysis using *Arabidopsis thaliana* Affymetrix gene chips determined plant responses to phosphate deprivation. *Proceedings of the National Academy of Sciences of the United States of America* 102, 11934–11939.

Mittler, R. and Blumwald, E. (2010) Genetic engineering for modern agriculture: challenges and perspectives. *Annual Review of Plant Biology* 61, 443–462.

Miyasaka, S., Bute, J., Howell, R. and Foy, C. (1991) Mechanisms of aluminum tolerance in snapbeans. Root exudation of citric acid. *Plant Physiology* 96, 737–746.

Moftah, A.E. and Michel, B.E. (1987) The effect of sodium chloride on solute potential and proline accumulation in soybean leaves. *Plant Physiology* 83, 238–240.

Mohammadi, M., Taleei, A., Zeinali, H., Naghavi, M.R., Ceccarelli, S. and Grando, B.M. (2005) QTL analysis for phenologic traits in doubled haploid population of barley. *International Journal of Agriculture and Biology* 7, 820–823.

Morphew, R., Arndt, R., Lauter, C., Tesfaye, M., Samac, D. and Temple, G. (2004) Characterization of altered metabolism in transgenic alfalfa overexpressing malate dehydrogenase to confer aluminum tolerance. *Abstracts of Papers of the American Chemical Society* 227, 148.

Muchero, W., Diop, N.N., Bhat, P.R., Fenton, R.D., Wanamaker, S., Pottorff, M. *et al.* (2009) A consensus genetic map of cowpea [*Vigna unguiculata* (L) Walp.] and synteny based on EST-derived SNPs. *Proceedings of the National Academy of Sciences of the United States of America* 106, 18159–18164.

Munns, R. and Tester, M. (2008) Mechanisms of salinity tolerance. *Annual Review of Plant Biology* 59, 651–681.

Neale, D.B. and Ingvarsson, P.K. (2008) Population, quantitative and comparative genomics of adaptation in forest trees. *Current Opinion in Plant Biology* 11, 149–155.

Nord, E.A. and Lynch, J.P. (2009) Plant phenology: a critical controller of soil resource acquisition. *Journal of Experimental Botany* 60, 1927–1937.

Ochoa, I.E., Blair, M.W. and Lynch, J.P. (2006) QTL analysis of adventitious root formation in common bean under contrasting phosphorus availability. *Crop Science* 46, 1609–1621.

Pantalone, V.R., Kenworthy, W.J., Slaughter, L.H. and James, B.R. (1997) Chloride tolerance in soybean and perennial *Glycine* accessions. *Emphytica* 97, 235–239.

Park, Y.J., Lee, J.K. and Kim, N.S. (2009) Simple sequence repeat polymorphisms (SSRPs) for evaluation of molecular diversity and germplasm classification of minor crops. *Molecules* 14, 4546–4569.

Peng, H., Chenga, H.Y., Yua, X.W., Shic, Q.H., Zhangc, H., Lic, J.G. *et al.* (2009) Characterization of a chickpea (*Cicer arietinum* L.) NAC family gene, CarNAC5, which is both developmentally and stress-regulated. *Plant Physiology and Biochemistry* 47, 1037–1045.

Pérez-Torres, C.A., López-Bucio, J., Cruz-Ramírez, A., Ibarra-Laclette, E., Dharmasiri, S., Estelle, M. *et al.* (2008) Phosphate availability alters lateral root development in *Arabidopsis* by modulating auxin sensitivity via a mechanism involving the TIR1 auxin receptor. *Plant Cell* 20, 3258–3272.

Phang, T.-H., Shao, G. and Lam, H.-M. (2008) Salt tolerance in soybean. *Journal of Integrative Plant Biology* 50, 1196–1212.

Popelka, J.C., Terryn, N. and Higgins, T.J.V. (2004) Gene technology for grain legumes: can it contribute to the food challenge in developing countries? *Plant Science* 167, 195–206.

Proite, K., Leal-Bertioli, S.C., Bertioli, D.J., Moretzsohn, M.C., da Silva, F.R., Martins, N.F. *et al.* (2007) ESTs from a wild *Arachis* species for gene discovery and marker development. *BMC Plant Biology* 7, 7.

Rafalski, J.A. (2010) Association genetics in crop improvement. *Current Opinion in Plant Biology* 13, 174–180.

Raju, N.L, Gnanesh, B.N., Lekha, P., Jayashree, B., Pande, S., Hiremath, P.J. *et al.* (2010) The first set of EST resource for gene discovery and marker development in pigeonpea (*Cajanus cajan* L.). *BMC Plant Biology* 10, 45.

Ramaekers, L., Remans, R., Rao, I.M., Blair, M.W. and Vanderleyden, J. (2010) Strategies for improving phosphorus acquisition efficiency of crop plants. *Field Crops Research* 117, 169–176.

Ramírez, M., Graham, M.A., Blanco-López, L., Silvente, S., Medrano-Soto, A., Blair, M.W. *et al.* (2005) Sequencing and analysis of common bean ESTs. Building a foundation for functional genomics. *Plant Physiology* 137, 1211–1227.

Rangel, A.F., Mobin, M., Rao, I.M. and Horst, W.J. (2005) Proton toxicity interferes with the screening of common bean (*Phaseolus vulgaris* L.) genotypes for aluminum resistance in nutrient solution. *Journal of Plant Nutrition and Soil Science* 168, 607–616.

Rangel, A.F., Rao, I.M. and Horst, W.J. (2007) Spatial aluminium sensitivity of root apices of two common bean (*Phaseolus vulgaris* L.) genotypes with contrasting aluminium resistance. *Journal of Experimental Botany* 58, 3895–3904.

Rangel, A.F., Rao, I.M. and Horst, W.J. (2009) Intracellular distribution and binding state of aluminium in root apices of two common bean (*Phaseolus vulgaris*) genotypes in relation to Al toxicity. *Physiologia Plantarum* 135, 162–173.

Raynolds, M. and Tuberosa, R. (2008) Translational research impacting on crop productivity in drought prone environments. *Current Opinion in Plant Biology* 11, 171–179.

Rubio, G., Sorgonà, A. and Lynch, J.P. (2004) Spatial mapping of phosphorus influx in bean root systems using digital autoradiography. *Journal of Experimental Botany* 55, 2269–2280.

Rubio, V., Linhares, F., Solano, R., Martin, A.C., Iglesias, J., Leyva, A. *et al.* (2001) A conserved MYB transcription factor involved in phosphate starvation signaling both in vascular plants and in unicellular algae. *Genes and Development* 15, 2122–2133.

Salekdeh, G.H., Reynolds, M., Bennett, J. and Boyer, J. (2009) Conceptual framework for drought phenotyping during molecular breeding. *Trends in Plant Science* 14, 488–496.

Sanger, F., Nicklen, S. and Coulson, A.R. (1977) DNA sequencing with chain-terminating inhibitors. *Proceedings of the National Academy of Sciences of the United States of America* 74, 5463–5467.

Sato, S. and Tabata, S. (2006) *Lotus japonicus* as a platform for legume research. *Current Opinion in Plant Biology* 9, 128–132.

Schmutz. J., Cannon, S.B., Schlueter, J., Ma, J., Mitros, T., Nelson, W. *et al.* (2010) Genome sequence of the palaeopolyploid soybean. *Nature* 463, 178–183.

Schneider, M., Bairoch, A., Wu, C.H. and Apweiler, R. (2005) Plant protein annotation in the UniProt Knowledgebase. *Plant Physiology* 138, 59–66.

Seki, M., Umezawa, T., Urano, K. and Shinozaki, K. (2007) Regulatory metabolic networks in drought stress responses. *Current Opinion in Plant Biology* 10, 296–302.

Shao, H.B., Chu, L.Y., Jaleel, C.A., Manivannan, P., Panneerselvam, R. and Shao, M.A. (2009) Understanding water deficit stress-induced changes in the basic metabolism of higher plants – biotechnologically and sustainably improving agriculture and the ecoenvironment in arid regions of the globe. *Critical Reviews in Biotechnology* 29, 131–151.

Sharma, K.K. and Lavanya, M. (2002) Recent developments in transgenics for abiotic stress in legumes of the semi-arid tropics. JIRCAS Working Report, pp. 61–73.

Shi, H., Quintero, F.J., Pardo, J.M. and Zhu, J.-K. (2002). The putative plasma membrane Na+/H+ antiporter SOS1 controls long-distance Na+ transport in plants. *Plant Cell* 14, 465–477.

Shinozaki, K. and Yamaguchi-Shinozaki, K. (2007) Gene networks involved in drought stress response and tolerance. *Journal of Experimental Botany* 58, 221–227.

Streeter, J.G., Lohnes, D.G. and Fioritto, R.J. (2001) Patterns of pinitol accumulation in soybean plants and relationships to drought tolerance. *Plant, Cell and Environment* 24, 429–438.

Sun, Y.X., Wang, D., Bai, Y.L., Wang, N.N. and Wang, Y. (2006) Studies on the overexpression of the soybean *GmNHX1* in *Lotus corniculatus*: the reduced Na^+ level is the basis of the increased salt tolerance. *Chinese Science Bulletin* 51, 1306–1315.

Tesfaye, M., Liu, J., Allan, D.L. and Vance, C.P. (2007) Genomic and genetic control of phosphate stress in legumes. *Plant Physiology*144, 594–603.

Tester, M. and Langridge, P. (2010) Breeding technologies to increase crop production in a changing world. *Science* 327, 818–822.

Thi Lang, N. and Chi Buu, B. (2008) Fine mapping for drought tolerance in rice (*Oryza sativa* L.). *Omonrice* 16, 9–15.

Tian, J., Venkatachalam, P., Liao, H., Yan, X. and Raghothama, K. (2007) Molecular cloning and characterization of phosphorus starvation responsive genes in common bean (*Phaseolus vulgaris* L.). *Planta* 227, 151–165.

Timko, M.P., Rushton, P.J., Laudeman, T.W., Bokowiec, M.T., Chipumuro, E., Cheung, F. *et al.* (2008) Sequencing and analysis of the gene-rich space of cowpea. *BMC Genomics* 9, 103.

Tran, L.S, Quach, T.N., Guttikonda, S.K., Aldrich, D.L., Kumar, R., Neelakandan, A. *et al.* (2009) Molecular characterization of stress-inducible *GmNAC* genes in soybean. *Molecular Genetics and Genomics* 281, 647–664.

Tsai, S.N. (2003) Cloning and characterization of ion transporter genes from a salt-tolerant soybean variety. M.Phil. thesis, The Chinese University of Hong Kong, China.

Udvardi, M.K., Tabata, S., Parniske, M. and Stougaard, J. (2005) *Lotus japonicus*: legume research in the fast lane. *Trends in Plant Science* 10, 222–228.

Uhde-Stone, C., Zinn, K.E., Ramirez-Yáñez, M., Li, A., Vance, C.P. and Allan, D.L. (2003) Nylon filter arrays reveal differential gene expression in proteoid roots of white lupin in response to phosphorus deficiency. *Plant Physiology* 131, 1064–1079.

Uhde-Stone, C., Liu, J., Zinn, K.E., Allan, D.L. and Vance, C.P. (2005) Transgenic proteoid roots of white lupin: a vehicle for characterizing and silencing root genes involved in adaptation to P stress. *Plant Journal* 44, 840–853.

Valdés-López, O. and Hernández, G. (2008) Transcriptional regulation and signaling in phosphorus starvation: what about legumes? *Journal of Integrative Plant Biology* 50, 1213–1222.

Valdés-López, O., Arenas-Huertero, C., Ramírez, M., Girard, L., Sánchez, F., Vance, C.P. *et al.* (2008) Essential role of MYB transcription factor: PvPHR1 and microRNA: PvmiR399 in phosphorus-deficiency signalling in common bean roots. *Plant, Cell and Environment* 3, 1834–1843.

Vance, C.P. (2001) Symbiotic nitrogen fixation and phosphorus acquisition. Plant nutrition in a world of declining renewable resources. *Plant Physiology* 127, 390–397.

Varshney, R.K., Close, T.J., Singh, N.K., Hoigston, D.A. and Cook, D.R. (2009a) Orphan legume crops enter the genomics era! *Current Opinion in Plant Biology* 12, 202–210.

Varshney, R.K., Nayak, S.N., May, G.D. and Jackson, S.A. (2009b) Next-generation sequencing technologies and their implications for crop genetics and breeding. *Trends in Biotechnology* 27, 522–530.

Varshney, R.K., Hiremath, P.J., Lekha, P., Kashiwagi, J., Balaji, J., Deokar, A.A. *et al.* (2009c) A comprehensive resource of drought- and salinity-responsive ESTs for gene discovery and marker development in chickpea (*Cicer arietinum* L.). *BMC Genomics* 10, 523.

Verdoy, D., Coba De La Pena, T., Redondo, F.J., Lucas, M.M. and Pueyo, J.J. (2006) Transgenic *Medicago truncatula* plants that accumulate proline display nitrogen-fixing activity with enhanced tolerance to osmotic stress. *Plant, Cell and Environment* 29, 1913–1923.

Wei, W., Huang, J., Hao, Y.-J., Zou, H.-F., Wang, H.-W., Zhao, J.-Y. *et al.* (2009) Soybean GmPHD-type transcription regulators improve stress tolerance in transgenic *Arabidopsis* plants. *PLoS ONE* 4, e7209.

White, P.J. and Brown, P.H. (2010) Plant nutrition for sustainable development and global health. *Annals of Botany* 105, 1073–1080.

Winicov, I. (2000) Alfin1 transcription factor overexpression enhances plant root growth under normal and saline conditions and improves salt tolerance in alfalfa. *Planta* 210, 416–422.

Wu, P., Ma, L., Hou, X., Wang, M., Wu, Y., Liu, F. *et al.* (2003) Phosphate starvation triggers distinct alterations of genome expression in *Arabidopsis* roots and leaves. *Plant Physiology* 132, 1260–1271.

Xue, R.-G., Zhang, B. and Xie, H.-F. (2007) Overexpression of a *NTR1* in transgenic soybean confers tolerance to water stress. *Journal of Plant Cell, Tissue and Organ Culture* 89, 177–183.

Yan, F., Zhu, Y., Müller, C., Zörb, C. and Schubert, S. (2002) Adaptation of H^+-pumping and plasma membrane H^+ ATPase activity in proteoid roots of white lupin under phosphate deficiency. *Plant Physiology* 129, 50–63.

Yan, X., Liao, H., Beebe, S.E., Blair, M.W. and Lynch, J.P. (2004) QTL mapping of root hair and acid exudation traits and their relationship to phosphorus uptake in common bean. *Plant and Soil* 265, 17–29.

Yang, X.J. and Finnegan, P.M. (2010) Regulation of phosphate starvation responses in higher plants. *Annals of Botany* 105, 513–526.

Young, N.D. and Udvardi, M. (2009) Translating *Medicago truncatula* genomics to crop legumes. *Current Opinion in Plant Biology* 12,193–201.

Yu, B.J., Lam, H.M., Shao, G.H. and Liu, Y.L. (2005) Effects of salinity on activities of H+-ATPase, H+-PPase and membrane lipid composition in plasma membrane and tonoplast vesicles isolated from soybean (*Glycine max* L.) seedlings. *Journal of Environmental Science* 17, 259–262.

Zhang, X.-H., Zhong, W.Q. and Widholm, J.M. (2005) Expression of a fungal cyanamide hydratase in transgenic soybean detoxifies cyanamide in tissue culture and in planta to provide cyanamide resistance. *Journal of Plant Physiology* 162, 1064–1073.

Zhang, Y.J., Lynch, J.P. and Brown, K.M. (2003) Ethylene and phosphorus availability have interacting yet distinct effects on root hair development. *Journal of Experimental Botany* 54, 2351–2361.

Zhou, G.A., Chang, R.Z. and Qiu, L.J. (2010) Overexpression of soybean ubiquitin-conjugating enzyme gene *GmUBC2* confers enhanced drought and salt tolerance through modulating abiotic stress-responsive gene expression in *Arabidopsis*. *Plant Molecular Biology* 72, 357–367.

Zhou, Q.Y., Tian, A.G., Zou, H.F., Xie, Z.M., Lei, G., Huang, J. *et al.* (2008) Soybean WRKY-type transcription factor genes, *GmWRKY13*, *GmWRKY21*, and *GmWRKY54*, confer differential tolerance to abiotic stresses in transgenic *Arabidopsis* plants. *Plant Biotechnology Journal* 6, 486–503.

19 Trait Mapping and Molecular Breeding

S.K. Chamarthi, A. Kumar, T.D. Vuong, M.W. Blair, P.M. Gaur, H.T. Nguyen and R.K. Varshney

19.1 Introduction

The role of legumes in agricultural development has been that of providing long-term stability to agricultural systems. Legumes and cereals have co-evolved since ancient times. They have acted as a major contributory factor in sustaining agricultural production throughout the millennia. Grain and forage legumes are grown on ~190 million ha, and their production is about 300 million metric t worldwide (ICRISAT, 2009). Unfortunately, yield improvements in legume crops have not kept pace with those of cereals.

The majority of legumes, apart from soybean, have literally been termed 'orphan crops' in the sense that they are devoid of a well-developed infrastructure (both knowledge and physical capacity) for genetic and genomic analysis or molecular breeding. This lack of infrastructure has restricted the biotechnological crop improvement strategies available for these crops. In this context, there is a need to increase the availability of genomic data and resources in key species and also to decrease the barriers that limit the adoption of complex genomic data sets by crop improvement specialists. Part of the solution lies in training the next generation of scientists to navigate both basic and applied plant science. This in turn, will improve the capacity for the uptake of new biotechnologies and reduce the 'gap' between genomics and traditional versus modern molecular breeding. This chapter provides general concepts of trait mapping and molecular breeding in food legumes, citing the examples of soybean, common bean and chickpea where development and use of genetic and genomic resources are at an advanced stage.

19.2 Challenges in Legume Production

Legume production is greatly challenged by numerous biotic and abiotic stresses, which result in severe losses to agricultural production on a yearly basis. Most legume crops are affected by common insect pests, diseases and a range of abiotic stresses, including adaptation to acid, saline or low-fertility soils as well as adverse weather conditions such as drought, cold temperature or heat stress. However, the occurrence and severity of biotic and abiotic stresses differ from crop to crop, and by pathogen and the environmental conditions to which the crop is exposed, requiring crop-specific breeding approaches and management practices. These stresses are discussed in detail in Chapters 15 and 16.

The interaction between biotic and abiotic stresses is likely to be especially complex and damaging to legume crops in arid and semi-arid regions of the world, and de-convoluting such interactions is an important long-term challenge for legume improvement and molecular/physiological research. Although several breeding strategies ranging from classical breeding to more directed physiological and molecular genetic approaches have been implemented to cope with the threats of these stresses, a better understanding of the mechanisms underlying specific stresses will make molecular breeding truly feasible. The availability of tolerant and resistant cultivars to biotic and abiotic stresses is one of the most effective management practices when irrigated under a comprehensive integrated management approach, resulting in cost savings (i.e. insecticides, pesticides) and environmental protection.

Biotechnological approaches, such as marker-assisted breeding, tissue culture, *in vitro* mutagenesis and genetic engineering, can contribute to the speeding up of classical breeding and in overcoming major problems, such as lack of natural sources of genetic resistance to biotic and abiotic stresses and sexual incompatibility (Cook and Varshney, 2010). In the near future, great success in crop improvement will be possible by combining genomic tools with rational selection of germplasm and precise phenotyping for traits of interest, an approach termed 'genomics-assisted breeding' (Varshney *et al.*, 2005).

Improvement in agronomic/phenological traits of legumes is crucial in order to improve their use as human food, especially in developed countries. In the current scenario, legumes have become an increasingly important concern in marketing and profitability. Therefore, different quality characteristics of legumes such as seed size, mass and shape, storability, etc. are receiving greater attention in regard to genetic improvement. There is also an increasing interest in improving nutritional characteristics of legumes with enhanced contents of beta carotene, leutin, isoflavones and other nutraceuticals.

19.3 Molecular Breeding Approaches

The use of molecular markers for improving breeding efficiencies in plant breeding was first suggested in 1989 (Tanksley *et al.*, 1989; Melchinger, 1990). Today, plant breeding is rapidly evolving as more molecular genetic tools are being applied to commonly accepted field techniques (Kulwal *et al.*, 2010); recent advances in genomics have allowed identification of molecular markers associated with traits of interest to breeders. In this context, initially a linkage between a gene responsible for a trait of interest and a molecular marker is established and confirmed, validated using breeders' materials and subsequently used in DNA diagnostic tests to guide plant-breeding selection efforts (Morgante and Salamini, 2003; Gupta and Varshney, 2004) (Fig. 19.1). The process of indirect selection in crop improvement can be expedited by using molecular markers, which help in alleviating several time-/cost-consuming and labour-intensive aspects of direct screening under greenhouse and field conditions.

Trait mapping

Molecular breeding includes the identification of genotypically and phenotypically polymorphic plant genotypes, development of segregating mapping populations, genotyping of the mapping population, phenotyping of trait(s) of interest and marker–trait association analysis. Subsequently, mapped gene(s) or quantitative trait loci (QTL) can be introgressed individually or combined (pyramided) in an improved cultivar (Gupta *et al.*, 2010a). Two main approaches can be used to identify marker–trait associations: (i) linkage mapping; and (ii) association mapping (Fig. 19.1).

In general, linkage mapping-based gene/QTL studies involve: (i) development of an appropriate mapping population from contrasting parental genotypes for the trait of interest; (ii) identification of polymorphic markers; (iii) genotyping of the mapping population with polymorphic markers;

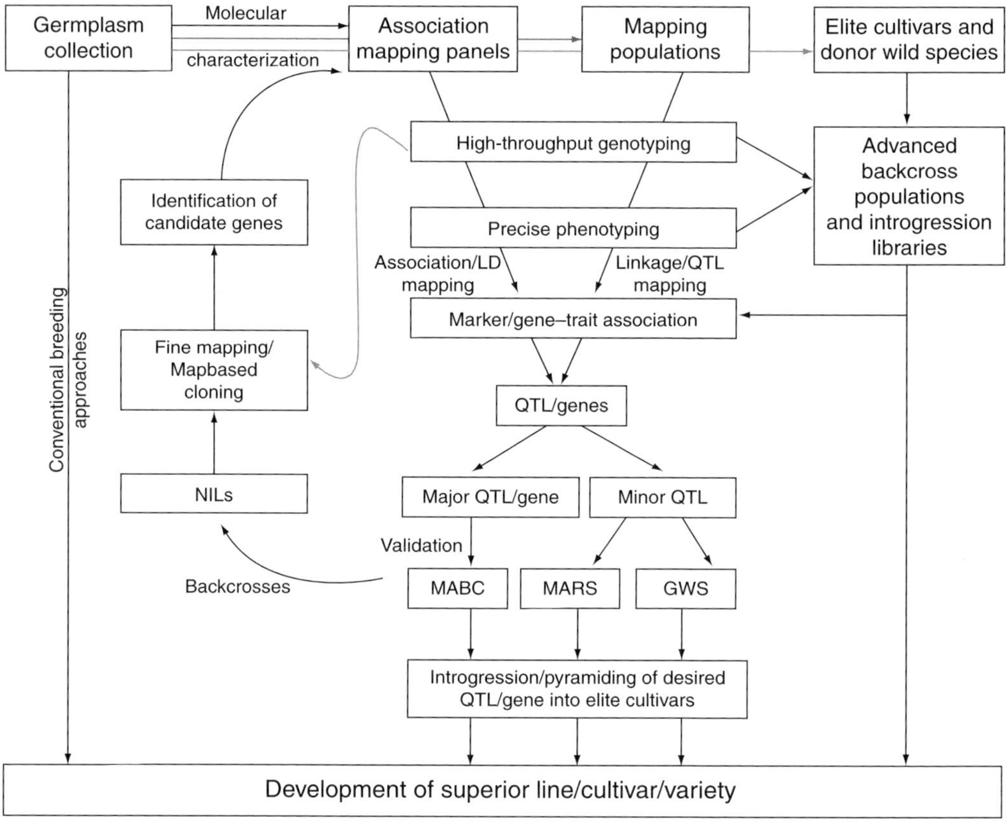

Fig. 19.1. A scheme showing integration of modern genetics and breeding approaches (trait mapping and molecular breeding) in crop improvement programmes. GWS, genome-wide selection; LD, linkage disequilibrium; MABC, marker-assisted backcrossing; MARS, marker-assisted recurrent selection; NILs, near-isogenic lines; QTL, quantitative trait loci.

(iv) construction of the genetic map based on genotyping data; (v) precise phenotyping of the mapping population in different environments; and (vi) marker–trait association using suitable genetic linkage and QTL analysis programmes (Varshney *et al.* 2009b). In legume species, linkage mapping-based approaches have been extensively used for mapping genes/QTL for resistance to diseases, nematodes and insects and for tolerance to abiotic stresses and several agronomic traits (Table 19.1).

Association mapping (also known as linkage disequilibrium mapping) has also been used in plant species for trait mapping. It has several advantages over conventional linkage mapping approach: (i) it takes less time as there is no need to develop a specialized mapping population – rather, an existing natural population is used; (ii) it is less expensive as the same association mapping panel and genotyping data generated can be used for mapping of different traits; (iii) resolution of mapping is high because of the use of a natural population that has several meiotic recombinants, unlike the few in mapping populations; and (iv) as compared with linkage mapping where a maximum of two alleles are obtained, a higher numbers of alleles are obtained to find trait associations.

Association mapping comprises the following two broad categories based on the scale and goal of a particular study: candidate gene-based association mapping and whole-genome association mapping. The former relates to

Table 19.1. Examples of gene/QTL mapping in soybean, common bean and chickpea.

Crop	Trait	Gene/QTL (*n*)	Marker type(s)	Reference(s) and related reference(s) cited
Disease resistance				
Soybean	Sclerotinia stem rot	16 QTL	SSRs	Guo *et al.* (2008); Vuong *et al.* (2008); Huynh *et al.* (2010)
	Phytophthora root rot	3–8 QTL/genes	AFLPs, SSRs, RAPDs, SCARs	Han *et al.* (2008); Li, X. *et al.* (2010); Wang *et al.* (2010)
	Brown stem rot	4 QTL/genes	RFLPs, AFLPs, SSRs	Patzoldt *et al.* (2005)
	Asian soybean rust	5 loci/QTL	SSRs, SNPs	Garcia *et al.* (2008); Silva *et al.* (2008); Chakraborty *et al.* (2009); Hyten *et al.* (2009)
	Soybean mosaic virus	Rsv1, Rsv3	SSRs	Shi *et al.* (2008)
	Sudden death syndrome	4 QTL	SSRs	Kazi *et al.* (2008)
Common bean	Anthracnose	Co-genes	RAPDs, AFLPs	Rodríguez-Suárez *et al.* (2007)
	Halo blight	Pse-1	SCARs	Miklas *et al.* (2009)
	Charcoal rot	1 QTL	AFLP	Hernández-Delgado *et al.* (2009)
	Fusarium wilt	PvPR2, PvPR1	RAPDs	Schneider *et al.* (2001)
	White mould	Fin, Phs	RAPD, SSR	Kolkman and Kelly (2003)
	Rust	UR-6, UR-13	SCAR	Mienie *et al.* (2005)
	Common bacterial blight	6 QTL	SSRs, STSs, SCARs	Liu *et al.* (2008); Vandemark *et al.* (2008)
	Bacterial brown spot	2 QTL	RAPDs	Jung *et al.* (2003)
	Bean common mosaic virus	bc-3; I	RAPD; SCAR	Johnson *et al.* (1997)
	Bean golden yellow mosaic geminivirus	bgm-1	SCAR	Blair *et al.* (2007b)
Chickpea	Ascochyta blight	AR2, ar1, ar1a, ar1b, ar2a, ar2b, Ar19	SSRs, RAPDs, DAF	Cho *et al.* (2004)
		QTL_{AR1}, QTL_{AR2}	SCARs, SSRs, RAPDs	Iruela *et al.* (2006, 2007)
		13 QTL	SSRs, RAPDs	Kottapalli *et al.* (2009)
	Fusarium wilt	foc-0, foc-1, foc-2, foc-3, foc-4, foc-5	SSRs, STSs, RAPDs	Cobos *et al.* (2005); Iruela *et al.* (2007)
	Rust	1 QTL	SSR	Madrid *et al.* (2008)
Nematode and insect resistance				
Soybean	Corn earworm	3 QTL	RFLPs	Rector *et al.* (1998)
	Cyst nematode	rhg1, rhg4	SSRs, SCARs, SNPs	Wu *et al.* (2009); Vuong *et al.* (2010)
Common bean	Leaf hopper	1 QTL	SSR, RFLP	Murray *et al.* (2004)
	Thrips	Tpr6.1	SSRs	Frei *et al.* (2005)

Continued

Table 19.1. Continued.

Crop	Trait	Gene/QTL (*n*)	Marker type(s)	Reference(s) and related reference(s) cited
	Bean pod weevil	4 QTL	SSRs, SCARs	Blair *et al.* (2006, 2007a)
	Bruchids	Arc gene	SSRs	Blair *et al.* (2010b, c)
Abiotic stress tolerance traits				
Soybean	Waterlogging	6 QTL	SSRs, RFLPs	Githiri *et al.* (2006)
	Chilling tolerance in seed yield	3 QTL	SSRs	Funatsuki *et al.* (2005)
	Salt stress	1 QTL/gene	SSR	Lee *et al.* (2009); Tuyen *et al.* (2010)
	Manganese toxicity	3 QTL	SSRs, RAPDs	Kassem *et al.* (2004)
	Phosphorus deficiency	fsw1, fsw2, rp1, fsw3, rp2, lp1, lp2	SSRs, RFLPs	Li *et al.* (2005)
	Iron deficiency chlorosis	3–19 QTL	SSRs, RFLPs	Lin *et al.* (2000)
	Aluminum tolerance	11 QTL	RFLPs	Qi *et al.* (2008)
	Sulphur-containing amino acids	7 QTL	SSRs	Panthee *et al.* (2006)
Common bean	Drought	2 QTL	RAPDs	Schneider *et al.* (1997)
	Phosphorus uptake	Pup4.1, 10.1 and 2.1	RAPDs	Beebe *et al.* (2006)
Agronomic/phenological traits				
Soybean	Specific leaf weight, leaf size	3–6 QTL	RFLPs	Mian *et al.* (1998)
	Seed weight	15 QTL	SSRs	Csanadi *et al.* (2001)
	Flowering time	4–9 QTL	SSRs	Su *et al.* (2010)
	Sprout yield	4 QTL	RFLPs	Lee *et al.* (2001)
	Seed isoflavones (genistein, daidzein, glycitein)	35 QTL	SSRs	Zeng *et al.* (2009); Gutierrez-Gonzalez *et al.* (2010)
	Seed size	7 QTL	SSRs	Hyten *et al.* (2004)
	Seed flooding tolerance	4 QTL	–	Sayama *et al.* (2009)
	Ability, frequency and efficiency of somatic embryogenesis	5 QTL	SSRs, AFLPs	Choi *et al.* (2010); Song *et al.* (2010)
	Early maturity	E_8	SSRs	Cober *et al.* (2010)
	Oligosaccharides and sucrose	6 QTL	SSRs	Kim *et al.* (2006)
	Vitamin E content	21 QTL	SSRs	Li, H. *et al.* (2010)
	Chlorophyll *a* fluorescence parameter	4–13 QTL	–	Yin *et al.* (2010)
	Developmental behaviour	15 QTL	SSRs	Sun *et al.* (2006)
	Browning in soybean seed coat	5 QTL	AFLPs, SSRs	Githiri *et al.* (2007)
	Domestication	2 QTL	SSRs	Liu *et al.* (2007)
	Seed composition	52 QTL	SSRs	Li *et al.* (2007)
	Seed shape	19 QTL	SSRs	Salas *et al.* (2006)
	Photoperiod insensitivity	2 QTL	SSRs	Liu and Abe (2010)
	Net-like cracking of seed coat	2 QTL	SSRs	Oyoo *et al.* (2010)
	Cleistogamy	4 QTL	SSRs, AFLPs	Khan *et al.* (2008)

Continued

Table 19.1. Continued.

Crop	Trait	Gene/QTL (*n*)	Marker type(s)	Reference(s) and related reference(s) cited
Bean	Seed mass, calcium, iron, zinc, tannin content	3–26 QTL	AFLPs	Guzman-Maldonado *et al.* (2003)
	Plant height, climbing ability, internode length, branch number	1–9 QTL	SSRs, RAPDs, SCARs	Checa and Blair (2008)
	Phenological traits, seed size traits, seed quality traits	31 QTL	SSRs, AFLPs, SCARs, ISSRs	Pérez-Vega *et al.* (2010)
	Nutritional traits (iron, zinc, tannins, phytate)	3–26 QTL	SSRs, RAPDs, AFLPs	Blair *et al.* (2009a, b, 2010a); Caldas and Blair (2009); Cichy *et al.* (2009)
	Nodulation number	4 QTL	RFLP	Nodari *et al.* (1993)
Chickpea	Seed size traits	2 QTL	SSRs	Hossain *et al.* (2010)
	Double podding	s	SSRs	Rajesh *et al.* (2002)
	Time to flowering	2 QTL	SSRs	Lichtenzveig *et al.* (2006)
	Beta carotene, leutin, seed weight	1–4 QTL	SSRs	Abbo *et al.* (2005)
	Flower colour	B/b	SSR	Cobos *et al.* (2005)

polymorphisms in selected candidate genes that appear to have roles in controlling phenotypic variation for specific traits, while the latter surveys genetic variation in the whole genome to find marker–trait associations for various complex traits (Zhu *et al.*, 2008). In taking the decision as to which is the method of choice, one has to consider the extent of linkage disequilibrium (LD) in the organism of interest.

Although the association mapping approach has been used recently in several cereals like maize, barley, wheat, etc. (Ersoz *et al.*, 2007), only a few examples have become available in legume species. The candidate gene-based approach has been successfully used to map different loci for iron deficiency chlorosis in soybean (Wang *et al.*, 2008). Similarly, several candidate genes implicated in oleate biosynthesis were mapped and their co-segregation with oleate and linoleate QTL investigated (Bachlava *et al.*, 2009). Other examples of trait mapping are shown in Table 19.1.

Next-generation sequencing and high-throughput genotyping technologies are becoming popular in legumes such as chickpea, common bean and soybean (Varshney *et al.*, 2009a, c, 2010b), accelerating their use in association mapping. For example, a high-throughput SNP genotyping platform (Illumina GoldenGate assay) developed in soybean (Hyten *et al.*, 2008) has been used for mapping soybean rust resistance (*Rpp3*) (Hyten *et al.*, 2009), SCN (soybean cyst nematode) (Vuong *et al.*, 2010), flooding and fatty acids (Vuong *et al.*, unpublished data).

Molecular breeding

Once markers are identified for a trait, they can be used for a variety of applications such as enhancing biological knowledge of the inheritance and genetic architecture of the trait, in addition to their use in breeding programmes. When molecular markers are used in breeding programmes, it is important to take into account the statistical power to identify QTL numbers, QTL effect, percentage of phenotypic variation explained, major and minor QTL through use of appropriate marker density on the genetic map and reasonable population sample size. Furthermore, markers identified in one population need to be validated in other population/germplasm collections, and closely linked markers flanking the QTL should be used for indirect selection of the trait.

Figure 19.1 shows a few molecular breeding approaches commonly used in breeding programmes, which are discussed in the following sections. Soybean is the legume crop in which these approaches have been most successful, and where the use of markers in breeding programmes is routine. Several improved lines/varieties for resistance to different SCN races (also known as HG types) (Concibido *et al.*, 1996; Cahill and Schmidt, 2004; Arelli and Young, 2009); phytophthora root rot and brown stem rot (Cahill and Schmidt, 2004); insect resistance (Narvel *et al.*, 2001; Walker *et al.*, 2002; Warrington *et al.*, 2008); low linolenic acid content (Sauer *et al.*, 2008); yield (Concibido *et al.*, 2003); and mosaic virus (Saghai Maroof *et al.*, 2008; Shi *et al.*, 2009) have been developed and released. In the case of common bean the use of molecular markers in breeding programmes is intermediate between that of soybean and chickpea, and marker-assisted selection (MAS) has been used principally to deploy single genes in large-scale programmes at CIAT for resistance to quarantined viruses (Miklas *et al.*, 2006a, b; Blair *et al.*, 2007a). More specifically, MAS has been successfully used for enhanced resistance to anthracnose in the bean cultivar Perola in Brazil (Raganin *et al.*, 2003), pinto beans in the USA (Miklas *et al.*, 2003a) and Andean climbing bean in Mexico/Colombia (Garzon *et al.*, 2008). On the other hand, molecular breeding activities have only just been initiated in chickpea. Several successful examples targeting the development of superior lines or released cultivars through molecular breeding are listed in Table 19.2.

Marker-assisted backcrossing

Marker-assisted backcrossing (MABC) is the simplest and most widely used molecular breeding approach in plant breeding. MABC has become a fast-track approach for increasing the genetic gain of plants, resulting in the development of improved varieties with better yield potential, improved quality and resistance against insects, pests and diseases (Collard and Mackill, 2008; Moose and Mumm, 2008; Ribaut *et al.*, 2010). Basically, this approach incorporates desirable major genes/QTL from an agronomically inferior source (the donor parent) into an elite cultivar or breeding line (the recurrent parent) without transfer of undesirable or deleterious genes from the donor (linkage drag).

The desired outcome of MABC is a line/cultivar containing only the major genes/QTL from the donor parent, while retaining the whole genome of the recurrent parent (Hospital and Charcosset, 1997; Varshney and Dubey, 2009; Varshney *et al.*, 2009b, Gupta *et al.*, 2010a). Three types of selection

Table 19.2. Examples of development/release of improved lines/cultivars in soybean and common bean using molecular breeding approaches.

Crop	Cultivar/breeding line	Trait	Country and year of release	Reference
Soybean	JTN-5503	Disease resistance	USA, 2005	Arelli *et al.* (2006)
	JTN-5303	Disease resistance	USA, 2005	Arelli *et al.* (2007)
	JTN-5109	Soybean cyst nematode resistance	USA, 2009	Arelli and Young (2009)
	DS-880	Soybean cyst nematode resistance	USA, 2010	Smith *et al.* (2010)
Bean	USPT-ANT-1	Disease resistance	USA, 2004	Miklas *et al.* (2003b)
	ABCP-8	Disease resistance	USA, 2005	Mutlu *et al.* (2005)
	ABC-Weihing	Disease resistance	USA, 2006	Mutlu *et al.* (2008)

can be exercised in MABC: foreground, recombinant and background. Foreground selection involves the selection of target genes/QTL on the carrier chromosome with the help of two flanking markers (Hospital and Charcosset, 1997). It can be used to select for laborious or time-consuming traits and it allows selection of heterozygous plants at the seedling stage and therefore identifies plants desirable for backcrossing. Furthermore, recessive alleles can be identified and selected, which is difficult to perform using conventional methods.

Recombination events between the target locus and linked flanking markers can also be identified in backcross (BC) progeny. This can be used to reduce linkage drag, which is difficult to overcome through the use of conventional backcrossing (Frisch *et al.*, 1999b). For this purpose, Hospital and Decoux (2002) have developed a statistical programme called 'Popmin' (http://moulon.inra.fr/~fred/programs/popmin) for calculating the minimum population size.

Background selection involves selection of BC progeny with highest proportion of recurrent parent (RP) genome, using unlinked markers present on 'non-carrier' chromosomes (Hospital and Charcosset, 1997; Frisch *et al.*, 1999b). The use of background selection during MABC to accelerate the development of a RP genome with additional genes has been referred to as complete line conversion (Ribaut *et al.*, 2002). While conventional backcrossing takes a minimum of six BC generations to recover the RP genome, the use of markers enables the similar degree of progress in two BC generations (Visscher *et al.*, 1996; Hospital and Charcosset, 1997; Frisch *et al.*, 1999a, b; Varshney and Dubey, 2009). Studies have also shown that the use of a limited number of markers on non-carrier chromosomes can be sufficient to recover more than 95% of the recurrent parent genome in three BC generations (Visscher *et al.*, 1996; Kumar *et al.*, 2010).

The MABC approach has also been used to construct near-isogenic lines (NILs) or chromosome segment substitution lines (CSSLs), which are often used for genetic analysis of genes/QTL (Peleman and van der Voort, 2003; Lorieux, 2005; Varshney *et al.*, 2010b). NILs are developed in the same way as advanced backcross (AB) lines by crossing a donor parent with a recurrent parent. After several generations of backcrossing, the advanced backcross lines are expected to contain all of the recurrent parent genome except for the chromosomal region containing a gene or QTL of interest. NILs have been utilized for validation of QTL, for fine mapping and can also be used directly in breeding programmes (Stuber *et al.*, 1999). NILs containing different genes affecting the same trait are very useful for comparing the effectiveness of these genes in different locations or environments (Fig. 19.1).

Another use of MABC is to pyramid various genes for multiple traits within the same cultivar (Koebner and Summers, 2003; Sharma *et al.*, 2004; Saghai Maroof *et al.*, 2008; Li, X. *et al.*, 2010). Several excellent reviews have documented the use of MABC for pyramiding genes/QTL resulting in the development of superior lines/varieties/hybrids in crop plants (Gupta *et al.*, 2010a, b; Kumar *et al.*, 2010; Varshney *et al.*, 2010a), and examples in soybean and common bean are presented in Table 19.2. However, the use of MABC has now been initiated in elite lines of chickpea at ICRISAT, in collaboration with its partners, for the introgression of QTL/genes for drought-related traits and resistance to diseases (fusarium wilt and ascochyta blight). In addition, the introgression of root trait QTL is in progress in collaboration with the Indian Institute of Pulses Research (IIPR) and the India and Ethiopia Institute of Agricultural Research (EIAR), Ethiopia. Molecular breeding for development of superior lines with enhanced resistance to fusarium wilt and ascochyta blight has been initiated recently in collaboration with several Indian partners, including IIPR, Jawaharlal Nehru Krishi Vishwavidyalaya, Mahatma Phule Krishi Vidyapeeth and the Agricultural Research Station, Gulburga.

Marker-assisted recurrent selection

One of the limitations of MABC is that only a limited number of desirable alleles can be introgressed at a time. To overcome this limitation, particularly in the case of complex traits like drought tolerance, the marker-assisted recurrent selection (MARS) approach has been proposed for transferring/pyramiding of superior QTL/gene alleles for trait(s) of interest in one genetic background (Bernardo and Charcosset, 2006; Varshney and Dubey, 2009; Gupta *et al.*, 2010a, b; Ribaut *et al.*, 2010). The genetic gain feasible through MARS has been estimated as being higher than that via MABC (Bernardo and Charcosset, 2006).

In principle, MARS is a forward breeding approach combining MAS (Stam, 1995) with increase in the frequency of favourable alleles/QTL at multiple loci (Edwards and Johnson, 1994; Koebner and Summers, 2003; Eathington, 2005). This involves multiple cycles of marker-based selection that include improvement of F_2 progeny by one cycle of MAS based on marker data and phenotypic data, followed by three recombination cycles of the selected progenies based on marker data only and repetition of these cycles to develop the population for multi-location phenotyping (Ribaut *et al.*, 2010; Tester and Langridge, 2010). In MARS, a selection index is used that gives weights to markers according to the relative magnitude of their estimated effects on the trait (Lande and Thompson, 1990; Edward and Johnson, 1994). For the successful use of MARS, a number of factors including heritability of the target traits, marker coverage in the genome, reliability of marker–trait associations, family size, number of families and type of population should be considered (Mayor and Bernardo, 2009). Moreover, knowledge of the quantitative traits can be very useful in enhancing the selection response through MARS with the help of candidate gene markers, or tightly linked markers, each having a relatively large effect. The response to MARS decreases as the knowledge of the number of minor QTL associated with the trait decreases (Charcosset and Moreau, 2004; Bernardo and Charcosset, 2006).

The MARS approach has been/is being used extensively in maize breeding in both the private and public sectors. For instance, it has been employed to fix six marker loci in two different F_2 populations that showed an increase in the frequency of marker alleles, from 0.50 to 0.80 (Edward and Johnson, 1994). Several multinational companies, such as Syngenta and Monsanto, are using MARS in their breeding programmes in several crops, including soybean (Ribaut *et al.*, 2010).

Recently, some international agricultural research centres (IARCs), such as ICRISAT and CIAT, in collaboration with the Generation Challenge Program (GCP), have initiated the use of MARS in chickpea and common bean, respectively, for pyramiding favourable drought-tolerant alleles. Therefore, the potential of MARS is yet to be demonstrated in legume breeding for the development of superior lines/genotypes.

Genome-wide selection

A new approach based on genome-wide marker profiling, called 'genome-wide selection (GWS)' or 'genomic selection (GS)', has been proposed for complex traits that are controlled by many genes/QTL, each of small effect. Basically, this method predicts genomic estimated breeding values (GEBVs) of progenies, which are calculated for progenies based on both phenotyping and genotyping data. These GEBVs are then used to select the superior progeny lines for advancement in the breeding cycle (Heffner *et al.*, 2009; Jannink *et al.*, 2010). Several computational tools are available or are being developed to calculate GEBVs, such as BLUPs (best linear unbiased prediction) programmes, and the geostatistical mixed model has recently been developed as a tool in GS (Robinson, 1991; Streeck and Piepho, 2010). This approach is not required to elucidate marker–trait associations by QTL mapping (Bernardo, 2010a, b; Tester and Langridge, 2010). Furthermore, it has been shown that double-haploid (DH) populations are very useful in GWS compared with F_2 populations, when many QTL control a trait (Mayor and Bernardo, 2009). Currently, however, there is little information available on the use of GWS in crop plants, although recent developments in plant genomics make it feasible to generate genome-wide marker data (using SNPs) to start GWS in breeding programmes. In the next few years GWS is expected to be used in legumes, at least in soybean.

Introgression of superior alleles from wild species

Plant breeders mostly use existing germplasm and landraces to develop new varieties for desirable agronomic traits. However, yields have remained stagnant partly because sufficient genetic diversity is missing for progress in some of the traits, due to genetic bottlenecks that occurred during the domestication process (Tanksley and McCouch, 1997; Gur and Zamir, 2004). It is well known that wild species/relatives are the reservoirs for resistance genes to many biotic and abiotic stresses. However, their transfer from wild species to elite cultivars through conventional breeding has been limited, mainly due to the associated transfer of undesired alleles (linkage drag). However, it is now feasible to recover/transfer the favourable alleles in elite germplasm left behind by the domestication process more efficiently, using innovative genomics-assisted breeding strategies such as molecular maps and integrative QTL analysis. In this context, several methods for transferring superior alleles from wild species have been suggested and some of these are discussed below.

One approach is the construction of introgression libraries using the genetic background of elite lines by introgressing small wild species segments in a systematic manner. Introgression libraries are made up of introgression lines (ILs) that are produced by successive backcrossing (generally three to four generations) to the recurrent parent. The introgressed fragments can be monitored using molecular markers, either in each generation or at chosen stages. Fixation of the materials is obtained by either selfing or using double-haploid methodology. As a result, each line possesses one or several homozygous chromosomal fragments of the donor genotype, introgressed into a recurrent background genome. These fragments should be arranged continuously from the first to the last chromosome, either manually or using a computer software-aided process (graphical genotyping). The whole donor genome is thus represented by a set of small, contiguous overlapping fragments. This differs from the more traditional approach of introducing resistance genes from wild species into elite cultivars through genetically balanced mapping populations of progeny recombinant inbred lines (RILs) derived from an early generation (e.g. F_2 plants or families or F_1-derived doubled haploids). Such populations contain an equal proportion of exotic and elite genotypes, and deleterious effects of exotic alleles may mask the desired target gene effect. Therefore the development of introgression lines represents a significant advantage over the previously used RIL-type populations. In regard to legumes, some reports on the development of introgression libraries have become available in soybean using wild soybean species (*Glycine soja*) (Concibido *et al.*, 2003), and in groundnut from synthetic tetraploids (Foncéka *et al.*, 2009).

Another important approach used in the transfer of superior alleles from wild species into cultivated germplasm is based on the advanced-backcross QTL (AB-QTL) analysis proposed by Tanksley and Nelson (1996). This method proved effective in detecting additive, dominant, partially dominant and over-dominant QTL. This approach uses repeated backcrossing with the elite parent but decreases the number and size of the exotic introgressions, thereby reducing the burden of linkage drag. During backcrossing cycles, the transfer of desirable genes/QTL is monitored by molecular markers. The segregating BC_2F_2 or BC_2F_3 population generated during backcrossing (F_2 or F_3 stages) is then used not only for recording phenotyping data for the trait of interest, but also for genotyping with polymorphic molecular markers. These data are then used for QTL analysis, leading to simultaneous discovery of QTL and the generation of introgression lines. Once favourable QTL alleles are identified, only a few additional marker-assisted generations are required to develop full NILs that can be field tested and used for variety development.

The AB-QTL approach has been used in common bean and soybean. For instance, in the case of common bean, Blair *et al.* (2005) used a cross between a wild Colombian accession and an Andean cultivar to develop $BC_2F_{3:5}$-derived lines for AB-QTL analysis of yield traits, and finding that segregation distortion was minimal except at a few domestication syndrome genes. Similar populations have been developed for: (i) introgression of high seed iron content from wild Mexican accessions into the Andean and Meso-American background cvs Cerinza or Tacana; and (ii) introgression of drought tolerance between the common bean gene pools from Meso-American sources to Andean cultivars as part of the Tropical Legumes project on adaptation to drought-prone marginal regions of eastern and southern Africa. Population sizes in these AB-QTL mapping populations have ranged from 157 to 300 genotypes, with various experimental designs used for analysis. In the case of soybean, for instance, Chaky *et al.* (2003) generated 296 $BC_2F_{4:6}$ backcross introgression lines (BILs) from the cross *Glycine max* (Dunbar) × *Glycine soja* (PI 326582A). This study provided several QTL for seed yield, seed protein and oil, in addition to some late-maturing and taller BILs.

19.4 Conclusions

It is evident that several success stories on both trait mapping and molecular breeding are available in soybean. The availability of the soybean genome sequence (Schmutz *et al.*, 2010) and the establishment of high-throughput SNP genotyping platforms (Hyten *et al.*, 2008) are expected further to accelerate molecular breeding in soybean improvement.

In the case of common bean, efforts aimed at trait mapping and molecular breeding are extensive, as compared with chickpea. However, molecular breeding activities in common bean have remained focused mainly on simple inherited traits like disease resistance. In the case of chickpea, although several examples on trait mapping are available, the use of molecular breeding activities as compared with soybean and common bean is still at the preliminary stage. As both of these legume crops are very important in sub-Saharan Africa and South America (common bean) and South Asia (chickpea), CIAT and ICRISAT have initiated molecular breeding programmes in these crops to improve complex traits like drought tolerance, through the use of MABC and MARS approaches as a part of the Tropical Legume (TL-I) project of the Generation Challenge Programme (GCP) in collaboration with the Bill and Melinda Gates Foundation (BMGF) (http://www.generationcp.org/gcptli/). With the goal of sustainable crop production in these legume crops, it is essential that national agricultural research programmes in the developing countries of sub-Saharan Africa, South Asia and South America should lead or actively participate in the molecular breeding of these legumes. Shortage of appropriate human resources and physical infrastructure in developing countries, however, are challenging issues. The establishment of the Integrated Breeding Platform (IBP) as a one-stop shop for accessing genotyping services, information and data management, decision support, statistical tools and technical support will help in overcoming some of above-mentioned limitations.

In summary, due to advances in sequencing, genotyping, biometrics and bioinformatics, the future of molecular breeding in legume crops is promising, not only in soybean, common bean and chickpea but also in other crops like lentil, faba bean and pigeon pea, which are still considered 'orphan legume crops'.

Acknowledgements

The authors are grateful to the Generation Challenge Programme (GCP) of the Consultative Group on International Agriculture Research (CGIAR); to the Tropical Legumes projects (TL-I and TL-II) of the Bill and Melinda Gates Foundation; and to the Department of Biotechnology (DBT), Government of India for providing financial support for funding legume genomics research at ICRISAT. The authors also wish to express their thanks to M. Isabel Vales, Mahendar Thudi, Reyazul Rouf Mir and Manish Pandey from ICRISAT for discussions and useful suggestions while preparing the manuscript.

References

Abbo, S., Molina, C., Jungmann, R., Grusak, M.A., Berkovitch, Z., Reifen, R. *et al.* (2005) Quantitative trait loci governing carotenoid concentration and weight in seeds of chickpea (*Cicer arietinum* L.). *Theoretical and Applied Genetics* 113, 1357–1369.

Arelli, P.R. and Young, L.D. (2009) JTN-5109 soybean germplasm resistant to nematode population infecting cv. Hartwig. Agronomy Society of America, *The Abstracts* 268–218, p.133 (http://a-c-s.confex.com/crops/2009am/).

Arelli, P.R., Young, L.D. and Mengistu, A. (2006) Registration of high yielding and multiple disease resistant soybean germplasm JTN-5503. *Crop Science* 46, 2723–2724.

Arelli, P.R., Pantalone, V.R., Allen, F.L. and Mengistu, A. (2007) Registration of soybean germplasm JTN-5303. *Journal of Plant Registrations* 1, 69–70.

Bachlava, E., Dewey, R.E., Burton, J.W. and Cardinal, A.J. (2009) Mapping candidate genes for oleate biosynthesis and their association with unsaturated fatty acid seed content in soybean. *Molecular Breeding* 23, 337–347.

Beebe, S.E., Rojas, M., Yan, X., Blair, M.W., Pedraza, F., Muñoz, F. *et al.* (2006) Quantitative trait loci for root architecture traits correlated with phosphorus acquisition in common bean. *Crop Science* 46, 413–423.

Bernardo, R. (2010a) Potential and practicality of genome-wide selection in plants. *Plant and Animal Genomes XVIII Conference*, 9–13 January, San Diego, California.

Bernardo, R. (2010b) Genome-wide selection with minimal crossing in self-pollinated crops. *Crop Science* 50, 624–627.

Bernardo, R. and Charcosset, A. (2006) Usefulness of gene information in marker-assisted recurrent selection: A simulation appraisal. *Crop Science* 46, 614–621.

Blair, M.W., Astudillo, C., Restrepo, J., Bravo, L.C., Villada, D. and Beebe, S.E. (2005) Análisis multilocacional de líneas de fríjol arbustivo con alto contenido de hierro en el departamento de Nariño. *Fitotecnia Colombiana* 5, 20–27.

Blair, M.W., Iriarte, G. and Beebe, S. (2006) QTL analysis of yield traits in an advanced backcross population derived from a cultivated Andean x wild common bean (*Phaseolus vulgaris* L.) cross. *Theoretical and Applied Genetics* 112, 1149–1163.

Blair, M.W., Fregene, M.A., Beebe, S.E. and Ceballos, H. (2007a) Marker assisted selection in common beans and cassava. In: Guimaraes, E.P., Ruane, J., Scherf, B.D., Sonnino, A. and Dargie, J.D (eds) *Marker-Assisted Selection: Current Status and Future Perspectives in Crops, Livestock, Forestry and Fish.* Food and Agriculture Organization of the United Nations, Rome, pp. 81–116.

Blair, M.W., Rodriguez, L.M., Pedraza, F., Morales, F. and Beebe, S. (2007b) Genetic mapping of the bean golden yellow mosaic geminivirus resistance gene *bgm-1* and linkage with potyvirus resistance in common bean (*Phaseolus vulgaris* L.). *Theoretical and Applied Genetics* 114, 261–271.

Blair, M.W., Astudillo, C., Grusak, M., Graham, R. and Beebe, S. (2009a) Inheritance of seed iron and zinc content in common bean (*Phaseolus vulgaris* L.). *Molecular Breeding* 23, 197–207.

Blair, M.W., Sandoval, T.A., Caldas, G.V., Beebe, S.E. and Páez, M.I. (2009b) Quantitative trait locus analysis of seed phosphorus and seed phytate content in a recombinant inbred line population of common bean (*Phaseolus vulgaris* L.). *Crop Science* 49, 237–246.

Blair, M.W., Medina, J.I., Astudillo, C., Rengifo, J., Beebe, S.E., Machado, G. *et al.* (2010a) QTL for seed iron and zinc concentrations in a recombinant inbred line population of Mesoamerican common beans (*Phaseolus vulgaris* L.). *Theoretical and Applied Genetics* 121, 237–248.

Blair, M.W., Muñoz, C., Buendía, H.F., Flower, J., Bueno, J.M. and Cardona, C. (2010b) Genetic mapping of microsatellites markers around the arcelin bruchid resistance locus in common bean. *Theoretical and Applied Genetics* 121, 393–402.

Blair, M.W., Prieto, S., Diaz, L.M., Buendía, H.F. and Cardona, C. (2010c) Linkage disequilibrium at the APA-Arcelin insecticidal seed storage protein locus of common bean (*Phaseolus vulgaris* L.). *BMC Plant Biology* 10, 79.

Cahill, D.J. and Schmidt, D.H. (2004) Use of marker-assisted selection in a product development breeding program. In: Fischer, T. (ed.) *New Directions for A Diverse Planet. Proceedings of 4th International Crop Science Congress,* 26 September–1 October, Brisbane, QLD, Australia (www.cropscience.org.au/icsc2004/).

Caldas, G.V. and Blair, M.W. (2009) Inheritance of condensed tannin content and relationship with seed color and pattern genes in common bean (*Phaseolus vulgaris* L.). *Theoretical and Applied Genetics* 119, 131–142.

Chakraborty, N., Curley, J., Frederick, R.D., Hyten, D.L., Nelson, R.L., Hartman, G.L. *et al.* (2009) Mapping and confirmation of a new allele at *Rpp1* from soybean PI 594538A conferring RB lesion type resistance to soybean rust. *Crop Science* 49, 783–790.

Chaky, J.M., Specht, J.E. and Cregan, P.B. (2003) Advanced backcross QTL analysis in a mating between *Glycine max* and *Glycine soja*. *Plant and Animal Genome Abstracts*, pp. 545.

Charcosset, A. and Moreau, L. (2004) Use of molecular markers for the development of new cultivars and the evaluation of genetic diversity. *Euphytica* 137, 81–94.

Checa, O.E. and Blair, M.W. (2008) Mapping QTL for climbing ability and component traits in common bean (*Phaseolus vulgaris* L.). *Molecular Breeding* 22, 201–215.

Cho, S.H., Chen, W.D. and Muehlbauer, F.J. (2004) Pathotype-specific genetic factors in chickpea (*Cicer arietinum* L.) for quantitative resistance to *Aschochyta* blight. *Theoretical and Applied Genetics* 109, 733–739.

Choi, P., Mano, Y., Ishikawa, A., Odashima, M., Umezawa, T., Fujimura, T. *et al.* (2010) Identification of QTLs controlling somatic embryogenesis using RI population of cultivar × weedy soybean. *Plant Biotechnology Reports* 4, 23–27.

Cichy, K.A., Caldas, G.V., Snapp, S.S. and Blair, M.W. (2009) QTL analysis of seed iron, zinc, and phosphorus levels in an Andean bean population. *Crop Science* 49, 1742–1750.

Cober, E.R., Molnar, S.J., Charette, M. and Voldeng, H.D. (2010) A new locus for early maturity in soybean. *Crop Science* 50, 524–527.

Cobos, M.J., Fernandez, M., Rubio, J., Kharrat, M., Moreno, M.T., Gil, J. *et al.* (2005) A linkage map of chickpea (*Cicer arietinum* L.) based on populations from Kabuli x Desi crosses: location of genes for resistance to *Fusarium* wilt race 0. *Theoretical and Applied Genetics* 110, 1347–1353.

Collard, B.C.Y. and Mackill, D.J. (2008) Marker-assisted selection: an approach for precision plant breeding in the twenty-first century. *Philosophical Transactions of the Royal Society B* 363, 557–572.

Concibido, V.C., Denny, R.L., Lange, D.A., Orf, J.H. and Young, N.D. (1996) RFLP mapping and molecular marker-assisted selection of soybean cyst nematode resistance in PI 209332. *Crop Science* 36, 1643–1650.

Concibido, V.C., La Vallee, B., Mclaird, P., Pineda, N., Meyer, J., Hummel, L. *et al.* (2003) Introgression of a quantitative trait locus for yield from *Glycine soja* into commercial soybean cultivars. *Theoretical and Applied Genetics* 106, 575–582.

Cook, D.R. and Varshney, R.K. (2010) From genome studies to agricultural biotechnology: Closing the gap between basic plant science and applied agriculture. *Current Opinion in Plant Biology* 13, 115–118.

Csanadi, G., Vollmann, J., Stift, G. and Lelley, T. (2001) Seed quality QTLs identified in a molecular map of early maturing soybean. *Theoretical and Applied Genetics* 103, 912–919.

Eathington, S. (2005) Practical applications of molecular technology in the development of commercial maize hybrids. In: *Proceedings of the 60th Annual Corn and Sorghum Seed Research Conferences*. American Seed Trade Association, Washington, DC.

Edwards, M. and Johnson, L. (1994) RFLPs for rapid recurrent selection in analysis of molecular marker data. *Joint Plant Breeding Symposium Series, American Society for Horticultural Science*, CSSA, Madison, Wisconsin, pp. 33–40.

Ersoz, E.S., Yu, J. and Buckler, E.S. (2007) Applications of linkage disequilibrium and association mapping in crop plants. In: Varshney, R.K. and Tuberosa, R. (eds) *Genomics-Assisted Crop Improvement: Vol. 1: Genomics Approaches and Platforms*. Springer, Dordrecht, The Netherlands, pp. 97–119.

Foncéka, D., Hodo-Abalo, T., Rivallan, R., Faye, I., Sall, M.N., Ndoye, O. *et al.* (2009) Genetic mapping of wild introgressions into cultivated peanut: a way toward enlarging the genetic basis of a recent allotetraploid. *BMC Plant Biology* 9, 103.

Frei, A., Blair, M.W., Cardona, C., Beebe, S.E., Gu, H. and Dorn, S. (2005) QTL mapping of resistance to *Thrips palmikarny* in common bean. *Crop Science* 45, 379–387.

Frisch, M., Bohn, M. and Melchinger, A.E. (1999a) Comparison of selection strategies for marker-assisted backcrossing of a gene. *Crop Science* 39, 1295–1301.

Frisch, M., Bohn, M. and Melchinger, A.E. (1999b) Minimum sample size and optimal positioning of flanking markers in marker assisted backcrossing for transfer of a target gene. *Crop Science* 39, 967–975.

Funatsuki, H., Kawaguchi, K., Matsuba, S., Sato, Y. and Ishimoto, M. (2005) Mapping of QTL associated with chilling tolerance during reproductive growth in soybean. *Theoretical and Applied Genetics* 111, 851–861.

Garcia, A., Calvo, E.S., Kiihl, R.A., Harada, A., Hiromoto, D.M. *et al.* (2008) Molecular mapping of soybean rust (*Phakopsora pachyrhizi*) resistance genes: discovery of a novel locus and alleles. *Theoretical and Applied Genetics* 117, 545–553.

Garzon, L.N., Ligarreto, G.A. and Blair, M.W. (2008) Molecular marker-assisted backcrossing of anthracnose resistance into Andean climbing beans (*Phaseolus vulgaris* L.). *Crop Science* 48, 562–570.

Githiri, S.M., Watanabe, S., Harada, K. and Takahashi, R. (2006) QTL analysis of flooding tolerance in soybean at an early vegetative growth stage. *Plant Breeding* 125, 613–618.

Githiri, S.M., Yang, D., Khan, N.A., Xu, D., Komatsuda, T. and Takahashi, R. (2007) QTL analysis of low temperature induced browning in soybean seed coats. *Journal of Heredity* 98, 360–366.

Guo, X., Wang, D., Gordon, S.G., Helliwell, E., Smith, T., Berry, S.A. *et al.* (2008) Genetic mapping of QTLs underlying partial resistance to *Sclerotinia sclerotiorum* in soybean PI 391589A and PI 391589B. *Crop Science* 48, 1129–1139.

Gupta, P.K. and Varshney, R.K. (2004) Cereal genomics: an overview. In: Gupta, P.K. and Varshney, R.K. (eds) *Cereal Genomics*. Kluwer Academic Publishers, Dordrecht, The Netherlands, pp. 1–18.

Gupta, P.K., Kumar, J., Mir, R.R. and Kumar, A. (2010a) Marker assisted selection as a component of conventional plant breeding. *Plant Breeding Reviews* 33, 145–217.

Gupta, P.K., Langridge, P. and Mir, R.R. (2010b) Marker-assisted wheat breeding: present status and future possibilities. *Molecular Breeding* 26, 145–161.

Gur, A. and Zamir, D. (2004) Unused natural variation can lift yield barriers in plant breeding. *PLOS Biology* 2, 1610–1615.

Gutierrez-Gonzalez, J.J., Wu, X., Gillman, J.D., Lee, J.D., Zhong, R., Yu, O. *et al.* (2010) Intricate environment-modulated genetic networks control isoflavones accumulation in soybean seeds. *BMC Plant Biology* 10, 105 (doi: 10.1186/1471-2229-10-105).

Guzman-Maldonado, S.H., Martinez, O., Acosta-Gallegos, J.A., Guevara-Lara, F. and Paredes-Lopez, O. (2003) Putative quantitative trait loci for physical and chemical components of common bean. *Crop Science* 43, 1029–1035.

Han, Y., Teng, W., Yu, K., Poysa, V., Anderson, T., Qiu, L. *et al.* (2008) Mapping QTL tolerance to *Phytophthora* root rot in soybean using microsatellite and RAPD/SCAR derived markers. *Euphytica* 162, 231–239.

Heffner, E.L., Sorrells, M.E. and Jannink, J.L. (2009) Genomic selection for crop improvement. *Crop Science* 49, 1–12.

Hernández-Delgado, S., Reyes-Valdés, M.H., Rosales-Serna, R. and Mayek-Pérez, N. (2009) Molecular markers associated with resistance to *Macrophomina phaseolina* (Tassi) Gold. in common bean. *Journal of Plant Pathology* 91, 163–170.

Hospital, F. and Charcosset, A. (1997) Marker assisted introgression of quantitative trait loci. *Genetics* 147, 1469–1485.

Hospital, F. and Decoux, G. (2002) Popmin: a program for the numerical optimization of population sizes in marker assisted backcross programs. *Journal of Heredity* 93, 383–384.

Hossain, S., Ford, R., McNeil, D., Pittock, C. and Panozzo, J. F. (2010) Inheritance of seed size in chickpea (*Cicer arietinum* L.) and identification of QTL based on 100-seed weight and seed size index. *Australian Journal of Crop Science* 4, 126–135.

Huynh, T.T., Bastien, M., Iquira, E., Turcotte, P. and Belzile, F. (2010) Identification of QTLs associated with partial resistance to white mold in soybean using field-based inoculation. *Crop Science* 50, 969–979.

Hyten, D., Pantalone, V., Sams, C., Saxton, A., Landau-Ellis, D., Stefaniak, T. *et al.* (2004) Seed quality QTL in a prominent soybean population. *Theoretical and Applied Genetics* 109, 552–561.

Hyten, D.L., Song, Q., Choi, I.Y., Yoon, M.S., Specht, J.E., Matukumalli, L.K. *et al.* (2008) High-throughput genotyping with the GoldenGate assay in the complex genome of soybean. *Theoretical and Applied Genetics* 116, 945–952.

Hyten, D.L., Smith, J.R., Frederick, R.D., Tucker, M.L., Song, Q. and Cregan, P.B. (2009) Bulked segregant analysis using the GoldenGate assay to locate the *Rpp3* locus that confers resistance to soybean rust in soybean. *Crop Science* 49, 265–271.

ICRISAT (2009) Food security and diversification in the drylands. *Annual Report 2009*, ICRISAT, Patancheru, India, pp, 1–76.

Iruela, M., Rubio, J., Barro, F., Cubero, J.I., Millan, T. and Gil, J. (2006) Detection of two quantitative trait loci for resistance to Ascochyta blight in an intra-specific cross of chickpea (*Cicer arietinum* L.): development of SCAR markers associated with resistance. *Theoretical and Applied Genetics* 112, 278–287.

Iruela, M., Castro, P., Rubio, J., Cubero, J.I., Jacinto, C., Millan, T. *et al.* (2007) Validation of a QTL for resistance to *Ascochyta* blight linked to resistance to *Fusarium* wilt race 5 in chickpea (*Cicer arietinum* L.). *European Journal of Plant Pathology* 199, 29–37.

Jannink, J.L., Lorenz, A.J. and Iwata, H. (2010) Genomic selection in plant breeding: from theory to practice. *Briefings in Functional Genomics and Proteomics* 9, 166–177.

Johnson, W.C., Guzman, P., Mandala, D., Mkandawire, A.B.C., Temple, S., Gilbertson, R.L. and Gepts, P. (1997) Molecular tagging of the *bc-3* gene for introgression into Andean common bean. *Crop Science* 37, 248–254.

Jung, G., Ariyarathne, H.M., Coyne, D.P. and Nienhuis, J. (2003) Mapping QTL for bacterial brown spot resistance under natural infection in field and seeding stem inoculation in growth chamber in common bean. *Crop Science* 43, 350–357.

Kassem, M.A., Meksem, K., Kang, C.H., Njiti, V.N., Kilo, V., Wood, A.J. *et al.* (2004) Loci underlying resistance to manganese toxicity mapped in a soybean recombinant inbred line population of 'Essex' x 'Forest'. *Plant Soil* 260, 197–204.

Kazi, S., Shultz, J., Afzal, J., Johnson, J., Njiti, V.N. and Lightfoot, D.A. (2008) Separate loci underlie resistance to root infection and leaf scorch during soybean sudden death syndrome. *Theoretical and Applied Genetics* 116, 967–977.

Khan, N.A., Githiri, S.M., Benitez, E.R., Abe, J., Kawasaki, S., Hayashi, T. *et al.* (2008) QTL analysis of cleistogamy in soybean. *Theoretical and Applied Genetics* 117, 479–487.

Kim, M.S., Park, M.J., Jeong, W.H., Nam, K.C. and Chung. J.I. (2006) SSR marker tightly linked to the Ti locus in soybean [*Glycine max* (L.) Merr.]. *Euphytica* 152, 361–366.

Koebner, R.M.D. and Summers, R.W. (2003) 21st century wheat breeding: Plot selection or plate detection? *Trends in Biotechnology* 21, 59–63.

Kolkman, J.M. and Kelly, J.D. (2003) QTL conferring resistance and avoidance to white mold in common bean. *Crop Science* 43, 539–548.

Kottapalli, P., Gaur, P.M., Katiyar, S.K., Crouch, J.H., Buhariwalla, H.K., Pande, S. *et al.* (2009) Mapping and validation of QTLs for resistance to an Indian isolate of *Ascochyta* blight pathogen in chickpea. *Euphytica* 165, 79–88.

Kulwal, P.L., Thudi, M. and Varshney, R.K. (2010) Genomics interventions in crop breeding for sustainable agriculture. In: *Encyclopedia of Sustainability Science and Technology* <in press>.

Kumar, J., Mir, R.R., Kumar, N., Kumar, A., Mohan, A., Prabhu, K.V. *et al.* (2010) Marker assisted selection for pre-harvest sprouting tolerance and leaf rust resistance in bread wheat. *Plant Breeding* 129, 617–621.

Lande, R. and Thompson, R. (1990) Efficiency of marker-assisted selection in the improvement of quantitative traits. *Genetics* 124, 743–756.

Lee, J.D., Shannon, J.G., Vuong, T.D. and Nguyen, H.T. (2009) Inheritance of salt tolerance in wild soybean (*Glycine soja* Sieb. and Zucc.) accession PI 483463. *Journal of Heredity* 100, 798–801.

Lee, S.H., Park, K.Y., Lee, H.S., Park, E.H. and Boerma, H.R. (2001) Genetic mapping of QTL conditioning soybean sprout yield and quality. *Theoretical and Applied Genetics* 103, 702–709.

Li, H., Liu, H., Han, Y., Wu, X., Teng, W., Liu, G. *et al.* (2010) Identification of QTL underlying vitamin E contents in soybean seed among multiple environments. *Theoretical and Applied Genetics* 120, 1405–1413.

Li, W., Sun, D., Du, Y., Chen, Q., Zhang, Z., Qiu, L. *et al.* (2007) Quantitative trait loci underlying the development of seed composition in soybean [*Glycine max* (L.) Merr.]. *Genome* 50, 1067–1077.

Li, X., Han, Y., Teng, W., Zhang, S., Yu, K., Poysa, V. *et al.* (2010) Pyramided QTL underlying tolerance to *Phytophthora* root rot in mega-environments from soybean cultivars Conrad and Hefeng 25. *Theoretical and Applied Genetics* 121, 651–658.

Li, Y.D., Wang, Y.J., Tong, Y.P., Gao, J.G., Zhang, J.S. and Chen, S.Y. (2005) QTL mapping of phosphorus deficiency tolerance in soybean (*Glycine max* L. Merr). *Euphytica* 142, 137–142.

Lichtenzveig, J., Bonfil, D.J., Zhang, H.B., Shtienberg, D. and Abbo, S. (2006) Mapping quantitative trait loci in chickpea associated with time to flowering and resistance to *Didymella rabiei* the causal agent of *Ascochyta* blight. *Theoretical and Applied Genetics* 113, 1357–1369.

Lin, S.F., Grant, D., Cianzio, S. and Shoemaker, R.C. (2000) Molecular characterization of iron deficiency chlorosis in soybean. *Journal of Plant Nutrition* 23, 1929–1939.

Liu, B. and Abe, J. (2010) QTL Mapping for photoperiod insensitivity of a Japanese soybean landrace sakamotowase. *Journal of Heredity* 2, 251–256.

Liu, B., Fujita, T., Yan, Z.H., Sakamoto, S., Xu, D. and Abe, J. (2007) QTL mapping of domestication-related traits in soybean (*Glycine max*). *Annals of Botany* 100, 1027–1038.

Liu, S., Yu, K. and Park, S.J. (2008) Development of STS markers and QTL validation for common bacterial blight resistance in common bean. *Plant Breeding* 127, 62–68.

Lorieux, M. (2005) CSSL Finder: a free program for managing introgression lines (2005http://mapdisto.free.fr/).

Madrid, E., Rubiales, D., Moral, A., Moreno, M.T., Millan, T., Gil, J. *et al.* (2008) Mechanism and molecular markers associated with rust resistance in a chickpea interspecific cross (*Cicer arietinum* × *Cicer reticulatum*). *European Journal of Plant Pathology* 121, 45–53.

Mayor, P.J. and Bernardo, R. (2009) Genomewide selection and marker-assisted recurrent selection in doubled haploid versus F_2 populations. *Crop Science* 49, 1719–1725.

Melchinger, A.E. (1990) Use of molecular markers in breeding for oligogenic disease resistance. *Plant Breeding* 104, 1–19.

Mian, M.A.R., Ashley, D.A. and Boerma, H.R. (1998) An additional QTL for water use efficiency in soybean. *Crop Science* 38, 390–393.

Mienie, C.M., Liebenberg, M.M., Pretorius, Z.A. and Miklas, P.N. (2005) SCAR markers linked to the common bean rust resistance gene Ur-13. *Theoretical and Applied Genetics* 111, 972–979.

Miklas, P.N., Delorme, R. and Riley, R. (2003a) Identification of QTL conditioning resistance to white mold in snap bean. *Journal of the American Society for Horticultural Sciences* 128, 564–570.

Miklas, P.N., Kelly, J.D. and Singh, S.P. (2003b) Registration of anthracnose-resistant pinto bean germplasm line USPT-ANT-1. *Crop Science* 43, 1889–1890.

Miklas, P.N., Kelly, J.D., Beebe, S.E. and Blair, M.W. (2006a) Common bean breeding for resistance against biotic and abiotic stresses: From classical to MAS breeding. *Euphytica* 147, 105–131.

Miklas, P.N., Hu, J., Grunwald, N.J. and Larsen, K.M. (2006b) Potential application of TRAP (Targeted Region Amplified Polymorphism) markers for mapping and tagging disease resistance traits in common bean. *Crop Science* 46, 910–916.

Miklas, P.N., Fourie, D., Wagner, J., Larsen, R.C. and Mienie, C.M.S. (2009) Tagging and mapping *Pse-1* gene for resistance to halo blight in common bean differential cultivar UI-3. *Crop Science* 49, 41–48.

Moose, S. and Mumm, R.H. (2008) Molecular plant breeding as the foundation for 21st century crop improvement. *Plant Physiology* 147, 969–977.

Morgante, M. and Salamini, F. (2003) From plant genomics to breeding practice. *Current Opinion in Biotechnology* 14, 214–219.

Murray, J.D., Michaels, T.E., Cardona, C., Schaafsma, A.W. and Pauls, K.P. (2004) Quantitative trait loci for leafhopper (*Empoasca fabae* and *Empoasca kraemeri*) resistance and seed weight in the common bean. *Plant Breeding* 123, 474–479.

Mutlu, N., Miklas, P.N., Steadman, J.R., Vidaver, A.V., Lindgren, D., Reiser, J. and Pastor-Corrales, M.A. (2005) Registration of pinto bean germplasm line ABCP-8 with resistance to common bacterial blight. *Crop Science* 45, 806.

Mutlu, N., Urrea, C.A., Miklas, P.N., Steadman, J.R., Pastor Corrales, M.A., Lindgren, D.T. *et al.* (2008) Registration of common bacterial blight, rust and bean common mosaic resistant great northern bean germplasm line ABC – Weighing. *Journal of Plant Registrations* 2, 120–124.

Narvel, J.M., Walker, D.R., Rector, B.G., All, J.N., Parrott, W.A. and Boerma, H.R. (2001) A retrospective DNA marker assessment of the development of insect resistant soybean. *Crop Science* 41, 1931–1939.

Nodari, R.O., Tsai, S.M., Guzman, P., Gilbertson, R.L. and Gepts, P. (1993) Towards an integrated linkage map of common bean. III. Mapping genetic factors controlling host bacteria interactions. *Genetics* 134, 341–350.

Oyoo, M.E., Githiri, S.M., Benitez, E.R. and Takahashi, R. (2010) QTL analysis of net-like cracking in soybean seed coats. *Breeding Science* 60, 28–33.

Panthee, D.R., Pantalone, V.R., Sams, C.E., Saxton, A.M., West, D.R., Orf, J.H. *et al.* (2006) Quantitative trait loci controlling sulfur containing amino acids, methionine and cysteine, in soybean seeds. *Theoretical and Applied Genetics* 112, 546–553.

Patzoldt, M.E., Grau, R., Stephens, P.A., Kurtzweil, N.C., Carlson, S.R. and Diers, B.W. (2005) Localization of a quantitative trait locus providing brown stem rot resistance in the soybean cultivar Bell. *Crop Science* 45, 1241–1248.

Peleman, J.D. and van der Voort, J.R. (2003) Breeding by design. *Trends in Plant Science* 7, 330–334.

Pérez-Vega, E., Paneda, A., Rodríguez-Suárez, C., Campa, A., Giraldez, R. and Ferreira, J.J. (2010) Mapping of QTLs for morpho-agronomic and seed quality traits in a RIL population of common bean (*Phaseolus vulgaris* L.). *Theoretical and Applied Genetics* 120, 1367–1380.

Qi, B., Korir, P., Zhao, T., Yu, D., Chen, S. and Gai, J. (2008) Mapping quantitative trait loci associated with aluminum toxin tolerance in NJRIKY recombinant inbred line population of soybean (*Glycine max*). *Journal of Integrative Plant Biology* 50, 1089–1095.

Raganin, V.A., Sanglard, D.A., de Souza, T.L.P.O., Moreira, M.A. and de Barros, E.G. (2003) Simultaneous transfer of resistance genes for rust, anthracnose, and angular leaf spot to cultivar Perola assisted by molecular markers. *Annual Report of the Bean Improvement Cooperative* 46, 159–160.

Rajesh, P.N., Tullu, A., Gil, J., Gupta, V.S., Ranjekar, P.K. and Muehlbauer, F.J. (2002) Identification of an STMS marker for the double-podding gene in chickpea. *Theoretical and Applied Genetics* 105, 604–607.

Rector, B.G., All, J.N., Parrott, W.A. and Boerma, H.R. (1998) Identification of molecular markers linked to quantitative trait loci for soybean resistance to corn earworm. *Theoretical and Applied Genetics* 96, 786–790.

Ribaut, J.M., Jiang, C. and Hoisington, D. (2002) Simulation experiments on efficiencies of gene introgression by backcrossing. *Crop Science* 42, 557–565.

Ribaut, J.M., de Vicente, M.C. and Delannay, X. (2010) Molecular breeding in developing countries: Challenges and perspectives. *Current Opinion in Plant Biology* 13, 1–6.

Robinson, G.K. (1991) That BLUP is a good thing: the estimation of random effects. *Statistical Science* 6, 15–32.

Rodríguez-Suárez, C., Méndez-Vigo, B., Paneda, A., Ferreira, J.J. and Giraldez, R. (2007) A genetic linkage map of *Phaseolus vulgaris* L. and localization of genes for specific resistance to six races of anthracnose (*Colletotrichum lindemuthianum*). *Theoretical and Applied Genetics* 114, 713–722.

Saghai Maroof, M.A., Jeong, S.C., Gunduz, I., Tucker, D.M., Buss, G.R. and Tolin, S.A. (2008) Pyramiding of soybean mosaic virus resistance genes by marker assisted selection. *Crop Science* 48, 517–526.

Salas, P., Oyarzo-Llaipen, J.C., Wang, D., Chase, K. and Mansur, L. (2006) Genetic mapping of seed shape in three populations of recombinant inbred lines of soybean [*Glycine max* L. Merr.]. *Theoretical and Applied Genetics* 113, 1459–1466.

Sauer, M.L.A., Scott, R.A. and Cheesbrough, T.M. (2008) Marker assisted selection for low linolenic acid content in soybean. *Journal of Crop Improvement* 21, 139–155.

Sayama, T., Nakazaki, T., Ishikawa, G., Yagasaki, K., Yamada, N., Hirota, N. *et al.* (2009) QTL analysis of seed-flooding tolerance in soybean (*Glycine max* [L.] Merr.). *Plant Science* 4, 514–521.

Schmutz, J., Cannon, S.B., Schlueter, J., Ma, J., Mitros, T., Nelson, W. *et al.* (2010) Genome sequence of the palaeopolyploid soybean. *Nature* 463, 178–183.

Schneider, K.A., Brothers, M.E. and Kelly, J.D. (1997) Marker-assisted selection to improve drought resistance in common bean. *Crop Science* 37, 51–60.

Schneider, K.A., Grafton, K.F. and Kelly, J.D. (2001) QTL analysis of resistance to Fusarium root rot in bean. *Crop Science* 41, 535–442.

Sharma, K.D., Winter, P., Kahl, G. and Muehlbauer, F.J. (2004) Molecular mapping of *Fusarium oxysporum* f. sp. *ciceris* race 3 resistance gene in chickpea. *Theoretical and Applied Genetics* 108, 1243–1248.

Shi, A., Chen, P., Li, D.X., Zheng, C., Hou, A. and Zhang, B. (2008) Genetic confirmation of 2 independent genes for resistance to soybean mosaic virus in J05 soybean using SSR markers. *Journal of Heredity* 99, 598–603.

Shi, A., Chen, P., Li, D., Zheng, C., Zhang, B. and Hou, A. (2009) Pyramiding multiple genes for resistance to soybean mosaic virus in soybean using molecular markers. *Molecular Breeding* 23, 113–124.

Silva, D.C.G., Yamanaka, N., Brogin, R.L., Arias, C.A.A., Nepomuceno, A.L., Di Mauro, A.O. *et al.* (2008) Molecular mapping of two loci that confer resistance to Asian rust in soybean. *Theoretical and Applied Genetics* 117, 57–63.

Smith, J. (2010) USDA, ARS, National Genetic Resources Program. *Germplasm Resources Information Network – (GRIN).* National Germplasm Resources Laboratory, Beltsville, Maryland (http://www.ars-grin.gov/cgi-bin/npgs/acc/search.pl?accid=PI+659348).

Song, X., Han, Y., Teng, W., Sun, G. and Li, W. (2010) Identification of QTL underlying somatic embryogenesis capacity of immature embryos in soybean (*Glycine max* (L.) Merr.). *Plant Cell Reports* 29, 125–131.

Stam, P. (1995) Marker-assisted breeding. In: Van Ooijen, J.W. and Jansen, J. (eds) Biometrics in plant breeding: applications of molecular markers. In: *Proceedings of the Ninth Meeting of the EUCARPIA Section Biometrics in Plant Breeding*, Wageningen, The Netherlands, pp. 32–44.

Streeck, T.S. and Piepho, H.P. (2010) Genome-wide selection by mixed model ridge regression and extensions based on geostatistical models. *BMC Proceedings* 4 (Suppl, 1), S1–S8.

Stuber, C.W., Polacco, M. and Senior, M. L. (1999) Synergy of empirical breeding, marker-assisted selection, and genomics to increase crop yield potential. *Crop Science* 39, 1571–1583.

Su, C.F., Lu, W.G., Zhao, T.J. and Gai, J.Y. (2010) Verification and fine-mapping of QTL conferring days to flowering in soybean using residual heterozygous lines. *Chinese Science Bulletin* 55, 499–508.

Sun, D., Li, W., Zhang, Z., Chen, Q., Ning, H., Qiu, L. *et al.* (2006) Quantitative trait loci analysis for the developmental behavior of soybean [*Glycine max* (L.) Merr.]. *Theoretical and Applied Genetics* 112, 665–673.

Tanksley, S. and McCouch, S. (1997) Seed banks and molecular maps: unlocking genetic potential from the wild. *Science* 277, 1063–1066.

Tanksley, S.D. and Nelson, J.C. (1996) Advanced backcross QTL analysis: a method of the simultaneous discovery and transfer of valuable QTLs from unadapted germplasm into elite breeding lines. *Theoretical and Applied Genetics* 92, 191–203.

Tanksley, S.D., Young, N.D., Paterson, A.H. and Bonierbale, M.W. (1989) RFLP mapping in plant breeding: new tools for an old science. *Nature Biotechnology* 7, 257–264.

Tester, M. and Langridge, P. (2010) Breeding technologies to increase crop production in a changing world. *Science* 327, 818–822.

Tuyen, D.D., Lal, S.K. and Xu, D.H. (2010) Identification of a major QTL allele from wild soybean (*Glycine soja* Sieb. & Zucc.) for increasing alkaline salt tolerance in soybean. *Theoretical and Applied Genetics* 121, 105–115.

Vandemark, G.J., Fourie, D. and Miklas, P.N. (2008) Genotyping with real-time PCR reveals recessive epistasis between independent QTL conferring resistance to common bacterial blight in dry bean. *Theoretical and Applied Genetics* 117, 513–522.

Varshney, R.K. and Dubey, A. (2009) Novel genomic tools and modern genetic and breeding approaches for crop improvement. *Journal of Plant Biochemistry and Biotechnology* 18, 127–138.

Varshney, R.K., Graner, A. and Sorrells, M.E. (2005) Genomics-assisted breeding for crop improvement. *Trends in Plant Sciences* 10, 621–630.

Varshney, R.K., Close, T.J., Singh, N.K., Hoisington, D.A. and Cook, D.R. (2009a) Orphan legume crops enter the genomics era! *Current Opinion in Plant Biology* 12, 202–210.

Varshney, R.K., Hoisington, D.A., Nayak, S.N. and Graner, A. (2009b) Molecular plant breeding: methodology and achievements. In: Somers, D.J., Langridge, P. and Gustafson, J.P. (eds) *Methods in Molecular Biology, Plant Genomics*. Humana Press, Totowa, New Jersey, pp. 283–304.

Varshney, R.K., Nayak, S.N., May, G.D. and Jackson, S.A. (2009c) Next-generation sequencing technologies and their implications for crop genetics and breeding. *Trends in Biotechnology* 27, 522–530.

Varshney, R.K., Thudi, M., May, G.D. and Jackson, S.A. (2010a) Legume genomics and breeding. *Plant Breeding Reviews* 33, 257–304.

Varshney, R.K., Glaszmann, J.C., Leung, H. and Ribaut, J.M. (2010b) More genomic resources for less studied crops! *Trends in Biotechnology* 28, 452–460.

Visscher, P.M., Haley, C.S. and Thompson, R. (1996) Marker assisted introgression in backcross breeding programs. *Genetics* 144, 1923–1932.

Vuong, T.D., Diers, B.W. and Hartman, G.L. (2008) Identification of QTL for resistance to Sclerotinia stem rot (*Sclerotinia sclerotiorum*) in soybean plant introduction 194639. *Crop Science* 48, 2209–2214.

Vuong, T.D., Sleper, D.A., Shannon, J.G. and Nguyen, H.T. (2010) Novel quantitative trait loci for broad-based resistance to soybean cyst nematode (*Heterodera glycines Ichinohe*) in soybean PI 567516C. *Theoretical and Applied Genetics* 121, 1253–1266.

Walker, D.R., Boerma, H.R., All, J.N. and Parrott, W.A. (2002) Combining *cry1Ac* with QTL alleles from PI 229358 to improve soybean resistance to lepidopteran pests. *Molecular Breeding* 9, 43–51.

Wang, H., Waller, L., Tripathy, S., St. Martin, S.K., Zhou, L., Krampis, K. *et al.* (2010) Analysis of genes underlying soybean quantitative trait loci conferring partial resistance to *Phytophthora sojae*. *The Plant Genome* 3, 23–40.

Wang, J., McClean, P.E., Lee, R., Goss, R.J. and Helms, T. (2008) Association mapping of iron deficiency chlorosis loci in soybean (*Glycine max* L. Merr.) advanced breeding lines. *Theoretical and Applied Genetics* 116, 777–787.

Warrington, C.V., Zhu, S., Parrot, W.A., All, J.N. and Boerma, H.R. (2008) Seed yield of near-isogenic soybean lines with introgressed quantitative trait loci conditioning resistance to corn earworm (Lepidoptera: Noctuidae) and soybean looper (Lepidoptera: Noctuidae) from PI 229358. *Journal of Economic Entomology* 101, 1471–1477.

Wu, X., Blake, S., Sleper, D.A., Shannon, J.G., Cregan, P.B. and Nguyen, H.T. (2009) QTL, additive, and epistatic effects for SCN resistance in PI 437654. *Theoretical and Applied Genetics* 118, 1093–1105.

Yin, Z., Meng, F., Song, H., He, X., Xu, X. and Yu, D. (2010) Mapping quantitative trait loci associated with chlorophyll *a* fluorescence parameters in soybean (*Glycine max* (L.) Merr.). *Planta* 231, 875–885.

Zeng, G., Li, D., Han, Y., Teng, W., Wang, J., Qiu, L. *et al.* (2009) Identification of QTL underlying isoflavone contents in soybean seeds among multiple environments. *Theoretical and Applied Genetics* 118, 1455–1463.

Zhu C., Gore, M., Buckler, E.S. and Yu, J. (2008) Status and prospects of association mapping in plants. *The Plant Genome* 1, 5–20.

20 Improving Protein Content and Nutrition Quality

J. Burstin, K. Gallardo, R.R. Mir, R.K. Varshney and G. Duc

20.1 Introduction

Legumes have been part of the human diet since the early ages of agriculture. Many legume species are still an irreplaceable source of dietary proteins and other nutrients for humans, especially in vegetarian diets of the developing countries (Wang *et al.*, 2003). Legume seeds contain from 16 to 50% protein and provide one third of all dietary protein nitrogen (Graham and Vance, 2003). Thus legumes, as a complement to cereals, make one of the best solutions to protein-calorie malnutrition, particularly in developing countries. Legumes constitute the main component of traditional dishes throughout the world, where maize and beans, rice and lentils, barley and peas, wheat and chickpeas are eaten together. The carbon energy supply that is needed upon germination is stored in grain legume seeds either in the form of oil (soybean, groundnut) or as starch (common bean, pea, faba bean, lentil, chickpea, cowpea, mung bean). In addition, these are also an important source of the 15 essential minerals required by man (Wang *et al.*, 2003), of complex carbohydrates, of soluble fibres and of other compounds that are alternatively considered anti-nutritional or health-promoting: trypsin inhibitors, tannins, phytate, saponins and oligosaccharides have recently been associated with various health benefits, such as protective effects against cardiovascular diseases, cancers and diabetes (Champ, 2002; Clemente *et al.*, 2009). Since the main challenge for grain legumes in human nutrition is linked to their role as a source of protein, the genetic improvement made for protein content, bioavailability and nutritional quality in food legume crops is discussed in this chapter.

20.2 Improving the Protein Content of Grain Legume Seeds

Genetic and environmental variability and heritability

A good amount of genetic variability is required for the genetic improvement of a trait. In the recent past, studies have been conducted in estimating genetic variability for protein content among a number of accessions of various food legume crops, and it has been observed that ample amount of variability is present for this trait (Table 20.1); this varied from 26.5 to 57% in soybean, 20.9–29.2% in common bean, 15.8–32.1% in pea, 22–36% in faba bean, 19–32% in lentil, 16–28% in chickpea, 16–31% in cowpea, 21–31% in mung bean and 16–24% in pigeon pea. These results were compiled from various studies, and thus variability among these could also in part reflect the environmental variability of protein content.

Table 20.1. Principal constituents of grain legume seeds: range of variation (% seed weight).

Content (%)					
Protein	Oil	Starch	Fibre	Sucrose	Reference(s)
Soybean					
35.1–42.0	17.7–21.0	1.5	20.0	6.2	Hedley (2001)
34.7–55.2	6.5–28.7	–	–	–	NGRP (2001)
40.0–45.0	19.0–21.5	–	–	–	Hyten *et al.* (2004)
41.8–49.4	15.2–20.7	–	–	–	Chung *et al.* (2003)
40.4–50.6	13.4–21.2	–	–	–	Brummer *et al.* (1997)
31.7–57.4	–	–	–	–	Jun *et al.* (2008)
26.5–47.6	–	–	–	–	Vollman *et al.* (2000)
Groundnut					
–	44.0–50.0	–	–	–	Lord and Wakelam (1950)
20.7–28.1	–	–	–	–	Dwivedi *et al.* (1990)
16.0–34.0	–	–	–	–	Jambunathan *et al.* (1985)
Common bean					
20.9–27.8	0.9–2.4	41.5	10.0	5.0	Hedley (2001)
23.0–29.2	–	–	–	–	Coelho *et al.* (2009)
Pea					
18.3–31.0	0.6–5.5	45.0	12.0	2.1	Hedley (2001)
24.0–32.4	–	45.5–54.2	8.9–11.9	–	Gabriel *et al.* (2008b)
21.9–34.4	1.4–4.7	18.6–54.5	5.9–12.7	1.3	Bastianelli *et al.* (1998)
20.6–27.3	–	–	–	–	Burstin *et al.* (2007)
15.8–32.1	–	–	–	–	Blixt (1978)
Faba bean					
26.1–38.0	1.1–2.5	37.0–45.6	7.5–13.1	0.4–2.3	Duc *et al.* (1999)
22.4–36.0	1.2–4.0	41.0	12.0	3.3	Hedley (2001)
26.0–29.3	–	42.2–51.5	–	–	Avola *et al.* (2009)
Lentil					
23.0–32.0	0.8–2.0	46.0	12.0	2.9	Hedley (2001)
25.1–29.2	–	46.0–49.7	13.1–14.7	2.1–3.2	Wang *et al.* (2009)
18.6–30.2	–	–	–	–	Hamdi *et al.* (1991)
Chickpea					
15.5–28.2	3.1–7.0	44.4	9.0	2.0	Hedley (2001)
18.7–21.1	–	42–45.1	–	–	Frimpong *et al.* (2009)
17.1–19.8	–	48.0–54.9	–	–	Frimpong *et al.* (2009)
–	–	–	2.7–11.7	–	Cho *et al.* (2002)
12.4–31.5	–	–	–	–	Hulse (1975)
Cowpea					
23.5	1.3	–	–	–	Hedley (2001)
24.8	1.9	–	6.3	–	Kabas *et al.* (2006)
20.9–36.0	2.6–4.2	–	–	–	Oluwatosin (1997)
16.0–31.0	2.4–4.3	–	–	–	Adekol and Oluleye (2007)
23.1–27.3	–	–	–	–	Bliss *et al.* (1973)
Mung bean					
22.9–23.6	1.2	45.0	7.0	1.1	Hedley (2001)
21.0–31.3	1.2–1.6	–	8.9–12.9	–	Anwar *et al.* (2007)
23.7–31.4	–	–	–	–	Lawn and Rebetzke (2006)
Pigeon pea					
19.5–22.9	1.3–3.8	44.3	10.0	2.5	Hedley (2001)
15.9–24.1	–	–	–	–	Upadhyaya *et al.* (2007)

The heritability of a trait is an important factor for efficient selection. It has been shown that environmental conditions significantly influence the seed protein content in most legumes (Dwivedi *et al.*, 1990; Hamdi *et al.*, 1991; Oluwatosin, 1997; Saxena *et al.*, 2002; Frimpong *et al.*, 2009). However, environmental effects in seven different environments had a similar magnitude of effects on protein content to genetic effects when studied in 255 genotypes of pea (Matthews and Arthur, 1985). Gueguen and Barbot (1988) found protein content varying from 18.1 to 27.8% for the cultivar 'Amino', depending on the environment. Karjalainen and Kortet (1987) showed that protein content was positively associated with the sum of temperature from sowing to maturity, and temperature during flowering and beginning of seed filling, while it was negatively associated with July precipitation. Larmure *et al.* (2005) further specified the effect of temperature during seed filling on seed protein content through its effect on the nitrogen:carbon (N:C) ratio. All environmental factors affecting nitrogen nutrition, such as drought stress, soil compactness, root diseases and pests may also influence seed protein content through their impact on nitrogen availability (Biarnès *et al.*, 2000). Foroud *et al.* (1993) described a variable effect of the level and timing of water stress on the protein content of soybean. Aerial disease could also have negative effects by increasing the N:C ratio of assimilates reaching the seeds (Garry *et al.*, 1996). Several authors also reported intraplant variability resulting from fluctuating environmental factors and N:C availability during seed filling of different fruiting nodes (Escalante and Wilcox, 1993; Atta *et al.*, 2004).

Genotypic/environmental effects are also usually significant, even though often of lower magnitude (Matthews and Arthur, 1985; Oluwatosin, 1997; Lawn and Rebetzke, 2006; Burstin *et al.*, 2007; Frimpong *et al.*, 2009). As a result, seed protein content heritability values are variable across studies, depending on the extent of genetic variability analysed, unpredictable environmental variation and experimental design. None the less, seed protein content heritability is generally moderate to high among accessions (20–80%), suggesting that selection for protein can be successful. In the case of soybean, it is interesting to note that the selection of varieties with significantly increased protein content was achieved quite rapidly using backcrossing or recurrent selection (Brim and Burton, 1978; Wilcox and Cavins, 1995; Cober and Voldeng, 2000). This was achieved due to: (i) the large variability present for protein content in germplasm; (ii) the sufficiently high heritability values; and (iii) the mostly additive inheritance (Chung *et al.*, 2003).

To study the genetics of protein content in soybean quantitative trait loci (QTL) analysis was conducted, which resulted in the identification of two major QTL controlling seed protein content in a population derived from a cross between a *Glycine soja* accession from China and a *Glycine max* breeding line (Diers *et al.*, 1992). The *G. soja* parent possessed positive alleles of the QTL-LGI and QTL-LGE. Sebolt *et al.* (2002) investigated the stability of these QTL alleles in the *G. max* genetic background by developing the near-isogenic lines having homozygous alleles of *G. soja* QTL responsible for high protein content. Among the two QTL, only the QTL-LGI allele showed a significant effect in the *G. max* background, although it increased plant height and reduced yield and oil, seed size and maturity. Subsequently, a genetic association was shown between protein content QTL-LGI and other QTL controlling oil content, maturity and yield in different lines recombining within the target regions under introgress (Nichols *et al.*, 2006). These results have confirmed that the linkage between the protein content QTL and yield QTL can be broken. The effect of this QTL-LGI was further validated by marker-assisted selection (MAS) involving improvement in protein content of soybean lines carrying homozygous alleles from the high-protein parent (Yates *et al.*, 2004). Many other soybean seed protein content QTL have been identified in a range of environments and in several genetic backgrounds, which are presented elsewhere (Vuong *et al.*, 2007). QTL controlling seed protein content were investigated on 17 soybean mapping populations and found to be located on all the linkage groups of soybean genome, except for LG B1, D1b, D2, J and O. The identified QTL may be efficiently utilized for developing future soybean varieties with desirable seed components through MAS (http://www.SoyBase.org).

In pea, the selection for yield has led to a rapid and undesirable decrease in protein content. Burstin *et al.* (2007) analysed the variation of protein content in five environments using a recombinant inbred line (RIL) mapping population segregating for *afila* controlling leaf tendril formation, *le* controlling internode length and plant height, and *rms6* controlling plant branching. In this population, eight QTL controlling seed protein content variation were observed. Among these, five were stable across at least two environments, of which two were located in the same genomic region, where the above three genes were located suggesting their pleiotropic effects on several traits. Tar'an *et al.* (2004) also reported two (of three) seed protein QTL in pea showing consistency in many environments. Irzykowska and Wolko (2004) reported five QTL in a cross segregating for the *r* gene controlling starch synthesis and wrinkled seed phenotype. Genetic variability for seed protein content was studied in pigeon pea using wild relatives and improved varieties (Saxena *et al.*, 2002). The results indicated the possibility of developing genotypes possessing high protein content similar to their wild relatives and seed characters similar to cultivated types.

Seed protein content, yield and related traits

Highly negative correlations between protein and oil are well documented in soybean and other food legume crops (Dwivedi *et al.*, 1990; Oluwatosin, 1997; Hyten *et al.*, 2004). Correlation between starch and protein content has also been reported as negative when the gene pool was considered in pea (Bastianelli *et al.*, 1998) and chickpea (Frimpong *et al.*, 2009). There are often negative correlations between protein content and yield, but it has also been reported variously as either non-significant or positive (Bliss *et al.*, 1973; Hamdi *et al.*, 1991; Oluwatosin, 1997; Cober and Voldeng, 2000; Lawn and Rebetzke, 2006; Burstin *et al.*, 2007; Frimpong *et al.*, 2009). Similarly negative correlations have been reported between seed size and protein content in pigeon pea (Saxena *et al.*, 2002), but some promising lines with high protein content and large seed size have been obtained at ICRISAT, suggesting the possibility of simultaneous improvement in protein content and yield-contributing traits.

Potential limitations in protein content improvement

The rate of protein accumulation depends on both the sink strength capacity of seeds and the source strength of vegetative parts for nitrogenous assimilates resulting from nitrogen acquisition, assimilation, transport and mobilization (Salon *et al.*, 2001; Munier-Jolain *et al.*, 2008). The rate of dry matter accumulation depends on the accumulation of all constituents, including carbohydrates and oil, and relates mostly to carbon supply through efficient photosynthesis and effective biosynthetic pathways. De-podding and defoliation studies were conducted in several legume species in order to analyse the effect of source:sink ratio variation on seed constituents' accumulation. Burstin *et al.* (2007) analysed the genetic variability of the effect of de-podding on the seed protein content of eight pea genotypes. The effects of genotype, de-podding and genotype × de-podding on seed protein content were all significant. For all genotypes, seed protein increased dramatically when the source:sink ratio increased. However, there was still a significant variation for seed protein content among the eight genotypes once de-podded. This suggests that N source capacity is the major limiting factor for seed protein content in pea, but also that the maximal rate of protein accumulation in the seed is significant. Similar results were obtained for soybean (Proulx and Naeve, 2009; Rotundo *et al.*, 2009). Three types of genes/QTL could be identified for seed protein content (Burstin *et al.*, 2007; Gallardo *et al.*, 2008): (i) major genes controlling developmental processes and having pleiotropic effect on the whole plant phenotype and on the source–sink structure; (ii) genes of plant metabolism controlling source–sink relationships at the plant metabolism level; and (iii) genes solely controlling the capacity of seeds to accumulate storage compounds. The impact of these different types of effectors on yield will probably be different.

Improving nitrogen supply to the seed

The establishment and optimal functioning of symbiosis in food legumes, together with efficient mobilization of assimilates from vegetative parts to the seeds, controls the availability of nitrogen to the growing seeds. Many genes involved in the control of nodulation have recently been identified (Ferguson *et al.*, 2010). Pea mutants with absence of N_2 fixation activity produce lower seed yield and protein content, which can be alleviated by adequate mineral fertilization, whereas an autoregulation mutant of pea displaying a supernodulating phenotype has a reduced shoot biomass and seed yield, associated with higher seed protein content (Sagan *et al.*, 1993). A reduced root development was detected in supernodulating mutants of soybean or pea (Olsson *et al.*, 1989; Bourion *et al.*, 2007), which may be explained by a competition effect of nodules for C, with a secondary effect of lower access to soil resources. The importance of fine tuning between root and nodule establishment and functioning for final C:N equilibrium in seeds is illustrated by these extreme mutant phenotypes. The efficiency of N-fixing symbiosis relies on the carbon supply from aerial parts to the root parts. In a recent study, Bourion *et al.* (2010) located QTL for root development in the region of QTL for seed protein content. However, QTL should be refined and further work is needed in order to define the ideotype of root/nodule/shoot development.

Improving seed sink strength

Functional interactions exist among different seed constituents: for example, the disruption of the *r* gene abolishes starch synthesis in pea seeds, leading to a wrinkled seed phenotype having a profound impact on seed metabolism, where elevated sucrose content impacts the accumulation of storage protein families (Wang and Hedley, 1993). By knocking down the accumulation of one of the constituents, the percentage of the others will increase. However, this may have a detrimental effect on seed yield. This strategy is viable if it allows production of specific seed products for market purposes. Strategies to increase seed sink strength have been tested through the manipulation of amino acids and sucrose flux to the developing embryo (Weber *et al.*, 2005). It has been shown that seed-specific overexpression of an amino acid permease in pea increases amino acid supply to, and the level of protein in, the seed (Weigelt *et al.*, 2008). This indicates a stimulation of storage protein synthesis by increased amino acid availability. Seed-specific overexpression of a bacterial phosphoenol pyruvate carboxylase in *Vicia narbonensis* increased seed protein content with a compensatory effect on seed number and seed weight (Rolletschek *et al.*, 2004). In general, the genetics of seed protein content largely remain a mystery. However, with the advent of high-throughput genotyping and phenotyping tools, we think that two directions should be pursued in order to gain on efficiency in breeding: whole genome selection and plant modelling of interacting processes.

20.3 Improving Seed Protein Composition for Better Digestibility and Nutrient Balance

Seed storage proteins are synthesized during seed development and confined in membrane-bound organelles until they are hydrolysed upon germination to provide carbon and nitrogen skeletons for the developing seedlings. Grain legume storage proteins include two major classes of salt-soluble globulins: the 7/8S vicilins and 11/12S legumins, each of which consists of a family of closely related molecules (Boulter and Croy, 1997). Proteome reference maps have been developed for soybean (Hadjduch *et al.*, 2005), pea (Bourgeois *et al.*, 2009) and lentil (Scippa *et al.*, 2010) revealing a complex composition of grain legume globulins. Other proteins (albumins,

Table 20.2. Range of variation (g/100 g protein) in four amino acids of grain legume seeds.

Species	Lysine	Methionine	Cysteine	Tryptophan	Reference
Soybean	22.4–24.1	4.4–8.8	5.1–7.3	4.4–5.1	Panthee *et al.* (2006)
Pea	15.5–19.7	2.0–2.4	2.9–3.6	2.0–2.7	Gabriel *et al.* (2008b)
	14.8–23.0	2.1–3.3	2.9–4.2	1.6–3.2	Bastianelli *et al.* (1998)
Faba bean	17.3–21.6	2.3–2.9	2.9–4.3	2.0–3.2	Duc *et al.* (1999)
Lentil	4.5–12.6	1.2–1.7	0.4–0.5	–	Rozan *et al.* (2001)
Cowpea	–	1.9–2.8	1.6–2.1	3.0–3.7	Bliss *et al.* (1973)
	4.9–9.0	0.52–2.05	0.84–2.24	0.72–1.91	Oluwatosin (1997)

glutelins) complete the protein fraction of the seed, and the composition in the different protein fractions depends on the species (Boulter and Croy, 1997; Gallardo *et al.*, 2008; Montoya *et al.*, 2010).

The various storage proteins have different *in vitro* and *in vivo* digestibility depending on their structural characteristics (Crévieu *et al.*, 1997; Gabriel *et al.*, 2008a, b; Montoya *et al.*, 2010). Although studies on grain legume seed protein digestibility in the human are scarce (Mariotti *et al.*, 2001), some relevant information can be found in digestibility surveys on monogastric animals. Generally, β-sheet structures are less digestible than α-helix structures and glycosylated proteins are less susceptible to hydrolysis. Several attempts have been made in order to improve protein digestibility through the suppression or overexpression of a particular storage protein family (Burow *et al.*, 1993), but these did not generally yield the expected outcome. Therefore, searching variability among germplasms for protein composition patterns favourable to digestibility has been suggested as an alternate strategy for improvement (Montoya *et al.*, 2010). Several authors reported intra-specific variability for seed protein composition in food legume crops such as pea (Tzitzikas *et al.*, 2006), lentil (Scippa *et al.*, 2010) and soybean (Natarajan *et al.*, 2006). This variability was also observed for other minor proteins, which may also have a role in protein digestibility (Vigeolas *et al.*, 2008).

Essential amino acids are important in human nutrition, as they cannot be synthesized and there is therefore dependency on dietary sources of these amino acids. Among these, tryptophan and the sulfur-containing amino acid methionine are the most limiting in legume seeds (Table 20.2). By contrast, legume seed proteins are rich in lysine while cereal seed proteins are low in this amino acid (Wang *et al.*, 2003).

Genetic manipulations have been used in attempts to improve seed quality, in particular towards increasing methionine levels. The strategy employed in improving amino acid balance in legumes is to modify storage protein composition in favour of the accumulation of sulfur-rich proteins. To this end, transgenic plants expressing the sulfur-rich 2S albumin genes from Brazil nut and sunflower in the seeds of soybean and lupin (*Lupinus angustifolius* L.), respectively, were developed (Altenbach *et al.*, 1989; Molvig *et al.*, 1997). Although these transgenic plants increased seed methionine levels, the introduced sulfur-rich sink proteins generally had allergenic properties (Pastorello *et al.*, 2001). Importantly, the accumulation of such foreign proteins in seeds occurred at the expense of other sulfur compounds, such as free sulfur amino acids and glutathione (Tabe and Droux, 2002), which thus limited the rate of synthesis of sulfur amino acids during seed development. Activating the synthesis of essential amino acids might therefore be a possible route for improvement of amino acid balance in legume seeds.

Lysine regulates the flux of carbon and nitrogen towards methionine synthesis (see Jander and Joshi, 2010). A similar feedback inhibition was also observed for the tryptophan biosynthetic pathway, where tryptophan itself inhibits its own biosynthetic pathway by negatively regulating anthranilate synthase, which catalyses the conversion of chorismate to anthranilate (Ufaz and Galili, 2008). Interestingly, the modulation of feedback inhibition in these pathways allowed an increase in the synthesis of some amino acids. For example, the introduction of genes encoding anthranilate synthase forms insensitive to feedback inhibition enhanced the accumulation of tryptophan in seeds (Ufaz and Galili, 2008; Ishimoto *et al.*, 2010).

These findings open perspectives towards modification of the synthesis of essential amino acids in legume seeds through the identification of feedback-insensitive natural allelic variants in genes of amino acid biosynthetic pathways. However, the up-accumulation of amino acids under free forms in seeds could have negative effects on agronomic traits. For example, the germination ability of transgenic seeds containing very high levels of free lysine or tryptophan was reduced (Zhu and Galili, 2003; Wakasa *et al.*, 2006).

Availability of sulfur and nitrogen in the environment determines the amino acid balance in mature seeds. Legume seeds produced in sulfur-limiting conditions, but with adequate nitrogen, generally contained reduced levels of sulfur-rich storage proteins and accumulated more sulfur-poor proteins (Tabe and Droux, 2002). In soybean, Paek *et al.* (1997) reported an increase in the proportion of sulfur-poor protein as protein concentration increased. Wilcox and Shibles (2001), by contrast, found a constant sulfur:nitrogen ratio in a population segregating for seed protein content, but seed yield was not high in this population and thus sulfur was probably not limiting in this context. In cowpea, Bliss *et al.* (1973) found a positive correlation between seed protein content and protein methionine content. In chickpea, the application of nitrogen, phosphorus and sulfur fertilizers improves the levels of protein and essential amino acids (Gupta and Singh, 1982; Williams and Singh, 1987). In pea seeds, reduced levels of sulfur-rich proteins in conditions of limited sulfur availability were shown to be primarily a consequence of reduced levels of their mRNA (Higgins *et al.*, 1986). *O*-acetylserine and free methionine, but not free cysteine, were implicated as signalling molecules controlling the expression of genes for sulfur-rich storage proteins in legume seeds (Tabe *et al.*, 2010 and references therein).

These findings indicate that the capacity of legume plants to regulate the flux of sulfur and nitrogen compounds to the seeds should be considered if the accumulation of sulfur-rich storage proteins is to be increased. Sulfate is one of the dominant forms of sulfur found in the phloem supplying pods during legume seed development (Tabe and Droux, 2001). In plants, sulfate can either be reduced to sulfide leading to the synthesis of cysteine, the precursor for methionine synthesis, or it can be stored in the vacuoles. Considering the importance of sulfate for the synthesis of sulfur compounds, one limiting step for accumulation of sulfur-rich proteins could be the uptake of sulfate by the root and its distribution within the plant by membrane-localized sulfate transporters. In several species strongly regulated by sulfur deficiency, sulfate transporters of high affinity have been identified that facilitate the uptake of sulfate by the root (SULTR1-1 and SULTR1-2) or its translocation from source to sink (SULTR1-3) (see Hawkesford and De Kok, 2006). Other transport forms of sulfur in the phloem include glutathione and *S*-methylmethionine, which can reconverted to cysteine and methionine (Bourgis *et al.*, 1999). Interestingly, a characterization of knockdown *Arabidopsis* mutants for isozyme 2 of homocysteine methyltransferase, which converts *S*-methylmethionine to methionine, suggests that increasing the transport of *S*-methylmethionine from vegetative tissues to seeds could increase seed methionine levels (Lee *et al.*, 2008).

20.4 Breeding for Minor Compounds (Seed Protein Bioavailability for Humans)

Minor seed compounds in grain legumes exert either positive or negative influence on protein bioavailability by impacting digestibility or acceptability. Studies have documented the possibility of improving the nutritional values of grain legumes as animal feeds, mainly for monogastric animals. However, these findings cannot be easily extrapolated to humans because, even if monogastric, human beings have their own physiology varying with age, and the human diet is composed of a diversity of ingredients generating high dilutions and complex interactions. This is why seed compounds termed anti-nutritionals in feeds have been removed by breeding, even though some of them may have a positive role in human chronic disease prevention, i.e. cancer, cardiovascular disease, diabetes and obesity. The genetic variability available for these compounds, including trypsin inhibitors, lectins, α-galactosides and phytic acids, may thus help breeders significantly improve the protein component of human diets (Table 20.3).

Trypsin inhibitors are present in most grain legume seeds, but these inhibitory activities in soybean seeds are usually reduced by processing. However, in soybean, null alleles have been identified for both Bowman–Birk and Kunitz trypsin inhibitors, facilitating the development of low-trypsin-inhibitor cultivars (Clarke and Wiseman, 2000). In pea, large genetic variability is available for the activity of Bowman–Birk trypsin/chymotrypsin inhibitor proteins (TIAs) (Bastianelli *et al.*, 1998). The polymorphism in coding and promoter sequences of genes at the *Tri* locus accounts for most of the variation in TIAs, and this allows the initiation of MAS (Page *et al.*, 2002). Even if low TIA activity is of benefit in pig or poultry feed digestibility, its high content in foods should be of positive value because the presence of trypsin inhibitors in pea reduced levels of HT29 colonic cancer cells *in vitro* (Domoney *et al.*, 2009). If validated *in vivo*, this would encourage breeding for high trypsin inhibitor content in food or nutraceutical applications.

Most grain legume cotyledons contain lectins (haemagglutinins), polysaccharide-binding proteins that bind to glycoprotein on the epithelial surface of the small intestine, interfering with nutrient absorption and increasing the production of mucins and a loss of plasma proteins in the intestinal lumen (Pusztai, 1989). In plants, lectins are very diverse and are involved in plant defences (Etzler, 1985) or symbiosis with *Rhizobia* (van Rhijn *et al.*, 2001). Although some natural variability exists for lectin haemagluglutinin activity in germplasm (Valdebouze *et al.*, 1980), its low content and toxicity do not allow for the definition of a breeding target for this trait.

The glycosides vicine and convicine (VC) are in abundance in the cotyledons of faba bean seeds; in general, their concentration varies from 6 to 14 g/kg DM in mature seeds of released cultivars. The presence of VC causes favism, which is an acute haemolytic problem caused by the ingestion of faba beans in G6PD-deficient human individuals (Arese and De Flora, 1990). A mutant allele, *vc-* has been identified that reduces VC contents by ten- to 20-fold (Duc *et al.*, 1989). Since the determination of VC content by chemical analysis in seeds is costly, molecular markers have been proposed to assist in selection for genotypes of low VC content (Gutierrez *et al.*, 2006).

Flavonoids are major phenolic compounds involved in the determination of seed coat colour and tannin synthesis (Nozzolillo *et al.*, 1989, Caldas and Blair, 2009), tannins are binding to proteins and reducing their digestibility. In pea and faba bean, a single gene mutation has a pleiotropic effect in eliminating tannins from the seed coat and determining the white flower trait. The zero tannin trait increases protein digestibility in pigs or poultry by about 10% when compared with tannin-containing lines (Grosjean *et al.*, 1999; Crepon *et al.*, 2010). This quality trait is economically valuable for feed efficiency, and zero-tannin varieties have been bred in Europe. In common bean, the genetics of seed coat colour and tannin content was shown to be under the control of 12 QTL (Caldas and Blair, 2009), but limited data are available for individual phenolic compounds. The removal of tannins from the human diet may have positive nutritional effects, but it has a certain impact on the level of astringency, with positive or negative consumer reactions according to dietary habits. The health benefits of proanthocyanidins may deserve some attention. The diverse colours of common beans are suggested to be important sources of dietary antioxidants (Beninger and Hosfield, 2003).

The anti-nutritional effect of phytic acid is associated with mineral complexing (especially Zn, Ca and Fe) and the inactivation of digestive enzymes; it induces a reduction in bioavailability of minerals and proteins in foods and hence may be of nutritional concern (Frossard *et al.*, 2000). On the other hand, phytic acid may have protective effects such

Table 20.3. Levels of some minor constituents of grain legume seeds.

Species	TIA (TIU/mg)	Tannins (g/kg)	Saponin (g/kg)	Total α-galactosides (% DM)	Phytic acid (g/kg)	Reference
Soybean	–	–	6.5	–	–	Kadlec *et al.* (2001)
	–	–	–	–	6.2–20.5	Saghai Maroof *et al.* (2009)
	43–83	–	–	–	–	Guillamon *et al.* (2008)
Common bean	–	–	2.3–3.5	0.4–8.0	–	Kadlec *et al.* (2001)
	–	0–38.5	–	–	–	Caldas and Blair (2009)
	–	–	–	–	2.9–17.8	Blair *et al.* (2009)
	17–51	–	–	–	–	Guillamon *et al.* (2008)
Pea	–	–	1.1	2.3–9.6	–	Kadlec *et al.* (2001)
	1.0–14.6	0.04–7.4	0.3–1.0	3.6–10	1.3–10.2	Bastianelli *et al.* (1998)
	1.9–6.8	–	–	–	–	Gabriel *et al.* (2008)
	6.0–15.0	–	–	–	–	Guillamon *et al.* (2008)
Faba bean	0.3–5.3	0.1–10.4	–	1.4–6.2	3.8–13.4	Duc *et al.* (1999)
	–	–	0.1	1.0–4.5	–	Kadlec *et al.* (2001)
	–	2.1–3.2	–	–	–	Avola *et al.* (2009)
	0.8–3.6	–	–	–	–	Filipetti *et al.* (1999)
	5.0–10.0	–	–	–	–	Guillamon *et al.* (2008)
Lentil	–	–	1.1	1.8–7.5	–	Kadlec *et al.* (2001)
	1.9–2.8	3.4–6.1	–	–	6.2–8.8	Wang *et al.* (2009)
	3.0–8.0	–	–	–	–	Guillamon *et al.* (2008)
Chickpea	–	–	2.3	2.0–7.6	–	Kadlec *et al.* (2001)
	12.7	–	–	–	–	Singh and Jambunatham (1981)
	10.3	–	–	–	–	Singh and Jambunatham (1981)
	15.0–19.0	–	–	–	–	Guillamon *et al.* (2008)
Cowpea	12.0–16.6[a]	–	–	–	–	Vasconcelos *et al.* (2010)
	–	0.3–6.9	–	–	–	Plahar *et al.* (1997)
	–	–	–	–	9.9–16.4	Singh (1999)

TIA, trypsin/chymotrypsin inhibitor activity. [a] De-hulled seeds.

as a decrease in the risk of iron-mediated colonic cancer and lowering of serum cholesterol and triglycerides (Champ, 2002). In common bean, five QTL were identified that controlled total and net seed phytate content (Blair *et al.*, 2009).

Lipoxygenase activity can cause unpleasant tastes and aromas when reacting with seed lipids. In soybean and pea, null mutants were found for three and two LOX genes, respectively. Their molecular characterization (Forster *et al.*, 1999; Lenis *et al.*, 2010) has been accomplished and offers the possibility of breeding for lipoxygenase-free varieties. On the other hand, a large number of different saponins also exist in legumes (Heng *et al.*, 2006). These compounds, although contributing to the bitterness of pea as well as of soybean, have positive hypocholesterolaemic and anti-carcinogenic effects that have also been studied (Champ, 2002). Some genetic diversity has already been described for both the quantity and quality of seed saponins (Table 20.3; Heng *et al.* 2006), but its genetic basis is unknown.

20.5 Conclusion

There is an urgent need to develop new references on the health-promoting and nutritional values of grain legumes. Determining the value of particular fractions in nutraceutic applications may provide new markets with higher added value. Although the effects and cost of the technological treatment of bioactive components have not been calculated, this may help in choosing between genetic strategies and technological processes. Several studies have demonstrated the effectiveness of proteins in protection against parasitic insects or fungi. Attempts to modify the contents of minor bioactive compounds will involve an appraisal of their consequences on plant behaviour in regard to biotic or abiotic stresses.

References

Adekola, O.F. and Oluleye, F. (2007) Influence of mutation induction on the chemical composition of cowpea *Vigna unguiculata* (L.) Walp. *African Journal of Biotechnology* 6, 2143–2146.

Altenbach, S.B., Pearson, K.W., Meeker, G., Staraci, L.C. and Sun, S.M. (1989) Enhancement of the methionine content of seed proteins by the expression of a chimeric gene encoding a methionine–rich protein in transgenic plants. *Plant Molecular Biology* 13, 513–522.

Anwar, F., Latif, S., Przybylski, R., Sultana, B. and Ashraf, M. (2007) Chemical composition and antioxidant activity of seeds of different cultivars of mungbean. *Journal of Food Science* 72, 503–510.

Arese, P. and De Flora, A. (1990) Pathophysiology of hemolysis in glucose-6-phosphate dehydrogenase deficiency. *Seminars in Hematology* 27, 1–40.

Atta, S., Maltese, S. and Cousin, R. (2004) Protein content and dry weight of seeds from various pea genotypes. *Agronomie* 24, 257–266.

Avola, G., Gresta, F. and Abbate, V. (2009) Diversity examination based on physical, technological and chemical traits in a locally grown landrace of faba bean (*Vicia faba* L. var major). *International Journal of Food Science and Technology* 44, 2568–2576.

Bastianelli, D., Grosjean, F., Peyronnet, C., Duparque, M. and Régnier, J.M. (1998) Feeding value of pea (*Pisum sativum* L.) 1. Chemical composition of different categories of pea. *Animal Science* 67, 609–619.

Beninger, C.W. and Hosfield, G.L. (2003) Antioxidant activity of extracts, condensed tannin fractions, and pure flavonoids from *Phaseolus vulgaris* L. seed coat color genotypes. *Journal of Agriculture and Food Chemistry*, 51, 7879–7883.

Biarnès, V., Faloya, V., Chambenoit, C. and Roullet, G. (2000) Facteurs de variabilité: Teneur en protéines des graines de pois protéagineux: les effets du milieu et de la variété. *Perspectives Agricoles* 261, 32–34.

Blair, M.W., Sandoval, T.A., Caldas, G.V., Beebe, S.E. and Páes, M.I. (2009) Quantitative trait locus analysis of seed phosphorus and seed phytate content in a recombinant inbred line population of common bean. *Crop Science* 49, 237–246.

Bliss, F.A., Barker, L.N., Franckowiak, J.D. and Hall, T.C. (1973) Genetic and environmental variation of seed yield, yield components, and seed protein quantity and quality of cowpea. *Crop Science* 13, 656–660.

Blixt, S. (1978) Problems relating to pea breeding. *Agri Hortique Genetica* 36, 56–87.
Boulter, D. and Croy, R.R.D. (1997) The structure and biosynthesis of legume seed storage proteins: a biological solution to the storage of nitrogen in seeds. *Advances in Botanical Research* 27, 1–84.
Bourgeois, M., Jacquin, F., Savois, V., Sommerer, N., Labas, V., Henry, C. *et al.* (2009) Dissecting the proteome of pea mature seeds reveals the phenotypic plasticity of seed protein composition. *Proteomics* 9, 254–271.
Bourgis, F., Roje, S., Nuccio, M.L., Fisher, D.B., Tarczynski, M.C., Li, C. *et al.* (1999) S-methylmethionine plays a major role in phloem sulfur transport and is synthesized by a novel type of methyltransferase. *Plant Cell* 11, 1485–1498.
Bourion, V., Laguerre, G., Depret, G., Voisin, A.S., Salon, C. and Duc, G. (2007) Genetic variability in nodulation and root growth affects nitrogen fixation and accumulation in pea. *Annals of Botany* 100, 589–598.
Bourion, V., Rizvi, S.M.H., Fournier, S., de Larambergue, H., Galmiche, F., Marget, P. *et al.* (2010) Genetic dissection of nitrogen nutrition in pea through a QTL approach of root, nodule, and shoot variability: *Theoretical and Applied Genetics* 121, 71–86.
Brim, C.A. and Burton, J.W. (1978) Recurrent selection in soybeans. II. Selection for increased percent protein in seeds. *Crop Science* 19, 494–498.
Brummer, E.C., Graef, G.L., Orf, J., Wilcox, J.R. and Shoemaker, R.C. (1997) Mapping QTL for seed protein and oil content in eight soybean populations. *Crop Science* 37, 370–378.
Burow, M.D., Ludden, P.W. and Bliss, F.A. (1993) Suppression of phaseolin and lectin in seeds in common bean, *Phaseolus vulgaris* L.: increased accumulation of 54 kDa polypeptides is not associated with higher seed methionine concentrations. *Molecular and General Genetics* 241, 431–439.
Burstin, J., Marget, P., Huart, M., Moessner, A., Mangin, B., Duchene, C. *et al.* (2007) Developmental genes have pleiotropic effects on plant morphology and source capacity, eventually impacting on seed protein content and productivity in pea. *Plant Physiology* 144, 768–781.
Caldas, G.V. and Blair, M.W. (2009) Inheritance of seed condensed tannins and their relationship with seed coat color and pattern genes in common bean (*Phaseolus vulgaris* L.). *Theoretical and Applied Genetics* 119, 131–142.
Champ, M. (2002) Non-nutrient bioactive substances of pulses. *British Journal of Nutrition* 3, S307–S319.
Cho, S.H., Kumar, J. and Shultz, J.L. (2002) Mapping genes for double podding and other morphological traits in chickpea. *Euphytica* 128, 285–292.
Chung, J., Babka, H.L., Graef, G.L., Staswisk, P.E., Lee, D.J., Cregan, P.B. *et al.* (2003) The seed protein, oil, yield QTL on linkage group I. *Crop Science* 43, 1053–1067.
Clarke, E.J. and Wiseman, J. (2000) Developments in plant breeding for improved nutritional quality of soya beans II. Anti-nutritional factors. *Journal of Agricultural Science* 134, 125–136.
Clemente, A., Moreno, F.J., Marín-Manzano, M.C., Jiménez, E. and Domoney, C. (2009) The cytotoxic effect of Bowman–Birk isoinhibitors, IBB1 and IBBD2, from soybean (*Glycine max*) on HT29 human colorectal cancer cells is related to their intrinsic ability to inhibit serine proteases. *Molecular and Nutrition Food Research* 54, 396–405.
Cober, E.R. and Voldeng, H.D. (2000) Developing high-protein, high-yield soybean populations and lines. *Crop Science* 40, 39–42.
Coelho, R.C., Faria, M.A., Rocha, J., Reis, A., Oliveira, M.P.P.P. and Nunes, E. (2009) Assessing genetic variability in germplasm of *Phaseolus vulgaris* L. collected in Northern Portugal. *Scientia Horticulturae* 122, 333–338.
Crepon, K., Marget, P., Peyronnet, C., Carrouee, B., Arese, P. and Duc, G. (2010) Nutritional value of faba bean (*Vicia faba* L.) seeds for feed and food. *Field Crops Research* 115, 329–339.
Crévieu, I., Carré, B., Chagneau, A.M., Quillien, L., Guéguen, J. and Bérot, S. (1997) Identification of resistant pea (*Pisum sativum* L.) proteins in the digestive tract of chickens. *Journal of Agricultural and Food Chemistry* 45, 1295–1300.
Diers, B.W., Keim, P., Fehr, W.R. and Shoemaker, R.C. (1992) RFLP analysis of soybean seed protein and oil content. *Theoretical and Applied Genetics* 83, 608–612.
Domoney, C., Charlton, A., Chinoy, C., Vigeolas, H., Geigenberger, P. and Clemente, A. (2009) Manipulating seed quality traits in pea (*Pisum sativum* L.) for improved feed and food. *Grain Legumes* 52, 12–13.
Duc, G., Sixdenier, G., Lila, M. and Furstoss, V. (1989) Search of genetic variability for vicine and convicine content in *Vicia faba* L. A first report of a gene which codes for nearly zero-vicine and zero-convicine contents. In: Huisman, J., Van der Poel, A.S.M.J.F.B. and Liener, I.E. (eds) *Recent Advances of Research in Fanti-nutritionnal Factors in Legume Seeds*. Pudoc, Wageningen, The Netherlands, pp. 305–313.
Duc, G., Marget, P., Esnault, R., Le Guen, J. and Bastianelli, D. (1999) Genetic variability for feeding value of faba bean seeds (*Vicia faba* L.). Comparative chemical composition of isogenics involving zero-tannin and zero-vicine genes. *Journal of Agricultural Science* 133, 185–196.

Dwivedi, S.L., Jambunathan, R., Nigam, S.N., Raghunath, K., Shankar, K.R. and Nagabhushanam, G.V.S. (1990) Relationship of seed mass to oil and protein contents in peanut (*Arachis hypogaea* L.). *Peanut Science* 17, 48–52.

Escalante, E.E. and Wilcox, J.R. (1993) Variation in seed protein among nodes of normal-protein and high-protein soybean genotypes. *Crop Science* 33, 1164–1166.

Etzler, M.E. (1985) Plant lectins: molecular and biological aspects. *Annual Review of Plant Physiology* 36, 209–234.

Ferguson, B.J., Indrasumunar, A., Hayashi, S., Lin, M.H., Lin, Y.H., Reid D.E. *et al.* (2010) Molecular analysis of legume nodule development and autoregulation. *Journal of Integrative Plant Biology* 52, 61–76.

Filipetti, A., Azadegan, G.H. and De Pace, C. (1999) Breeding strategies for seed protein content and trypsin inhibitors inferred from combining ability and heterosis in test-crosses of *Vicia faba*. *Plant Breeding* 118, 411–416.

Foroud, N., Mûndel, H.H., Saindon, G. and Entz, T. (1993) Effect of level and timing of moisture stress on soybean yield, protein, and oil responses. *Field Crops Research* 37, 195–209.

Forster, C., North, H., Afzal, N., Domoney, C., Hornostaj, A., Robinson, D.S. *et al.* (1999) Molecular analysis of a null mutant for pea (*Pisum sativum* L.) seed lipoxygenase-2. *Plant Molecular Biology* 39, 1209–1220.

Frimpong, A., Sinha, A., Tar'an, B., Warkentin, T.D., Gossen, B.D. and Chibbar, R.N. (2009) Genotype and growing environment influence chickpea (*Cicer arietinum* L.) seed composition. *Journal of the Science of Food and Agriculture* 89, 2052–2063.

Frossard, E., Bucher, M., Machler, F., Mozafar, A. and Hurrell, R. (2000) Potential for increasing the content of bio-availability of Fe, Zn and Ca in plant for human nutrition. *Journal of the Science of Food and Agriculture* 80, 861–869.

Gabriel, I., Lessire, M., Juin, H., Burstin, J., Duc, G., Quillien, L. *et al.* (2008a) Variation in seed protein digestion of different pea (*Pisum sativum* L.) genotypes by cecectomized broiler chickens: 1. endogenous amino acid losses, true digestibility and *in vitro* hydrolysis of proteins. *Livestock Science* 113, 251–261.

Gabriel, I., Quillien, L., Cassecuelle, F., Marget, P., Juin, H., Lessire, M. *et al.* (2008b) Variation in seed protein digestion of different pea (*Pisum sativum* L.) genotypes by cecectomized broiler chickens: 2. Relation between *in vivo* protein digestibility and pea seed characteristics, and identification of resistant pea polypeptides. *Livestock Science* 113, 262–273.

Gallardo, K., Thompson, R. and Burstin, J. (2008) Reserve accumulation in legume seeds. *CR Biologies* 331, 755–762.

Garry, G., Tivoli, B., Jeuffroy, M.H. and Citharel, J. (1996) Effects of Ascochyta blight caused by *Mycosphaerella pinodes* on the translocation of carbohydrates and nitrogenous compounds from the leaf and hull to the seed of dried-pea. *Plant Pathology* 45, 769–777.

Graham, P.H. and Vance, C.P. (2003) Legumes. Importance and constraints to greater use. *Plant Physiology* 131, 872–877.

Grosjean, F., Barrier-Guillot, B., Bastianelli, D., Rudeaux, F., Bordillon, A. and Peyronnet, C. (1999) Feeding value of three categories of pea (*Pisum sativum* L.) for poultry. *Animal Science* 69, 591–599.

Gueguen, J. and Barbot, J. (1988) Quantitative and qualitative variability of pea (*Pisum sativum* L.) protein composition. *Journal of the Science of Food and Agriculture* 42, 209–224.

Guillamon, E., Pedrosa, M.M., Burbano, C., Cuadrado, C., de Cortez Sanchez, M. and Muzquiz, M. (2008) The trypsin inhibitors present in seed of different grain legume species and cultivar. *Food Chemistry* 107, 68–74.

Gupta, N. and Singh, R.S. (1982) Effect of nitrogen, phosphorus and sulfur nutrition on protein and amino acids in Bengal gram (*Cicer arietinum* L.). *Indian Journal of Agricultural Research* 16, 113–117.

Gutierrez, N., Avila, C.M., Duc, G., Marget, P., Suso, M.J., Moreno, M.T. *et al.* (2006) CAPs markers to assist selection for low vicine and convicine contents in faba bean (*Vicia faba* L.). *Theoretical and Applied Genetics* 114, 59–66.

Hajduch, M., Ganapathy, A., Stein, J.W. and Thelen, J.J. (2005) A systematic proteomic study of seed filling in soybean. Establishment of high-resolution two-dimensional reference maps, expression profiles, and an interactive proteome database. *Plant Physiology* 137, 1397–1419.

Hamdi, A., Erskine, W. and Gates, P. (1991) Relationships among economic characters in lentil. *Euphytica* 57, 109–116.

Hawkesford, M.J. and De Kok, L.J. (2006) Managing sulfur metabolism in plants. *Plant Cell Environment* 29, 382–395.

Hedley, C.L. (2001) *Carbohydrate in Grain Legume Seeds*. CAB International, Wallingford, UK, pp. 1–11.

Heng, L., Vincken, J.P., van Koningsveld, G., Legger, A., Gruppen, H., van Boekel, T. *et al.* (2006) Bitterness of saponins and their content in dry peas. *Journal of the Science of Food and Agriculture* 86, 1225–1231.

Higgins, T.J., Chandler, P.M., Randall, P.J., Spencer, D., Beach, L.R., Blagrove, R.J. *et al.* (1986) Gene structure, protein structure, and regulation of the synthesis of a sulfur-rich protein in pea seeds. *Journal of Biological Chemistry* 261, 11124–11130.

Hulse, J.H. (1975) Problems of nutritional quality of pigeonpea and chickpea and prospectus of research. In: *International Workshop on Grain Legumes*, 13–16 January, ICRISAT, Hyderabad, India, pp. 189–208.

Hyten, D.L., Pantalone, V.R., Sams, C.E., Saxton, A.M., Landau-Ellis, D., Stefaniak, T.R. *et al.* (2004) Seed quality QTL in a prominent soybean population. *Theoretical and Applied Genetics* 109, 552–561.

Irzykowska, L. and Wolko, B. (2004) Interval mapping of QTLs controlling yield-related traits and seed protein coat in *Pisum sativum*. *Journal of Applied Genetics* 45, 297–306.

Ishimoto, M., Rahman, S.M., Hanafy, M.S., Khalafalla, M.M., El-Shemy, H.A., Nakamoto, Y. *et al.* (2010) Evaluation of amino acid content and nutritional quality of transgenic soybean seeds with high-level tryptophan accumulation. *Molecular Breeding* 25, 313–326.

Jambunathan, R., Raju, S.M. and Barde, S.P. (1985) Analysis of oil content of groundnuts by nuclear magnetic resonance spectrometry. *Journal of the Science of Food and Agriculture* 36, 162–166.

Jander, G. and Joshi, V. (2010) Recent progress in deciphering the biosynthesis of aspartate-derived amino acids in plants. *Molecular Plant* 3, 54–65.

Jun, T.-H., Van, K., Kim, M.Y., Lee, S.-H. and Walker, D.R. (2008) Association analysis using SSR markers to find QTL for seed protein content in soybean. *Euphytica* 162, 179–191.

Kabas, O., Yilmaz, E., Ozmerzi, A. and Akinci, I. (2006) Some physical and nutritional properties of cowpea seed (*Vigna sinensis* L.) *Journal of Food Engineering* 79, 1405–1409.

Kadlec Bjergegaard, C., Gulewicz, K., Horbowicz, M., Jones, A., Kadlec, P., Kintia, P. *et al.* (2001) Carbohydrate chemistry. In: Hedley, C.L. (ed.) *Carbohydrates in Grain Legume Seeds*. CAB International, Wallingford, UK, pp. 15–59.

Karjalainen, R. and Kortet, S. (1987) Environmental and genetic variation in protein content of peas under northern conditions and breeding implications. *Journal of Agriculture Science Finland* 59, 1–9.

Larmure A., Salon, C. and Munier-Jolain, N.G. (2005) How does temperature affect C and N allocation to the seeds during the seed-filling period in pea? Effect on seed nitrogen concentration. *Functional Plant Biology* 32, 1009–1017.

Lawn, R.J. and Rebetzke, G.J. (2006) Variation among Australian accessions of wild mungbean (*Vigna radiate* ssp. *sublobata*) for traits of agronomic, adaptive, or taxonomic interest. *Australian Journal of Agricultural Research* 57, 119–132.

Lee, M., Huang, T., Toro-Ramos, T., Fraga, M., Last, R.L. and Jander, G. (2008) Reduced activity of *Arabidopsis thaliana* HMT2, a methionine biosynthetic enzyme, increases seed methionine content. *The Plant Journal* 54, 310–320.

Lenis, J.M., Gillman, J.D., Lee, J.D., Shannon, J.G. and Bilyeu, K.D. (2010) Soybean seed lipoxygenase genes: molecular characterization and development of molecular marker assays. *Theoretical and Applied Genetics* 120, 1139–1149.

Lord, J.W. and Wakelam, J.A. (1950) Changes occurring in the proteins as a result of processing groundnuts under selected industrial conditions. *British Journal of Nutrition* 4, 154–160.

Mariotti, F., Pueyo, M.E., Tome, D., Bérot, S., Benamouzig, R. and Mahé, S. (2001) The influence of the albumin fraction on the bioavailability and postprandial utilization of pea protein given selectively to humans. *Journal of Nutrition* 131, 1706–1713.

Matthews, P. and Arthur, E. (1985) Genetic and environmental components of variation in protein content of peas. In: Hebblethwaite, P.D., Heath, M.C. and Dawkins, T.C.K. (eds) *The Pea Crop: A Basis for Improvement*. Butterworth, London, p. 486.

Molvig, L., Tabe, L.M., Eggum, B.O., Moore, A.E., Craig, S., Spencer, D. *et al.* (1997) Enhanced methionine levels and increased nutritive value of seeds of transgenic lupins (*Lupinus angustifolius* L.) expressing a sunflower seed albumin gene. *Proceedings of the National Academy of Sciences U.S.A.* 94, 8393–8398.

Montoya, C.A., Lallès, J.P., Beebe, S. and Leterme, P. (2010) Phaseolin diversity as a possible strategy to improve the nutritional value of common beans (*Phaseolus vulgaris*). *Food Research International* 43, 443–449.

Munier-Jolain, N., Larmure, A. and Salon, C. (2008) Determinism of carbon and nitrogen reserve accumulation in legume seeds. *CR Biologies* 331, 780–787.

Natarajan, S.S., Xu, C., Bae, H., Caperna, T.J. and Garrett, W.M. (2006) Characterization of storage protein in wild (*Glycine soja*) and cultivated (*Glycine max*) soybean seeds using proteomic analysis. *Journal of Agricultural and Food Chemistry* 54, 3114–3120.

NGRP (2001) National Genetic Resources Program, USDA–ARS. Germplasm Resources Information Network (GRIN). Available at http://www.ars–grin.gov/npgs/descriptors/soybean

Nichols, D.M., Glover, K.D., Carlson, S.R., Specht, J.E. and Diers, B.W. (2006) Fine mapping of a seed protein QTL on soybean linkage group I and its correlated effects on agronomic traits. *Crop Science* 46, 834–839.

Nozzolillo, C., Ricciardi, L. and Lattanzio, V. (1989) Flavonoid constituents of seed coats of *Vicia faba* (Fabaceae) in relation to genetic control of their color. *Canadian Journal of Botany* 67, 1600–1604.

Olsson J., Nakao, P., Bohlool, B.B. and Gresshoff, P.M. (1989) Lack of systemic suppression of nodulation in split root systems of supernodulating soybean (*Glycine max*. L. Merr.) mutants. *Plant Physiology* 90, 1347–1352.

Oluwatosin, O.B. (1997) Genetic and environemental variation for seed yield, protein, lipid and amino acid composition in cowpea (*Vigna unguiculata* (L) Walp). *Journal of Science and Food Agriculture* 74, 107–116.

Paek, N.C., Imsande, J., Shoemaker, R.C. and Shibles, R. (1997) Nutritional control of soybean seed storage protein. *Crop Science* 37, 498–503.

Page, D., Aubert, G., Duc, G., Welham, T. and Domoney, C. (2002) Combinatorial variation in coding and promoter sequences of genes at the *Tri* locus in *Pisum sativum* accounts for variation in trypsin inhibitor activity in seeds. *Molecular Genetics and Genomics* 267, 359–369.

Panthee, D.R., Pantalone, V.R., Sams, C.E., Saxton, A.M., West, D.R., Orf, J.H. *et al*. (2006) Quantitative trait loci controlling sulfur containing amino acids, methionine and cysteine, in soybean seeds. *Theoretical and Applied Genetics* 112, 546–553.

Pastorello, E.A., Pompei, C., Pravettoni, V., Brenna, O., Farioli, L., Trambaioli, C. *et al*. (2001) Lipid transfer proteins and 2S albumins as allergens. *Allergy* 56 (Suppl. 67), 45–47.

Plahar, W.A., Annan, N.T. and Nti, C.A. (1997) Cultivar and processing effects on the pasting characteristics, tannin content and protein quality and digestibility of cowpea (*Vigna unguiculata*). *Plant Foods for Human Nutrition* 51, 343–356.

Proulx, R.A. and Naeve, S.L. (2009) Pod removal, shade and defoliation effects on soybean yield, protein and oil. *Agronomy Journal* 101, 971–978.

Pusztai, A. (1989) Biological effects of dietary lectins. In: Huisman, J., Van der Poel, A.S.M.J.F.B. and Liener, I.E. (eds) *Recent Advances of Research in Anti-nutritionnal Factors in Legume Seeds*. Pudoc, Wageningen, The Netherlands, pp. 17–29.

Rolletschek, H., Borisjuk, L., Radchuk, R., Miranda, M., Heim, U., Wobus, U. *et al*. (2004) Seed-specific expression of a bacterial phosphoenolpyruvate carboxylase in *Vicia narbonensis* increases protein content and improves carbon economy. *Plant Biotechnology Journal* 2, 211–219.

Rotundo, J.L, Borrás, L., Westgate, M.E. and Orf, J.H. (2009) Relationship between assimilate supply per seed during seed filling and soybean seed composition. *Field Crops Research* 112, 90–96.

Rozan, P., Kuo, Y.H. and Lambein, F. (2001) Amino acids in seeds and seedlings of the genus *Lens*. *Phytochemistry* 58, 281–289.

Sagan, M., Ney, B. and Duc, G. (1993) Plant symbiosis as a tool to analyse nitrogen nutrition and yield relation ship in field-grown peas (*Pisum sativum* L.). *Plant and Soil* 153, 33–45.

Saghai Maroof, M.A., Glover, N.M., Biyashev, R.M., Buss, G.R. and Grabau, E.A. (2009) Genetic basis of the low-phytate trait in the soybean line CX1834. *Crop Science* 49, 69–76.

Salon, C., Munier-Jolain, N.G., Duc, G., Voisin, A.S., Grandgirard, D., Larmure, A. *et al*. (2001) Grain legume seed filling in relation to nitrogen acquisition: a review and prospects with particular reference to pea. *Agronomie* 21, 539–552.

Saxena, K.B., Kumar, R.V. and Rao, P.V. (2002) Pigeonpea nutrition and its improvement. In: Basra, A.S. and Randhawa, I.S. (eds) *Quality Improvement in Field Crops*. Food Products Press, New York, pp. 227–260.

Scippa, G.S., Rocco, M., Ialicicco, M., Trupiano, D., Viscosi, V., Di Michele, D. *et al*. (2010) The proteome of lentil (*Lens culinaris* Medik.) seeds: Discriminating between landraces. *Electrophoresis* 31, 497–506.

Sebolt, A.M., Shoemaker, R.C. and Diers, B.W. (2002) Analysis of a qualitative trait locus allele from wild soybean that increases seed protein concentration in soybean. *Crop Science* 40, 1438–1444.

Singh, U. (1999) Cooking quality of pulses. *Journal of Food Science and Technology* 36, 1–14.

Singh, U. and Jambunatham, R. (1981) Studies on desi and kabuli chickpea (*Cicer arietinum* L.) cultivars – levels of protease inhibitors, levels of polyphenolic compounds and in vitro digestibility. *Journal of Food Science* 46, 1364–1367.

Tabe, L.M. and Droux, M. (2001) Sulfur assimilation in developing lupin cotyledons could contribute significantly to the accumulation of organic sulfur reserves in the seed. *Plant Physiology* 126, 176–187.

Tabe, L.M. and Droux, M. (2002) Limits to sulfur accumulation in transgenic lupin seeds expressing a foreign sulfur-rich protein. *Plant Physiology* 128, 1137–1148.

Tabe, L., Wirtz, M., Molvig, L., Droux, M. and Hell, R. (2010) Overexpression of serine acetlytransferase produced large increases in O-acetylserine and free cysteine in developing seeds of a grain legume. *Journal of Experimental Botany* 61, 721–733.

Tar'an, B., Warkentin, T., Somers, D.J., Miranda, D., Vandenberg, A., Blade, S. *et al.* (2004) Identification of quantitative trait loci for grain yield, seed protein concentration and maturity in field pea (*Pisum sativum* L.). *Euphytica* 136, 297–306.

Tzitzikas, E.N., Vincken, J.P., De Groot, J., Gruppen, H. and Visser, R.J.F. (2006) Genetic variation in pea seed globulin composition. *Journal of Agricultural and Food Chemistry* 54, 425–433.

Ufaz, S. and Galili, G. (2008) Improving the content of essential amino acids in crop plants: goals and opportunities. *Plant Physiology* 147, 954–961.

Upadhyaya, H.D., Reddy, K.N., Gowda, C.L.L. and Silim, S.N. (2007) Patterns of diversity in pigeonpea (*Cajanus cajan* (L.) Millsp.) germplasm collected from different elevations in Kenya. *Genetic Resources in Crop Evolution* 54, 1787–1795.

Valdebouze, P., Bergeron, E., Gaborit, T. and Delort-Laval, J. (1980) Content and distrbution of trypsin inhibitors and hemagglutinins in some legume seeds. *Canadian Journal of Plant Sciences* 60, 695–701.

Van Rhijn, P., Fijishige, N.A., Lim, P.O. and Hirsch, A.M. (2001) Sugar binding activity of pea lectin enhances heterologous infection of transgenic alfalfa plants by *Rhizobium leguminosarum* biovar. *viciae. Plant Physiology*, 126, 133–144.

Vasconcelos, I., Machado Maia, F.M., Farias, D.F., Campello, C.C., Urano Carvalho, A.F. *et al.* (2010) Protein fractions, amino acid composition and anti-nutritonal constituents of high-yielding cowpea cultivars. *Journal of Food Composition and Analysis* 23, 54–60.

Vigeolas, H., Chinoy, C., Zuther, E., Blessington, B., Geigenberger, P. and Domoney, C. (2008) Combined metabolomic and genetic approaches reveal a link between the polyamine pathway and albumin 2 in developing pea seeds. *Plant Physiology* 146, 74–82.

Vollman, J., Fritz, C.N., Wagentristl, H. and Ruckenbauer, P. (2000) Environmental and genetic variation of soybean seed protein content under Central European growing conditions. *Journal of the Science of Food and Agriculture* 80, 1300–1306.

Vuong, T.D., Wu, X., Pathan, M.D.S., Valliyodan, B. and Nguyen, H.T. (2007) Genomics approaches to soybean improvement. In: Varshney, R.K. and Tuberosa, R. (eds) *Genomics Assisted Crop Improvement: Vol. 2: Genomics Applications in Crops*. Springer-Verlag, New York, pp. 243–279.

Wakasa, K., Hasegawa, H., Nemoto, H., Matsuda, F., Miyazawa, H., Tozawa, Y. *et al.* (2006) High-level tryptophan accumulation in seeds of transgenic rice and its limited effects on agronomic traits and seed metabolite profile. *Journal of Experimental Botany* 57, 3069–3078.

Wang, N., Hatcher, D.W., Toews, R. and Gawalko, E.J. (2009) Influence of cooking and dehulling on nutritional composition of several varieties of lentils (*Lens culinaris*). *Food Science and Technology*, 42, 842–848.

Wang, T.L. and Hedley, C.L. (1993) Genetic and developmental analysis of the seed. In: Casey, R. and Davies, D.R. (eds) *Peas: Genetics, Molecular Biology and Biotechnology*. CAB International, Wallingford, UK, pp. 83–120.

Wang, T.L., Domoney, C., Hedley, C.L., Casey, R. and Grusak, M.A. (2003) Can we improve the nutritional quality of legume seeds? *Plant Physiology* 131, 886–891.

Weber, H., Borisjuk, L. and Wobus, U. (2005) Molecular physiology of legume seed development. *Annual Review of Plant Biology* 56, 253–279.

Weigelt, K., Küster, H., Radchuk, R., Müller, M., Weichert, H., Fait, A. *et al.* (2008) Increasing amino acid supply in pea embryos reveals specific interactions of N and C metabolism, and highlights the importance of mitochondrial metabolism. *The Plant Journal* 55, 909–926.

Wilcox, J.R. and Cavins, J.F. (1995) Backcrossing high seed protein to a soybean cultivar. *Crop Science* 35, 1036–1041.

Wilcox, J.R. and Shibles, R.M. (2001) Interrelationship among seed quality attributes in soybean. *Crop Science* 41, 11–14.

Williams, P.C. and Singh, U. (1987) Nutritional quality and the evaluation of quality in breeding programmes. In: Saxena, M.C. and Singh, K.B. (eds) *The Chickpea*. CAB International, Wallingford, UK, pp. 329–356.

Yates, J.L., Harris, D.K. and Boerma, H.R. (2004) Marker-assisted selection around a major QTL on LG-I increases seed protein content in backcross-derived lines of soybean. In: *Soy2004 the 10th Biennial Conference of the Cellular and Molecular Biology of the Soybean*, Columbia, Missouri, pp. 67.

Zhu, X. and Galili, G. (2003) Increased lysine synthesis coupled with a knockout of its catabolism synergistically boosts lysine content and also transregulates the metabolism of other amino acids in *Arabidopsis seeds. Plant Cell* 15, 845–853.

21 Underutilized Food Legumes: Potential for Multipurpose Uses

Nazmul Haq

21.1 Introduction

Global food security is becoming uncertain, with increasing dependence on a few major staple crops, and this is also the case for food legumes. Many food legumes are underutilized, although these could provide even better food products, more balanced natural diets, better pharmaceutical products, natural insecticides and flavouring, dyes, as well as their usage for the preparation of beverages (Haq, 2004). Several authors (NAS, 1979; Haq, 1983; van der Maesen and Somaatmadja, 1989; Chomchalow *et al.*, 1993; Bhag Mal, 1995; Graham and Vance, 2003; Aletor and Aladetimi, 2006; Smartt, 2008; Gowda *et al.*, 2009) have highlighted the potential of several underutilized food legumes and the need for research to improve their utilization and nutrition security. In this chapter, the potential of selected legume species with diverse uses is discussed. These were selected because of recent research interest and their potential suitability in various cropping systems and adaptation to a range of climates in sustainable agricultural production systems. However, this chapter excludes legume trees, which are also important for food, feed and environmental maintenance (see Haq *et al.*, 2007; Bhat and Karim, 2009).

21.2 Jack Bean (*Canavalia ensiformis* (L.) DC.)

Canavalia ensiformis is an annual or short-lived (perennial) plant native to the West Indies and Central America, and which is now cultivated throughout the tropics and subtropics (Table 21.1). The pods are flat and sword-shaped, 5–12 cm long, and contain a variable number of white seeds. It may grow under an annual rainfall of 640–4290 mm, in a temperature range of 14.4–27.8°C and at soil pH 4.5–8.0. It is relatively drought tolerant, and also tolerant to waterlogging and salinity to some extent, and it is a short-day type (Table 21.2).

The protein content of the seed is in the range of 23.8–27.6%. Seeds are used in foods, and have potential use in the production of protein concentrates. Akapapunam and Sefa-Dedah (1997) have stated that the crop's nutritional characteristics open up the possibilities of new and highly nutritious food products. Young leaves, flowers, pods and immature seeds are eaten as vegetables; dry seeds are eaten after long cooking as they require detoxification on account of their content of HCN and other toxins. The crop is primarily cultivated for forage (Akinlade *et al.*, 2007) and green manure. It is also grown as a cover crop to control soil erosion.

The information on genetic resources and crop improvement research on this species is

Table 21.1. Distribution of underutilized food legumes and their biological characteristics.

Species	Common name	Distribution	Life form	Growth habit
Canavalia ensiformis	Jack bean	Tropics, China, Japan	Annual, perennial	Climbing, bushy erect, semi-erect
Lablab purpureus	Lablab bean	Tropics and sub-tropics	Annual/ perennial	Climbing, erect
Lupinus mutabilis	Tarwai/pearl bean	Andean region, Europe, Africa, Australia	Annual	Erect
Macrotylema uniflorum	Horse gram	SE Asia, tropical Asia, Africa, Australia	Annual	Herbs with twining branches
Mucuna pruriens	Velvet bean	Tropics and subtropics, Asia, Africa to western hemisphere via Mauritius	Annual, perennial	Climbing, bushy, erect, semi-erect
Phaseolus acutifolius	Tepary bean	USA, Mexico, Costa Rica	Annual	Vine, erect
Phaseolus lunatus	Lima bean	Andean, subtropics	Annual	Erect, vine
Psophocarpus tetragonolobus	Winged bean	Asia, Africa, Pacific & Indian Ocean Islands, C & S America	Perennial, annual	Climbing
Sphenostylis stenocarpa	African yam bean	Tropical Africa	Annual	Climbing
Vigna angularis	Adzuki bean	Far East, India, SE Asia, China, USA, S. America, Angola, Zaire, New Zealand	Annual	Erect, bushy
V. aconitofolius	Moth bean	S & SE Asia, USA, Australia, China	Annual	Erect
V. subterranean	Bambara groundnut	Africa, S & SE Asia, N Australia, S & C America	Annual	Short-creeping trailing stem
V. umbellata	Rice bean	Asia, Pacific Islands, E Africa, Mauritius, N Australia, USA	Annual. Perennial	Climbing, erect and semi-erect

limited; however, there has been some collection in India, Indonesia and Africa. One of the priorities for genetic improvement is to produce high-yielding and low-toxic varieties. Long maturity period, pod shattering, presence of toxins, non-acceptability of flavour, texture and cooking problems are drawbacks of the species. High-yielding varieties with better flavour and texture would encourage farmers to grow this crop.

Agronomic information is not well documented. Seeds are sown in rows 75 cm apart with 4–60 cm between plants in the rows. The seed rate is 25–30 kg/ha when planted in rows and 40–60 kg/ha when broadcast for green manure. Climbing types need support. Green pods are ready for harvesting after 3–4 months and ripe seeds from 6–10 months. At present the forage yield is 6000 kg/ha and dry bean yield is 1500 kg/ha. Pests and diseases are few, but include armyworm and pod weevil. Diseases include *Colletotrichum*, mosaic (asparagus bean), and green and yellow viruses. Jack bean is widely used in

Table 21.2. Cultivation conditions of underutilized food legumes.

Species	Soil	Temperature (°C)	Day length	Rainfall (mm)	Altitude (m)	Strength
Canavalia ensiformis	Depleted, leached soil; pH 4.0–8.0	14.4–27.8	Short day	64–4290	1800	Resistance to drought, salinity, waterlogging; multiple uses; possibilities of new and highly nutritious food products
Lablab purpureus	Poor, well-drained sandy loam, clay; pH 5.0–6.5	18–30	Short and long day	700–900	1250	Multiple uses; suitable for dry areas; nutritious
Lupinus mutabilis	Wide range of soils, sandy, loam, moderately acidic, coarse texture Soil, pH 5.0–7.0	15–25	Short day	450–1000	4000	Low alkaloid types; compact inflorescences makes the crop suitable for large-scale cultivation because of ease of harvesting
Macrotylema uniflorum	Wide range of soils, sandy to loam, stony, gravel, upland; pH 5.0–7.5	20–30	Short day	700–4300	1800	High seeding habit; tolerant to drought; dry-area, short duration crop
Mucuna pruriens	Wide range of well-drained soil including heavy clays; prefer sandy loam for optimum yield; pH 5.0–6.5	20–30	Short day	1200–1500	2100	Wide range of variability in germplasms provide opportunities for further exploitation; food, feed and industrial uses
Phaseolus acutifolius	Poor, shallow soils of pH 5.0–7.0	17–26	Short day	500–1700	–	Tolerates drought, heat, slopes. Resistance to diseases. Many uses, can be used as catch crop
Phaseolus lunatus	Well-drained soils	8.7–27.5	Short day/ day/neutral	310–4290	2400	Successful in semi-arid, humid and sub-tropics. Deep rooting, multiple products to exploit
Psophocarpus tetragonolobus	Prefers well-drained, sandy loam. Wide range of soil, pH 5.0–6.8	18–30	Short day	1500–2500	2200	High protein in seeds and tubers; high nodulating ability. Potential for multi-nutritional products

Continued

Table 21.2. Continued.

Species	Soil	Temperature (°C)	Day length	Rainfall (mm)	Altitude (m)	Strength
Sphenostylis stenocarpa	Wide range of well-drained soils, clay, acid & highly leached sandy soils, rocky soils	30–40	–	900–2000	1950	Seeds and tubers contain high protein, wider adaptation, higher seed and tuber in the systems.
Vigna angularis	Light to heavy clay, acidic pH 5.0–7.5	15–30	Sensitive to day length	500–1700	420	Suitable for subtropics and high altitudes in the tropics; tolerance to drought, heat, frost, viral diseases; multi-purpose nutrition crop
V. aconitofolius	Poor sandy, well-drained loam; pH 5.0–8.1	22–41	Short day	250–500	1500	Hardy, very drought resistant; 2–3 months to mature seeds; grows on different soil textures, multiple uses
V. subterranean	Wide range: loam, poor soil where groundnut cannot be grown; pH 5.0–6.5	20–28	Short day	600–1200	1600	Grows in dry areas; taste & flavour liked by consumers; fits traditional farming systems
V. umbellata	Wide range of types. Prefers fertile loam	18–30	Short day	1000–1500	1800	Multi-purpose nutrition crop; short duration; low susceptibility to pests, diseases

intercropping with sugarcane, coffee, tobacco, rubber and sisal. It is also used as a cover crop for cocoa, coconut, citrus and pineapple.

21.3 Lablab Bean (*Lablab purpureus* (L.) Sweet)

Lablab bean is indigenous to South-east Asia and has been introduced in Africa and other tropical and subtropical countries. It has now spread throughout the tropics and is cultivated in warmer regions of the world. It is perennial but normally grown as an annual or biennial, with a dwarf or bushy erect and climbing habit. The stem is usually 2–3 m but can be up to 10 m in length. Lablab bean thrives well in temperatures of 18–30°C and, although it is highly drought resistant, it requires irrigation in the early growth stage. It can grow in poor soils, but sandy loam and clay soils are ideal at a pH range of 5.0-6.5 (Table 21.3), provided these are well drained.

Seeds are sown at the end of the wet season in rows 60–70 cm apart, with 30–45 cm between plants and 30–40 kg seeds planted per hectare. Trellises or stakes are needed to support climbing plants. Farmers use few fertilizers to grow this crop. Harvesting for vegetables continues for 70–120 days after sowing, which can be continued until seeds mature. Seed and green pod yield ranges from 1460 to 2200 kg/ha and 2600 to 4500 kg/ha, respectively depending on variety and location. In mixed cropping the yield of seed is about 450 kg/ha, while fodder yield is 5000–10,000 kg/ha.

Lablab bean grows well intercropped with finger millet, pigeon pea, maize and coffee; the yield of the bean crop increases when it is intercropped with maize. Haque *et al.* (2003) reported lablab bean-based intercropping systems in north-west Bangladesh. The plant develops a thick cover on the soil and forms a good mulch in orchards and plantations. Its production in multiple cropping systems is an added bonus. Few diseases and pests have been reported for lablab bean. However, diseases due to *Cercospora*, *Colletotrichum*, *Leveilla*, *Rhizoctonia*, *Xanthomonas*, *Pseudomonas* and mosaic virus have been observed in various habitats, these being controlled by seed treatment with chemicals. Insect pests include pod borer (*Adisura* spp.), bean leaf beetle (*Ceratoma*), red gram moth (*Exelastis*), bollworm (*Helicoverpa*) and spotted pod borer (*Maruca*); root-knot nematodes also attack the crop. Insect pests can be controlled by common insecticides.

Young leaves, flowers and pods of lablab are used as vegetables. Mature seeds are consumed after cooking as *dhal* and sometimes used as a substitute for broad beans in the preparation of the fried bean cake *tanniah*. Protein concentrates can be made from seeds. Plants are used as forage for livestock (Ajayi *et al.*, 2009) and also grown for grazing (NAS, 1979; Murphy and Colucci, 1999). It also makes good silage and is used as green manure in soil improvement and often grown as a second crop in rice fields. Seeds and leaves are also used for medicinal purposes (Bhag Mal, 1995).

Germplasm collection and evaluation of lablab bean have been carried out in many countries in Asia, with the aim of selection and improvement of the crop. India, Indonesia, Australia and the ILRI (International Livestock Research Institute) and some national institutions in Asia have been storing collections. The University of Bangalore, India has been involved in a systematic improvement programme for the crop (Gowda, 2010). A large variation in plant height, number of pods/plant, number of seeds/pod, length of pod and seed weight/plant has been observed (Ayisi *et al.*, 2004; Gowda, 2010; Islam, 2010). Small-podded types are common in India and Papua New Guinea, while longer-pod types are found in Indonesia and West Africa. Genetic variability for pod and seed was very high, as detected by RAPD markers and RFLP studies (Liu, 1996; Sultana *et al.*, 2000; Singh *et al.*, 2006; Rai *et al.*, 2010). This information has provided an opportunity for breeding strategies to improve the crop. Studies at Bangalore University have identified and improved several superior types, both for grain and vegetable use, and results on heterosis have shown that improvement of this crop for seed and vegetables is possible. Although this crop is sensitive to photoperiod, photo-neutral,

Table 21.3. Chemical composition of potential food legumes.

Species	Protein (%)	Fat (%)	Carbohydrate (%)	Fibre (%)	Ash (%)	Calcium mg/100g	Phosphorus mg/100 g	Iron mg/100 g
Canavalia ensiformis	23.8–27.6	2.6	45.2–56.9	7.4	3.2	158.0	298.0	7.0
Lablab purpureus	21.5–29.0	1.2–1.0	60.1	6.8–8.6	3.8	98.0	345.0	3.9
Lupinus mutabilis	32.0–46.0	13.0–23.0	30.4	7.0–11.0	3.6	28.0	168.0	1.9
Macrotylema uniflorum	22.0–25.1	0.5–2.0	57.3–60.0	3.0–5.3	2.8–6.3	0.28–0.34	0.39	0.27–7.6
Mucuna pruriens	15.1–23.4	3.6	59.2	5.7	3.9	0.18	0.99	–
Phaseolus acutifolius	22.0–25.0	1.5	60.0–65.0	3.0–4.0	4.0	112.0	310.0	–
Phaseolus lunatus	14.4–26.4	1.5	58.0	3.7	3.4	133.0	445.0	5.6
Psophocarpus tetragonolobus								
seeds:	21.1	0.12	74.1	5.7	3.2	6.1	437.0	2.2
tubers:	3.0–20.0	0.4–1.1	27.2–30.5	1.6–17	1.7	25.0	30.0	0.05
Sphenostylis stenocarpa								
seeds:	21.1–29.0	0.12	74.1	5.7	3.2	6.1	437.0	–
tubers:	12.0–19.0	0.6	86.3	1.1	2.3	28.0	257.0	–
Vigna angularis	19.9–25.3	0.6	57.1–64.4	5.7–9.8	3.9–4.3	136.0–353.0	260.0	7.6–9.8
V. aconitofolius	23.6–28.6	1.1	56.5	4.5	3.5	0.2–0.3	0.1–0.7	0.009
V. subterranean	14.0–24.0	6.0	62.0	5.0	3.0	94.0	293.0	4.7
V. umbellata	16.0–25.8	0.9	64.9	3.8–4.8	3.3–4.8	315.0–450.0	197.0–393.0	5.8

short- and long-day cultivars have been identified (Dutta *et al.*, 2007; Gowda, 2010). Several varieties have also been identified as being resistant to yellow mosaic virus and pod borer (Duke, 1987; Bhag Mal, 1995).

21.4 Tarwai or Pearl Lupin (*Lupinus mutabilis*)

Lupinus mutabilis, commonly known as tarwai or pearl lupin, is a New World domesticate and is significant in the Andean region (Peru, Bolivia, Ecuador and Chile). This species is cultivated at both high altitudes (1800–4000 m) and low latitudes (0–22°S). It can be grown on sandy, loamy and moderately acidic soils with pH 5.0–7.0, and is tolerant to drought and frost; *L. mutabilis* is grown as a grain crop, as well as for green manuring.

Tarwai has high protein (32–46%) and fat content (13–23%) (Table 21.3). Seeds and vegetative parts contain an alkaloid (0.007–3.970%), but toxicity of the species varies seasonally and geographically and it often increases after flowering. Seeds have been used for human consumption. The nutritional protein value of the species is significantly increased when supplemented by synthetic methionine at a level of 2% of total protein. Tarwai seeds are used for animal feed and are increasingly being substituted for soyameal, groundnut cake and fish meal in the production of high-quality livestock rations. A bitter compound, lupinex is extracted from seeds and can be used to protect plants from insects and diseases (Kahnt and Hijazi, 2008).

The germplasm collection of *L. mutabilis* is very limited and needs to be explored. Some collections are maintained in gene banks in Bolivia, Germany, France, Peru, Spain and USA (Haq, 1993). The species has a wide genetic variation in plant form, vegetative growth, inflorescence types, susceptibility to frost and diseases, and protein, oil and alkaloid content, indicating adaptation to micro-habits, typical of the diverse Andean environment. Some progress on improving this crop has been made in Australia, Europe and in Latin America, reviewed by Haq (1993). Some selections have been made for low alkaloid content (Sweetingham and Clements, 2005), early maturing, compact inflorescence with long axis, and high protein and oil content. Chaudhary (1993) reported a positive correlation between higher plant density and seed yield.

It is also been grown in Europe, including the UK (Haq, 1993) on a trial basis, although Hardy and Huyghe (1997) believe that the existing genotypes have limited suitability for adoption under European conditions. It has also been tried in Pakistan under rainfed conditions and found to have potential for further trials (Chaudhary and Cheema, 1998).

The seeding rate is 120–180 kg/ha, depending on variety. A spacing of 15 cm between plants and 20 cm between rows is normally used. Harvesting for seeds is carried out at about 4–5 months after sowing. The application of 20–30 kg/ha nitrogen and 300–500 kg/ha of phosphates is recommended for raising a good crop of tarwai. In poor, sandy soils, 28–56 kg/ha potash is also applied. Seed yield varies considerably in different growing environments, and in Bolivia it has been reported to be 1200 kg/ha, in Ecuador 600–700 kg/ha and in Peru 300–2000 kg/ha. In the UK, a seed yield of 2360 kg/ha has obtained from experimental plots (Chaudhary, 1993).

In regard to diseases attacking *L. mutabilis*, anthracnose in particular causes serious problems, while wilting and damping-off diseases are also recorded. Bean yellow mosaic causes damage to seed stocks, and mycoplasma has been reported from Peru. Pests include aphids, *Heliothis* spp. and cutworms, while birds and rodents may cause considerable damage. The species has potential use in crop diversification programmes. Rotation of tarwai with potato, maize and quinoa in the Andes is beneficial. The effect of rotation in breaking disease cycles has also been reported (Chaudhary, 1993). The species can be used to control soil erosion and in agroforestry.

21.5 Horse Gram (*Macrotylema uniflorum* (Lam.) Verdc)

Horse gram originated in South-east Asia, probably native to the centre of Hindustan, and is distributed throughout tropical Asia, tropical Africa, the West Indies and Australia. It is extensively grown in dry areas. It is an annual herb with 30–50 cm twining branches, and pods are 3–5 cm long containing 5–7 small seeds. Horse gram is a short-day crop and grows in a wide variety of soils, from sandy to red loam and even in stony and gravelly upland soils with poor fertility. However, it does not tolerate waterlogging and strong alkalinity conditions. It grows at a temperature of 20–30°C and is successfully grown as a dryland crop with less than 30 cm rainfall. It grows at pH 5.0–7.5 and at an elevation of 1800 m (Table 21.2).

The seeds contain 0.5–2.0% fat and 22–25% crude protein and are rich in lysine, tyrosine and arginine but deficient in cystine and tryptophan. The hay contains 16.2% crude protein and 1.8% fat. The seed is used like *dhal* and it is fermented to produce a sauce similar to soy sauce, while seeds are also boiled for animal feed. Stems, leaves and husks are used as forage. It is also grown for green manure and a cover crop on eroded hilly slopes and red laterite soils. The seeds and fresh plants are used in medicines (Bhag Mal, 1995).

There are a few scattered collections of germplasm in various countries of Asia and the Pacific region, although information on actual collections in countries other than India is not available. Germplasm evaluation of horse gram has, to date, shown wide variation for plant height, pod number/plant, pod length seed weight and seed yield. These variations provide a basis for considerable improvement of the crop (Bhag Mal, 1995; Sunil *et al.*, 2008; Ravindra and Sundar, 2009). Genotypes have been identified for resistance to bean beetle (*Collosobruchus chinensis*), nematodes (*Rotylenchulus reniformis*) and yellow mosaic virus. Attempts to improve the crop have been confined to selection for high yield and a few varieties have been developed. Molecular and biochemical investigations are in progress to improve the quality of horse gram (Dutta *et al.*, 2007).

Horse gram is cultivated with minimal land preparation, but good agronomic practice increases yields. The seed is sown in 30 cm rows 7.5–10.0 cm apart. The crop matures in 4–6 months and yield varies based on genotype and management practices (Keshava *et al.*, 2007). This ranges from 130–1200 kg/ha in India to 1120–2200 kg/ha in Australia. Green forage yield in India varies from 5000 to 14,000 kg/ha, and is 4400 kg/ha in Australia (Jansen, 1989). The main diseases are rhizoctonia and anthracnose; pests affecting the crop include pod borers, grasshoppers, caterpillars and aphids. Horse gram is cultivated as a sole crop, in both intercropping and rotation. It is intercropped with cereals (e.g. pearl millet, sorghum, maize), oilseeds (e.g castor, niger) and many other legumes, in particular pigeon pea. Witcombe *et al.* (2008) reported that the variety AK-42 in India yielded more grain (> 60%) when it was intercropped with maize. It has potential to be exploited in different cropping systems depending on ecotypes and the purpose of growing. Virk *et al.* (2006) have recommended the crop for use in crop diversification programmes.

21.6 Velvet Bean (*Mucuna pruriens* (Wall, ex Wight) Bak. ex Burck)

Commonly known as velvet bean, this crop originated in Asia and has now been distributed throughout the western hemisphere. It is now cultivated in many tropical and subtropical countries for food, forage and cover. The velvet bean is annual or perennial, with a mainly climbing habit although bushy forms also exist. The crop can be grown on a wide range of soil types, including heavy clay, and is tolerant of fairly acidic soil (pH 5.0–6.5). The crop grows profusely in areas with an average annual rainfall of 1200–1500 mm (Tables 21.1, 21.2).

The protein content of the seed ranges from 15.1 to 23.41% (Table 21.3) and some genotypes of velvet bean are high in methionine, which is low in other legumes. In Southeast Asia the immature pods and leaves are used as a vegetable, while in parts of Asia and Africa the seeds are roasted and eaten. Seeds are also fermented after removal of the coat to produce bean cake and *tempe*. In Africa, soups and stews are prepared after the seeds have been boiled and toxic substances removed. The use of the seed as a source of high-viscosity starch as a thickening agent for food products, and as an adhesive in the paper and textile industries, has been investigated with promising results. Furthermore, L-dopa has been extracted from the seed to provide a commercial drug for the treatment of Parkinson's disease. Crop use is primarily as a forage for ruminants (Vedivel and Janardhanan, 2000) and as a cover crop in India, the USA, Australia, Malaysia and parts of Africa (Klassen *et al.*, 2006).

Genetic diversity has been recorded in germplasm (Pugalenthi *et al.*, 2005; Lorrenzetti *et al.*, 2010). Types with drought resistance and varieties with a wide range of maturity periods were observed. The average seed yield of velvet bean in Australia is 560 kg/ha, whereas in the USA average seed yield is 1680–3360 kg/ha. The crop is fairly free from pests and diseases, although some bacterial and fungal diseases and insect pests have been reported. The free L-dopa in the velvet bean resists attack from insects and facilitates seed storage; it is used as a mixed crop with sugar cane and maize and in rotation with sugar cane in Burma. Armstrong *et al.* (2008) reported that when the crop was intercropped with maize and other legumes (lablab and runner bean) it produced significantly higher yield.

21.7 Tepary Bean (*Phaseolus acutifolius* A. Gray)

Tepary bean is native to the south-western USA and Mexico and has been grown since pre-Columbian times. It is a drought-resistant legume and grows in desert and semi-desert conditions extending from Arizona to Costa Rica. The crop grows well where annual rainfall is less than 400 mm. It has both bushy and semi-vining forms, with an average height of 75 cm. The pods are about 8 cm in length and contain 5–6 seeds with diverse colours, but are commonly buff.

Tepary beans are high in protein (23–25%) and are used as dry beans. They are eaten like other beans after soaking, but some Native Americans use them in soups, stews and, when ground, as a meal. Tepary beans have a sweet, nutty flavour, different from other beans in taste. The bean cooks faster than other beans and has a creamier texture when cooked. Lectins and other compounds in tepary bean are thought to be useful in cancer treatment. It is grown as a catch crop and requires little weeding.

Germplasm collection has been carried out primarily in Mexico and the USA. There is a wide variation in yield and yield-related characters. Zambre *et al.* (2006) have developed a protocol for *Agrobacterium*-mediated transformation to incorporate desired genes.

This species is drought tolerant but needs plentiful rain for germination. Short day length is favourable to raising a good crop. The crop matures in 75–85 days, depending on cultivar and location. It is mostly grown in small plots but has a wider potential for cultivation in arid or semi-arid conditions. Seeds are sown in rows at a rate of 11–17 kg/ha, and seed yields of 500–800 kg/ha without irrigation and 900–1700 kg/ha with irrigation have been reported (Jansen, 1989). However, a yield of 5000 kg/ha has also been reported from experimental plots. In Kenya increase in yield was reported as significant when a *Rhizobium* strain was used (Shisanaya, 2002). The plant is mostly disease free but is susceptible to root rot attack and *Rhizoctonia* spp. Leaf borers are the major pests. Tepary bean has been grown in association with maize. An improved yield of tepary bean was observed when intercropped with maize, but unfortunately maize yield was decreased.

21.8 Lima Bean (*Phaseolus lunatus* L.)

Lima bean originated in the central Andean region of America, spreading throughout America, the Pacific region, Asia and parts of Africa. It is an annual or perennial herb, the bush form attaining a height of 60 cm and the climbing form growing up to 4m. It is generally self-fertilizing but has a variable degree of outcrossing.

Large-seeded types are more sensitive to temperature than small-seeded. Day-neutral types of lima bean occur which require an optimum temperature of 16–27°C; the plant does not flower if the temperature is above 30°C. It is grown in tropical and subtropical lowlands but will also grow at up to 2500 m. The crop can survive in as little as 500–600 mm annual rainfall once established, but thrives better at an annual rainfall of 900–1500 mm. It prefers well-aerated, well-drained soils of pH 6.0–6.8, but can tolerate acidity as low as pH 4.4.

The use of lima bean seeds for food has been reported since 5000 BC. Immature seeds, leaves and pods have use as a vegetable. Green lima beans are also canned or frozen. Mature dry seeds are used as a pulse and also mixed with other ingredients to make food preparations – its flour is added to bread flour to make noodles and is also used as bean paste. Seeds and leaves are also used in traditional medicine. The plant can be used as cattle fodder. The silage contains 27.3% dry matter and up to 14.2% digestible nutrients. It is also grown as either cover or a green manure crop.

Germplasm of lima bean has been collected by CIAT, Colombia and IITA, Nigeria, and some collections exist in Latin America, North America, Europe, Africa and Asia (Bettencourt *et al.*, 1989). Considerable genetic variability exists for plant and seed characters (Martinez-Castilo *et al.*, 2004; Akande and Balogun, 2007; Arsene *et al.*, 2007; Castifieiras *et al.*, 2007), which has also been studied using molecular markers (Caicedo *et al.*, 1999; Asante *et al.*, 2008). Genetic studies have shown that there are around 22 characters that are monogenic and 10 in linked combinations that can be used for further genetic studies (Bhag Mal, 1995). Recent work has shown success in transferring genes through interspecific hybridization, and the method allows avenues for transferring desirable traits from other *Phaseolus* group members, in particular from *P. vulgaris*. However, research on improvement of lima bean is limited to the tropics, although research has shown that lima bean has broad adaptability and potential for good yield in the tropics. The knowledge already acquired for *Phaseolus* beans and cowpea can assist in the development of high-yielding, disease-/pest-resistant types.

Lima bean plants are spaced at 10–15 cm and 60–75 cm between rows. Cultural practices are similar to those for common bean (*P vulgaris*) in relation to fertilizers and management. Harvesting is carried out at 65–90 days after sowing, depending on variety. Seed yield ranges from 1000 to 1500 kg/ha as a monocrop and 200–600 kg/ha as a mixed crop. However, the experimental yield is much higher and ranges from 2000 to 2500 kg/ha for the bush type and 3000–4000 kg/ha for the climbing type. This shows the potential of higher yield through germplasm exploitation.

Damaging diseases include web blight, anthracnose, root rot, downy mildew, bacterial spot, common blight and mosaic viruses. Pests affecting lima bean include root-knot nematode, pod borers and bruchids. Lima bean is grown in small plots in the tropics, but also as a companion with other crops including tree crops. It has good potential for agroforestry systems, as do other legumes.

21.9 Winged Bean (*Psophocarpus tetragonolobus* (L.) DC)

Winged bean is distributed throughout the Asia and Pacific regions, in the Caribbean and in Africa, and is now being grown in the USA. The origin of this species has not yet been determined. The crop is mainly grown as a green vegetable, but it is also grown widely on a large scale as a tuber crop in Papua New Guinea and Burma. It is a climbing plant, both perennial and annual, growing 3–4 m tall, and at present trellises are needed to produce a heavy crop. The plant grows well in hot, humid areas with 2500 mm annual rainfall. It appears to tolerate a wide range of soil conditions, the optimum being sandy loam or heavy clay of pH 5.0–5.7. It is a short-day plant and temperature is important as photoperiod controls flowering.

Immature pods (containing 1.9–4.3% protein) and unripe seeds (containing

4.6–10.7% protein) are used as vegetables. The seed is the most important part of the plant, containing high protein (up to 46%) and oil (26%) levels, and various plant products can be made that are similar to those of soybean. The tuber contains up to 20% (dry weight) protein which is much more than that of conventional root crops. However, further work is needed on anti-nutritional factors in tubers. The haulm can be used in animal feeds.

The germplasm resources, genetics and breeding of the crop have been reported by Haq (1982). A large number of germplasm collections, mainly from Asia and Papua New Guinea, have been evaluated in diverse climatic conditions in various parts of the world. Considerable variation has also been found in both quantitative and qualitative characters (Patel and Loknathan, 1998). Variation in morphology, physiology, chemical compositions, maturity (110–180 days), reaction to pests and diseases, resistance to drought and nodule formation has been reported.

Seeds are sown at different spacings for seed and tuber production, the two primary products varying from country to country (Bhag Mal, 1995). Fresh pod yields of up to 55,700 kg/ha, grain yields of up to 5000 kg/ha and tuber yields of 17,700 kg/ha fresh weight have been reported, although the tuber yield depends on the duration of the crop. These yields are projected from small experimental plots with good management. However, one yield trial in a 0.5 ha plot with selected lines produced 2000–2700 kg/ha of seeds consistently in two crop seasons in Sri Lanka. The diseases false rust, leaf spot and yellow mosaic virus and the insects/pests such as maruca, lady bug and root-knot nematode cause damage to the crop. The crop is used in parts of Asia as an intercrop with sweet potato, sugar cane and other grain legumes. Its use in relay, and mixed cropping has shown promise in Thailand and Sri Lanka. However, a yield of 2000 kg/ha was achieved with selected lines as a catch crop in an old coconut plantation in Sri Lanka (Samranayake and Gunasena, 2010). High nodulating ability has also been observed, and this makes the crop suitable as a cover crop.

21.10 African Yam Bean (*Sphenostylis stenocarpa* Hochst. Ex A. Rich)

African yam bean is widely distributed throughout tropical Africa. Although this crop can be used for seed and tubers, in Nigeria it is grown solely for seeds. It is believed that it originated in Ethiopia (NRI, 1987). However, Potter and Doyle (1992) have argued that the origin of the species might be in West and Central Africa; both cultivated and wild types are found in tropical Africa as far south as Zimbabwe, throughout West Africa from Guinea to southern Nigeria, particularly in Togo, Côte d'Ivoire, Ghana and Cameroon, in Central Africa and in East Africa from northern Ethiopia to Mozambique, including Tanzania and Zanzibar. It grows on a wide range of soils, including acidic and highly leached sandy soils with a pH of 5.0–6.5. It can be grown at up to 1800 m in climates ranging from savannah to rainforest (1000 mm rainfall during the growth stage). It is an annual with a climbing habit, grows up to 3 m and needs staking for good cropping. The pods are slightly woody, containing 20–30 seeds and are up to 30 cm long.

The seed contains crude protein (21–29%), with amino acid levels of lysine (9.28 g/16.0 g N) and methionine comparable to those of other legumes such as soybean (Eromosole *et al.*, 2008). However, Akande (2010) has reported that it is rich in minerals such as phosphorus, iron and potassium, although it also contains some anti-nutrients (Fasoyiro *et al.*, 2006, mentioned by Akande, 2010). The seeds are processed with water and some condiments, then wrapped in plantain leaves and boiled and eaten. Flour is mixed with cassava to make soups, and the beans are also used as a paste and sauce. Seeds are previously soaked in water for 12 h. The tubers (containing 12.5–19.0% protein) are eaten like potatoes; resembling sweet potatoes, they are 5.0–7.5 cm long and weigh on average 50–150 g. The tubers are an important source of starch (65–70%) and protein in tropical Africa. It can also be used as green manure for soil restoration and as a feed for livestock.

Germplasm has been collected, particularly by IITA, and evaluated by various institutions. Amoatey *et al.* (2000), Adewale *et al.* (2008) and Akande (2010) have carried out germplasm characterization and found variability in days to maturity, number of seeds per pod, seed size and seed weight. Significant genetic variability was also observed for nutrient and anti-nutrient contents. Potter and Doyle (1992) and Aletor and Aladetimi (2006) found variation for both seed and tuber characters through morphometric and isozyme analyses. High genetic diversity was reported in Nigerian accessions using RAPD primers by Moyib *et al.* (2008).

Both seeds and tubers are used for planting at the beginning of the rainy season. Tubers are ready for harvesting 5–10 months after planting. Seed and tuber yield is reported to be 3000 kg/ha and 1800 kg/ha, respectively. Diseases reported include powdery mildew, leaf spot and stem rust, while virus mosaics are also recorded. Pests, such as leaf-rolling caterpillars and leaf miners, cause damage to the foliage, and thrips damage the flowers. Nematodes attack the roots. The crop is also grown in association with yam, cassava, maize and okra in traditional farming systems (Klu *et al.*, 2001). It can be excellent in rotation for ground cover, as it has a high nitrogen-fixing ability.

21.11 Adzuki Bean (*Vigna angularis* (Willd.))

Adzuki bean is thought to have originated in the Far East but is now grown in India, South-east Asia, China, the USA, South America, Angola, Zaire and New Zealand. It is adapted to both temperate and sub-temperate climates. It is annual, erect and/or bushy with a determinate growth habit, usually 30–90 cm tall. Its flowers (6–12 in cluster) are bright yellow; pods are cylindrical and 6–12 cm long with 4–12 seeds. It can grow in a temperature range of 15–30°C and is reported to grow under an annual precipitation of 500–17,000 mm, on all types of soil from light to heavy clay but does not grow well on extremely acidic soils of pH 5.0–7.5.

Seeds are used as human food; they contain protein (19.9–25.3%) and minerals, and their nutritive value is comparable to that of rice bean. They are reported to contain trypsin and chymotrypsin inhibitors. The beans are boiled or fried and often eaten with rice and ground flour. They are also used in the preparation of cakes, sweetmeats and can be candied. In Japan, it is used largely for human consumptions in the form of meal or paste. Young beans are used as a vegetable. It is also grown for forage and green manure in Japan and China. The seeds possess medicinal properties and are reported to be used in various treatments.

Germplasm collection and evaluation have been carried out by various institutions, such as AVRDC in the Far East and also in South and South-east Asia. Variation in growth habit, time of maturity and seed colours is observed and wide variation has also been noted in yield-related characters (Bhag Mal, 1995; Xu *et al.*, 2000; Zong *et al.*, 2003; Yoon *et al.*, 2007; Kaga *et al.*, 2008; Redden *et al.*, 2009). Xu *et al.* (2008) found genetic differences when they evaluated accessions from eight Asian countries using AFLP and SSR markers. They found that the most diverse accessions were from Japan, China and Korea. Variations in susceptibility to phytopthera stem rot, powdery mildew and *Ascochyta* are also reported (Bhag Mal, 1995). Selections have been made at institutions in Asia, and AVRDC has released some varieties for wider cultivation. Breeding work to improve resistance to brown stem rot has also been carried out. Gene transfer from other related species to adzuki bean is possible, as reported by many researchers (Rashid and Haq, 1993). In India superior lines with high yield are selected and recommended for cultivation (Dutta *et al.*, 2007).

The standardization of agronomic practices for the successful cultivation of adzuki bean has not been determined. Seeds are planted 2.5 cm deep, spaced 30 cm apart in rows 60–90 cm apart depending on variety and local habitat. Seeding rate is usually 10–20 kg/ha. Fertilization is recommended at a rate of 300, 100 and 100 kg/ha of superphosphate, potassium sulfate and ammonium sulfate, respectively. The seed yield ranges from 1000 to 2699 kg/ha. Several diseases and pests and are reported, including leaf blight, rust, charcoal rot, anthracnose, leaf spot, stem rot and powdery mildew, while the pests include pod worm, butter bean borer maggots, aphids and cyst nematodes. The adzuki bean is a short-duration crop and is commonly grown with rice bean in mixed cropping systems. In Japan, the crop is grown in rotation with rice. Like other legumes intercropping with cereal crops offers promise for commercial cultivation and the crop can be made more attractive economically.

21.12 Moth Bean (*Vigna aconitifolius* (Jacq.))

Moth bean is indigenous to the Indian subcontinent and is now widely distributed through the semi-arid areas of South and Southeast Asia; it is also grown in the USA and Australia. It is 10–40 cm tall with 60–130 cm long, trailing branches. Pods are 2.5–5.0 cm long with 4–9 seeds. It can grow at a height of up to 1300 m above sea level. Cultivation of the crop is reported from the drier tracts of Sri Lanka, Myanmar, Malaysia, southern China and western USA. It is an annual crop with a spreading growth habit. Moth bean is drought resistant and mostly grown in arid and semi-arid regions. It performs well in poor, sandy soils and in drylands. The crop tolerates a range of soil pH of 5.0–8.1.

The seed is nutritious and contains 21.0–26.7% protein. It is used in foods, mainly as *dhal* in rural areas of arid regions of Asia. The seeds are reported to be a good source of lysine, leucine and vitamin A. Its immature and mature pods are used as a vegetable. Seeds may be processed for starch and have some medicinal value as well. Leaves, pod shells and branches are used as fodder for animals. It can be used as green manure and also as a cover crop.

Germplasm collection and evaluation have been carried out in some Asian countries, and a wide range of variation for growth characters and yield has been reported (Bhavsar and Birari, 1991; Gowda, 2010). Varietal differences in quality characters are reported and differences in disease and insect pest resistance were also observed. High-yielding varieties have also been developed using chemical mutagen; breeding and biotechnology methods are detailed by Dutta *et al.* (2007).

Seeds are sown 10 cm apart with rows spaced at 30–50 cm. Seeding rate is 12–15 kg/ha for a sole crop and 8–10 kg/ha as an intercrop. Fertilization with 10 kg N/ha, 20 kg P_2O_5/ha is normally practised to achieve a satisfactory crop. Seed yield ranges from 200 to 1500 kg/ha. The green fodder yield is about 2500 kg/ha and dry matter yield is 800 kg/ha (Gill and Yadav, 1989). Bacterial wilt, leaf spot, anthracnose and yellow mosaic virus are the major diseases that threaten, and hairy caterpillar, galerucid beetles, jassids and bean weevil are the major insect pests attacking the crop. It can be grown as an intercrop with other legumes and cereals, and is also rotated as a green manure crop with cotton. Intercropping with cereals can produce good crop yields. Further studies are needed to determine suitable crop combinations and ideotypes.

21.13 Bambara Groundnut (*Vigna subterranea* (L.) Verdc)

Bambara groundnut is indigenous to tropical Africa, but it is also grown in Asia, North Australia, and South and Central America. Recently, the importance of this crop was highlighted by Mkandawire (2010). It is an annual herb of up to 30 cm in height, with creeping and multi-branched lateral stems. Pods are about 2.5 cm in diameter containing up to four seeds. It can be grown in sub-humid to dry regions where the growing of other food legumes is risky. It grows best in a temperature range of 20–28°C and can be cultivated in savannah areas with a seasonal rainfall of 600–750 mm, although for optimum yields an annual rainfall of 750–900 mm is suitable (Linnemann and Azam-Ali, 1993). It grows well on poor soils and thrives on light, sandy, well-drained loam with a pH of 5.0–6.5.

Immature seeds are normally eaten fresh, boiled or grilled and the young pods are used in soup. Mature seeds contain 14–24% protein and 6–12% oil; they are nutritionally valuable as the protein has a high lysine content and exhibits beneficial complementary effects when consumed with cereals that are low in lysine. Flour from mature seeds is mixed with oil or butter to form a porridge, and is also used in other dishes. Bambara groundnut can also be canned. The haulm is also useful for animal feeding. The processing and marketing status has been reported by NAS (2006).

Germplasm collection and evaluation have been carried out throughout Africa (Heller *et al.*, 1997), the largest collection being held by IITA while smaller collections are stored by the national institutions on that continent. More exploration and collections are needed from high-rainfall areas. A wide variation is reported by Linnemann and Azam-Ali (1993) and Begemann *et al.* (2003) from the evaluation of germplasm at different locations. Genetic differences were also studied using AFLP and RAPD markers by IITA (Begemann *et al.*, 2003). Some evaluation was carried out in other ecological systems to evaluate the potential for cultivation and use (NAS, 2006). Variation in yield-related characters of diverse ecotypes has provided the basis for crop improvement. However, the underground flowers make cross-pollination difficult but, once the desired types are selected, it is possible to use them for improvement as the plants are self-compatible and largely self-pollinated. Schenkel *et al.* (2003) suggested that single-seed descent and pure-line selection in breeding programmes would be useful in the improvement of this crop.

Maturity time of the crop depends on cultivar and climatic conditions. Bunch and spreading types mature in 90–120 and 120–150 days, respectively. Crop yield varies from 742kg/ha to 4400kg/ha. Chomchalow (1993) reported the yield of fresh pods to be 2970–3325kg/ha when grown as an intercrop in Thailand, while in West Java a yield of 5000–6000kg/ha was obtained when it was grown as a monocrop. Yields of shelled nuts of 2600kg/ha have been reported from Tanzania (NAS, 2006) and 3580kg/ha from Zimbabwe (Haq, 1993). A few diseases affect the crop – cercospora (leaf spot) and erysiphe (mildew) are reported to cause damage. Among the pests, rats, red ants, cutworms and grasshoppers affect crop growth. Lists of diseases and pests are given by Linnemann and Azam-Ali (1993). The crop is grown as an intercrop in association with millet, sorghum, maize, cassava, groundnut, cowpea, okra, pumpkin and sugar cane. In many countries in Africa it grows in rotation with cereals, yams and legumes (Andika *et al.*, 2008). The residual effects of groundnut, bambara groundnut, fallow and a previous maize crop on grain yield have been shown by Linnemann and Azam-Ali (1993).

21.14 Rice Bean (*Vigna umbellata* (Thumb.) Ohwi and Ohashi)

Rice bean is native to Indo-China and it is now grown in Asia and other areas of the tropics such as Mauritius, East Africa, West Indies, the Pacific islands, Brazil, Australia and the USA. The crop is adapted to conditions of high temperature and humidity, and is suitable for the lowland tropics where other crops are difficult to grow. It can thrive under an annual rainfall of 700–1730mm and annual mean temperature of 18–30°C, with soil pH ranging from 6.8 to 7.5. It grows at an altitude of up to 2000m. It may be annual or perennial, with a shot-stemmed, erect, semi-erect or twining growth habit. Rice bean is a short-day legume and the day length threshold for this species is less than 12h.

The seed contains 16–25% protein and is a good source of calcium. It is reported to contain the vitamins thiamine, niacin and riboflavin, and large amounts of iron and phosphorus. Rice bean is a multi-purpose species like other underutilized food legumes. Immature pods and leaves are used as both a vegetable and a pulse in Asia. Its use as bean sprouts has also been noted in Asia. The seed is made into soups and stews, for production of bean sprouts and is also processed into *dhal*. It is used as a fodder crop, green manure and a cover crop. A large number of germplasm collections exist in Asia, particularly in India and AVRDC, Taiwan. The evaluation of germplasm has shown considerable variation in agronomic characters (Joshi *et al.*, 2008; Muthusamy *et al.*, 2008). Variation has also been reported in the biochemical composition of seeds, crude protein, amino acid profile, iron, calcium and phosphorus.

Several institutes have produced a large number of strains through selection and breeding. The yield of seed and forage has been increased through crossing, and interspecific hybridization can be used to produce desirable high-yielding varieties for wider cultivation (Rashid *et al.*, 1987). Mutation breeding has also shown promise in developing desirable varieties (Singh and Tomar, 1989). Dutta *et al.* (2007) reported a linkage map for rice bean using RFLP, RAPD and RAPA markers. This could help to select breeding materials for improvement of the crop. Growth, maturity and yield of rice bean vary depending on cultivar, climatic conditions and the time of sowing. For example, the crop matures within 60 days in Angola, but in Eastern India and Bangladesh it takes about 130 days to produce an economic seed yield. Seed yield varies from 600 to 2100 kg/ha and green fodder from 20,000 to 22,000 kg/ha 70–80 days after sowing. The crop is remarkably free from pests and diseases, which makes it a potential source of disease resistance for other species in the genus. Rice bean can be intercropped with sorghum, pearl millet, maize, minor millets and grasses, as well as with other legumes. Because of the crop's duration, rice bean has been grown in rotation with rice in Asia, where it increases the fertility of paddy fields. It is also grown as a mixed crop with maize.

21.15 Conclusions

The situation of some underutilized crops for production has improved over the last few years, as several funding agencies and organizations have given impetus to the improvement of underutilized food legumes. This has increased knowledge on some species, but even so we have not seen any specific so-called underutilized legume species become a major species in farming systems. The research on these potential crops is still minimal in comparison with major crops. This may be due to limited research funds, but also to disjointed research carried out without proper market research and possible outlets. However, some underutilized fruit tree legumes have been penetrating the global market because of their diverse uses and potential for added value (Haq *et al.*, 2007).

The crops discussed in this chapter are already playing a role in traditional cropping systems in subsistence agriculture. The demand for food, and also increasing input costs, necessitate the use of these underutilized food legumes in existing systems as rotation crops, intercrops and catch crops in various combinations with cereals and industrial crops, including plantation crops. Haq (1983) mentioned that multiple cropping of rice and the introduction of the groundnut to Africa have replaced rice bean and bambara groundnut. The situation is similar even in major food legumes (i.e. lentil, mung bean) that were replaced by wheat. The large-scale cultivation of underutilized legumes is limited because of the lack of improved genotypes. However, recent research has shown that large genetic diversity and diverse ecotypes of these crops exist, making them suitable for crop diversification systems; and their use could be extended beyond present subsistence agriculture. For example, rice bean can be grown during the fallow period for three to four months after transplanted *aman paddy* is harvested in north-eastern Bangladesh. This will not only earn extra cash but will also help in increasing soil fertility. Similarly, suitable genotypes of bambara groundnut can be grown where other crops cannot be grown due to poor soil or adverse climate conditions. The velvet bean, too, has wider adaptation and could emerge as a food and industrial crop where cowpeas cannot thrive. Similarly, adzuki bean is a popular food in Japan and Korea and this, combined with its medicinal value, merits consideration for serious cultivation, at least in Asia and the Pacific region. The potential of horse gram is enormous, as it is a short-duration crop and grows well under low rainfall and high temperature; it can grow in poor soils and in semi-arid and arid areas. The crop needs minimal inputs and management and can be cultivated commercially for grain, fodder and green manure. Lablab bean has established regional and international markets for its fresh and processed products.

The above examples have shown that there is a need for continuous research with special emphasis on market chains. The improvement in these high-protein crops with multi-purpose uses by interdisciplinary research will not only develop crops for smallholders but will also enable these to be established more widely in agricultural systems, where it will help generate a balanced diet for farmers and produce extra cash to meet their other needs.

Acknowledgement

I thank Dr. J. Smartt for reading this manuscript and for his valuable comments and suggestions.

References

Adewale, B.D., Kehinde, B.O., Odu, O.B. and Dumet, D.J. (2008) The potential of African yam bean (*Sphenostylis stenocarpa* Hochst, Ex. A. Rich) in Nigeria: character and genetic diversity. In: Smartt, J. and Haq, N. (eds) *New Crops and Uses: Their Role in A Rapidly Changing World*. CUC, UK, pp. 265–276.

Ajayi, F.T., Akande, S.R., Adegbite, A.A. and Idowu, B. (2009) Assessment of seven underutilized grain legume foliage as feed resources for ruminants. *Livestock Research for Rural Development* 21, www.lrrd.org/irrd21/1/ajay 21149.htm (accessed 19 March 2010).

Akande, S.R. (2010) Germplasm characterization of African yam bean (*Sphenostylis stenocarpa*) from south-west *Nigeria. PGR Newsletter (FAO-Bioversity)* 154, 25–29.

Akande, S.R. and Balogun, M.O. (2007) Evaluation and heritability studies of local lima bean (*Phaseolus lunatus* L.) cultivars from south-west Nigeria. *Revista UDO Agricola* 7, 22–28.

Akapapunam, M.A. and Sefa-Dedah, S. (1997) Jack bean (*Canavalia ensiformis*): Nutrition related aspects and needed nutrition research. *Plant Foods Human Nutrition* 50, 93–99.

Akinlade, J.A., Farinu, G.O., Taiwo, A.A., Aderinola, O.A., Adebayo, T.A., Ojebiyi, O.O. *et al.* (2007) Agronomic and nutritive evaluation of jack bean (*Canavalia ensiformis*) for fodder in the derived Savannah Zone in Nigeria. *International Journal of Agricultural Research* 2, 1059–1063.

Aletor, V.A. and Aladetimi, O.O. (2006) Compositional evaluation of some cowpea varieties and some underutilised legumes in Nigeria. *Molecular Nutrition and Food Research* 33, 999–1007.

Amoatey, H.M., Klu, G.Y.P., Bansa, D., Kumaga, F.K., Aboagye, L.M., Benett-Larty, S.O. *et al.* (2000) The African yam bean (*Sphenostylis stenocarpa*): a neglected crop in Ghana. *West African Journal of Applied Ecology* 1, 53–59.

Andika, D.O., Onyango, M.O.A. and Onyango, J.C. (2008) Role of bambara groundnut in cropping systems in western Kenya. In: Smartt, J. and Haq, N. (eds) *New Crops and Uses: Their Role in A Rapidly Changing World*. CUC, UK, pp. 204–212.

Armstrong, K.L., Albrecht, A.L., Laur, J.G and Riday, H. (2008) Intercropping corn with lablab bean, velvet bean, and Scarlet runner bean for forage. *Crop Science* 48, 371–379.

Arsene, I.Z.B. and Baudoin, J.P (2007) Geographical allozymes differentiation in wild *Phaseolus lunatus* L. of the central valley of Costa Rica and its implications for conservation and management populations. *Biotechnology and Agronomy Society of the Environment* 1, 287–297.

Asante, I.K., Offei, S.K., Addy, R. and Carson, A.G. (2008) Phenotypic and seed protein analysis in 31 lima bean (*Phaseolus lunatus*) accessions in Ghana. *West African Journal of Applied Ecology* 12 (www.ajol.info/index.php/wajae/issue/view/6415).

Ayisi, K.K., Bopape, M.P. and Pengelly, B.C. (2004) Assessment of the variation in growth and yield of diverse lablab (*Lablab purpureus*) germplasm in Limpopo province, South Africa. *ACIAR Proceedings* No. 115, 44–49.

Begemann, F., Mukeme, I. and Obel-Lawson, E. (2003) *Promotion of Bambara Groundnut*. IPGRI, Rome, pp. 122.

Bettencourt, E., Konopka, J. and Damania, A.B. (1989) *Directory of Crop Germplasm Collections. 1. Food Legumes*. International Board for Plant Genetic Resources, Rome.

Bhag Mal (1995) *Underutilized Grain Legumes and Pseudocereals – Their Potentials in Asia*. FAO (RAPA), Bangkok, Thailand, pp. 162.

Bhat, R. and Karim, A.A. (2009) Exploiting the nutritional potential of wild and underutilized legumes. *Comprehensive Reviews in Food Science and Food Safety* 8, 305–331.

Bhavsar, V.V. and Birari, S.P. (1991) Variability, correlation and path analysis in moth bean. *Journal of Maharashtra Agriculture University* 14, 148–150.

Caicedo, A.L., Gaitan, E., Duque, M.C., Chica, O.T., Debouck, D.G. and Tohme, J. (1999) AFLP fingerprinting of *Phaseolus lunatus* L. and related wild species from South America. *Crop Science* 39, 1497–1507.

Castifieiras, L., Guzman, F.A., Dugue, M.C., Shagarpdsky, T., Cristobal, R and Vicente, M.C. (2007) AFLPs and morphological diversity of *Phaseolus lunatus* L. in Cuban home gardens: approaches to recovering the lost *ex situ* collection. *Biodiversity and Conservation* 16, 2847–2865.

Chaudhary, G.A. (1983) Studies on the population density and root development in *Lupinus mutabilis* in relation to yield and yield components. PhD thesis, University of Southampton, UK.

Chaudhary, G.A. and Cheema, N.M. (1998) Prospects of *Lupin* cultivation in Pakistan. In: Haq, N., Anthony, K., Sarwar, M. and Ahmed, Z. (eds) *Underutilized Crops of Pakistan*. PARC, Islamabad, Pakistan, pp. 41–46.

Chomchalow, N., (1993) Winged bean. In: Chomchalow, N., Gowda, C.L.L. and Laosuwan, P. (eds) *Underexploited and Potential Food Legumes in Asia*. FAO, Bangkok, Thailand, pp. 109–115.

Chomchalow, N., Gowda, C.L.L. and Laosuwan, P. (1993) *Underexploited and Potential Food Legumes in Asia*. FAO, Bangkok, Thailand, p. 150.

Duke, J.A. (1987) *Handbook of Legumes. World Economic Importance*. Plenum Press, New York, pp. 345.

Dutta, M., Phogar, B.S. and Dillon, B.S. (2007) Genetic improvement and utilization of major underutilized crops in India. In: Ochatt, S. and Jain, S.M. (eds) *Breeding of Neglected and Underutilized Crops, Spices and Herbs*. Science Publishers, Enfield, New Hampshire, pp. 251–298.

Eromosole, C.O., Argoundade, I.C., Eromosole, J.C. and Ademuyiwa, O. (2008) Extractability of African yam bean (*Spenostylis stenocarpa*) protein in acid salt and aqueous media. *Food Hydrocollide* 22, 1622–1628.

Fasoyiro, S.B., Ajbede, S.R., Omole A.J., Adeniyan, O.W. and Farinde, E.O. (2006) Proximate minerals and anti-nutritional factors of underutilized grain legumes in south-western Nigeria. *Nutrition and Food Science* 26, 18–23.

Gill, A.S. and Yadav, C.L. (1989) Performance of moth varieties for forage yields under rainfed conditions. *Agriculture Science Digest* 9, 105–106.

Gowda, C.L.L., Upadhya, H.D. and Ghaffar, M.A. (2009) Importance of underutilized indigenous legumes in Asia-Pacific region. *Acta Horticulturae* 752, 95–98.

Gowda, M.B. (2010) Genetic enhancement of dolichos bean through integration of conventional breeding and molecular approaches. Final Report to the Kirkhouse Trust, UK, pp. 56.

Graham, P.H. and Vance, P.C. (2003) Legumes: Importance and constraints of greater use. *Plant Physiolology* 131, 872–877.

Haq, N. (1982) Germplasm resources, breeding and genetics of the winged bean. *Z. Pflanzanzücht* 88, 1–12.

Haq, N. (1983) New food legume crops for the tropics. Better crops for the future. *Ciba Foundation Symposium* 97, 144–160.

Haq, N. (1993) Lupins (*Lupinus* species). In: Williams, J.T. (ed.) *Pulses and Vegetables*. Chapman & Hall, London, pp. 103–130.

Haq, N. (2004) New plant-derived ingredients. In: Smith, J.M. (ed.) *Technology of Reduced Additive Foods*. Blackwell Science, Oxford, UK, pp. 106–124.

Haq, N. (2007) Underutilized fruits: A resource for sustainable livelihoods. In. Ochatt, S. and Jain, S.M. (eds) *Breeding of Neglected and Underutilized Crops*. Spices and Herbs, Science Publishers, Enfield, New Hampshire, pp. 1–28.

Haq, N., Bowe, C. and Dunsiger, Z. (2007) Challenges to stimulating the adoption and impact of indigenous fruit trees in tropical agriculture. In: Akinnifesi, F.K., Leakey, R.R.B, Ajayi O.C., Sileshi, G., Tchoundjeu, Z., Matakala, P. *et al.* (eds) *Indigenous Fruit Trees in the Tropics: Domestication, Utilization and Commercialization*. CAB International, Wallingford, UK, pp. 51–69.

Haque, M.E., Rahman, M., Rahman, M.A., Roy, A.K. and Sikdar, B. (2003) Lablab bean-based intercropping systems in northwest region of Bangladesh. *Pakistan Journal of Biological Science* 6, 948–951.

Hardy, A. and Huyghe, C. (1997) Physiological bases of the poor adaptation of current *Lupinus mutabilis* genotypes to European conditions. *Grain Legumes* 15, 9–10.

Heller, J., Begemann, F. and Mushonga, J. (1997) *Bambara Groundnut*. IPGRI, Rome, pp. 166.

Islam, T. (2010) Morpho-agronomic diversity of hyacinth bean (*Lablab purpureus* (L.) Sweet) accessions from Bangladesh. *PGR Newsletter* 156, 73–78.

Jansen, P.C.M. (1989) *Phaseolus acutifolius* A. Gray In: Van der Maesen, L.J.G. and Somaatmadja, S. (eds) *Plant Resources of South-East Asia*. (No.1). *Pulses*. Pudoc, Wageningen, The Netherlands, pp. 53–55.

Joshi, K.D., Bhaduri, B., Gautam, R., Bajracharya, J. and Hollington, P.B. (2008) Rice bean: a multi-purpose underutilized legume. In: Smartt, J. and Haq, N. (eds) *New Crops and Uses: their Role in a Rapidly Changing World*. CUC, UK, pp. 234–248.

Kaga, A., Isemura, T., Tomooka, N. and Vaughan, D.A. (2008) The genetics of domestication of the azuki bean (*Vigna angularis*). *Genetics* 178, 1013–1036.

Kahnt, G. and Hijazi, L.A. (2008) Use of lupinex to increase crop yield and improve harvest quality with lesser nitrogen fertilization. *Journal of Agronomy and Crop Science* 166, 228–237.

Keshava, B.S., Halepyati, A.S., Puiari, B.T. and Desai, B.K. (2007) Yield and economics of horse gram (*Macrotyloma uniflorum* Lam. Verde) as influenced by genotypes, plant densities and phosphorus levels. *Karanatka Journal of Agricultural Science* 20, 589–591.

Klassen, W., Codallo, M., Zasada, I.A. and Abdul-Baki, A.A. (2006) Characterisation of velvet bean (*Mucuna pruriens*) lines for cover crop use. *Proceedings of Florida State Horticulture Society* 119, 258–262.

Klu, G.Y.P., Amoatey, H.M., Bansa, D. and Kumaga, F.K. (2001) Cultivation and use of African yam bean (*Sphenostylis stenocarpa*) in Volta region of Ghana. *Journal of Food Technology in Africa* 6, 74–77.

Linnemann, A.R. and Azam-Ali, S. (1993) Bambara groundnut (*Vigna subterranea*). In: Williams, J.T. (ed.) *Pulses and Vegetables*. Chapman & Hall, London, pp. 103–130.

Liu, C.J. (1996) Genetic diversity and relationships among *Lablab purpureus* genotypes evaluated using RAPD as markers. *Euphytica* 90, 115–119.

Lorrenzetti, F., MacIsaac, S., Arnason, J.T., Awang, D.V.C. and Buckles, D. (2010) The phytochemistry, toxicology, and food potential of velvet bean (*Mucuna* Adans. Spp., Fabacea). Available at www.idrc.ca/en/ev-31916-201-1-DO_TOPIC (accessed 13 May 2010).

Martinez-Castilo, J., Zizumbo-Villarreal, D. and Colunga-Garciamarin, P. (2004) Intraspecific diversity morpho-phenological value of *Phaseolus lunatus* L. from Yucatan Peninsula, Mexico. *Economic Botany* 58, 354–380.

Mkandawire, C.H. (2010) Rediscover some of the underutilised and neglected crops of the world with a view to broaden our food resource base: Bambara groundnut. Available at www.ecotravelpage.info/article/re-discover

Moyib, O.K., Gbadegesin, M.A., Aina, O.O. and Odunola, O.A. (2008) Genetic variation within a collection of Nigerian accessions of African yam bean (*Sphenostylis stenocarpa*) revealed by RAPD primers. *African Journal of Biotechnology* 7, 1839–1846.

Murphy, A.M. and Colucci, P.E. (1999) A tropical forage solution to poor quality ruminant diets: a review of *Lablab purpureus*. *Livestock Research for Rural Development* 11. www.lrrd.cipan.org.co/lrrdH/2/colu112.htm (accessed 20 May 2010).

Muthusamy, S., Kanagarjan, S. and Ponnusamy, S. (2008) Efficiency of RAPD and ISSR markers systems in accessing genetic variation of rice bean (*Vigna umbellata*) landraces. *Electronic Journal of Biotechnology* 11, 1–10 (available online).

NAS (1979) *Tropical Legumes: Resources for the Future*. National Academy of Sciences, Washington, DC, pp. 331.

NAS (2006) *Bambara Bean. Lost Crops of Africa Vol. 11. Vegetables*. National Academy of Sciences, Washington, DC, pp. 52–73.

NRI (1987) African yam bean (*Sphenostylis stenocarpa*). *Root Crops* 6, pp. 308.

Patel, D.P and Loknathan, T.R. (1998) Evaluation of winged bean (*Pshopocarpus tetragonolobus*) germpalsm. *Indian Journal of Plant Genetic Resources* 12. Available at www.indianjournals.com/ijor.aspx?target (accessed 16 June 2010).

Potter, D. and Doyle, J.J. (1992) Origins of the African yam bean (*Spenostylis stenocarpa* Leguminosae) evidence from morphology, isozymes, chloroplast DNA and linguistics. *Economic Botany* 46, 276–292.

Pugalenthi, M., Vedivel, V. and Siddhuraju, P. (2005) Alternative food/feed perspectives of an underutilized legume. *Mucuna puriens* var. *utilis* – a review. *Plant Foods and Human Nutrition* 60, 201–218.

Rai, N.A.K., Singh, P.K., Singh, M., Durra, D. and Rai, M. (2010) Genetic relationship among hyacinth bean (*Lablab purpureus*) genotypes cultivars from different races based on quantitative traits and random amplification polymorphic DNA marker. *African Journal of Biotechnology* 9, 137–144.

Rashid, K.A. and Haq, N. (1993) *In vitro* culture of the parents, interspecific hybrids in colchiploid embryos of *Vigna* spp. *Journal of Science (Mal)* 1, 1–12.

Rashid, K.A., Smartt, J. and Haq, N. (1987). Hybridization in the genus *Vigna*. In: Shanmugasundaram, S. and Mclean, B.T. (eds) *Mungbean*. AVRDC, Taipei, Taiwan.

Ravindra, R. and Sundar, S.T.B. (2009) Nutrition value of horse gram for egg type chicks and growers. *Tamilnadu Journal of Veterinary and Animal Science* 5, 125–131.

Redden, J.R., Basford, K.E., Kroonenberg, P.M., Islam, A.F.M., Ellis, R., Wang, S. *et al.* (2009) Variation in adzuki bean (*Vigna angularis*) germplasm grown in China. *Crop Science* 49, 771–782.

Samranayake, R.G.A.R. and Guansena, H.P.M. (2010) Studies on winged bean-maize intercrop systems. Available at www.Goviya/lk/agri_learning/maize

Schenkel, W., Marrure, F.J., Saywan, R.S. and Ng, Q. (2003) Crop improvement and breeding In: Begemann, F., Mukeme, I. and Obel-Lawson, E. (eds) *Promotion of Bambara Groundnut.* IPGRI, Nairobi, Kenya, pp. 4–7.

Shisanaya, C.A. (2002) Improvement of drought adapted tepary bean (*Phaseolus acutifolius* A. Gray va. Latifolius) yields through biological nitrogen fixation in semi-arid S.E.-Kenya. *European Journal of Agronomy* 16, 13–24.

Singh, S.K., Chakraborty, S., Singh, A.K. and Pandey, P.K. (2006) Cloning, restriction mapping and phylogentic relationship of genomic compounds MYMIV from lablab bean. *Bioresource Technology* 97, 1807–1814.

Singh, V.P. and Tomar, Y.S. (1989) A photosensitive promising mutant of rice bean (*Vigna umbellata*). *Legume Research* 12, 47–48.

Smartt, J. (2008) *Grain Legumes: Evolution and Genetic Resources.* Cambridge University Press, Cambridge, UK.

Sultana, N., Ozaki, Y. and Okubo, H. (2000) The use of RAPD markers in lablab bean (*Lablab purpureus* (L.) Sweet) phylogeny. *Bulletin of the Institute of Tropical Agriculture Kyushu University* 23, 45–51.

Sunil, N., Sivraj, N., Pandravada, S.R., Kamala, V., Reddy, P.R. and Varaprasad, K.S. (2008) Genetic and geographical divergence in horse gram germplasm for Andraha Pradesh. *Plant Genetic Resources Characterization and Utilization* 7, 84–87.

Sweetingham, M. and Clements, J. (2005) Prospects for commercial pearl *Lupin. Beanstalk* 6, 5.

Van der Maesen, L.J.G. and Somaatmadja, S. (1989) *Plant Resources of South-East Asia. (1). Pulses.* Pudoc, Wageningen, The Netherlands, pp. 105.

Vedivel, V. and Janardhanan, K. (2000) Nutritional and anti-nutritional composition of velvet bean: an underutilized food legume in South India. *International Journal of Food Science and Nutrition* 51, 279–287.

Virk, D.S., Charkraborty, M., Ghosh, J. and Harris, D. (2006) Participatory evaluation of horse gram (*Macrotylema uniflorus*) varieties and their on-station responses to on-farm seed priming in eastern India. *Experimental Agriculture* 42, 411–425.

Witcombe, J.R., Billore, M., Singhal, H.C., Patel, N.B., Tikka, S.B.S., Saini, D.P. *et al.* (2008) Improving the food security of low-resource farmers: Introducing horse gram into maize-based cropping systems. *Experimental Agriculture* 44, 339–348.

Xu, H.X., Jing, J., Tomooka, N., Kaga, A., Isemura, T. and Vaughan, D.A. (2008) Genetic diversity of the azuki bean (*Vigna angularis* (Wild.) Ohwi & Oshahi) gene pool as assessed by SSR markers. *Genome* 51, 728–738.

Xu, R.Q., Tomooka, N. and Vaughan, D.A. (2000) AFLP markers for characterizing the azuki bean complex. *Crop Science* 40, 808–815.

Yoon, M.S., Lee, J., Kim, C.Y. and Baek, H.J. (2007) Genetic relationships among cultivated and wild *Vigna angularis* (Wild.) Ohwi et. Ohashi and relatives from Korea based on AFLP markers. *Genetic Resources and Crop Evolution* 54, 875–883.

Zambre, M., Montagu, M.V., Angenon, G. and Terryn, N. (2006) Tepary bean. *Methods in Molecular Biology* 343, 407–414.

Zong, X.X., Kaga, Tomooka, N., Wang, X.W., Han, O.K. and Vaughn, D.A. (2003) The genetic diversity of the *Vigna angularis* complex in Asia. *Genome* 46, 647–658.

22 Legumes as a Model Plant Family

S.B. Cannon, Shusei Sato, Satoshi Tabata, N.D. Young and G.D. May

22.1 Introduction

The human population derives the majority of its nutrition either directly or indirectly (via animal protein) from two plant families: the grasses and the legumes. Grain legumes alone contribute 33% of human protein nutrition (Vance *et al.*, 2000). Thus, it is critical for genetic improvement of legume crop species that we make good use of information in this plant family. The term 'model' in biology connotes an organism that exemplifies processes or characteristics that apply broadly to many other organisms. In this sense, the legumes contain numerous species that have been used as important models in plant biology. The first of these was Gregor Mendel's pea (*Pisum sativum* L.), which he used to investigate the basic principles of trait heritability and genetic linkage. More recent models include *Medicago truncatula* and *Lotus japonicus*, both used to investigate the genetic basis of nodulation and symbiotic nitrogen fixation (SNF), among many other processes. In this sense, the legume family contains several prominent biological models.

More broadly, every legume species that has been used to uncover new biological knowledge can be thought of as a potential model for that characteristic, applicable to some other set of species at greater or lesser taxonomic distances. In this sense, the legumes can be thought of as a network (or, indeed, tree) of related traits, developmental patterns, biochemistries and so on. More simply, the legumes comprise a coherent genetic system. This broader sense of the term 'model' is worth noting because several technologies and trends make it possible to consider relating and integrating virtually all discoveries across the legume family. As more genomes and near-complete transcriptomes are sequenced, it becomes increasingly possible to determine orthology relationships among all genes and notable genomic features from such data sets; and to associate those features with traits and developmental patterns – and then to compare traits and patterns for homology and for underlying sequence similarities and differences.

The concept of 'legumes as a model plant family' can be broadened further still: the legumes family can be thought of as a model for other taxonomic families. The kinds of general processes to be investigated at the family level include the pace, timing and geography of speciation; the role of polyploidy in speciation; and the processes of evolution of functions and forms. The legumes have numerous characteristics that make them well suited as a 'model family'. The family is remarkably large and diverse, with more than 20,000 species (Doyle and Luckow, 2003) on nearly all terrestrial biomes. The family contains clades with unusual patterns of diversification (for example, with little species radiation in some of the early-diverging clades, and prodigious speciation in the papilionoid clade and some sub-clades). The family contains great morphological diversity, exemplified by the dramatically different floral structures in various basal lineages (including some with radially symmetric flowers), the mimosoid clade (with reduced petals and a 'pompom' of anthers) and the papilionoid clade (with its characteristic fused and asymmetrical pea- or bean-like floral structures). And of course the family includes many species with nodules, housing symbiotic nitrogen-fixing rhizobial bacteria. With access to more nitrogen, the legumes have also evolved elaborated nitrogen-rich chemistries, including a wide range of alkaloids, non-protein amino acids and protein-rich seeds.

These three senses of the term 'model' might be summarized as: 'a few bright lights (a few selected species)' or 'a thousand points of light (many informative species)' or 'a model family (a family as model for other families)'.

22.2 Development of Legumes as a Model Plant Family: Recent Progress

There have been, over the last decade, numerous reviews of legume genomics and cross-legume research (Graham and Vance, 2003; Young *et al.*, 2003; Gepts *et al.*, 2005; Zhu *et al.*, 2005; Cannon *et al.*, 2009; Varshney *et al.*, 2009; Young and Udvardi, 2009; Sato *et al.*, 2010). It is informative to trace the progress in legume research in this set of reviews, and particularly, relative to the 10-year roadmap laid out in a white paper from 2005, reporting on the Cross-Legume Advances through Genomics (CATG) planning conference in Santa Fe, New Mexico, in December 2004 (Gepts *et al.*, 2005). The CATG white paper (a report on the conference) presented a vision and plan for development of 'legumes as a model plant family'. The broad goals of this plan were: '(i) knowledge about the legume family as a whole; (ii) understanding about the evolutionary origin of legume-characteristic features such as rhizobial symbiosis, flower and fruit development, and its nitrogen economy; and (iii) pooling of genomic resources across legume species to address issues of scientific, agronomic, environmental, and societal importance' (Gepts *et al.*, 2005).

The strategy described in the CATG white paper focused on developing one or two reference systems for the two primary crop-containing clades: the Hologalegina, represented by *L. japonicus* and *M. truncatula*; and the phaseolid/millettioid clade, represented by *Glycine max* (soybean) (Fig. 22.1 and discussed below). Besides these 'nodal' species, *Phaseolus* and *Arachis* would also be targeted for development of a range of genomic resources, including physical maps and BAC-end sequencing, molecular markers, EST sequencing, microarray sequencing and ultimately, 'sequencing of the *Phaseolus* genome and gene-rich regions for *Arachis*' (Gepts *et al.*, 2005). Additionally, cross-legume markers would be developed for species including 'pea, lentil, chickpea, faba bean, lucerne, clover, cowpea, pigeonpea, and lupin (*Lupinus* spp.)'. Here, the need for species in other basal clades of legumes was mentioned. Interestingly, *Chamaecrista* has, in fact, since been developed as a model in its own right (Singer *et al.*, 2009).

The progress since the 2005 meeting has been significantly more rapid than had been anticipated at that time. We now have three mostly-complete genome sequences (soybean, *M. truncatula* and *L. japonicus*), and several more under way (common bean, pigeon pea, narrow-leafed lupin and mung bean). Beyond these, extensive transcriptome sequence sets, and sizeable genetic maps, are available for leguminous crops (pea, lentil, chickpea, faba bean, clovers, cowpea, pigeon pea, groundnut and lupin) that had received little attention, in terms of modern molecular breeding techniques, until recently (Varshney *et al.*, 2009). Development can also be seen beyond the core nodal/model species (*Medicago*, *Lotus*, *Glycine*) to the next tier of legume crops and forages, such as pea, common bean, faba bean, lentil, chickpea, lucerne, etc. These 'second-tier' species include many important crops, but have not – at least until very recently – had extensive

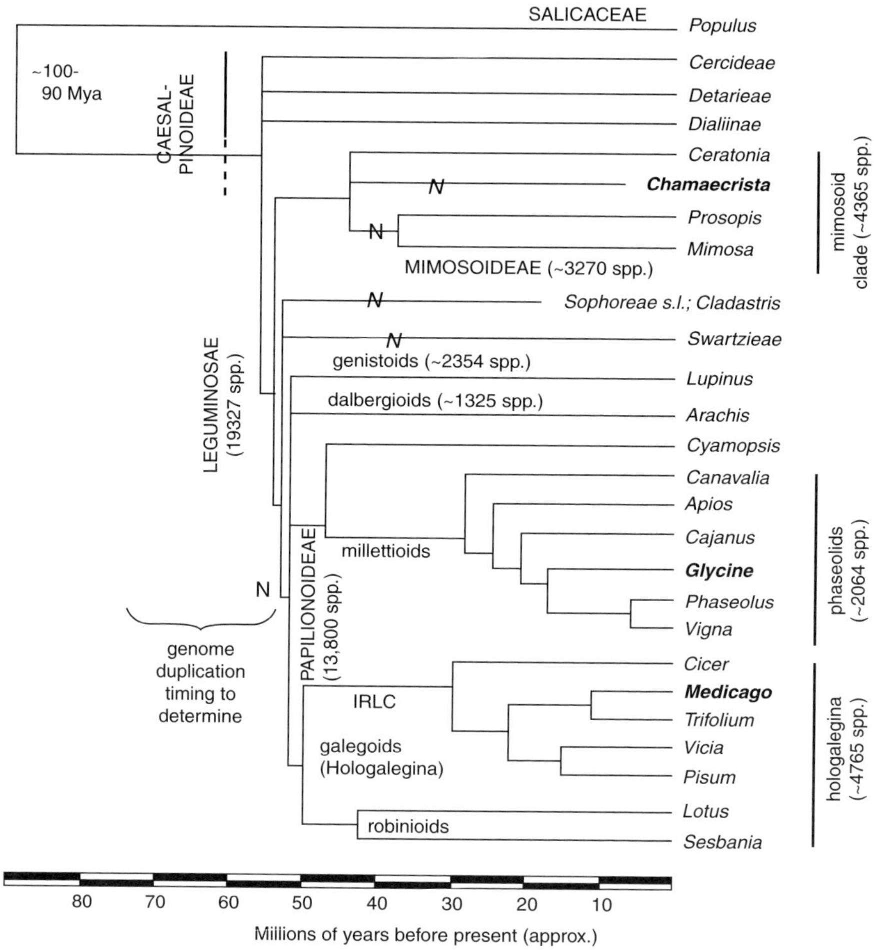

Fig. 22.1. Taxonomic relationships among selected legume genera. Each genus contains one or more food crops, or a genomic model (*Medicago*, *Lotus*, *Chamaecrista*). IRLC, 'inverted-repeat-loss clade'. Approximate inferred speciation dates follow the timings in Lavin *et al.* (2005). The phylogeny is after Lavin *et al.*, (2005) and Lewis *et al.* (2005). Common names and uses are given in Table 22.1. Figure redrawn from Cannon *et al.* (2005), with permission.

genetic or genomic resources. A 'third tier' would include those crops that are currently marginal with limited breeding programmes, and few to no genomic resources.

The legume family contains at least four dozen domesticated or partly domesticated food and forage species (Cannon *et al.*, 2009), including many 'third-tier' species that are only partially domesticated or are important in particular limited geographical regions (lablab, winged bean, Hausa groundnut, tarwi lentil, etc.). These may prove increasingly important if they are able to fill particular agroecological niches, and if various agronomic deficiencies or marketing constraints can be overcome. Additionally, numerous potential forage and green-manure crops are important in particular agroecosystems, but have received little formal breeding attention.

As examples of several legume crops arguably in the 'second-tier,' faba bean and pea can take advantage of winter and spring moisture where they are useful, both for fodder and seed production and in crop rotations.

Although they have broad global adoption, the funding and development of genomic resources for these plants lag behind that of the nodal genomic species. Further, their adoption and use is surprisingly limited in some regions: while faba bean is very important in China and the Mediterranean region, it is a minor crop in the USA, despite suitable climate in many areas (Monfreda *et al.*, 2008).

Consider next, several examples of 'third-tier' crops – defining these as crops with few genomic resources and with limited geographic use, and perhaps agronomic problems to be solved. *Lathyrus sativus* (cicerchia, chickling vetch or grass pea), is a short-cycle crop that is able to tolerate both waterlogged and dry soil (Campbell, 1997). It is used, therefore, to take advantage of remnant moisture at the end of rice culture in India, and as an early spring crop in the Mediterranean. The chief deficiency in *Lathyrus* as a crop is its production of a non-protein amino acid, β-oxalyl-diamino-propionic acid (ODAP) that is toxic if consumed in quantity; however, low-ODAP lines exist and may be useful as the basis for further breeding programmes (Campbell, 1997). Winter moisture and otherwise inaccessible nutrients and moisture are accessible to deep-rooted and cold- and drought-hardy perennials such as *Lupinus polyphyllus* (Washington lupin). Although most lupin species produce toxic alkaloids, there are low-alkaloid lines of *Lupinus angustifolius*, *Lupinus albus* and *Lupinus polyphyllus* (Kurlovich *et al.*, 2008). Numerous legumes produce edible tubers, including *Pachyrhizus* edulis (jicama), *Vigna vexillata* (tuber cowpea) (Karuniawan *et al.*, 2006) and *Apios Americana* (groundnut, Indian potato or potato bean) (Blackmon and Reynolds, 1986). Among these, *Apios*, in particular, is promising, as it was a staple of Eastern and Midwestern North American Indians, and it produces a palatable, nutritious tuber, with approximately 15% protein by dry weight (Wilson *et al.*, 1986, 1987). Furthermore, *Apios* tolerates waterlogging and acid soils, so may be a useful crop where soil acidity has increased due to long-term intensive agriculture with high nitrogen use – a problem now evident in much of China's cropland, for example (Guo *et al.*, 2010). In arid environments, long-lived, extremely drought-tolerant shrubs or trees are locally important, including *Prosopis glandulosa* (honey mesquite) and the African *Cordeauxia edulis* (yeheb nut) (Graham and Vance, 2003).

A confluence of factors may lead to rapid crop improvement in species that are currently minor or niche-specific crops. First, the extensive development of the existing pre-eminent models in the legumes *M. truncatula*, *L. japonicus* and *G. max* (soybean) means that a great deal of information about gene function is available for use in other crop species – providing there are sufficient technologies for making use of this information through breeding methods or biotechnology. Secondly, the extent of similarity across several groups of legumes that contain the largest number of agronomic species suggests that gene functions and locations will frequently be conserved across many legume species. Thirdly, radical decreases in the cost of sequencing and other genomic technologies mean that impressive resources can now be relatively inexpensively developed for virtually any crop.

22.3 Legume Taxonomy in Relation to Agronomic Species

The legumes comprise approximately 20,000 species (Doyle and Luckow, 2003). The majority of these (approx. 13,800 species) are in the Papilionoideae subfamily. Of the remaining scant third (~5530), about two-thirds (~3270) are in the Mimosoideae subfamily and the remainder are in a collection of early-diverging clades, traditionally placed in the Caesalpinoideae.

Most domesticated legume species are in the papilionoid subfamily (Table 22.1; Fig. 22.1). These include the various beans in the millettioid clade; and the peas, vetches (such as faba bean) and clovers in the 'hologalegina' clade (also called 'galegoid' or 'cool-season legume' clade). Nevertheless, a large number of other economically important species are found in the Mimosoideae and early-diverging clades, including a vast number of little-studied species (many of them tropical).

Table 22.1. Selected food and model legumes. Crop legume species are grouped in the three classical legume subfamilies: Caesalpinioideae, Mimosoideae and Papilionoideae; and then by clade and tribe.

Clade	Tribe	Binomial	Common name	Uses	Characteristics
Cercicdae	Cercicdeae	*Tylosema esculentum*	Marama bean	s,t	D,P
Detarieae	Detarieae	*Detarium senegalense*	Sweet detar	s	P
Detarieae	Detarieae	*Tamarindus indica*	Tamarind	p	P
Umtiza	Caesalpinieae	*Ceratonia siliqua*	Carob	s,p	P
Caesalpinieae	Caesalpinieae	*Chamaecrista fasciculata*[a]	Partridge pea	m	D,P
Caesalpinieae	Caesalpinieae	*Cordeauxia edulis*	Yeheb-nut	s	D,P
Mimosoid	Mimoseae	*Parkia speciosa*	Petai	s,p,f	D,P
Mimosoid	Mimoseae	*Prosopis glandulosa*	Honey mesquite	s,p,f	D,P
Mimosoid	Mimoseae	*Desmanthus illinoensis*	Illinois bundle flower	s,f	D,P
Mimosoid	Mimoseae	*Inga edulis*	Ice cream bean	p	P
Indigoferoid	Indigofereae	*Cyamopsis tetragonoloba*	Guar/cluster bean	s,p,f	
Genistoid	Genisteae	*Aspalathus linearis*	Rooibos tea	l	D,P
Genistoid	Genisteae	*Lupinus albus*	White lupin[a]	s	[c]
Genistoid	Genisteae	*Lupinus angustifolius*	Narrow-leaved lupin	s	[c]
Genistoid	Genisteae	*Lupinus luteus*	Yellow lupin[a]	s	[c]
Genistoid	Genisteae	*Lupinus mutabilis*	Andean lupin; tarwi	s	C[c]
Genistoid	Genisteae	*Lupinus polyphyllus*	Washington lupin	s,f	C,P[c]
Dalbergioid	Aeschynomeneae	*Arachis hypogaea*	Peanut/ groundnut[a]	s	
Galegoid	Galegeae	*Glycyrrhiza glabra*	Licorice	t	P
Galegoid	Hedysareae	*Caragana arborescens*	Pea shrub	s,p	D,C,P
Galegoid	Cicereae	*Cicer arietinum*	Chickpea	s	
Galegoid	Trifolieae	*Trigonella f.-graecum*	Fenugreek	s	
Galegoid	Trifolieae	*Medicago truncatula*[ab]	Barrel medic	f,m	
Galegoid	Vicieae/Fabeae	*Lathyrus sativus*	Chickling vetch	s,f	D[c]
Galegoid	Vicieae/Fabeae	*Lens culinaris*	Lentil[a]	s	
Galegoid	Vicieae/Fabeae	*Pisum sativum*[a]	Pea[a]	s,p,f,m	
Galegoid	Vicieae/Fabeae	*Vicia faba*	Faba bean[a]	s	
Robinioid	Loteae	*Lotus tetragonolobus*	Asparagus pea	p	

Continued

Table 22.1. Continued.

Clade	Tribe	Binomial	Common name	Uses	Characteristics
Robinioid	Loteae	*Lotus japonicus*[ab]	Birdsfoot trefoil	f,m	P
Robinioid	Sesbanieae	*Sesbania spp.*	Agati	f,l,s,p	F,P
Millettioid	Phaseoleae	*Pediomelum esculentum*	Breadroot	t	D,P
Millettioid	Phaseoleae	*Apios americana*	Potato bean	t	P
Millettioid	Phaseoleae	*Cajanus cajan*	Pigeon pea[a]	s,p	D,P
Millettioid	Phaseoleae	*Canavalia ensiformis*	Jack bean	s,p,f	[c]
Millettioid	Phaseoleae	*Lablab purpureus*	Hyacinth bean	s,p,f	
Millettioid	Phaseoleae	*Glycine max*[a]	Soybean[a]	s,m	
Millettioid	Phaseoleae	*Pachyrhizus erosus*	Jicama/yam bean[a]	t	
Millettioid	Phaseoleae	*Phaseolus coccineus*	Scarlet runner bean	s,p	
Millettioid	Phaseoleae	*Phaseolus lunatus*	Lima bean	s	[c]
Millettioid	Phaseoleae	*Phaseolus vulgaris*[a]	Common bean[a]	s,p	
Millettioid	Phaseoleae	*Phaseolus acutifolius*	Tepary bean	s,p	D
Millettioid	Phaseoleae	*Macrotyloma geocarpum*	Hausa groundnut	s	D
Millettioid	Phaseoleae	*Psophocarpus spp.*	Winged bean	p,t	
Millettioid	Phaseoleae	*Vigna angularis*	Adzuki bean[a]	s	
Millettioid	Phaseoleae	*Vigna aconitifolia*	Moth bean	s	
Millettioid	Phaseoleae	*Vigna mungo & radiata*	Mung bean[a]	s	
Millettioid	Phaseoleae	*Vigna subterranea*	Bambara groundnut	s	D
Millettioid	Phaseoleae	*Vigna unguiculata*	Cowpea[a]	s,p	

Primary uses: s, seed; t, tuber or root; p, pod or pod wall; m, model; l, leaf; f, forage. Characteristics: D, drought-tolerant; C, cold-tolerant; P, perennial; F, flooding-tolerant.
[a] Model or 'major' crop legumes;
[b] Sequenced legume genomes;
[c] Varieties that may contain toxins (alkaloids or cyanogenic glycosides) removable in preparation. Source: Cannon *et al.* (2005), with permission.

The millettioid clade is represented by the models *G. max* (soybean), with a completely sequenced genome, and by *Phaseolus vulgaris* (common bean), with a genome sequencing project under way. Some other genera with species of interest in this clade include *Vigna* (cowpea), *Cajanus* (pigeon pea), *Lablab* (hyacinth bean, used throughout Asia), *Apios* ('potato bean', an edible North American tuber crop), *Cyamopsis* (used as a vegetable and for guar gum), *Pachyrrhiza* (jicama root), *Macrotyloma* (Hausa groundnut; a drought-tolerant African bean with growth and pod characteristics similar to groundnut (peanut)) and *Phosphocarpus* (winged bean; a large bean common in south Asia and the Pacific islands).

The galegoid clade is represented by the models *M. truncatula* and *L. japonicus*, both with genome sequences that are substantially complete in the euchromatic regions. This clade can be further subdivided into the 'inverted repeat loss clade' (IRLC) and the robinioid clade. The IRLC clade contains *Medicago*, *Cicer* (chickpea), *Lathyrus* (grass pea), *Trifolium* (clovers), *Vicia* (vetches and faba bean), *Pisum* (several pea species), *Glycyrrhiza* (liquorice) and *Lens* (lentil). The robinioid clade includes *Lotus* L. *japonicus*, *tetragonolobus* (asparagus pea), *Sesbania* (a forage and green manure used in flooded rice fields) and *Robinia* (containing the black locust tree, used ornamentally and for durable timber).

There are several early-diverging clades within the papilionoid subfamily. The dalbergioid clade includes *Arachis* (groundnut) and more than a thousand other species. The genistoid clade inlcudes several species of domesticated lupins (e.g. *Lupinus angustifolius*, narrow-leafed lupin; and *Lupinus mutabilis*, Andean lupin or 'tarwi'), as well as more than 2300 other species (Lewis *et al.*, 2005). Clades in this subfamily diverging even earlier include several small clades with incompletely resolved taxonomic relationships. Intriguingly, several of these early genera contain no nodulating species (e.g. *Sophora* and *Cladastris*), raising the possibility that nodulation in the papilionoids may have arisen after the origin of the papilionoid subfamily. The nodulation and systematic data are also, however, compatible with multiple losses of nodulation.

The mimosoid subfamily is taxonomically well defined, and large in terms of species composition (with approximiately 3270 species) (Lewis *et al.*, 2005). Some of the plants in this subfamily have been used locally for food, including *Prosopis* (e.g. honey mesquite in the American south-west), *Inga edulis* (ice cream bean in South America), *Desmanthus illinoensis* (a perennial grain producer in the North American great plains) (Vail *et al.*, 1992) and *Parkia speciosa* (petai bean, in South-east Asia).

Beyond the papilionoid and mimosoid subfamilies is a collection of smaller clades, generally early diverging in the family and sometimes with poorly resolved molecular systematics, traditionally called the Caesalpinioideae subfamily (Fig. 22.1). Molecular systematics results place some caesalpinoid clades basally in the papilionoid subfamily, some along a grade leading to the mimosoid subfamily and some in separate lineages. Some caesalpinoid legumes of agronomic value include *Ceratonia* (carob), *Tamarindus* (tamarind), *Detarium* (sweet detar), *Cordeauxia edulis* (yeheb nut) (Wickens and Storey, 1984) and *Tylosema esculentum* (marama bean) (Fox and Norwood-Young, 1982).

22.4 Chromosomal and Gene Order Conservation

Orthology and synteny relationships are important because they mean that positional relationships determined around a locus of interest in one species can serve as a guide (when synteny holds) for relative gene positions in orthologous loci in other species. This means that, to the extent that synteny holds, hard-earned QTL and positional cloning information from one species has a good chance of being applied in related species.

Synteny is extensive within the Phaseoleae, and also within the Hologalegineae. Even in species between these two clades (e.g. between *Medicago* and *Glycine*, separated by ~55 million years (Lavin *et al.*, 2005), synteny often extends as far as whole chromosome arms (unpublished results). Between *M. truncatula* and *L. japonicus*, separated by more than 50 million years, most genes occur in approximately ten large blocks of synteny (Cannon *et al.*, 2006). Even at greater evolutionary distances, conservation of chromosomal arm-scale blocks is seen between *Medicago* and *Arachis* (groundnut) (Bertioli *et al.*, 2009) and between *Medicago* and lupin (Nelson *et al.*, 2006).

In the Phaseoleae, another indicator of chromosomal stability between species is given by chromosome counts. Most genera in the Phaseoleae have 11 chromosomes in the haploid genome (in 42 of 57 genera with chromosome counts in the IPCN database) (Goldblatt and Johnson, 2008). Agronomic species with $n=11$ include *Amphicarpa* (American hogpeanut), *Apios* (groundnut, Indian potato or potato bean), *Cajanus* (pigeon pea), *Canavalia* (jack bean and sword bean), *Lablab*, *Macrotyloma* (Hausa groundnut and horse gram), *Pachyrhizus* (jicama tuber), *Phaseolus* (common bean, lima bean and tepary bean), *Vigna* (cowpea, yard-long bean, mung bean, moth bean, etc.) and *Voandzeia* (Bambara groundnut). The exact extent of synteny remains to be established for these species, but conservation of chromosome numbers across all of these phaseolid species suggests that relative gene orders within corresponding chromosomes will generally be conserved across whole chromosomes across all of these species. Conservation of synteny at the chromosome scale is supported, so far, at least between *Vigna radiata*, *Vigna unguiculata* and *P. vulgaris* (Boutin *et al.*, 1995; Choi *et al.*, 2004).

The only prominent exception in chromosome numbers among agronomic Phaseoleae is *Glycine* (soybean), with $n=20$, as a result of an episode of polyploidy estimated to have occurred between ~5 million years ago (Mya) (Doyle and Egan, 2009) and 10–13 Mya (Schmutz *et al.*, 2010). Another species in this clade, with limited agronomic use (in Asia, for its root starch and young leaves), is *Pueraria montana* (kudzu), with $n=22$. Chromosomal instabilities have been observed after polyploidy; this may explain the relatively large numbers of rearrangements observed in *Glycine* when the chromosomes are compared among one another (Schmutz *et al.*, 2010). Some soybean chromosomes correspond across most of their lengths: 4 and 6, 3 and 19, 10 and 20; however, others are substantially rearranged. Chromosome 13, for example, matches parts of at least eight others (Schmutz *et al.*, 2010). Similar extents of rearrangements are also seen in comparisons with *Phaseolus* markers (P. McClean, North Dakota, 2009, personal communication). It appears that most of the rearrangements are in the *Glycine* lineage, which is suggested by the fact that the *Phaseolus* and *Vigna* genetic maps are almost entirely collinear (Boutin *et al.*, 1995; Choi *et al.*, 2004). The rearrangements are also evident in comparisons between *Vigna* markers and *Glycine* chromosomal sequences, where chromosomes in *Vigna* map to multiple chromosomes in *Glycine* (Muchero *et al.*, 2009).

In the hologalegina clade, chromosome numbers of 7 and 8 are most common. Among food species the following have $n=7$ (all in tribe Fabeae): *Vicia* (faba bean), *Pisum* (pea), *Lens* (lentil) and *Lathyrus* (cicerchia). In tribe Cicereae, *Cicer* (chickpea) has $n = 8$. The various clover species (typified by *M. truncatula*) are predominantly $n = 8$. Although *Medicago* and *Pisum* are in separate tribes within the hologalegina, they share almost complete chromosome-scale synteny (Kalo *et al.*, 2004) – with the exception of *Medicago* chromosome 6, which has limited synteny to pea linkage groups VII and VI (Kalo *et al.*, 2004). *Medicago* chromosome 6 is atypical in that it is small and unusually repeat-dense (Cannon *et al.*, 2006).

22.5 Genome Conservation and Change from Sequenced Legume Genomes

The complete, high-quality draft genome sequence of the soybean genome has now been published, and the majority of genomic sequence from euchromatic regions have been published for *L. japonicus* and *M. truncatula* (Cannon *et al.*, 2006; MGSC, 2007; Sato *et al.*, 2008). Taking the more complete of the genome sequences as a point of reference, we can consider ways in which other legume genomes may be similar or different.

Most soybean genes (about 78%) occur in the chromosome ends, which make up less than half of the genome sequence (Schmutz *et al.*, 2010). The remaining gene-poor genomic sequence, in the large pericentromeres, is primarily composed of transposable elements and satellite repeats. The pattern of substantially lowered gene density (and frequency of genetic recombination) in pericentromeres appears to be a general feature of plant genomes, observed in all plant genomes sequenced to date. However, the extent of pericentromeric repeats differs between genome: pericentromeres in *Arabidopsis* and rice are much smaller, comprising roughly 15% of the genomes, while those in sorghum are large, comprising ~62% of the heterochromatin. The pattern of pericentromeric repeats in soybean is also somewhat different than in *Arabidopsis* or rice, with the latter having relatively long gradients of increasing transposon density, up to the location of the centromeric satellite repeats, while the euchromatin-heterochromatin boundaries in soybean are sharper, occurring over approximately a megabase on most chromosomes (compared with a typical soybean chromosome size of 47.5 Mb). It can be inferred by genome size and sequenced euchromatic sequence-space in *L. japonicus* and *M. truncatula* that the pericentromeres are relatively smaller in these genomes than in soybean. The exact nature of the euchromatin/heterochromatin boundaries, and the content of the pericentromeric sequences, are not yet known in *Medicago* and *Lotus*, as those genome sequencing projects have explicitly focused on generating sequence for the gene-rich euchromatic regions.

While most genes in the soybean genome are found in the euchromatic regions, approximately 21% occur in the pericentromeric regions (Schmutz *et al.*, 2010). This figure depends on determination of pericentric boundaries – which are strikingly clear in most soybean chromosomes, and are indicated by transition from gene-rich, recombinogenic DNA to gene-poor regions with severe suppression of recombination. Genes within the pericentromeres typically occur in roughly the same order as homoeologous genes in euchromatic regions of the soybean genome, but spread out in the pericentromeres by inserted transposons. While the average density across the genome is 8.9 genes/100 kb in the euchromatin, it is 1.9 genes/100 kb in the heterochromatin (about 4.7-fold lower). It is not yet known what the comparable densities are in the heterochromatin of *Medicago* or *Lotus* (or other legume genomes). It is possible that in these genomes (and in phaseolid genomes) that have not undergone recent polyploidy, the pericentromeres have been more stable for longer periods of time and that genes have gradually been lost from or have migrated away from the pericentromeres of these genomes. This remains to be seen however, as the genomes for these and other legumes are completed.

Recombination in the large pericentromeric regions is severely suppressed. Only 7% of recombination is seen in these regions, which comprise approximately 57% of the genome (Schmutz *et al.*, 2010). One consequence is that a substantial number of genes in the low-recombination pericentromeres are in very low-recombination regions; agronomic traits in these areas will typically be subject to linkage drag that may span virtually the entire pericentromere. Similarly, QTLs for traits in the pericentromeres may be represented by high-density, high-resolution maps, yet still span many megabases. This was the case with QTLs for seed protein in soybean, identified in several backgrounds (Bolon *et al.*, 2010). The major QTL reported in these studies spans only 3 cM on linkage group I (Gm20), but encompasses 8.4 Mb on the chromosome. On the other hand, the number of candidate genes within this region is lessened by the low gene density in the region. Other legume species will also face this kind of challenge, to a greater or lesser degree.

The timing of polyploidy in soybean remains uncertain, estimated at ~13 Mya in Schmutz *et al.* (2010) and between 5 and 10 Mya in Doyle and Egan (2009). The wide range of estimated dates is due partly to uncertainties in rates of change in the *Glycine* lineage (particularly after a dramatic genomic event such as polyploidy), and partly due to the possibility that the *Glycine* polyploidy may have involved an alloploidy event, combining genomes of species separated by an unknown evolutionary distance – perhaps up to several million years. About 75% of genes occur in multiple copies (from any source, including local duplications or older whole-genome duplications). Just looking at the recent episode of polyploidy, 43.4% of soybean genes have matches in the corresponding region. These consequences of polyploidy do not affect other agronomic species in the Phaseoleae (which lack clear signs of recent polyploidy). However, outside the Phaseoleae, species with relatively recent polyploid histories include several clover species, groundnut, lucerne and lupins.

22.6 Translating Trait Information between Species

Several examples illustrate the use of information derived from one species to identify similar traits in other species. Common to these examples is flexibility in the choices of species to make best use of genomic tools and information where it is available: perhaps using map-based cloning in the target species; and synteny information and sequence with a model species to identify candidate genes; and transformation systems in another species to test function.

A gene identified in *Arabidopsis* has been used to identify orthologous genes and homologous traits in common bean, and tested for complementation in soybean. A mutant in inflorescence architecture identified in *Arabidopsis*, called 'terminal flower 1' (TFL1) was identified in common bean, and showed complete linkage with the determinacy trait (PvDT1), scorable as determinate bush or indeterminate vine forms (Kwak *et al.*, 2008). Orthologues to the bean PvDT1 were then identified in soybean (GmDT1 and GmDT2). Complementation tests in *Arabidopsis* established similar functions of these genes in soybean (Tian *et al.*, 2010).

A gene identified in maize has been used to identify orthologues with similar function in soybean. Low-phytate mutants have been generated in several crops as a result of gamma irradiation or chemical mutagenesis. Several responsible genes have been identified in rice, wheat and maize, including genes in the families MIPS, myo-inositol kinases, inositol polyphosphate kinases and multidrug resistance-associated protein ATP-binding cassettes (MRP ABC) transporters (Saghai Maroof *et al.*, 2009). Genes in two of these gene families occur where the QTLs for the low-phytate trait have been mapped in soybean. These candidate genes were evaluated for polymorphisms in the low- and high-phosphate mapping population, leading to the identification of two homoeologous MRP ABC transporters as causative genes (a double mutant) in the soybean low-phytate line (Saghai Maroof *et al.*, 2009).

The cross-species functionality of genes has also been listing a transformation approach. Yang *et al.* (2008) used *M. truncatula* to clone the RCT1 gene that confers resistance to *Colletotrichum trifolii* (anthracnose). The resistant allele from *M. truncatula*, when transformed into *M. sativa*, confers resistance to multiple anthracnose strains in lucerne.

Two loci identified as mutants in pea, affecting floral morphology (keeled wings (K) and lobed standard 1 (LST1)), were tested and used to identify candidate genes in *L. japonicus* (Wang *et al.*, 2008).

A trait described by Mendel has been traced through four species (Armstead *et al.*, 2007). A trait similar to Mendel's 'I' locus, for green or yellow seed, was observed and mapped in meadow fescue. A candidate gene was identified in rice by synteny analysis. The trait was then fine-mapped in pea, leading to the gene orthologous to the candidate identified in rice. Function of the gene was then tested in *Arabidopsis*.

Genes involved in nodule development and responses have been identified in one species and then others. The deduction, over the last decade, of genes underlying nodule initiation and early development has made extensive use of multiple and reciprocal models (reviewed in Oldroyd *et al.*, 2009; Sato *et al.*, 2010). One example is the set of orthologous receptor kinase genes responsible for perception of nod factor and arbuscular mycorrhiza: SYMRK in *L. japonicus*, DMI2/NORK in *M. truncatula* and NORK in *M. sativa* (Gherbi *et al.*, 2008).

22.7 New Technologies and Directions

A rationale for encouraging international focus on a single or a small number of intensively studied models is that if the cost of developing genomic tools is high, then as much value as possible should be squeezed from the initial investments in tool development. Additionally, methods of reverse genetics have enabled generation of tagged genotypic changes that can be evaluated for phenotypic effects. In effect, methods such as transposon tagging or mutagenesis screens cheaply generate new phenotypes linked with known genotypic changes. In the case of development of *Arabidopsis* as a model, phenotype generation was therefore relatively inexpensive, while sequencing (both genomic and EST) was expensive.

Biological models have clearly been spectacularly successful and undoubtedly will continue to be vital, but several factors may shift the balance towards research on a larger number of species and, indeed, to different concepts of 'model'.

The cost of sequencing has plummeted by at least two orders of magnitude over the last decade: from tens of millions of US dollars for complete genome sequencing to tens or hundreds of thousands (at the time of writing – at least for deep-coverage genome resequencing). Low-coverage genome resequencing and full transcriptome profiling can be carried out almost trivially in single 'runs' of next-generation sequencing reactions.

Similarly, the costs of developing large genetic maps have dropped significantly (though less rapidly than sequencing costs have dropped). It is now feasible to develop a genetic map of over 1000 markers in the low tens of thousands of US dollars. For larger budgets, haplotype projects can determine allele states of tens of thousands of markers over an entire germplasm collection. There are currently large-scale haplotyping and resequencing projects under way for soybean and *Medicago*. These open the door to genome-wide association and genomic selection studies.

It seems likely that for half a dozen of the most agronomically important legumes, the following resources will probably be available in the next few years: high-quality reference genome sequences; extensive genetic maps; genome resequencing or haplotype data; complete transcriptome sequence and transcriptome expression profiling. Indeed, genome sequencing projects are either under way or complete for soybean, *M. truncatula*, *L. japonicus*, common bean, cowpea, mung bean, pigeon pea, narrow-leafed lupin and diploid lucerne. And, for most legume species of agronomic interest, the following resources will probably be generated: more-or-less complete (though fragmentary) genome sequences; near-complete transcriptome sequences; moderately sized genetic maps; and at least partial expression profiling data.

It is conceivable that most positional cloning work, to determine gene function, will continue to be done in a small number of model species (soybean, *Medicago* and *Lotus*). However, if extensive genomic resources are soon available for numerous other species (e.g. groundnut, pea, chickpea, pigeon pea, cowpea, common bean), research communities working on these species will be empowered to identify gene function (or at least positional trait associations) in these species first, rather than in the models. After all, the traits of interest are often available (and segregating) in these species rather than in the models. For example, numerous leaf and growth-habit and flower-morphology phenotypes and mutants are available in pea; and seed-coat and growth-habit and pod characteristic traits are available in common bean. These QTLs and mutants will probably not be reproduced easily in the current primary legume models. For traits such as winter-hardiness or perenniality, even *M. truncatula* will probably not be a replacement for *M. sativa* (lucerne). Similarly, soybean may be an intractable model for various traits in the other beans, particularly as soybean frequently contains duplicated genes left over from the *Glycine* polyploidy event.

There are numerous websites devoted to data sets in particular legume crops. Portals such as comparative-legumes.org and grain-legumes.com provide links to additional specialized legume websites, and sites such as phytozome.net, legoo.org and symapdb.org provide resources and analytical tools for multiple data sets. A focus of comparative-legumes.org is to help integrate services across numerous sites. For example, sequence searches there currently provide results linking to instances of four genome browsers and to the *Medicago* Gene Atlas.

As the genetic bases of functions are determined in any given species, that species becomes the *de facto* model for that trait. Two classes of technology would further amplify this trend, and greatly accelerate development of orphan crops (as well as more peripheral crops, still on the verge of domestication).

The first game-changing 'technology' consists of the filling out and integration of genetic and genomic information across the legumes: essentially the completion of sequence and map data, and concomitant bioinformatics resources across a wide range of species. This means, for example, deep transcriptomic profiles for all species of interest, showing expression patterns of each gene across a wide range of tissues and conditions; and high-quality gene annotations, applied across all species and updated and propagated as new information is added from any species. This would allow facile identification of orthologues and candidate genes in QTL regions of interest.

The second game-changing technology would be dramatically improved transformation technologies, more widely applicable (i.e. to more species). This may not merely be wishful thinking; some promising technologies available now include zinc-finger nucleases, capable of introducing targeted lesions or single-base changes (Maeder *et al.*, 2008; Townsend *et al.*, 2009); and engineered viruses, capable of delivering RNAi knockdown across a range of legume species (Zhang *et al.*, 2010). The ability to efficiently knock out or modify genes in a wide range of species would enable researchers in chickpea (for example) to test gene function in a handful of candidate genes under a drought QTL. While QTL regions often encompass hundreds or even thousands of genes, identifying promising candidate genes should be made possible by the combination of better functional annotations (generally transferrable across orthologues in the legumes) and better information about the expression patterns of each gene across a wide range of tissues and conditions. However, it is also worth emphasizing that while the revolution in sequencing will make some things possible in less-studied species, the one-gene-at-a-time tools such as RNAi and zinc-finger have a long way to go before they will enable the kind of functional genomics that are expected of the models in soybean, *Medicago* and *Lotus*.

The combination of mature, complete data sets with bioinformatics resources to enable translation of information easily across species, and better plant transformation technologies applicable to many crop species, promises to shift plant biology in the legumes from 'a few bright lights' to 'a thousand points of light'. This gives some hope for more rapid improvement of many orphan crops, and perhaps domestication of species with great potential that have been intractable to date.

References

Armtead, I., Donnison, I., Aubry, S., Harper, J., Hortensteiner, S., James, C. *et al.* (2007) Cross-species identification of Mendel's I locus. *Science* 315, 73–74.

Bertioli, D.J., Moretzsohn, M.C., Madsen, L.H., Sandal, N., Leal-Bertioli, S.C., Guimaraes, P.M. *et al.* (2009) An analysis of synteny of *Arachis* with *Lotus* and *Medicago* sheds new light on the structure, stability and evolution of legume genomes. *BMC Genomics* 10, 45–55.

Blackmon, W.J. and Reynolds, B.D. (1986) The crop potential of *Apios americana* – preliminary evaluations. *HortScience* 21, 1334–1336.

Bolon, Y.T., Joseph, B., Cannon, S.B., Graham, M.A., Diers, B.W., Farmer, A.D. *et al.* (2010) Complementary genetic and genomic approaches help characterize the linkage group I seed protein QTL in soybean. *BMC Plant Biology* 10, 41.

Boutin, S.R., Young, N.D., Olson, T.C., Yu, Z.H., Vallejos, C.E. and Shoemaker, R.C. (1995) Genome conservation among three legume genera detected with DNA markers. *Genome* 38, 928–937.

Campbell, C.G. (1997) *Grass pea. Lathyrus sativus L. Promoting the Conservation and Use of Underutilized and Neglected Crops*. 18. Institute of Plant Genetics and Crop Plant Research, Rome.

Cannon, S.B., Crow. JA., Heuer, M.L., Wang X., Cannon, E.K., Dwan, C. *et al.* (2005) Databases and information integration for the *Medicago truncatula* genome and transcriptome. *Plant Physiology* 138, 1–3.

Cannon, S.B., Sterck, L., Rombauts, S., Sato, S., Cheung, F., Gouzy, J. *et al.* (2006) Legume genome evolution viewed through the *Medicago truncatula* and *Lotus japonicus* genomes. *Proceedings of the National Academy of Sciences U.S.A.* 103, 14959–14964.

Cannon, S.B., May, G.D. and Jackson, S.A. (2009) Three sequenced legume genomes and many crop species: rich opportunities for translational genomics. *Plant Physiology* 151, 970–977.

Choi, H.K., Mun, J.H., Kim, D.J., Zhu, H., Baek, J.M., Mudge, J. *et al.* (2004) Estimating genome conservation between crop and model legume species. *Proceedings of the National Academy of Sciences U.S.A.* 101, 15289–15294.

Doyle, J.J. and Egan, A.N. (2009) Dating the origins of polyploidy events. *New Phytologist* 186, 73–85.

Doyle, J.J. and Luckow, M.A. (2003) The rest of the iceberg. Legume diversity and evolution in a phylogenetic context. *Plant Physiology* 131, 900–910.

Fox, F.W. and Norwood-Young, M.E. (1982) *Food from the Veld*. Delta Books, Johannesburg, Republic of South Africa.

Gepts, P., Beavis, W.D., Brummer, E.C., Shoemaker, R.C., Stalker, H.T., Weeden, N.F. *et al.* (2005) Legumes as a model plant family. Genomics for food and feed report of the cross-legume advances through genomics conference. *Plant Physiology* 137, 1228–1235.

Gherbi, H., Markmann, K., Svistoonoff, S., Estevan, J., Autran, D., Giczey, G. *et al.* (2008) SymRK defines a common genetic basis for plant root endosymbioses with arbuscular mycorrhiza fungi, rhizobia, and Frankiabacteria. *Proceedings of the National Academy of Sciences U.S.A* 105, 4928–4832.

Goldblatt, P. and Johnson, D.E. (2008) *Index to Plant Chromosome Numbers*. Missouri Botanical Garden, St. Louis, Missouri.

Graham, P.H. and Vance, C.P. (2003) Legumes: importance and constraints to greater use. *Plant Physiology* 131, 872–877.

Guo, J.H., Liu, X.J., Zhang, Y., Shen, J.L., Han, W.X., Zhang, W.F. *et al.* (2010) Significant acidification in major Chinese croplands. *Science* 327, 1008.

Kalo, P., Seres, A., Taylor, S.A., Jakab, J., Kevei, Z., Kereszt, A. *et al.* (2004) Comparative mapping between *Medicago sativa* and *Pisum sativum. Molecular Genetics Genomics* 272, 235.

Karuniawan, A., Iswandi, A., Kale, P.R., Heinzemann, J. and Gruneberg, W.J. (2006) *Vigna vexillata* (L.) A. Rich. cultivated as a root crop in Bali and Timor. *Genetic Resources and Crop Evolution* 53, 213–217.

Kurlovich, B.S., Stoddard, F.L. and Earnshaw, P. (2008) Potential and problems of *Lupinus polyphyllus* Lindl. domestication. In: Palta, J.A. and Berger, J.B. (eds) *12th International Lupin Conference*, Freemantle, Western Australia, pp. 304–307.

Kwak, M., Velasco, D. and Gepts, P. (2008) Mapping homologous sequences for determinacy and photoperiod sensitivity in common bean (*Phaseolus vulgaris*). *Journal of Heredity* 99, 283–291.

Lavin, M., Herendeen, P.S. and Wojciechowski, M.F. (2005) Evolutionary rates analysis of Leguminosae implicates a rapid diversification of lineages during the tertiary. *Systematic Biology* 54, 575–594.

Lewis, G., Schrire, B., Mackind, B. and Lock, M. (2005) *Legumes of the World*. Royal Botanic Gardens, Kew, UK.

Maeder, M.L., Thibodeau-Beganny, S., Osiak, A., Wright, D.A., Anthony, R.M., Eichtinger, M. *et al.* (2008) Rapid 'open-source' engineering of customized zinc-finger nucleases for highly efficient gene modification. *Molecular Cell* 31, 294–301.

MGSC (2007) *Medicago truncatula* genome 'Mt2.0' release whitepaper. Available at http://medicago.org/genome/downloads/Mt2/

Monfreda, C.N., Ramankutty, N. and Foley, J.A. (2008) Farming the planet: 2. Geographic distribution of crop areas, yields, physiological types, and net primary production in the year 2000. *Global Biogeochemical Cycles* 22, GB1022.

Muchero, W., Diop, N.N., Bhat, P.R., Fenton, R.D., Wanamaker, S., Pottorff, M. *et al.* (2009) A consensus genetic map of cowpea [*Vigna unguiculata* (L) Walp.] and synteny based on EST-derived SNPs. *Proceedings of the National Academy of Sciences U.S.A.* 106, 18159–18164.

Nelson, M.N., Phan, H.T., Ellwood, S.R., Moolhuijzen, P.M., Hane, J., Williams, A. *et al.* (2006) The first gene-based map of *Lupinus angustifolius* L. – location of domestication genes and conserved synteny with *Medicago truncatula. Theoretical and Applied Genetics* 113, 225–238.

Oldroyd, G.E., Harrison, M.J. and Paszkowski, U. (2009) Reprogramming plant cells for endosymbiosis. *Science* 324, 753–754.

Saghai Maroof, M.A., Glover, N.M., Biyashev, R.M., Buss, G.R. and Grabau, E.A. (2009) Genetic basis of the phytate trait in the soybean line CX1834. *Crop Science* 49, 69–76.

Sato, S., Nakamura, Y., Kaneko, T., Asamizu, E., Kato, T., Nakao, M. *et al.* (2008) Genome structure of the legume, *Lotus japonicus. DNA Research* 15, 227–339.

Sato, S., Isobe, S. and Tabata, S. (2010) Structural analyses of the genomes in legumes. *Current Opinion in Plant Biology* 13, 146–152.

Schmutz, J., Cannon, S.B., Schlueter, J., Ma, J., Mitros, T., Nelson, W. *et al.* (2010) Genome sequence of the paleopolyploid soybean. *Nature* 463, 178–183.

Singer, S.R., Maki, S.L., Farmer, A.D., Ilut, D., May, G.D., Cannon, S.B. *et al.* (2009) Venturing beyond beans and peas: what can we learn from Chamaecrista? *Plant Physiology* 151, 1041–1047.

Tian, Z., Wang, X., Lee, R., Li, Y., Specht, J.E., Nelson, R.L. *et al.* (2010) Artificial selection for determinate growth habit in soybean. *Proceedings of the National Academy of Sciences U.S.A.* 107, 8563–8676.

Townsend, J.A., Wright, D.A., Winfrey, R.J., Fu, F., Maeder, M.L., Joung, J.K. *et al.* (2009) High-frequency modification of plant genes using engineered zinc-finger nucleases. *Nature* 459, 442–445.

Vail, J., Kulakow, P. and Benson, L. (1992) Illinois Bundleflower: prospects for a perennial seed crop. In: Smith, D.D. and Jacobs, C.A. (eds) *Recapturing a Vanishing Heritage, Proceedings of the Twelfth North American Prairie Conference,* Cedar Rapids, Indiana.

Vance, C.P., Graham, P.H. and Allan, D.H. (2000) Biological nitrogen fixation. Phosphorus: a critical future need. In: Pedrosa, F.O., Hungria. M., Yates, M.G. and Newton, W.E. (eds) *Nitrogen Fixation: from Molecules to Crop Productivity.* Kluwer Academic Publishers, Dordrecht, The Netherlands, pp. 506–514.

Varshney, R.K., Close, T.J., Singh, N.K., Hoisington, D.A. and Cook, D.R. (2009) Orphan legume crops enter the genomics era! *Current Opinion in Plant Biology* 12, 202–210.

Wang, Z., Luo, Y., Li, X., Wang, L., Xu, S., Yang, J. *et al.* (2008) Genetic control of floral zygomorphy in pea (*Pisum sativum* L.). *Proceedings of the National Academy of Sciences U.S.A.* 105, 10414–10419.

Wickens, G.E. and Storey, I.N.J. (1984) *Cordeauxia edulis* Hemsley. *Survey of the Economic Plants of the Arid and Semi-Arid Tropics.* Royal Botanic Gardens, Kew, UK.

Wilson, P.W., Gorny, J.R., Blackmon, W.J. and Reynolds, B.D. (1986) Fatty acids in the American groundnut (*Apios americana*). *Journal of Food Science* 51, 1387–1388.

Wilson, P.W., Pichardo, F., Blackmon, W.J. and Reynolds, B.D. (1987) Amino acids in the American groundnut (*Apios americana*). *Journal of Food Science* 52, 224–225.

Yang, S., Gao, M., Xu, C., Gao, J., Deshpande, S., Lin, S. *et al.* (2008) Alfalfa benefits from *Medicago truncatula*: the RCT1 gene from *M. truncatula* confers broad-spectrum resistance to anthracnose in alfalfa. *Proceedings of the National Academy of Sciences U.S.A.* 105, 12164–12169.

Young, N.D. and Udvardi, M. (2009) Translating *Medicago truncatula* genomics to crop legumes. *Current Opinion in Plant Biology* 12, 193–201.

Young, N.D., Mudge, J. and Ellis, T.H. (2003) Legume genomes: more than peas in a pod. *Current Opinion in Plant Biology* 6, 199–204.

Zhang, C., Bradshaw, J.D., Whitham, S.A. and Hill, J.H. (2010) The development of an efficient multipurpose bean pod mottle virus viral vector set for foreign gene expression and RNA silencing. *Plant Physiology* 153, 52–65.

Zhu, H., Choi, H.K., Cook, D.R. and Shoemaker, R.C. (2005) Bridging model and crop legumes through comparative genomics. *Plant Physiology* 137, 1189–1196.

23 Plant Genetic Resources and Conservation of Biodiversity

S. Sardana, Mohar Singh, S.K. Sharma and Neha Rajan

23.1 Introduction

Plant genetic resources are reservoirs of genes and gene complexes, which contribute enormously towards genetic improvement of crop plants. Legume crops are an important component of the human diet, complementing cereal-based diets with minerals, vitamins and, in particular, proteins; provide nutritious green fodder and feed to livestock, restore and maintain soil fertility and also fit well within cropping patterns. However, the narrow genetic base of most of the released varieties and the non-availability of appropriate genetic material have been major constraints for their spectacular growth worldwide, with average productivity showing only a marginal increase during the last four decades. In the present global scenario with the continuously expanding need for varietal improvement, there is an urgent need systematically to collect/assemble, evaluate, utilize and conserve the genetic resources of food legumes, for both the present and posterity.

23.2 Collection and Conservation of Germplasm Diversity

To enrich food legume germplasm, it is clear that germplasm must be collected from diversity-rich areas of the world. Genetic variability, accumulated and conserved safely, is immensely valuable to mankind both now and in the future. The management of crop genetic resources includes their conservation or maintenance in a particular state, and two different complementary approaches: namely, *ex situ* and *in situ* conservation. The establishment of *ex situ* germplasm collections has been the result of several decades of global efforts to conserve plant biodiversity; and collected diversity is conserved almost entirely *ex situ* as seeds, including wild and domestic annuals as well as perennials. The international institutions have a global mandate for research on particular crops and thus now have the global collection of major pulse crops, namely chickpea, pigeon pea, lentil, common bean, cowpea and faba bean (Table 23.1). The majority of collections at these centres represent variability from both the centre of origin and centre of primary diversity. For example, the International Centre for Agricultural Research in the Dry Areas (ICARDA) collection (48%) consists of accessions from Central and West Asia and North Africa (Furman *et al.*, 2009); the International Crops Research Institute for the Semi-Arid Tropics (ICRISAT) have collected/assembled 19,187 active accessions of chickpea and 13,632 of pigeon pea from over 40 countries (Gowda and Upadhyaya, 2006). Legume crops were also collected from

Table 23.1. Legume collections held in trust for the World Collection by international centres (2006–2007).

Centre	Crop	Accessions (*n*)
ICRISAT (India)	Chickpea	17,117
	Pigeon pea	13,389
ICARDA (Syria)	Chickpea	11,986
	Faba bean	5,454
	Lentil	9,487
CIAT (Colombia)	Bean	35,225
IITA (Nigeria)	Cowpea	15,003
	Wild *Vigna*	1,632
AVRDC (Taiwan)	*Vigna*	10,821

CIAT, International Center for Tropical Agriculture; IITA, International Institute of Tropical Agriculture; AVRDC, Asian Vegetable Research and Development Center.
Source: www.Singer.Cgiar.org

Russia, Mali, Nigeria, Malawi and Zambia between 1977 and 1980. Exploration for and collection of wild *Vigna* species have been conducted by Japan in collaboration with Thailand, Sri Lanka and Vietnam (Tomooka *et al.*, 2000). Introduction is one important aspect of germplasm collection that can meet the varietal needs for different growing situations and overcoming the narrow genetic diversity mainly observed in legume crops.

The role of NBPGR in the collection and conservation of food legume diversity in India

The Indian subcontinent is an important source of diversity for pulse crops. The regions that are rich in legume crops are: (i) the western Himalayas, which include cold, arid tracts (field pea, lentil and French bean); (ii) the north-eastern regions and eastern Himalayas (rice bean, winged bean, adzuki bean, black gram, lablab bean, sword bean and pea); (iii) the eastern peninsular region (rice bean, chickpea, pigeon pea and horse gram/*kulthi*); (iv) the western arid/semi-arid region (moth bean, cluster bean/guar, cowpea, black gram and green gram); (v) the central tribal region (cowpea, chickpea, pigeon pea, black gram and green gram); and (vi) the western peninsular region (cowpea, horse gram, mung bean, pigeon pea and black gram).

Wild Vigna species (*V. bourneae, V. capensis, V. dalzelliana, V. khandalensis, V. grandis, V. hainiana, V. minima, V. mungo* var. *sylvestris, V. radiata* var. *sublobata, V. aconitifolia* var. *silvestris, V. radiata* var. *setulosa, V. trilobata* and *V. vexillata*) are widely distributed in the western and eastern Ghats, north-eastern hills, north-western plains, the peninsular region and the northern Himalayas.

Cajanus and Atylosia, two former genera, together constitute a single taxon (Smart, 1990), and a high diversity of wild species of Cajanus (*C. Albicans, C. cajanifolius, C. elongatus, C. goensis, C. grandiflorus, C. heynei, C. lineatus, C. mollis, C. platycarpus, C. pubescens, C. rugosus, C. scarabaeoides, C. sericeus, C. trinervius, C. villosus* and *C. volubilis* var. *volubilis*) occurs in India, with the maximum concentration in the western Ghats, the north-eastern region and the eastern peninsular tract. In chickpea only one species, *Cicer microphyllum*, the Indian wild relative, occurs in the high hills of the western Himalayas. The wild form of *Lathyrus aphaca* occurs as a weed in the northern plains/Gangetic plains, and also in the temperate zone alongside other species (Arora, 1988).

For enrichment of the germplasm of food legumes, it is important that this is collected from the diversity-rich areas of the world. The earliest record of exploration and collection of pulses in India was reported by Shaw and Khan (1931) and Bose (1932). During the 1960s, under the PL 480 programme, a large number of accessions of legumes were collected by the Plant Introduction Division of the Indian Agricultural Research Institute (now NBPGR). Explorations under the Regional Pulse Improvement Project of the US Department of Agriculture during the early 1970s resulted in the collection of 7064 accessions of chickpea. Since the inception of NBPGR (1976), a large number of crop-specific and multi-crop explorations have been conducted independently and in coordination with the Indian Council of Agricultural Research (ICAR) institutes, State Agricultural Universities (SAUs), State Agricultural Departments and other research institutes of the country. This strategy was further strengthened with the National Agricultural Technological Project (NATP) from 1999 to 2005; under this project, several explorations were conducted and over 11,800 accessions of legume crops from different parts of the country were collected: chickpea (657), pigeon pea (1268), urd bean (1658), mung bean (1155), lentil (552), pea (638), Lathyrus (311), French bean/common bean (1201), cowpea (1545), moth bean (198), horse gram (1125), rice bean (469), faba bean (113) and wild Vigna species (186).

Besides cultivated forms, wild species are also a rich source of resistance to several biotic and abiotic stresses (Harlan, 1984). Therefore, NBPGR have also made efforts to collect wild species of *Vigna* from various parts of the country: *Cicer microphyllum* from the higher hills of Lahul and Spiti of Himachal Pradesh; *Cajnus cajanifolius* from Kurda, Orissa and wild forms of rice bean (*Vigna umbellata*) from the natural/disturbed habitats in the Khasi and Jaintia hills of Meghalaya. Recent crop-specific explorations (2006–2007) focusing on *Vigna* spp. have resulted in the collection of *V. acnotifolia* (weedy form), *V. dalzelliana*, *V. khandalensis*, *V. minima*, *V. mungo* var. *sylvestris*, *V. radiata* var. *sublobata* and *V. vexillata* from the Western Ghats of Maharashtra; and from Rajasthan the wild species *Vigna trilobata*, *V. radiata* var. *sublobata* and *V. hainiana*. Dana (1998) also collected wild *Vigna* species, namely, *V. aconitifolia* var. *sylvestris*, *V. dalzelliana*, *V. hainiana*, *V. khandalensis*, *V. mungo* var. *sylvestris*, *V. radiata* var. *setulosa*, *V. radiata* var. *sublobata* and *V. trilobata*, from the states of Gujarat, Rajasthan, Maharashtra, Madhya Pradesh, Bihar, Orissa and West Bengal during the period 1974–1994.

Over the past three decades, NBPGR has contributed to introducing over 1,06,150 accessions (including trials) of legume crops from 56 countries under strict quarantine measures. Some of the promising germplasm introduced possesses resistance to diseases, pests and nematodes; tolerance to drought, heat, cold, chilling and salt stresses; bold seeds, high yield potential, wider adaptability, low neurotoxin content and desirable quality traits; and these are now being used in legume improvement programmes across the country (Table 23.2). The main contributors of these accessions are CIAT (French bean, pigeon pea); ICARDA, (chickpea, faba bean, *Lathyrus*, lentil, pea); ICRISAT (chickpea, pigeon pea); IITA (cowpea, *Vigna* spp.); AVRDC (mung bean, *Vigna* spp.); Bogor Research Institute for Food (BORIF), Bogor, Indonesia (mung bean, *Vigna* spp, pigeon pea); Commonwealth Scientific and Industrial Research Organization (CSIRO), Canberra, Australia (cowpea, *Vigna* spp., winged bean); and the United States Department of Agriculture (USDA) (pea, *Vigna* spp.).

The *ex situ* seed gene bank at NBPGR comprises 12 long-term modules (total capacity: 1 million accessions) maintained at –20°C for housing the base collections. Most of the food legumes have orthodox seeds that can be dried and stored for a long period with minimum loss of viability. The national gene bank at NBPGR, New Delhi, which is primarily responsible for conservation of germplasm on long-term basis holds 56,841 accessions of food legumes belonging to 23 genera and 61 species as the base collections (Table 23.3), and 10,235 duplicate safety samples of pigeon pea and lentil received from ICARDA and ICRISAT. Besides, 554 released varieties of 18 legume crops have also been conserved as the base collection. The germplasm conserved as a base collection is assigned a national identify number, dried to seed moisture of around 5 ± 2 at 15°C and 15% relative humidity (RH). The accessions meeting international standards (IBPGR, 1994), with seed viability of more than 85% and 2000–4000 seeds, are transferred to long-term storage. Base collections are regularly monitored for their seed viability, quantity and health at 10-year intervals. For conservation of active collections, seeds are kept under medium–term storage (5°C and 40% RH). The Indian Institute of Pulses Research, Kanpur serves as the National Active Germplasm Site (NAGS) and maintains over 59,500 working collections of pulse crops in coordination with AICRP (All-India Coordinated Research Project) centres. A total of 632 accessions of 12 pulse

Table 23.2. Promising trait-specific germplasm of legume crops.

Trait	Accessions	Country
Adzuki bean (*Vigna angularis*)		
Long pods; good yield	EC87899-1	
Chickpea (*Cicer arietinum*)		
Ascochyta blight	EC 382754-56, EC383763-64, EC 381867-81, EC 554848-57, EC599949	Syria
Fusarium wilt	EC 381886-88	Syria
Large-seeded *kabuli* type; resistance to ascochyta blight	EC 499759	USA
Large-seeded *kabuli* landraces; resistant to ascochyta blight and fusarium wilt	EC556541-42	Spain
Resistance to leaf miner	EC-381899-90, EC 3811892-93, EC 381895	Syria
Cold tolerance	EC 381903-06, EC 519355-381, EC 595376, EC 566900-909	Syria, Australia
Chilling tolerance	EC 583236	Australia
Heat tolerance	EC 565197-214	USA
Salt tolerance	EC 566910-919	Australia
Drought tolerance	EC 389382-85, EC 381895, EC381903-06, EC 381909	Syria
Large-seeded *kabuli*	EC 431475, EC 571855- 572003	Bangladesh, Syria
Extra-large-seeded *kabuli*	EC 583236	USA
Isogenic lines	EC 44104-105	Spain
High yield; resistance to ascochyta blight	EC 600092-93	Israel
Cowpea (*Vigna unguiculata*)		
Resistance to aphids, bruchids, thrips, striga; adaptation to tropical conditions	EC 336589-90	Nigeria
Earliness; drought tolerance	EC 384918-934	Nigeria
High yield; tolerance to drought	EC 391006-09	Taiwan
Heat tolerance; resistance to fusarium wilt, root-knot nematode	EC 496737	USA
High yield	EC 514421	USA
Day-neutral types	EC 566356-357	Taiwan
Suitable for rainfed conditions	EC 582501	USA
Resistance to cowpea aphid-borne mosaic virus, bacterial blight; early maturity	EC 587822	Senegal
Faba bean (*Vicia faba*)		
High yield	EC 284356-375	Syria
Resistance to chocolate leaf spot	EC 303650-71	Syria
Resistance to ascochyta blight	EC 303672-91	Syria
Home garden cultivar; small-seeded	EC 466695	USA
Heat tolerance; resistance to fusarium wilt, root-knot nematode	EC 466882	USA
Resistance to seed-borne potyvirus, BCMV, ashy stem blight	EC 502154	USA

Continued

Table 23.2. Continued.

Trait	Accessions	Country
Large-seeded; resistance to curly top virus	EC565673-74	USA
French bean (*Phaseolus vulgaris*)		
High yield	EC 397815-39	
Resistance to bacterial blight	EC 398478-574, EC 398575-612	USA
Resistance to BCMV	EC 406066-070 and EC 402962-85	Colombia
Dwarf; high yielding; resistance to diseases	EC 421101-108	Zambia
Virus resistance	EC 463964-966	USA
High yielding; early maturing	EC 467266	Canada
Tolerance to common bacterial blight; resistance to BCMV, lodging	EC 498445	Canada
Large-seeded; resistance to curly top virus	EC 565673-74	USA
High yield; resistance to all known strains of BCMV and root rot	EC 589388	USA
High yield; resistance to lodging, BCMV, common tungro virus	EC 589468	USA
Resistance to BCMV; tolerance to white mould; semi-determinate growth habit	EC 590327	Canada
Non-nodulating genetic stock	EC 590328	Canada
Resistance to common bacterial blight	EC 592938	USA
Resistance to yellow and orange strains of bacterial wilt	EC 599950	Canada
Small-seeded; resistance to common bacterial blight	EC 593020	USA
Grass pea (*Lathyrus sativus*)		
Low neurotoxin contents	EC 322620-31	Syria
High yield; drought tolerance	EC 389370-73	Syria
Lablab bean (*Lablab purpureus*)		
High yield; bruchid resistance	EC 382814-15	Costa Rica
Lentil (*Lens culinaris*)		
Earliness	EC 397814	Syria
Good plant vigour	EC 299646, EC 399648	Syria
Bold seeds	EC 267626,EC 267659, EC 267673, EC 397814	Syria
Resistance to vascular wilt	EC382684-714	Syria
Winter-hardy; multiple resistant lines	EC 382715-25	Syria
High yield; drought tolerant	EC 389386-89	Syria
High yield	EC 397781-814	Slovakia
High yield; yellow cotyledons; seed coat light green; lack of seed coat mottling	EC 499760	USA
Large seeds; high yield	EC 550084	USA
Winter-hardy	E 608175	USA
High yield; high level of winter-hardiness	EC 631332	Turkey

Continued

Table 23.2. Continued.

Trait	Accessions	Country
Moth bean (*Vigna aconitifolia*)		
High yield	EC 390994-97	Taiwan
Mung bean (*Vigna radiata*)		
Wide adaptability; earliness; resistance to tungro mosaic virus	EC 118889, EC 118894, EC 118895, EC162584, EC158782, EC159734	Taiwan
Resistance to charcoal rot, leaf crinkle; tolerance to drought, flood; photoperiod insensitive	EC 318985-319057	Taiwan
High yielding	EC 391170-75	Indonesia
Large-seeded; long-podded with shiny green seed coat	EC 393407-10	Bangladesh
Heat tolerant; short and long duration	EC 397138, EC 396394-396423	Thailand
High yielding	EC 390990-93	Taiwan
High yielding	EC 428862	Nepal
Resistance to MYMV	EC 564801-818, EC 565626-633	Taiwan
Early maturity	EC 512780-793	USA
Resistance to powdery mildew	EC 605445	Australia
Pea (*Pisum sativum*)		
Earliness	EC 324118, EC 328751, EC 328754, EC 328789	
Long pods	EC 342007, EC 398588, EC 398596	USA
Powdery mildew resistance	EC 322745, EC 381853-66	Canada, USA
Fusarium wilt resistance	EC 384889-91	USA
Drought tolerance	EC 389374-77	Syria
Sugary pods	EC 334160-63	USA
Large-seeded; resistance to race 1 of fusarium wilt, powdery mildew; high yield	EC 499761-62	USA
Multiple disease resistance	EC 507770-71	USA
Stiff stem (lodging resistance)	EC 548807-811	Russia
Vegetable type; high-podded	EC 564801-818	Russia
Ascochyta blight resistance	EC 595959	USA
Pigeon pea (*Cajanus cajan*)		
High yield	EC 284065	Australia
Early and medium duration	EC 215296-97	Malawi
Landraces with desirable traits	EC 377961-79	Kenya
Rice bean (*Vigna umbellata*)		
High yield	EC 93452, EC 101887, PI 247685, PI 247693	USA
High yield; drought tolerance	EC 3901002-05	Taiwan
Urd bean (*Vigna mungo*)		
Resistance to beanfly, bruchids	EC 245976-77	Taiwan
Resistance to yellow mosaic virus	EC 390998-391001	Taiwan

BCMV, bean common mosaic virus; MYMV, mungbean yellow mosaic virus.

crops belonging to different species, namely *Cajanus* (7), *Cicer* (9), *Cymopsis* (1), *Lens* (1), *Macrotyloma* (2), *Phaseolus* (1), *Pisum* (1), *Rhynchosia* (7), *Vicia* (1) and *Vigna* (27) have been cryo-stored at −180°C in liquid nitrogen, and one accession of *Cicer microphyllum* has been conserved through *in vitro* culture at NBPGR.

Table 23.3. Base collections maintained at the national gene bank of NBPGR (as of February 2010).

Crop	Total accessions	Wild species
Adzuki bean	167	0
Chickpea	13,990	72
Cluster bean	3,822	32
Cowpea	3,440	3
Faba bean	579	1
French bean	3,107	9
Horse gram	2,709	6
Khesari	2,599	16
Lablab bean	1,153	6
Lentil	2,374	28
Lima bean	35	1
Moth bean	1,540	36
Mung bean	3,723	11
Parkia	3	3
Pea	2,862	1
Pigeon pea	10,385	135
Rice bean	1,978	89
Scarlet runner bean	36	0
Sword bean	111	24
Urd bean	1,595	61
Velvet bean	59	4
Vigna	346	321
Winged bean	228	0
Total	56,841	859

23.3 Characterization, Evaluation and Documentation

For effective utilization of germplasm, it is important that it is characterized and evaluated for both important agro-morphological traits and biotic and abiotic stresses. For systematic characterization and evaluation of various legume crops, NBPGR has developed minimal descriptors (Mahajan *et al.*, 2000). A large number of accessions were characterized and evaluated by NBPGR, IIPR, other ICAR institutes, SAUs and at IARC, namely ICRISAT, ICARDA, IITA and CIAT. This has resulted in the identification of genetic stocks of various pulse crops by various workers, as reported by Sardana *et al.* (2005). Multiplicational evaluation of chickpea (2792 accessions) and pigeon pea (2142 accessions) during the period 2004–2007 has resulted in identification of some germplasm with good agronomic traits and resistance/tolerance to various biotic and abiotic stresses.

Registration of germplasm

The organization NBPGR has been recognized as a nodal agency for the registration of genetic stocks/unique germplasm of crops and its effective utilization. A large amount of legume germplasm has been characterized and evaluated, and unique/promising germplasm for early maturity, bold seeds, resistance/tolerance to biotic and abiotic stresses, stable cytoplasmic genetic male sterile (CGMS) lines, and fertility restorer lines of stable CMS lines were identified and are now registered with NBPGR, Among those registered include the pigeon pea elite germplasm line ICPL 87162 for high protein (Reddy *et al.*, 1997); ICP 9145, ICP 8863, ICP 112922, ICP 11299 and ICO 12745 for fusarium wilt resistance (Reddy *et al.*, 1995a, b); and in chickpea, ICC 4958 for drought tolerance (Saxena *et al.*, 1993) and ILC 3800, ILC 5901 and ILC 7738 for leaf minor resistance (Singh and Weigand, 1996).

Development of core collections

Over the last four decades there has been increased emphasis on germplasm exploration and collection, resulting in huge collections in most gene banks. Initially, however, the germplasm in the gene banks could not be characterized effectively and remained underutilized, until the formation of core sets and mini-core sets and their evaluation enhanced their utilization (Frankel, 1984). A core collection represents about 10% of the entire collection that captures most of the available diversity of species (Brown, 1989), while a mini-core collection represents about 1% but captures most of the useful variation of the crop (Upadhyaya and Ortiz, 2001). Using passport information the characterization and evaluation data generated over a period of time, scientists at various institutes, including CG centres, have developed global core and mini-core collections in important grain legumes (Table 23.4). The optimal and convenient size of these collections has led to increased demand for germplasm use by researchers. The utilization of such diverse accessions in crop breeding programmes should be enhanced to develop improved cultivars with a broad genetic base.

At NBPGR, a core collection of 1532 mung bean accessions has been created (Bisht *et al.*, 1998). ICRISAT has developed a core collection of chickpea from the global collection of 16,991 accessions from 44 countries (Upadhyaya *et al.*, 2001) and a core collection of pigeon pea from 12,153 accessions assembled from 56 countries (Reddy *et al.*, 2005). USDA has developed core collections of chickpea from 3873 accessions, lentil from 2390 accessions and pea from 2888 accessions (Simon and Hannan, 1995). Although cores may not always enhance access to germplasm with unique or extremely rare characteristics, the legume cores have been useful for directing users at the preliminary stages of germplasm evaluation. A mini-core subset consisting of 211 chickpea accessions from 1956 core collection accessions was also developed and this subset, due to drastically reduced size, will enable a point of entry to the proper exploitation of chickpea genetic resources (Upadhyaya and Ortiz, 2001).

The core collection of lentil, comprising 287 accessions, has been evaluated for variation in phenological and morphological characters that indicated sufficient variation to warrant their use in breeding programmes (Tullu *et al.*, 2001). Germplasm resistant to fusarium wilt race 2 has been identified in the *Pisum* core collection (McPhee *et al.*, 1999). Significant favourable variation was observed in the USDA core collection of *Pisum* germplasm, which could be used to increase both seed yield and total biomass production of adapted lines (McPhee and Muehlbaur, 2001). The chickpea mini-core developed at ICRISAT was evaluated at IIPR Kanpur during in 2003/2004, when 12 accessions were selected for use in the crop improvement programme, which will help to broaden the genetic base of the cultivars (Gowda and Upadhyaya, 2006).

Knowledge gained on heritable variation for anti-nutritional factors among germplasm accessions in *Lathyrus* has made it possible to develop varieties with low ODAP (β-N-oxaly-L-α, β-diaminopropionic acid). Screening of horse gram germplasm has resulted in the identification of one wild germplasm (IC212722) with 38% protein content (Yadav *et al.*, 2004).

Molecular characterization

The documentation and dissemination of information on genetic resources of legume crops are important for their effective utilization. As a result, various national and international centres have published a number of catalogues on various food legumes. In India, the characterization and evaluation of various food legume crops at NBPGR has led to the publishing of 31 catalogues for 14 legume crops describing 24,205 accessions; IIPR, Kanpur has also published a catalogue

Table 23.4. Core and mini-core collections developed at international centres.

Crop	Accessions	Traits	Type of collection	Accessions in subset	Centre
Chickpea	3,350	NA	Core	505	ICRISAT
	16,991	13	Core	1956	ICRISAT
	1,956	22	Mini-core	211	ICRISAT
Pigeon pea	12,153	14	Core	1290	ICRISAT
	1,290	33	Mini-core	146	ICRISAT
Lentil	1,000	17	Core	234	ICARDA
Mung bean	1,153	23	Core	203	NBPGR
Phaseolus	1,117	14	Core	147	CIAT

on chickpea germplasm (Singh and Kumar, 2004) describing 1097 accessions. ICARDA has published four catalogues, two on *kabuli* chickpea (Singh *et al.*, 1983, 1991) and one each on lentil (Erskine and Witcombe, 1984) and faba bean (Robertson and Sherbeeny, 1988); ICRISAT has published one catalogue on the global collection of chickpea (Pundir *et al.*, 1988) and two on pigeon pea (Ramanandan *et al.*, 1988a, b). AVRDC has published a catalogue on mung bean (Tay *et al.*, 1989) and CSIRO, Australia has published a catalogue on urd bean (Imrie *et al.*, 1981), while IITA, Nigeria has published a catalogue on cowpea (IITA, 1974). ICARDA has also documented a catalogue on 268 accessions of wild *Cicer* species for various morphological traits. Efforts have also been made to publish a monograph (Chandel *et al.*, 1988) and research bulletin (Sarma *et al.*, 1995) on rice bean.

Molecular techniques such as random amplified polymorphic DNA (RAPD), restriction fragment length polymorphism (RFLP), amplified fragment length polymorphism (ALFP), sequence tagged microsatellite (STMS) and simple sequence repeats (SSR) have also been used in different centres for the characterization of germplasm. For example, at NBPGR in India 571 varieties and 633 elite landraces of 11 pulse crops have been characterized using molecular markers, which revealed moderate to low levels of polymorphism. Principle component analysis has also shown a high degree of genetic similarity among cultivars, which is due to similarity in their pedigree and narrow genetic base (Karihaloo *et al.*, 2001).

23.4 Genetic Enhancement using Wild Relatives

The genetic base of pulse varieties is quite narrow and needs immediate corrective measures by involving unadapted accessions, exotics and wild relatives in hybridization programmes (Kumar *et al.*, 2004). Wild relatives of legumes are known as valuable sources for resistance to several biotic and abiotic stresses beside yield components (Table 23.5). The real challenge, therefore, is to convert the collected PGR through pre-breeding into parental lines, which will be more acceptable to plant breeders. Wide diversity was observed in 45 morphological characters for 206 accessions of 14 wild *Vigna* species, namely *Vigna khandalensis*, *V. radiata* var. *sublobata*, *V. radiata* var. *setulosa*, *V. mungo* var. *silvestris*, *V. hainiana*, *V. umbellata*, *V. dalzelliana*, *V. bourneae*, *V. minima*, *V. trilobata*, *V. aconitifolia*, *V.vexillata*, *V. pilosa* and *V. glabrescens* (Bisht *et al.*, 2005).

Harlan and de Wet (1971) initiated the concept of the gene pool, which has proved useful to plant breeders in the creation of useful variation in crop improvement. Wild species can enrich the gene pool, where the main aim over recent decades has been to improve yields in different legume crops. The wild relative of *Cicer*, namely, *C. judaicum*, has been reported to be resistant to botrytis grey mould (Meeta and Bedi, 1987), fusarium wilt (Nene and Haware, 1980) and has high methionine content, while *C. pinnatifidum* is resistant to botrytis grey mould and has high tryptophan content. In efforts at improvement of mung and urd bean, the sub-gene pool of wild types in accession PLN 5 of *V. radiata* var. *sublobata* (Singh and Ahuja, 1977) and IW 3390 of *V. mungo* var. *silvestris* (Reddy and Singh, 1993) have already been identified as potential sources of MYMV (mung bean yellow mosaic virus) resistance, and TC 1966 from *V. radiata* var. *sublobata* was identified as carrying a gene for bruchid tolerance (Tomooka *et al.*, 1992). Ferguson and Robertson (1999) observed wide variation for phenological and agro-morphological traits in 310 accessions of wild lentils (*Lens culinaris* subsp. *orientalis*, *L. odomensis*, *L. ervoides*, *L. nigricans* and *L. lamottei*); and Gupta and Sharma (2006) in 70 wild accessions (*L. culinaris*, subsp. *orientalis*, *L. odomensis*, *L. ervoides* and *L. nigricans*). These latter authors suggested that valuable variation existing among wild accessions should be exploited following introgression with cultivated lentils.

23.5 Utilization of Germplasm

Plant genetic resources are the most valuable and essential raw material for crop

Table 23.5. Sources of resistance in wild species of legume crops.

Species	Accessions	Traits	Reference
Cicer judaicum	EC 382438-39, EC-382451, EC541557-558	Cold tolerance; resistance to leaf miners, bruchids, ascochyta blight	Anonymous (1997) Anonymous (2005)
C. bijugum	EC541549-50	Resistance to leaf miners, ascochyta blight	Anonymous (2005)
C. chorassanicum	EC541551-52	Resistance to leaf miners	Anonymous (2005)
C. cuneatum	EC541553-54	Resistance to *Callosobruchus chinensis*	Anonymous (2005)
C. echinospermum	EC 382414, EC541555-56	Resistance to fusarium wilt	Anonymous (1997) Anonymous (2005)
C. pinnatifidum	EC 382450, EC541557-558	Resistance to fusarium wilt,	Anonymous (1997) Anonymous (2005)
C reticulatum	EC541561-62	Resistance to bruchids, cyst nematodes	Anonymous (2005)
C. yamashitae	EC541563-64	Resistance to leaf miners	Anonymous (2005)
Cajanus scarobaeoides	ICPW 111, ICPW 128	Resistance to cyst nematode (*Heterodera cajani*)	Sharma *et al.* (1993)
C. pemingia	ICPW 194, ICPW 202, ICPW 203	Resistance to *Heterodera cajani*	Sharma *et al.* (1993)
Lens nigricans	ILWL 138	Resistance to wilt, ascochyta blight	Bayaa *et al.* (1994)
L. nigricans	ILWL 37	Resistance to rust, wilt, powdery mildew	Gupta and Sharma (2006)
L. ervoides	ILWL 40, ILWL 41, ILWL 42, ILWL 251	Resistance to wilt, rust, powdery mildew	Gupta and Sharma (2006)
Vigna radiata var. *sublobata*	TC 1966	Resistance to bruchids (*Callosobruchus* sp.)	Tomooka *et al.* (1992)
V. mungo var. *silvestris*	IW 3390	Resistance to MYMV	Reddy and Singh (1993)
Pisum sativum var. *arvense*	EC548807-011	Resistance to lodging	Anonymous (2005)
P. sativum var. *abyssinicum*	EC548872	Resistance to lodging	Anonymous (2005)
P. sativum var. *elatius*	EC548813	Resistance to lodging	Anonymous (2005)

MYMV, mung bean yellow mosaic virus.

improvement. In fact, they provide a reservoir of genes to tailor newer plant types and insurance against nature's vagaries. Before 1960, most of the improved varieties were acquired by direct selection from natural variability collected through explorations in different agro-climatic regions within and outside the countries. Later, the efforts were directed towards diversification of genetic base by evolving high-yielding and widely adapted cultivars by different breeding methods. Over a period, a large number of superior genotypes of legumes has been released for general cultivation in different parts of the India (Singh *et al.*, 2005), as well as in other countries of the world, utilizing the useful and desirable traits from germplasm. To date 31 exotics have contributed in the development of 45 cultivars, namely, chickpea (8), pigeon pea (2), field pea

(5), mung bean (11), lentil (1) and rajmash (4) released in India (Ali *et al.*, 2006). Of the 520 legume varieties released to date, 225 are selections either directly or indirectly from landraces and germplasm. This indicates the importance of direct utilization of germplasm in legume improvement. Some germplasm lines of Indian origin were also utilized to release varieties of chickpea in (Nepal (Sita), Bangladesh (Nabin, Barichhola 3), Myanmar (Schwe Kyemon), Ethiopia (Mariye) and Kenya (ICCL83110) (Sethi and van Rheenen, 1994); pigeon pea in Australia (Haunt and Quest), Indonesia (Megha) and Fiji (Kamica) (Ariyanayagam and Jain, 1994); mung bean in Vietnam (DX113, DX102A), Costa Rica (ASVEG78), Ecuador (Boliche 457), Fiji (Station 25, Station 27, Station 46), Indonesia (Manyar, Nuri, Gelatik, Walet), Philippines (BPIMG 2, BPIMG 4), Sri Lanka (T 77) and Thailand (Kamphanegsaen 2); and urdbean in Thailand (U Thong 2 and Phitsanulok 2).

23.6 Conclusions

It is obvious that the need for introduction of legume genotypes will continue, since new diseases and new niches are continually appearing. Preferred traits for introduction/collection of legume germplasm from diversity-rich areas are: short duration, bold seeds, temperature tolerance, salinity tolerance, resistance against wilt and ascochyta blight for chickpea; short duration, CMS sources, resistance to wilt and phytophthora blight for pigeon pea; short duration, bold seeds, photo-thermal insensitivity, resistance against bruchids, powdery mildew and cercospora leaf spot for mung bean and urd bean; bold seeds, resistance against rust and vascular wilt for lentil; dwarfness, afila types, short duration, resistance to powdery mildew for field pea; cold tolerance (< 5°C) and resistance to BCMV (bean common mosaic virus) for rajmash; and low neurotoxin for *Lathyrus*. Efforts need to be directed towards the collection and conservation of endangered and wild germplasm from threatened areas of diversity. This should be complemented by safe duplication of germplasm at other locations, preferably near the place of collection, and multi-location evaluation of food legume germplasm for agro-morphological traits, disease and pest resistance; tolerance to abiotic stresses should take priority. The molecular characterization of unique/useful genetic stocks should also be established to assert their sovereign rights. Germplasm enhancement in the past has made little progress; future efforts should be aimed at transferring useful genes from exotic or wild types to more adapted germplasm/varieties by the crop improvement institutes using the biotechnology tools now available.

References

Ali, M., Gupta, S., Singh, B.B. and Kumar, S. (2006) Role of plant introduction in varietal development of pulses in India. *Indian Journal of Plant Genetic Resources* 19, 346–352.

Anonymous (1997) *Annual Report of Plant Genetic Resources 1996–1997*, NBPGR, Pusa Campus, New Delhi, India.

Anonymous (2005) *Annual Report of Plant Genetic Resources 2004–2005*, NBPGR, Pusa Campus, New Delhi, India.

Ariyanayagam, R.P. and Jain, K.C. (1994) Pigeonpea germplasm management and enhancement. In: Bantilan, M.C.S. and Joshi P.K. (eds) *Summary Proceedings of a Workshop on Research Evaluation and Impact Assessment*. ICRISAT, Patancheru, India, pp. 32–37.

Arora, R.K. (1988) The Indian gene centre – Priorities and prospects of collection. In: Paroda, R.S., Arora, R.K. and Chandel, K.P.S. (eds) *Plant Genetic Resources: Indian Perspective*. NBPGR Publications, New Delhi, India, pp. 66–75.

Bayaa, B., Erskine, W. and Hamdi, A. (1994) Response of wild lentil to *Ascochya fabae* f.sp. *Lentis* from Syria. *Genetic Resources and Crop Evolution* 41, 61–65.

Bisht, I.S., Mahajan, R.K. and Patel, D.P. (1998) The use of characterisation data to establish the Indian mungbean core collection and assessment of genetic diversity. *Genetic Resources and Crop Evolution* 45, 127–133.

Bisht, I.S., Bhat, K.V., Lakhanpaul, S., Latha, M., Jayan, P.K., Biswas, B.K. *et al.* (2005) Diversity and genetic resources of wild *Vigna* species in India. *Genetic Resources and Crop Science* 52, 53–68.

Bose, R.D. (1932) Studies in Indian pulses. No.4. Mung or mungbean (*Phaseolus radiatus* (Linn.). *Indian Journal of Agricultrul Sciences* 2, 607–624.
Brown, A.H.D. (1989) Core collections: A practical approach to genetic resources management. *Genome* 31, 818–824.
Chandel, K.P.S., Arora, R.K. and Pant, K.C. (1988) *Rice Bean: A Potential Grain Legume. NBPGR Science Monograph* No. 12. NBPGR, New Delhi, India.
Dana, S. (1998) Collection of wild *Vigna* in seven states in India. *Green Journal* 1, 9–12.
Erskine, H.W. and Witcombe, J.R. (1984). *Lentil Catalogue*. International Centre for Agricultural Research in the Dry Areas (ICARDA), Aleppo, Syria.
Ferguson, M.E. and Robertson, L.D. (1999) Morphological and phenological variation in the wild relatives of lentil. *Genetic Resources and Crop Evolution* 46, 3–12.
Frankel, O.H. (1984) Genetic perspectives of germplasm conservation In: Arber, W. (ed.) *Genetic Manipulations: Impact on Man and Society*. Cambridge University Press, Cambridge, UK, pp. 161–170.
Furman, B.J., Coyne, C., Redden, B., Sharma, S.K. and Vishnyakova, M. (2009) Genetic resources: collection, characterization, conservation and documentation, In : Erskine W., Muehlbauer, F.J., Sarker, A. and Sharma, B. (eds) *The Lentil: Botany, Production and Uses*. CABI International, Wallingford, UK, pp. 64–75.
Gowda, C.L.L. and Upadhyaya, H.D. (2006) International crop germplasm exchange at ICRISAT. *Indian Journal of Plant Genetic Resources* 19, 418–427.
Gupta, D. and Sharma, S.K. (2006) Evaluation of wild *Lens* taxa for agro-morphological traits, fungal diseases and moisture stress in North Western Indian Hills. *Genetic Resources and Crop Evolution* 53, 1233–1241.
Harlan, J.R. (1984) Evaluation of wild relatives of crop plants. In: Holden, J.H.W. and Willium, S.J.T. (eds) *Crop Genetic Resources: Conservation and Evaluation*. George Allen and Unwin, London, pp. 212–222.
Harlan, J.R. and de Wet, J.M.J. (1971) Toward a rational classification of cultivated plants. *Taxonomy* 20, 509–517.
IBPGR (1994) *Genebank Standards*. International Board for Plant Genetic Resources, FAO, Rome.
IITA (1974) *Cowpea Germplasm Catalog*. International Institute of Tropical Agriculture, Ibadan, Nigeria.
Imrie, B.C., Beech, D.F., Blogg, D. and Thomas, B. (1981) A catalogue of the *Vigna radiata* and *V. mungo* germplasm collection of division of tropical crops and pastures. *Genetic Resources Communication* 2, CSIRO, Australia.
Karihaloo, J.L., Bhat, K.V., Lakhanpaul, S., Mahapatra, T. and Randhawa, G.J. (2001) Molecular characterzation of germplasm. In: Dhillon, B.S., Varaprasad, K.S., Singh, M., Archak, S., Srivastava, U. and Sharma, G.D. (eds) *National Bureau of Plant Genetic Resources: A Compendium of Achievements*. National Bureau of Plant Genetic Resources, New Delhi, India, pp. 166–182.
Kumar, S., Gupta, S., Chandra, S. and Singh B.B. (2004) How wide is the genetic base of pulse crops. In: Ali, M., Singh, B.B., Kumar, S. and Vishwa, D. (eds) *Pulses in New Perspective*. Indian Society of Pulses Research and Development, IIPR, Kanpur, India, pp. 211–221.
Mahajan, R.K., Sapra, R.L., Srivastava, U., Singh, M. and Sharma, G.D. (2000) *Minimal Descriptors (for Characterization and Evaluation) of Agri-horticultural Crops (Part 1)*. National Bureau of Plant Genetic Resources, New Delhi, India.
McPhee, K.E. and Muehlbaur, F.J. (2001) Biomass production and related characters in the core collection of *Pisum* germplasm. *Genetic Resources and Crop Evolution* 48, 195–203.
McPhee, K.E., Tullu, A., Kraft, J.M. and Muehlbauer, F.J. (1999) Resistance to *Fusarium* wilt race 2 in the *Pisum* core collection. *Journal of the American Society of Horticultural Sciences* 124, 28–31.
Meeta, M. and Bedi, P.S. (1987) Evaluation of gram germplasm against *Botrytis cnerea. Indian Phypathology* 39, 296.
Nene, Y.L. and Haware, M.P. (1980) Screening chickpea for resistance to wilt. *Plant Disease* 64, 379–380.
Pundir, R.P.S., Reddy, K.N. and Mengesha, M.H. (1988) *ICRISAT Chickpea Germplasm Catalogue: Evaluation and Analysis*. International Research in the Semi-Arid Tropics (ICRISAT), Patancheru, AP, India.
Ramanandan, P., Shastri, D.V.S.S.R. and Megesha, M.H. (1988a) *ICRISAT Pigeonpea Germplasm Catalogue: Evaluation and Analysis*. International Research in the Semi-Arid Tropics (ICRISAT), Patancheru, AP, India.
Ramanandan, P., Shastri, D.V.S.S.R. and Megesha, M.H. (1988b) *ICRISAT Pigeonpea Germplasm Catalogue: Passport Information*. International Research in the Semi-Arid Tropics (ICRISAT), Patancheru, AP, India.

Reddy, K.R. and Singh, D.P. (1993) Inheritance of resistance to mungbean yellow mosaic virus. *Madras Agricultural Journal* 80, 199–201.

Reddy, L.J., Saxena, K.B., Jain, K.C., Singh, U., Green, J.M. *et al.* (1997) Registration of high protein pigeonpea elite germplasm ICPL 87162. *Crop Science* 37, 294.

Reddy, L.J., Upadhyaya, H.D., Gowda, C.L.L. and Sube, S. (2005) Development of core collection in pigeonpea [*Cajanus cajan* (L.) Millspaugh] using geographic and qualitative morphological descriptors. *Genetic Resources and Crop Evolution* 52, 1049–1056.

Reddy, M.V., Nene, Y.L., Raju, T.N., Kannaiyan, J., Remanandan, P., Mangesha, M.H. *et al.* (1995a) Registration of pigeonpea germplasm line ICP 9145 resistant to *Fusarium* wilt. *Crop Science* 35, 1231.

Reddy, M.V., Nene, Y.L., Raju, T.N., Kannaiyan, J., Remanandan, P., Mangesha, M.H. *et al.* (1995b) Registration of four pigeonpea germplasm lines resistant to *Fusarium* wilt: ICP 8863, ICP 11292, ICP 11299 and ICP 12745. *Crop Science* 35, 595.

Robertson, L.D. and Sherbeeny, M.E.L. (1988) *Faba Bean Germplasm Catalog*. International Centre for Agricultural Research in Dry Areas (ICARDA), Aleppo, Syria.

Sardana, S., Dhillon, B.S., Mahendra, S. and Mishra, S.K. (2005) Pulses germplasm: collection, conservation and utilization. In: Singh G., Singh, S.H. and Singh, K.J. (eds) *Pulses*. Agrotech Publishing Academy, Udaipur, India, pp. 95–144.

Sarma, B.K., Singh, M., Gupta, H.S., Singh, G. and Srivastava, L.S. (1995). *Studies in Rice Bean Germplasm. Research Bulletin No. 34*. ICAR Research Complex for NEH Region, Barapani, Meghalaya, India.

Saxena, N.P., Krishnamurthy, L. and Johansen, C. (1993) Registration of a drought-resistant chickpea germplasm. *Crop Science* 33, 1424.

Sethi, S.C. and van Rheenen, H.A. (1994) Genetic improvement of chickpea. In: Bantilan, M.C.S. and Joshi P.K. (eds) *Summary Proceedings of a Workshop on Research Evaluation and Impact Assessment*. International Research in the Semi-Arid Tropics (ICRISAT), Patancheru, AP, India, pp. 26–31.

Sharma, S.B., Ramanandan, P. and Jain, K.C. (1993) Resistance to cyst nematode (*Heterodera cajani*) in pigeonpea cultivars and in wild relatives of *Cajanus*. *Annals of Applied Biology* 123, 75–81.

Shaw, F.J.F. and Khan, A.R. (1931) Studies in Indian pulses. Some varieties of Indian gram (*Cicer arietinum L.*). *Indian Agriculture Botanical Survey* 19, 27–47.

Simon, C.J. and Hannan, R.M. (1995) Development and use of core subsets of cool-season food legume germplasm collections. *Horticulture Research* 30, 907.

Singh, B.V. and Ahuja, M.R. (1977) *Phaseolus subobatus* Rox. – A source of resistance to yellow mosaic virus for cultivated mung. *Indian Journal of Genetics* 37, 130–132.

Singh, D.P., Singh, B.B. and Kumar, S. (2005) Varietal imrovement in pulses. In: Singh G., Singh, S.H. and Singh, K.J. (eds) *Pulses*. Agrotech Publishing Academy, Udaipur, India, pp. 145–172.

Singh, K.B. and Weigand, S. (1996) Registration of three leaf minor resistant chickpea germplasm lines: ILC 3800, ILC 5901 and ILC 7738. *Crop Science* 36, 472.

Singh, K.B., Malhotra, R.S. and Witcombe, J.R. (1983) *Kabuli Chickpea Germplasm Catalog*. ICARDA, Aleppo, Syria.

Singh, K.B., Holly, L. and Bejiga, G. (1991) *A Catalog of Kabuli Chickpea Germplasm*. ICARDA, Aleppo, Syria.

Singh, N. and Kumar, S. (2004) *Chickpea Germplasm Catalogue*. Indian Institute of Pulses Research, Kanpur, India.

Smartt, J. (1990) *Grain Legumes: Evolution and Genetic Resources*. Cambridge University Press, Cambridge, UK, pp. 379.

Tay, D.C.S., Huang, V.K. and Chen, C.Y. (1989) *Germplasm Catalogue of Mungbean and Other Vigna* sp. AVRDC Publication No. 89, AVRDC, Taiwan.

Tomooka, N.C., Lairungreang, P., Nakeeraks, Y.E. and Thavarasook, C. (1992) Development of bruchid resistant mungbean line using wild mung bean germplasm in Thailand. *Plant Breeding* 109, 60–66.

Tomooka, N., Egawa, Y. and Kaga, A. (2000) Biosystematics and genetic resources of the genus *Vigna* subgenus *Ceratotropis*. In: Oono, K., Vaughen, D., Tomooka, N., Kaga, A. and Miyazaki, S. (eds) *The Seventh MAFF International Workshop on Genetic Resources, Part 1. Wild Legumes*. Ministry of Agriculture, Forestry and Fisheries and National Institute of Agrobiological Resources, Japan, pp. 37–62.

Tullu, A., Kushmenoglu, I., McPhee, K.E. and Muehlbauer, F.J. (2001) Characterization of core collection of lentil germplasm for phenology, morphology, seed and straw yields. *Genetic Resources and Crop Evolution* 48, 143–162.

Upadhyaya, H.D. and Ortiz, R. (2001) A mini core subset for capturing diversity and promoting utilization of chickpea genetic resources. *Theoretical and Applied Genetics* 102, 1292–1298.

Upadhyaya. H.D., Bramel, P.J. and Singh, S. (2001) Development of a chickpea core subset using geographic distribution and quantitative traits. *Crop Science* 41, 200–210.

Yadav, Sangita, Negi, K.S. and Mandal, S. (2004) Protein and oil rich wild horsegram. *Genetic Resources and Crop Evolution* 51, 629–633.

24 Seed Dormancy and Viability

J.Y. Asibuo

24.1 Introduction

The seed encloses the embryo as the new plant in miniature; its structure and physiology ensures dispersal and is well provided with food reserves to nourish the growing seedling until it establishes itself as a self-sufficient, autotrophic organism. In some scientific publications, the term germination is used loosely and sometimes incorrectly, and so it is important to clarify its meaning. Germination begins with water uptake by the dry seed and ends with the start of elongation by embryonic axis, usually the radicle (Bewley and Black, 1994). Water uptake by the mature dry seed is triphasic, the first phase involving rapid water uptake followed by a plateau phase; the final phase involves increase in water uptake after germination is completed, as the radicle elongates (Bewley, 1997; Manz *et al.*, 2005). One of the first observations upon imbibition is the resumption of respiratory activity, which can be detected within minutes. After a steep initial increase in oxygen consumption, the rate declines until the radicle penetrates the surrounding structures. This is followed by increased respiratory activity (Salon *et al.*, 1988; Botha *et al.*, 1992).

Many viable seeds fail to germinate under favourable environmental growing conditions. Such seeds are referred to as dormant. It may not be advantageous for a seed to germinate freely, even in seemingly favourable conditions. For example, germination of annuals in the spring allows time for vegetative growth and the subsequent production of offspring, whereas germination in similar conditions in the autumn could lead to the extinction of the vegetative plant during the winter. Seed dormancy is generally an undesirable characteristic in agricultural crops, where rapid germination and growth are required. However, some level of dormancy is desirable, at least during seed development. Therefore, dormancy is an adaptive trait that optimizes the distribution of germination over time in a population of seeds.

Defining seed dormancy is difficult because dormancy can only be measured by the absence of germination. There is therefore no agreement about the definition of seed dormancy. The array of ideas about dormancy is revealed by the number of classifications of dormancy employed by various authorities. Many authors have different views on dormancy, and in some instances there are contradictions. Seed dormancy has been defined as a block to the completion of germination of an intact viable seed under favourable conditions (Hilhorst, 1995; Bewley, 1997; Li and Foley, 1997). Vleeshouwers *et al.* (1995) defined dormancy as a seed characteristic, the degree of which defines what conditions should be met to make the seed germinate. Baskin and Baskin

(2004) defined a dormant seed as one that does not have the capacity to germinate in a specified period of time under any combination of normal physical environmental factors that are otherwise favourable for its germination.

Seed dormancy is an innate seed property that sets limits to the environmental conditions that must be met before the seed can germinate. It maximizes seedling survival by preventing germination under unfavourable conditions. Seed dormancy and germination in higher plants are controlled by complex adaptive traits that are influenced by a large number of genes, environmental factors such as light and temperature, the duration of seed storage (after ripening) and the plant hormones abscisic acid (ABA) and gibberellin (GA) (Koornneef *et al.*, 2002).

Dormancy has evolved differently across species through adaptation to the prevailing environment, so that germination occurs when conditions for establishing a new plant generation are likely to be suitable (Baskin and Baskin, 2004). A varied range of dormancy mechanisms have evolved due to the diversity of climates and environments in which they are sited. Dormancy should not just be associated with the absence of germination; rather, it is a characteristic of the seed that determines the conditions required for germination. Studies of genetics and physiology have shown the important roles of the plant hormones ABA and GA in the regulation of dormancy and germination. The use of quantitative genetics and the mutant approach has allowed further genetic elucidation of these traits and identification of previously unknown components (Koornneef *et al.*, 2002). Seed dormancy is often confused with seed persistence in the soil or on the plant, though even in scientific publications dormancy and persistence have been used interchangeably (Baskin and Baskin, 2004).

24.2 Classification of Seed Dormancy

Primary versus secondary dormancy

Many classification systems for seed dormancy have been suggested and used. Some of these classifications are complex while others are simple. A generally accepted distinction made in dormancy studies is that of primary versus secondary dormancy, which is based on the timing of dormancy onset rather than on the cause. Seeds that are released from the plant in a dormant state are said to exhibit primary dormancy; seeds that are released from the plant in a non-dormant state but which become dormant if the conditions for germination are unfavourable exhibit secondary dormancy (Bewley and Black, 1994; Hilhorst, 1995). Once primary dormancy is lost in response to prevailing environmental conditions, secondary dormancy will soon start to be induced if the conditions required to terminate dormancy and induce germination are absent. Freshly harvested mature, water-permeable dormant seeds are said to have primary dormancy, which has been induced with the involvement of ABA during seed maturation on the mother plant (Hilhorst, 1995; Kucera *et al.*, 2005).

Coat-imposed and embryo dormancy

The second system is seed coat-imposed and embryo dormancy, or both. This classification is based on the mechanism or location of constraints to germination.

Hierarchical system of dormancy

A more comprehensive and complex system was devised by Nikolaeva (1969), and adopted by Baskin and Baskin (1998, 2004). This hierarchical system includes five classes of dormancy, each subdivided into levels and types: physiological, morphological, morphophysiological, physical and combinational.

Physiological dormancy (PD)

Physiological dormancy is the most abundant form and is found in seeds of gymnosperms and all major angiosperms. It is the most prevalent dormancy form in temperate seed banks and the most abundant dormancy class in the field. Physiological dormancy is also the major form of dormancy in most seed model species in the laboratory. Seeds with physiological dormancy may only re-enter dormancy, that is, secondary dormancy, after primary dormancy has been interrupted (Baskin and Baskin, 1998). Physiological dormancy can be divided into three levels: deep, intermediate and non-deep (Baskin and Baskin, 2004). In deep dormancy, the embryos excised from these seeds either do not grow or will produce abnormal seedlings. Treatment with GA does not interrupt dormancy, and several months of cold or warm stratification are required before germination can take place (Baskin and Baskin, 2004). The great majority of seeds have non-deep physiological dormancy. Embryos excised from these seeds produce normal seedlings; GA treatment can interrupt this dormancy and, depending on species, dormancy can also be broken by scarification, after-ripening in dry storage, and cold or warm stratification.

Morphological dormancy (MD)

Morphological dormancy is evident in seeds with embryos that are underdeveloped (in terms of size), but differentiated into cotyledons and hypocotyl-radical. These embryos are not (physiologically) dormant, but simply need time to grow and germinate.

Morpho-physiological dormancy (MPD)

Morpho-physiological dormancy is also evident in seeds with underdeveloped embryos, but in addition these have a physiological component to their dormancy (Baskin and Baskin, 2004). These seeds therefore require a dormancy-interrupting treatment, for example a defined combination of warm and/or cold stratification which, in some cases, can be replaced by GA application.

Physical dormancy (PY)

Physical dormancy is caused by water-impermeable layers of palisade cells in the seed or fruit coat that control water movement. Mechanical or chemical scarification can interrupt PY dormancy.

Combinational dormancy (PY + PD)

Combinational dormancy is evident in seeds with water-impermeable coats (as in PY) combined with physiological embryo dormancy (Baskin and Baskin, 2004). Only a few species exhibit some degree of physiological dormancy once the seed or fruit coat becomes permeable to water, and seed can germinate over a wide range of temperatures in both light and darkness (Baskin and Baskin, 1998).

24.3 Dormancy in Legumes

The family Leguminosae has been found to have fully developed embryos (Baskin and Baskin, 1998), with some having an impermeable seed coat. These include the three subfamilies Caesalpinioideae, Mimosoideae and Papilionoideae. According to Corner (1951), the hardness and impermeability of the dried testa of the seeds of Leguminosae is caused mainly by the contraction of the walls of the palisade layer as the seed ripens. Egley (1989) also observed that heavily lignified cell walls may make the palisade layer impermeable to water. For seeds with physical dormancy to germinate, the water-impermeable layer(s) must become permeable, thereby allowing passage of water to the embryo. Contrary to many reports that the lens is the site of water entry into legume seeds, Egley (1979) found that covering the lens of hot water-pretreated seeds of the legume *Crotalaria spectabilis* with petroleum jelly did not prevent germination, implying that water uptake occurs in region(s) of the seed coat other than the lens. Morrison *et al.* (1998) showed that dry-heating caused disruption of the seed coat only at the lens in some legumes; in others an area on the seed coat, in addition to the lens region, was disrupted by dry heat. Seeds disrupted only at the lens had a thinner testa, thicker palisade layer and a thinner mesophyll layer (Morrison *et al.*, 1998).

A hard seed coat contributes to the viability of stored seeds. Dormancy and viability can be maintained for long periods in hard-seeded soybean accessions because their seed coats are impermeable to water (Rolston, 1978). In soybean, while some authors concluded that water reaches the embryo mainly through the testa (Noodén *et al.*, 1985; Chachalis and Smith, 2000), others suggested that water entrance occurs mainly through the hilar area (McDonald *et al.*, 1988a, b). On the other hand, in comparison with non-black seed-coated cultivars, black-coated soybean seeds have slower initial imbibition rates (Kuo, 1989; Chachalis and Smith, 2000), higher resistance to field deterioration (Tully *et al.*, 1981; Mugnisjah *et al.*, 1987), a tougher testa (Tully *et al.*, 1981), higher lignin contents and fungicidal properties (Krzyzanowski *et al.*, 1999).

In legumes, white seeds imbibe water more rapidly than coloured seeds and then germinate earlier. White seeds also suffer greater imbibition damage, as measured by higher solute leakage, which affects their vigour and viability (Powell, 1989; Kantar *et al.*, 1996). It was hypothesized that during dehydration of seeds, an enzymatic oxidation of phenolic compounds in the presence of oxygen might render the seed coat impermeable to water (Marbach and Mayer, 1975). Several mechanisms may explain how the chemical and structural composition of the testa determines the germination capacity of seeds. The oxidized flavonoid polymers may play a major role in limiting not only water entry, as seen in legumes, but also oxygen supply to the embryo, for example, as reported by Corbineau and Côme (1993) for cereals, and by contributing to the mechanical resistance of the testa. They may also inhibit the leaching of germination inhibitors out of the seed.

Sefa-Dedeh and Stanley (1979) concluded that during the first 3–12 h, hilar size was the most important controller of imbibition in cowpea (*Vigna unguiculata*) seeds, while the percentage of protein in the cotyledon was important between 12 and 24 h of imbibition. However, since the seeds absorbed nearly 80% in the first 3 h, it was concluded that thickness of the seed coat was the most important factor.

The hilum, the micropyle and the raphe have variously been suggested as paths of water imbibition in *Phaseoulus lunatus* (Korban *et al.*, 1981) and *Phaseolus vulgaris* (Agbo *et al.*, 1987); Korban *et al.* (1981) associated these differences with different cultivars. In a large number of legume species, seed coat colour changes during storage according to prevailing environmental conditions; in these cases, changes occur along with seed physiological deterioration and increase in seed coat permeability (Silva *et al.*, 1988). Seeds with unpigmented seed coat deteriorate more rapidly and are more susceptible to imbibition damage (Abdullah *et al.*, 1991; Asiedu and Powell, 1998). In fact, an association between rapid imbibition and white or partially white-coated seeds has been observed in cultivars of a large number of legume species; seeds of other colours tend to absorb slowly. This property has been attributed to seed coat permeability, adherence of the seed coat to cotyledons and thickness of the testa (Legesse and Powell, 1996). The seed coat of lima bean is impermeable to water, and hydration in wild lima bean is restricted to the hilum (Degreef *et al.*, 2002). These authors found that lima bean lacks innate dormancy, but that dormancy can be induced by high temperatures and low humidity. Moreover, germination and interruption of dormancy are controlled by the passage of water through the hilum.

Seed dormancy in groundnut has been shown to be affected by several factors. The seed coat of some groundnut genotypes may prevent germination by preventing water uptake or gas exchange, mechanically restraining the growth of the embryo, chemically inhibiting germination or acting as a barrier to photoreception (Hull, 1937). Groundnut has an indeterminate flowering pattern and therefore pods of the same plants vary in their maturity. Toole *et al.* (1964) observed that immature groundnut seeds have a long dormancy period, which declined as maturity progresses. Removal of the seed coat in groundnut has been found to improve seed germination. Toole *et al.* (1964) demonstrated that removal of the seed coat resulted in the loss of seed dormancy. Patil (1967) confirmed these results when he found that removal of the seed coat after 60 days of flowering slightly improved seed germination. Hammons (1973), however, contends that seed dormancy in groundnut is an inherent property of the seed and does not depend on an impervious or protective seed coat. Hormonal balance between ABA, which acts as germination inhibitor, and ethylene, which acts as a germination activator, is produced by the embryo through the action of cytokinin during seed imbibition, and the release of these chemicals varies for different genotypes of groundnut (Ketring and Morgan, 1971, 1972). Depending on the genetic constitution, different seed parts (coat, cotyledon and embryo) have also variously been reported to have a role in imparting dormancy in groundnut (Nautiyal *et al.*, 1994).

24.4 Plant Growth Regulators

In plants, ABA is synthesized through the cleavage and oxidation of carotenoids, and is catabolized through hydroxylation or by conjugation to glucose (Nambara and Marion-Poll, 2005). Dormant seeds treated with fluridone (a compound that inhibits carotenoid and, thus, ABA synthesis) have been found to have similar germination characteristics to non-dormant seeds, demonstrating that the continued production of ABA is required for dormancy maintenance in imbibed seeds of several species (Ali-Rachedi *et al.*, 2004; Kusumoto *et al.*, 2006; Feurtado *et al.*, 2007). There is substantial inferred evidence that ABA is involved in regulating the commencement of dormancy and in sustaining the dormant state, although there is a scarcity of unambiguous evidence that, physiologically, ABA really is an important controlling factor in the dormancy of most seeds (Bewley, 1997). Proof supporting the role of ABA in seed dormancy of many species is provided by a number of studies (Gubler *et al.*, 2005; Kermode, 2005; Kucera *et al.*, 2005). These studies concluded that exogenous ABA delays or blocks germination of seeds and embryos; in the immature seed, ABA maintains the embryo in a developing rather than germinating programme, so that seeds do not germinate while still attached to the mother plant; differences in susceptibility to preharvest sprouting have been linked to ABA content; the capability of the seed to synthesize ABA is necessary to acquire dormancy, so that dormancy is not established in seeds that are deficient in ABA because of mutation, transgenic modification or chemical inhibition of ABA synthesis; chemical inhibition of ABA synthesis also enhances the germination of previously dormant seeds; on the other hand, overexpression of genes that increase ABA content also delays germination; soon after incubation at germination temperatures, a greater decrease of ABA occurs in non-dormant seeds as opposed to dormant ones.

However, the direct contribution of ABA to the physiological modulation of the dormancy level is debatable because for the following reasons: (i) vivipary is a phenomenon distinct from lack of dormancy in the mature seed and, in fact, it also occurs in non-dormant species such as maize (Bewley and Black, 1994; Gianinetti and Vernieri, 2007); (ii) a number of plants have high ABA levels in the seed but show no dormancy (Bewley and Black, 1994); and (iii) in species with seed dormancy, inconsistent relationships between dormancy intensity in the mature grain and its ABA content have been observed (Kermode, 2005; Millar *et al.*, 2006). Studies on many species indicate that the synthesis of ABA after imbibition is a feature of the dormant seed (Kermode, 2005), although it can also occur in non-dormant seed (Bewley and Black, 1994). The change in ABA turnover can be seen as a consequence of dormancy removal by after-ripening, and does not appear to support the view that ABA is the cause of dormancy (Gianinetti and Vernieri, 2007).

The ratio of ABA:GA is an important determinant of germination (Finch-Savage and Leubner-Metzger, 2006). Although GA can stimulate germination of dormant seeds in some species, there are many instances where GA alone is ineffective, and it has been suggested that GA is necessary but not sufficient for dormancy release (Finkelstein *et al.*, 2008). There is also evidence that GA mediates the metabolism of ABA, and vice versa (Gonai *et al.*, 2004; Gubler *et al.*, 2008).

24.5 Release of Dormancy

Seed dormancy can be overcome by germination-promoting factors such as after-ripening, light and imbibed seed treatments such as chilling, warm stratification, light, gibberellins and other hormones (Kucera *et al.*, 2005) and smoke substances such as butenolide (Krock *et al.*, 2002). Furthermore, several compounds are known as being important stimulants of germination, a number of which are nitrogen (N)-containing compounds, including nitric oxide gas (NO), nitrite (NO_2^-) and nitrate (NO_3^-) (Alboresi *et al.*, 2005; Bethke *et al.*, 2007a). Bethke *et al.* (2007a) suggested that all N compounds affect germination through conversion to NO. Enzymatic NO production occurs mainly via nitrate reductase as a by-product of lipid catabolism or nitric oxide synthase (Crawford and Guo, 2005). Non-enzymatic conversion of nitrite to NO has also been demonstrated and was suggested to have special significance for seeds (Bethke *et al.*, 2007b).

Moreover, applied chemicals such as gibberellins also have germination-promoting effects (Ali-Rachedi *et al.*, 2004; Alboresi *et al.*, 2005). Auxins are known to play important roles in embryogenesis, although their role in the regulation of germination and seedling establishment remains unclear (Kucera *et al.*, 2005). Auxin alone is generally not considered to be important in the control of seed germination, but interaction between auxin, ABA, GA and ethylene was suggested to affect both germination and seedling establishment (Ogawa *et al.*, 2003; Carrera *et al.*, 2008).

None of these environmental factors are an absolute requirement for germination because the need for one factor depends on the other factors, as shown in the interaction between light and temperature by Cone and Spruit (1983). This requirement for exogenous factors depends very much on the genotype. It appears that the sensitivity of seeds to cold, nitrate and light is dependent upon the length of time over which they have been dry after-ripened. The seeds first become sensitive to nitrate, then to cold and finally to light (Finch-Savage *et al.*, 2007). Furthermore it has been shown that the rate of increase of sensitivity to environmental signals is not constant, seeds produced in different years having different responses. This is due to the fact that the depth of dormancy is determined not only by the genetic constitution of the crop, but also by the surrounding environment during seed formation (Donohue, 2005).

Several studies have analysed the expression of the genome in *Arabidopsis* seeds, at both the transcriptome and proteome level (Holdsworth *et al.*, 2008); these studies have identified a major role for translation in germination and dormancy release. Other studies have shown that important transcript and protein changes are happening in dry seeds during storage and that these changes might be targeted to release of dormancy as seeds after-ripen. These studies also indicate that the accumulation of both specific proteins (Chibani *et al.*, 2006) and gene transcripts (Bove *et al.*, 2005; Leubner-Metzger, 2006) can occur. The dry state, as described by Holdsworth *et al.* (2008), is mature seeds containing 5–10% water, depending on the species. This residual water is not uniformly distributed in the seed tissues, indicating that some areas may contain enough water to support gene expression (Leubner-Metzger, 2006).

References

Abdullah, W.D., Powell, A.A. and Matthews, S. (1991) Association of differences in seed vigour in long bean (*Vigna sesquipedalis*) with testa colour and imbibition damage. *Journal of Agricultural Science* 116, 259–264.

Agbo, G.N., Hosfield, M.A., Uebersax, M.A. and Klomparens, K. (1987) Seed microstructure and its relationship to water uptake in isogenic lines and a cultivar of dry beans (*Phaseolus vulgaris* L.). *Food Microstructure* 6, 91–102.

Alboresi, A., Gestin, C., Leydecker, M.T., Bedu, M., Meyer, C. and Truong, H.N. (2005) Nitrate, a signal relieving seed dormancy in Arabidopsis. *Plant Cell Environment* 28, 500–512.

Ali-Rachedi, S., Bouinot, D., Wagner, M.H., Bonnet, M., Sotta, B., Grappin. P. *et al.* (2004) Changes in endogenous abscisic acid levels during dormancy release and maintenance of mature seeds: studies with the Cape Verde Islands ecotype, the dormant model of *Arabidopsis thaliana*. *Planta* 219, 479–488.

Asiedu, E.A. and Powell, A.A. (1998) Comparisons of the storage potential of cultivars of cowpea (*Vigna unguiculata*) differing in seed coat pigmentation. *Seed Science and Technology* 26, 211–221.
Baskin, C.C. and Baskin, J.M. (1998) *Seeds: Ecology, Biogeography, and Evolution of Dormancy and Germination*. Academic Press, San Diego, California.
Baskin, J.M. and Baskin, C.C. (2004) A classification system for seed dormancy. *Seed Science Research* 14, 1–16.
Bethke, P.C., Libourel, I.G. and Jones, R.L. (2007a) Nitric oxide in seed dormancy and germination. In: Bradford, K. and Nonogaki, H. (eds) *Seed Development, Dormancy and Germination*. Blackwell Publishing, Oxford, UK, pp. 153–171.
Bethke, P.C., Libourel, I.G., Aoyama, N., Chung,Y.Y., Still, D.W. and Jones, R.L. (2007b) The *Arabidopsis* aleurone layer responds to nitric oxide, gibberellin, and abscisic acid and is sufficient and necessary for seed dormancy. *Plant Physiology* 143, 1173–1188.
Bewley, J. (1997) Seed germination and dormancy. *Plant Cell* 9, 1055–1066.
Bewley, J.D. and Black, M. (1994) *Seeds: Physiology of Development and Germination* (2nd edn). Plenum, New York.
Botha, F.C., Potgieter, G.P. and Botha, A.M. (1992) Respiratory metabolism and gene expression during seed germination. *Journal of Plant Growth Regulators* 11, 211–224.
Bove, J., Lucas, P., Godin, B., Oge, L., Jullien, M. and Grappin, P. (2005) Gene expression analysis by cDNA-AFLP highlights a set of new signaling networks and translational control during seed dormancy breaking in *Nicotiana plumbaginifolia*. *Plant Molecular Biology* 57, 593–612.
Carrera, E., Holman, T., Medhurst, A., Dietrich, D., Footitt, S., Theodoulou, F.L. *et al.* (2008) Seed after-ripening is a discrete developmental pathway associated with specific gene networks in *Arabidopsis*. *Plant Journal* 53, 214–224.
Chachalis, D. and Smith, M.L. (2000) Imbibition behavior of soybean (*Glycine max* (L.) Merril) accessions with different testa characteristics. *Seed Science and Technology* 28, 321–331.
Chibani, K., Ali-Rachedi, S., Job, C., Job, D., Jullien, M. and Grappin, P. (2006) Proteomic analysis of seed dormancy in *Arabidopsis*. *Plant Physiology* 142, 1493–1510.
Cone, J.W. and Spruit, C.J.P. (1983) Imbibition conditions and seed dormancy of *Arabidopsis thaliana*. *Physiology of Plants* 59, 416–420.
Corbineau, F. and Côme, D. (1993) The concept of dormancy in cereal seeds. In: *Proceedings of the 4th International Workshop on Seeds, Basic and Applied Aspects of Seed Biology*, 20–24 July, Angers, France.
Corner, E.J.H. (1951) The leguminous seed. *Phytomorphology* 1, 117–150.
Crawford, N.M. and Guo, F.Q. (2005) New insights into nitric oxide metabolism and regulatory functions. *Trends in Plant Science* 10, 195–200.
Degreef, J., Rocha, O.J., Vanderborght, T. and Baudoin, J.P. (2002) Soil seed bank and seed dormancy in wild populations of lima (*Fabaceae*): Considerations for *in situ* and *ex situ* consideration. *American Journal of Botany* 89, 1644–1650.
Donohue, K., Dorn, L., Griffith, C., Kim, E., Aguilera, A., Polisetty, C.R. *et al.* (2005) Environmental and genetic influences on the germination of *Arabidopsis thaliana* in the field. *Evolution* 59, 740–757.
Egley, G.H. (1979) Seed coat impermeability and germination of showy crotolaria (*Crotolaria spectabilis*) seeds. *Weed Science* 27, 355–361.
Egley, G. H. (1989) Water-impermeable seed coverings as barriers to germination. In: Taylorson, R.B. (ed.) *Recent Advances in the Development and Germination of Seeds*. Plenum Press, New York, pp. 207–223.
Feurtado, J.A., Yang, J., Ambrose, S.J., Cutler, A.J., Abrams, S.R. and Kermode, A.R. (2007) Disrupting abscisic acid homeostasis in western white pine (*Pinus monticola* Dougl. Ex D. Don) seeds induces dormancy termination and changes in abscisic acid catabolites. *Journal of Plant Growth Regulation* 26, 46–54.
Finch-Savage, W.E. and Leubner-Metzger, G. (2006) Seed dormancy and the control of germination. *New Phytologist* 171, 501–523.
Finch-Savage, W.E., Cadman, C.S., Toorop, P.E., Lynn, J.R. and Hilhorst, H.W. (2007) Seed dormancy release in *Arabidopsis* Cvi by dry after-ripening, low temperature, nitrate and light shows common quantitative patterns of gene expression directed by environmentally specific sensing. *The Plant Journal* 51, 60–78.
Finkelstein, R., Reeves, W., Ariizumi, T. and Steber, C. (2008) Molecular aspects of seed dormancy. *Annual Review of Plant Biology* 59, 387–415.
Gianinetti, A. and Vernieri, P. (2007) On the role of abscisic acid in seed dormancy of red rice. *Journal of Experimental Biology* 58, 3449–3462.
Gonai, T., Kawahara, S., Tougou, M., Satoh, S., Hashiba, T., Hirai, N. *et al.* (2004) Abscisic acid in the thermoinhibition of lettuce seed germination and enhancement of its catabolism by gibberellin. *Journal of Experimental Botany* 55, 111–118.

Gubler, F., Millar, A.A., Jacobsen, J.V. (2005) Dormancy release, ABA and pre-harvest sprouting. *Current Opinion in Plant Biology* 8, 183–187.

Gubler, F., Hughes, T., Waterhouse, P. and Jacobsen, J. (2008) Regulation of dormancy in barley by blue light and after-ripening: effects on abscisic acid and gibberellin metabolism. *Plant Physiology* 147, 886–896.

Hammons, R.O. (1973) Genetics of *Arachis hypogaea*. In: Wilson, C.T. (ed.) *Peanut: Culture and Uses*. Peanut Research and Education Association, Stillwater, Oklahoma, pp. 135–173.

Hilhorst, H.W.M. (1995) A critical update on seed dormancy. I. Primary dormancy. *Seed Science Research* 5, 61–73.

Holdsworth, M.J., Finch-Savage, W.E., Grappin, P. and Job, D. (2008) Post-genomics dissection of seed dormancy and germination. *Trends in Plant Science* 13, 7–13.

Hull, F.H. (1937) *Inheritance of Rest Periods of Seeds and Certain Other Characteristics in the Peanut*. Technical Bulletin No. 314. Florida Agricultural Experiment Station, Gainesville, Florida, pp. 317.

Kantar, F., Pilbeam, C.J. and Hebblethwaite, P.D. (1996) Effect of tannin content of faba bean (*Vicia faba*) seed on seed vigour, germination and field emergence. *Annals of Applied Biology* 128, 85–93.

Kermode, A.R. (2005) Role of abscisic acid in seed dormancy. *Journal of Plant Growth Regulation* 24, 319–344.

Ketring, D.L. and Morgan, P.W. (1971) Physiology of oilseed. II. Dormancy release in Virginia-type peanut seeds by plant growth regulators. *Plant Physiology* 47, 488–492.

Ketring, D.L. and. Morgan, P.W (1972) Physiology of oilseed. IV. Role of endogenous ethylene and inhibitory regulators during natural and induced after ripening of dormant Virginia-type peanut seeds. *Plant Physiology* 50, 382–387.

Koornneef, M., Bensink, L. and Hilhorst, H. (2002) Seed dormancy and germination. *Current Opinion in Plant Biology* 5, 33–36

Korban, S.S., Coyne, D. and Weihing, J.L. (1981) Rate of water uptake and sites of water entry in seeds of different cultivars of dry bean. *Horticultural Science* 16, 545–546.

Krock, B., Schmidt, S., Hertweck, C. and Baldwin, I.T. (2002) Vegetation derived abscisic acid and four terpenes enforce dormancy in seeds of the post-fire annual, *Nicotiana attenuata. Seed Science Research* 12, 239–252.

Krzyzanowski, F.C., Franca Neto, J.B., de Kaster, M. and Mandarino, J.M.G. (1999) Metodologia para seleção de genótipos de soja com semente resistente ao dano mecânico – relação com o conteúdo de lignina. In: *Resultados de Pesquisa da Embrapa Soja*. 1998. EMBRAPA/CNPSo, Londrina, Brazil, pp. 213–214.

Kucera, B., Cohn, M.A. and Leubner-Metzger, G. (2005) Plant hormone interactions during seed dormancy release and germination. *Seed Science Research* 15, 281–307.

Kuo, W.H.J. (1989) Delayed permeability of soybean seeds: characteristics and screening methodology. *Seed Science and Technology* 17, 131–142.

Kusumoto, D., Chae, S.H., Mukaida, K., Yoneyama, K., Yoneyama, K., Joel, D.M. *et al.* (2006) Effects of fluridone and norflurazon on conditioning and germination of *Striga asiatica* seeds. *Plant Growth Regulation* 48, 73–78.

Legesse, N. and Powell, A.A. (1996) Relationship between the development of seed coat pigmentation, seed coat adherence to the cotyledons and the rate of imbibition during the maturation of grain legumes. *Seed Science and Technology* 24, 23–32.

Leubner-Metzger, G. (2006) Hormonal interactions during seed dormancy release and germination. In: Basra, A.S (ed.) *Handbook of Seed Science and Technology*. Haworth's Food Products Press, Binghamton, New York, pp. 303–341.

Li, B. and Foley, M.E. (1997) Genetic and molecular control of seed dormancy. *Trends in Plant Science* 2, 384–389.

Manz, B., Muller, K., Kucera, B., Volke, F. and Leubner-Metzger, G. (2005) Water uptake and distribution in germinating tobacco seeds investigated in vivo by nuclear magnetic resonance imaging. *Plant Physiology* 138, 1538–1551.

Marbach, I. and Mayer A.M. (1975) Changes in catechol oxidase and permeability to water in seed coats of *Pisum elatius* during seed development and maturation. *Plant Physiolology* 56, 93–96.

McDonald Jr., M.B., Vertucci, C.W. and Roos, E.C. (1988a) Soybean seed imbibition: water absorption by seed parts. *Crop Science* 28, 993–997.

McDonald Jr., M.B., Vertucci, C.W. and Roos, E.C. (1988b) Seed coat regulation of soybean imbibition. *Crop Science* 28, 987–992.

Millar, A.A., Jacobsen J.V., Ross, J.J, Helliwell, C.A., Poole, A.T., Scofield, G. *et al.* (2006) Seed dormancy and ABA metabolism in *Arabidopsis* and barley: the role of ABA 8'-hydroxylase. *The Plant Journal* 45, 942–954.

Morrison, D.A., McClay, K., Porter, C. and Rish, S. (1998) The role of the lens in controlling heat-induced breakdown of testa-imposed dormancy in native Australian legumes. *Annals of Botany* 82, 35–40.
Mugnisjah, W.Q., Shimano, I. and Matsumoto, S. (1987) Studies on the vigor of soybean seeds. II. Varietal differences in seed coat quality and swelling components on seed during moisture imbibition. *Journal of the Faculty of Agriculture, Kyushu University* 3, 227–234.
Nambara, E. and Marion-Poll, A. (2005) Abscisic acid biosynthesis and metabolism. *Annual Review of Plant Biology* 56, 165–185.
Nautiyal, P.C., Ravindra, V. and Bandyopadhyay, A. (1994) Peanut seed dormancy. *ACIAR Food Legume Newsletter* 21, 2.
Nikolaeva, M.G. (1969) *Physiology of Deep Dormancy in Seeds*. National Science Foundation, Washington, DC.
Nooden, L.D., Blakley, K.A. and Grzybowski, J.M. (1985) Control of seed coat thickness and permeability in soybean. *Plant Physiology* 79, 543–545.
Ogawa, M., Hanada, A., Yamauchi, Y., Kuwahara, A., Kamiya, Y. and Yamaguchi, S. (2003) Gibberellin biosynthesis and response during *Arabidopsis* seed germination. *Plant Cell* 15, 1591–1604.
Patil, V.V. (1967) Dormancy studies in rice and groundnut. MSc (Agriculture) thesis, University of Poona, Poona, India.
Powell, A.A. (1989) The importance of genetically determined seed coat characteristics to seed quality in grain legumes. *Annals of Botany* 63, 169–175.
Rolston, M. (1978) Water impermeable seed dormancy. *Botanical Review* 44, 365–396.
Salon, C., Raymond, P. and Pradet, A. (1988) Quantification of carbon fluxes through the tricarboxylic acid cycle in early germinating lettuce embryos. *Journal of Biological Chemistry* 263, 12278–12287.
Sefa-Dedeh, S. and Stanley, D.W. (1979) The relationship of microstructure of cowpeas to water absorption and dehulling properties. *Cereal Chemistry* 56, 379–386.
Silva, A. A., Caemello, S.M. and Nakagawa, J. (1988) Germinação e vigor sementes de *Crotalaria lanceolata* E.Mey. I. Influência da cor do tegumento e da posição dos frutos na infrutescência. *Revista Brasileira de Sementes* 10, 67–73.
Toole, V.K., Bailey, W.K. and Toole, E.H. (1964) Factors influencing dormancy of peanut seeds. *Plant Physiology* 39, 822–832.
Tully, R.E., Musgrave, M.E. and Leopold, A.C. (1981) The seed coat as a control of imbibitional chilling injury. *Crop Science* 21, 312–317.
Vleeshouwers, M., Bouwmeester, H.J. and Karssen, C.M. (1995) Redefining seed dormancy: an attempt to integrate physiology and ecology. *Journal of Ecology* 83, 1031–1037.

25 Postharvest Technology

A.P. Rodiño, J. Kumar, M. De La Fuente, A.M. De Ron and M. Santalla

25.1 Introduction

The grain legumes constitute an important dietary constituent for humans and animals. They are considered as 'vegetarian's meat' because of the high protein content (15–34%, depending on species), which is double that of wheat and three times that of rice. The grain legumes are generally consumed as natural food products in form of whole grains or de-hulled or split grains. Therefore, size and shape of seeds, seed coat appearance, colour, cotyledon colour and uniformity are important for markets. Besides postharvest maintenance of grain legumes during storage, processing and marketing help to determine quality and minimize yield losses. Most of the commercial technologies available for this purpose are either obsolete or inadequate and result in heavy losses due to breakage and powdering of the grain. Successful efforts have been made to develop improved technologies to reduce losses and improve product quality. Postharvest technologies help agro-industries in making legume grains more preferable to consumers. Moreover, development of this industry would provide additional rural employment, improve nutritional standards, bring a better price to the grower and ensure supplies at lower prices to the consumer.

Postharvest processing of grain legumes necessitates a series of mechnical separations, milling operations and modifications for preparing the processed foods for consumption. The grain legume processing industry incorporates three major components.

1. Primary processing, i.e. cleaning, drying, storage, packaging, etc. Raw materials are purified by removal of foreign matter and immature grain, and are then prepared for grading and secondary processing.
2. The secondary processing stage mostly involves de-hulling, splitting, sorting and polishing of grains, i.e. processing of the raw material into products suitable for food uses or consumption after cooking, roasting, frying, etc.
3. Tertiary processing, which involves further processing of legume grains into useful food products, i.e. value addition and creation of ready-to-eat forms.

In this chapter, the first two components of postharvest processing are dealt with while the third is discussed in detail in Chapter 26.

25.2 Basic Considerations for Postharvest Processing

Preharvest management

Environmental conditions experienced during the growing season may change the physical properties of grains, leading to loss of quality and quantity during postharvest processing. Preharvest management of crops can minimize these losses and improve the grain quality. For example in lentil, alternating wet and dry periods during late stages of maturity can cause wrinkling, colour loss and brittle seed coat which, in turn, contribute to damage during storage and postharvest handling. Chemical desiccation of the crop prior to harvest may also be a factor in the finding of brittle seeds coats that chip and spilt more easily (Vandenberg, 2009). Quality in cowpea becomes a serious problem if rainfall occurs soon after maturity, and therefore preharvest fungicide sprays can be of benefit in preventing this. For avoidance of the grain damage, this crop should be harvested well before too much drying can occur and the drum speed kept low (250–300 rpm) in mechanized harvesting, to avoid splitting and cracking of grains (Cameron, 1999). Thus preharvest management of legume grain crops should be taken into consideration for obtaining improved quality and quantity in postharvest products.

Threshing

Primary processing starts with pod threshing and de-hulling of the whole seeds. In India, the threshing of food legumes is usually carried out manually using long, wooden sticks, thereby increasing grain damage. Advances in postharvest technology have made available threshing machinery that separates the seeds from the pods. However, it is necessary that crops have optimum moisture content in their grains in order to minimize threshing damage. It has been observed in harvested soybean of <12% moisture that losses through mechanical damage may be high, while in chickpea the degree of dryness determines its susceptibility to cracking. Therefore, the careful setting of drum speeds, conveyors and augers on threshing machines is required (http://www2.dpi.qld.gov.au/fieldcrops/3251.html).

Storage conditions

Before primary processing, the harvested grain legumes are stored in steel bins, jute bags, earthen pots, mud bins, bamboo baskets or in other types of receptacles at farmer, trader and industry levels. Appropriate technologies for handling and storage have been developed (Ali and Srivastava, 1993). It has been observed that the initial conditions of the storage environment may have an impact on processing quality of legume grains: high temperature, humidity, light and moisture content are all involved in the changing the seed coat colour (an important factor in marketing of grain legumes for consumption). For example, when the seed coat of green lentil is exposed to moderately high temperatures (20–30°C) at high relative humidity (RH) (100%), it slowly turns from green to an oxidized brown colour in 3 weeks or less. Studies have been conducted to optimize moisture levels during storage (Cenkowski *et al.*, 1989; Tang *et al.*, 1994; Juak-Knieval *et al.*, 2008), and it has been observed that a cool temperature (5°C) with 100% RH can prevent browning for up to 5 weeks (Nordstorm and Sistrunk, 1979; Nozzolillo and De Bezada, 1984). Storage of beans (*Phaseolus vulgaris* L.) at 1°C and 30% RH retained their original colour for one year, while 24°C and 80% RH accelerated darkening (Hughes and Sandsted, 1975; Edmister *et al.*, 1990; Gunes and Lee, 1997). Even at a moderately low temperature (10°C), darkening was slow in adzuki bean (*Vigna angularis*; Yousif *et al.*, 2003).

In faba bean (*Vicia faba* L.), postharvest darkening reduces its market value (Hughes and Sandsted, 1975); it has been observed that faba bean seeds darken rapidly and phenolic content falls when stored at higher temperature, moisture and light intensity (Nasar-Abbas *et al.*, 2009). In chickpea also, temperatures ranging from 33 to 35°C and 75% RH for 160 days caused darkening of the testa (Reyes-Moreno *et al.*, 1987). The moisture content during storage is one of the important factors affecting the quality of grain for primary and secondary processing; serious losses in milling quality may result during shelling of groundnut when kernels are dried below 7% moisture content (by weight) or stored at a temperature below 7°C. Therefore, for bulk storage of unshelled groundnuts, it is important to maintain at least 7.5% kernel moisture content and store them at 10°C and 65% RH. Under these conditions, unshelled groundnuts can be stored without significant loss in quality for about 10 months (http://www.vasat.icrisat.org/?q=node/190). In North America, lentil matures during the cooler season, and therefore it becomes necessary to reduce the moisture content immediately after harvest using aeration fans in the bins. In general, storage of pulses (legume grain crops) is safe at 10–20% moisture (Ali, 2004).

Mechanical dryers using biomass, solar energy and electrical energy are now available, but sun drying is the most common practice. Farmers expose their legume grains to sunshine at various intervals after storage to reduce moisture content (Giga *et al.*, 1992). Alternatively, solarization by solar heaters can be used where lethal temperatures are required to kill all insects. For example, 65°C is required to kill all life stages of *Callosobruchus maculatus*, the cowpea storage insect (Murdock and Shade, 1991). Solar heaters are now available for this purpose, being suitable for treating small quantities of cowpea (25–50 kg), but it is imperative they be used correctly in order to heat the grains sufficiently and for long enough to kill all insects without losing quality.

Grain legumes are more prone to insect and pest attack during storage than cereals, and are hence more difficult to store. Milling losses in insect-damaged grain are even higher, as more breakage and powdering occurs with such grains. In Ghana, a study demonstrated that the level of damage due to insects during storage reached up to 60% in cowpea and 15–20% in bambara nut. It has been demonstrated through trials that traditional methods such as the application of either shea nut butter or a mixture of ash and chilli powder can reduce damage levels to 5% (equivalent to a weight loss of only 1%); these methods make a significant contribution to household food security. Fumigation with phosphine before storage of cowpea grains has been shown more useful in treating sacks of grain than traditionally used chemicals, including rat poison, pesticide sprays and phosphine tablets, which are neither safe nor effective for maintaining grain quality.

25.3 Primary Processing

Primary processing involves the cleaning of legume grains and their separation into desired quality classes on the basis of physical properties such as diameter, thickness, density and grain colour. Therefore, important physical properties including surface area and volume have been studied in several crops, including chickpea (Konak *et al.*, 2002), cowpea (Ige, 1977; Yalcin, 2007), soyabean (Deshpande *et al.*, 1993), pea (Yalcin *et al.*, 2007), lentil (Carman, 1996) and beans (Koutsika and Traka Mavrona, 2008).

The cleaning phase uses various screens, air flows, separators and de-stoning mechanisms to remove unwanted organic or inorganic matter. Specific machinery for large-scale primary processing has been developed, and modifications used for processing include those to elevators, conveyors and storage systems in order to reduce damage to both grain and seedcoat, and to maintain high visual quality. For example, in most of the food legumes including lentil, the primary processs begins with storage and with bagging of cleaned grains (Vandenberg, 2009). The operations required between these processes are highly variable and flexible, in accordance with the economics of meeting consumer demand. Since the level of grain quality is based on specific diameter, thickness, colour, damage and admixture specifications, shape, density and hardness, these physical parameters are used when setting machine parameters (Sahay, 2003). The general procedure used in primary processing is outlined in Fig. 25.1 and discussed briefly below.

Cleaning

The threshed grains are generally impure, due to the presence of straw, stones, inert matter, etc. Traditionally in rural areas, sieves are used for small-scale primary processing. The simplest cleaning method in soybean

Operations	Activities
Stored uncleaned grain ↓	Storage of legume grains in specific conditions required for crop
Pre-cleaning ↓	Removal of coarse and fine materials based on stationary screens
Air cleaning ↓	Use of a combination of air, gravity and screens to remove chaff, straw and dust and to separate grains on the basis of size, shape and density
Indent cleaning ↓	Grains uniform in diameter are separated
Gravity separation ↓	Separation of grains based on specific weight
De-stoning ↓	Removal of heavier materials like stones, glass and metal
Grading ↓	Grains are graded on the basis of size and colour as per customer demands
Packaging	The cleaned and graded grains are packaged into bags, containers, bulk containers, etc. for delivery

Fig. 25.1. General outline of steps involved in the primary processing of legume grains (modified from Vandenberg, 2009).

involves tossing the beans into the air and letting the wind carry off the lightest impurities; however, this cleaning method does not eliminate the heavier impurities. Improving legume quality is best achieved mechanically, principally with the air-screen separator that uses a combination of air, gravity and screens to separate seeds based on size, shape and density. Models equipped with vibrating screens and replaceable sieves are now commercially available. The mechanical cleaning of lentil grains has been discussed in detail by Vandenberg (2009). In general, the basic principle used in cleaning grains can be similary applied to other legumes, although the size of the screen mesh will vary with the size and shape of the grain. Adjustments can be made to settings to achieve optimum quality in the final product. For example, for soyabean, adjustments to air speed, chaffer and sieve settings can be made as required, depending upon weather conditions. The setting of the cleaning fan to provide maximum air exposure without blowing the beans into the return elevator or out from the back end by adjustment of the chaffer allows the fan to separate pods and stalk fragments from the beans.

Following air-screening, indent cleaning is generally used for separating grains by diameter into different sizes, for which mechanical separators are commercially available. Some machines have inbuilt air-screen separators, which use a combination of air, gravity and screens to separate grains based on size, shape and density.

The primary aspiration drains off chaff, straw, dust or decreased grains, then the gravity separator cleans grains of similar size but different weight. It can be used effectively to remove partially eaten (for example by storage insects), immature or broken seeds to ensure maximum quality of the final product. De-stoning machines are used to separate heavier materials like stones, glass and metal (Sahay, 2003; Vandenberg, 2009). Cowpea grown as a dried pea product can be directly combined using a platform head or a row crop head. Adjustments to combined settings, and possibly screen/sieve sizes, should be made for the cowpea seed (Quinn, 1999).

25.4 Secondary Processing

The cleaned and graded grains are processed further into various forms by secondary processing, of which de-hulling or decortication is one of the main steps. De-hulled grains are consumed as whole or split grains in the form *dhal*. The recovery of de-hulled grain after milling is often called milling efficiency, defined in pulses in various ways. Ehiwe and Reichert (1987) defined it as percentage recovery of hull and de-hulled grain yield during decortication. The terms milling and dehulling efficiency have been used by Wang (2005), who described these as the percentage of whole and split de-hulled grains and percentage recovery of de-hulled whole grains, respectively, during the process of decortication. A more precise definition of de-hulling efficiency is total percentage recovered of split and unsplit cotyledons with less than 2% hull during decortication (Vandenberg, 2009).

De-hulling efficiency and loss are deterimed by several parameters that have been optimized in several pulse crops. The de-hulling index was studied in pigeon pea using response surface methodology (RSM; Phirke *et al.*, 1996; Goyal *et al.*, 2008), and it has been reported that hot milling (heat treatment after urea application) reduces de-hulling time by 50% (Phirke *et al.*, 1996). Process parameters and enzymatic pretreatment were found to be optimized using RSM in this crop (Verma *et al.*, 1993). Machine conditions for milling of pigeon pea were optimized without considering the time required for de-hulling (Mandhyan and Jain, 1992). The carborundum number of de-husking rollers has also been shown to affect the recovery of de-hulling pigeon pea grain (Sahay and Bisht, 1987).

The moisture content of grain plays an important role in de-hulling, as efficiency decreases with increases in moisture content (Ramakrishnaiah and Kurien, 1983). The milling of *dhal* is the main secondary processing procedure carried out in South Asia, especially in India where around 75% of total pulses are milled into *dhal*, mostly by domestic, cottage and small- to medium-scale industries. It represents the third largest food grain processing industry in India after rice and wheat. Ali (2004) observed that in India there are around 11,000 *dhal* mills with an average capacity of 10–20 t/day. However, these mills do not follow standard processes because there is variation in pulse crops, varieties, environments and treatments. In other countries too, the milling of *dhal* is becoming popular – Turkey, Sri Lanka, Egypt, Syria, Canada, Australia and the United Arab Emirates (Vandenberg, 2009). The use of the *chakki*, a vertical stone, is common in India, but it involves a high level of grain damage (up to 20–45%) and poor-quality *dhal* (Ali, 2004). Thus efforts have been made at several research centres in India to develop small-capacity mills to achieve better recovery of *dhal*. These low-cost machines are usually of 1–2 horsepower (HP), with de-hulling, splitting and aspiration capability; use of these requires pretreatment of the grains with various agents including water and oil, salts, chemicals or heat alone or in combination, in order to remove the pericarp (see below for details). However, these mills have not attained popularity among villagers, despite their low cost and refined pretreatment techniques.

The commercial de-hulling or *dhal* mills comprise cleaning and grading units and pitting, pretreatment and drying units. The cost of these commercial mills depends upon the capacity and degree of automation

Operations	Activities
Cleaning and grading legume grains	Product of primary processing is used
Pitting	Abrasive roller machine is used to scratch grains for facilitating the entry of oil/water in to the grains during pretreatments. This process is also part of the de-hulling procedure
Pretreatment with oil and water or other agents	Pretreatments are given with edible oil/water/salt for loosening of husk. Sometimes water and oil are applied simultaneously. This process reduces the total time required in processing
Tempering and steeping	Legume grains are covered and left for 12–18 h for penetration of oil/water into the cotyledons after oil/water mixing
Drying	For equilibrium of grain temperature and moisture following sun-drying, generally for 1–5 days depending on weather conditions
De-hulling and splitting	The dried and pretreated grains are milled in an abrasive mill for removing the seed coat. Complete splitting needs many passes (generally 3–9) and hence de-husking is preferably achieved by subjecting the grains to abrasive forces and splitting by attrition and/or impact. This process results in whole grains, split grains and hulls
Sieving	For separating the fine hull from whole and split grains. This process is also required to obtain the whole de-hulled grains
Aspiration	Husk is separated with aspiration and sold as livestock feed
Grading	Unsplit and split de-hulled grains are graded on the basis of size and colour as per costumer demand
Polishing	Split/*dhal*/whole de-hulled grains are polished to add lustre and shine to the product. *Dhal* is polished in different ways such as nylon polishing, oil/water polishing, colour polishing, but sometimes consumers prefer unpolished *dhal*
Packaging	The graded split grains are packed into bags, containers or bulk containers for delivery

Fig. 25.2. Outline of general steps involved in secondary processing for de-hulling or *dhal* milling in legume grains.

introduced for handling of materials. The general procedure of secondary processing for de-hulling is outlined in Figure 25.2. The process of de-hulling begins with the cleaned and graded grain product derived from primary processing, and requires various additional and optional steps that vary greatly by plant.

In the decortication process of legume grains, loosening of the binding between the seed coat and cotyledon is required – usually a gum (such as galactomannan) or lignin layer (Siegel and Fawcett, 1976). The variablity in thickness and depth of this layer influences the binding strength (Muller, 1967), the thickness being reported as low as 5% of the seed in some varieties of cowpea and as high as 30% in some lupin species (Kurien, 1984). Lentil, chickpea, pea and lathyrus crops are considered to be in the category of pulses that are relatively easy to de-hull compared with pigeon pea, green gram and back gram (Muller, 1967). Therefore, pretreatment is generally suggested to loosen the seed coat. Singh (1995) reviewed various pretreatments used in pigeonpea; commercial mills use ~1% edible oil as pretreatment for loosening the husk of those legume grains that are difficult to mill (Singh, 1995; Sokhansanj and Patil, 2003). In black gram, the use of 0.85% oil treatment with 90°C drying temperature has been reported for maximum recovery of *dhal* (Tiwari *et al.*, 2005). The hot milling treatment is optimal for obtaining maximum recovery of de-hulled grains in pigeon pea (Phirke *et al.*, 1996).

For *dhal* milling, grains are given an initial 'pitting' in the roller mill before pretreatment, to crack the husk and thereby improve the absorption of pretreatment agents. In the case of black gram, the coating of wax and dust is removed by initial scouring in a roller mill, facilitating the absorption of water or oil. Thus the steps of pitting, addition of water, heating and cooling are designed to reduce the natural binding of the seed coat to the cotyledon.

The pretreated grains are then tempered and dried, which aids penetration of oil/water into the cotyledons, and grain temperature is equilibrated after drying, natural sun-drying is most commonly practised. The drying period varies from one to five days, depending upon weather conditions. Some *dhal* mills are equipped with dryers for continuous operation, especially in the rainy season and/or unfavourable weather conditions.

De-husking and splitting are the most important unit operations of any *dhal* milling process; these are carried out either in a single operation or, more advantageously, as independent operations. The pretreated grains are subjected to abrasive/scouring forces for removal of husks and splitting of cotyledons into two equal halves. The *chakki* is used in some operations to de-husk and split the grains. For the splitting of de-husked and moistened grains, either the vertical disk burr mill is used or grains are allowed to fall on a hard or cemented surface from sufficient height (Sahay, 2003). For processing of split lentils, it is also possible to use a second set of horizontal stone-milling machinery that can be adjusted to produce 100% split products from de-hulled grains (Vanderberg, 2009). Although moisture addition adversely affects husking, it helps in splitting the grain, and thus water addition prior to husking often leaves patches of husk on the split cotyledons (*dhal*). These patches have to be removed by scouring in polishing machines. Husk separated by aspiration may be used for livestock feed.

After milling the split and whole grains are polished to remove fine dust, as a means of improving the visual quality of the product. Polishing with water is common practice, and is accomplished in same facilities by adding small amounts to the product stream as it passes in a horizontally mounted screw-conveyer prior to bagging. Adding a small amount of vegetable oil to the product stream is also used for certain markets, the Middle East in particular. In some cases, lentils are double-oiled for specific markets. Oiling is also a traditional method of preparation of de-hulled pulses in South Asia. The practice is commonly used in Africa and India at the household level, and to control grain storage insects such as *Bruchus* spp. occuring in Europe, North Africa and South-west Asia.

Thermal processing in glass jars or cans and infrared drying are other forms of secondary processing (Cenkowski *et al.*, 1989; Vandenberg, 2009). European and North American consumers are generally more familiar with thermally processed pulse products in ready-to-eat forms such as soup and side dishes. Variations on the thermal process include packaging of prepared dishes in foil or plastic envelopes or bags. Thermal processing may alter the nutritional profile, depending to some extent on the temperature used in the process and the chemical composition of the brine solution used in the canning procedure (Singh and Diwakar, 1993). Infrared drying is applied to rehydrate whole grains to a moisture content of 19–39%; for this pupose, an infrared wave length of 1000–1500 nm can be used for rapid reduction of moisture content. Dehydration triggers changes to cooking properties, reducing cooking time and altering starch gelatinization and solubilization through changes in porosity and water dispersibility (Scanlon *et al.*, 2005). This process has been commercialized for whole lentils and other grain products by Infraready Products Ltd, to produce quick-cooking whole grains (Vandenberg, 2009).

25.5. Grading and Packaging

Both cleaned grains with hull and whole or split de-hulled grains are graded and packed for consumers. Therefore, grading and packaging operations are common steps in both primary and seconday processing. On the basis of consumer demand, cleaned grains are graded for uniformity of specific diameter or thickness of grain. This grading is based on grain size, which can incur additional cost to the final product, and demand for sizing can be very specific. As in the case of lentil, consumers in Spain prefer to pay extra for extra-large green lentils and large beans, and in Bangladesh or other part of South Asia for extra-small red lentils. The size of faba and chickpea grains also determines consumer preference, as large-seeded varieties require longer cooking times than small-seeded (Hsieh *et al.*, 1998). To this end, efforts have been made to develop sizing and grading machines and now both, hand- and power-operated graders are available commercially (Ali, 2004). In chickpea, separation of *dhal* is carried out according to size and soundness using rotating-reel graders, the separate compartments of which have different screen mesh pore sizes (Sahay, 2003).

The grading of legume grains is on the basis of colour; machinery that sorts by colour is available, having been adapted from other industries. A computerized image analysis system has now been developed that can grade the legume grains on the basis of size, shape and colour, and this has been utilized to evaluate the volume and surface area of lentil, bean and pea (Kadlec *et al.*, 2006; Fıratıgil-Durmus *et al.*, 2010). The basic principle here is to compare the grain colour by passage through sensors that are adjusted to an acceptable predominant colour range, with colours lying outside the selected range being rejected by air-jetting (Vandenberg, 2009). In Canadian lentil production this system is often used, depending on the market price of colour-sorted and -unsorted lentil.

After grading, legume food grains are packed in polypropylene or jute bags as whole grains or spilt forms. In Canada, the packaging of whole lentil grains is done as per local or export specifications. For this purpose, various types of bagging, containerization and shipping methods are used. Exported lentils are generally packaged in polypropylene bags of 25–100 kg. Consumer-ready packages are typically sold in 500 or 1000 g polyethene bags, or possibly up to 15 kg, particularly for the restaurant and institutional sector (Vandenberg, 2009). Private firms in Turkey supply all types of dry legumes in 1 kg packets or 25–50 kg polythene bags. In Spain, the beans are supplied in plastic and cloth bags. In India, food legume grains are consumed as whole grains as well as in split *dhal* form; the cleaned and graded lentils are packed in 50 or 100 kg jute bags for wholesale and retail traders. Packaging of split *dhal* is available in 500 or 1000 g polythene bags in market. Various types of packaging machines are now available, and commercial *dhal* mills have bagging and packaging units as part of both primary and seconday processing (Ali, 2004). Bagging systems for grain legumes are flexible and can be influenced by packaging trends in other crops or can be changed according to consumer requirements and market trends.

25.6 Conclusions

Some traditional pest control methods can significantly reduce postharvest losses of legume grains, allowing longer storage and thus increasing household food security. Farmers can also increase income by selling their surplus grain later in the season, when market prices are higher. Significant levels of postharvest losses affect the economy and the welfare of farmers, consumers and traders, and therefore some key socio-economic and technological aspects of postharvest processing and preservation need to be considered in regard to reducing these postharvest losses. A comparative study on various postharvest practices among small, medium and large farms would enable researchers and policy makers to identify postharvest loss-reducing technologies for specific groups of farmers. Frequently farmers are required to harvest immature crops – mainly fruits and vegetables – during those times when they need immediate cash and/or when prices are high. The relationship between large capital investment and postharvest losses, and information on postharvest practices, should be collated through research. Provision of policy guidelines for increasing postharvest research and development activities carried out by research institutions and universities to determine the technologies best suited at the farmer and grain-processor levels is also needed. Production of cereals (rice and wheat) has almost doubled since 1970, but in the case of food legume crops no such improvements have been attained. On the other hand, the demand for vegetarian food within an expanding population is increasing. Postharvest losses due to inadequate processing and preservation facilities must be given due attention to ensure food security at both the macro and micro level.

Acknowledgements

This work was supported by the project INCITEOTPXI1403088ES from the Galician Government. The first and the last three authors are grateful to the Diputación Provincial de Pontevedra for farm facilities.

References

Ali, N. (2004) Postharvest technology and value addition in pulses. In: Ali, M, Singh, B.B., Kumar, S. and Dhar V (eds) *Pulses in a New Perspective*. Indian Society of Pulses Research and Development, IIPR, Kanpur, India, pp. 530–543.

Cameron, A.G. (1999) Forage and cowpeas. Agnote, northern territory of Australia. Agnote No. 304 (E34) (www.ntigfgtpa.com/pdf/304/pdf).

Cenkowski, S., Sokhansanj, S. and Sosulski, F.W. (1989) Equilibrium moisture content of lentils. *Canadian Agricultural Engineering* 31, 159–162.

Edmister, J.A., Breene, W.M., and Serugendo, A. (1990) Influence of temperature, water activity and time on cookability and colour of a stored Rwandan dry bean (*Phaseolus vulgaris*) mixture. *Journal of Stored Products Research* 26, 121–126.

Ehiwe, A.O.F. and Reichert, R.D. (1987) Variability in de-hulling quality of cowpea, pigeon pea and mung bean cultivars determined with TADD. *Cereal Chemistry* 64, 81–93.

Fıratlıgil-Durmus, E., Šarka, E., Bubnik, Z., Schejbal, M. and Kadlec, P. (2010) Size properties of legume seeds of different varieties using image analysis. *Journal of Food Engineering* 99, 445–451

Goyal, R.K., Vishwakarma, R.K. and Wanjari, O.D. (2008) Optimization of the pigeon pea de-hulling process. *Biosystems Engineering* 99, 56–61.

Gunes, G. and Lee, C.Y. (1997) Colour of minimally processed potatoes as affected by modified atmosphere packaging and antibrowning agents. *Journal of Food Science* 62, 572–575.

Hughes, P.A. and Sandsted, R.F. (1975) Effect of temperature, relative humidity and light on the colour of 'California Light Red Kidney' bean seeds during storage. *Hortscience* 10, 421–423.

Juak-Knieval, D.C., Vandenberg, A. and Bett, K.E. (2008) Slow darkening in pinto bean (*Phaseolus vulgaris* L.) seed coats is controlled by a single major gene. *Crop Science* 48, 189–193.

Kadlec, P., Skulinova, M., Šarka, E. and Fort, I. (2006) Microwave and vacuum microwave drying of germinated pea seeds. In: *Proceedings of 17th International Congress of Chemical and Process Engineering CHISA* (CD ROM), Prague.

Konak, M., Arman, K.C. and Aydin, C. (2002) Physical properties of chick pea seeds. *Biosystems Engineering* 82, 73–78.

Koutsika-Sotiriou, M. and Traka-Mavrona, E. (2008) Snap Bean. In: Prohens, J. and Nuez, F. (eds) *Vegetables II. Handbook of plant breeding*. Springer, New York, Volume 2, pp. 27–83.

Kurien, P.P. (1984) De-hulling technology of pulses. *Research and Industry* 29, 207–214.

Mandhyan, B.L. and Jain, S.K. (1992) Optimization of machine conditions for milling of pigeon pea. *Journal of Food Engineering* 18, 91–96.

Milan-Carrillo, J. (2000) Hard-to-cook phenomenon in chickpeas (*Cicer arietinum* L.): effect of accelerated storage on quality. *Plant Foods for Human Nutrition* 55, 229–241.

Muller, F.M. (1967) Cooking quality of pulses. *Journal of the Science of Food and Agriculture* 18, 292–295.

Nasar-Abbas, S.M., Siddique, K.H.M., Plummer, J.A., White, P.F., Harris, D., Dods, K. *et al.* (2009) Faba bean (*Vicia faba* L.) seeds darken rapidly and phenolic content falls when stored at higher temperature, moisture and light intensity. *Food Science and Technology* 42, 1703–1711.

Nordstorm, C.L. and Sistrunk, W.A. (1979) Effect of type of bean, moisture level, blanch treatment and storage time on quality attributes and nutrient content of canned dry beans. *Journal of Food Science* 44, 392–395.

Nozzolillo, C. and De Bezada, M. (1984) Browning of lentil seeds, concomitant loss of viability, and the possible role of soluble tannins in both phenomena. *Canadian Journal of Plant Science* 64, 815–824.

Phirke, P.S., Bhole, N.G. and Adhaoo, S.H. (1996) Response surface modeling and optimization of de-hulling of pigeon pea with different pre-treatments and conditions. *Journal of Food Science Technology* 33, 47–52.

Quinn, J. (1999) *Cowpea, a Versatile Legume for Hot, Dry Conditions.* Thomas Jefferson Institute, Springfield, Virginia (available at: www hort.purdue.edu/newcrop articles/ji-compea html).

Ramakrishnaiah, N. and Kurien, P.P. (1983) Variabilities in the de-hulling characteristics of pigeon pea (*Cajanus cajan* L.) cultivars. *Journal of Food Science Technology* 20, 287–291.

Reyes-Moreno, C., Okamura-Esparza, J., Armienta-Rodelo, E., Gomez-Garza, R.M., Sahay, K.M. and Bisht, B.S. (1987) Optimization of grades of carborundum for best recovery of dhal in milling of different pulses. *Agricultural Engineering Today* 9, 11–22.

Sahay, K.M. (2003) Postharvest technology and value addition in chickpea. In: Ali, M and Kumar S. and Singh N.B. (eds) *Chickpea Research in India.* IIPR, Kanpur, India, pp. 315–325.

Siegel, A. and Fawcett, B. (1976) *Food Legume Processing and Utilization (with Special Emphasis on Application in Developing Countries).* International Development Research Centre, Ottawa, Canada, pp. 88.

Singh, F. and Diwakar, B. (1993) *Nutritive Value and Uses of Pigeonpea and Groundnut.* Skilled Development Series 14, IRCISAT, Patancheru, AP, India, pp. 39.

Singh, U. (1995) Methods of de-hulling of pulses: a critical appraisal. *Journal of Food Science Technology* 32, 81–93.

Singh, U., Santosa, B.A.S. and Rao, P.V. (1992) Effect of de-hulling methods and physical characteristics of grains on dhal yield of pigeon pea (*Cajanus cajan*) genotypes. *Journal of Food Science and Technology* 29, 350–353.

Sokhansanj, S. and Patil, R.T. (2003) De-hulling and splitting pulses. In: Chakraverty, A., Mujumdar, A.S., Raghavan, G.S.V. and Ramaswamy, H.S. (eds) *Handbook of Post Harvest Technology.* Marcel Dekker, New York.

Tang J., Sokhansanj, S. and Sosulski, S. (1994) Moisture absorption characteristics and hard seed development in Laird lentils. *Cereal Chemistry* 3, 428.

Tiwari, B.K., Jaganmohan, R. and Vasan, B.S. (2005) Effect of heat processing on milling of black gram and its end product quality. *Journal of Food Engineering* 88, 553–560.

Vandenberg, A. (2009) Postharvest processing and value addition. In: Erskine, W., Muehlbauer F.J., Sarker, A. and Sharma, B. (eds) *The Lentil: Botany, Production and Uses.* CAB International, Wallingford, UK, pp 391–407.

Verma, P., Saxena, R.P., Sarkar, B.C. and More, P.K. (1993). Enzymatic pretreatment of pigeon pea (*Cajanus cajan* L.) grain and its interaction with milling. *Journal of Food Science and Technology* 30, 368–370.

Wang, N. (2005) Optimization of a laboratory de-hulling process for lentil (*Lens culinaris*). *Cereal Chemistry* 82, 671–676.

Yalcın, I. (2007) Physical properties of cowpea (*Vigna sinensis* L.) seed. *Journal of Food Engineering* 79, 57–62.

Yalcın, I., Ozarslan, C. and Akbas, T. (2007) Physical properties of pea (*Pisum sativum*) seed. *Journal of Food Engineering* 79, 731–735.

Yousif, A.M., Kato, J. and Deeth, H.C. (2003) Effect of storage time and conditions on the seed coat colour of Australian adzuki beans. *Food Australia* 55, 479–484.

26 Value Addition and International Trade

M. Gupta, B.K. Tiwari and T. Norton

26.1 Introduction

Grain food legumes are cultivated all over the world, and the dried seeds of harvested food legume crops (i.e. pulses) make an important contribution to the diet of all cultures and play a particularly important role in the diets of Asian countries (Singh and Singh, 1992). Apart from their role in common diets, grain food legumes have many advantages in the gluten-free market. First, they are high in iron, and consequently go towards alleviating iron deficiency anaemia and vitamin B deficiency, which are regular concerns of coeliacs. Secondly, they help with constipation, which is another problem that arises as a result of a diet low in whole wheat fibre, wheat bran and other dietary fibres. Lastly, because the gluten-free market is growing rapidly in North America and Europe, a wide variety of specialty products, e.g. ready-to-eat baked goods, mixes for breads, buns, muffins, cakes, cookies, pastries, pizza crusts, pastas, cereals, soups, sauces and other foods, can be developed using pulses. Such a wide range of attributes make grain food legumes an essential ingredient for adding value to a large range of gluten-free products.

26.2 Food Value of Grain Legumes

Pulses contain more protein than any other plant type, and have many other important nutritional properties that meet the dietary needs of many cultures and social classes. They have, therefore, been justifiably described as 'the poor man's meat'. While the nutritive properties of pulses resemble those of whole cereal grains, there are important differences. First, the pulse protein is low in sulfur-containing amino acids, but rich in lysine, an amino acid that is deficient in many cereals. A combination of pulses and cereal proteins may, therefore, have a nutritive quality on a par with animal proteins. Secondly, pulses, as a class, are good sources of the B group vitamins apart from riboflavin, with the greater part of these vitamins retained in the dried product, as no losses comparable to those caused during the milling and cooking of cereals occur. Consequently, pulses act as an excellent prevention against beriberi. Thirdly, although pulses, like cereal grains, are devoid of vitamin C, a large amount of ascorbic acid is formed during their germination. Sprouted pulses are, therefore, an important food for protection against scurvy (Morrow, 1991). Owing to this fortunate attribute, Asian and African dieticians make beneficial use of sprouted pulses for their menus, especially when fresh vegetables and fruits are scarce or too expensive. Finally, the digestion of pulses and the absorption of their principal nutrients are practically complete in the gut for people without gastrointestinal disorders, and nearly as effective as the assimilation of cereals. However, due to the potential development of complications, only small quantities of well-cooked pulses should be included in the diets of patients with stomach disorders (Sharma and Garry, 1995).

Most grain growers will be familiar with pulses, but are probably less familiar with the nutritional properties of pulses and products made from them. Pulses provide a similar level of energy to grains. Like grains, most of the energy is provided by carbohydrate rather than fat (about 50–60%). All pulses have a low glycaemic index (that is, the carbohydrate is digested slowly). They are also low in fat, containing only 2–6%, most of which is provided in the form of polyunsaturated and monounsaturated fatty acids. Pulses are also a good source of fibre. Those that are consumed with the seed coat included, such as green lentils, are much higher in dietary fibre than those that have been de-hulled before consumption (such as split peas or red lentils). Figure 26.1 shows the different uses of food legumes; these are used as starters, main dishes, side dishes or in salads. The de-husked or decorticated and whole seeds are used as *dhal* and taken with *chapatis* and cooked rice. Whole seeds take longer to cook than the de-husked and decorticated varieties, which have a better digestibility. They are also commonly used in the form of flour such as that of Bengal gram, green gram and black gram, known as *besan*; and for mixing with cereal flour in various proportions for *chapatis* and other preparations (Cai *et al.*, 2001). The low moisture content and hard seed coat of dry pulses permit their storage over long periods.

The practice of utilizing sprouted young pulses as a fresh vegetable is widespread in the Orient. The storage of dried seed followed by sprouting enables a continuous supply of fresh vegetable material to be produced as required. There is a large increase in nutrients in sprouted pulses when compared with their dried embryos. During the sprouting process, vitamin, mineral and protein levels increase substantially, with a corresponding decrease in calorie and carbohydrate content, which leads to an improvement in nutritive value and digestibility of the pulse. Starch is then broken down to dextrin and maltose, and proteins are broken down to polypeptides, peptides and amino acids. Some of the bound iron is converted to a more readily assimilated form, and phosphorus is liberated from phytate. The ascorbic acid or vitamin C content rises from negligible levels in the seed to 12 mg/100 g after 18 h of germination. Riboflavin and niacin contents increase significantly. All these changes are brought about by enzymes that become active during germination (Singh and Singh, 1992).

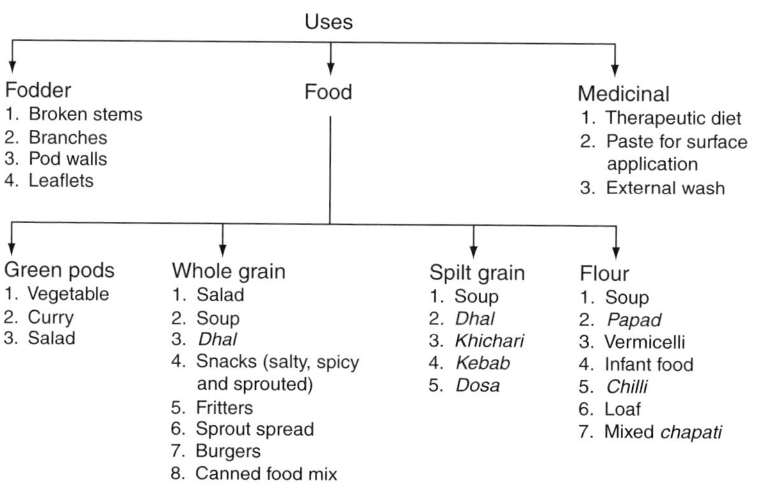

Fig. 26.1. Uses of food legumes (modified from Raghuvanshi and Singh, 2009).

26.3 Value Addition

World legume production is around 58.9 t, grown in an area of about 70.5 Mha (FAO, 2008). Asia accounts for about 50% of global pulse production, while India's share is about 24.6% of this (Fig. 26.2), but its production trends are not very satisfactory. During recent decades production of pulses has somewhat stagnated, and this has put a lot of pressure on their per capita availability. As per international standards, the recommended personal dietary requirement is 55 g/day, but current consumption is now only 30 g/day, against 67g/day in India four decades ago; this suggests that the per capita consumption of all pulses is declining in South Asia. In rural India the pulse has been the major protein source, with its decline in production unfortunately now causing several health issues. Consumption could be increased by introducing the value-added products of less preferred pulses, resulting in higher investment return from those crops with relatively high availability and low price. Value-addition processing can also increase the market value of those food legumes that has reduced their market value as a whole food.

'Value added' is a term frequently mentioned when discussing the future profitability of agriculture. Its popularity rose substantially during the 1990s to the point that it has become one of today's buzzwords. In general, adding value is the process of changing or transforming a product from its original state to a more valuable state that is preferred in the marketplace. As a specific example, a more narrow definition would be to economically add value to an agricultural product (such as pulses) by processing it into a product (such as flour) desired by customers (such as baked products). It is important to identify the value-added activities that will support the necessary investment in research, processing and marketing. With the continuous movement towards a global economy, the international market for value-added products is growing. Market forces have led to greater opportunities for product differentiation and added value to raw commodities because of: (i) increased consumer demands regarding health, nutrition and convenience; (ii) efforts by food processors to improve their productivity; and (iii) technological advances that enable producers to produce what consumers and processors desire (Siebert *et al.*, 1997).

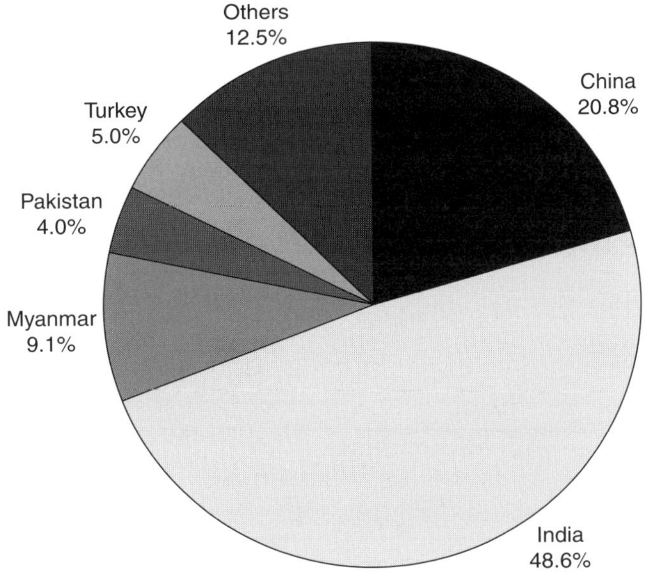

Fig. 26.2. Production of pulses in Asia (FAO, 2008).

Snack foods

Pulse-based snack foods are popular in Southeast Asia. In Thailand, snacks made from peas are popular. Pulses offer a good base for making extruded snack foods because they easily allow for flavour addition and they retain a good crunch. New snack foods are being launched based on pulses, e.g. chickpea chips. Pulse flour, such as chickpea flour, has been used to make a variety of snack foods. In India, and Pakistan there is a big market for deep-fried de-hulled whole or split pulses. A good amount of chickpea, field pea and mung bean goes toward this type of value addition in these countries.

Breakfast foods

Breakfast bars are the biggest growth area in the breakfast food industry. Pulses can be treated to remove any beany flavours and then softened for easy eating and added to cereals or bakery products. Due to their high protein content, peas and lentils could be added to breakfast cereals in the same way as soy is already being used.

Pre-prepared meals

Consumer demand for fast and easy meal solutions, such as microwaveable or partly prepared, pulse-based meals, is common to both Australian and Asian time-poor consumers. Consequently, pre-prepared meals are a growing segment of the retail market. The food service sector has also expressed a preference for prepared or quick-cooking pulse products. In developing pre-prepared meals, consideration of flavour, appearance and ability to retain texture and colour when reheated is important. The packaging should have an attractive appearance, small convenient size and adequate shelf life. Pre-prepared meals include microwaveable meals, partly prepared meals (which require simple addition and heating) and home meal replacements. Examples of pulse-based pre-prepared meals include:

- patties and cold meat replacements;
- vegetarian meals;
- prepared salads and soups; and
- bean-based packaged foods.

26.4 International Trade

For pulses, which are primarily grown in rainfed conditions and therefore subject to seasonal and yearly variation in supply, market information is essential and costly. To some extent, the same complexities in obtaining market information and adjusting the scale of the lots seem to influence the international market. Import prices of regularly traded pulses, such as chickpea, seem to track fairly well, but for many other pulses price variations seem common. A few observations may be made on the frequency and basic structure of international trade flows. First, it is probable that import transactions are seasonally structured, whereby import prices fall below national wholesale prices. It seems likely that wholesale prices in major consumer centres are also seasonally structured. To some extent this observation is borne out by the participation of countries that produce pulses at different times of the calendar year. Their participation in South Asian trade would mean that storage time and costs are minimized. This reasoning stands and falls with the dominance of seasonal price fluctuation in major producer and consumer centres of the world. Although it is evident that supplies from many directions, produced at different times of the year, participate, it still needs to be proved that importers effectively minimize costs through inter-seasonal sourcing from a policy point of view, where there seems to be no specific hindrance to international trade. Normally, some types of staple trading markets emerge at locations of concentrated demand and/or supply. It should be noted that the expansion in international trade in pulses in the subregion seems to have taken place without the inclusion of pulses in the international mainstream commodity futures. Such markets usually include cash trading prices and also futures.

The rationale of entering trade in an international marketplace is to stabilize prices and to spread risk over larger geographical areas as well as over time. The advantage of derived and option markets is that the costs of information and transactions decrease, and that the resulting price formation may incorporate risk. Conditions for such trade are many: (i) some continuity in trade; (ii) sufficiently large turnover to support the information and transactional infrastructure; (iii) accountability and credibility of firms; and (iv) constant market demand channelled through industry. On the surface it seems that because of the absence of a structured futures market in pulses in South Asia these conditions are not met, and the accompanying trade mechanisms are not in place (Akridge *et al.*, 1997). However, there are many reasons to assume that most of the trading mechanisms are in place; in fact some pulses and their products, notably soybean, meal and oil and red bean are listed commodities in Chicago and Tokyo.

Importers and exporters of pulses most likely use risk-spreading transactional options through their trading networks; collection and local wholesale traders and possibly also processors also use risk-spreading transactions on the basis of price expectations. The point is that trade in pulses is primarily regional. The total volume of trade in South Asia is quite considerable, at almost one third of world trade; it is thus not the size of the market, but the spatial structure of the pulse trade that constrains size and information.

For information improvement, one has to focus on domestic wholesale markets as well as the international market. The current efforts in India to improve the local collection market are most interesting and need careful monitoring and analysis. It is most important to recognize that an international trading system can never make up for weaknesses in the domestic market, whether these are located in production, transport or processing, or in all three fields, as is usually the case. A mature domestic market would also offer traders the opportunity of time-bound options and transactions; in other words, the same options and market information are accessible in international markets. These options are sometimes, rather misleadingly, called parallel trade – a better term would be direct trade. Among the recommendations for improvement of the domestic market are contract farming and other ways to secure supply in a more or less fixed time frame. Nevertheless, it is most likely that current trade already includes the purchase of farm products in advance as well as contract farming. Such transactions may well cause shifts in land allocation of producers if prices justify, or in general investment in pulse production. Experiments with auction-type trade in production centres may be expanded with direct trade information, linking producer prices in import prices in trade centres. Likewise, direct trade information links with production centres may strengthen the efficiency of major consumer centre markets (Soe, 1994).

The idea of expanding the geographic span of trade through improved access to trade information is the bread and butter of private traders. One must realize that traders in agricultural produce tend to stabilize and even define their business by limiting their exposure to information and to concentrate on – and to some extend capture – a market they know well. Innovation in terms of information is likely to be connected to generations of people, education and exposure to information and, most importantly, to investment in rural information and transport infrastructure. The process of expanding the accessibility of market information will depend on technical innovation, improvement in infrastructure and a healthy agricultural industry. The issue for follow-up is therefore the functioning of the domestic market and its prospects for innovation, both technical and in terms of information.

26.5 Trade Scenario

The total volume of world pulse production almost increased by 50% during the years 1980–2004, surpassing for the first time in history the 60 million t mark in 2004. Although global production showed an overall upward trend over this period, there was a variation in growth. While it grew relatively fast between 1980 and 1990 (3.6% p.a.), production stayed almost stagnant afterwards. Growth in the 1980s was driven by the developed countries, who expanded their output by 8% annually from 10.7 million t in 1980 to 20.8 million t in 1990. Further analysis shows that production growth in the developed countries over the period 1980–2004 would have been much higher had it not been for decline in the transition economies; aggregate pulse production in these countries contracted by 12% annually during the 1990s, in line with the general trend in their agricultural production following the reforms in their economies. The data show that industrialized countries doubled their share in world pulse production, up from 13% in 1980–1982 to 26% in 2002–2004 (FAO, 2004).

The fast growth of pulse production in the developed relative to the developing countries, as a whole, can partly be explained by the large yield differential between the two country groups. While yields in the developed countries grew by 2% annually from 1980 to 2004, yield growth in the developing countries was substantially smaller (0.4% p.a.). As a result, the yield gap between the two groups has widened, rising from 0.5 t/ha in 1980–1982 to 1.1 t/ha in 2002–2004.

The lagging behind of pulse productivity in developing countries can be explained by several factors, including: (i) production in the developing countries is largely of a subsistence nature, while in developed countries it is commercial; (ii) lack of investment because pulse cultivation is generally a small-scale activity that is not viewed as a sector capable of generating economic returns; (iii) the expansion of irrigated land has pushed pulses into marginal zones with the better land used to grow cereals; (iv) an agricultural policy focusing on cereals for food security purposes; and (v) limited research and lack of technology and availability to farmers of improved cultivars (Hofer, 2004).

Per capita consumption of pulses in the developing countries stagnated overall and registered a drastic decline in some regions, especially in Asia and sub-Saharan Africa. These trends reflect changing dietary patterns and consumer preferences but, in several countries, also the failure of domestic production to keep pace with population growth. Often this was the result of government preference for increasing production and self-sufficiency in cereals. As a result, per capita pulse production, reflecting availability from domestic sources, has declined. Simple graphical analysis shows that pulse consumption in the developing countries very closely follows movements in domestic production. In the industrialized countries, by contrast, per capita food consumption of pulses has increased. A plausible explanation for this is the increased consumer awareness of the health benefits of dry legumes. High levels of animal protein in the diets of industrialized countries have stimulated consumers to look for alternative sources, and with good levels of protein and fibre along with low fat content, pulses represent an excellent alternative. Another factor that may have contributed to this is international migration, which has accelerated during the period of study.

Developing countries have often met their growing pulse requirements through increased imports, reflected in a growing overall trade deficit. A large portion of this, however, was due to larger imports by India, the world's largest producer and consumer of pulses. Rising disposable incomes in this country, especially among the poorer segments of the population, are spurring demand. Another important market is the Near East/North Africa region, where imports are sustained by population growth (Tyagi *et al.*, 1993).

26.6 Trade Expansion

Global trade of pulses in 2001–2003 averaged about 9 million t/year, with a total value of some US$3 billion. Trade in pulses grew rapidly between 1980 and 2003 (5% p.a.), much faster than output, and as a result the proportion of pulse production traded increased significantly, from 7% in the early 1980s to 16% in 2001–2003. Nevertheless, pulse trade remains a relatively thin market, especially when compared with other food commodities, namely grains. The expansion in international trade of pulses has provided a good opportunity for several countries to expand their exports. It is noteworthy that, despite their growing trade deficit, the developing countries, as a group, increased their pulse shipments by 150% between 1980 and 2003 and more than doubled their export earnings from pulse sales (Fig. 26.3).

In regard to outlook, per capita pulse consumption in the Near East/North Africa and Latin America and Caribbean regions is projected to stay at current levels, while it is likely to fall further in South Asia because of a shift to consumption of higher-value livestock products and fruits and vegetables. By contrast, an increase is expected in sub-Saharan Africa, reversing the decline experienced in recent years. It is also expected that net imports by developing countries will grow in order to meet their growing demand. On the supply side, constraints to productivity growth and production in the developing countries are expected to persist, unless corrective measures are taken. As such, production growth is expected to lag behind demand. Consequently, the recent trend in pulse imports by this group of countries will most probably continue.

Since pulses play an important role in the food security of a relatively large proportion of the population in the developing countries, it is recommended that the development of new pulse varieties and cultivation technologies be reinforced by adequate policies, support programmes in education and training of farmers, supply of input and credit and the development of appropriate marketing channels (Vishwanath, 1993). In other words, governments are recommended to create an enabling environment in order for this sector to develop. The import levels of pulses (mainly by developing countries) are largely determined by shortfalls in domestic production. For example, India is the largest producer and consumer of pulses; however, the majority of pulses are produced on marginal, rainfed regions resulting in wide variability in yield and production. Overall trade and price levels are therefore dependent on the success or otherwise of the importers' own domestic crops. The international trade of pulses over the last 40 years has grown significantly; the annual average growth rate in the international trade of all key pulse commodities ranged between 5% and 8% per annum compared with wheat and rice at 0.81% and 1.49% per annum, respectively. The level of pulses traded will continue to expand as traditional producing/consuming nations continue to reduce their pulse production in favour of other, higher-paying commodities such as wheat and rice. Competitive pressures will build in the industry between Canada, the USA, Australia, Turkey and Syria over the short to medium term, with China

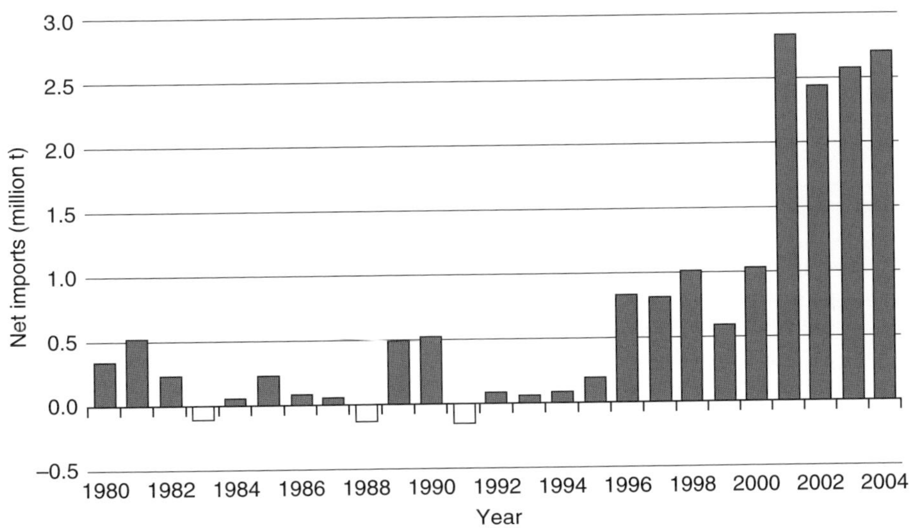

Fig. 26.3. Net imports of pulses in developing countries.

over the medium term and the former Soviet republics over the long term. Increased international competition will drive the search for a technological edge among traders to maintain supplies of cost-competitive pulses. It will also drive the entrepreneurial efforts of traders looking to deliver to precise customer requirements, efficiently and profitably (Soe, 1994). This will necessitate a more strategic approach for long-term industry development, positioning and planning. There are strong niche market opportunities for higher-quality, higher-value-added product for the growing middle classes of the Indian subcontinent and the Middle East. Demand for pulses in stock feeds is likely to increase with the forecast expansion in meat consumption driven by rising per capita incomes in developing nations.

There is momentum developing for the use of pulse fractions (proteins, starches and fibres) as food and industrial ingredients, although these are considered as potential opportunities for the longer term. Recent high prices for pulses have attracted many suppliers and supplies to the market – for example, note the phenomenal growth of the Canadian and Australian pulse industries over the last 10 years. However, the consumer market is able to absorb increased supplies but at a lower price. Producers and exporters need to have an effective risk management strategy in place and a cost/price profile able to withstand the vagaries of this market condition (Jensen, 2002).

Research, development and extension will continue to play a key role in the production of cost-competitive pulse products and in developing sustainable competitive advantage in the international market. Australia's key competitor, Canada, is seeking to improve its market positioning through R&D outcomes in enhancing productivity and production reliability. There are also concerns that India's recent R&D drive to stabilize pulse production may affect the longer-term position of the subcontinent as a key importer of pulses. Price movements that rise too high will see the market engage in product substitution (as the bulk of the pulse market is extremely price conscious); for example, food constitutes about 50% of household expenditure in India (MOFPI, India). Therefore even a slight movement in price will see a switch to more affordable food items. Canada is currently a major exporter of green lentils but is now looking at becoming a leading exporter of split red lentils. The key enablers they are looking at in regard to this strategy are increased disease-resistant varieties, the ability to multiply seed at a faster rate than 'Mediterranean' climates and the existing processing/splitting capacity in the state of Saskatchewan (Vishwanath, 1993).

References

Akridge, J., Downey, D., Boehlje, M., Harling, K., Barnard, F. and Baker, T. (1997) *Agricultural Input Industries. Food System 21. Gearing Up for the New Millennium*. Purdue University Cooperative Extension Service, West Lafayette, Indiana.

Cai, R., Klamczynska, B. and Baik, B.K. (2001) Preparation of bean curds from protein fractions of six legumes. *Journal of Agricultural and Food Chemistry*, 49, 3068–3073.

FAO (2004) Pulse production data sheets. Available at http://faostat.fao.org/faostat/form?collection =Production. Crops

FAO (2008) *Crop Prospects and Food Situation*. FAO, Rome.

Hofer, J. (2004) Legume crops and their origin. *Grain Legumes* 40, 9–40.

Jensen, E.S. (2002) The contribution of grain legumes, currently underutilised in the EU, to a more environmentally-friendly and sustainable European Agriculture. In: *Grain Legumes for Sustainable Agriculture*, 26 September. Org Print, Strasbourg, France.

Morrow, B. (1991) The rebirth of legumes. *Food Technology* 45, 96–121.

Raghuvanshi, R.S. and Singh, D.P. (2009) Food preparation and use. In: Erskine, W., Muehlbauer, F.J., Sarker, A. and Sharma, B. (eds) *The Lentil: Botany, Production and Uses*. CABI, Wallingford, UK, pp. 408–424.

Sharma, A. and Garry, P. (1995) *Foodgrains, Pulses, Oilseeds and Cotton in India. The Potential Impact of Unilateral Trade Liberalization* (mimeo). NCAER, New Delhi, India.

Siebert, J.W., Jones, R. and Sporleder. T.L. (1997) The VEST model: an alternative approach to value added. *Agribusiness* 13, 561–568.

Singh, U. and Singh, B. (1992) Tropical grain legumes as important human foods. *Economic Botany*, 46, 310–321.

Soe, T. (1994) Policies and institutions related to grain trade: problems and prospects both in the short-run and the medium-term. Paper presented at the Training Seminar of Developing an Efficient Marketing System for Food Grains, 21–25 November,Yangon, Burma.

Tyagi, V.P., Kumar, S., Pandey, R.K. and Bhalerao, M.M. (1993) Marketing of oilseeds and pulses. A sample study. *India Journal of Agricultural Economics* 48, 374.

Viswanath, N.B. (1993) Performance in production and marketing. A case study of pulses in Karnataka, *India Journal of Agricultural Economics* 48, 415.

Index

Note: Page references in *italic* refer to tables; those in **bold** refer to figures